© VCH Verlagsgesellschaft mbH, D-6940 Weinheim (Federal Republic of Germany), 1985

Vertrieb:
VCH Verlagsgesellschaft, Postfach 12 60/12 80, D-6940 Weinheim (Federal Republic of Germany)
USA und Canada: VCH Publishers, 303 N.W. 12th Avenue, Deerfield Beach FL 33442-1705 (USA)

ISBN 3-527-26225-3

Instrumentelle Multielementanalyse

Herausgegeben von
Bruno Sansoni

Prof. Dr. Bruno Sansoni
Zentralabteilung f. Chemische Analysen
Kernforschungsanlage Jülich GmbH
D-5170 Jülich 1
Bundesrepublik Deutschland

1. Auflage 1985

Lektorat: Dr. Hans-F. Ebel
Herstellerische Betreuung: Ellen Böckmann

CIP-Kurztitelaufnahme der Deutschen Bibliothek

Instrumentelle Multielementanalyse / hrsg. von Bruno Sansoni
– 1. Aufl. – Weinheim; Deerfield Beach, FL: VCH, 1985.
  ISBN 3-527-26225-3
NE: Sansoni, Bruno [Hrsg.]

© VCH Verlagsgesellschaft mbH, D-6940 Weinheim (Federal Republic of Germany), 1985

Alle Rechte, insbesondere die der Übersetzung in andere Sprachen, vorbehalten. Kein Teil dieses Buches darf ohne schriftliche Genehmigung des Verlages in irgendeiner Form – durch Photokopie, Mikroverfilmung oder irgendein anderes Verfahren – reproduziert oder in eine von Maschinen, insbesondere von Datenverarbeitungsmaschinen, verwendbare Sprache übertragen oder übersetzt werden.

All rights reserved (including those of translation into other languages). No part of this book may be reproduced in any form – by photoprint, microfilm, or any other means – nor transmitted or translated into a machine language without written permission from the publishers.

Satz: Graphische Betriebe der KFA, Jülich
Fotographie der ersten Umschlagseite: W. Hilgers, ZCH
Druck: betz-druck gmbh, D-6100 Darmstadt 12
Bindung: Buchbinderei K. Schaumann, D-6100 Darmstadt
Printed in the Federal Republic of Germany

Gewidmet

dem Andenken

unserer beiden bedeutenden Kollegen

Prof. Dr. Hans Wolfgang Nürnberg

Institut für Angewandte Physikalische Chemie

Kernforschungsanlage Jülich GmbH, Jülich

1930 — 1985

und

Prof. Dr. Gordon F. Kirkbright

Department of Instrumentation and Analytical Science

Institute of Science and Technology

University of Manchester, England

1939 — 1984

die uns für immer verlassen haben

VORWORT

Beliebige Materialproben lassen sich durch eine Analyse der in ihnen enthaltenen Elemente chemisch charakterisieren. Dies kann qualitativ durch Nachweis der Art der Elemente und quantitativ durch Bestimmung der zugehörenden Gehalte oder Konzentrationen erfolgen. Dies sind zwar nicht die einzigen Parameter zur Charakterisierung eines Materials, aber bis heute die am weitaus häufigsten verwendeten. Wegen ihrer in den letzten Jahrzehnten stark gestiegenen Bedeutung interessieren dabei nicht nur Haupt- und Nebenbestandteile, sondern in zunehmendem Maße auch Elementspuren. Dies bringt eine Vielzahl der in einer Probe zu analysierenden Elemente mit sich.

Multielementanalysen erlauben es, in der gleichen Probe in einem Arbeitsgang eine Vielzahl von Elementen, entweder gleichzeitig oder kurz hintereinander, zu bestimmen. Der Arbeitsaufwand ist nur wenig größer als bei der Analyse eines einzelnen Elementes nach einer Monoelementmethode. Die Instrumentelle Multielementanalyse versucht dabei, chemische Arbeitsschritte so weit wie möglich zu vermeiden und allein physikalische Methoden anzuwenden. Das wird durch Einsatz verschiedenartiger Spektrometriearten ermöglicht und durch weitgehende Automation und Computerisierung des Instrumentariums begünstigt.

Die heutigen Methoden der Instrumentellen Multielementanalyse können dank extremen Auflösungsvermögens, höchst empfindlicher Detektoren, hoch gezüchteter Elektronik zur Kompensation des Untergrundes und von Interelementeffekten, weitgehender Computerisierung mit angeschlossener anspruchsvoller Datenauswertung und Automatisierung, bei vergleichbarem Zeit-, Betriebsmittel- und Personalaufwand zehn bis achtzig (!) Elemente nebeneinander bestimmen. Eine entsprechende Monoelementmethode bewältigt aus der gleichen Probenmenge im allgemeinen jeweils nur ein einzelnes Element.

Es liegt auf der Hand, daß dies einen Umbruch sowohl der Forschungsstrategien als auch der Organisation und Einrichtung des Analysenlaboratoriums bewirkt. Dem Auftraggeber aus Naturwissenschaft, Medizin, Technik und Umweltschutz, der zunächst nur ein oder einzelne wenige Elemente in wertvollen, schwer zugänglichen oder vorzubereitenden Proben bestimmt haben möchte, werden jetzt bei nur wenig größerem Personalaufwand 10, 20 und mehr Elementgehalte angeboten. Das muß zu einem Überdenken der Strategie des Forschungsprojektes führen. In gleichwertiger Partnerschaft mit dem Analytiker können die zusätzlichen Elementgehalte sinnvoll in die Untersuchung einbezogen werden. In der Massenspektrometrie wird für metallische Proben der Traum des Analytikers wahr, „fast alle Elemente in einer Probe" oberhalb von 0,1 bis 1 atom-ppm in einem Meßgang zu erhalten. Bei zu geringen Gehalten werden zumindest die Nachweisgrenzen angegeben. Im Einzelfall ist zu diskutieren, ob die relativ geringen Mehrkosten für die Multielementanalyse deren Vorteile aufwiegen oder nicht. Ausgangspunkt sollte dabei aber nicht das Forschungsprojekt in seiner ursprünglichen Form, sondern in der durch Hinzunahme weiterer Elemente erweiterten Fassung sein.

Es darf allerdings nicht übersehen werden, daß jede einzelne Multielementmethode meistens nur für eine bestimmte Gruppe von Elementen hinsichtlich Nachweisstärke, Störungsfreiheit und Bestimmbarkeit optimale Eigenschaften besitzt. Sie bedeutet immer einen Kompromiß bei der Auswahl optimaler Versuchsparameter, welche alle zu bestimmenden Elemente berücksichtigen muß. Es liegt in der Natur eines solchen Kompromisses, daß man mit der gleichen Methode für nur ein einzelnes zu bestimmendes Element gegebenenfalls günstigere Parameter finden und damit Nachweisstärke, Reproduzierbarkeit und Richtigkeit verbessern könnte. Es wird außerdem der Einsatz mehrerer, sich ergänzender Multielementmethoden sinnvoll. Sie ermöglichen die Analyse möglichst vieler Elemente nebeneinander mit einer mittleren erreichbaren Nachweisstärke und Richtigkeit. Ein Beispiel hierfür findet sich auf dem Umschlag dieses Buches. Dort handelt es sich um die Instrumentelle Multielementanalyse einer uranhaltigen Bodenprobe.

Weniger offenkundig ist, daß ein vollautomatisiertes Großgerät zwar häufig durch einen Techniker betrieben werden kann, aber andererseits Überwachung, Probenvorbereitung, Eichung und Qualitätskontrolle der erhaltenen Analysenergebnisse, insbesondere im Spurenbereich, an den Analytiker höhere Anforderungen stellen als bisher. Das trifft insbesondere für die Beherrschung von spektralen Interferenzen und Matrixstörungen zu.

Dieser Gesichtspunkt wird oft übersehen, da moderne Großgeräte dank weitgehender Computerisierung und Automation immer irgendwelche Elementgehalte, Kurven oder schöne Diagramme liefern; es fragt sich nur, ob es die richtigen sind. Der Analytiker ist heute als Wissenschaftler noch mehr als früher von der bedauerten „Dienstmagd" zum echten Partner des Forschers aufgerückt.

Die Zentralabteilung für Chemische Analysen betreibt in der Kernforschungsanlage Jülich einen Chemischen Analysendienst für anspruchsvolle Grundlagenforschung und Technik. Sie verfügt zur Analyse von Elementen und Radionukliden über eines der umfangreichsten Methodenspektren. Es konnte daher seit Jahren das Zusammenwirken verschiedenartiger Multielementmethoden studiert werden. Im Beispiel des erwähnten Umschlagbildes zu diesem Band waren es Funkenmassenspektrometrie, Bogen- und ICP-Atomemissionsspektrometrie, Neutronenaktivierungsanalyse und Röntgenfluoreszenzanalyse im Vergleich zur Atomabsorptionsspektrometrie als Monoelementmethode. Thema und Durchführung des Symposiums gingen von der Erfahrung aus, daß der durch diese modernen Großgeräte der Multielementanalyse bewirkte Umbruch bei Auftraggebern und Administration häufig noch nicht erkannt und ins Bewußtsein aufgenommen worden ist.

Bei der Rechtfertigung der Existenz Analytischer Zentrallaboratorien und bei der Entscheidung über die Notwendigkeit der Anschaffung teurer Großgeräte tauchen immer wieder als Gegenargumente die Fragen auf: braucht man überhaupt ein so teures Großgerät; wenn schon eines, warum dann gleich mehrere; kann man sich nicht auf die Analyse nur eines Teiles aller Elemente des Periodensystems beschränken und den anderen Teil nach außerhalb in Auftrag geben; soll man gleichbleibende Routineaufträge, die von einem Analysenautomaten lösbar sind, nicht besser vor Ort ausführen und zentral nur schwierige Einzelanalysen, die dezentral nicht lösbar sind? Auch in der Forschungspolitik wird bei der Bewilligung von Großgeräten wegen der Höhe der in Frage stehenden Geldbeträge eine Rechtfertigung gefordert.

Das über Erwarten große Interesse am Symposium, das sich in über 120 Vortragsanmeldungen sowie 320 Teilnehmern äußerte, unterstreicht die Aktualität des gewählten Themas, welches erstmals den Begriff „Instrumentelle Multielementanalyse" zusammenfaßt. In einer angeschlossenen Industrieausstellung zeigten 25 Gerätehersteller zum Teil neueste Entwicklungen von Multielement-Analysengeräten. Diese sind im Anhang abgebildet und kurz beschrieben. Erstmals konnten aus gezielten Spenden der Aussteller Stipendien für Aufenthaltskosten an ausgewählte Doktoranden und Diplomanden vergeben werden.

Der vorliegende Band ist mehr als ein normaler Symposiumsband. Das Gebiet war durch 46 eingeladene Vorträge mehr oder weniger systematisch abgedeckt worden. Es fehlen allerdings das Manuskript mit dem Überblick über die ICP-Spektrometrie sowie über die Multielementanalyse in der Werk- und Reinststofforschung. Hier sei deshalb auf Übersichtsartikel in der Originalliteratur verwiesen.

Die folgenden Beiträge behandeln als Analysenmethoden die Funken-, Laser-, ICP-Plasma- und Sekundärionen-Massenspektrometrie sowie die Ionenmikrosonde; wellenlängen- und energiedispersive, totalreflektierende, protonen- und teilcheninduzierte Röntgenfluoreszenzanalyse; Funken-, Bogen-, Glimmentladungs-, ICP- und MIP-Plasma-Atomemissionsspektrometrie; Atomabsorptions- und ICP-Plasma-Atomfluoreszenzanalyse; Voltammetrie; Aktivierungsanalyse mit Reaktor-, epithermischen, thermischen, 14 MeV-Neutronen und geladenen Teilchen, Monostandard- und Indikator-Neutronenaktivierungsanalyse; hochauflösende Gamma- und Alpha-Spektrometrie. Besondere Schwerpunkte bilden die ICP-Atomemissions-, ICP-Atomfluoreszenzanalyse und ICP-Massenspektrometrie. Neu sich abzeichnende Methoden sind außer der bereits kommerziell gewordenen ICP-Massenspektrometrie, Atomfluoreszenz- und totalreflektierenden Röntgenfluoreszenzanalyse die Vorwärtsstreuung sowie die protoneninduzierte Röntgenfluoreszenzanalyse mit Synchrotronstrahlung.

Anwendungsgebiete sind vor allem die Geochemie, Kosmochemie, Lagerstättenkunde, Archäometrie, Bildende Kunst, Gläser, Reinst- und Werkstoffe, Elektronikmaterialien, Wasser- und Abwasser, Umwelt-, Lebens- und Lebensmittelwissenschaften, hoch radioaktive Materialien und radioaktive Abfälle.

Das Symposium „Instrumentelle Multi-Element-Analyse" wurde vom Arbeitskreis Mikro- und Spurenanalyse von Elementen (A.M.S.E.L.) in der Fachgruppe Analytische Chemie der Gesellschaft Deutscher Chemiker (GDCh) veranstaltet, gemeinsam mit dem GDCh-Arbeitskreis Radioanalytik, der Arbeitsgemeinschaft zur Förderung der Radionuklidtechnik (AFR), dem Verein Deutscher Eisenhüttenleute (VDEh), der Arbeitsgemeinschaft Massenspektrometrie der DPG, GDCh, und Deutschen Bunsengesellschaft für Physikalische Chemie sowie der Sektion Geochemie der Deutschen Mineralogischen Gesellschaft. Es wurde gemeinsam mit dem 11. Seminar für Aktivierungsanalyse vom 2. bis 5. April 1984 in der Kernforschungsanlage Jülich GmbH veranstaltet.

Mitglieder des Wissenschaftlichen Komitees waren B. Sansoni, Jülich (Vorsitzender); K. Bächmann, Darmstadt; M. Grasserbauer, Wien; B. Grauert, Münster; K.H. Koch, Dortmund; K. Laqua, Dortmund; F. Lux, München; J. Möller, Berlin; H.W. Nürnberg, Jülich; K.H. Schmitz, Duisburg; G. Tölg, Dortmund; K.H. Wedepohl, Göttingen. Das Ortskomitee bestand aus B. Sansoni, H. Heckner, U. Herpers, B. Krahl-Urban (Tagungsbüro der KFA), H. Petri, L. Radermacher und G. Wolff. Die Textverarbeitung lag in Händen von Frau A. Schorn, Hilfe beim Lesen der Korrekturen leisteten C. Freiburg, H. Heckner, H. Petri, L. Radermacher, G. Wolff, Frau R. Kurth und Frau A. Sansoni. Die Verantwortung für die Originalbeiträge liegt bei den Autoren, die ihre Beiträge druckfertig geliefert haben.

Als damaliger Vorsitzender des veranstaltenden Arbeitskreises AMSEL dankt der Herausgeber dem Vorstand der Kernforschungsanlage Jülich GmbH, insbesondere den Herren Dr. R. Theenhaus und A.W. Plattenteich, für die Förderung von Tagung und Geräteausstellung.

Jülich, im Juni 1985 *Bruno Sansoni*

Inhalt

Vorwort VII
Autorenverzeichnis XVII

EINFÜHRUNG 1

Einführung in die Instrumentelle Multielementanalyse 3
B. Sansoni, Jülich

Selected opportunities in analytical chemistry 57
George H. Morrison, Ithaca, USA

Einige physikalische Betrachtungen zur zweckmäßigen Auswahl von
Multi-Element-Methoden 65
K. Laqua, Dortmund

Instrumentelle Multielementbestimmung in der täglichen Praxis 75
K. Ohls, Dortmund

BESTIMMUNGSMETHODEN 85

KERNSTRAHLUNGSSPEKTROMETRIE 87

Review of the present state of gamma-spectrometry for multi-radionuclide analysis 87
U. Herpers, Köln

Multi-element-analysis by means of alpha particle spectrometry 98
M. Helmbold, Jülich

AKTIVIERUNGSANALYSE 109

Review on the monostandard method in multielement neutron activation analysis 109
A. Alian, B. Sansoni, Jülich

Comparison of different neutron activation analysis methods for multielement analysis
of geological material 123
A. Alian, R.G. Djingova, B. Sansoni, Jülich

Single comparator method in 14 MeV neutron activation analysis 133
J. Janczyszyn, Cracov

Simultaneous determination of the halogens Cl, Br, I in biological materials with
epithermal neutron activation; results and comparison with conventional methods 137
A. Wyttenbach, L. Tobler, V. Furrer, Würenlingen

The ratio method as the basis for a prediction-optimization INAA- program 141
R. Gwozdz, Copenhagen

Detection limits in multielement neutron activation analysis – some observations on
signal-to-noise ratio and counting statistics 145
M. Sankar Das, S. Yegnasubramanian, Bombay

Minimierung der Blindwerte bei der Instrumentellen Multi-Element-Bestimmung in
biologischen Proben 153
F. Lux, T. Bereznai, S. Knobl, München

Reduction of the background in the γ-spectra of neutron activated biological samples
by separation of ^{32}P 163
S. Knobl, F. Lux, München

A simple dead time correction for routine instrumental neutron activation analysis of
short-lived isotopes 169
A. Faanhof, Pretoria

Analytical investigations using the 14 MeV neutron activation facility Korona 173
R. Pepelnik, B. Anders, E. Bössow, H.-U. Fanger, Geesthacht, Hamburg

Indicator activation with a 14 MeV neutron generator 177
V. Cercasov, Stuttgart

Oberflächenprobleme bei Aktivierungsanalysen mit geladenen Teilchen 179
H. Weniger, K. Bethge, G. Wolf, Frankfurt am Main

Some remarks on volatility losses of several sample components during activation
with high energy Bremsstrahlung 183
C. Segebade, Berlin

MASSENSPEKTROMETRIE 185

Massenspektrometrie zur Multi-Element-Analyse in Festkörpern. Möglichkeiten
durch optimale Wahl der Ionisierungsmethode 185
H.E. Beske, Jülich

High precision spark source mass spectrometry by multielement isotope dilution 195
K.P. Jochum, Mainz

Laser mass spectroscopy, a useful instrument for the multi-element analysis of solids ... 201
A.W. Witmer, Eindhoven

In situ multi element trace analysis with SIMS 211
M. Grasserbauer, Wien

ICP-MASSENSPEKTROMETRIE 227

The potential of ICP source mass spectrometry 227
A.L. Gray, Guildford, England

RÖNTGENFLUORESZENZANALYSE 237

Advances in X-ray fluorescence analysis of soft and ultrasoft X-rays 237
T. Arai, Osaka

Die Röntgenbox – RFA im Rasterelektronenmikroskop 255
R. Eckert, Stuttgart

Energy dispersive X-ray fluorescence analysis with multiple total reflection – an
improvement of detection limits 257
K. Freitag, Ahrensburg

Applications of total reflection X-ray fluorescence in multi-element analysis 269
W. Michaelis, A. Prange, J. Knoth, Geesthacht, Hamburg

PIXE analysis – physical basis and examples of applications 291
B. Gonsior, M. Roth, Bochum

Röntgenfluoreszenzanalyse mit Hilfe der Synchrotronstrahlung (SYRFA) 301
P. Ketelsen, A. Knöchel, W. Petersen, G. Tolkiehn, Hamburg

ATOMEMISSIONSSPEKTROMETRIE 311

High-resolution ICP spectroscopy using a computer-controlled Echelle spectrometer
with predisperser in parallel slit arrangement 311
P.W.J.M. Boumans, J.J.A.M. Vrakking, Eindhoven

Low-consumption ICP emission spectrometry 329
L. de Galan, Delft

Spectral interferences and matrix effects in optical emission spectroscopy 337
J.A.C. Broekaert, Dortmund

Sample introduction, signal generation and noise characteristics for argon
inductively-coupled plasma optical emission spectroscopy 347
G.F. Kirkbright, Manchester

A contribution to the direct analysis of solid samples by spark erosion combined
to ICP-OES 359
J.A.C. Broekaert, F. Leis, K. Laqua, Dortmund

Three-filament and hollow-cylinder microwave induced plasmas as excitation
sources for emission spectrometric trace analysis of solutions and gaseous samples ... 363
D. Kollotzek, P. Tschöpel, G. Tölg, Schwäbisch Gmünd

Erfahrungen mit der Lichtleiterkopplung bei einem ICP-Spektrometer 373
A. Golloch, H.-M. Kuß, R. Rütjes, K.H. Schmitz, Duisburg

Development of an optimization criterion for ICP atomic emission spectrometers 377
H. Friege, P. Werner, Düsseldorf

Simultane oder sequentielle Messungen mit der ICP-Atomemissionsspektrometrie? 381
G. Drews, Mainz

VORWÄRTSSTREUUNG 385

Application of forward scattering to simultaneous multielement determination 385
H. Debus, S. Ganz, W. Hanle, G. Hermann, A. Scharmann, Giessen

Intensity and spectral distribution of the resonance radiation of sodium in forward
scattering measured with a cw dye laser 391
S. Ganz, M. Gross, W. Hanle, G. Hermann, A. Scharmann, Giessen

ATOMFLUORESZENZSPEKTROMETRIE 397

Recent developments and applications in hollow cathode lamp-excited ICP atomic
fluorescence spectrometry 397
D.R. Demers, Bedford, *E.B.M. Jansen*, Zoeterwoude

ATOMABSORPTIONSSPEKTROMETRIE 411

One set of conditions: Prerequisite for multi-element determination using
electrothermal atomization 411
G. Schlemmer, B. Welz, Überlingen

VOLTAMMETRIE 415

Potentialities of voltammetry in environmental oligo-element analysis of trace metals 415
H.W. Nürnberg

IONENCHROMATOGRAPHIE 445

Multi-Element-Analyse mittels Ionen-Chromatographie 445
G. Schwedt, Stuttgart, *B. Rössner, Da-ren Yan*, Göttingen

SONSTIGE ANALYSENSCHRITTE 455

VORKONZENTRIERUNG 457

Chemische Multi-Elementanreicherung – Probleme der Anpassung an Probenmaterial
und Bestimmungsmethode 457
E. Jackwerth, Bochum

Anreicherung von Seltenen Erden, Uran und Thorium und Bestimmung durch
Röntgenfluoreszenzanalyse 485
G. Hartmann, B. Sarx, H. Klenk, K. Bächmann, Darmstadt

STANDARDS 491

Multi-Element Standards 491
O. Suschny, Seibersdorf

Multi-element standards from oxide powders for sequential X-ray spectrometers 501
C. Freiburg, W. Reichert, Jülich, *A. Solomah*, Pinawa Manitoba

DATENVERARBEITUNG UND -BEURTEILUNG 505

Datenbeurteilung in der industriellen Multielementanalytik 505
K.-H. Koch, Dortmund

Analytische und geochemische Kontrolle der Multielementanalytik geologischen
Materials auf statistischer Grundlage 517
D. Sauer, Wien

Generation of accuracy in multielement systems by reconstitution of the sample (II) 519
G. Staats, Dillingen

Multi-element trace analysis of geothermal waters: Problems, characteristics
and applicability 523
R. Vandelannoote, W. Blommaert, L. Van't dack, R. Van Grieken, R. Gijbels, Antwerpen

Interdependence of selectivity and precision in multielement analysis 529
M. Otto, Freiberg, W. Wegscheider, Graz

Ein neues Datenbank-, Informations- und Auswertungssystem Chemometrie 531
J. Bürstenbinder, Berlin

ANWENDUNGEN 533

GEOWISSENSCHAFTEN 535

Vor- und Nachteile der ICP-Atomemissionsspektroskopie und der Atomabsorptionsspektroskopie bei der Analyse geochemischer und biologischer Materialien 535
H. Heinrichs, H.J. Brumsack, K.H. Wedepohl, Göttingen

Möglichkeiten der Instrumentellen Multi-Element-Analyse in der Wasserchemie 539
K.-E. Quentin, München

Analyse von Seltenerdelement-Mineralen mit Hilfe der ICP-OES 549
J. Luck, Berlin

Verfahren zur Bestimmung der Seltenen Erden (SE) in geologischen Matrizes mit
Hilfe der ICP-OES... 551
J. Erzinger, Giessen

Determination of antimony, indium, rhenium, selenium, tellurium and tin in
geochemical samples by RNAA 553
J. Nonaka, Mainz

Feasibility of beta-ray spectrometry in INAA: Applications in geo- and cosmochemistry .. 558
G. Weckwerth, Mainz

Optimierung der Multielementanalytik geologischen Materials in Pulverform für
wellenlängendispersive Röntgenfluoreszenz 563
N. Müller, Wien

Multi-Spurenelement-Analyse in Gneisen und deren Schwermineralfraktionen
mittels INAA 565
W. Kiesl, F. Kluger, Wien

Optimierung des ICP-OES-Spektrometers für die Anwendung auf geochemische
Multielementanalysenverfahren 570
P. Dolezel, Wien

Instrumental neutron activation analysis of small spheres of various origins 573
H.T. Millard Jr., Denver, P. Englert, U. Herpers, Köln

UMWELTFORSCHUNG 577

Application of multi-elemental neutron activation analysis in environmental research 577
R. Dams, Gent

Instrumentelle Analyse von Luftstaub durch Aktivierung mit Photonen und
Photoneutronen 594
B.F. Schmitt, C. Segebade, Berlin

Fly ash of a waste incineration facility as a reference material for instrumental
multielementanalysis 596
C. Segebade, B.F. Schmitt, Berlin

Multi-element analysis of single dust grains in the µg-mass range 598
K. Thiel, J. Peters, Köln, W. Schröder, Jülich

MATERIALWISSENSCHAFTEN 603

Neutron activation analysis for the modern electronics industry 603
M.L. Verheijke, J. Hanssen, H. Jaspers, L. Steuten, P. Wijnen, Eindhoven

LEBENS- UND LEBENSMITTELWISSENSCHAFTEN 607

Current aspects of multielement analysis in the life science 607
R. Michel, Köln, *G.V. Iyengar*, Jülich, *R. Zeisler*, Gaithersburg

Instrumentelle Multi-Element-Analyse in der Lebensmittelanalytik 621
G. Schwedt, Stuttgart

Simultane Multi-Element-Bestimmung in Wein durch
ICP-Plasma-Emissionsspektralanalyse 629
H. Eschnauer, H. Meierer, R. Neeb, Mainz

RADIO- UND KERNCHEMIE 635

Studies on the chemical nature of highly radioactive microparticles by
SEM-EDX, AES, ESCA 635
F. Baumgärtner, A. Huber, R. Henkelmann, München

Probleme bei der Analyse von simuliertem hochaktivem Waste (HAW) 641
W. Coerdt, F. Geyer, E. Mainka, H.G. Müller, S. Weis, Karlsruhe

Multi-element analysis of waste water using AES-ICP and XRF methods 643
P. Hoffmann, K.H. Lieser, R. Speer, R. Pätzold, Darmstadt

In-line determination of U, Np, and Pu in process streams by energy-dispersive XRF 645
P. Hoffmann, T. Hofmann, K.H. Lieser, Darmstadt

SONSTIGE ANWENDUNGEN 647

Einsatz eines simultanen ICP-Spektrometers im Analysendienst eines
Forschungszentrums 647
G. Wolff, H. Nickel, H. Lippert, Jülich

Instrumental analysis and provenance of archaeological artifacts 657
E. Pernicka, Heidelberg

Instrumentelle Element-Analyse – ein Werkzeug bei der Untersuchung von
Kunstwerken 667
F. Mairinger, Wien

Multi element analysis in archaeometry 669
J. Riederer, Berlin

METHODENVERGLEICH 675

Kombinierte Anwendung spektrochemischer Analysenmethoden bei der
Multielementanalyse geologischen Materials an Großserien 677
E. Schroll, D. Sauer, Wien

Intercomparison of the multielement analytical methods TXRF, NAA and ICP with
regard to trace element determinations in environmental samples 693
W. Michaelis, H.-U. Fanger, R. Niedergesäß, H. Schwenke, Geesthacht

Multi-Element-Analyse von vier geologischen Standard-Referenz-Proben mit Hilfe
der ICP-OES, NAS und SSMS 711
P. Dulski, J. Luck, W. Szacki, Berlin

Atomemissionsspektrometrie mit ICP- und DC-ARC-Anregung: Ein Vergleich 713
G. Drews, Mainz

Comparative analysis of natural rock samples by mass spectrometry,
X-ray fluorescence, atomic absorption spectrometry and gamma-ray spectrometry 717
B.T. Hansen, Münster, *F. Henjes-Kunst*, Karlsruhe, *A. Baumann*, Münster,
U. Jecht, Karlsruhe

ANHANG ... 721

Großgeräte zur Instrumentellen Multielementanalyse ... 723
L. Radermacher, B. Sansoni, Jülich

Abkürzungen für die Bezeichnung von Analysenmethoden ... 751
B. Sansoni, Jülich

Sachverzeichnis ... 768

Autorenverzeichnis

Alian, A. 109,123	Friege, H. 377	Klenk, H. 485
Anders, B. 173	Furrer, V. 137	Kluger, F. 565
Arai, T. 237		Knobl, S. 153,163
		Knöchel, A. 301
	de Galan, L. 329	Knoth, J. 269
Bächmann, K. 485	Ganz, S. 385,391	Koch, K.H. 505
Baumann, A. 717	Geyer, F. 641	Kollotzek, D. 363
Baumgärtner, F. 635	Gijbels, R. 523	Kuß, H.-M. 373
Bereznai, T. 153	Golloch, A. 373	
Beske, H.E. 185	Gonsior, B. 291	
Bethge, K. 179	Grasserbauer, M. 211	Laqua, K. 65,359
Blommaert, W. 523	Gray, A.L. 227	Leis, F. 359
Bössow, E. 173	van Grieken, R. 523	Lieser, K.H. 643,645
Boumans, P.W.J.M. 311	Gross, M. 391	Lippert, H. 647
Brumsack, H.J. 535	Gwozdz, R. 141	Luck, J. 549,711
Broekaert, J.A.C. 337,359		Lux, F. 153,163
Bürstenbinder, J. 531		
	Hanle, W. 385,391	
	Hansen, B.T. 717	Mainka, E. 641
Cercasov, V. 177	Hanssen, J. 603	Mairinger, F. 667
Coerdt, W. 641	Hartmann, G. 485	Meierer, H. 629
	Heinrichs, H. 535	Michaelis, W. 269,693
	Helmbold, M. 98	Michel, R. 607
van't Dack, L. 523	Henjes-Kunst, F. 717	Millard, H.T. 573
Dams, R. 577	Henkelmann, R. 635	Morrison, G.H. 57
Da-ren Yan 445	Hermann, G. 385,391	Müller, H.G. 641
Das, M.S. 145	Herpers, U. 87,573	Müller, N. 563
Debus, H. 385	Hoffmann, P. 643,645	
Demers, D.R. 397	Hofmann, Th. 645	
Djingova, R.G. 123	Huber, A. 635	Neeb, R. 629
Dolezel, P. 570		Nickel, H. 647
Drews, G. 381,713		Niedergesäß, R. 693
Dulski, P. 711	Iyengar, G.V. 607	Nonaka, J. 553
		Nürnberg, H.W. 415
Eckert, R. 255	Jackwerth, E. 457	
Englert, P. 573	Janczyszyn, J. 133	Ohls, K. 75
Erzinger, J. 551	Jansen, E.B.M. 397	Otto, M. 529
Eschnauer, H. 629	Jaspers, H. 603	
	Jecht, U. 717	
	Jochum, K.P. 195	Pätzold, R. 643
Faanhof, A. 169		Pepelnik, R. 173
Fanger, H.-U. 173,693		Pernicka, E. 657
Freiburg, C. 501	Ketelsen, P. 301	Peters, J. 598
Freitag, K. 257	Kiesl, W. 565	Petersen, W. 301
	Kirkbright, G.F. 347	Prange, A. 269

Quentin, K.-E.	539	

Radermacher, L.	723	
Reichert, W.	501	
Riederer, J.	669	
Rössner, B.	445	
Roth, M.	291	
Rütjes, R.	373	

Sansoni, B.	3, 109, 123, 723, 751
Sarx, B.	485
Sauer, D.	517, 677
Scharmann, A.	385, 391
Schlemmer, G.	411
Schmitt, B.F.	594, 596
Schmitz, K.H.	373
Schröder, W.	598
Schroll, E.	677

Schwedt, G.	445, 621
Schwenke, H.	693
Segebade, C.	183, 594, 596
Solomah, A.	501
Speer, R.	643
Staats, G.	519
Steuten, L.	603
Suschny, O.	491
Szacki, W.	711

Thiel, K.	598
Tobler, L.	137
Tölg, G.	363
Tolkien, G.	301
Tschöpel, P.	363

Vandelannoote, R.	523
Verheijke, M.L.	603
Vrakking, J.J.A.M.	311

Weckwerth, G.	558
Wegscheider, W.	529
Wedepohl, K.H.	535
Weis, S.	641
Welz, B.	411
Weniger, H.	179
Werner, P.	377
Wijnen, P.	603
Witmer, A.W.	201
Wolf, G.	179
Wolff, G.	647
Wyttenbach, A.	137

Yegnasubramanian, S.	145

Zeisler, R.	607

Einführung

EINFÜHRUNG
IN DIE INSTRUMENTELLE MULTIELEMENTANALYSE
Bruno Sansoni
Zentralabteilung für Chemische Analysen (ZCH)
Kernforschungsanlage Jülich GmbH
D-5170 Jülich 1

1. BEGRIFFSBILDUNG

Das Thema "Instrumentelle Multielementanalyse" (Instrumental multi element analysis) beinhaltet vier verschiedene Begriffe.

Analyse (analysis, assay) bedeutet soviel wie Chemische Analyse oder Materialanalyse. Der Zusatz "chemisch" weist nicht mehr wie früher auf die Herkunft der verwendeten Untersuchungsmethode hin. Er bezieht sich heute vielmehr auf die Ermittlung der "chemischen" Zusammensetzung der zu untersuchenden Materialprobe (sample). Die Analyse erlaubt es, entweder qualitativ die Art der einzelnen Materialkomponenten (components) zu erkennen oder deren Gehalte (pro Gewicht) (content) und Konzentrationen (pro Volumen) (concentration) zu bestimmen. Die Strukturanalyse ermittelt darüber hinaus die geometrische Anordnung und Art der chemischen Verknüpfung der Bausteine einer Materialprobe.

Der Gang einer chemischen Analyse (analytical scheme) besteht nach Abbildung 1 im allgemeinen und kompliziertesten Fall aus 11 bis 13 Einzelschritten (steps) (64, 61), von denen nur einer der Physik, zwei der Mechanik, fünf der Chemie und zwei bis drei der Angewandten Mathematik entnommen sind. Dabei ist die häufig ausschließlich betrachtete physikalische Messung des Analysensignales (23, 50, 80) nur einer dieser Teilschritte. Weiterhin wird häufig übersehen, daß dessen Hauptschwierigkeit weniger in der eigentlichen physikalischen Messung an sich, als vielmehr in der Umrechnung der gemessenen Einheiten des Analysensignales in die gesuchten Gehalte oder Konzentrationen, also im Problem der Kalibrierung (calibration) (36) liegt.

Es ist das strategische Konzept (strategy) (64, 61), so viele Teilschritte wie möglich zu vermeiden. Dies sind vor allem die zeitraubende und experimentelle Erfahrung erfordernde chemische Veraschung (65), Auflösung, Vorkonzentrierung (34), Trennung (49, 41) und Meßpräparateherstellung. Man wird also zunächst versuchen, die Analyse rein instrumentell durchzuführen. Erst wenn Schwierigkeiten auftreten, müssen chemische Schritte hinzugezogen werden. Vor allem bei Spurenanalysen in schwierigen Grundmaterialien (matrix), aber auch bei Störungen durch Begleitelemente (interference) ist dies häufig der Fall.

SCHRITTE EINER SPURENANALYSE VON ELEMENTEN UND RADIONUKLIDEN IN BELIEBIGEN MATERIALIEN

FORMULIERUNG EINER FRAGE	„Chemische" Analyse	„Instrumentelle" Analyse
1. Probenahme 2. Probenvorbereitung	① ②	① ②
3. Veraschung 4. Auflösung	③ ④	
5. Vorkonzentrierung 6. Trennung 7. Meßpräparateherstellung	⑤ ⑥ ⑦	(⑦)
8. Messung des Analysensignales – integral – spektral – Eichung	⑧	⑧
9. Datenverarbeitung 10. Datenbeurteilung 11. Qualitätskontrolle	⑨ ⑩ ⑪	⑨ ⑩ ⑪
BEANTWORTUNG DER FRAGE		

Abbildung 1:
Teilschritte des allgemeinen Analysenganges für alle Elemente in allen Materialien, nach (64, 61)

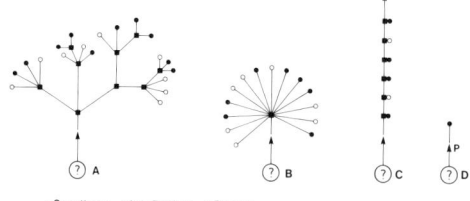

■ Operationen   ○ kein Ergebnis   ● Ergebnis

Abbildung 2:
Struktur chemischer Analysenmethoden, nach H. Kaiser (36)

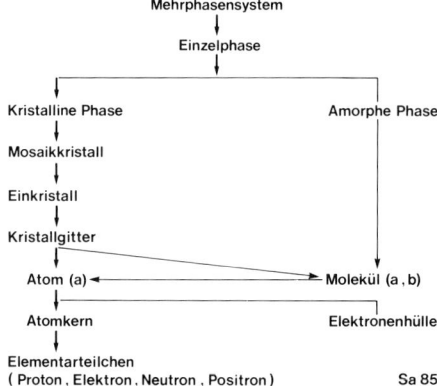

Abbildung 3: Baustufen der Materie
In Anlehnung an (42) erweitert.
(a) Nach Elektronenabgabe als Kation, nach Elektronenaufnahme als Anion
(b) Komplexe aus zentralem Metallkation und Liganden als Sonderfall von Molekülen

Elementanalyse bedeutet, daß es sich um den Nachweis oder die Gehaltsbestimmung der in der Probe enthaltenen chemischen Elemente handelt. Die übrigen Baustufen der Materie werden dabei nicht berücksichtigt. Der im angelsächsischen neuerdings verwendete Begriff "speciation" bezieht sich dagegen auf die zusätzliche Ermittlung der Art der Verknüpfung der gefundenen Elemente zu Molekülen, Atom- oder Molekülionen.

Multi- im Gegensatz zu "Mono"elementanalyse bringt zum Ausdruck, daß in einem Analysenschritt in der gleichen Probe jeweils nicht nur ein, sondern möglichst viele Elemente nebeneinander nachgewiesen oder bestimmt werden sollen. Es ist nicht festgelegt, ab wieviel Elementen man von "Multi"analyse spricht. Bei der Vorbereitung zu diesem Symposium wurde von einer unteren Grenze von zehn Elementen ausgegangen. Bei weniger Elementen kann man von einer "Oligo"elementanalyse sprechen. "Mono"elementanalyse bezieht sich auf nur ein einzelnes zu analysierendes Element. Zu den Oligoelementmethoden gehört die hier nur kurz behandelte Voltammetrie, Polarographie und gelegentlich die elektrochemisch oder photometrisch indizierte Titration. Ausgesprochene Monoelementmethoden sind die klassische Titration, Gravimetrie, Spektralphotometrie und Atomabsorptionsspektrometrie. In Sonderfällen der Atomabsorptionsspektrometrie lassen sich allerdings auch einige wenige Elemente oder Komponenten gleichzeitig bestimmen.

Instrumentell weist darauf hin, daß die Messung des Analysensignales mit physikalischen Instrumenten erfolgt und chemische Analysenschritte ganz oder zumindest weitgehend entfallen. Im Gegensatz dazu verwendet die "chemische" Analyse neben physikalischen bzw. instrumentellen zusätzlich auch chemische Methoden. Früher sprach man bei der Kombination von chemischen und physikalischen Methoden von "Verbundverfahren". Dieser Begriff erübrigt sich nach der allgemeinen Auffassung, welche Abbildung 1 zugrunde liegt. Der Begriff Verbundanalyse hatte sich historisch in einer Zeit entwickelt, in der man entweder nur chemische oder nur physikalische Analysenmethoden in entsprechend organisierten Laboratorien anwendete. Heute ist diese Unterscheidung nicht mehr erforderlich.

Zusammenfassend beschreibt das Thema Instrumentelle Multielementanalyse die qualitative und quantitative Analyse möglichst vieler chemischer Elemente nebeneinander in der gleichen Probe, mit physikalischen Methoden und unter weitgehendem Verzicht auf chemische Arbeitsverfahren. Diese sollen möglichst nur für eine etwaige Probenvorbereitung verwendet werden.

Bleibt eine feste oder flüssige Probe während der Analyse unverändert, so arbeitet die Methode "zerstörungsfrei" (non destructive). Erfolgt die

Multielementbestimmung gleichzeitig im selben Probenvolumen, so ist sie "simultan" (simultaneous), werden die Elemente hintereinander bestimmt, so wird sie als "sequentiell" (sequential) bezeichnet (Abbildung 2). Erwünscht sind simultane Methoden, jedoch kommen ihnen schnelle sequentielle sehr nahe. Allerdings verbrauchen sequentielle Methoden häufig größere Probenvolumina für die Bestimmung der gleichen Anzahl von Elementen.

Ausgesprochene Simultanmethoden sind die Funken-Massenspektrometrie (37, 12, 22, 20), klassische Flammen- (16), Bogen-, Funken- (47, 55, 68, 69, 70) und Plasma- (73) Emissionsspektralanalyse, energiedispersive Röntgenfluoreszenzanalyse (31, 72), sowie Gamma- und Alphaspektrometrie (5, 39) mit Vielkanalanalysatoren. Ausgesprochen sequentiell arbeiten dagegen die wellenlängendispersive Röntgenfluoreszenzanalyse (31, 72) in ihrer Normalausführung, die Atomabsorptionsspektrometrie (75, 77, 15, 38) mit mehreren Hohlkathodenlampen, ein Teil der ICP- Emissions- und der ICP-Atomfluoreszenzspektrometrie (38) sowie die Voltammetrie (44, 27, 54, 74, 53).

Die gleichzeitige Analyse von vielen Elementen nebeneinander haben die klassische Emissionsspektral-, die Röntgenfluoreszenz- und die gammaspektrometrische Neutronenaktivierungsanalyse (39, 19) schon seit langem ermöglicht. Daher sind Instrumentelle Multielementmethoden an sich nicht neu. Man kann aber erst seit dem Zusammentreffen mehrerer Neuentwicklungen der letzten Jahre von einem eigenen Schwerpunkt oder Arbeitsgebiet der Multielementanalyse sprechen.

2. DAS UNTERSUCHUNGSMATERIAL

Alle chemischen Stoffe dieser Erde und eines großen Teiles des Weltalls, auch Materie oder Material (material) genannt, lassen sich durch ihre "chemische" Zusammensetzung beschreiben und ausreichend charakterisieren. Die Zusammensetzung einer gegebenen Materialprobe ist vielschichtiger als allgemein angenommen. Der folgende Überblick hält sich an die verschiedenen Baustufen der Materie (42) (Abbildung 3).

Danach besteht die Mehrzahl aller chemischen Stoffe oder Materialien zunächst aus den unmittelbar mit dem Auge oder Lichtmikroskop wahrnehmbaren Mehrphasensystemen (multi phase system) (z.B. Granit). Diese setzen sich ihrerseits aus den Einzelphasen (phase) (Quarz, Feldspat, Glimmer) zusammen. Die Zahl reiner Phasen ist naturgemäß sehr viel kleiner als diejenige der daraus zusammengesetzten Kombinationen von Mischphasen. In reinen Phasen haben deren Oberflächen (surface) häufig andere physikalische, chemische und kristallographische Eigenschaften als der verbleibende Hauptanteil (bulk), trotz praktisch gleicher chemischer Bruttozusammensetzung. Deshalb werden beide getrennt betrachtet und analysiert. Man unterscheidet kristalline (crystalline) und amorphe (amorphous) Phasen. Die kristallinen Phasen bestehen aus Kristallgittern (crystal lattice), in denen sich Moleküle oder Atome unter bestimmten Symmetriebedingungen regelmäßig zusammengelagert haben. Die kristallinen Phasen setzen sich häufig aus Mosaikkristallbereichen zusammen, die ihrerseits Einkristallcharakter haben. Moleküle bestehen aus Atomen. Kristalline und amorphe Phasen unterscheiden sich wesentlich durch den Typ der chemischen Bindung bei der Verknüpfung der einzelnen Atomarten. Im Kristallgitter haben Atome und Moleküle häufig ein oder mehrere Elektronen abgegeben und bilden Atom- und Molekülionen als Bausteine. Die Ionenladungen ermöglichen aufgrund starker elektrostatischer Kräfte einen besonders festen Zusammenhalt im Gitter. Komplexionen entstehen durch Anlagerung von Ligandenmolekülen oder -anionen an zentrale Metallkationen und spielen eine besondere Rolle unter den Ionen.

Atome und Atomionen besitzen eine Elektronenhülle (electron sheath), die für das chemische Verhalten des Atoms maßgebend ist. Der Atomkern (atomic nucleus) ist Träger des Hauptteiles der Masse des Atomes. Atomkerne ihrerseits bauen sich aus Elementarteilchen (elementary particle) auf. Die für den Chemiker wichtigsten sind Protonen, Elektronen, Neutronen und Positronen. Mit ihnen hat der Analytiker jedoch nur im Falle der von den schwersten und instabilen Atomarten ausgesandten radioaktiven Strahlung zu tun oder bei der Aktivierungsanalyse mit geladenen Teilchen (charged particle).

Aus nur einer Atomart aufgebaute reine Phasen nennt man chemische Elemente (element), aus mehreren Atomarten oder Molekülen zusammengesetzte chemische Verbindungen (compound).

Die nachfolgenden Aufsätze behandeln Analysenmethoden für Elemente und betrachten daher vornehmlich nur die Baustufen Atom und dessen Elektronenhülle. Es ist wichtig darauf hinzuweisen, daß zur vollständigen Charakterisierung eines Materials darüber hinaus noch die Kenntnis der höher organisierten Baustufen, manchmal außerdem auch der darunter liegenden des Atomkernes (Isotope) notwendig ist. Hierüber sagt die Elementanalyse im allgemeinen nichts aus. Das chemische, biochemische und ökologische Verhalten von Metallen kann von deren Einbindung in Ionen, Moleküle, Komplexionen, Molekülionen, Phasen oder Mischphasen wesentlich stärker bestimmt werden als von der Art des Elementes selbst. Daher erlangen Analysenmethoden zur Untersuchung dieser Baustufen zunehmende Bedeutung. Sie werden hier nicht behandelt.

Trotzdem tragen Elementanalysen auch heute noch den größten Anteil zur Charakterisierung verschiedenartigster Materialien bei.

Vereinfachend ist, daß nicht alle chemischen Stoffe Mehrphasensysteme, sondern oft nur reine Phasen sind, die aus einer einzigen Atom- oder Molekülart bestehen (Gold, Schwefel, Zucker, Feldspat). Weiterhin kann man bei der Untersuchung eines Elementes durch Messung einer charakteristischen Eigenschaft der Elektronenhülle seines Atome gleichzeitig auch die Art des zugehörenden Atomkernes erkennen. Bei Konstanz der Isotopenverhältnisse eines Elementes in der Natur kennt man dann indirekt außerdem auch die zugehörenden Isotopengehalte.

Ein gegebenes Material enthält Haupt-, Neben- und Spurenbestandteile (major, minor, trace components). Bei Gehalten im Bereich von 100 bis 1 % spricht man von Haupt-, zwischen 1 und 0,1 bis 0,01 % von Neben- und unterhalb von 0,01 % von Spurenbestandteilen. Gehalte an Spurenbestandteilen gibt man häufig in "Teilen pro Million" (ppm), "Teilen pro Milliarde" (ppb) und gelegentlich noch in "Teilen pro Trillion" (ppt) an. Es bedeuten dabei ppm = $\mu g/g$, ppb = ng/g, ppt = pg/g; 0,01 % sind 100 ppm.

Tabelle 1 enthält als Beispiel für die Vielfalt der Materialproben des Analytikers eine Auswahl der im Laboratorium des Verfassers innerhalb weniger Jahre analysierten chemischen Stoffe. Die Analysenproben konnten, ohne Anspruch auf Vollständigkeit, grob in folgende Klassen eingeteilt werden: Reinstmetalle, Reinstlegierungen und verwandte Systeme; metallische, oxidische, silikatische und karbidische Werkstoffe; Chemikalien, Komplexsalze; Materialien für die Nukleartechnik, radioaktiver Abfall und Reaktorüberwachung; sonstige technische Betriebsprodukte; Gesteine, Minerale, Böden, Wässer, biomedizinisches Material und andere Umweltproben sowie sonstige Materialien.

Tabelle 2 gibt eine Zusammenstellung über die in diesen Materialien analysierten Komponenten.

Abbildung 4 bringt eine Übersicht über die Elementzusammensetzung ausgewählter wichtiger Materialien an Haupt-, Neben- und Spurenbestandteilen. Es handelt sich dabei teilweise um gut untersuchte Standardreferenzmaterialien, in denen besonders viele Elementgehalte zuverlässig analysiert worden sind. Bei weiterer Verfeinerung der Analysenmethoden dürften in den Gehaltsbereichen von unterhalb 10 ppb noch weitere Elemente gefunden werden. In der Geochemie besagt der Satz von der Allgegenwart der Elemente, daß bei ausreichend empfindlichen Analysenmethoden in geeigneten geologischen Proben nahezu alle stabilen chemischen Elemente des Periodensy-

Analysierte Materialien

Es wurden über 220 verschiedenartige Materialarten mit einer ungewöhnlich breiten Stoffauswahl untersucht.

1. **Reinstmetalle:** Ag, Al, Au, Cd, Cr, Cu, Fe, Gd, La, Li, Mg, Mo, Nb, Ni, Pd, Pr, Pt, Rb, Rh, Sb, Sm, Ta, Th, Ti, U, V, W, Yb, Zr.
2. **Reinstlegierungen und verwandte Systeme:** Ag-Al, Al-Ce-La, Al-Ce-Se, Al-Cu, Al-Cu-Mg, Al-Dy, Al-La-Tm, Al-Li, Al-Mg, Al-Mg-Ni, Al-Mg-Si, Al-Ni, Al-V, Al-Zn, Al-Sm, Al-Tm, Al-U, Al-Y, Au-Fe, Au-Ga, Au-Cu, Au-Ni, Be-Ti, Ca-Hg, Ce-P, Ce-Co, Ce-Cu-Si, Ce-La-Pd, Ce-Pd, Ce-Pr, Co-P, Co-Pt, Co-Fe-Mo, Cr-Fe-Mn, Cs-Mn, Cu-Ga, Cu-Mn, Cu-Ni, Cu-Pd, Cu-Te, Er-La-Pd, Er-Pd, Eu-Pd-Y, Eu-Sr-O, Eu-Sr-S, Fe-Ni, Fe-P, Fe-Pd, Fe-Ti, Gd-S, Hg-In, Mg-Al, Ho-Pd-Si, Ho-Sn, La-Pd, Li-Pb, Mg-Ni, Mo-Pr, Mo-S, Mo-Te, Mo-Zr, Nb-V-H, Ni-Pt, Ni-Si, Ni-Sr-Y, Ni-Zn, Pd-Tb, Pd-Tm-Y, Pd-Y, Sn-Y, Zr-Zr.
3. **Metallische Werkstoffe:** Ag-beschichtetes Al, Amalgame, Bleilegierungen, Chromnickelstähle, Edelstähle, Hochtemperaturlegierungen, Incoloy, Molybdänbasislegierungen, Monel, Nickelbasislegierungen, niedrig legierte und synthetische Stähle, Stahlgranulate, Ra- neynickel, TZM, UA-Stähle, Zircaloy, Zirkon. - Cadmiumbleche, Eisenspäne, korrodierte Telefonkabel, Kupferfolie, Messingteile, Metallglasproben, Metallspäne, Metallfolien, Nickelblech, Niobpulver, Reintitan-Werkstoffteile, Schweißnähte, Schweißstähle, Targetmaterialien aus Massenseparator auf Xyklotron, Thermoelemente, Turbinenschaufelmaterial, Titannickel-Sinterstoffe, Ventile, Verzunderungsschichten, Woodsche Legierung, Zahnräder.
4. **Oxidische Werkstoffe:** Al-Oxid, Aluminium, Aluminiumoxidhydrate, Aluminium-Zirkonium-Mischoxide, Abscheidungen auf AUR-Brennelementkugeln, Berylliumoxid, Calziumoxid, Eisenoxidhydrat, Europiumoxid, Keramiken auf Aluminiumoxid-Basis, Lithiumaluminat, Magnesiumoxidsteine, Nickelchloridkatalysatoren, Titandioxid, Titandioxidhydrat, Strontiumaluminate, Strontiumtitanate, simulierte Wasserglaser, Zirkonoxideinkristalle und Pulver.
5. **Silikatische Werkstoffe:** Bruchstücke von Diaphragmen, Siliziumdioxid, Strontiumalumosilikate.
6. **Karbidische Werkstoffe:** Chromkarbid, Siliziumkarbid.
7. **Chemikalien:** Ammoniumnitrat, Bariumbromid, Bariumkarbonat, Borsäurepräparate, Calzium[fluorid, Kaliumkarbonat, Kupfer]sulfat, Lithiumkarbonat, Ruteniumnitrat, verschiedene hochreine Säuren und Reagenzien, Silbernitrat, Zinn-, und Zinn-Rheniumkomplexsalze. Atzmittel [für Elektrotechnik, Beizspülwasser, Cyanidlaugen, Elektrolytlösungen, Zinkfarben.
8. **Komplexsalze:** $2Ce(NO_3)_2 \cdot 3Mg(NO_3)_2 \cdot 24H_2O$, K_2SnCl_6, $(NH_4)_2TiF_6$, $(NH_4)_2SnCl_6$, $Ni(NH_3)_6J_2$.
9. **Brennelemententwicklung:** Abscheidungen auf AUR-Brennelementkugeln, Bornitrid in Graphitkugeln, Carburierungsmittel, coated particles, Graphitmaterial, Kernbrenn[lösung], Reaktorgraphit, Spaltproduktgemische.
10. **Radioaktiver Abfall:** Brennelementeluate, Rückstände auf Walzentrocknern, simulierte Wastegläser und deren Eluate, Wastelösungen, Abscheidungen von Wasserjveraigung, Gasproben aus Strahlrohr FRJ 2.
11. **Reaktorüberwachung FRJ-1 und FRJ-2:** Leichtwasserüberwachung (AVR) 1 und -2 (Primär- und Sekundärkreislauf, Absetzblock, Beckenanlage, Hilfskreisläufe, Rückkühlkreislauf), Wasserauf[bereitungsanlage], Schwerwasserüberwachung (Primärkreislauf), He- liumkondensat), Fortluft.
12. **Sonstige technische Betriebsprodukte:** Ablagerungen aus NFE-Versuchsreaktoren, - auf Klima[filtern - auf Magnetankern von Relais, - auf PVC-Schlauch (ZFR), - auf Thermoelementen und Wärmeaustauschern (AVR); Abriebe; Filterrückstände, Kaowool, Molybdän- Ölproben vom HHU-Störfall; Kathoden- und Anodenraum technischer Elektrolysierzellen; Beläge auf Deckel, Graphitfilz, Zirkonoxidkugeln und Pulver von HT-Versuchsstand; Berylliumoxidrückstände auf Graphitkugeln, Deckel und Rezipientenwandung (IRB); Betonproben, Harzabscheidungen, Glasfaser- und Kunststoff[beschichtung] (AP); Katalysatoren verschiedener Arten; Kohle; Rückstände von Produktwasser EVA; Ru-103-Standards; Schlackenp[lastersteln]; Schlacken; Ventile; Wasserkreislauf[rückstände]; Wasserproben aus Zyklotron, ZnSn[salfe]; Si-Scheiben, Abwasserrückstand auf Filter, Behälter, Dichtungen, Düngemittel, Druckfarben, Eiführmasse [für Rohrölen, Elektrolyt, Feuerlöschpulver, Formteile, Graphit, Hautschutzcremes, Ionenaustauscher (anorganische und organische), Lösungsmittel, Metallschutzspray, Metallspritzpulver, Methanol, Molybdänchalkogenide, Nylon, Polyacetylen, Polypropylen- und Stahlgewebe, Polyvinylalkohol, Schwereester, Steinwolle, Turbinenöl, Getrocknete Farbstoffe, Glührückstände, Katalysatorabrieb, Korrosionsprodukte, Metallkatalysatoren, Verascnungsrückstände und Aufschlußlösung von Papierherstellung, Ölrückstände, Produktgas (EVA), Siliconschichten auf Beschichtungspapier, Wischproben -.
13. **Gesteine, Minerale, Böden, Wässer:** Cordierit, Ilmenit, Magnetit, Muskovit, Nontronit, Pyrit-Andesit, Braunkohle, Dazit, Erze, Granitite, Ölschiefer, Posidonienschiefer, Tone; verschiedenartige Böden, Bodenextrakte, Fluß- und Meeressedimente, Gletschereis, Meersalz, synthetisches Meer- und Flußwasser, Salzlauge.
14. **Biologisches Material:** Algen, Blutserum, Cellulose mit Mikroorganismen, Fingernägel, Fleisch, Haare, Hirn, Humisäuresalze, Leber, Nieren, Nierensteine, Tinten[aschgen, Urin, Vollblut, Zähne, Zahnstein.
15. **Sonstige Umweltproben:** Braunkohlenasche, Fischaschen, Flugstaub, Getreide, Gras, Klärschlamm, Luftstaub, Meersalz, Pflanzenaschen, Sedimente, Steinkohleasche, Wasserrückstände.

Tabelle 1: Beispiele für Analysenproben aus dem Analysendienst eines Forschungszentrums
In der ZCH innerhalb von drei Jahren im Analysendienst analysierte Proben

Analysierte Komponenten

85 von 89 natürlich vorkommenden Elementen

H, He, Li, Be, B, C, N, O, F, Ne, Mg, Al, Si, P, S, Cl, Ar, K, Ca, Sc, Ti, V, Cr, Mn, Fe, Co, Ni, Cu, Zn, Ga, Ge, As, Se, Br, Kr, Rb, Sr, Y, Zr, Nb, Mo, Ru, Rh, Pd, Ag, Cd, In, Sn, Sb, Te, I, Xe, Cs, Ba, La, Ce, Pr, Nd, Sm, Eu, Gd, Tb, Dy, Ho, Er, Tm, Yb, Lu, Hf, Ta, W, Re, Os, Ir, Pt, Au, Hg, Tl, Pb, Bi, Po, Rn, Ra, Th, U. Die fehlenden vier Elemente sind ebenfalls bestimmbar

146 von über 2200 bekannten Radionukliden:

H-3, Be-7, C-11, F-18, Na-24, Mg-27, Al-28, Si-31, Cl-38, Ar-41, K-40, K-42, Ca-47, Ca-49, Sc-46, Sc-47, Ti-51, V-49, V-52, Cr-51, Mn-54, Mn-56, Fe-59, Co-55, Co-56, Co-58, Co-60, Ni-57, Ni-65, Cu-61, Cu-64, Zn-65, Zn-69m, Ga-66, Ga-67, As-76, Se-75, Br-80, Br-82, Kr-79, Rb-86, Sr-85, Zr-95, Zr-97, Nb-95, Nb-97, Mo-93m, Mo-99, Tc-95, Tc-96, Tc-99, Ru-103, Rh-105, Pd-109, Ag-110m, Cd-115, Cd-117, Sn-113, In-114m, Sb-122, Sb-124, Sb-125, Te-131, I-125, I-131, I-133, Xe-125, Xe-133, Xe-135, Cs-134, Cs-136, Cs-137, Ba-131, Ba-133, Ba-137m, Ba-139, Ba-140, La-140, Ce-141, Ce-143, Ce-144, Pr-142, Pr-144, Nd-147, Sm-145, Sm-153, Eu-152, Eu-154, Eu-155, Gd-153, Gd-159, Tb-160, Dy-165, Ho-166, Tm-170, Yb-169, Yb-175, Yb-177, Lu-177, Lu-176m, Lu-177, Hf-175, Hf-181, Ta-182, W-181, W-187, Hf-192, Pt-197, Pt-199, Au-198, Au-199, Hg-197, Hg-197m, Hg-203, Tl-208, Pb-210, Pb-212, Po-214, Bi-214, Po-210, Pa-216, Rn-220, Ra-224, Ra-228, Th-228, Th-233, Pa-233, U-234, U-236, U-238, Np-239, Pu-239, Pu-240, Pu-242, Am-241, Cm-242, Cm-244.

Darüber hinaus können die meisten übrigen Gamma- und Alphastrahler in Konzentrationen oberhalb der Bestimmungsgrenze gemessen werden

261 von etwa 287 bekannten stabilen Isotopen:

H-1, H-2, Be-9, B-10, B-11, C-12, C-13, N-14, N-15, O-16, O-18, F-19, Na-23, Mg-24, Mg-25, Mg-26, Al-27, Si-28, Si-29, Si-30, P-31, S-32, S-33, S-34, Cl-35, Cl-37, K-39, K-41, Ca-40, Ca-42, Ca-43, Ca-44, Ca-46, Ca-48, Sc-45, Ti-46, Ti-47, Ti-48, Ti-49, Ti-50, V-50, V-51, Cr-50, Cr-52, Cr-53, Cr-54, Mn-55, Fe-54, Fe-56, Fe-57, Fe-58, Co-59, Ni-58, Ni-60, Ni-61, Ni-62, Ni-64, Cu-63, Cu-65, Zn-64, Zn-66, Zn-67, Zn-68, Zn-70, Ga-69, Ga-71, Ge-70, Ge-72, Ge-73, Ge-74, Ge-76, As-75, Se-74, Se-76, Se-77, Se-78, Se-80, Se-82, Br-79, Br-81, Kr-80, Kr-82, Kr-83, Kr-84, Kr-86, Rb-85, Rb-87, Sr-84, Sr-86, Sr-87, Sr-88, Y-89, Zr-90, Zr-91, Zr-92, Zr-94, Zr-96, Nb-93, Mo-92, Mo-94, Mo-95, Mo-96, Mo-97, Mo-98, Mo-100, Ru-96, Ru-98, Ru-99, Ru-100, Ru-101, Ru-102, Ru-104, Rh-103, Pd-102, Pd-104, Pd-105, Pd-106, Pd-108, Pd-110, Ag-107, Ag-109, Cd-106, Cd-108, Cd-110, Cd-111, Cd-112, Cd-113, Cd-114, Cd-116, In-113, In-115, Sn-112, Sn-114, Sn-115, Sn-116, Sn-117, Sn-118, Sn-119, Sn-120, Sn-122, Sn-124, Sb-121, Sb-123, Te-120, Te-122, Te-123, Te-124, Te-125, Te-126, Te-128, Te-130, Cs-133, Ba-130, Ba-132, Ba-134, Ba-135, Ba-136, Ba-137, Ba-138, La-138, La-139, Ce-136, Ce-138, Ce-140, Ce-142, Pr-141, Nd-142, Nd-143, Nd-144, Nd-145, Nd-146, Nd-148, Nd-150, Sm-144, Sm-147, Sm-148, Sm-149, Sm-150, Sm-152, Sm-154, Eu-151, Eu-153, Gd-152, Gd-154, Gd-155, Gd-156, Gd-157, Gd-158, Gd-160, Tb-159, Dy-156, Dy-158, Dy-160, Dy-161, Dy-162, Dy-163, Dy-164, Ho-165, Er-162, Er-164, Er-166, Er-167, Er-168, Er-170, Tm-169, Yb-168, Yb-170, Yb-171, Yb-172, Yb-173, Yb-174, Yb-176, Lu-175, Lu-176, Hf-174, Hf-176, Hf-177, Hf-178, Hf-179, Hf-180, Ta-180, Ta-181, W-180, W-182, W-183, W-184, W-186, Re-185, Re-187, Os-184, Os-186, Os-187, Os-188, Os-189, Os-190, Os-192, Ir-191, Ir-193, Pt-190, Pt-192, Pt-194, Pt-195, Pt-196, Pt-198, Au-197, Hg-196, Hg-198, Hg-199, Hg-200, Hg-201, Hg-202, Hg-204, Tl-203, Tl-205, Pb-204, Pb-206, Pb-207, Pb-208, Bi-209, Th-232, U-233, U-234, U-235, U-236, U-238.

Von über 40 000 berücksichtigten kristallinen Phasen etwa 152:

CaHg, CaHg$_2$, CaHg$_3$, CaHg$_4$, CaHg$_5$, Ca$_3$Hg, Ca$_5$Hg, Ca$_3$Hg$_2$, Mg$_2$Hg, Mg$_3$Hg, Mg$_4$Hg, Mg$_5$Hg, Mg$_5$Hg$_3$, Cr$_2$C, Cr$_3$C$_2$, Cr$_7$C$_3$, Fe$_3$C, Mo$_2$C, MoS$_2$, Fe$_2$Mo$_2$C$_4$, Zr$_3$C, ZrN, ZrO$_2$, MgCl$_2$, LiCl, NiAl, Ni$_3$Al, Al$_2$O$_3$, Na$_2$O, Quarz, Dolomit, Calcit, Anhydrit, Albit, Anorthoklas, Muskovit, Kalifilit, Orthoklas, Mikroklin, Polyxydomen, Polyphenyten, CePt$_5$, Mg$_2$S, J$_2$SMg, Fe$_3$P, MnSeO$_4$, NaAsi, NaS, Na$_2$S, γN, N$_2$Pr, Au$_2$Ce$_3$, EfPt$_2$, TbPd$_5$, CeCoCu$_2$Si$_2$, Mo$_3$Si, Mo$_3$Si, GeO$_2$, T$_2$NA$_4$, T$_2$NA, CsMo$_2$, Nd$_3$, N$_3$P$_2$, Nd$_3$S$_4$, NdSi$_3$, Y$_3$N$_4$, Dy$_3$N$_4$, Er$_2$O$_3$, Yb$_3$N$_4$, Yb$_3$Co$_4$Ge$_1$, Er$_3$Co$_3$Ge$_2$, Tm$_2$Pt$_5$, HoPt$_3$Ge$_6$, Pd$_2$Er$_5$, Pd$_4$Ce$_3$, Pd$_3$Pr, CoCl$_2$, Ni$_2$P, CoP, CoWP, BaMoO$_4$, Ag$_2$O, Cu$_2$O, Ag$_2$SO$_4$, Mo$_2$O$_5$, SnO$_2$, PbO$_2$, Cu$_2$Pt, La$_4$Pd$_3$, Yb$_3$Pd$_5$, BaMo$_2$, SrMoO$_4$, NaYbF$_4$, YbMoO$_4$, TbSeO$_4$, CeLaPd, NH$_4$NO$_3$, SrMoO$_4$, (NH$_4$)$_2$MoO$_4$, $5Fe_2O_3 \cdot MoO_3$, Li$_4$MoO$_4$, MoO$_2$, MoO$_3$, Mg$_{13}$, Na$_2$Mo$_2$O$_7$, MgO, MgO$_2$, Na$_3$Mo$_5$, CaMoO$_4$, (UO)$_2$(MoO$_4$), Mg$_2$O$_7$, Na$_2$MoO$_4$, UO$_2$, UO$_3$, UOH$_2$, SrMoO$_4$, Na$_2$Mo$_2$O$_7$, Ag$_2$O, Na$_3$OCl, NaOHH$_2$O, ZnMo$_2$O$_7$, (OH)$_2$Mn, ZnMoO$_4$, Dy$_4$As$_2$, YSb, Tm$_2$Ni$_5$, α-Pt, CoMoO$_4$, CoWO$_4$, Ca$_5$(WO$_4$), H$_4$O$_3$, Pd$_3$Y, Pr$_2$Fe$_3$, U$_3$O$_8$, Pd$_5$Sm, Sm$_2$O$_3$, Sb$_2$O$_3$, Sb$_2$O$_5$, Sb$_2$O$_3$, Cu, Te, Ag$_2$Te, Pb$_2$Cu$_2$Mo$_{14}$, Sr$_3$, Ag$_3$, SrO$_4$, ThO$_2$, $2BaO \cdot 3ZnO$, BaO, $S_2O_4^{2-}$, ZnO, ZnO$_2$, ZrO$_2$.

Tabelle 2: Beispiele für analysierte Komponenten in Proben aus dem Analysendienst eines Forschungszentrums (ZCH)

stems nachgewiesen werden können. Nehmen wir als Beispiel den Standardgranit USGS-G2 (26 b) in Abbildung 4. Darin sind bisher etwa 80 Elemente gefunden und bestimmt worden. Das sind fast alle stabilen Elemente des Periodensystems.

Tabelle 3 enthält als Beispiel für biologisches Material die chemische Elementzusammensetzung zahlreicher Probenarten aus dem menschlichen Organismus (61). Tabelle 4 gründet darauf eine Einteilung nach Hauptbestandteilen (61). Tabelle 5 gibt eine biologisch-toxikologische Klassifizierung.

Zwangsläufig erhebt sich die Frage, wie weit es berechtigt ist, die Spurenanalyse zu immer geringeren Spurengehalten zu entwickeln. Die Antwort liegt darin, daß dies so lange sinnvoll ist, wie die Spurengehalte noch einen nachweisbaren Einfluß auf die Eigenschaften des untersuchten Materiales haben. In diesem Sinne interessieren in der Geochemie in Böden und Gesteinen noch Gehalte von etwa 1 bis 0,1 ppb; in der Materialforschung metallischer Werkstoffe 10 bis 1 ppb, in ausgewählten Reinststoffen (3, 83) 1 bis 0,1 ppb; in biologischem Material und insbesondere in Blutserum für diagnostische Zwecke (11) noch 0,5 - 1 ppb. Die zur Zeit allgemein sinnvolle untere Grenze für die Analyse von Elementspuren dürfte bei etwa 0,1 ppb liegen, bei Meerwasser entsprechend tiefer, im ppt-Bereich. Die Tendenz wird teilweise zu noch niedrigeren Gehalten gehen.

3. KLASSISCH-CHEMISCHE MULTIELEMENTANALYSE

Im Zeitalter der Instrumentellen Multielementanalyse wird leicht vergessen, daß es bereits seit langem eine entsprechende chemische Multielementanalyse gibt. Es sind dies die klassischen Trennungsgänge für Kationen und teilweise auch Anionen (Abbildung 5). Der klassische Kationentrennungsgang geht auf Heinrich Rose (1795 - 1864) zurück und wurde 1829 in Berlin veröffentlicht (58). Er wurde 1841 von Carl Remigius Fresenius in der "Anleitung zur qualitativen chemischen Analyse" (25) in eine viele Jahrzehnte gültige Form gebracht. Im ersten Viertel dieses Jahrhunderts hat ihn A.A. Noyes modifiziert und in den Dreißiger Jahren Werner Fischer durch Einführung von Urotropin als weiterem Fällungsreagenz verbessert.

Die nachzuweisenden Elemente liegen in wässrigen Lösungen als hydratisierte Kationen ($(Me.aq)^+$, $Me(H_2O)_6^{2+}$ etc.) und hydratisierte Anionen ($Cl^-.aq$, SO_4^{2-}, AsO_4^{3-}) vor. Es gibt kaum spezifische chemische Einzelelementnachweise. Daher muß man das Multielementgemisch durch mehrere hintereinandergeschaltete quantitative Gruppentrennungen so lange in immer

Einführung 11

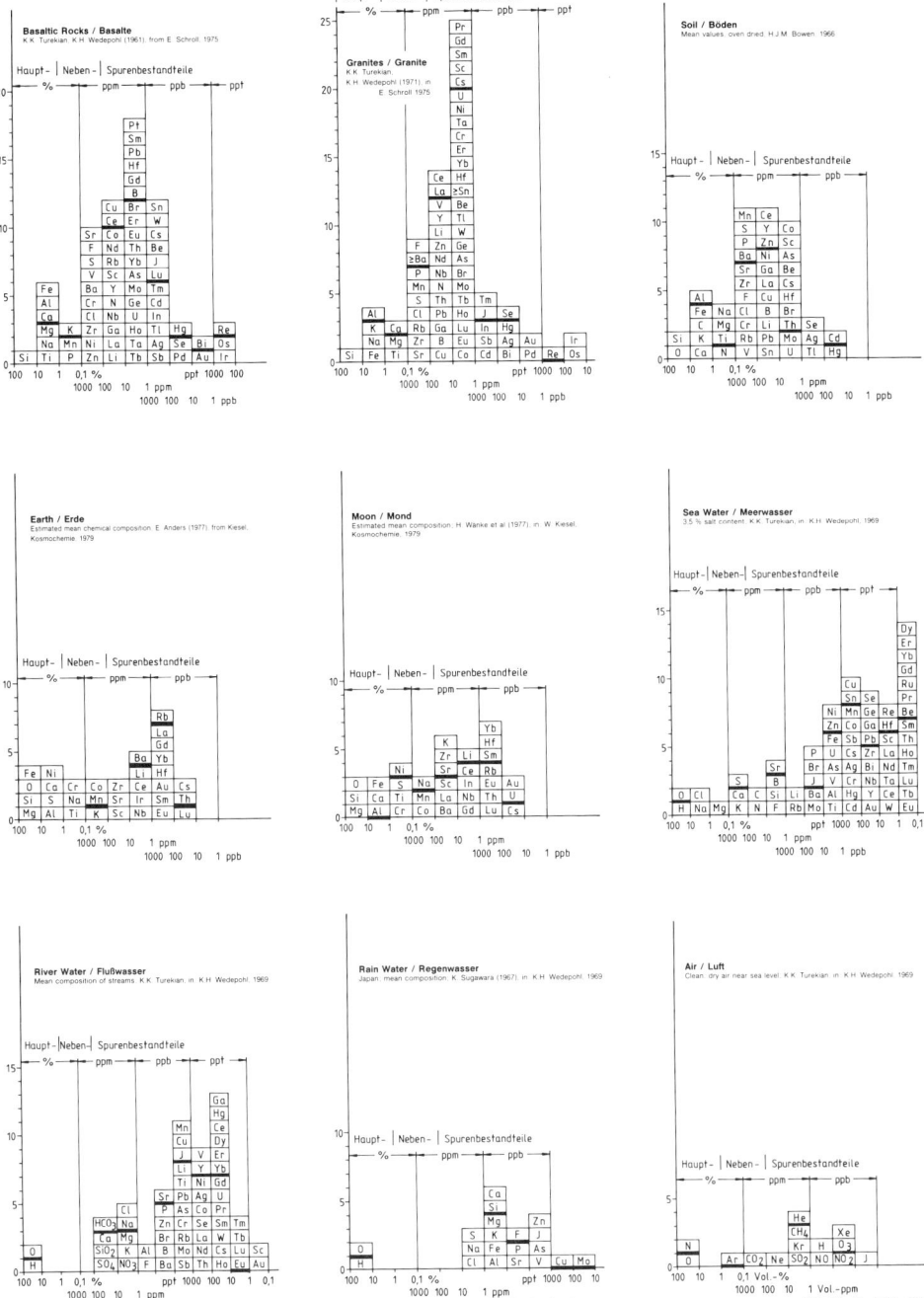

Abbildung 4: Chemische Zusammensetzung der Materie: Haupt-, Neben- und Spurenbestandteile
Gehalt (Gewicht/Gewicht); Konzentration (Gewicht/Volumen)
Abszisse: Gehalt bzw. Konzentration, Größenordnung von links nach rechts abnehmend
Ordinate: Anzahl Elemente pro Größenordnung, Gehalt bzw. Konzentration von unten nach oben zunehmend;
Trennlinie bei halber Größenordnung

12 Sansoni

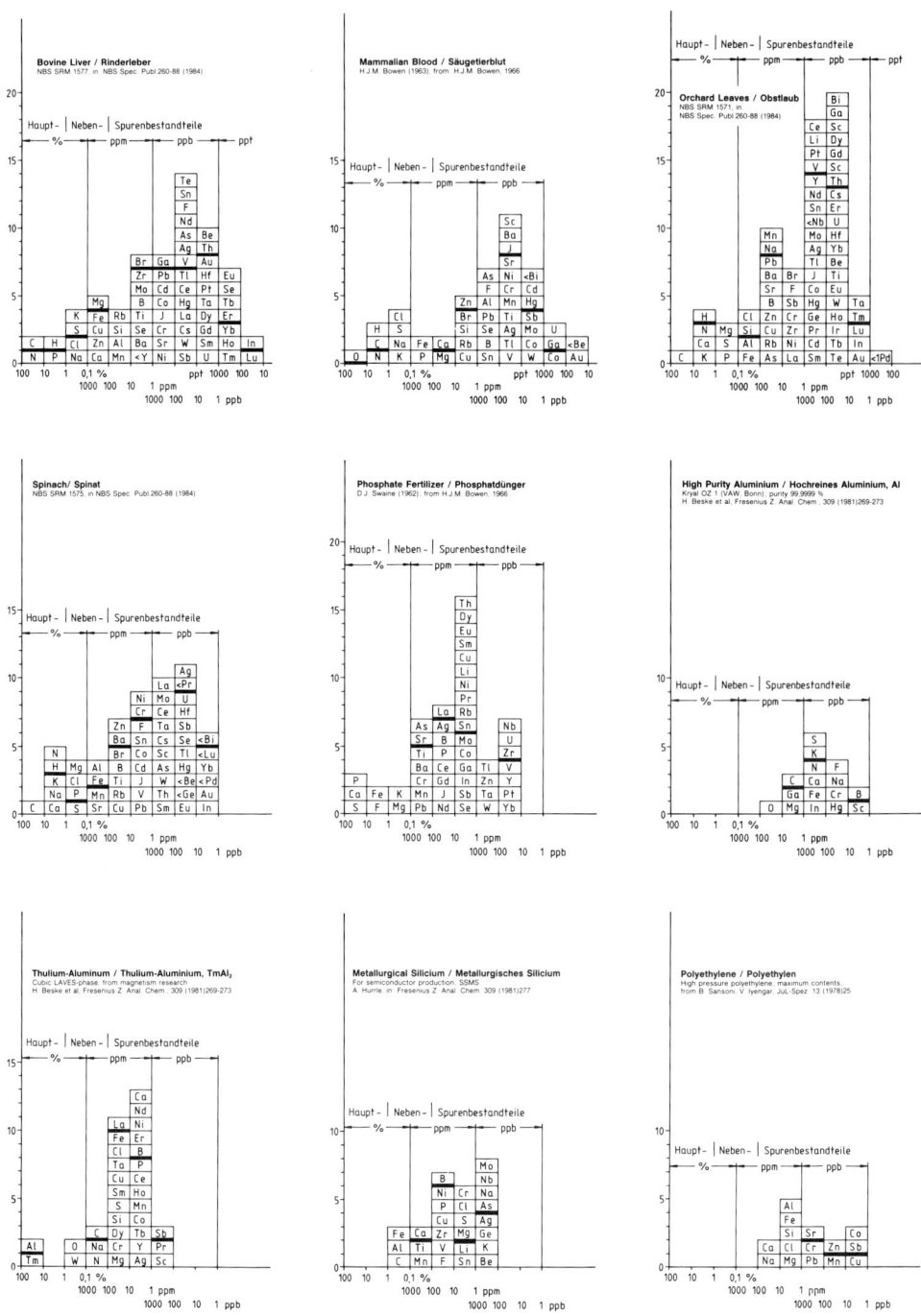

Abbildung 4: Fortsetzung

Contents: xxx = >.10mg/g dry weight, resp. >.17
Contents: xx = 1–10mg/g dry weight, resp. 0,1–17
Contents: x = 0,1–1mg/g dry weight, resp. 100–1000ppm
without x = <0,1mg/g dry weight, resp. <100ppm

Samples	Ca	Cl	Fe	K	Mg	Na	P	S	Si	Zn	Additional elements
Solids:											
Adrenal	x	xx	x	xx	x	xx	xx	xx	x		
Aorta	xx	xx	x	xx	x	xx	xx	xx	x		
Bone (composite)	xxx	x	x?	xx	xx	xx	xxx	x	–	x	
Brain (composite)	x	xxx?	x	xx	x	xx	xxx?	xx	x		
Breast	x	xx	–	xx		xx	xx	xx			
Esophagus	x	xx	x	xx	x	xx	xx	xx			Cu:x; Mn:x; Sn:x; Zr:x; (ICRP)
Feces (a)	xx	xxx		xxx	xx	xx	xxx	xx	x	x	
Gall bladder	x	xx	–	xx	x	xx	xx	xx	–		
Hair	x	xx		x		xx?	x	xxx	x?	x	
Heart	x	xx	x	xxx?	x	xx	xx	xx		x?	
Intestine	x	xx	x	xx	x	xx	xx	xx	x?		
Kidney (composite)	x	xxx?	x	xx	x	xx	xx	xx	x	x	
Larynx	x	xx	x	xx	x	xx	xx	xx	–		
Liver	x	xx	xx	xx	x	xx	xxx?	xx	x?	x	
Lung	x	xx	x	xx	x	xx	xx	xx	x		Al:x
Lymph	x	xx	x	xx	x	xx	xx	xx		xx	Al:x
Muscle (skeletal)	x	xx	x	xxx?	x	xx	xx	xx		x	
Nail	x	xx		xx		xx	x	xx	x	x	Al:x
Ovary	x	xx		xx	x	xx	xx	xx			
Pancreas	x	xx	x	xx	x	xx	xx	xx		x	
Placenta	xx	xx	xx	xx	x	xx	xx	xx			
Prostate	x	xx	x	xx	x	xx	xx	xx		x	
Skin (composite)	x	xx		xx	x	xx	xx?	xx	x		
Spleen	x	xx	xx	xx	x	xx	xxx?	xx		x?	
Stomach	x	xx	x	xx	x	xx	xx	xx	x	x	
Testis	x	xx	x	xx	x	xx	xx	xx			
Thymus		x	x	xx	x	xx	xx	xx			
Thyroid	xx?	xx	x	xx		xx	xx	xx		x?	I:xx
Tongue	x	xx	x	xxx	x	xx	xx	xx		x?	
Tooth (composite)	xxx	x		xx	xx	xx	xxx	x	x?	x	Ba:x; F:x; Sr:x?;
Trachea	x	xx	x	xx	x	xx	xx	xx	–		
Urinary bladder	x	xx	x	xx	x	xx	xx	xx			
Uterus	x?	xx	x	xx	x	xx	xx	xx		x?	
Liquids:											
Blood (composite)		xx	x	xx		xx	x	xx			
Erythrocytes		xx	xx	xx		x	x	?			
Serum (human)	x?	xx				xx	x	xx			
Cerebrospinal fluid	xx	xx		x		xx					
Gastric juice			x		xx						
Bile (gallbladder)	xx	xx	x	xxx	xx	xxx	xx	–		x	
Milk (mature) (a)	x	x		x		x	x	x			
Pancreatic juice		xx		x		xx					
Prostatic fluid	xx	xx	–	xx	–	xx	–	–		xx	
Seminal fluid	x	xx	–	xx	x	xx	–	–		x	
Sweat	xx	xx		x	xx	xx	–				
Urine	x?	??	xx	xx	x						

– = contents not available or uncertain;
? = borderline cases; (all values according to ICRP reference man)

(V. Iyengar, B. Sansoni, 1976)

Tabelle 3: Elementzusammensetzung biomedizinischer Proben
aus dem menschlichen Organismus Erwachsener (61)

Element content	Ca	Cl	Fe	K	Mg	Na	P	S	Si	Zn
xxx >1 %	bone tooth	brain feces kidney		feces heart muscle tongue bile*	bile*	bone brain feces liver spleen tooth		hair		
xx 0,1–1 %	aorta feces placenta thyroid CFS bile prostatic fluid sweat	all others, excepting bone and milk	liver placenta, spleen, excepting erythro- cytes	all others excepting hair, serum CSF gastric juice milk pancreatic juice	bone feces tooth bile* sweat urine	all others excepting erythro- cytes milk urine	all others excepting hair lung nail; all body fluids (without bile)	all others excepting bone tooth; among fluids only blood and serum	lymph nodes	prostatic fluid
x 100– 1000 ppm	**Remarks:** 1. Feces also: Cu (x); Mn (x); Sn (x); 2. Al (x) in lung, lymph nodes, nail 3. I (x) in thyroid 4. tooth also: Ba (x); F (x); Sr (x?) 5. bone also: F (xx) 6. dry urine also: Cl (xxx); K (xxx); Na (xxx); P (xxx); S (xxx); Ca (xx); 6. dry urine also: Mg (xx); Br (x); Si (x); *Bile taken from gall bladder							(x): adrenal aorta brain feces hair intestine kidney liver lung nail skin stomach tooth	(x): bone feces hair heart kidney liver muscle nail pancreas prostate spleen stomach thyroid tongue tooth uterus	

Tabelle 4: Einteilung von biomedizinischem Material nach der Elementzusammensetzung an Hauptbestandteilen (61)

Structural elements: C, Ca(*), H, N, O, P, S

Electrolyte elements: Ca, Cl, K, Mg, Na

Trace Elements:

1. **Essential:**
   1.1 Biologically important: Co, Cr, Cu, F, Fe, I, Mg, Mn, Mo, Ni,
   1.1 Biologically important: Se, Si, Sn, V, Zn
   1.2 Clinically significant: (Co), (Cr), Cu, Fe, Mg, (Se), Zn

2. **Suspected to be essential:** As, Ge, Rb

3. **Regularly found in the tissue:** Al, B, Br, Ga, Li, Sc, Sr, Ti

4. **Toxic:**
   4.1 Potentially toxic: As, Cd, Hg, Pb, Sb, Se
   4.2 Major environmental contaminants: Cd, Hg, Pb
   4.3 Industrial hazards: Be, Bi, Cr, Mn, Ni, Sb, Si, Th, Te, U, V, W

5. **Others:** Ag, Au, Ba, Ce, Cs, Nb, Pt, rare earths, Te, W, Zr

6. **Radioactive contaminants(**):** Po, Ra, Rn, Th, U; Am, Cm, Np

*) Bone only.
**) if not listed as elements before.

(V. Iyengar, B. Sansoni, 1976)

Tabelle 5: Biologisch-toxikologische Einteilung von Elementen in biologischem und medizinischem Material (61)

Einführung 15

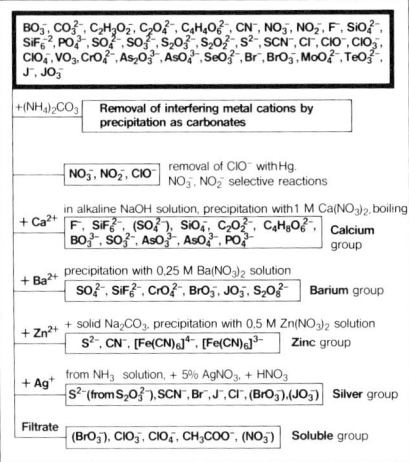
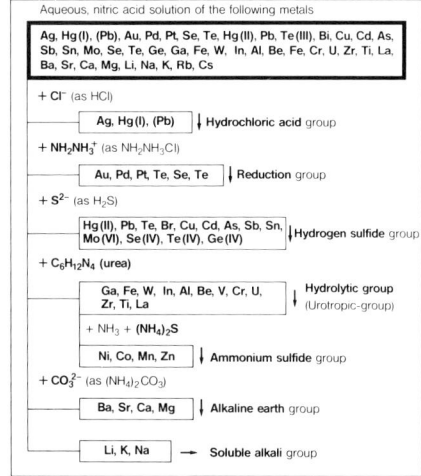

Abbildung 5: Klassisch-chemische Multielementanalyse durch Kationen- und Anionentrenngänge.
Diese haben meist Baumstrukturen nach Abbildung 2.

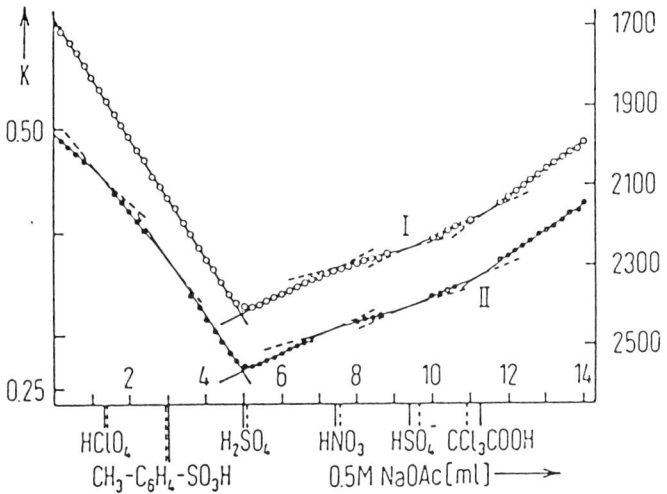

Abbildung 6: Sequentielle Oligo-Anionenbestimmung von 6 Anionen durch klassisch-chemische Titration
von Brønsted'schen Säuren mit der Base Azetat im nichtwässrigen Lösungsmittel wasserfreie Essigsäure (Eisessig).
Konduktometrische Titration eines Gemisches von $HClO_4/p-CH_3-C_6H_4-SO_3H/H_2SO_4$ (97%)/HNO_3 CCl_3COOH in 99 ml Eisessig + 1 ml H_2O mit 0,5 M CH_3COONa (in Eisessig).
——— gefunden, ----------- theoretische Äquivalenzpunkte.
Kurve I: (linke Ordinate) K ($10^4 \Omega^{-1} \cdot cm^{-1}$);
Kurve II: (rechte Ordinate) in Skalenteilen (Umrechnungsfunktion in K ist nicht linear).

kleinere Elementgruppen auftrennen, bis spezifische Nachweise durch charakteristische Fällung, Farbänderung oder Gasentwicklung möglich werden. Gegebenenfalls erfolgt die Auftrennung bis zu den Einzelelementen. Diese Analysenmethode hat eine typische "Baumstruktur" nach Abbildung 2, Symbol B. Im klassisch-chemischen Trennungsgang wird die abzutrennende Elementgruppe in eine während der Fällung neu gebildete feste Phase überführt und diese anschließend durch Filtrieren und Zentrifugieren von der in Lösung bleibenden Elementgruppe mechanisch abgetrennt (Abbildung 5). Die Kunst besteht darin, die unterschiedlichen Versuchsbedingungen so auszugleichen, daß jeweils alle Elemente einer Gruppe entweder quantitativ gefällt werden oder vollständig in Lösung bleiben, nicht jedoch teilweise.

Der Kationentrennungsgang von H. Rose (58) enthält bereits die Gruppenfällungen mit Salzsäure (Ag, Hg, Pb), Schwefelwasserstoff in schwach saurer Lösung (Ag, Hg, Bi, Pb, Cd, As, Sn, Sb, Au), Ammoniumsulfid (Fe, Al, Ni, Co, Zn, Mn), Ammoniumcarbonat (Ba, Sr, Ca) und Natriumhydrogenphosphat (Mg). In Lösung bleibt die Gruppe der Alkalien (Li, Na, K). Die Verwendung nichtwässriger Lösungsmittel hat neuartige Trenneffekte ermöglicht, die erst wenig genutzt sind (66).

Eine der Hauptschwierigkeiten liegt in dem bei der Auswahl optimaler Versuchsbedingungen zu treffenden Kompromiß. Die sorgfältig ausgearbeiteten Bedingungen müssen bei der praktischen Anwendung genauestens eingehalten werden.

Der klassische Kationentrennungsgang hat aus heutiger Sicht mehr Nachteile als Vorteile. Er eignet sich nur für Haupt- und Nebenbestandteile, meist jedoch nicht mehr für Elementspuren. Dazu sind die Trennungen im allgemeinen nicht ausreichend vollständig. Für Spurenkonzentrationen eignen sich diese Trennungsgänge nur in Verbindung mit Fremdträgern. Für die quantitative Multielementanalyse sind sie nur bedingt brauchbar. Die Ausführung des Trennungsganges ist sehr zeitraubend, erfordert viel persönliche Erfahrung und hohe Experimentierkunst. Die sequentielle Analyse von 40 Elementen kann bis zu 30 Arbeitsstunden erfordern. Im Vergleich dazu liefert die simultane ICP-Atomemissionsspektralanalyse im günstigen Fall das Ergebnis bereits nach fünf Minuten. Bei geringer Abweichung von den optimalen Arbeitsbedingungen werden die Trennungen unvollständig und es verschlechtern sich die Nachweisgrenzen.

Diese Trennungsgänge waren für das vorige und die erste Hälfte dieses Jahrhunderts eine große Leistung. Sie ermöglichten erstmals überhaupt qualitative Multielementanalysen und in günstigen Fällen halbquantitative Abschätzung von Konzentrationen. Mit ihrer Hilfe wurden zahlreiche neue chemische Elemente entdeckt. Für die meisten Elementkombinationen ließen sich geeignete Gruppen- und Einzeltrennungen finden. Es waren Absolutmethoden, die keine Vergleiche mit Standards benötigten. Darüber hinaus erlaubt es die klassisch-chemische Elementanalyse häufig, gleichzeitig auch die chemische Form des zu analysierenden Elementes, also dessen Oxidationsstufe, seine Ionenform und/oder Komplexzusammensetzung (Speciation) zu ermitteln.

In Sonderfällen erlaubt auch die klassische Titration, nicht jedoch die Gravimetrie, die sequentielle Bestimmung mehrerer Elemente und darüber hinaus Anionen nebeneinander in der gleichen Lösung. Abbildung 6 zeigt die Oligo-Anionenbestimmung von ClO_4^-, $CH_3 \cdot C_6H_4 \cdot SO_3^-$, HSO_4^-, NO_3^-, SO_4^{2-}, $C \cdot Cl_3 \cdot COO^-$ durch konduktometrische Titration im Lösungsmittel wasserfreie Essigsäure (Eisessig) mit der BRØNSTEDschen Base Acetat an einem Beispiel des Autors (60).

4. INSTRUMENTELLE MULTIELEMENTANALYSE

Im Gegensatz zur chemischen, beruht die Instrumentelle Elementanalyse (1, 6, 50, 55, 76, 80, 36) auf der quantitativen Messung von physikalischen Analysensignalen (analytical signal), welche mit geeigneten Meßinstrumenten erfolgt. Das Signal geht von einer Baustufe des zu untersuchenden Elementes aus und ist für dieses charakteristisch. Es ist jedoch bereits hier darauf hinzuweisen, daß die Messung nach Abbildung 1 nur einer von mehreren notwendigen Analysenschritten ist, auf die später eingegangen wird.

4.1 Analysensignal

Das Analysensignal ist Träger der Information, deren Quelle eine Baustufe des zu messenden Atomes in der Analysenprobe. Zwischen der Intensität I des Analysensignales und dem gesuchten Gehalt oder der Konzentration c des Elementes in der Materialprobe muß die zugehörige Analysenfunktion (36)

$$I = f(c)$$

vollständig bekannt sein. Das betrifft nicht nur die Abhängigkeit von der Konzentration c der zu bestimmenden Elemente, sondern auch den Einfluß des Grundmateriales, der Matrix (matrix) und der übrigen, möglicherweise störenden Komponenten (matrix und interelement interferences). In reinen Lösungen ist dieser Zusammenhang häufig linear und das Analysensignal einfach proportional dem gesuchten Gehalt. Bei Nichtlinearität sollte die Analysenfunktion zumindest gut reproduzierbar sein. Bei nicht genügend bekannter Analysenfunktion hilft eine Relativmessung gegen das Analysensignal eines geeigneten Standards (24, 26, 48, 51, 36). Dieser muß jedoch grundsätzlich genau die gleiche Zusammensetzung der Matrix und Begleitkomponenten haben. Der Gehalt des Standards an den gesuchten Elementen wird in umfangreichen Zertifizierungsanalysen (certification) von anerkannten Laboratorien in Ringanalysen (intercomparison run; round robin) ermittelt. In diesem Falle liegen Referenz-Standardmaterialien (reference standard materials) vor. Stehen diese nicht zur Verfügung, was häufig der Fall ist, so müssen synthetische Mischungen mit simulierten ähnlichen Zusammensetzungen herangezogen werden. Die Gammaspektrometrie (5, 39, 43) der Neutronenaktivierungsanalyse ist wegen der geringen Selbstabsorption in der Probe eine der wenigen Methoden, bei denen Mischungen wässriger Elementlösungen als Standard ausreichen und die Matrix, sofern sie aus leichteren Elementen besteht, weggelassen werden kann. Auch die Funken-Massenspektrometrie kann als Absolutmethode dienen, wenn man einen Fehler um den Faktor bis zu 2 zuläßt.

Die qualitative Art des gesuchten Elementes in der Probe läßt sich häufig aus der Energie (Wellenzahl, Wellenlänge, Massenzahl, Potential) des Analysensignales ermitteln. Dies erfolgt in der Spektrometrie durch Messung geeigneter Spektren.

Es stehen zahlreiche Arten von Analysensignalen und dementsprechend viele verschiedene Spektrometrien für die Messung zur Verfügung (Tabelle 6) (6, 80, 17).

Das Analysensignal kann entweder aus (a) elektromagnetischer Wellenstrahlung, (b) ionisierender Kernstrahlung, (c) Atommassenstrahlung oder (d) einem Strom-Spannungssignal bestehen.

Nur in Sonderfällen wird das Signal ohne Anregung spontan von den Atomen des zu untersuchenden Elementes ausgesandt. Dies ist nur bei der aus Elementarteilchen bestehenden ionisierenden Kernstrahlung (5, 39, 43) der Fall, die von instabilen, radioaktiven Atomkernen ausgestrahlt wird. Alphastrahlung besteht aus Alphateilchen (Helium-4-Kernen), Betastrahlung aus Elektronen, Gammastrahlung aus hoch energetischen Photonen bzw. extrem kurzwelliger elektromagnetischer Strahlung und niederenergetische Photonenstrahlung (Low energy photons) aus etwas längerwelliger elektromagnetischer Strahlung. Da die ionisierende Strahlung durch Umwandlungen im Atomkern entsteht, ist sie sehr energiereich und nachweisstark. Diese Signale ermöglichen mit die empfindlichsten Nachweise und Bestimmungen überhaupt.

Einführung 19

Analysensignal, Δ Bausteile	Energie, ΔE (e.V.)	Wellenlänge, λ (nm)
Neutronenstrahlung:		
Schnelle Neutronen	≦ 10.000.000	
Spontanspaltung von Cf-252	1–10.000.000	
Pu-239/Be-Quelle	5.140.000	
thermische Neutronen	< 0,1	
Alphastrahlung		
Po-212	8.780.000	
Po-210 (RAF)	5.305.000	
U-238	4.200.000	
Betastrahlung		
P-32	1.710.000	
S-37	3.100.000	
Ba-137 m	662.000	
Gammastrahlung		
S-37	3.100.000	
Co-60	1.332.000	
	1.173.000	
Ba-137 m	662.000	
Dy-165 m	108.000	
Dy-165	95.000	
Röntgenstrahlung (Niedrigenergetische Photonenstrahlung)		0,001 – 10 nm
– innere Umwandlungen im Kern		
U-238	112.000	
Co-60 m	48.000	
	59.000	
– K-Einfang in Elektronenhülle		
Fe-55 (Mn$_{Kα}$-Strahlung)	5.900	
Röntgenfluoreszenzstrahlung		
– K-Strahlung		
Na	≈ 1.000	
Ga	≈ 10.000	
Ra	≈ 100.000	
– L-Strahlung		
Ni	≈ 1.000	
W	≈ 10.000	
Atomemission und -absorption:		
Cu, 4s ⟶ 4p Elektron	6	200
Molekülabsorption		
– o-Chinon, Elektronenübergänge zwischen Molekülorbitalen	5	300
– Ligandenfeld, Trennung von e_{2g} und t_{2g} in (Ti(H$_2$O)$_6^{3+}$)	2	50.000
Chemische Reaktionen		
Dissoziationsenergie von O$_2$	5	300
Infrarotstrahlung:		
Streckschwingungen von C = O	0,2	5.000
Elektronenspinresonanz (ESR)		
Umwandlung Borscher Magnetonen bei 100 Gauss	0,000.001	1.000.000.000
Kernmagnetische Resonanz (NMR)		
Umwandlung von Kernmagnetonen bei 1000 Gauss	0,000.000.006	20.000.000.000

Daten teilweise nach B. Magyar und K. H. Lieser

Tabelle 7: Die Energie von Analysensignalen der Kernstrahlungs- und elektromagnetischen Spektrometriearten

Änderung in der Baustufe	Anregung der Baustufe durch	von der Baustufe ausgesandtes Analysensignal	Spektrometrieart
instabiler Atomkern ⟶ anderer instabiler oder stabiler Atomkern	Spontan	α-Strahlung (Helium-4)	Alphaspektrometrie
	Spontan	β-Strahlung (Elektronen)	Betaspektrometrie
	Spontan	γ-Strahlung (Photonen)	Gammaspektrometrie
	Spontan	Röntgenstrahlung durch Kernisomerieänderung	Röntgenspektrometrie
		Röntgenstrahlung durch innere Konversion	Niederenergiephotonenspektrometrie
Stabiler Atomkern ⟶ Radionuklid	**Neutronen** – thermische – epithermische – Reaktorneutronen	wie ① zusätzlich: prompte γ-Strahlung	wie ① zusätzlich: prompte Gammaspektrometrie ②
	Geladene Teilchen: – Elektron – Proton – Deuteron – He-3, He-4	wie ① und ②	Gammaspektrometrie Röntgenfluoreszenzspektrometrie (RFA) protoneninduzierte Röntgenfluoreszenzspektrometrie (PIXE) Deuteronen-, Alphateilcheninduzierte RFA ③
	Photonen, intensiv (γ-Strahlung)	wie ①, ② jedoch nicht universell	Gamma- und/oder Röntgenspektrometrie ④
Innere Elektronenhülle von Atomen (K-, L-, Schale)	Kathoden- oder äußere Elektronenstrahlung	Röntgenstrahlung	Röntgenfluoreszenzspektrometrie ⑤
Mittlere Elektronenhülle von Atomen	VUV-Strahlung (Vakuumultraviolett)	Harte Ultraviolettstrahlung	Vakuum-Ultraviolettspektrometrie ⑥
Äußere Elektronenhülle des Atoms		Optische Strahlung (elektromagnetische Strahlung von UV bis NIR Bereich)	Atomspektrometrie, jeweils in Emission, Fluoreszenz, Absorption, Streuung, Reflexion etc. ⑦
	Flamme	dsgl.	Flammen-Atomemissionsspektrometrie Flammen-Atomfluoreszenzspektrometrie Flammen-Atomabsorptionsspektrometrie
	Elektrothermal (ET)		ET-Atomemissionsspektrometrie ET-Atomfluoreszenzspektrometrie ET-Atomabsorptionsspektrometrie
	Gleichstrombogen	dsgl.	Bogen-Atomemissionsspektrometrie
	Hochfrequenzfunken	dsgl.	Funken-Atomemissionsspektrometrie
	Glimmentladung	dsgl.	Glimmentladungs-Atomemissionsspektrometrie
	Hochfrequenzplasma (ICP)		ICP-Plasma-Atomemissionsspektrometrie ICP-Atomfluoreszenzspektrometrie
	Laser	dsgl.	Laser-Atomemissionsspektrometrie Laser-Atomfluoreszenzspektrometrie (Lasermikrosonde)
Molekülorbitale	Sichtbares Licht	Absorption	Optische Molekülabsorptionsspektrometrie (Spektralphotometrie)

Tabelle 6: Die verschiedenen Arten von Analysensignalen und daraus entwickelte Spektrometriearten für die Instrumentelle Multielementanalyse

Dies macht sich die Aktivierungsanalyse dadurch zunutze, daß sie nicht-radioaktive (inactive) Elemente vor der Messung durch Bestrahlung mit Neutronen oder geladenen Teilchen in geeignete Radionuklide überführt und als solche meßbar macht.

Eine andere Möglichkeit, Analysensignale von nicht-radioaktiven Elementen zu erhalten, ist die Anregung zur Aussendung längerwelliger elektromagnetischer Strahlung. Durch Bestrahlung mit Elektronen, Protonen oder schwereren geladenen Teilchen aus Kathodenstrahlröhre, Beschleuniger, Zyklotron oder Synchrozyklotron können Elektronen der innersten K- und L-Schalen der Elektronenhülle des Atoms angeregt werden. Sie werden dadurch auf ein energiereicheres, instabiles, Energieniveau (excited level) angehoben, verlieren nach kurzer Zeit die Anregungsenergie und fallen auf das Ausgangsniveau (ground state) zurück. Dabei wird harte bis weiche Röntgenstrahlung als K_α-, K_β-, L_α- , L_β- usw. Strahlung ausgesandt (31, 72). Die Wellenlänge nimmt nach dem Moseleyschen Gesetz mit steigender Ordnungszahl des Elementes gesetzmäßig zu und kann umgekehrt zur Ermittlung der Ordnungszahl und damit zum qualitativen Nachweis des Elementes herangezogen werden.

Durch Anregung mittlerer Elektronenhüllen des Atomes kommt es zur Aussendung von elektromagnetischer Strahlung im Vakuum-Ultraviolett-Gebiet (VUV), bei Elektronen der äußeren Elektronenhüllen zu ultravioletter (UV), sichtbarer (VIS) und naher infraroter (NIR) Strahlung (6, 80, 7, 1, 16, 23, 38, 46, 47, 50, 55, 68, 69, 70, 73, 75, 77, 17). Es ist dies das Gebiet der klassischen Optik. Die Anregung kann, mit zunehmender Intensität durch Flamme, Hohlkathodenlicht, Gleichstrombogen, Wechselstromfunken, Glimmentladung, Radiofrequenzplasma oder Laserstrahl erfolgen. Bei Molekülen wird das Analysensignal im nahen, mittleren und fernen Infrarotgebiet durch Elektronenübertragung, Molekülschwingungen, Ober- und Grundschwingungen sowie Rotationen von Molekülen erzeugt. Mikrowellen entstehen durch Rotation in Flüssigkeiten und Feststoffen. Änderungen Bohrscher Magnetonen oder von Kernmagnetonen führen zur kernmagnetischen- bzw. Kernquadrupolresonanz.

In dieser Reihenfolge nimmt nach Tabelle 7 die Energie des Analysensignales stark ab und seine Wellenlänge entsprechend zu. Danach ist der Energieunterschied zwischen der Gammastrahlung aus dem Atomkern einerseits und Änderungen von Kernmagnetonen in der kernmagnetischen Resonanz andererseits gewaltig. Er beträgt etwa 14 Größenordnungen in Elektronenvolt! Es sieht so aus, als ob in der Elementanalyse Analysensignale mit den höchsten Energien die niedrigsten Nachweisgrenzen ermöglichen. So lassen sich aus der jeweiligen Halbwertszeit eines gammastrahlenden Radionuklides des zu bestimmenden Elementes folgende absolute Nachweisgrenze berechnen (43):

Halbwertszeit	Zahl der Atome	Konzentration (Mol)
1 h	5.200	$8,6 \cdot 10^{-21}(!)$
1 d	125.000	$2,1 \cdot 10^{-19}$
1 a	$4,55 \cdot 10^7$	$7,6 \cdot 10^{-17}$
10^5 a	$4,55 \cdot 10^{12}$	$7,6 \cdot 10^{-12}$
10^9 a	$4,55 \cdot 10^{16}$	$7,6 \cdot 10^{-8}$

Diese Angaben sind für 1 Umwandlung/Sekunde berechnet.

Diese Aussage hat jedoch keine allgemeine Gültigkeit. Denn ein wesentlicher Faktor für die Empfindlichkeit eines Analysensignales ist die Intensität I der Strahlung. So haben die Analysensignale der klassischen Atomemissionsspektrometrie zwar um etwa 5 Größenordnungen niedrigere Energien als diejenigen der Gammaspektrometrie, der Median der Nachweisgrenzen für Elemente ist jedoch nach Abbildung 15 nur um 2 Größenordnungen verschieden.

Grundsätzlich anderer Natur sind die Signale der Massenspektrometrie (22, 12, 37). Hier handelt es sich um Strahlung aus gasförmigen Atomionen der betreffenden Elemente, die nicht nur in alle Elemente, sondern in alle Isotope nach unterschiedlicher Massenzahl aufgetrennt werden. Das Analysensignal wird hier also nicht nach Energien, sindern nach Massen zerlegt. Nachweis und Bestimmung dieser Signale erfolgen anschließend sehr unspezifisch entweder photographisch mit der Photoplatte oder photoelektrisch.

Bedingt damit vergleichbar sind die Strom-Spannungssignale der Voltammetrie (44). Sie entstehen in wässrigen Lösungen reduzier- und oxidierbarer Atom- und Molekülionen, welche bei angelegter steigender elektrischer Spannung charakteristische Stromspitzen ergeben.

Analysensignale werden auf zwei verschiedene Weisen gemessen: (a) integral durch Bruttomessung, oder (b) spektral durch Spektrometrie.

Die integrale Bruttomessung ist meßtechnisch am einfachsten, liefert jedoch einen erheblich geringeren Informationsgehalt. Zur Multielementanalyse eignen sich einfache Bruttomessungen im allgemeinen nicht.

4.2 Spektrometrie

Die Instrumentelle Multielementanalyse beruht überwiegend auf der Messung von Spektren geeigneter Analysensignale. Diese erlauben Aussagen nicht nur über die Intensität des Signales, sondern auch über seine Energie. Die dazu verwendete Methode heißt Spektrometrie. Wird das Spektrum auf einer Photoplatte aufgenommen, so spricht man von Spektrographie (spectrography), wird es photometrisch gemessen, von Spektrometrie (spectrometry).

Als Spektrometrie im allgemeinen Sinne bezeichnet man die nach Energie oder Masse aufgelöste Messung der in Tabelle 7 zusammengefaßten Analysensignale. Dementsprechend kann man eine Kernstrahlen-, eine elektromagnetische sowie eine Massenspektrometrie unterscheiden. Die klassische optische ist ein Sonderfall der elektromagnetischen Spektrometrie. Voraussetzung für diese Methoden ist, daß die zu messenden Energien charakteristisch von den zu untersuchenden Komponenten (Elementen) der Probe sowie die Intensitäten in definierter Weise und möglichst linear von deren Gehalten bzw. Konzentrationen abhängen.

Zur Messung werden die Energie auf der X-Achse und die Intensität I auf der Y-Achse aufgetragen. Je nach Natur des Analysensignales kann das Spektrum entweder aus einzelnen wenigen oder vielen, deutlich voneinander getrennten Einzellinien (Linienspektrum), eng aufeinander folgenden Bändern (Bandenspektrum) oder aus einem Kontinuum (Kontinuumspektrum) bestehen. Das Spektrum kann in unterschiedlichen Bereichen auch mehr als einen dieser Typen beinhalten.

Grundsätzlich kann man mit allen in Tabelle 7 aufgeführten Analysensignalen eine Spektrometrie aufbauen. Hinzu kommen zahlreiche weitere physikalische Effekte, welche in Zukunft noch neue Spektrometriearten erwarten lassen. Besonders groß ist die Vielfalt auf dem Gebiet der Spektrometrie von Festkörperoberflächen, wo man heute über 60 bis 70 Varianten unterscheidet (vergl. auch Anhang 7.2.1.8).

4.3 Spektrometer

Spektren werden mit einem Spektrographen oder Spektrometer (6, 23, 80, 39) gemessen. Ersterer verwendet eine photographische Aufzeichnung, letzterer die photoelektrische Messung. Da letztere heute überwiegt, spricht man meist von Spektrometer. Bei dem klassischen Spektroskop wurde das Spektrum mit dem Auge beobachtet.

Der zentrale Teil aller Spektrometer ist das Gerät, welches das Analysensignal nach Energien oder Massen auftrennt. Es ist im Falle der optischen Emissionsspektralanalyse ein Prisma (Glas oder Quarz), Gitter oder Interferometer. In der Kernstrahlenspektrometrie (5, 39, 43) erfolgt die Umwandlung der Kernstrahlungsimpulse in elektrische im Detektor. Im anschließenden Impulshöhenanalysator wandelt der Analog-Digital-Wandler die aus dem Detektor kommenden elektronischen Impulse um. Die Impulse werden nach Energien geordnet, sortiert und gespeichert. Die Ausgabe der Spektren erfolgt numerisch über Drucker oder graphisch über Plotter und Schreiber. Sie kann auf Magnetband oder Plattenspeicher gespeichert werden.

In der Massenspektrometrie (20) trennt der Massenseparator die gasförmigen und ionisierten Atome und Isotope durch magnetische, elektrische oder doppelt fokussiert durch eine Kombination von elektrischen und magnetischen Feldern nach Atommassen-/Ionenladungsverhältnissen.

Alle Spektrometrien, die mit gasförmigen Atomen oder Ionen arbeiten, benötigen für die Feststoffanalyse zur "Anregung" noch eine Verdampfungs-, Atomisierungs- und Ionisierungsquelle. Dies entfällt bei der Gammaspektrometrie. In der Alpha- und Röntgenspektrometrie für weiche Röntgenspektrometrie entfällt zwar die Verdampfungs- und Atomisierungs-/Ionisationsquelle, es ist jedoch eine Abscheidung der zu analysierenden Komponenten in möglichst trägerfreier Schicht durch umfangreiche chemische Probenvorbereitung erforderlich.

Zu dieser Kombination von Anregungsquelle /Dispersionseinrichtung/ Detektor kommen je nach Spektrometrieart noch optische und elektronische Zusatzeinrichtungen.

Die Variationsmöglichkeit für die Entwicklung von Spektrometrien wird erhöht, indem man sie entweder auf Emission oder Absorption, Streuung, Fluoreszenz, Phosphoreszenz oder Polarisation der zu messenden Strahlung gründet. Außerdem kann man die verschiedenen Verdampfungs- und Anregungsquellen mit den unterschiedlichen Zerlegungs- bzw. Trenneinrichtungen für die Strahlung kombinieren.

5. INSTRUMENTELLE MULTIELEMENTANALYSE DURCH SPEKTROMETRIE

Die zur Instrumentellen Multielementanalyse eingesetzten Spektrometrien sollen (a) hohes Auflösungsvermögen haben, um auch eng nebeneinander liegende Spektrallinien auflösen zu können, (b) einen breiten Spektralbereich überdecken, (c) linienarme Spektren mit möglichst wenig unnötigen und störenden Linien liefern, (d) geringe Matrix-, Interelement- und spektrale Störungen haben und sich (e) für Simultanbetrieb, zumindest aber schnelle sequentielle Arbeitsweise eignen.

Wegen dieser Anforderungen bleibt nur eine kleinere Anzahl von Spektrometriearten übrig. Es eignen sich vor allem die gut aufgelösten Spektren der Gammastrahlung (Abbildung 7) und Röntgenstrahlung (Abbildung 8), die optische Strahlung der Atomemission aus Bogen- (Abbildung 9), Funken-, Glimmentladungs-, Plasma- und Laseranregung sowie die in der Funken-, Niedervoltbogenentladung (Abbildung 10) oder mit der Laserquelle des Massenspektrometers erzeugte Atom- bzw. Atomionenstrahlung. Durch besondere Linienarmut und hervorragende Trennung der Elementlinien zeichnen sich die Linienspektren der Atomabsorptions- (77, 78, 79) und Atomfluoreszenzspektrometrie (38) aus. Die Alpha-Spektrometrie (Abbildung 11) (5, 39, 43) eignet sich zur Multiradionuklidanalyse dann, wenn sie scharfe und getrennte Peaks ergibt. Dazu muß das zu analysierende Radionuklidgemisch im allgemeinen von der Matrix getrennt und trägerfrei in dünnster Schicht als Meßpräparat vorliegen. Aber auch endliche Schichtdicken mit stark verbreiterten Peaks können gelegentlich noch zur simultanen Bestimmung von maximal 5 bis 8 Alphastrahlern herangezogen werden, wie dies Abbildung 11 zeigt. Betaspektren eignen sich wegen ihres breiten Kontinuums im allgemeinen nicht zur Multielementanalyse.

5.1 Optische Atomspektrometrie (1, 6, 7, 16, 23, 38, 46, 47, 55, 68, 69, 70, 73, 75, 77, 80, 83, 17)

Ein Spektrometer für elektromagnetische Strahlung im optischen Bereich besteht im allgemeinen aus einer (a) Verdampfungs- und Anregungseinrichtung, einer (b) Zerlegungseinheit und einem (c) Detektor. Dazwischen befindet sich ein (d) Eintrittsspalt mit einer Optik, die den zu zerlegenden Lichtstrahl durch die zerlegende Einrichtung führt und danach (e) den Eingangsspalt im Detektor als Spektrallinie abbildet.

Der Monochromator ist ein Teil des Spektrometers, ohne Verdampfungs- und Anregungsquelle, aber mit zusätzlichem Austrittsspalt. Man kann damit über den ganzen Spektralbereich hinweg ein ausgewähltes schmales Wellenlängenband isolieren und abbilden.

Verdampfung und Anregung werden häufig in der gleichen Geräteanordnung ausgeführt. Als Verdampfungs- und Anregungsquelle stehen Flamme, Gleichstrombogen, Wechselstrom- und Hochfrequenzfunke, Glimmentladung, Hochfrequenzplasma und Laser zur Verfügung. Die Einrichtung zur spektralen

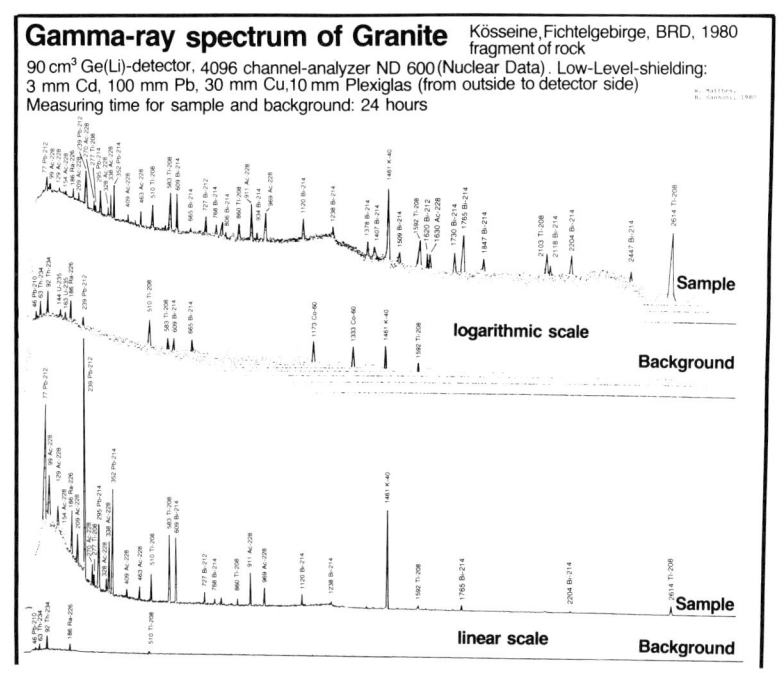

Abbildung 7: Beispiel für ein Low-level Gammaspektrum hoher Auflösung (Kösseine-Granit), mit Ge(Li)-Detektor und Vielkanalanalysator in mit 10 cm Blei und 6 cm Kupfer abgeschirmter Kammer gemessen

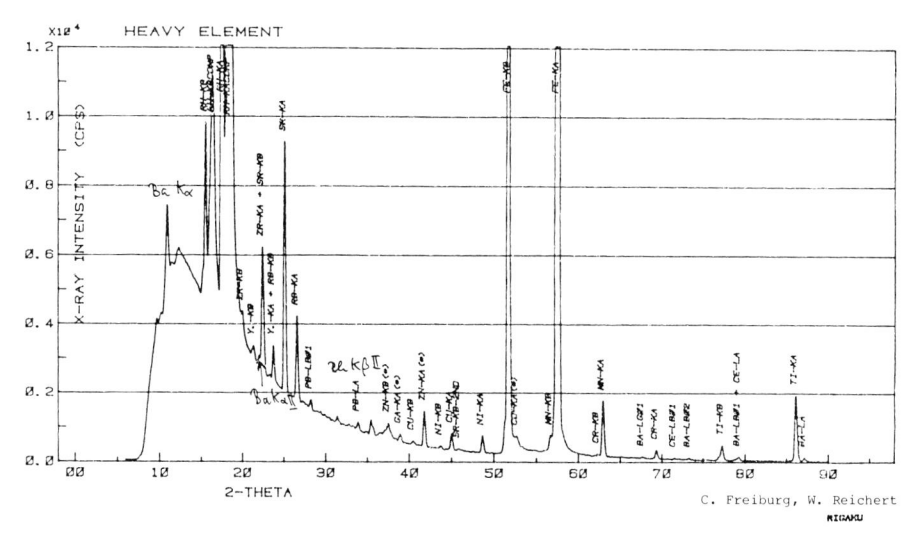

Abbildung 8: Beispiel für Röntgenfluoreszenzspektrum, uranhaltige Bodenprobe

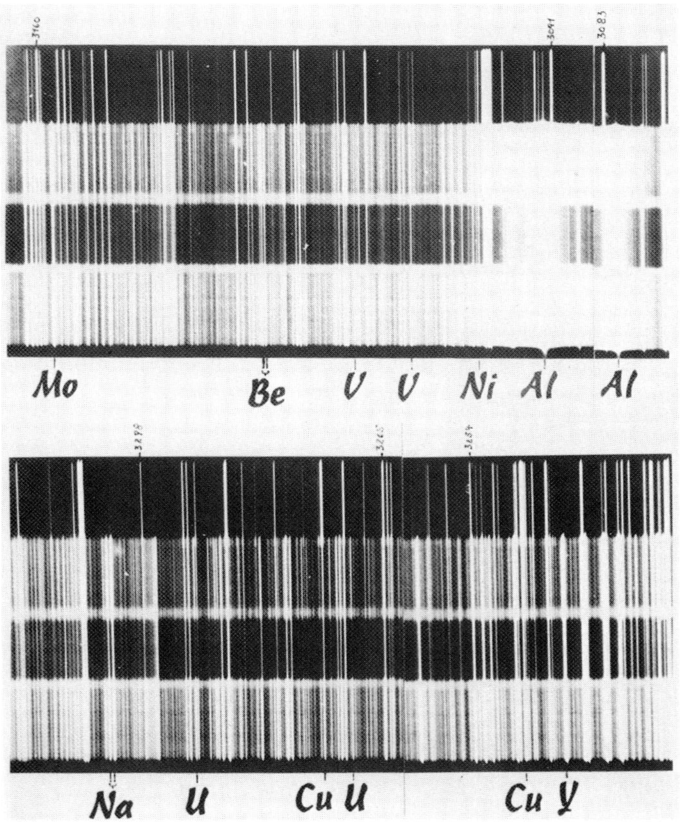

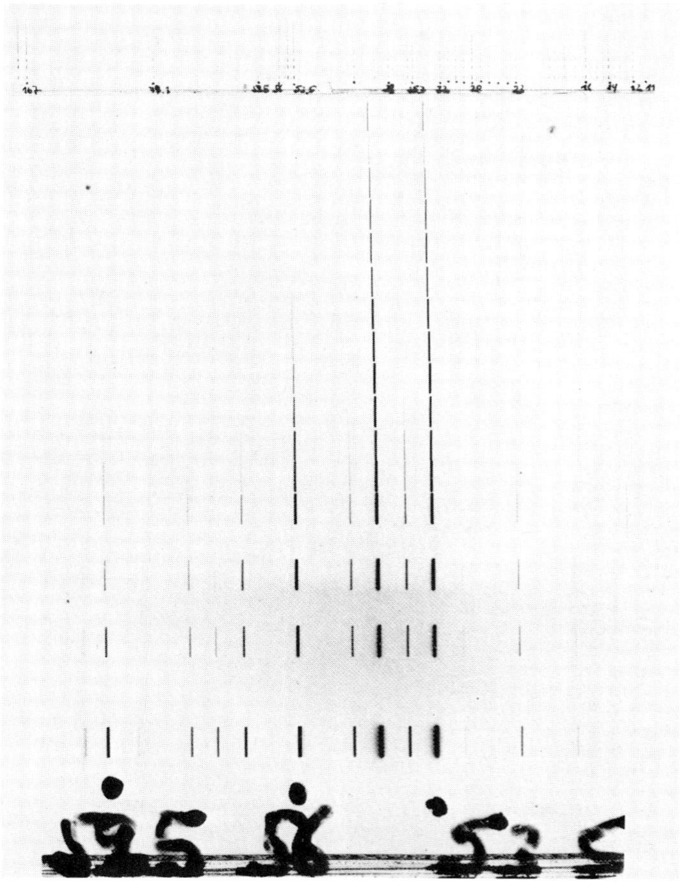

Abbildung 10: Beispiel für ein hochauflösendes Funktenmassenspektrum mit Niedervolt-Bogenentladungsionenquelle (uranhaltiger Boden), Gesamtspektrum und Teilausschnitt

Abbildung 9: Beispiel für ein optisches Gleichstrombogen-Atomemissionsspektrum (uranhaltiger Boden), Teilausschnitt aus dem UV-Gebiet bei 308–300 nm

Alpha-ray spectrum of Granite

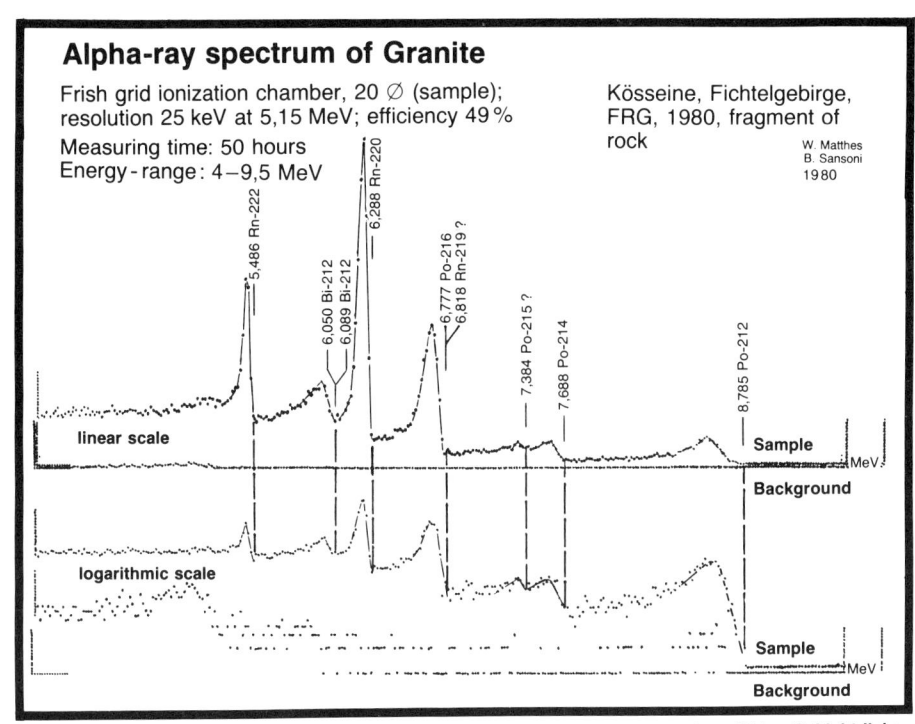

Frish grid ionization chamber, 20 ⌀ (sample);
resolution 25 keV at 5,15 MeV; efficiency 49 %
Measuring time: 50 hours
Energy-range: 4–9,5 MeV

Kösseine, Fichtelgebirge, FRG, 1980, fragment of rock

W. Matthes
B. Sansoni
1980

Peaks: 5,486 Rn-222; 6,050 Bi-212; 6,089 Bi-212; 6,288 Rn-220; 6,777 Po-216; 6,818 Rn-219 ?; 7,384 Po-215 ?; 7,688 Po-214; 8,785 Po-212

linear scale — Sample / Background
logarithmic scale — Sample / Background

Abbildung 11: Beispiel für ein Low-level Alphaspektrum von Kösseine-Granit mit endlicher Schichtdicke, in Großflächen-Gitterionisationskammer mit 20 cm ⌀ Probenkammer.

Zerlegung des Lichtes, also das Spektrometer im engeren Sinne, kann, wie bereits erwähnt, entweder ein Glas- oder Quarzprisma, Gitter oder Interferometer sein.

Als Detektor dient im einfachsten Falle die Fotoplatte. Die weiterführende photoelektrische Aufzeichnung erfolgt durch Photozelle, Photomultiplier, Photoleiter oder Phototransistor. Mit ihrer Hilfe wird der aufgetrennte Lichtstrahl in elektrische Stromimpulse verwandelt, die mit Hilfe von Gleichstromverstärkern und -integratoren, Wechselstromverstärkern oder über Vielkanalmittelung gemessen werden können. Die Anzeige erfolgt analog oder digital. – Im Spektrometer ist die Verdampfung und Anregung der Analysenprobe der komplexeste, daher unübersichtlichste und am schlechtesten reproduzierbare Teilschritt. Hierbei müssen die feste oder flüssige Probe gleichzeitig verdampft, in freie gasförmige Atome dissoziiert, diese ionisiert und schließlich angeregt werden. In Zukunft wird daher eine stärkere Trennung von Verdampfungs- und Anregungsschritt zu übersichtlicheren und besser reproduzierbaren Verhältnissen führen. Diese Entwicklung hat erst begonnen.

Die Flammenanregung kann wegen der relativ niedrigen Flammentemperatur nur eine begrenzte Zahl von Elementen erfassen. Die Bogenanregung hingegen ermöglicht ausgesprochene Multielementanalysen. Der Schwerpunkt liegt bei qualitativen und halbquantitativen Übersichtsanalysen, es werden aber auch quantitative Bestimmungen durchgeführt. Die Hochfrequenzfunkenanregung ist besonders für die halbquantitative bis quantitative Spurenelementbestimmung geeignet, wird aber weniger häufig angewendet. Hochfrequenzplasmen regen wegen ihrer extrem hohen Temperaturen sehr viele Elemente an, zerstören außerdem fast alle chemische Bindungen und eliminieren dadurch chemische Interferenzen weitgehend. Alle drei genannten Anregungsarten haben den Nachteil, daß sie extrem linienreiche Spektren liefern und daher leicht spektralen Störungen ausgesetzt sind. Die erwähnte Atomabsorptionsspektrometrie ist im Gegensatz dazu linienarm, eignet sich aber nicht ohne weiteres zur Multielementbestimmung (77-79). Extrem linienarm ist die Atomfluoreszenz, die bereits zur schnellen sequentiellen Multielementanalyse von bis zu 12 Elementen Anwendung gefunden hat (18).

5.2 Kernstrahlungsspektrometrie (5, 39, 43)

Spektrometer für Alpha-, Beta-, Gamma- und weiche Röntgenstrahlung (Low energy photon spectrometry) sind unterschiedlich aufgebaut. Allen gemeinsam ist der Wegfall der Verdampfungs- und Anregungsquelle der optischen Spektrometrie. Für die Multielement- bzw. Multiradionuklidanalyse ist vor allem die hochauflösende Gammaspektrometrie bis in den Bereich der Röntgenstrahlung von Bedeutung, gefolgt von der Alphaspektrometrie. In der Reihe: Alpha-, Beta-, Gammaspektrometrie nimmt der Strahlungsuntergrund stark zu und dementsprechend die Nachweisstärke ab. Für die extreme Spurenanalyse von Radionukliden hat daher die Alphaspektrometrie günstigere Nachweisgrenzen. Andererseits nimmt in umgekehrter Reihenfolge die Selbstabsorption in der Analysenprobe stark zu und die Durchdringungsfähigkeit entsprechend ab. Infolgedessen ist die Geometrie der Analysenprobe bei der Gammaspektrometrie wenig kritisch. Die Alphaspektrometrie hingegen verlangt extrem dünne, meist sogar trägerfreie Meßpräparate, um scharfe, deutlich getrennte und quantitativ auswertbare Peaks zu ergeben.

5.2.1 Hochauflösende Gammaspektrometrie

Das Gammaspektrometer besteht aus einem mechanischen Probenhalter, einem hochauflösenden Detektor mit angeschlossenem Vor- und Hauptverstärker, Vielkanalanalysator, Datenspeicher, Computerauswertung und Datenausgabe mit Plotter, Schreiber oder Magnetband (39).

Als Halbleiterdetektoren dienen der mit Lithium dotierte Germanium- und neuerdings auch der Reinstgermaniumdetektor. Beide haben den Vorteil eines hohen Auflösungsvermögens. Der Reinstgermaniumdetektor kann noch weit ins Röntgengebiet verwendet werden. Im Vergleich zum NaJ(Tl)-Szintillationsdetektor haben jedoch beide den Nachteil geringerer Zählerausbeute und einer aufwendigeren Elektronik. Vielkanalanalysatoren mit etwa 4000 bis 8000 Kanälen und weitgehender Mikrocomputerisierung und hohem Bedienungskomfort sind heute als stationäre oder sogar tragbare Geräte sehr preiswert im Handel. Halbleiterdetektoren sind allerdings noch relativ teuer. Der Ge(Li)-Detektor erfordert ständige Kühlung mit flüssigem Stickstoff auch während der Aufbewahrung. Der Reinstgermaniumdetektor hingegen nur während der Spektrometrie. Reinstgermaniumdetektoren werden in Zukunft Ge(Li)-Detektoren weitgehend ersetzen.

5.2.2 Hochauflösende Röntgenspektrometrie (39)

Sie ist der vorausgehend behandelten Gammaspektroskopie sehr ähnlich. Als Detektoren sind der Reinstgermaniumdetektor bis herab zu etwa 20 - 40 keV und der Si(Li)-Detektor sowie das Proportional-Zählrohr (46) bis herunter zu etwa 1 keV anwendbar. Als Fenster dienen dünnste Berylliumfolien. Die energiedispersive Röntgenspektrometrie erlaubt im Gegensatz zur sequentiellen wellenlängendispersiven über einen Vielkanalanalysator schnelle Simultanbestimmungen.

5.2.3 Hochauflösende Alphaspektrometrie (5, 39, 43)

Die Meßanordnung muß hier die hohe Selbstabsorption nicht nur in der Analysenprobe, sondern auch in Luft berücksichtigen. Erstere erfordert die aufwendige Herstellung von dünnsten und daher möglichst trägerfreien Präparateschichten. Als Detektoren eignen sich vor allem die Ionisationskammer und Halbleiterdetektoren, flüssige Szintillatoren dagegen nur bedingt. Bei den festen Szintillatoren haben CsJ und ZnS Auflösungen in der gleichen Größenordnung wie bei der Gammaspektroskopie (etwa 8 %). Bei Flüssigkeitsszintillatoren ist die Auflösung mit etwa 20 % wesentlich schlechter, jedoch beträgt die Zählausbeute fast 100 %. Die mit Abstand höchste Auflösung ist auch hier mit Halbleiterdetektoren, vor allem mit dem Si(Li)-Detektor zu erreichen. Die Auflösung geeigneter Gitterionisationskammern kann in günstigen Fällen mit etwa 25 keV für die Co-60 Linie fast an diejenige der Si(Li)-Detektoren herankommen (Abbildung 11). Dabei darf der Probendurchmesser bis zu 20 cm ⌀ betragen. Die Zählausbeute ist jedoch in der Ionisationskammer wesentlich größer als beim Halbleiterdetektor, dementsprechend lassen sich die Meßzeiten verkürzen.

5.3 Massenspektrometrie (20)

Das Massenspektrometer besteht aus einer (a) Verdampfungs- und Ionisierungsquelle, (b) einer Ionentrennungseinrichtung und (c) dem Ionennachweis. Alle drei Schritte erfordern Hochvakuum.

(a) Verdampfung und Ionisierung der Probe erfolgen entweder durch thermische Ionisation auf Metalloberflächen (W, Re), Vakuumentladungsquellen oder Teilchenbeschuß aus Ionenquellen. Zur ersten Gruppe gehört die Thermionenquelle, zur zweiten die Niederspannungsentladungs- sowie Hochfrequenzfunken-Ionenquelle und zur dritten die Laseranregungs- sowie Ionenbeschuß(SIMS)-Ionenquelle.

(b) Im Ionenseparator werden die erzeugten gasförmigen Ionen nach Massen und Energien soweit aufgetrennt, daß alle Isotope eines Elementes nebeneinander als deutlich getrennte Linien vorliegen.

(c) Die Detektion der getrennten Isotope erfolgt entweder mit der Fotoplatte oder elektrisch. Letzteres nach der Auffängermethode mit anschließender Verstärkung oder mit offenem Sekundärionenvervielfacher als Vorverstärker.

Zahlreiche Elektronikeinrichtungen sind zur extremen Stabilisierung der elektrischen und magnetischen Felder, zur Steuerung der Ionenzelle wie zum Erhalt des Hochvakuums erforderlich.

Im einzelnen eignet sich die Thermionenquelle besonders zur Isotopenhäufigkeitsanalyse in Feststoffen. Vakuumentladungsquellen werden am häufigsten zur Multielement-Spurenanalyse von Feststoffen verwendet. Darin werden alle Elemente der Probe ionisiert. Die Elementkonzentration im Ionenstrahl ist proportional der Elementkonzentration der Ausgangsprobe. Während des Verdampfens sind nur geringe Fraktionierungseffekte zu beobachten. Dabei bildet die Hochfrequenzfunken-Ionenquelle aus den Metallen der Probe positiv geladene Kationen, dagegen negative Ionen aus Nichtmetallen. Beides kann zur Analyse genutzt werden. Es entstehen zahlreiche Ionen eines Elementes und Isotopes. Es wurden bis zu 15-fach geladene Atomionen beobachtet. Infolgedessen ist das Spektrum sehr linienreich. Die Niederspannungsentladungs-Ionenquelle ermöglicht Multielementspurenanalysen mit relativ hoher Reproduzierbarkeit. Sie ist gekennzeichnet durch intensives Auftreten von Linien mehrfach geladener Ionen, die alle zur Auswertung herangezogen werden. Mit größter Intensität entstehen Linien der 2- und 3-fach geladenen Ionen. Bei der Ionenbeschuß-Ionenquelle werden Argon- oder Stickstoffionen beschleunigt und auf die Oberfläche der zu untersuchenden Metallprobe geschossen. Durch Sputtern in den obersten Schichten bilden sich positiv geladene Sekundärionen (SIMS). Die Laser-Ionenquelle erzeugt sehr hohe Temperaturen. Sie kann daher alle Elemente der Probe, Metalle und Nichtmetalle, verdampfen und ionisieren. Für die Ionentrennung werden meist doppelt fokussierende Massenspektrometer oder schnelle dynamische Massenspektrometer mit meist fotografischem Ionennachweis eingesetzt. Es bauen sich relativ hohe Ionenströme auf, die zu starken störenden Raumladungen führen können. Die Glimmentladungsquelle ist erst neuerdings zur Massenspektrometrie angewendet worden. Sie läßt für metallische Proben relativ hohe Reproduzierbarkeiten erwarten.

Der eigentliche Ionenseparator trennt alle erzeugten Ionen der Isotope eines Elementes nach ihrem Verhältnis Masse/Ionenladung. Im statischen Ionentrennsystem werden entweder ein magnetisches, ein elektrisches Sektorfeld oder ein kombiniert magnetisch/elektrisches Feld (Doppelfokussierung) verwendet. Zu den dynamischen Trennsystemen gehören das Energiebilanz-Spektrometer, Flugzeit-Streckspektrometer (Ionen unterschiedlicher Masse haben unterschiedliche Fluggeschwindigkeit) und Bahnstabilitäts-Spektrometer (die Ionen müssen zum Durchlaufen des Spektrometers bestimmten Stabilitätsbedingungen genügen). Von dieser Klasse ist das Quadrupol-Massenspektrometer besonders wichtig. Hier erfolgt die Massentrennung durch Schwingungen der Ionen in einem elektrischen Quadrupolfeld.

Das Massenspektrum hat an sich eine einfache Struktur (Abbildung 9). Jedes Isotop bzw. Ion des Isotopes bildet eine deutlich getrennte Linie. Die Bestimmung eines Elementes wird allerdings kompliziert durch die Überlagerung mit Linien der Isotope anderer Elemente mit gleicher Massenzahl, die Bildung von mehr- bis vielfach geladenen Atomionen, Molekül- und Komplexionen sowie Clustern. Im ersteren Falle erlaubt Höchstauflösung eine feinere Unterscheidung.

Ohne Zweifel sind die Funken- und Niedervoltbogenentladungs-Massenspektrometrie die extremsten Multielementmethoden. Sie ermöglichen Simultanbestimmungen nahezu aller stabilen Elemente des Periodensystems gleichzeitig in einem Arbeitsgang und aus einer Analysenprobe. Allerdings sind die Geräte- und Wartungskosten sehr hoch. Die Auswertung erfordert einigen Aufwand. Umfangreiche Vorkenntnisse sind erforderlich. Richtigkeit und Reproduzierbarkeit bedürfen weiterer Verbesserungen.

5.4 Voltammetrie

Die zu analysierende Lösung reduzier- und oxidierbarer Elemente befindet sich in einer elektrochemischen Zelle. Über eine Quecksilbertropf- sowie Bezugselektrode wird an die wässrige Lösung eine von Null beginnende und ansteigende Spannung gelegt und die resultierende Stromstärke gemessen. Während des Spannungsanstieges erfolgt jeweils nach Erreichung des charakteristischen Redoxpotentiales der einzelnen Redoxsysteme die elektrochemische Umsetzung und der entsprechende Stromausschlag. Die nach Abbildung 12 erhaltenen Stromspannungskurven ähneln formal optischen Absorptionsspektren. Damit lassen sich reduzier- und oxydierbare Elemente bzw. Ionen in der Lösung nachweisen und bestimmen. Die auf der X-Achse aufgetragene Spannung ist über das entsprechende Redoxpotential ein qualitatives Maß für die Art des zu bestimmenden Redoxsystemes bzw. Elementes in der Lösung. Die Intensität der zugehörenden Stromstärke auf der Y-Achse ermöglicht die quantitative Bestimmung der Konzentration.

Die Voltammetrie erlaubt die sequentielle Oligoelementbestimmung von Elementen, deren Halbstufen-Redoxpotentiale genügend weit auseinanderliegen. Im Beispiel der Differentialpulspolarographie von Abbildung 12 liegen die Nachweisgrenzen der Elemente Cu, Tl, Pb, Cd, Zn zwischen 1 und 10 ppb.

Bei der Inversvoltammetrie (anodic stripping voltammetry) geht der Polarographie eine elektrolytische Anreicherung im Quecksilbertropfen voraus. Das ermöglicht eine Erhöhung der Empfindlichkeit um den Faktor 10 bis 100. Es lassen sich etwa ein Dutzend Metalle durch Inversvoltammetrie bestimmen u.a. Mn, Co, Ni, Cu, Zn, Ga, In, Sn, Tl, Pb, Bi. Die Nachweisgrenzen liegen zwischen 0,1 und 0,5 ppb. Diese Methode ist für extrem niedrige Spurenelementkonzentrationen insbesondere von verschiedenen toxischen Metallen in der Umweltforschung besonders wichtig. Sie erlaubt die empfindlichsten Bestimmungsmethoden für Blei und Cadmium in diesen Matrices. Die Inversvoltammetrie ist jedoch eine Monoelementmethode.

6. BESTIMMBARE ELEMENTKOMBINATIONEN UND NACHWEISGRENZEN

Für die Anwendung einer Spektrometrie zur Multielementanalyse ist entscheidend wichtig, welche Elemente (a) überhaupt, (b) simultan oder (c) sequentiell nebeneinander und (d) mit welcher Nachweis- bzw. Bestimmungsgrenze sie analysiert werden können. Streng vergleichbare Daten unter gleichen Versuchsbedingungen gibt es für die hier besprochenen Spektrometriearten nur selten. Daher haben die folgenden Angaben nur orientierenden Charakter.

Die Zahl der mit der jeweiligen Methode bestimmbaren Elemente in Tabelle 8 bezieht sich auf praktische Erfahrungen im Analysendienst der Zentralabteilung für Chemische Analysen (ZCH) der KFA. Daher sind manche analysierbaren Elementkombinationen weniger vollständig und die Nachweisgrenzen konservativer angegeben als in der Übersichtsliteratur (17, 80). Nachweisgrenzen gelten meist nur für reine wässrige Lösungen des Einzelelementes. Störende Begleitelemente und Grundmaterialien können erhebliche Verschlechterungen dieser Nachweisgrenzen bewirken. Nach Tabelle 8 nimmt die Anzahl der mit Instrumentellen Multielelementmethoden bestimmbaren Elemente in folgender Reihe ab: Funkenmassenspektrometrie, Neutronenaktivierungsanalyse, Röntgenfluoreszenzanalyse, Atomemissionsspektrometrie mit Gleichstrombogen, mit ICP- Plasma und als Oligoelementmethode die Voltammetrie. Die angegebenen Elementzahlen bedeuten nicht, daß diese auch alle nebeneinander simultan bestimmbar sind.

Es folgen stichwortartige Angaben zu verschiedenen Methoden. Die angegebenen Nachweisgrenzen sind innerhalb einer Größenordnung nach Gruppen mit steigenden Werten angegeben. Ein Teil der allgemeinen Angaben zur Methode

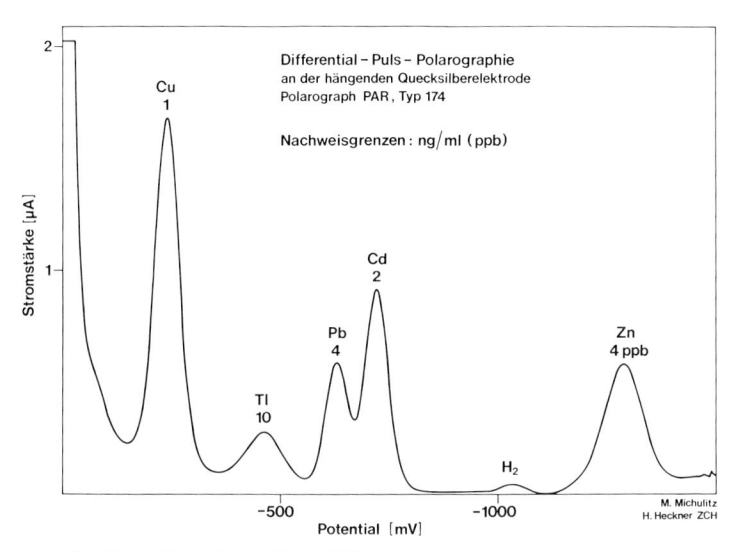

Abbildung 12: Beispiel für ein Differentialpulspolarogramm
Abszisse: Vorgegebene Spannung, Ordinate: Stromstärke
Polarograph PAR, Typ 174, hängende Hg-Elektrode Leitsalz: 2 m K_2CO_3;
Elementlösungen: je 1 ppm (als Chlorid)
Tropfzeit: 2 sec; Tropfengröße 2,4 mg/sec;
Ausgangspotential: 0 mV, Endpotential — 1500 mV;
Amplitude: 50 mV; Empfindlichkeit: 2 µA/volle Skala; Schreibergeschwindigkeit: 0,5 mV/sec.

Tabelle 8: Mit Multielementmethoden bestimmbare Anzahl Elemente und deren Nachweisgrenzen im Analysendienst (ZCH), unter realistischen Bedingungen

Multielementmethode	Anzahl bestimmbarer Elemente	Nachweisgrenzen (ng/g, ppb)	Bemerkung
Funkenmassenspektrometrie, Niedervolt-Bogenentladung			
a) in Reinstaluminium	84	10 - 700	at ppb!
b) in Reinstgold	84	10 - 800	at ppb!
Neutronenaktivierungsanalyse, mit Reaktorneutronen	72	0.01 - 10.000	100 mg Probe; $8 \cdot 10^{13}$ n · cm^{-2} · sec^{-1}
Röntgenfluoreszenzanalyse, wellenlängendispersiv			
a) Wässrige Lösungen	70	3.000 - 10.000	
b) Pulverpresslinge in SiO_2-Pulver	60	10.000 - 20.000	(ähnlich Böden)
Atomemissionsspektrometrie Gleichstrombogen-Anregung			
a) ZCH, Routinebetrieb	60	3.000 - 500.000	1 mg Probe
b) Nach Taylor und Ahrens	70	500 - 500.000	
Atomemissionsspektrometrie Anregung mit induktiv gekoppeltem Plasma, reine wässrige Lösungen Simultanbestimmung	50	1 - 200	
Voltammetrie, reine wässrige Lösungen	25		
a) Inversvoltammetrie, nach Anreicherung um ca. Faktor 100	10	0,1 - 0,5	
b) Potentiometrie mit ionensensitiven Elektroden	4	10 - 500	

sind der auf de Galan (17) zurückgehenden Zusammenstellung von Winefordner (80) entnommen.

6.1 Funkenmassenspektrometrie (FMS)

Es lassen sich feste, flüssige und gasförmige Proben untersuchen. Zur Elementbestimmung eignen sich besonders Festproben. Es ist keine chemische Probenvorbereitung erforderlich, nur eine mechanische Formgebung. Die Methode ist zunächst nur für elektrisch leitende Materialien geeignet. Die verbrauchte Probenmenge beträgt bei Metallen etwa 5 mg. Aus dem Probenmaterial muß ein Stiftpaar mit 1 - 2 mm Ø und 15 mm Länge geschnitten werden. Nichtleitende Stoffe können durch Zumischen von Metallpulver elektrisch leitend gemacht und mäßig gut analysiert werden. In metallischen Proben lassen sich nahezu alle stabilen Elemente des Periodensystems nachweisen und bestimmen, darüber hinaus auch deren Isotope. Die allgemeine Nachweisgrenze liegt an diesem speziellen Funkenmassenspektrometer der ZCH zwischen 1 ppb - 100 ppb. Die Methode hat relativ große systematische Fehler mit einem Faktor 2, arbeitet jedoch standardfrei. Die Niedervoltbogenentladungs-Ionenquelle erreicht Reproduzierbarkeiten von ± 20 %. Messung und Auswertung dauern bei fünffacher Wiederholung etwa 2 Manntage.

Folgende Nachweisgrenzen werden mit der Niedervolt-Bogenentladung in einem doppeltfokussierenden Massenspektrometer mit Fotoplattenregistrierung erreicht (H.E. Beske, F.G. Melchers, G. Frerichs, ZCH):

a) in Reinstgold
1 - 10 ppb: Li
10 - 100 ppb: Be, B, C, N, F, Na, Mg, Al, P, S, Cl, K, Ca, Sc, V, Cr, Mn, Fe, Co, Cu, As

100 - 1000 ppb: O, Si, Ti, Ni, Zn, Ga, Ge, Se, Br, Rb, Sr, Y, Zr, Nb, Mo, Ru, Rh, Pd, Ag, Cd, In, Sn, Sb, Te, J, Cs, Ba, La, Ce, Pr, Nd, Sm, Eu, Gd, Tb, Dy, Ho, Er, Tm, Yb, Lu, Hf, Ta, W, Re, Os, Ir, Pt, Hg, Tl, Pb, Bi, Th, U

b) in Reinstaluminium
1 - 10 ppb: Li, B
10 - 100 ppb: Be, F, Na, K, Sc, Cr, Fe, Cu, As
100 - 1000 ppb: C, N, O, Mg, Si, P, S, Cl, Ca, Ti, V, Mn, Co, Ni, Zn, Ga, Ge, Se, Br, Rb, Sr, Y, Zr, Nb, Mo, Ru, Rh, Pd, Ag, Cd, In, Sn, Sb, Te, J, Cs, Ba, La, Ce, Pr, Nd, Sm, Eu, Gd, Tb, Dy, Ho, Er, Tm, Yb, Lu, Hf, Ta, W, Re, Os, Ir, Pt, Au, Hg, Tl, Pb, Bi, Th, U

Es sind jeweils ca. 80 Elemente simultan bestimmbar.

6.2 Neutronenaktivierungsanalyse (NAA)

Die Neutronenaktivierungsanalyse eignet sich zur Multielementanalyse in Lösungen und Feststoffen. Diese können organischer oder anorganischer Natur sein. Die Probenmengen betragen im allgemeinen 0,1 bis 1 g. Die quantitative Analyse der erhaltenen Gammaspektren erfolgt über Computerauswertung. Systematische Fehler sind durchaus möglich. Die Reproduzierbarkeit auch im Spurenbereich ist häufig gut (etwa $\pm$ 0,5 bis 20 %). Bei Spurenanalysen ist die Gefahr einer Einschleppung von Verunreinigungen nach der Aktivierung gering. Im Gegensatz zur Multistandardmethode (relative method) eignet sich besonders die Monostandardmethode (comparator method) für Multielementanalysen. Sie benötigt keine Multielementstandards und kommt mit nur einem Einelementstandard (comparator) aus (2).

Unter optimalisierten Bedingungen lassen sich mit Reaktorneutronen (Neutronenfluß 8.10^{13} n.cm^{-2}.sec^{-1}; Ge(Li)-Detektor mit P/C = 25; 100 mg Probeneinwaage) maximal bis zu etwa 70 Elemente mit Nachweisgrenzen zwischen 0,01 ppb - 10 ppm bestimmen. Die simultan bestimmbare Anzahl liegt jedoch erheblich darunter (G. Erdtmann, ZCH).

10 - 100 ppt:	Eu, Ir, Au
100 - 1000 ppt:	Sm, Sc, Re, Np, Lu, Ho, In
1 - 10 ppb:	Mn, As, La, Tb, Dy, Tm, Pu; Cu, Ga, W, Hg, Th; Sb, Hf, Os, U; Na, Br, Cs, Ce; Xe; Ta, Er, Yb, Pa; Co
10 - 100 ppb:	Ca, Rh; Ar, V, Cr, Se, Ru, Ag, Te, J, Pt; Pd, Pr, Nd, Gd; Kr; Mo
100 - 1000 ppb:	Ba; K, Zn, Rb, Sr, Cd; Al, Ge; Cl
1 - 10 ppm:	Ni, Sn; Y, Zr; Ti; Fe; Mg

6.3 Röntgenfluoreszenzanalyse (RFA)

Sie eignet sich für feste und flüssige Proben mit anorganischer oder organischer Matrix. Für Feststoffe sind Einwaagen von etwa 1 g, bei Flüssigproben 8 - 10 ml, in der Filtertechnik nur ca. 0,1 ml erforderlich. Bestimmbar sind Elemente oberhalb der Ordnungszahl Z 9, mit weiterem Geräteaufwand außerdem B, C, O. Auch die qualitative Analyse gelingt für Elemente mit Z > 9. Realistische Nachweisgrenzen liegen zwischen 10 bis 1 ppm. Systematische Fehler sind wegen störender Selbstabsorption, Korngrößeneffekten, Sekundäranregungen häufig. Die Präzision kann bei $\pm$ 10 % liegen. Eine Sonderstellung nimmt die hoch empfindliche energiedispersive RFA mit Totalreflexion ein, die auch zur Spurenanalyse von Lösungen, Emulsionen oder Suspensionen im unteren ppb-Bereich geeignet ist. Die Meßzeit beträgt simultan 100 - 500 sec.

Die beiden Beispiele beziehen sich auf die sequentielle wellenlängendispersive RFA (C. Freiburg, W. Reichert).

a) Pulverpreßlinge in SiO_2-Matrix oder Sandböden. Es sind etwa 70 Elemente bestimmbar. Angaben ungefähr geschätzt, Meßzeiten 20 - 100 sec je Element.

1 - 10 ppm: Cr, Mn, Fe, Co, Ni, Cu, Zn, Ga, Ge, As, Se, Br, Rb, Sr, Y, Zr, Nb, Mo, Ru, Pd, Ag, Cd, In, Sn, Sb, Te, J, Cs, Ba, La, Ce, Pr, Nd, Sm, Eu, Gd, Tb, Dy, Ho, Er, Tm, Yb, Lu, Ta, W, Re, Os, Ir, Pt, Au, Hg Tl, Pb, Bi, Th, U

10 - 50 ppm: F, Na, Mg, Al, Si (Matrix), P, S, Cl, K, Ca, Sc, Ti, V

b) Wässrige Lösungen, 100 sec Meßzeit. Etwa 70 Elemente bestimmbar.

1 - 10 ppm: Al, Si, P, S, Cl, K, Ca, Sc, Ti, V, Cr, Mn, Fe, Co, Ni, Cu, Zn, Ga, Ge, As, Se, Br, Rb, Sr, Y, Zr, Nb, Mo, Ru, Pd, Ag, Cd, In, Sn, Sb, Te, J, Cs, Ba, La, Hf, Ta, W, Re, Os, Ir, Pt, Au, Hg, Tl, Pb, Bi; Ce, Pr, Nd, Sm, Eu, Gd, Tb, Dy, Ho, Er, Tm, Yb, Lu, Th, U

10 - 100 ppm: F, Na, Mg

6.4 Atomemissionsspektrometrie (AES)

6.4.1 Flammenanregung (F-AES)

Flüssige Proben, meist anorganisch; 1 - 5 ml Lösung, in Sonderfällen einige Mikroliter. Nur Metalle analysierbar. Nachweisgrenzen bei der quantitativen Analyse zwischen 1 ppb und 1000 ppm. Abhilfe bei systematischen Fehlern durch simulierte Standardlösung, Zufügen von Ionisationspuffern oder chemische Trennungen. Reproduzierbarkeit $\pm$ 0,5 - 5 %.

6.4.2 Gleichstrom-Bogenanregung (DC-AES)

Hauptsächlich anorganische Festproben; Probeneinwaagen etwa 1 mg. Es sind viele Elemente, vor allem Metalle und einige wenige Nichtmetalle analysierbar. Die Methode wird viel zur qualitativen, aber auch zur quantitativen Analyse eingesetzt (1). Die Nachweisgrenzen liegen zwischen etwa 1 - 100 ppm, die Reproduzierbarkeiten in der Größenordnung von $\pm$ 10 bis 20 %. Häufig kleinere systematische Fehler. Abhilfe durch Puffern mit leicht ionisierbaren Substanzen, z.B. Lithiumsalzen.

a) 1,5 m Gitterspektrograph (Typ Wadsworth), Gitter mit 600 Strichen pro mm; 1 mg Probe, normale Analysendienstbedingungen (W. Hilgers, G. Wolff, ZCH). Etwa 60 Elemente bestimmbar.

1 - 10 ppm:	Be, Mn, Mo; B, Yb; Li, Na, Fe, Cu, Rb, Ag, Ba, Tl, Pb; Mg, V, Sr
10 - 100 ppm:	Ca, Ni, In, Sn, Au, Bi, Gd; Al, Ti, Cr, Co, Pd, Sb; Si; Ge, Nb, Sc, Y, Zr, Eu, Tm; Pt; Hg, Ho, Er
100 - 1000 ppm:	Ru, La, Re, Dy; Zn; Sm; P, As, Te, Hf, W, Tb; K; Ga; Ta, Ce

b) Optimalisierte Bedingungen nach Ahrens und Taylor (1) im Spektralbereich etwa 220 bis 900 nm. Die angegebenen Nachweisgrenzen für etwa 70 Elemente sind für die jeweils empfindlichste Spektrallinie geschätzt:

100 - 1000 ppb:	Li, Na, Cu, Ag
1 - 10 ppm:	Cr, Rb, In, Tl; Mg, Al, K, Ca, Sc, Cs; Zn, Ga; V, Fe, Ge, Sr, Mo, Ba, Pb; Be, B, Ti, Mn, Co, Y, Zr, Ru, Rh, Pd, Cd, Sn, La, Ag, Pr, Nd, Eu, Tb, Dy, Ho, Er, Tm, Yb, Lu
10 - 100 ppm:	Si, Sb, W, Bi; Os, Ir, Pt; F, P, As, Hf(?), Ta(?), Re, Hg, Th, U
100 - 1000 ppm:	Te, Gd(?); Ce, Sm(?)

6.4.3 Wechselstrom-Funkenanregung (AC-AES)

Feste oder flüssige Proben, elektrisch leitend, wenige mg Einwaage. Vor allem für quantitative Elementanalyse. Nachweisgrenzen zwischen 10 - 1000 ppm oder 10 - 1000 ng für viele Elemente. Reproduzierbarkeit $\pm$ 5 - 10 %.

6.4.5 ICP-Plasmaanregung (ICP-AES)

Die Methode ist für Flüssigkeiten und nur in Ausnahmefällen für elektrisch leitende anorganische Festproben geeignet. Es sind etwa 2 - 5 ml Lösung erforderlich. Selten zur qualitativen, meist zur quantitativen Multielementbestimmung verwendet. Im allgemeinen wenig systematische Fehler durch chemische Störungen, wegen des extrem hohen Linienreichtums jedoch häufig spektrale Störungen. Auch Interferenzen durch Ionisierung bedingen spektrale Störungen. Abhilfe durch Änderung der Plasmabedingungen über elektrische Parameter, Regulierung der Gaszufuhr und Zusatz von Ionisierungspuffern. Reproduzierbarkeit $\pm$ 0,2 - 5 %.

Simultanspektrometer ARL, Typ 34.000; Gitter mit 1080 Strichen pro mm, betrieben in 1. bis 3. Ordnung; Ar/Ar-Plasma (27,1 MHz; 1,25 kW); Meinhard-Zerstäuber; Ansauggeschwindigkeit 2 ml/min. Simultangerät für 46 Elemente (G. Wolff, W. Hilgers, H. Nickel, H. Lippert, ZCH):

30 - 100 ppt:	Be, Ca
100 - 1000 ppt:	Mg, Sr; Ba; Mn
1 - 10 ppb:	Zn, Zr, Cd; Li, Cu, Ti, V, Co, Mo, Ag, La; Nb, Gd; B; Si, Ru
10 - 100 ppb:	Cr, Ni, Sn, Ta, Bi, Th; Na, Al, Fe, Sb, Te; W, Ce; Hg; C, P, As; S, K, Tl, U; Se; Pb

100 - 1000 ppb: Rb
10 - 100 ppm: Cs

6.4.6 Mikromethode in heißer Graphit-Hohlkathode

Probeneinwaagen bis herab zu 50 bis 100 ug Festprobe. 2 m-Gitterspektrograph, Czerny-Turner-Anordnung; Gitter etwa 1200 Striche/nm. Damit sind etwa 25 Elemente bestimmbar (H. Nickel, G. Wolff, ZCH).

300 ppt: Li, Mg
1 - 10 ppb: Be, Ag; Na, Mn, Cu, Sr, Cd; Cr, Fe; Pb
10 - 100 ppb: Al, Co, Ni, Zn, Mo, In, B, Pd, Tl; V; Ca
100 - 1000 ppb: Rh; Pt

6.5 Simultane Atomfluoreszenzspektrometrie (ICP-AFS)

Sequentielles Multielementspektrometer Baird (82) (ICP-HCL-AFS) mit ICP-Plasma-Anregung und 12 kreisförmig darum angeordneten Hohlkathodenlampen im Pulsbetrieb; Argongas. Jedes Element erfordert eigene Hohlkathodenlampe. In einem Meßzyklus sind jeweils 12 Elemente bestimmbar. Nachweisgrenzen für ca. 30 Elemente (nach Firmenangabe):

0,1 - 1 ppb: Li, Se, Zn; Ca, Cd, Mg; Be
1 - 10 ppb: Ag, As, Na; Cu, Si; Mn; Cr, K, Rh; Ni; Co, Pd
10 - 100 ppb: Fe; Al; Au, Ba; Mo, Pb
100 - 1000 ppb: Sn
1 - 10 ppm: B

6.6 Differential-Puls-Polarographie (DPP)

Sie kann im Analysendienst für maximal 25 Elemente als Oligoelementmethode für jeweils etwa 1 bis 5 Elemente eingesetzt werden. Polarograph PAR, Typ 149, mit Hg-Tropfelektrode (H. Heckner, M. Michulitz):

Ti, V, Cr, Mn, Fe, Co, Ni, Cu, Zn, Ga, Ge, As, Mo, Cd, In, Sn, Sb, W, Tl, Pb, Bi, V, Pu

6.7 Inversvoltammetrie (DPASV)

Vorkonzentrierung am hängenden Hg-Tropfen mit Anreicherung um Faktor ca. 10 - 100). Es sind etwa 12 Elemente einzeln bestimmbar. Polarograph PAR, Typ 149 (M. Michulitz, H. Heckner, ZCH):

10 - 100 ppt: Ni, Co
100 - 1000 ppt: Zn, Cd, Pb, Bi; Cu, Sn, Tl; As, Se

6.8 Volumetrische Titration

Im Analysendienst häufiger verwendete Titrationen (A. Klinkmann, C. Schwarz, G. Wolff, ZCH):

1 - 10 ppm: C
10 - 100 ppm: P, S; Mn; N, Ca, Cr; Mg
100 - 1000 ppm: As; V, Fe, Cu, Zn, Sn, Sb; Ag

6.9 Gravimetrie

Im Analysendienst gelegentlich noch verwendete Methoden (A. Klinkmann, C. Schwarz, G. Wolff, ZCH) für:

10 - 100 ppm: Mg; Si, Ni; Si
100 - 1000 ppm: Co, Cd, Nb; Cu, Sn, Ta, W, Au, Pb

7. Die übrigen Analysenschritte

Wie aus Abbildung 1 hervorgeht, ist die physikalische Messung von Analysensignalen im allgemeinen Analysengang nur <u>ein</u> Teilschritt neben neun bis zwölf anderen.

Es besteht kein Zweifel darüber, daß auch die anderen Teilschritte mit Fehlern behaftet sind, die sich im Endergebnis summieren. Einige von ihnen führen zu erheblich größeren Fehlern als die eigentliche physikalische Messung. Wegen der Begrenzung des Themas dieses Buches kann hierauf nicht näher eingegangen werden. Im folgenden seien daher nur einige weiterführende Beispiele mit Literaturstellen, insbesondere aus dem Arbeitskreis des Verfassers, angegeben.

7.1 Probenahme (4, 56, 61, 62, 71)

Es ist jedem Analytiker geläufig, daß die Probenahme als erster Teilschritt zugleich die größten Fehlermöglichkeiten beinhaltet. Bei unsachgemäßer Ausführung können hier systematische Fehler von über Tausend Prozent auftreten! Aus Raumgründen sei beispielhaft nur die Probenahme von biologischem Material erwähnt (61) (Abbildung 13).

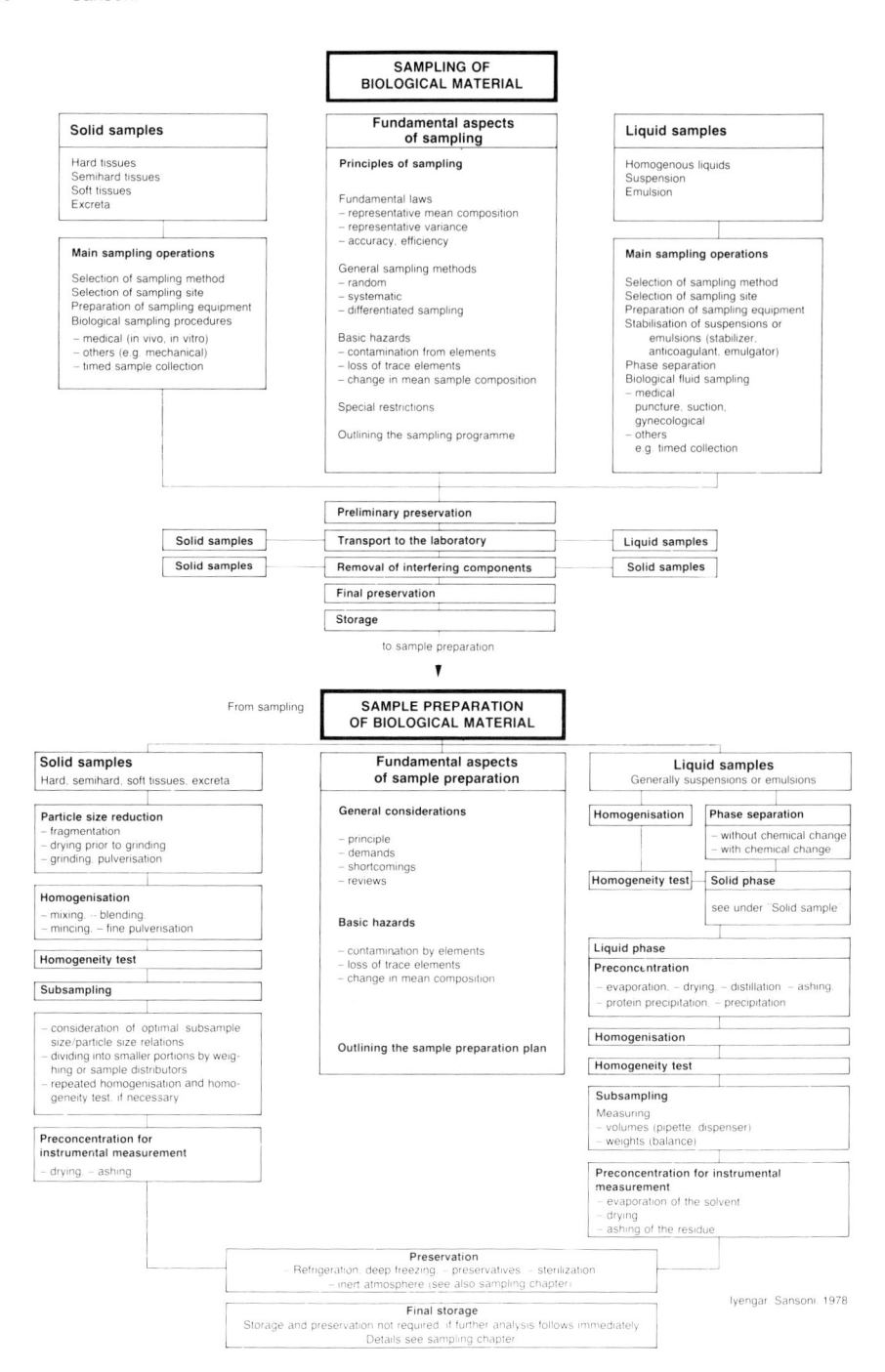

Abbildung 13: Schema der Probenahme und Probenvorbereitung von biologischem Material (61)

Drei Grundregeln gelten für jede Probenahme: 1) Repräsentative mittlere Zusammensetzung der Analysenproben im Vergleich zur zu beurteilenden Gesamtheit, 2) repräsentative Varianz der Elementgehalte in Analysenprobe und Gesamtheit, 3) die Richtigkeit aller Probenahmeoperationen zusammen soll zumindest die gleiche Größenordnung wie die Richtigkeit der nachfolgenden Analysenschritte haben. In der Praxis müssen Kompromisse zwischen diesen Anforderungen einerseits und Kosten- sowie Personalkapazität andererseits geschlossen werden.

Hauptgefahrenquellen sind Verunreinigungen durch zu bestimmende Elemente von außen, Verluste an zu bestimmenden Elementspuren in der Probe und eine Änderung der mittleren Zusammensetzung des Analysenmaterials (24, 61, 81, 82). Wegen vieler weiterer zu beachtenden Faktoren sei auf die Originalarbeit verwiesen (61).

7.2 Probenvorbereitung

Die Probenvorbereitung (9, 10, 21, 24, 28, 33, 40, 45, 51, 57, 61, 65, 34) ist der Analysenteilschritt mit den zweitgrößten Fehlermöglichkeiten. Abbildung 13 behandelt als Beispiel wiederum biologisches Material. Feste Proben müssen in den meisten Fällen zunächst zerkleinert werden (45). Dem schließt sich eine Homogenisierung mit nachfolgendem Homogenitätstest an. Dann wird die Probe in Unterproben geteilt (subsampling). Hierbei müssen wiederum die unter Probenahme (7.1) genannten Forderungen erfüllt sein. Eine teilweise und einfache Vorkonzentrierung für die anschließende instrumentelle Messung gelingt im Falle von biologischem Material durch Trocknen oder Veraschen. Die endgültige Aufbewahrung der Proben kann bei biologischem Material den Zusatz von Stabilisatoren einschließen. Schwer lösliche Materialproben müssen chemisch aufgeschlossen werden (9, 10, 21). Im Anschluß daran erfolgt bei organischen Proben die Veraschung (28, 40, 65). Eine Übersicht über mögliche Veraschungsmethoden gibt Tabelle 9. Hinsichtlich weiterer Einzelheiten sei hier auch auf die Originalarbeit (65) verwiesen.

Probenahme und Probenvorbereitung sind Teilschritte der Analyse, die bisher einer Automation am längsten widerstanden hatten. In den letzten Jahren ist jedoch eine zunehmende Anzahl automatisierter Geräte, zumindest für einzelne Teilschritte, kommerziell zugänglich geworden. Die Firma Zymark hat das erste speziell für chemische Probenvorbereitung entwickelte Robotersystem auf den Markt gebracht.

1. DRY ASHING	2. WET ASHING	3. OTHERS
1.1 High Temperatures	2.1 High Temperatures	3.1 High Temperatures
1.1.1 Combustion with air/oxygen	2.1.1 Normal Pressure	3.1.1 Oxidative fusion (21, 39, 70, 198)
1.1.1.1 Stationary system	2.1.1.1 Oxidising mineral acids	3.1.2 Oxidation in nitric acid vapour (101, 292)
A. Open muffle furnace, air under atmospheric pressure (59, 78, 154, 174, 289)	A. HNO_3 (31, 60, 80, 113, 126, 238) B. H_2SO_4 (51, 99)	3.1.3 Oxidation with ozone (158)
B. Oxygen flask (25, 117, 184, 262, 286, 287)	C. $HClO_4$ (105, 272, 335) D. HNO_3/H_2SO_4 (219, 2F2)	3.1.4 Halogenation (59, 85, 159, 306, 320)
C. Oxygen bomb, high pressure (30, 33, 91, 118, 192, 264, 265, 271)	E. $HNO_3/HClO_4$ (215, 273, 274) F. HNO_3/H_2SO_4 (36, 137, 210, 220, 245)	3.1.5 Reductive (32, 40, 44, 73, 175, 275, 307)
1.1.1.2 Streaming system	2.1.1.2 Hydrogen peroxide	
A. Oxygen/air stream in combustion tube (48, 107, 131, 195, 233, 298, 317)	A. 30 % H_2O_2 (52, 68) B. 50 % H_2O_2/H_2SO_4 (101)	
B. H_2/O_2-flame in a closed cooled system (33, 73, 153, 168, 169, 266, 328)	2.1.2 High pressure	
1.1.2 Pyrolytic decomposition	2.1.2.1 Oxidising mineral acids in teflon bomb (27, 161, 173)	
A. Heating under inert gas e.g. Ar, N_2 (187, 192)	2.1.2.2 Hydrogen peroxide in bomb (63)	
B. Reduction with H_2 at high temperature (295, 321)		
1.2 Low Temperatures	2.2 Low Temperatures	3.2 Low Temperatures
1.2.1 Plasma of oxygen gas at 100 - 125 °C	2.2.1 OH radicals from H_2O_2/Fe^{2+} at 100 - 220 °C (Fenton's reagent) (163, 164, 165, 253, 254, 255, 256, 257, 258) A. Open beaker B. Closed system	3.2.1 Enzymatic (116, 162, 185, 277)
A. Radiofrequency electrical fields (96, 97, 98, 176, 302, 303, 311)		3.2.2 Radiative (308, 312)
B. Microwave electrical fields (142, 143)		

Tabelle 9: **Übersicht über Veraschungsmethoden für organisches und biomedizinisches Material**
Die Literaturstellen beziehen sich auf die Originalarbeit (65, 61)

7.3 Chemische Fehlerquellen

Die vorstehend bereits genannten drei Hauptfehlerquellen jeder Element-Spurenanalyse sind nach (61): (a) die Verunreinigung der Probe durch Spuren der zu bestimmenden Elemente, (b) ein Verlust von zu bestimmenden Elementspuren und (c) Änderungen in der mittleren Zusammensetzung der Probe hinsichtlich der zu bestimmenden Elemente.

Damit ist grundsätzlich jeder Teilschritt der Spurenelementanalyse bis zur Beendigung der Messung des Analysensignales konfrontiert. Daher bedeuten die teilweise oder vollständige Einsparung der fehlerträchtigen chemischen Teilschritte bei der instrumentellen Elementanalyse eine grundsätzliche Verbesserung.

Die Kontamination einer Probe durch Elementspuren kann aus der Umgebung der Probe, aus dem Analysenteilschritt selbst und von der ausführenden Person stammen.

Einzelheiten über diese wichtigen Fehlerquellen enthält die Originalliteratur (3, 9, 10, 24, 33, 50, 51, 56, 57, 61, 62, 65, 76, 81, 82, 83, 34).

7.4 Datenverarbeitung und Datenbeurteilung

Der hohe Anfall von Analysendaten bei modernen Analysenautomaten erfordert eine angemessene Datenverarbeitung sowohl unmittelbar am Meßgerät, aber auch durch zusätzliche Rechner. Dies beinhaltet die Ausgabe von verarbeiteten Analysendaten als abgabefertige Tabellen, Diagramme, Grafiken, Analysenprotokolle einschließlich von statistischen Auswertungen (29, 30, 35, 48, 59, 63, 36).

Für die Datenverarbeitung und -auswertung steht heute eine umfangreiche Software zumeist als "IBM PC-kompatible" Programme kommerziell zur Verfügung. Das in der ZCH entwickelte Programm ZCH-3/2 zur Datenkonzentrierung enthält die in Tabelle 10 aufgeführten Statistikoperationen. Es führt wesentlich weiter als nur zur üblichen Berechnung von Mittelwert, Standardabweichung und Vertrauensbereich. Es beinhaltet eine anspruchsvollere Datenreduktion (besser: Datenkonzentrierung), die auch für schiefe (logarithmische) Datenkollektive anwendbar ist. Es folgen Korrelationsrechnung, pattern recognition und andere multivariante Statistiken, die in der "Chemometrie" behandelt werden.

Hierauf kann nicht weiter eingegangen werden. Es sei nur darauf hingewiesen, daß entgegen der landläufigen Meinung auch eine unsachgemäße Datenauswertung zu beträchtlichen Fehlern führen kann. An einem Beispiel der Überwachung von Luftstaub konnte gezeigt werden (63), daß Jahresmittel aus den gemessenen Tageswerten der Konzentration der Luft an verschiedenen toxischen Metallspuren (Pb, Cd, Zn, Ca) um bis zu etwa 40 % (!) falsch waren, wenn anstatt des hier anzuwendenden Medians x das arithmetische Mittel x genommen wurde. In Nähe der Bestimmungsgrenze (Cd) lag die Abweichung sogar bei etwa 70 %. Das liegt vor allem daran, daß die Gaußsche Normalverteilung des Datenkollektives nicht gegeben war. In Geochemie und Umweltforschung folgen die Daten für die Gehalte und Konzentrationen von Spurenelementen im Normalfall nicht einer Gaußschen, sondern einer logarithmischen Verteilung. Diese läßt sich häufig besser durch eine logarithmische Normalverteilung annähern.

Weiterführende Literatur über die hier nicht behandelte Definition, Festlegung und Messung von Nachweis- sowie Bestimmungsgrenzen, Reproduzierbarkeit und Richtigkeit findet sich bei (2, 24, 29, 40, 48, 63, 67, 36), über die Qualitätskontrolle der erhaltenen Analysendaten bei (29, 30, 35, 48) und über die Automation von Elementanalysenverfahren bei (13, 14).

Definition of the data sample and different measures for location of the data

1. **Data sample.**
   n values of the variable x are $x_1, x_2, x_3, \ldots, x_n$, arranged in increasing order with the rank number N: $X_1, X_2, X_3, \ldots, X_N$, where $X_i = x_i$. No particular type of distribution is assumed.

2. **Data location**

   2.1. Arithmetic mean
   $$\bar{x} = \frac{\sum_i x_i}{n}$$

   2.2. Mean of logarithms
   $$\bar{x}_{\log} = \frac{\sum_i \log x_i}{n}$$

   2.3. Geometric mean
   $$x_G = 10^{\bar{x}_{\log}}$$

   2.4. Harmonic mean
   $$\bar{x}_H = \frac{\sum_i 1/x_i}{n}$$

   2.5. Non-parametric estimation: median
   when n is odd
   $$\tilde{x} = X_{(N+1)/2}$$
   when n is even
   $$\tilde{x} = \frac{X_{N/2} + X_{(N/2)+1}}{2}$$

Skewness and kurtosis

Skewness
$$g_1 = \frac{\sum_i (x_i - \bar{x})^3}{n \cdot s^3}, \text{ where}$$
$$s = \sqrt{\frac{\sum_i (x_i - \bar{x})^2}{n}}$$

Kurtosis
$$g_2 = \frac{\sum_i (x_i - \bar{x})^4}{n \cdot s^4}, \text{ where}$$
$$s = \sqrt{\frac{\sum_i (x_i - \bar{x})^2}{n}}$$

Dispersion around the central value

1. Standard deviation
$$s = \sqrt{\frac{\sum_i (x_i - \bar{x})^2}{(n-1)}}$$

2. Relative coefficient of variation (Rel. $\%_o$)
$$V_r = s \cdot 100 / \bar{x}$$

3. Mean deviation from the arithmetic mean
$$DM = \frac{\sum_i |(x_i - \bar{x})|}{n}$$

4. Standard deviation of logarithms
$$s_{\log} = \sqrt{\frac{\sum_i (\log x_i - \bar{x}_{\log})^2}{(n-1)}}$$

5. Geometric standard deviation
$$= \text{antilog } s_{\log}$$

6. Non-parametric estimates:
   – Range $R = X_N - X_1$
   – 80 % Inter decile range $I_{80\%} = $ 9th decile – 1st decile

Tests for outliers

1. t-test:
$$t = \frac{(x_n - \bar{x})}{s \cdot \sqrt{n(n-1)}} \text{ or } \frac{(\bar{x} - x_1)}{s' \cdot \sqrt{n(n-1)}}$$
where $\bar{x}$ and s are calculated from $(n-1)$ observations, omitting the suspected observation. t is compared with the critical value for $(n-2)$ degrees of freedom.

2. Nalimov test:
$$r = \frac{(x_n - \bar{x})}{s \cdot \sqrt{(n-1)/n}} \text{ or } \frac{(\bar{x} - x_1)}{s \cdot \sqrt{(n-1)/n}},$$
for $n < 26$;
r is compared with the critical value for $(n-2)$ degrees of freedom.

Confidence intervals of central values

1. of arithmetic mean
$$= \bar{x} \pm \frac{t \cdot s}{\sqrt{n}}$$

2. of median
$$x_{(h)} \leq \tilde{x} \leq x_{(n-h+1)}, \text{ where}$$
$$h = \frac{n - t \cdot \sqrt{n-1}}{2}, \text{ reliable when } n < 50$$

3. of logarithmic mean
$$= \bar{x}_{\log} \pm \frac{t \cdot s_{\log}}{\sqrt{n}}$$

4. of geometric mean
$$= 10^{\left(\bar{x}_{\log} - \frac{t \cdot s_{\log}}{\sqrt{n}}\right)} \text{ and } 10^{\left(\bar{x}_{\log} + \frac{t \cdot s_{\log}}{\sqrt{n}}\right)}$$

Value t for $(n-1)$ degrees of freedom, corresponding to 95 % or 99 % probability.

3. Grubbs test:
$$T = \frac{(x_n - \bar{x})}{s} \text{ or } \frac{(\bar{x} - x_1)}{s}$$
for $n \leq 30$;
T is compared with the critical value corresponding to n.

4. Dixon test:
$$r_{i,j} = \frac{(x_n - x_{n-i})}{(x_n - x_{1+j})} \text{ or } \frac{(x_{1+i} - x_1)}{(x_{n-j} - x_1)}$$
where $i = 1, 2; j = 0, 1, 2$. i and j differ for different values of n. $r_{i,j}$ is compared with the critical value corresponding to n.

Tabelle 10: Datenreduzierung für normal und schief verteilte Datenkollektive
Routinerechenprogramm ZCH-3/2 der ZCH (63)

8. STRATEGIE DES EINSATZES VON INSTRUMENTELLEN MULTIELEMENTMETHODEN

8.1 Vor- und Nachteile

In der Routineanalytik wird der größte Teil der Spurenelementanalysen mit spektrometrischen Methoden ausgeführt. Für die zeitraubenden chemischen Methoden ist hier im allgemeinen nicht mehr genügend Personal- und damit Kostenkapazität vorhanden.

(1) Je mehr Elemente gleichzeitig zu bestimmen sind, desto lohnender wird der Einsatz Instrumenteller Multielementmethoden. Der Schwerpunkt liegt dabei zunächst in einer raschen Übersicht über eine Vielzahl von Elementgehalten in der Probe, bei denen es nicht immer auf höchste Richtigkeit und Reproduzierbarkeit ankommt. In einem zweiten Schritt lassen sich dann die Bestimmungen für einzelne herausgegriffene Elemente oder Elementgruppen verbessern. Für besonders präzise und richtige Bestimmungen im Bereich extrem niedriger Spurenelementgehalte werden sich allerdings zusätzliche chemische Analysenschritte (Auflösung, Vorkonzentrierung, Abtrennung sowie Meßpräparateherstellung) vielfach nicht vermeiden lassen.

Bei der Auswahl der geeignetsten Instrumentellen Multielementanalyse sind verschiedene Gesichtspunkte zu berücksichtigen.

(2) An erster Stelle steht die Frage, ob die zu bestimmenden Elementarten und deren Konzentrationen im optimalen Einsatzbereich der verfügbaren Methode liegen. Ist nur eine einzige Multielementmethode vorhanden, so wird man mit ihr arbeiten müssen, auch wenn sie für einzelne Elementkombinationen nicht optimal ist. Man wird sich umsehen, ob eine günstigere Multielementmethode im eigenen Laboratorium vorhanden oder in benachbarten Laboratorien und Instituten zugänglich ist.

(3) Für rasche Übersichtsanalysen sehr vieler Elemente in einer Probe ist es zweckmäßig, mehrere Instrumentelle Multielementanalysen nacheinander zu kombinieren. Jede Methode hat ihren optimalen Bereich, in dem bestimmte Elementgehalte zuverlässiger bestimmt werden können als andere, die unsicher sind. Ein Beispiel hierfür gibt Abbildung 14. Sie zeigt das Ergebnis des Einsatzes der Funkenmassenspektrometrie mit Niedervolt-Bogenentladungsquelle, Bogen-Atomemissionsspektrometrie, wellenlängendispersiven Röntgenfluoreszenzanalyse, ICP-Atomemissionsspektrometrie und Atomabsorption für die halbqualitative bis quantitative Gehaltsbestimmung von möglichst vielen Elementspuren in einem uranhaltigen Boden. Dieses Diagramm liegt auch dem Einbandentwurf dieses Buches zugrunde. Hier sieht man zum Beispiel sofort, daß der Urangehalt von drei Methoden nahezu über-

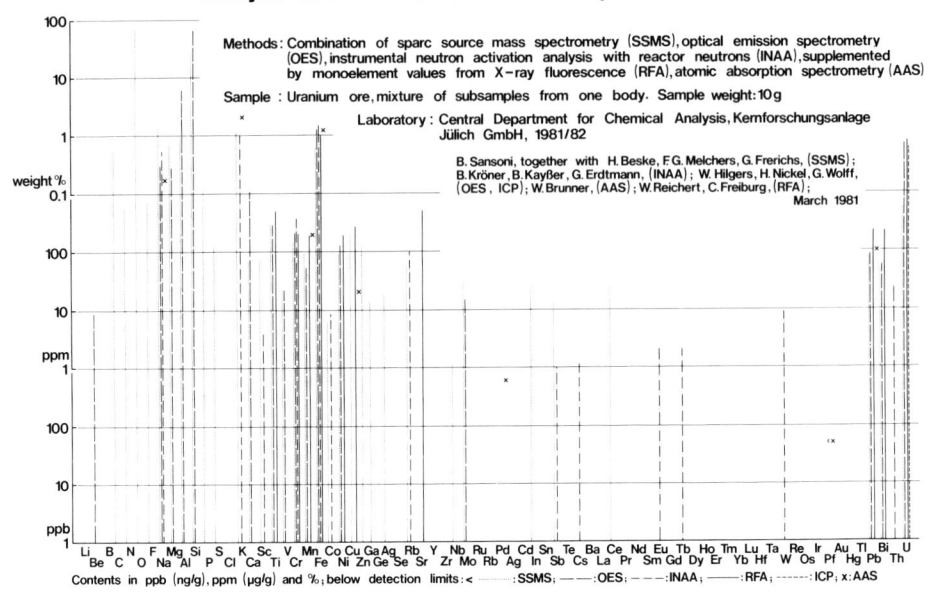

Semiquantitative Instrumental and Simultaneous Multielement Analysis of All Elements in One Sample

Methods: Combination of sparc source mass spectrometry (SSMS), optical emission spectrometry (OES), instrumental neutron activation analysis with reactor neutrons (INAA), supplemented by monoelement values from X-ray fluorescence (RFA), atomic absorption spectrometry (AAS)

Sample: Uranium ore, mixture of subsamples from one body. Sample weight: 10 g

Laboratory: Central Department for Chemical Analysis, Kernforschungsanlage Jülich GmbH, 1981/82

B. Sansoni, together with H. Beske, F.G. Melchers, G. Frerichs, (SSMS); B. Kröner, B. Kayßer, G. Erdtmann, (INAA); W. Hilgers, H. Nickel, G. Wolff, (OES, ICP); W. Brunner, (AAS); W. Reichert, C. Freiburg, (RFA);

March 1981

Contents in ppb (ng/g), ppm (µg/g) and %; below detection limits: < ; ———:SSMS; — · —:OES; ———:INAA; ———:RFA; ·······:ICP; x:AAS

Abbildung 14: Einsatz mehrerer Instrumenteller Multielementmethoden zur Analyse von Haupt-, Neben- und Spurenbestandteilen in einer uranhaltigen Bodenprobe.

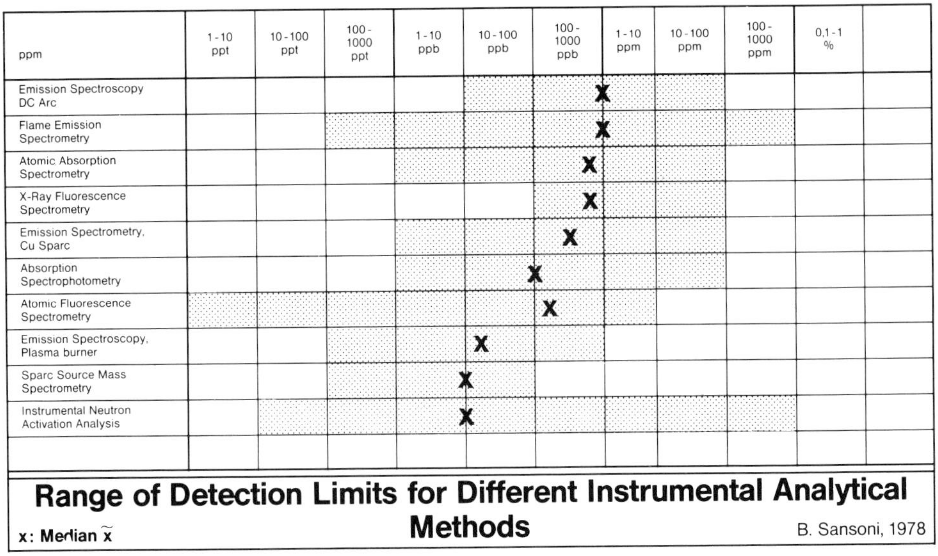

ppm	1-10 ppt	10-100 ppt	100-1000 ppt	1-10 ppb	10-100 ppb	100-1000 ppb	1-10 ppm	10-100 ppm	100-1000 ppm	0.1-1 %
Emission Spectroscopy DC Arc						x				
Flame Emission Spectrometry						x				
Atomic Absorption Spectrometry							x			
X-Ray Fluorescence Spectrometry							x			
Emission Spectrometry, Cu Sparc						x				
Absorption Spectrophotometry					x					
Atomic Fluorescence Spectrometry					x					
Emission Spectroscopy, Plasma burner				x						
Sparc Source Mass Spectrometry				x						
Instrumental Neutron Activation Analysis				x						

Range of Detection Limits for Different Instrumental Analytical Methods

x: Median $\tilde{x}$

B. Sansoni, 1978

Abbildung 15: Bereiche und Mediane der Nachweisgrenzen für verschiedene Methoden der instrumentellen Elementanalyse

einstimmend gefunden wird, die vierte Methode jedoch fast um eine Größenordnung darunter liegt. Mit hoher Wahrscheinlichkeit sind die drei übereinstimmenden Werte richtiger als der Ausreißer.

(4) Ein weiterer wichtiger Punkt ist der dynamische Bereich einer Multielementmethode. Er soll möglichst viele Konzentrationsbereiche aller zu bestimmenden Elementgehalte der Probe umfassen, um diese jeweils mit der gleichen, linearen Eichkurve ermitteln zu können. Andernfalls müssen mehrere Multielementbestimmungen bei unterschiedlicher Verdünnung ausgeführt werden. Dadurch kann viel vom Multielementcharakter der Methode verloren gehen. Extreme Beispiele sind die ICP-Atomemissionsspektrometrie mit einem sehr großen Konzentrationsbereich von 4 - 5 Größenordnungen und die Atomabsorptionsspektrometrie mit einem extrem kleinen dynamischen Bereich von nur etwa 2 Größenordnungen.

(5) Die Kostenkalkulation für die Neuanschaffung von Geräten zur Multielementanalyse wird naturgemäß günstiger, wenn sie auf die Einzelelementbestimmung und nicht auf die Analysenprobe bezogen wird. Die Anschaffungskosten amortisieren sich umso früher, (a) je größer die Serien gleichartiger Proben und b) je größer die in jeder Probe zu bestimmende Anzahl Elemente ist. Die Frage ist nur, ob alle von der Instrumentellen Multielementanalyse gelieferten Elementgehalte dem Auftraggeber so nützlich sind, daß er sie auch vergütet. Deshalb wird eine andere Kostenabschätzung auch davon ausgehen, daß ursprünglich nur einzelne der bestimmbaren Elemente vom Auftraggeber bestellt wurden und die übrigen Elementgehalte gewissermaßen eine kostenlose Dreingabe sind. Das Ergebnis dieser Betrachtung hängt vom Einzelfall ab.

(6) Welche Multielementanalysemethoden sind für ein modernes Analysenlaboratorien notwendig und einsatzbereit zu halten? Für Lösungen ist nach wie vor die Atomabsorptionsspektrometrie die am meisten verwendete Instrumentelle Analysenmethode. Sie hat allerdings ausgesprochen Monoelementcharakter. Als erste Multielementmethode für Lösungen kommz danach derzeit ohne Zweifel die ICP-Atomemissonsspektrometrie in Betracht. Zur Multielementanalyse in Feststoffen leistet die Röntgenfluoreszenzanalyse für Haupt- und Nebenbestandteile gute Dienste. Sie hat allerdings häufig mit systematischen Fehlern zu kämpfen, auch reicht sie nicht bis in den untersten ppm-Bereich. Sie eignet sich jedoch besonders für große Probenserien mit ähnlicher Zusammensetzung, für welche die Arbeitsbedingungen vorher sorgfältig ausgearbeitet worden sind. Für quantitative Übersichtsanalysen wird häufig die optische Bogen-Atomemissionsspektrometrie eingesetzt. Allerdings lassen hier Reproduzierbarkeit und Richtigkeit zu wünschen übrig. Für Multielementanalysen mit extrem vielen Elementen des Perioden-

systems bis herab zu etwa 0,1 atom-ppm in elektrisch leitenden Metallen oder Graphit ist die Funkenmassenspektrometrie die optimale Methode der Wahl. Für Nichtleiter gewinnt die Laser-Massenspektrometrie an Bedeutung. Beide Methoden sind jedoch sehr kosten- und personalaufwendig. Die Neutronenaktivierungsanalyse zeigt für bestimmte Elemente höchste Empfindlichkeiten, für andere ist sie jedoch unempfindlich. Einer ihrer Hauptvorteile ist die Zerstörungsfreiheit und eine geringe Gefahr der Einschleppung von Kontaminationen. Sie setzt jedoch die Verfügbarkeit eines Atomreaktors voraus, wenn schon nicht am gleichen Ort, so doch wenigstens in der Nähe.

Als Ergänzung kommt noch die eine oder andere Spektrometrieart hinzu. Aussichtsreich sind hierbei für einfachere Matrices wie zum Beispiel Wasser, die schnelle sequentielle ICP-Atomfluoreszenzspektrometrie und die Voltammetrie. Letztere erlaubt als Inversvoltammetrie die empfindlichsten Bestimmungen toxischer Metalle in Umweltproben bis in den sub-ppb-Bereich.

(7) Eine optimale Multielementanalyse sollte möglichst wenig störanfällig für wechselnde Matrix- und Elementzusammensetzung sein. Hier bietet die extrem hohe Temperatur der ICP-Plasmamethoden die Gewähr, daß nahezu alle chemischen Bindungen gespalten und dadurch die meisten chemischen Störungen eliminiert werden. Nachteilig ist andererseits der extrem große Linienreichtum, der häufig zu spektralen Interferenzen führt.

(8) Große Routineserien erfordern möglichst weitgehende Vollautomation. Diese kann aber auch schon bei kleineren Serien von 10 bis 20 Proben Personalkosten einsparen helfen.

(9) Die anschließende Computerauswertung muß dem großen Anfall von Meßdaten angepaßt sein. Multielementmethoden liefern häufig in kurzer Zeit viele Meßreihen, die durch den Rechner am Gerät häufig nicht vollständig aufgearbeitet werden können. Dann wird das Zusammenstellen von Tabellen, Berechnen von statistischen Daten, Zeichnen von Diagrammen und Kurven oder das Schreiben der Analysenprotokolle sehr viel zeitraubender als die eigentliche Messung. Hier geht der zukünftige Trend dahin, unmittelbar am Meßgerät Mikroprozessoren zu haben, die auch eine anspruchsvollere Datenverarbeitung, zum Beispiel multivariante Statistiken der Chemometrie unmittelbar am Gerät berechnen und als abgabefertige Protokolle mit Tabellen und graphischen Darstellungen ausgeben.

(10) Unverzichtbar ist eine ausreichende Qualitätskontrolle (29, 48) der erhaltenen Analysendaten. Das beinhaltet das Einschleusen von Kalibrierlösungen und Standards, die Berechnung von Kalibrierkurven und Richtig-

keiten sowie deren Verwendung zur anschließenden Korrektur von unrichtigen Meßwerten.

(11) Eine genaue Kenntnis von Matrix-, Interelement- und spektralen Störungen ist unerlässlich, um die Richtigkeit der bestimmten Elementgehalte sicherstellen oder zumindest beurteilen zu können.

(12) Ein großer Linienreichtum der Spektren ist besonders störend bei der Bogen-, Funken- und Plasma-Atomemissionsspektrometrie. Auch in der Funkenmassenspektrometrie stören die vielen Linien der bei der Elementanalyse an sich gar nicht gewünschten Isotope. Besonders linienarm sind die Spektren der Atomabsorption- und vor allem der Atomfluoreszenz.

(13) Da das Signal/Rausch-Verhältnis die Nachweisgrenzen wesentlich beeinflußt, sind möglichst rauscharme Spektren mit geringem Untergrund erwünscht. Einen sehr geringen Untergrund haben die Alpha- und Atomfluoreszenzspektren.

(14) Einen Vergleich der Nachweisgrenzen für die einzelnen Elementanalysemethoden gibt Abbildung 15. Dort ist jeweils der gesamte Bereich für alle bestimmbaren Elemente als Block angegeben. Das darin markierte Kreuz bezeichnet den Median. Die zugrundeliegenden Nachweisgrenzen sind allerdings nicht auf neuestem Stand (40). Danach nimmt die Nachweisstärke (Nachweisgrenze) der Instrumentellen Elementanalysemethoden in folgender Reihe zu:
Bogen- und Flammen-Atomemission < Röntgenfluoreszenz, Atomabsorption (Flamme) < Funken-Atomemission < Atomfluoreszenz < Plasma-Atomemission, Atomabsorption (elektrothermal) < Funkenmassenspektrometrie, Neutronenaktivierung.

8.2 Umbruch der Forschungsstrategie

Schon häufig hat in der Geschichte der angewandten Naturwissenschaften und Technik die Entdeckung neuer analytischer Methoden eine Änderung bisheriger Forschungsstrategien bewirkt. Der geschilderte Einsatz moderner instrumenteller Multielementanalysemethoden hat ebenfalls grundsätzliche und weitreichende Folgen für die Forschungsplanung des Auftraggebers aus Naturwissenschaft, Medizin, Technik und Umweltschutz. Diese sind jedoch noch keineswegs voll in das Bewußtsein gedrungen.

Aus mehreren Gründen wird hier ein Umbruch der Denkweise notwendig.

(1) In der Auftragsanalytik hat der Auftraggeber bisher nur das eine oder die wenigen Elemente in seinen Analysenproben bestimmen lassen, die für die ursprüngliche Aufgabe bzw. das Forschungsprojekt benötigt wurden. Weitere zusätzliche Elementgehalte interessierten nicht oder kaum und bildeten nur unerwünschte "Datenfriedhöfe". Jetzt aber kann der Auftraggeber mit nur unwesentlich höherem Aufwand und Kosten statt eines oder zweier Elementgehalte manchmal deren zehn, zwanzig oder mehr, zunächst ungewollt, mitgeliefert bekommen. Das führt logischerweise zu der Konsequenz, in echter Partnerschaft mit dem Analytiker das Forschungsziel neu zu überdenken und die zusätzlichen Elemente mit in die Zielsetzung des Projektes einzuarbeiten und gezielt zu berücksichtigen. Bei seltenen, komplizierten und schwer zugänglichen Proben sind Probenahme und Probenvorbereitung häufig unvergleichlich aufwendiger und kostspieliger als die zusätzliche Mitbestimmung durch eine Multielementanalyse. Dies ist bereits seit langem in der Geochemie eine übliche Denkweise. Dort werden heute in den oft mühsam zu beschaffenden und vorzubereitenden Proben grundsätzlich alle Elemente, die durch vernünftigen Aufwand mit Multielementmethoden analysierbar sind, bestimmt. Dabei spielt es keine Rolle, ob die erhaltenen Elementgehalte sofort benötigt werden oder nicht.

(2) Ein weiterer Grund liegt in der zunehmenden Vollautomatisation instrumenteller Analysengeräte. Bislang hatte man häufig nicht nur möglichst wenige Elemente bestimmen, sondern auch so wenig Proben wie möglich analysieren lassen, um Kosten und vermeintlichen Aufwand zu sparen. Hier schafft die zunehmende Vollautomation eine völlig neue Situation. Es bedeutet einen nur unwesentlichen Mehraufwand, mit einem Analysenautomaten statt zehn deren hundert Proben zu analysieren, soferne die Matrix und Elementzusammensetzung gleichartig sind. In diesem Fall wird man den Auftraggeber aufklären, nicht so wenig Proben wie möglich, sondern so viele wie für sein Projekt sinnvoll, zur Analyse zu geben.

(3) Eine dritte Änderung der Strategie bewirkt der Zwang zur Rationalisierung und Kosteneinsparung. Vollautomatisierte Multielementanalysemethoden stehen häufig vor dem Dilemma viel zu kleiner Analysenserien, die keinen wirklich vollen Nutzen mehr aus der Vollautomation ziehen können. Je größer die Probenserien, desto früher hat sich die teure Großgeräteanschaffung amortisiert, desto eher lohnt die Anschaffung. Es ist daher eine zwangsläufige Entwicklung, kleine und dezentrale Analyselaboratorien mit für sich jeweils zu kleinem Probenanfall zu einem größeren zentralen Laboratorium zusammenzulegen, um dadurch große Probenserien ähnlicher Zusammensetzung zu erhalten. Dabei wird man auch Analysenverfahren gegebenenfalls so ändern, daß verschiedenartige Elemente mit unterschiedlicher Konzentration und sogar Matrix mit dem gleichen Einheitsverfahren analy-

siert werden können. Ob die relativ geringen Mehrkosten für die Multielementanalyse durch den beträchtlichen Zuwachs an Information aufgewogen werden, hängt vom Einzelfall ab und läßt sich nicht allgemein vorhersagen.

(4) Auf der anderen Seite bringt die moderne Automation Gefahren mit sich. Diese können bei ausgesprochenen Instrumentellen Multielementmethoden besonders schwerwiegend sein. Die zugehörenden Großgeräte sind nach Festlegung der detaillierten Versuchsparameter durch einen Fachanalytiker zwar sehr leicht von einem Techniker, Laboranten oder sogar einer weniger gebildeten Hilfskraft zu bedienen. Der Analysenautomat liefert dann Zahlenreihen, Tabellen, Grafiken mit statistischer Auswertung, bis hin zu abgabefertigen Analysenprotokollen. Nur sind es leider nicht immer die richtigen Zahlen. Systematische Fehler des Analysenverfahrens wirken sich hier besonders verheerend aus. Nicht nur die Wartung der Maschine, sondern auch die Beurteilung der Richtigkeit der erhaltenen Analysenergebnisse erfordern jahrelange Erfahrung eines hoch qualifizierten Analytikers. Dies trifft vor allem im Bereich von Spurenkonzentrationen zu. Es ist daher unabdingbare Voraussetzung für die richtige Anwendung automatisierter Multielementanalysemethoden, durch ständige Kalibrierung, Überprüfung der Eichkurven, Vergleich mit synthetischen Standards und zertifizierten Standardreferenzmaterialien in ausreichend kurzen Abständen von zum Beispiel 10 bis 20 Analysenproben für eine ausreichende Qualitätskontrolle der erhaltenen Analysenergebnisse zu sorgen. Besonders störanfällig sind dabei die chemischen Methoden der Probenvorbereitung, Anreicherung und Trennung.

Zusammenfassend erfüllen die Multielementmethoden wie alle rein instrumentellen Analysenmethoden das eingangs erwähnte strategische Konzept, soviel überflüssige Teilschritte des allgemeinen Analysenganges (Abbildung 1) wie möglich wegzulassen. Es bleiben nur mehr Probenahme, mechanische Probenvorbereitung, gegebenenfalls Auflösung, Meßpräparateherstellung, Messung des Analysensignales, Datenverarbeitung, Datenbeurteilung und Qualitätskontrolle übrig. Gelegentlich ist der Wegfall der chemischen Analysenschritte allerdings mit einer Einbuße an Zuverlässigkeit und Richtigkeit des Ergebnisses erkauft. Dafür wird mit nur wenig größerem Aufwand ein vielfaches an Information geliefert, dem eine geänderte Forschungsplanung Rechnung zu tragen hat.

9. Literatur

1) L.H. Ahrens, S.R. Taylor, Spectrochemical Analysis, 2nd. Edition, Addison-Wesley Publishing Company, Inc., Reading, London, 1961; 454 S.

2) (a) A. Alian, B. Sansoni, Comparison of different methods for activation analysis of geological and pedological samples: reactor and epithermal neutron activation, relative and monostandard method; Berichte der Kernforschungsanlage Jülich, Nr. 75 (April 1980); 46 S.

   (b) A. Alian, B. Sansoni, Review on the monostandard method in multi-element neutron activation analysis, in: B. Sansoni, (Hrsg.)Instrumentelle Multielementanalyse, Verlag Chemie, Weinheim etc, 1985, im Druck;

3) I.P. Alimarin (Ed.), Analysis of High-purity Materials, Israel Program for Scientific Translations, Jerusalem 1968; 584 S.

4) H. Aly, R. Serrano (Hersg.), Probenahme, Theorie und Praxis, Verlag Chemie, Weinheim etc., 1980; 384 S.

5) K. Bächmann, Messung radioaktiver Nuklide, Verlag Chemie GmbH, Weinheim/Bergstr., 1970; 183 S.

6) E.J. Baird, Introduction to Chemical Instrumentation, Electronic Signals and Operations, McGraw-Hill Book Company, Inc., New York etc., 1962

7) R.M. Barnes, Emission Spectroscopy, Dowden, Hutchinson & Ross, Inc., 1976; 548 S.

8) F. Barwinek, W. Kreuzer, B. Sansoni, Zum ökologischen Verhalten von Fe-55 in Pflanzen sowie in Fleisch von Rind und Wild, Fleischwirtsch. **60** (1980), 123-132

9) R. Bock (revised by I.L. MARR), A Handbook of Decomposition Methods in Analytical Chemistry , International Textbook Company Ltd., Bishopbriggs, Glasgow, 1979; 444 S. (mehrere tausend Literaturhinweise)

10) R. Bock, Aufschlußmethoden der anorganischen und organischen Chemie, Verlag Chemie GmbH., Weinheim/Bergstr., 1972; 232 S.

11) H.J.M. Bowen, Trace Elements in Biochemistry, Academic Press, London - New York, 1966; 241 S.

12) C. Brunnée, H. Voshage, Massenspektrometrie, Verlag Karl Thiemig KG, München, 1964; 316 S.

13) W. Brunner, Automatisierung zweier handelsüblicher Atomabsorptions-Spektrometer mit Flamme und elektrothermaler Anregung, Spezielle Berichte der Kernforschungsanlage Jülich, Nr. 46 (Juli 1979); 32 S.

14) W. Brunner, H. Heckner, B. Sansoni, Vollautomatischer Simultanbetrieb mehrerer Atomabsorptionsspektrometer im Analysenauftragsdienst eines Forschungszentrums durch Einsatz eines Prozeßrechners, in: K.H. Koch, H. Massmann (Hrsg.), 13. Spektrometertagung, Walter de Gruyter & Co, Berlin, New York, 1981; S 99-119

15) G.D. Christian, F.J. Feldman, Atomic Absorption Spectroscopy; Applications in Agriculture, Biology and Medicine, Wiley-Interscience, New York, 1970

16) J.A. Deans, T.C. Rains (Ed.), Flame Emission and Atomic Absorption Spectrometry-Components and Techniques, Vol 2, Marcel Dekker, New York, 1971; 362 S.

17) L. de Galan, Analytical Spectrometry, Adam Hilger, London 1971

18) D.R. Demers, I.E.B.M. Jansen, in: (Hrsg.) Instrumentelle Multielementanalyse, B. Sansoni, Verlag Chemie, Weinheim etc., 1985, in Druck;

19) D. DeSoete, R. Gijbels, J. Hoste, Wiley-Interscience, London etc., 1972; 836 S.

20) H.-J. Dietze, Massenspektroskopische Spurenanalyse, Akademische Verlagsgesellschaft Geest & Portig, H.G., Leipzig 1975; 224 S.

21) J. Dolezal, P. Povondra, Z. Sulcek, Decomposition Techniques in Inorganic Analysis, London Iliffe Books Ltd., New York, American Elsevier Publ. Comp., Inc. 1968; 224 S.

22) H. Ewald, H. Hintenberger, Methoden und Anwendungen der Massenspektroskopie, Verlag Chemie GmbH, 1953; 288 S.

23) G.W. Ewing, Instrumental Methods of Chemical Analysis, 4. Aufl. MacGraw-Hill, Book Company, New York etc., 1975; 560 S.

24) P.D. La Fleur (Ed.), Accuracy in Trace Analysis: Sampling, Sample Handling, Analysis, Vol I und II, U.S. Department of Commerce, Nat. Bur. Stand. (U.S.), Spec. Publ. 422, I und II, 1976; 645 und 636 S.

25) R. Fresenius, Anleitung zur qualitativen chemischen Analyse, Braunschweig, 7. Ed., 1852

26) (a) E.S. Gladney, C.E. Burns, D.R. Perrin, I. Roelandts, T.E. Gills, 1982 Compilation of Elemental Concentration Data for NBS Biological, Geological and Environmental Standard Reference Materials, NBS Special Publication 260-88, U.S. Department of Commerce, National Bureau of Standards, 1984; 221 S.

(b) E.S. Gladney, E. Burns, 1982 Compilation of Elemental Concentrations of Eleven United States Geological Survey Rock Standards, Geostandard Newsletter, 7 (1983) 3 ff.

27) M. Geissler, Polarographische Analyse, Verlag Chemie, Weinheim etc., 1981; 194 S.

28) T.T. Gorsuch, The Destruction of Organic Matter, Pergamon Press, Oxford etc., 1970; 151 S.

29) G. Gottschalk, Einführung in die Grundlagen der chemischen Materialprüfung, S. Hirzel Verlag, Stuttgart, 1966

30) G. Gottschalk, R.E. Kaiser, Einführung in die Varianzanalyse und Ringversuche, Soforthilfe für die richtige statistische Auswertung von Datengruppen, Bibliographisches Institut Mannheim etc.; 1976; 165 S.

31) P. Hahn-Weinheimer, A. Hirner, K. Weber-Diefenbach, Grundlagen und praktische Anwendung der Röntgenfluoreszenzanalyse (RFA), Vieweg Verlag, Braunschweig etc., 1984

32) G.V. Iyengar, W.E. Kollmer, H.J.M. Bowen, The Elemental Composition of Human Tissues and Body Fluids, A Compilation of Values for Adults, Verlag Chemie, Weinheim etc., 1978; 151 S.

33) G.V. Iyengar, B. Sansoni, Sample preparation of biological materials for trace element analysis, in (45), S. 73 - 107

34) E. Jackwerth, Chemische Multi-Elementanreicherung - Probleme der Anpassung an Probenmaterial und Bestimmungsmethode, in: Instrumentelle Multielementanalyse (B. Sansoni, Hrsg.), Verlag Chemie, Weinheim etc., 1985, im Druck;

35) R. Kaiser, G. Gottschalk, Elementare Tests zur Beurteilung von Meßdaten, BI-Hochschultaschenbuch Nr. 774, Mannheim 1972; 165 S.

36) H. Kaiser, Guiding Concepts Relating to Trace Analysis Proceedings IUPAC Conference on Analytical Chemistry, Kyoto, 1976; Butterworth, 1977; S. 35 - 61

37) H. Kienitz (Hrsg.), Massenspektrometrie, Verlag Chemie GmbH, Weinheim, Bergstr., 1968; 883 S.

38) G.F. Kirkbright, M. Sargent, Atomic absorption and fluorescence spectroscopy, Academic Press, London etc., 1974; 798 S.

39) G.F. Knoll, Radiation Detection and Measurement, John Wiley & Sons, New York etc., 1979; 816 S.

40) O.G. Koch, G.A. Koch-Dedic, Handbuch der Spurenanalyse, 2. Auflage, Springer-Verlag, Berlin etc., Teil 1 und 2, 1974; ca. 1600 S.

41) E. Krell, H.-P. Frey, G. Gawalek, G. Werner, Einführung in die Trennverfahren, VEB Deutscher Verlag für Grundstoffindustrie, 1975; 338 S.

42) E. Lange, in: E. Lange, H. Göhr, Thermodynamische Elektrochemie, Dr. Alfred Hüthig Verlag GmbH., Heidelberg, 1962

43) K.H. Lieser, Einführung in die Kernchemie, Verlag Chemie GmbH., 1969; 717 S.

44) J.J. Lingane, Electroanalytical Chemistry, 2nd Ed., Interscience Publ., Inc., New York, 1958; 669 S.

45) G.H. Lowrison, Crushing and Grinding, The Size of Solid Materials, Butterworth & Co, Publ., Ltd., London, 1974; 286 S.

46) B. Magyar, Guide-lines to Planning Atomic Spectrometric Analysis, Elsevier Scientific Publishing Company, Amsterdam etc., 1982; 269 S.

47) R. Mannkopf, G. Friede, Grundlagen und Methoden der chemischen Emissionsspektralanalyse. Eine Einführung mit praktischen Arbeitshinweisen, Verlag Chemie GmbH, Weinheim/Bergstr., 1975; 218 S.

48) D.L. Massart, A. Dijkstra, L. Kaufman, Evaluation and Optimization of Laboratory Methods and Analytical Procedures, A survey of Statistical and Mathematical Techniques, Elsevier Scientific Publishing Company, Amsterdam etc., 1978; 596 S.

49) J. Minczewski, J. Chwastowska, R. Dybczynski, Separation and Preconcentration Methods in Inorganic Trace Analysis, Ellis Horwood Ltd., Publ., Chichester and John Wiley & Sons, New York etc., 1982; 543 S. (etwa 2500 Literaturhinweise)

50) G.H. Morrison (Ed.), Trace Analysis, Physical Methods, Interscience Publishers, New York etc., 1965; 582 S.

51) N.N., Elemental Analysis of Biological Materials, Current Problems and Techniques with Special Reference to Trace Elements, Technical Report Series No. 197, International Atomic Energy Agency, Vienna, 1980; 367 S.

52) T. Nakanishi, B. Sansoni, Low-energy photon spectrometry in nondestructive neutron activation analysis of environmental samples, J. Radioanalyt. Chem., 37 (1977), 945 - 955

53) R. Neeb, Inverse Polarographie und Voltammetrie, Neuere Verfahren zur Spurenanalyse, Verlag Chemie GmbH., Weinheim/Bergstr. 1969; 256 S.

54) H.W. Nürnberg (Ed.) Electroanalytical Chemistry, John Wiley and Sons, London etc., 1974; 609 S.

55) M. Pinta, Detection and Determination of Trace Elements. Absorption Spectrophotometry, Emission Spectroscopy, Polarography, Ann Arbor Science Publ., Inc., Ann Arbor, 1975; 588 S.

56) O. Proske, F. Ensslin, P. Dickens, K. Möhl, Analyse der Metalle, Band 3, Probenahme, Springer-Verlag Berlin etc., und Verlag Stahleisen m.b.H., Düsseldorf, 1956; 452 S.

57) R.D. Reeves, R.R. Brooks, Trace Element Analysis of Geological Materials, John Wiley & Sons, New York, etc., 1978; 421 S.

58) H. Rose, Ausführliches Handbuch der analytischen Chemie, Braunschweig, 1, 1851; Traité pratique d analyse chimique, Paris, 1 (1832) 446 - 452

59) L. Sachs, Angewandte Statistik. Planung und Auswertung, Methoden und Modelle, Springer-Verlag Berlin etc., 1974; 545 S.

60) B. Sansoni, Differenzierung starker Säuren durch konduktometrische Titration in Eisessig., Angew. Chem. 76 (1964) 184

61) B. Sansoni, V. Iyengar, Sampling and sample preparation methods for the analysis of trace elements in biological materials, Spezieller Bericht der Kernforschungsanlage Jülich, Nr. 13, Mai 1978 (ISSN 0343-7639); 79 S.

62) B. Sansoni, G.V. Iyengar, Sampling and storage of biological materials for trace element analysis, in (45), S. 57 - 71

63) B. Sansoni, R.K. Iyer, R. Kurth, Concentration of analytical data as part of data processing in trace element analysis, Fresenius Z. Anal. Chem. 306 (1981) 212-232

64) B. Sansoni, W. Kracke, R. Winkler, Rapid assay of environmental radioactive contamination with special reference to a new method of wet ashing, Berichte der Gesellschaft für Strahlen- und Umweltforschung m.b.H., Neuherberg bei München, GSF-Bericht S 68 (März 1969); 15 S.

65) B. Sansoni, V.K. Panday, Ashing in Trace Element Analysis of Biological Material, in: S. Fachetti (Ed.), Analytical Techniques for Heavy Metals in Biological Fluids, Elsevier Science Publishers B.V., Amsterdam; 1982 S. 91 - 131 (384 Literaturhinweise)

66) B. Sansoni, R. Stolz, Separation and Determination of Inorganic Ions in Glacial Acetic Acid, Angew. Chem. Int. Ed., 2 (1963), 615-616

67) E. Schroll, Analytische Geochemie, Band I: Methodik, Ferdinand Enke Verlag Stuttgart 1875: 292 S.

68) W. Seith, K. Ruthardt, Chemische Spektralanalyse. Eine Anleitung zur Erlernung und Ausführung von Emissions-Spektralanalysen, Springer-Verlag, Berlin etc., 5. Auflage, 1958; 162 S.

69) M. Slavin, Emission Spectrochemical Analysis, Wiley-Interscience, New York, 1971

70) K. Slickers, Automatic Emission Spectroscopy, Brühl Druck + Pressehaus Gießen, 1980; 243 S.

71) K. Sommer, Probenahme von Pulvern und körnigen Massengütern; Grundlagen, Verfahren, Geräte, Springer-Verlag, Heidelberg etc., 1979, 305 S.

72) R. Tertian, F. Claisse, Principles of Quantitative X-Ray Fluorescence Analysis, Heyden, London etc., 1982

73) M. Thompson, J.N. Walsh, A Handbook of Inductively Coupled Plasma Spectrometry, Blackie, Glasgow, London, 1983; 273 S.

74) H. Vassos, G.W. Ewing, Electroanalytical Chemistry, John Wiley & Sons, New York etc., 1983; 255 S.

75) B.V. L Vov, Atomic Absorption Spectrochemical Analysis, Adam Hilger, London, 1970

76) E. Wänninen, Trace Analysis and the Contamination Problem, in: L. Ninistö (Ed.), Euroanalysis IV, Reviews on Analytical Chemistry, Akademiai Kiadu, Budapest, und Association of Finnish Chemical Societies, Helsinki, 1982; p. 157-171

77) B. Welz, Atomabsorptionsspektrometrie, 3. Auflage, Verlag Chemie, Weinheim etc., 1983; 527 S.

78) B. Welz (Hrsg.), Atomspektrometrische Spurenanalytik, Verlag Chemie, Weinheim etc., 1982; 564 S.

79) B. Welz (Hrsg.), Fortschritte in der atomspektrometrischen Spurenanalytik, Band 1, Verlag Chemie, Weinheim etc., 1984; 673 S.

80) J.D. Winefordner, Trace Analysis, Spectroscopic Methods for Elements John Wiley & Sons, New York etc., 1976; 484 S.

81) M. Zief, J.W. Mitchell, Contamination Control in Trace Element Analysis, John Wiley & Sons, New York etc., 1976; 262 S.

82) M. Zief, R. Speights (Ed.), Ultrapurity, Methods and Techniques, Marcel Dekker, Inc., New York, 1972; 699 S.

83) K.I. Zil'bershtein, Spectrochemical Analysis of Pure Substances, Adam Hilger Ltd., Bristol, 1977; 435 S. (ca. 1500 Literaturhinweise)

SELECTED OPPORTUNITIES IN ANALYTICAL CHEMISTRY

George H. Morrison

Department of Chemistry, Cornell University,
Ithaca, New York   14853   USA

SUMMARY

Analytical chemistry is the science of measurement and characterization of chemical systems. Its methodology involves contributions from all areas of chemistry, as well as physics, engineering, instrumentation, and computer sciences. In turn, analytical chemical methods contribute to the development of these fields and others. For example, the amazing advances in integrated circuits and computers have led to the production of powerful analytical instruments. But these advances would not have been possible without the development of analytical methods for the detection of trace impurities in silicon and the examination of semiconductor surfaces. Thus, analytical chemistry is at the core of science and technology and contributes to the solution of problems in many areas, including biology and medicine, environmental and atmospheric studies, energy research, archeology, and materials science by providing essential information about the composition and nature of substances.

Periodically it is important to evaluate the status of a field, and a Committee to Survey Opportunities in the Chemical Sciences, under the guidance of the U.S. National Academy of Sciences/National Research Council has been actively engaged in evaluating progress in the chemical sciences. The present paper describes progress in analytical chemistry. Since 1965 tremendous strides have been made in the areas of computers, lasers, liquid chromatography, surface analysis, mass spectrometry, and ion selective electrodes, to name a few, and these will be described. Current methodology in a number of important areas of analytical chemistry will be demonstrated by vignettes of selected techniques. Finally, opportunities and the future of analytical chemistry will be discussed.

1 INTRODUCTION

It is generally recognized that science is essentially the study of quantitative relationships and that the unambiguous interpretation of factual information demands accurate and precise quantitative measurements. Chemistry is of particularly vital importance to all areas of scientific endeavor. Analytical chemistry, the science of measurement and characterization of chemical systems, plays a crucial role in providing the best possible support to keep pace with, and spur development of, all areas of science and technology.

Chemical measurements are an integral part of many scientific studies of natural phenomena. Analytical chemistry in particular, stands out as the study of "real world" problems, since analytical techniques developed by chemists are often fruitfully applied in such diverse fields as biology and medicine, geology, environmental science, materials science and metallurgy, agriculture, energy research and development, microelectronics and solid state physics, as well as basic chemical research. Advancement of knowledge in these fields is necessary to maintain the health, comfort, and economy of our society.

Periodically, it is important to evaluate the status of a field, and the Committee to Survey the Opportunities in the Chemical Sciences under the Guidance of the U.S. National Academy of Sciences/National Research Council has been actively engaged in evaluating progress in the chemical sciences. The chairman of the committee is Professor George Pimentel of the University of California at Berkeley. Thus, the report which is due to be completed this summer has come to be known as the Pimentel Report (1). The previous report on the status of chemistry was the Westheimer report published in 1965.

On the occasion of this symposium on instrumental multielement analysis it is the objective of this paper to acquaint the specialists at this conference with some of the developments in the broader field of analytical chemisty. Some of this material was prepared for use in the Pimentel Report. The contributors include Professor Allen Bard, the University of Texas; Professors R. G. Cooks and Fred Lyttle, both of Purdue University; Professor David Hercules, the University of Pittsburg; Professor G. H. Morrison, Cornell University; Professors Robert and Janet Osteryoung of the State University of New York at Buffalo; Professor L. B. Rogers of the University of Georgia; and Professor Charles Wilkins, the University of California at Riverside.

The field of analytical chemistry is as broad as chemistry itself. Additionally, there is considerable input from physics, engineering, computer science and mathematics, biological and life sciences. All of this knowledge is required to solve analytical problems posed by "real" samples. To meet this challenge requires a broad research program to explore the many different physical and chemical phenomena which provide the basis for modern chemical analysis. Analytical chemists, with their emphasis on the fundamental aspects of chemical measurement, have made considerable strides in the past decades to keep pace with the greater challenges posed by increasingly complex samples. Additionally, the kind of information required to solve problems has become more sophisticated. Many samples require multicomponent analysis or surface analysis or both. In situ analysis of hostile or inaccessible media is often necessary. As ever, there is the ultimate requirement that measurements be as accurate and precise as posssible. A few highlights of this effort will be examined in the remainder of this text.

Analytical chemists are undoubtedly the leaders among the chemical scientists when it comes to integrating advances in other disciplines into chemistry. In the last two decades technological developments from outside of chemistry have revolutionized analytical chemistry, in many cases influencing the way analytical chemists think about experiments. Principal breakthroughs which have profoundly influenced analytical chemistry come from computer science, laser physics and biotechnology to name a few. It is worthwhile to reflect on how these have impacted analytical chemistry.

2 COMPUTERS

The development of computer hardware and software has profoundly influenced the practice of analytical chemistry since the Westheimer report in 1965. Broadly speaking, the major influence of computers in analytical chemistry has been in instrumental control and data processing. Today, most analytical instruments have some form of cybernetic control ranging from dedicated microprocessors to fully interfaced general purpose mainframe computers. Analytical experiments which could not be done two decades ago are now routinely performed under computer control. For example, consider the revitalization of traditional electrochemical methods wrought by computer interfacing. Pulse and other voltammetric techniques have been devised which depend on recent developments in solid state circuitry and computers for their existence and popularity (2). Very fast repetitive sampling and background subtraction have driven detection limits of many important metal ions to the sub-ppb and upper-ppt concentration range (3).

The enhanced throughput and sensitivity of fourier domain spectroscopic techniques are now well known to the analytical community. Fourier transform infrared spectroscopy is by far the most popular. High resolution and multidimensional nuclear magnetic resonance is only possible through the use of transform techniques (4). Additionally, the fourier transform has appeared in other areas including mass spectrometry (5), two dimensional electron paramagnetic resonance (6) and electrochemistry (7). The principles of fourier analysis have been known to mathematicians for almost 150 years but practical implementation remained difficult even with the advent of computers. Until the fast fourier transform algorithm was rediscovered by Cooley and Tukey (8) this logjam remained. The proliferation of transform techniques in instrumental analysis can be traced directly to this development.

Data reduction, analysis and interpretation have benefitted enormously from the influx of computer technology. Digital filtering of analytical data, proposed in a classic paper by Savitsky and Golay in 1964 (9), remains an active field of analytical research twenty years later (10). Recently, Kalman filtering techniques have been applied to UV-Visible spectroscopy (11) and electrochemistry (12). Algorithms for extraction of multicomponent information from analytical data have been developed by many workers. Multicomponent methods often bear a considerable resemblance to image processing methods developed by NASA for interplanetary investigations. Indeed, there are many common mathematical tools used such as two dimensional integral transforms and multivariate statistical analysis. Image processing is another computationally intensive area which awaits its full implementation in analytical chemistry. However, image processing has already appeared in mass spectrometry (13), gel electrophoresis chromatography (14) and emission excitation fluorescence spectroscopy (15).

3 LASERS

Even two decades of intensive research and development have not brought the full power of laser light sources to bear on analytical problems. Recently, spectacular developments have been reported using lasers to analyze samples. The principles of resonance ionization spectroscopy and its ramifications in single atom detection have been described (16). Resonance ionization and other multiphoton methods are only possible with laser excitation sources. These have considerable potential, as the selection of several excitation wavelengths can be utilized to analyze for a specific atom or molecule in an extremely complex background.

The development of laser sources was a considerable boom to ordinary Raman spectroscopy, not only for traditional laboratory investigations, but for such applications as remote sensing of airborne pollutant gases (17).

The trend toward the development of higher energy lasers, such as the neodymium YAG laser, has opened the way for variations of the basic Raman scattering experiment. Coherent Anti Stokes Raman Spectroscopy (CARS) (18), a nonlinear technique, has considerable promise, expecially as a localized probe for hostile environments such as the internal chambers of working jet engines (19).

4 BIOSENSITIVE ELECTRODES

Another branch of electroanalytical chemistry which has seen remarkable advances has been ion selective electrode technology. These devices which are comparatively inexpensive and well suited to a variety of remote and hostile environment sensing applications have benefitted tremendously from advances in enzymology. Enzyme ion selective electrodes are now routinely constructed which can, in principle, analyze for a specific molecular entity in a complex biological matrix (20). Another attractive feature of ion selective electrodes is their amenability to miniaturization. Intracellular potassium and sodium ion distributions have been reported using miniature ISE's (21).

5 COMBINED TECHNIQUES

Two techniques are often better than one in analytical chemistry. The marriage of gas chromatography and mass spectrometry is one of the most successful unions in analytical instrumentation. This is as close as possible to a general purpose analytical instrument as there currently is. Other combinations include gas chromatography-fourier transform infrared (GC/FT-IR), liquid chromatography-mass spectrometry (LC/MS), and tandem mass spectrometry (MS/MS). The latter example has become very popular in recent years. By using a mass spectrometer instead of a chromatographic procedure to effect initial separation, the analysis of multicomponent mixtures is greatly speeded up (22). Another example of a multitechnique approach in instrumentation is spectroelectrochemistry (23) where the surface of an electrochemical cell is monitored spectroscopically during the course of an oxidation-reduction sequence. The future holds considerable promise for further development of these methods as well as new combinations of existing and perhaps yet to be developed methods. Additonally, new ideas in data manipulation, interpretation and display are needed to keep pace with the copious volume of results produced by these new generation instruments. Advances in computer graphics (24) have only scratched the surface in this area.

6 CONCLUSION

Many fascinating and important developments have obviously been omitted from this short report. To some extent, the choice of examples reflects the author's knowledge, background and intellectual approach to the challenges offered by contemporary analytical chemistry. However, one of the key themes of this report has been the fact that analytical chemisty like many other fields has been part of the information explosion which has taken place in recent decades. It is no longer sufficient for the analytical chemist to only know and appreciate chemisty. Depending on one's field of specialization, a nodding acquaintance or intimate involvement with computer science, mathematics, applied physics, engineering, biology, as well as the principles of management science are needed. This is necessary, not only to solve contemporary analytical problems, but to seek out the analytical problems of future years. Many areas await the attention of analytical chemistry. Some, but not all, may include further forays into novel and hostile media, analysis of planets, deep sea environments, high temperature and supercritical fluids, and noninvasive probes into the human body itself.

7 ACKNOWLEDGEMENT

The author wishes to thank Mark G. Moran for invaluable assistance in the preparation of this manuscript.

8 REFERENCES

1. Research Briefing, Report of the Research Briefing Panel on Selected Opportunities in Chemistry, National Academy Press, Washington, D. C. (1983).

2. J. B. Flato, Anal. Chem. $\underline{44}$ (1972) 75A.

3. S. D. Brown, B. R. Kowalski, Anal. Chem. $\underline{52}$ (1978) 2240.

4. B. C. Gerstein, Anal. Chem. $\underline{52}$ (1983) 781A.

5. J. D. Baldeschwieler, Science $\underline{159}$ (1968) 263.

6. G. Millhauser, J. H. Freed, J. Chem. Phys. (in press).

7. D. E. Smith, Anal. Chem. 48 (1976) 221A,.

8. J. W. Cooley, J. W. Tukey, Math. Comput. 19 (1965) 297.

9. A. Savitsky, M. J. E. Golay, Anal. Chem. 36 (1964) 1627.

10. M. H. A. Bromba, H. Ziegler, Anal. Chem. 53 (1981) 1583.

11. C. B. M. Didden and H. N. J. Poulisse, Anal. Lett. 13 (A11) (1980) 921.

12. P. F. Seelig, H. N. Blount, Anal. Chem. 48 (1976) 252.

13. B. K. Furman, G. H. Morrison, Anal Chem. 52 (1980) 2305.

14. P. H. O'Farrell, J. Biol. Chem. 250 (1975) 4007.

15. M. L. Gianelli, J. B. Callis, N. H. Andersen, G. D. Christian, Anal. Chem. 53 (1981) 1357.

16. G. S. Hurst, M. G. Payne, S. D. Kramer, J. P. Young, Rev. Mod. Phys. 51 (1979) 767.

17. J. G. Grasselli et al., Chemical Applications of Raman Spectroscopy, Wiley Interscience, 1981, pp. 160-161.

18. A. B. Hawey, Anal. Chem. 50 (1978) 905A.

19. A. Weber (Ed.), Topics in Current Physics--Raman Spectroscopy of Gases and Liquids, Springer-Verlag, Berlin 1979, p. 283.

20. G. A. Rechnitz, Science 214 (1981) 287.

21. H. M. Brown, J. D. Owen, Ion Sel. Electrode Rev. 1 (1979) 145.

22. F. W. McLafferty, Acc. Chem. Res. 13 (1980) 33.

23. A. T. Hubbard, Acc. Chem. Res. 13 (1980) 177.

24. T. E. Graedel, R. McGill, Science 215 (1982) 1191.

EINIGE PHYSIKALISCHE BETRACHTUNGEN ZUR ZWECKMÄSSIGEN AUSWAHL VON MULTI-ELEMENT-METHODEN

K. Laqua

Institut für Spektrochemie und angewandte Spektroskopie,
Bunsen-Kirchhoff-Straße 11, D-4600 Dortmund 1

ZUSAMMENFASSUNG

Die Multi-Element-Fähigkeit verschiedener instrumenteller Analysenverfahren unter Ausnutzung von optischen, Röntgen- und Massenspektren werden anhand von folgenden Kriterien diskutiert: Informationskapazität, dynamischer Bereich der Analysensignale, Selektivität, Spezifität und Eigenschaften der Meßempfänger. Weiter wird unterschieden nach Sequenz-, Simultan- und Quasisimultan-Verfahren, nach Probenart, -zustand und -verbrauch und nach der Art der spektralen Aussonderung. Die maßgebenden Rauschquellen werden aufgezeigt und ihr Einfluß auf das Nachweisvermögen, gegebenenfalls auch auf die Präzision und Richtigkeit beschrieben.

1. EINLEITUNG

Eine ideale Multi-Element-Methode erlaubt die genaue Bestimmung aller interessierenden Elemente in einer vorgegebenen Analysenprobe mit möglichst geringer Probenvorbereitung. In praktischen Fällen müssen oft erhebliche Abstriche gemacht werden. Es ist daher zweckmäßig, die Anforderungen an eine Multi-Element-Methode zusammenzustellen.

2. ANALYTISCHE EIGENSCHAFTEN

2.1 Dynamischer Bereich

Hierunter soll der nutzbare Konzentrationsbereich verstanden werden, wobei nähere Angaben unumgänglich sind. In der Literatur werden oft Angaben wie "fünf Zehnerpotenzen" gemacht, ohne daß ausgeführt wird, wie die Grenzen dieses Bereichs festgelegt werden. Der analytisch nutzbare Konzentrationsbereich hängt zunächst von der Analysenlinie und deren Anregung ab. Das untere Ende kann nicht tiefer als die Nachweisgrenze liegen; allerdings ist bei dieser die Präzision dann für die üblichen Anforderungen zu schlecht. Auch nach oben kann die Analysenlinie den Bereich begrenzen, z.B. durch Abnahme der Empfindlichkeit infolge von Selbstumkehr. Als Ausweg bietet sich an, mehrere Analysenlinien für die Bestimmung eines Elements zu verwenden. Der Strahlungsempfänger soll imstande sein, den spektralen Untergrund oder den spektralen Blindwert präzis zu messen. Im allgemeinen ist dafür die

spektrale Strahlungsleistung des Untergrundes hoch genug, so daß bei Meßzeiten die
für Durchschnittsanalysen typisch sind, z.B. 10 s bei Photovervielfachern (PV) mehr
als 15000 Photoelektronen (bei der optischen Emissionsspektralanalyse (OES)) erzeugt werden und der Dunkelstrom meist vernachlässigt werden kann. Nach oben wird
der lineare Bereich eines PV durch Sättigung der letzten Dynode begrenzt. Diese
hängt von der internen Verstärkung des PV ab. Es kann daher durchaus zweckmäßig
sein, diese niedrig zu halten und elektronisch nachzuverstärken. Wird der Photonstrom integriert, so können hierdurch Schranken gesetzt sein. Bei kontinuierlich
andauernden Signalen kann durch wiederholtes Entladen des Integrationskondensators
der nutzbare Bereich nach oben erweitert werden, nicht jedoch bei schnell veränderlichen, impulsförmigen, z.B. bei der Verdampfung und Anregung durch Laserimpulse.
Photonenzählung ist nur in besonderen Fällen ratsam; z.B. wenn dadurch eine
wirkungsvolle Unterdrückung des Untergrundes oder anderer spektraler Ordnungen
erreicht wird. In der optischen Spektroskopie ist diese Art der Strahlungsmessung
nur dann anzuraten, wenn der Dunkelstrom des PV und nicht der meist beträchtliche
spektrale Untergrund die entscheidende Rauschquelle darstellt, z.B. in der NichtResonanz-Atomfluoreszenz-Spektroskopie. Die obere Grenze eines modernen Photonenzählers ist etwa ebenso hoch wie bei einer Dauerstrommessung des PV (200 MHz entsprechen unter üblichen Betriebsbedingungen einem Anodenstrom von 1 µA). Im analytisch nutzbaren dynamischen Bereich sollte die Präzision, angegeben durch eine
relative oder absolute Standardabweichung bei Wiederholmessungen an den Grenzen ein
bestimmtes Vielfaches der letzteren nicht überschreiten.

Alle Methoden, bei denen analytische Signale in Emission beobachtet werden, also
z.B. Röntgen-Emission und -Fluoreszenz-Spektroskopie, optische Emissions- und
Fluoreszenz-Spektroskopie, Massenspektroskopie, kohärente Vorwärtsstreuung zählen
dazu, ebenso die auf den Spektroskopien der Atomkerne beruhenden Verfahren, die
nicht in diesem Zusammenhang weiter behandelt werden sollen. Nicht geeignet aus
dieser Sicht sind Absorptionsmethoden. Während mit Emissionsverfahren auch unter
hohen Anforderungen drei bis vier Zehnerpotenzen möglich sind, können mit den
letzteren nur eineinhalb Zehnerpotenzen erreicht werden. Bei der Bestimmung mehrerer Elemente, insbesondere in stark wechselnden Konzentrationen und in festen Proben
kommen also im allgemeinen nur Emissionsverfahren infrage. Besonders günstig sind
bei der OES Anregungsquellen, die mit vermindertem Druck betrieben werden.

2.2 Informationskapazität

Je mehr Informationen über die Zusammensetzung einer Probe eine Methode
vermitteln kann, umso besser sollte sie sein. H. Kaiser [1] hat das Rüstzeug der
Informationstheorie auf die Analytik angewandt. Danach besitzen spektroskopische
Verfahren eine besonders hohe Informationskapazität; es kann angenähert beschrieben werden durch:

$$P_{inf} = \bar{R} \log_2 \bar{S} \ln \frac{\nu_b}{\nu_a} \text{ bit},$$

worin $\bar{R}$ das mittlere praktische Auflösungsvermögen, $\bar{S}$ die mittlere Anzahl der unterscheidbaren Stufen bei der Intensitätsmessung und ν_a und ν_b die Grenzen des Spektralbereiches darstellen. Für die optische Spektralanalyse erhält man mit Beispielswerten von $\bar{R} = 10^5$, $\bar{S} = 10$, $\nu_b = 800$ nm, $\nu_a = 200$ nm ein $P_{inf} = 4,61 \cdot 10^5$ bit. Bei dem gewählten Auflösungsvermögen hätten in dem Wellenlängenintervall 150 000 äquidistant verteilte Linien Platz, die aber nur von etwa 90 Elementen stammen können. Man wird also auch, wenn es lohnend erscheint, mehrere Linien für eine Bestimmung heranzuziehen, diese Informationskapazität nicht ausnutzen; sie ist also gewissermaßen leer. Für eine nichtspektrale Methode, z.B. Einzelbestimmung nach Vortrennung, erhält man

$$P_{inf} = n \log_2 \bar{S} \text{ bit},$$

für eine Bestimmung aller Elemente, also

$$P_{inf} \approx 300 \text{ bit}.$$

Die nicht gleichförmige Verteilung der Spekrallinien führt trotz der hohen Informationskapazität zu Störungen durch Linieninterferenzen, ein Problem, das in der jüngsten Zeit wieder eingehend bearbeitet wird [2,3]. Obwohl eine Erhöhung der Meßgenauigkeit nur einen geringen Zuwachs an Informationskapazität bringt, lohnt es sich fast immer. Die anderen spektroskopischen Verfahren haben eine niedrigere "spektrale" Informationskapazität, sind aber leichter hinsichtlich der Linieninterferenzen zu beherrschen. Während sich bei der OES etwa 200 000 Linien auf ein Wellenlängengebiet von 3 Oktaven Umfang verteilen, sind diese bei der RFA etwa 10 000 Linien auf 10 Oktaven und bei der Elementmassenspektroskopie etwa 1000 auf 7 Oktaven. Es besteht kein Zweifel daran, daß sich auch aus der Informationstheorie die besonders gute Eignung von spektroskopischen Methoden für die Multielementanalyse ergibt.

2.3 Selektivität

In der Definition nach H. Kaiser [4] ist eine Methode vollkommen selektiv, wenn man mit ihr nebeneinander mehrere Komponenten unabhängig voneinander bestimmen kann. Dies ist ein großer Vorteil bei der Multi-Element-Analyse, da dann im allgemeinen keine aufwendige Probenvorbereitung (Abtrennung usw.) notwendig ist. Zwar hat sich die Vielfalt der Auswertemöglichkeiten seit [4] erhöht dank des mit Computern möglichen höherem Rechenaufwandes, so daß im Prinzip sogar die Intensität in jedem Meßkanal von den Konzentrationen aller Bestandteile abhängen kann; der Einfluß des Rauschens der unerwünschten Signalteile läßt sich jedoch nicht eliminieren. Präzision und Nachweisvermögen werden aber verschlechtert. Einige spektroskopische Verfahren mit besonders günstiger Anregung, z.B. mit Glimmentladungen oder mit ICP können sehr selektiv sein, sofern Linieninterferenzen vermieden und Einflüsse des Untergrundes berücksichtigt werden können.

2.4 Spezifität

Diese spielt bei der Multielementanalyse keine große Rolle, da sie nach [4] auf die Bestimmung eines einzigen Elementes (in Gegenwart anderer) zugeschnitten ist.

2.5 Präzision und Richtigkeit

Bei vielen analytischen Aufgaben wird aus zwei Gründen eine hohe Präzision als Gütezeichen für Wiederholbarkeit (in ein und demselben Laboratorium) oder für Vergleichbarkeit (bei verschiedenen Laboratorien) gefordert: zur Festlegung enger Fertigungstoleranzen (Vertrauensbereich) und zur Erkennung von systematischen Fehlern. Da die Präzision durch Zufallsschwankungen begrenzt wird, kommt ihr innerhalb eines ausgearbeiteten Verfahrens eine fundamentale Rolle zu. Mit sorgfältiger statistischer Auswertung, die heute stets mit multipler Regression erfolgen sollte, lassen sich auch geringe systematische Fehler erkennen und gegebenenfalls korrigieren, sofern nur die Präzision dafür ausreicht. Als besonders präzise haben sich die RFA und OES mit Funken und Glimmlampenanregung bewährt. Für die Zukunft sind auch große Fortschritte bei der Massenspektrometrie mit Glimmlampe und ICP zu erwarten. Die SNMS (secondary neutral particles MS) sollte eine große Zukunft als eine richtige Ergebnisse liefernde Massenspektroskopie haben. Der Begriff der Genauigkeit als übergeordnet der Präzision und Richtigkeit sollte nur dann benutzt werden, wenn es auf eine deutliche Trennung dieser Begriffe nicht ankommt.

2.6 Simultane und sequentielle Multielementanalyse

Die klassischen Verfahren der simultanen Multielementanalyse waren die OES und die Massenspektroskopie mit photographischer Registrierung. Bei beiden wird nur ein dispersives Element benutzt und die Strahlung, bzw. die Teilchen werden zu gleicher Zeit registriert. Die Photoemulsion als zweidimensionaler Strahlungsempfänger mit Abmessungen, die ein optimales spektrales Auflösungsvermögen über große spektrale Bereiche zulassen, ist in dieser Hinsicht unerreicht. Bei der Durchschnittsanalyse mit unbegrenzter Probenmenge lassen sich besonders niedrige Nachweisgrenzen erreichen, wobei die Möglichkeit, den Untergrund und Störlinien zuverlässig zu messen, besonders in Erscheinung tritt. Als Ersatz für die Photoemulsion mit begrenzten Einsatzmöglichkeiten werden nun die Kanalplatten, ein-und zweidimensionale Siliziumphotodiodenmatrizen und Fernsehaufnahmeröhren (z.B. Vidikon) benutzt. Ihnen allen ist gemeinsam, daß entweder sehr begrenzte Spektral-bereiche mit hoher Auflösung (notwendig für die Spurenbestimmung) oder größere Spektralbreiche mit niedriger Auflösung erfaßt werden können. In jedem Fall ist aber eine, wenn auch beschränkte simultane Multielementanalyse mit Untergrundkorrektur möglich. Besondere Arten der Simultananalyse findet man bei der RFA. Bei der wellenlängendispersiven RFA wird nur simultan mit einer Quelle angeregt, die Analysenstrahlung aber parallel mit einer Anzahl von kompletten Ein-Element-Spektrometern gemessen. Etwas ähnliches ist auch bei der optischen Atomfluoreszenzanalyse in einem kommerziellen Gerät verwirklicht. Echte Simultangeräte sind auch die energiedispersiven RFA-Spektrometer, die Flugzeitmassenspektrometer und schließlich die optischen Polychromatoren, die zwar nur ein dispersives Element besitzen, sonst aber auch nur Vielkanalinstrumente mit beschränkter Einsatzmöglichkeit und oft mangelnder Eignung für die Spurenbestimmung sind. Ein besonderes Problem stellt hier die verläßliche Untergrundmessung dar. Für die simultane Multielementanalyse mit der OES und photoelektrischer Strahlungsmessung sollten besondere Geräte entwickelt werden, um das photographisch erreichbare Nachweisvermögen zu erreichen. Sequentielle Geräte, z.B. Monochromatoren mit rascher Wellenlängeneinstellung können für Multielementanalysen eingesetzt werden, wenn die Analysendauer und der Probenverbrauch (bei probenverbrauchenden Methoden) tragbar ist. Beispiele aus anderen Bereichen sind die RFA Sequenzspektrometer und die Quadrupol-Massenfilter. Bei der OES werden außerordentliche Anforderungen an die Präzision der Wellenlängeneinstellung gestellt (± $0,0001^0$), wenn auf das zeitraubende Profilieren der Linien verzichtet werden soll. Die bei Elementspurenbestimmungen notwendige Untergrundmessung und -korrektur läßt sich bei zeitlich konstanten Signalen ebenfalls sequentiell durchführen. Bei zeitlich rasch veränderlichen Signalen dagegen ist eine simultane Untergrundmessung notwendig, die im allgemeinen mindestens einen zweiten Meßkanal erfordert. Es ist nicht allgemein bekannt, wie gut photographische Verfahren sein können (siehe Abb. 1 in [5]).

2.7 Kalibrierung

Bei der Multielementanalyse von Flüssigkeiten bereitet die Kalibrierung im allgemeinen keine besonderen Schwierigkeiten. Synthetische Proben oder in kritischen Fällen Zugabeverfahren erfüllen den Zweck. Neuerdings hat man sich auch bei der Plasmenspektroskopie (ICP u.a.) auf die Erfahrungen der Funkenspektroskopie und der photographischen Bogenmethoden besonnen und mißt Intensitätsverhältnisse (gegenüber einem vorhandenen oder zugesetzten Bezugselement (siehe z.B.[6]). Langzeitschwankungen und korrelierte Kurzzeitschwankungen können auf diese Weise eliminiert werden. Ein ideales Sequenzspektrometer für die Multielementanalyse sollte also drei frei einstellbare Meßkanäle haben: für die Analysenlinie, den Untergrund und die Referenzlinie. Die Kalibrierung bei der Analyse fester Proben für eine Multielementanalyse kann sehr aufwendig sein. Man sollte hier Methoden auswählen, die so selektiv sind, daß die Kalibrierung für das Element unabhängig von der für die anderen ist. Diesem Ideal kommen Glimmentladungen sehr nahe. Dann sind auch nur wenige Kalibrierungsproben notwendig. Mit Hilfe von multiplen Regressionsverfahren lassen sich jedoch, wenn auch mit sehr viel höherem Aufwand, auch anspruchsvolle Multielementverfahren zuverlässig kalibrieren.

2.8 Anzahl der zu bestimmenden Elemente

Mit keinem Verfahren und keinem Gerät lassen sich in der OES oder AFS alle Elemente bestimmen, wohl aber mit der Massenspektroskopie. Bei den OES und RFA Verfahren müssen je nach Wellenlängenbereich verchiedene dispersive Elemente und verschiedene Strahlungsempfänger benutzt werden. Außerdem muß für bestimmte Bereiche die Strahlung im Vakuum verlaufen. Vielkanalverfahren, z.B. bei der RFA lassen sich entsprechend instrumentell verwirklichen. In der OES müssen meist mehrere Spektralapparate benutzt werden, wenn stets die nachweisstärksten Linien gemessen werden sollen. Für viele Elemente gibt es allerdings auch genügend nachweisstarke Linien im UV und kürzerwelligem sichtbaren Spektralbereich. Die Kopplung mehrerer Spektrometer an eine Anregungsquelle ist heutzutage mit Hilfe von UV-durchlässigen optischen Fasern leicht möglich. Davon wird in kommerziellen Geräten Gebrauch gemacht.

Für die Bestimmung mehrerer Elemente gleichzeitig oder kurz nacheinander müssen die Analysenlinien entsprechend angeregt werden. Hierfür kommmt eigentlich nur Stoß- anregung in einem heißen Plasma infrage.
Für lokales thermisches Gleichgewicht LTE erhält man die spektrale Strahlungsdichte einer Linie aus dem Boltzmann Gesetz und der Einsteinschen Beziehung. Bei geringer Ionisation strebt sie mit größer werdender Temperatur einem Grenzwert zu. Bei höherer Temperatur nimmt aber auch die Ionisation zu. So ist die Normtemperatur, d.h. die Temperatur, bei der die Linie ein Maximum der spektralen Strahldichte erreicht, [7,8] in vielen Fällen etwa 10 000 K. Für die Analyse ist aber nicht nur die Strahldichte der Linie (diese ist besonders wichtig bei begrenzter Probenmenge), sondern das Verhältnis der Strahldichten von Linie und Untergrund maßgebend. Die spektrale Strahldichte des Untergrundes besteht aus Rekombinationsstrahlung und der hier nicht infragekommenden Bremsstrahlung [9]. Da der Untergrund mit der Temperatur steigt, ist die Normtemperatur meist zu hoch für ein gutes Nachweisvermögen. Falk und Mitarbeiter [10] haben gezeigt, daß in be- stimmten, nichtthermischen Plasmen (unter vermindertem Druck) ein besseres Ver- hältnis von Linienintensität und Untergrundintensität erreichbar ist als im Falle von LTE. Auch im ICP herrscht wohl kein LTE. Weitere Quellen von spektralem Untergrund sind nichtaufgelöste Bandenspektren (besonders bei Niederdruckentladun- gen und bei Hochfrequenzentladungen) und die Linienflügel von Spektrallinien mit Dispersionsprofil [12]. Diese spielen besonders bei ICP-Anregung eine große Rolle und erklären den Einfluß der Matrix auf das Nachweisvermögen.

2.9 Anregung, Spektralapparate und Strahlungsführung

Man kann leicht ausrechnen, daß in räumlich ausgedehnten Quellen wie in einem ICP in Volumina von der Größenordnung 1 cm^3 die Atome des Analysenelements viele Male, unter Umständen 10^3 mal, während des Aufenthaltes in der Beobachtungszone angeregt werden und Strahlung in einer Spektrallinie aussenden. Bei der Massenspektroskopie dagegen verschwindet ein Atom, nach der Ionisierung. Abgesehen von der Frage des Untergrundes, die in beiden Fällen verschieden ist, sollte man denken, daß die Anregung zur Strahlung für die OES sehr viel effizienter ist als die Ionisierung für die MS. Dies ist aber wegen Beschränkungen des Raumwinkels nicht der Fall. Für die optische Spektralanalyse ist zu hoffen, daß es gelingt, in Zukunft Ionen als Quelle von Strahlung für analytische Zwecke zu benutzen. Diese Ionen kann man fokussieren, festhalten und viele Male anregen. In der OES sollte stets und solange der spektrale Untergrund gemessen werden kann, für die Multielementanalyse ein Spektralapparat mit hohem theoretischen Auflösungsvermögen benutzt werden, wenn kleine Konzentrationen bestimmt werden sollen.

2.10 Probenverbrauch und wirksames Probenvolumen

Die Methoden der optischen Spektroskopie, bei der freie Atome im gasförmigen Zustand benötigt werden, und die Massenspektroskopie können auch nach dem Probenverbrauch beurteilt werden. Die Anforderungen an den Probenverbrauch hängen von der analytischen Fragestellung ab. Bei typischen Gleichstrombogenverfahren werden 10 bis 20 mg einer pulverförmigen Probe verbraucht, die repräsentativ für eine große Menge sein soll. Bei sehr inhomogenen Proben wird ein höherer Verbrauch angestrebt. Dies kann man z.B. mit einem magnetisch stabilisierten Bogen erreichen oder in einem Doppelbogen. Bei der industriellen Analyse von Metallen mit Funkenentladungen werden je Funken etwa 10^{-7} g abgebaut, d.h. bei typisch 2000 Funken für eine Analyse 2 · 10^{-4} g, d.h. es werden große Anforderungen an die Homogenität der Probe und an die Probenahme gestellt. Mit einem Impulslaser lassen sich ohne weiteres 10^{-6} g per Laserschuß verdampfen. Entgegen oft vorgebrachter Meinung ist der Abbau in einer Glimmlampe nach Grimm bei einer Meßzeit von 10 s mit etwa 3 · 10^{-4} g in derselben Größenordnung wie bei Funken. Die Strahlungsausbeute ist allerdings geringer (Niederdruckentladung!). Bei der RFA spielt die Tiefe, aus der Strahlung austritt, eine ähnliche Rolle wie die Erosionstiefe bei der optischen Spektralanalyse. Bei Oberflächen- und Teifenprofilanalysen hängen Präzision (Zählen von genügend Photoelektronen), Nachweisvermögen und Tiefenauflösung voneinander ab. Zu den etablierten Oberflächenmethoden ist unlängst die Oberflächenanalyse mit der Glimmlampe getreten. Sie eignet sich für rasche Übersichtsanalysen mit verhältnismäßig großem Abbau und Tiefenauflösungen von einigen nm. Wegen der völlig anderen Zerstäubung durch viele neutrale, durch Resonanzumladung entstandene Atome, die aus einem großen Raumwinkelbereich auf die **Analysenfläche** auftreffen, sind Matrix-und Struktureffekte sehr viel geringer als beim Abbau durch Zerstäubung im Hochvakuum.

2.11 Nachweisvermögen

Von den hier erwähnten Multielementmethoden hat die Massenspektrometrie im allgemeinen das beste Nachweisvermögen, sowohl mit Funken wie mit Laser- oder ICP Anregung. Auch ändert sich das Nachweisvermögen nur wenig von Element zu Element. Die optischen spektroskopischen Methoden dagegen unterscheiden sich um mehrere Zehnerpotenzen je nach Element. Dies liegt daran, daß bei der Massenspektroskopie meist eine nahezu völlige Ionisierung erfolgt und es nur wenige Ionenarten gibt. Dagegen wird bei der OES die Anregungsenergie auf viele Linien verteilt mit sehr unterschiedlichen Anregungswahrscheinlichkeiten. Besonders nachweisstark sind bei den herkömmlichen Methoden die Bogenanalyse und Niederdruckentladungen mit heißer Hohlkathode. Bei der RFA gibt es jeweils ein Maximum des Nachweisvermögens für die Anregung der K- und der L-Spektren [11]. Der maßgebende Untergrund rührt von der Streustrahlung der anregenden Strahlung, des Kristalls, der Matrix, usw. her.

2.12 Probenvorbereitung

Der Aufwand für die Probenvorbereitung wird in vielen Fällen die Wahl der Methode bestimmen. So ist es zu erklären, daß bei der Metallanalyse die eben geschliffene Scheibe erwünscht ist und die nachweisstarke Hohlkathode wegen der komplizierten Probenform sich nicht einbürgern konnte. Ähnliches gilt für Pulver.

3. NEUE ENTWICKLUNGEN

Es ist sicher, daß in der Zukunft eine Reihe von neuen spektroskopischen Methoden eine Rolle spielen wird, die mit den traditionellen in Konkurrenz treten werden.

3.1 Kohärente Vorwärtsstreuung (CFS). Dieses Verfahren kann leicht mit AAS Geräten, die für Zeeman-Korrektur mit transversalem Magnetfeld eingerichtet sind, durchgeführt werden. Der dynamische Bereich ist groß, trotz der quadratischen Abhängigkeit des Signals von der Konzentration. Für Multielementanalysen kommt hauptsächlich die Anregung durch die kontinuierliche Strahlung einer Blitzlampe infrage [13]. Es ist fraglich, ob die analytischen Leistungen ausreichen. Der Aufwand (Monochromator etc.) ist sehr viel größer als für die Einzelelementbestimmung mit einer Hohlkathodenlampe als Anregungsquelle.

3.2 Auf kohärenter monochromatischer Laserstrahlung beruhende Methoden. Hierzu zählen 'Laser-enhanced Ionisation Spectroscopy' (LEI) auch optogalvanische Spektroskopie genannt. Dieses ist sicher im Prinzip eine Einzelelementmethode und kann gegebenenfalls zu einer leistungsstarken sequentiellen Multielementmethode ausgebaut werden.

3.2 Laser-Atomfluoreszenzspektroskopie. Hier gilt ähnliches wie bei LEI.

3.3 Totalreflektierende RFA. Diese Methode hat zwar deutliche Beschränkungen hinsichtlich der analysierbaren Proben - sie sollten matrix-frei sein - ist aber eine ausgesprochene Multielementmethode, die sich besonders durch ein hohes absolutes Nachweisvermögen auszeichnet.

4. SCHLUSSWORT

Neben der Erforschung neuer Methoden können wohl alle beschriebenen Methoden nach den angegebenen Kriterien verbessert werden. Prinzipielle Grenzen, wie sie von Mandelstam [14] für den Lichtbogen und die Hohlkathodenlampe angegeben worden sind oder von Klockenkämper und Laqua [15] für die Analyse mit Verdampfung durch Laser sind bisher auch nicht annähernd erreicht worden. Es lohnt sich in vielen Fällen, den physikalischen Ursachen nachzugehen, z.B. ungenügende Atomisierung bei der Verdampfung, zu kurze Verweilzeit im Plasma, ungünstige Temperaturen und Elektronendrucke und gezielt Verbesserungen anzubringen.

LITERATUR:

[1] H. Kaiser in 'Methodicum Chimicum' Band 1, Teil 2, (1973).

[2] P.W.J.M. Boumans, 'Line Coincidence Tables for Inductively Coupled Plasma Atomic Emission Spectrometry', second edition, Pergamon Press, Oxford 1984.

[3] J.M. Mermet, G. Trassy, Spectrochim. Acta 36B (1981) 269.

[4] H. Kaiser, Z. Anal. Chem. 260 (1972) 252.

[5] U. Haisch, K. Laqua, W.-D. Hagenah, H. Waechter, Fresenius' Z. Anal. Chem. 316 (1983) 157.

[6] S.A. Myers, D.H. Tracy, Spectrochim. Acta 38B (1983) 1227.

[7] R.W. Larenz, Z. Phys. 129 (1951) 327.

[8] R. Diermeier, H. Krempl, Z. Phys. 200 (1967) 239.

[9] W. Lochte-Holtgreven, 'Plasma Diagnostics', North-Holland Publishing Company, Amsterdam 1968.

[10] H. Falk, E. Hoffmann, I. Jaeckel, Ch. Lüdke, Spectrochim Acta 34B (1979) 333.

[11] E.P. Bertin, 'Principles and Practice of X-Ray Spectometry Analysis', Plenum Press, New York, 1975.

[12] P.W.J.M. Boumans, J.J.A.M. Vrakking, Spectrochim. Acta 39B (1984) 1291.

[13] H. Debus, S. Ganz, M. Gross, G. Hermann, A. Scharmann, Optoelektronik Nr. 2 (1985) 123.

[14] S.L. Mandelstam, V.V. Nedler, Spectrochim. Acta 17 (1961) 885.

[15] R. Klockenkämper, K. Laqua, Spectrochim. Acta 32B (1977) 207.

INSTRUMENTELLE MULTIELEMENTBESTIMMUNG IN DER TÄGLICHEN PRAXIS

Knut Ohls
Hoesch Hüttenwerke AG, Chemische Laboratorien, D-4600 Dortmund 1

ZUSAMMENFASSUNG

Der Begriff "Instrumentelle Multielementbestimmung" ist in den letzten Jahren zu einem Schlagwort geworden. Jede Art von Verbundanalyse wird heute mit einbezogen, so daß hier an die klassische, simultane Multielementbestimmung erinnert werden soll, die in den letzten 25 Jahren zu einem wirtschaftlichen Faktor von großer Bedeutung geworden ist. Am Beispiel der Spektralanalyse werden die Vor- und Nachteile der simultan messenden Methoden gegenüber der sequentiellen Technik diskutiert.

Um die Entwicklung der ursprünglichen Multielementbestimmung zu veranschaulichen, werden die Simultanmethoden mit kompakten Originalproben beschrieben: die RFA als Material erhaltende, beliebig wiederholbare und die ICP- oder Funkenemissionsspektrometrie als Material verbrauchende Methoden. Die Probleme der Mikro- und Spurenanalyse in Industrielaboratorien werden als Grund für den zunehmenden Einsatz fester Proben gedeutet.

Zu der klassischen Multielementbestimmung gehört auch der Hinweis auf die bewährten Techniken der Eichung spektrometrischer Methoden. Die heutige Diskussion darf nicht so gedeutet werden, als ließen sich erst jetzt richtige Daten erstellen. Der über lange Jahre erfolgreiche Einsatz der Spektralanalyse, mit deren Daten ganze Prozeßabläufe gesteuert werden, beruhte letztlich auf der parallel ablaufenden chemisch-analytischen Eichkontrolle. Die Probleme durch die Einschränkung der chemischen Analyse in den Laboratorien und die Versuche, Ersatztechniken für die Eichung zu finden, müssen ständig angesprochen werden, um das Erreichte zu erhalten.

Wenn heute der Begriff "MULTIELEMENTANALYSE" genannt wird, dann fühlen sich derart viele Analytiker angesprochen, daß der Eindruck berechtigt erscheint, bisher sei eigentlich nur die Einzelelementbestimmung betrieben worden. Dies wird auch dadurch gestützt, daß genormte Analysenverfahren (DIN, EURONORM und ISO) bisher überwiegend nur die Bestimmung eines einzigen Elementes aus einer Probe beschreiben, obwohl dies der analytischen Praxis schon seit langer Zeit nicht mehr entspricht.

In unserem Industriezweig werden seit nunmehr 25 Jahren die verschiedenen Methoden der Spektralanalyse mit dem Zweck der MULTIELEMENTBESTIMMUNG eingesetzt (1. Bild), anfangs mehr aus zeitlichen Gründen, um den Produktionsprozeß beschleunigen zu helfen, heute mehr und mehr aus Gründen des rationellen Ablaufes analytischer Prozesse.

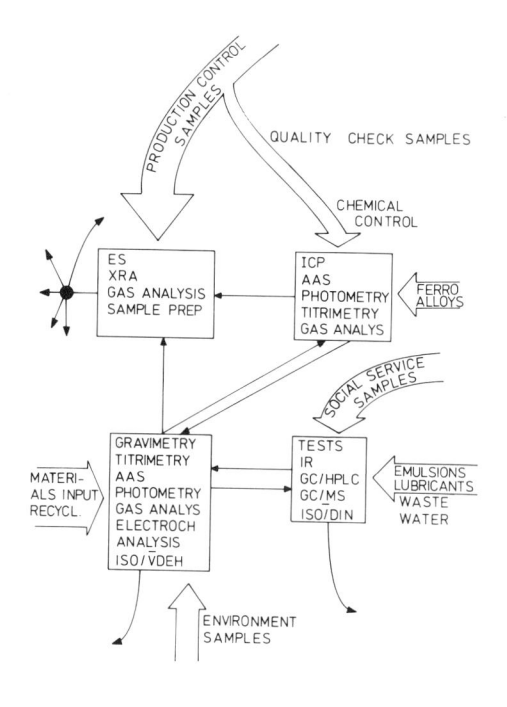

1. Bild: Informationssystem LABOR

Diese gesamte Richtung der schnellen, produktionsbegleitenden Analytik wurde bald mit dem Begriff "ROUTINEANALYSE" belegt, der ebenso falsch ist wie der Begriff "MULTIELEMENTANALYSE". Hier muß wesentlich genauer definiert werden. Die Routineanalytiker sind aufgrund der gesammelten Erfahrungen vorsichtig und hellhörig geworden, und es kann nicht ohne Kritik geschehen, wenn fast alle anderen in den oft geschmähten Kreis aufgenommen werden möchten.

So exakt wir uns bemühen, analytische Daten zu erstellen, so wenig exakt gehen wir mit der Sprache um. Es muß doch heißen: Stoffe werden analysiert, Elemente werden bestimmt und "multi" bedeutet mehr als eins oder ein paar Elemente.

Seit langer Zeit wird zwischen der Multikomponenten- oder MULTI-ELEMENTBESTIMMUNG und der VERBUNDANALYSE unterschieden. Erstere ist die simultane Messung und Bestimmung von vielen Bestandteilen (z. B. 4 - 20 Elemente) aus der Analysenprobe einer Stoffart, letztere ist die sequentielle Bestimmung von mehreren Bestandteilen aus einer Analysenprobe, z. B. die photometrische, atomabsorptionsspektrometrische oder voltammetrische Bestimmung von zwei oder mehr Elementen nacheinander aus einer einzigen Lösung.

Diese Unterscheidung halte ich für wichtig, weil sie die Art der einsetzbaren Analysenprinzipien beinhaltet. Im Gegensatz zur MULTIELEMENTBESTIMMUNG kann die VERBUNDANALYSE mit typischen Einzelelementbestimmungsverfahren durchgeführt werden. Das Aneinanderreihen von Einzelelementbestimmungen ist grundsätzlich verschieden von der eigentlichen MULTIELEMENTBESTIMMUNG, bei der das Vorliegen der gesamten Information vorausgesetzt wird. Das erfordert prinzipiell simultan messende Methoden, wie die Röntgenfluoreszenzspektrometrie und Emissionsspektralanalyse oder die γ-Spektrometrie und Festkörpermassenspektrometrie. Ein weiteres Kriterium der eigentlichen MULTIELEMENTBESTIMMUNG war durch den Einsatz der Originalprobe gegeben.

Am Beispiel der Emissionsspektroskopie läßt sich zeigen, daß die meßtechnischen und zeitlichen Unterschiede zwischen "simultan" und "sequentiell" geringer geworden sind, zumal eigentlich nur simultan gemessen und dann sequentiell ausgewertet wurde.

Gegenüber der Fotoplatten-Spektrographie ist nicht die Information sondern die Schnelligkeit stark verbessert worden. Seit mehr als 10 Jahren gibt es zwar Geräte, die simultan messen und registrieren können, so daß erst die Datenverarbeitung im Computer sequentiell erfolgt, doch hat sich diese von der RFA oder γ-Spektrometrie her bekannte

Technik nicht allgemein durchgesetzt. Der damalige "Marktführer" hatte dies nicht im Programm, was wieder einmal zeigt, daß auch die analytische Entwicklung weitgehend von den Geräteherstellern bestimmt wird.

Eine Entwicklung in die entgegengesetzte Richtung stellt die Verlagerung des sequentiellen Schrittes hin zum Messen dar. Sicherlich sind die Zeitunterschiede in manchen Anwendungsfällen nicht so entscheidend wie bei uns, und somit besteht im Fall der Emissionsspektralanalyse die Wahl der sequentiellen Arbeitsweise mit drehbarem Gitter oder beweglichem Multiplier.

Entscheidend für die Klassifizierung beider, der simultan und der sequentiellen Arbeitsweise, scheint mir zu sein, daß dies aufgrund der vorliegenden analytischen Aufgaben betrachtet wird. Die Vorteile der einen Technik sind die Nachteile der anderen: Die Optimierung der Elementbestimmung, die gleichzeitige Messung des spektralen Untergrundes oder die Wahl verschiedener Meßwellenlängen sind nur sequentiell möglich. Nur simultan liegt die gesamte erreichbare Information vor und Langzeitstabilität, Präzision oder Analysenzeit sind günstiger.

Vom allgemein analytischen Standpunkt aus betrachtet - wenn Analytik mit der Gewinnung von Information gleichgesetzt wird - ist die Simultanbestimmung vorzuziehen. In der täglichen Praxis sind deren Vorteile, wie Präzision der Daten, Langzeitstabilität der Apparatekonditionen und die Analysenzeit, wichtige Kriterien. Es wird häufig gesagt, daß bei wechselnder Probenmatrix oder unterschiedlichen Elementkombinationen der Einsatz sequentieller Methoden günstiger sei. Das ist nur teilweise richtig und muß differenzierter gesehen werden.

Für das Erstellen einer Analysenmethode, z. B. mit der ICP-Emissionsspektrometrie, ist es vorteilhaft, in sequentieller Arbeitsweise die Bedingungen und Analysenlinien festzulegen, den sie umgebenden Wellenlängenbereich und den spektralen Untergrund zu messen sowie aus dessen Schwankung und dem Linie/Untergrund-Verhältnis die Bestimmungsgrenzen abzuschätzen.

Wenn die Fragestellung es erfordert, sollte die Methode dann simultan messend weitergeführt werden. Sinnvoll wird dies bei einer zu bestimmenden Elementanzahl von ≥ 4 pro Analysenprobe. Ab dieser Grenze wird die simultan messende ICP-Spektrometrie wirtschaftlicher als z. B. die Atomabsorptionsspektrometrie, vorausgesetzt es fallen täglich große Probenanzahlen an (> 40). Bereits die beschriebene Arbeitstechnik an diesem Beispiel zeigt, daß die simultane MULTIELEMENTBESTIMMUNG den größeren Schwierigkeitsgrad insgesamt aufweist, denn hierbei müssen in vieler Hinsicht, wie z. B. für die Anregungsparameter, die optische Abbildung, die Meßverstärkung und Meßzeit, Kompromisse eingegangen werden. Dagegen ist es heute möglich, bei der sequentiellen Arbeitsweise zwischen den einzelnen Meßvorgängen in dieser Hinsicht zu optimieren, also Parameter in der Zeit zu verändern, die zur Suche der neuen Meßwellenlänge benötigt wird.

Den genannten Vorteilen der Sequenzbestimmung und der damit einfacheren Eichung der Analysenverfahren steht ein Verlust an Information gegenüber, was selten bedacht wird. In der täglichen Praxis werden zwar häufig die selben Elemente aus nahezu gleichbleibenden Matrices bestimmt. Hierbei ist es jedoch eine wichtige Aufgabe, sowohl die Veränderung der Matrix als auch das Vorhandensein unerwarteter Begleitelemente messend zu erkennen. Die programmierte Sequenztechnik ist hier überfordert; der komplette Spektrendurchlauf benötigt viel zu viel Zeit. Andererseits ist bei Material verbrauchenden Methoden die MULTIELEMENTBESTIMMUNG in Mikroproben (Mikroanalyse) sequentiell nicht durchführbar.

Bei richtiger Auslegung kann ein simultan messendes Analysensystem auch diese Veränderungen erkennen und in die Korrektur mit einbeziehen, wie das z. B. bei der RFA schon länger möglich ist. Hier und neuerdings auch bei ICP-Spektrometern sind kombinierte Geräte entwickelt worden, also Sequenz- und Simultanspektrometer in einem. Bei größeren Laboratorien ist die Anschaffung von beiden Typen nebeneinander zu empfehlen.

An einigen Beispielen soll im folgenden Teil gezeigt werden, wie sich die KLASSISCHE, SIMULTAN MESSENDE MULTIELEMENTBESTIMMUNG (oder einfach ausgedrückt: "Routineanalyse") in den letzten Jahren entwickelt hat.

Die Aufgaben sind klar vorgegeben worden:

1. Einschränkung der zeitaufwendigen chemischen Analyse durch Übernahme vieler Aufgaben in den Bereich der schnellen, produktionsbegleitenden Analytik.

2. Entwicklung von Kontrollmechanismen zur Prüfung der chemisch-analytischen Richtigkeit, z. B. durch kombinierten Einsatz von Atomabsorptions- und ICP-Spektrometrie zur

Erzeugung großer Datenmengen.

3. Auffangen aller neuen Anforderungen mit den bestehenden Analysensystemen, z. B. Mikro- und Spurenanalysen in normalen Industrielaboratorien.

Zahlreiche Aufgaben der chemischen Analyse sind in den letzten Jahren auf die RÖNTGENFLUORESZENZSPEKTROMETRIE (RFA) übertragen worden, was auch auf die Entwicklung der simultanen RFA seit 1967 zurückzuführen ist. Die Leistung und Stabilität der Anregung wurde erhöht und durch den geometrisch-optischen Aufbau gelang es, im Mittel etwa 30 Elemente simultan messend zu erfassen. Mit einer gezielt veränderten Probenvorbereitung wurde es möglich, sowohl die unterschiedlichsten Metalle, wie z. B. Ferrolegierungen nach Umschmelzen und Schleuderguß, als auch oxidische Materialien verschiedener Art nach Schmelzaufschluß zu analysieren. Der Massenanteilbereich der Elemente liegt bei uns für die RFA zwischen 80 - 0,01 %. Die Bestimmung von Anteilen nahe 100 %, z. B. bei Reinststoffen, ist quantitativ bzw. mit ausreichender Präzision nicht möglich. Die Spurenanalyse mit der RFA wird versucht, indem die Meßzeiten erheblich verlängert oder Anreicherungstechniken vorgeschaltet werden oder beides gleichzeitig erfolgt. Aus verschiedenen Gründen haben diese, sicherlich noch entwicklungsfähigen Methoden bisher keinen allgemeinen Eingang in die tägliche Praxis gefunden.

Die weiteren Anforderungen zur verstärkten analytischen Kontrolle großer Datenmengen und die spurenanalytischen Fragestellungen führten zum Einsatz der simultan messenden ICP-Emissionsspektrometrie. Als Lösungsmethode erfordert die ICP-MULTIELEMENTBESTIMMUNG die Herstellung von Analysenlösungen. Universelle Verfahren bestehen für Stähle und Eisenlegierungen oder für oxidische Stoffe unterschiedlicher Art, wobei kaum mehr als 8 - 10 Elemente in einer Lösung gehalten werden können. Der geschwindigkeitsbestimmende Schritt ist hierbei die Lösungsherstellung, so daß sich nach Erstellung der Meßmethoden die Entwicklungsarbeiten jetzt auf die Automatisierung des Lösens konzentrieren. Metallproben lassen sich relativ leicht in Säuregemischen lösen und filtrieren. Eine Variante zur Automatisierung dieses Vorganges wird überall vom Labor LUX (ARBED) vorgetragen. Weitaus schwieriger ist das vollständige Lösen oxidischer Materialien. Das IRSID (Metz) hat jetzt den eigenen Aufschlußautomaten dahingehend geändert, daß die Schmelze anschließend direkt gelöst wird.

Das ICP-Simultanspektrometer ist inzwischen mit der Möglichkeit ausgestattet worden, den Primärspalt mit Hilfe eines vom Computer gesteuerten Schrittmotors nach beiden Seiten hin exakt zu verstellen. Die resultierenden Linienprofile machen eventuelle Koincidenzen und den der Linie benachbarten spektralen Untergrund sichtbar.

Durch simultanes Messen auf z. B. 10 Stellungen des Primärspaltes über jeweils 1 s liegen in unserem Fall die Profilkurven für 40 Elemente vor (2. Bild) - und damit praktisch die q u a l i t a t i v e Zusammensetzung der Analysenprobe. Aus dem Vergleich der ausgegebenen Nettointensitäten mit bestehenden Eichfunktionen kann der Anteilbereich für jedes Element geschätzt werden. Gleichzeitig läßt sich diese Information nutzen, um gewisse Korrekturmaßnahmen aufzurufen.

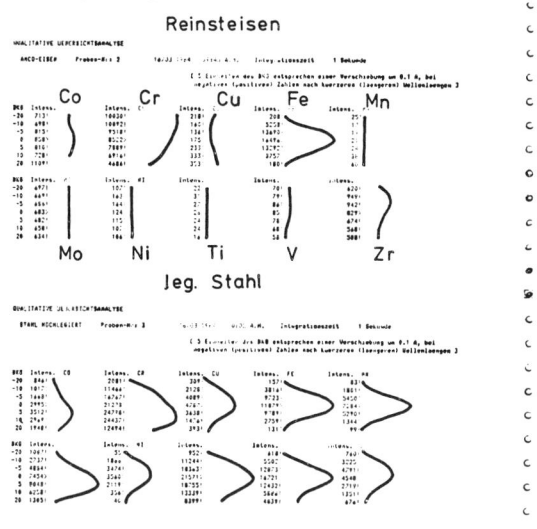

2. Bild: Profilkurven

In Zusammenhang mit der quantitativen Multielementbestimmung sind alle Techniken bevorzugt, die direkt von der kompakten Probe ausgehen. Das hat prinzipiell zwei Gründe: einmal den zeitlichen Ablauf, und zum anderen wird die Probe nicht verdünnt oder durch Kontaminationen bzw. Verluste verändert. Die Güte der analytischen Daten hängt dann jedoch von der Homogenität der kompakten Proben (Struktur, Phasen, Korngrenzenanreicherungen u. a.) ab, und die Eichung der spektralanalytischen Verfahren ist in dieser Hinsicht nicht befriedigend. Hier wird nämlich die chemische Bestimmung an Spänen der gleichen Probe durchgeführt, und die Daten werden auf die kompakte Probe übertragen, ohne die Struktur der Probe ausreichend gut zu kennen.

Im Fall der ICP-Spektrometrie lassen sich kompakte Proben auf zweierlei Weise analysieren:

Durch die Kombination von Abfunkkammer und ICP-Brenner kann im Bogen erzeugter Metalldampf in das Plasma geleitet werden. Die "Auflösung" der metallischen Probe erfolgt somit getrennt von der spektralen Anregung, so daß von der Probe verursachte Effekte keine Rolle mehr spielen. Es besteht in allen Fällen ein linearer Zusammenhang zwischen Meßwert und Konzentration. Oxidische (geologische) Materialien können nach Mischen mit z. B. Cu-Pulver ebenfalls in nur geringer Verdünnung analysiert werden.

Der zweite Weg ist die Verwendung eines Graphittiegels im Plasma, was wir SET genannt haben. Die Ausarbeitung spurenanalytischer Verfahren erfolgt an einem kombinierten Sequenzspektrometer, an das neben dem normalen Zerstäuberkammersystem/Brenner eine weitere Anregung für die SET angeschlossen ist. Durch den mechanischen "Fahrstuhl" wird die Position des Tiegels im Plasma reproduzierbar erreicht. Während dieser Operation bleibt das Plasma offen brennen. Der Tiegel wird leer zur Blindwertbestimmung und anschließend mit der Probe eingeführt. Mit dieser Technik werden für zahlreiche Elemente Bestimmungsgrenzen erreicht, wie das bisher nur mit der flammenlosen AAS möglich war.

Die einzige Hantierung mit der Probe im Laboratorium ist die Einwaage in den Graphittiegel.

Anhand der Bestimmung von Abriebmetallen in Schmierölen läßt sich auf eine weitere Entwicklung hinweisen, die die FUNKENEMISSIONSSPEKTRALANALYSE betrifft. Inhomogene Analysenproben, in diesem Fall: Partikel unterschiedlicher Größe in Lösung, lassen sich kaum mit ausreichender Präzision analysieren, wenn nicht die gesamte Probe verdampft und über die gesamte Zeit intergierend gemessen wird.

Die neue Elektrodenanordnung der Fa. SPECTRO, wobei die Radelektrode als cup-Elektrode benutzt und die gesamte Probe mit einem "gerichteten" Funken verdampft wird, führt darüber hinaus zu einer weiteren Information. Beim Einsatz eines Einzelfunkenanalysators, wie er auch zur Bestimmung der Al-Bindungsformen in kompakten Stahlproben benutzt wird, kann aus der Häufigkeitsverteilung auf die Größenverteilung der Partikel geschlossen werden.

Diese kurzen Hinweise sollen zeigen, daß auch die klassische Multielementbestimmung lebt und weiterentwicklungsfähig ist. Die wesentlichen Impulse hierfür sind in der Vergangenheit aus Industrielaboratorien gekommen. Ob das in Zukunft noch geschehen wird, wage ich zu bezweifeln. Es bleibt somit die Frage: Genügt uns eine kommerziell geprägte Entwicklung?

Selbst in der Hand behalten wir dagegen die Art und Durchführung der Eichung und des Kalibrierens der Meßgeräte. In der simultanen Multielementbestimmung ist ein hohes Maß an Präzision erreichbar, die Richtigkeit der Daten muß ständig überprüft werden. Da die chemischen Prüfmethoden eine oft schlechtere Präzision zeigen, besonders wenn es um kleinere Anteilbereiche geht, wird auch versucht, eine physikalische Eichung einzuführen: die Wägung von sogenannten Primärsubstanzen, die größtenteils keine Titersubstanzen sind. Diese Art der "Herstellung von Richtigkeit" ist nicht mit der chemisch-analytischen identisch, die sich nicht herstellen läßt. Es wird nur eine höhere Richtigkeit vorgetäuscht, die nicht real sein kann, wenn z. B. die spektralanalytischen Daten mit genormten, chemischen Analysenmethoden verglichen werden. Dies ist immer noch die tägliche Praxis.

Die chemisch-analytische Richtigkeit, die ich definiere als die Wiederfindungswahrscheinlichkeit des wahrscheinlichsten Wertes, wird allein durch die Standardabweichung eines Absolutverfahrens begrenzt, d. h. der wahrscheinlichste Wert liegt irgendwo innerhalb des gegebenen Streubereiches. Es sind Verfahren mit dem geringsten Streubereich auszuwählen - besser kann es nicht gemacht werden, und es ist auch nicht besser - wie die tägliche Praxis zeigt.

Wichtig für die Multielementbestimmung mit festen Proben ist die bessere Herstellung von Standardreferenzproben unterschiedlicher Zusammensetzung. Die Richtigkeit eines Relativverfahrens bezieht sich nur auf den wahrscheinlichsten Wert des Massenanteiles in der Eichprobe. Die Benutzung von Eichproben mit chemisch analysierten Massenanteilen kann nicht Daten hervorbringen, die richtiger sind, als diejenigen, die nur die globale Zusammensetzung der festen Proben beschreiben. In bezug auf die Spurenbestimmung kann nur empfohlen werden, Referenzverfahren hoher Güte zu installieren und zu verwenden. Die Eichung für die simultane Multielementbestimmung ist die schwierigste und somit immer noch verbesserungsfähig.

Bestimmungsmethoden

REVIEW OF THE PRESENT STATE OF GAMMA-SPECTROMETRY
FOR MULTI-RADIONUCLIDE ANALYSIS

U. Herpers
Institut für Kernchemie
der Universität zu Köln

SUMMARY

Some select gamma-ray spectroscopy systems with high efficiency and energy resolution for different applications are described. The salient performance characteristics, energy-range, efficiency and energy resolution of various semiconductor detectors are reported. Standardized computer controlled spectrometers of different design and a computer-based multi-purpose spectrometer system are presented.

1. INTRODUCTION

A nuclear radiation detection system consists of a detector, which proves the occurrence of a nuclear event and of an equipment of electronical units that provide information about events, such as the energy of each event. However, using high-resolution gamma-spectroscopy, one of the requirements for the simultaneous determination of different radionuclides is the knowledge of the appropriate detector and spectroscopy system needed for the relevant x- and gamma-rays.

2. DETECTOR SYSTEM

The peaks observed in the pulse height spectrum from a sodium iodide detector are very broad. It is therefore difficult to separate peaks generated by gamma-rays with closely lying energies. It is also sometimes difficult to distinguish very weak peaks from a broadly distributed background continuum on which they may be superimposed. For these reasons, there is often a preference for detectors with better energy resolution, particularly in the analysis of complex gamma-ray spectra with many closely lying peaks.

For these applications, germanium semiconductor detectors have become the instrument of choice. Many of these currently in use are of the lithium-drifted type, whereas nearly all newly fabricated germanium-detectors are now of the high purity germanium type.

The illustrations and charts in Fig.1 depict the various detector geometries that are commercially available, the energy range they cover and their salient performance characteristics. It is demonstrated that the Silicon(Li)-, Planar-Germanium and Low-Energy Germanium (LEGe) detectors are highly sensitive and versatile research tools over an energy range of approximately 1 keV to several MeV. Therefore, one of the interesting applications of these detectors and also of the Reverse-Electrode-Germanium detector (REGe) is the activation analysis, because most high energy, photon-emitting radioisotopes have corresponding low-energy photons in much greater abundances. Normally, the detectors are designed to take advantage of the yield of these low energy photons. Relatively small volumes are used to provide a low sensitivity to the higher energy gamma-rays of the radioisotopes and background. However, all detectors share very similar operational properties.

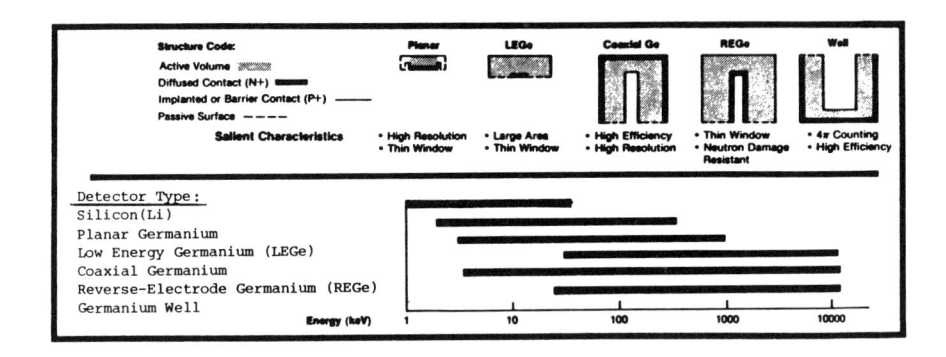

Fig.1: Structure, salient characteristics and energy ranges of various semiconductor detectors.

2.1 RESOLUTION

The resolution of a detector is a measure of its ability to distinguish between the closely spaced energies from a radiation source. Any resolution figure normally includes statistical and electronic contributions, and is defined as the full width of the recorded peak at one-half (FWHM) or one tenth (FWTM) maximum height. For semiconductor detectors, the system resolution is often specified in terms of FWHM in keV. Table 1 shows some typical system resolutions at various

Detector Type	Energy		
	5.9 keV	122 keV	1332 keV
Planar	185 eV	520 eV	-
LEGe	400 eV	600 eV	1.9 keV
Coaxial	-	0.8 keV	1.9 keV
REGe	0.7 keV	0.8 keV	1.9 keV

Table 1: Typical system resolutions afforded by the various detectors at various energy levels.

energy levels for different types of germanium detectors. As demonstrated, the energy resolution is excellent, typically a few tenths of a percent at moderate gamma-ray energies. This figure can be contrasted with a resolution of about 5% for a typical sodium iodide scintillation spectrometer.

2.2 EFFICIENCY

One of the important properties of a detector is its detection efficiency. Efficiencies are generally quoted as a fractional or percentage probability of recording a count under specified conditions and can be quoted as either "absolute" or "intrinsic". For absolute efficiencies, the probability of recording a count is normalized to the number of quanta emitted (generally in all directions) by the source.-Often it is only necessary to know the change in detector efficiency with respect to the given energy, so that just the relative efficiency is adequate.- On the other hand, intrinsic efficiencies are instead normalized to only the number of quanta striking the detector and producing a pulse.

In most cases of application of high-resolution gamma spectroscopy in the pure and applied fields of nuclear science the knowledge of the absolute efficiency of the detector of choice is essential. Typical absolute efficiency curves for various germanium detectors are given in Fig.2. However, if the efficiency is known in advance, it is possible to predict the number of recorded counts for a source of known intensity under specified geometric conditions and consequently, it is also possible to determine the absolute activity of a radioactive sample under the same specified geometric conditions.

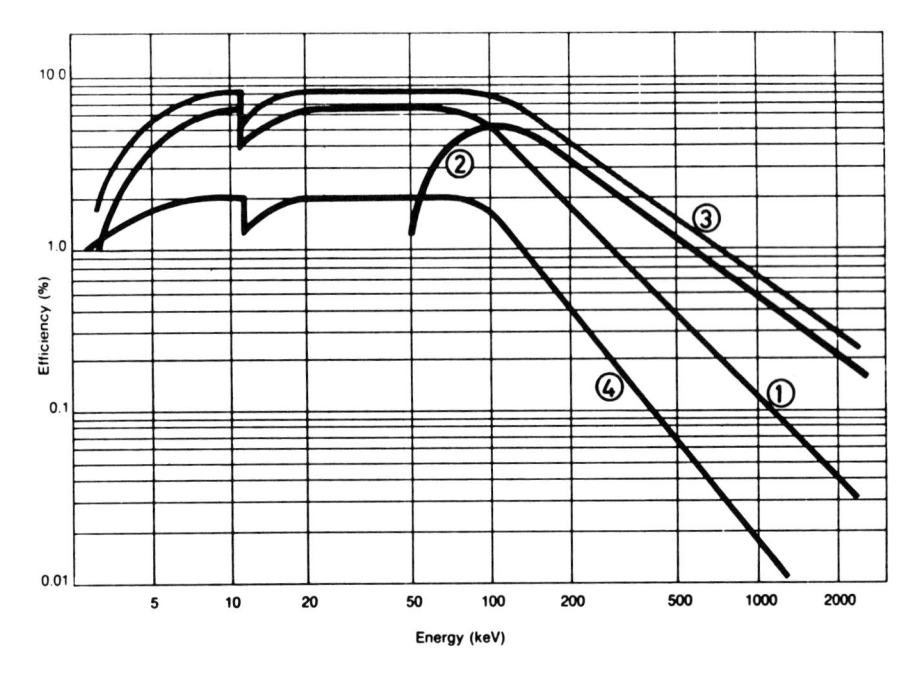

Fig.2: Typical absolute efficiency curves for various germanium-detectors with 2.5 cm source to end-cap spacing.
(Code: 1 - LEGe, 10 cm^2 x 15 mm thick;
2 - Coaxial Ge, 10% Relative Efficiency;
3 - REGe, 15% Relative Efficiency;
4 - Planar, 200 mm^2 x 10 mm thick.)

3. SPECTROSCOPY-SYSTEM

A typical gamma-ray spectrometer setup consists of a detector of choice with high-resolution and efficiency, high-voltage, preamplifier, amplifier and multichannel-analyzer with read-out. But, since the installation of computers in laboratories such a simple system is antiquated and uneffective in this traditional way. Therefore, computer-controlled gamma-ray spectrometers have been developed in manifold versions.

3.1 COMPUTER-BASED GAMMA SPECTROMETER

The development of high-resolution high-sensitivity semiconductor detectors and automatic multichannel analyzer systems has made possible the rapid acquisition of spectrographic data. This has created a need for computer spectral analysis systems. So the computer-based gamma-ray spectrometer is a natural extension of spectrometry, since the enormous number of output-data must be processed in some way. However, the standardized computer controlled gamma-ray spectrometer requires a reliable method of data accumulation remote from a central computer system, which can provide control of the data collection and read-out, as well as allow the analysis of complex gamma-ray spectra and the interpretation of the data.

In the designing of such a system three conditions must be satisfied: First, the data accumulation and the transfer of data must be completely automated and under control of the central processor. Second, the remote system must be able to monitor and verify its own operation. Third, the remote spectrometer system must be able to control external devices. Consequently, such a remote analyzer system can be achieved either by implementation of a commercial hard-wired analyzer that could be controlled remotely via commands received from a central computer, or by implementation of a microcomputer as the controlling device for the analyzer system, in which the computer receives and interprets commands from the central computer and performs the required functions. In the first version, the multichannel analyzer of the spectrometer branch is operated by a computer via an interface. The interface controls the functions of collection, stop, read-out, clear and display. Additionally, it can transfer data from the analyzer to the computer or vice versa. Normally, the multichannel analyzer has internal microprocessors that perform some analysis independently of the main computer, such as peak search, peak net area calculations, energy calibration, isotope identification, as well as spectrum manipulation functions (1-8). The second version is a more sophisticated system. This remote analyzer spectrometer consists of a detector system with preamplifier and linear pulse amplifier, an analog-to-digital converter interfaced to a microcomputer and a microcomputer working as a computer-based pulse height analyzer interfaced to a central computer. The advantage is that the data memory is separate from the central computer memory. It can be addressed directly by the central processing unit in the central computer via a special databus extender. Thus, the central computer memory is able to use a time-sharing monitor system, allowing more independent users, e.g. one user can be taking data, while another can be analyzing data or preparing computer programs. An excellent compilation of paper dealing with this subject is to be found in the Proceedings of the American Nuclear Society Topical Conference (9) and in the Proceedings of the 1981 International Conference on Modern Trends in Activation Analysis (10).

For selected research applications a CAMAC-based gamma-ray spectrometer may be more useful. This multichannel analyzer system is installed around a computer and uses CAMAC (Computer Automated Measurement And Control) as a primary interface (11,12). If the application demands single processing prior to the data acquisition, often NIM (Nuclear Instrumental Module) equipment is included. For its basic tasks, the multichannel analyzer acquires, analyzes, stores and displays pulse height data as required. But through the CAMAC-standard for computer-interfacing modules and the software, this system can easily be expanded to perform much more than a multichannel analyzer is expected to. This concept may be a framework for a general purpose laboratory automation system.

The design and implementation of very large data bases such as those associated with gamma-ray spectroscopy require the use of sophisticated software-techniques for data storage and handling. A data base management system must provide at least methods for construction of the data collection, methods to update, correct and modify values in the data collection, methods to retrieve selected data from the data collection and methods to reduce large quantities of data to a more usable form. Therefore, a software system for the routine analysis of gamma-spectra normally contains the following components: An automatic on-line spectrum analysis program for semiconductor detectors, an interpretation program coupled to the spectrum analyzer program to reduce the calculated gamma-ray data to isotopes present in the sample and corresponding element concentrations, software for the control of peripherical and special equipment, as e.g. automatic sample changers, software to compare and to combine results from different measurements, software for communications in the various components of the software system, software to generate status informations and analysis reports.

3.2 AUTOMATED MEASURING SYSTEM

Considering the different halflives of nuclides special measuring systems working automatically recently were developed, especially with respect to the applicability in the field of neutron activation analysis (see Workshop on Activation Analysis with Short-Lived Nuclides (13)). A high rate gamma-ray spectroscopy system combined with a very fast pneumatic transport system suited for measurements of short-lived isomeric transitions is extensively described in the literature (14-20). The spectrometer system consisting of a modified Ge(Li)-detector enables to collect spectra of high quality up to a very high input-rate. For digitization of pulse height special analog-digital-converters are needed (21), offering a constant conversion time at exceptional linearity and stability. The registration of spectra is performed by means of a minicomputer connected to a larger laboratory computer system.

For the activation analysis of nuclides with halflives in the region of seconds up to some minutes no special or modified equipments are necessary. Therefore, automatically working systems consisting of a pneumatic transport system and gamma-ray spectrometer branches controlled by a central computer are not extraordinary. Consequently, they are developed manifold from different institutions and worldwide installed at most of the nuclear research reactors (13).

Especially in the field of radiochemical neutron activation analysis and also in most cases of instrumental neutron activation analysis it is necessary to measure longer lived radionuclides of different energies from some keV up to some MeV. In those cases, an automated multipurpose measuring system is very useful. A blockscheme of a typical computer-based framework for a general purpose laboratory system is given in Fig. 3. The measuring system consists of two computer-based Ge(Li)-spectrometers with sample changers for automatic analysis; two heavy-shielded Ge(Li)-spectrometers and two gamma-ray spectrometers with well-type detectors for special problems; a spectrometer with a $Ge_{intrinsic}$-detector for the spectroscopy of low gamma-ray energies; a Si(Li)-spectrometer for the spectroscopy of x-rays and an anticompton-

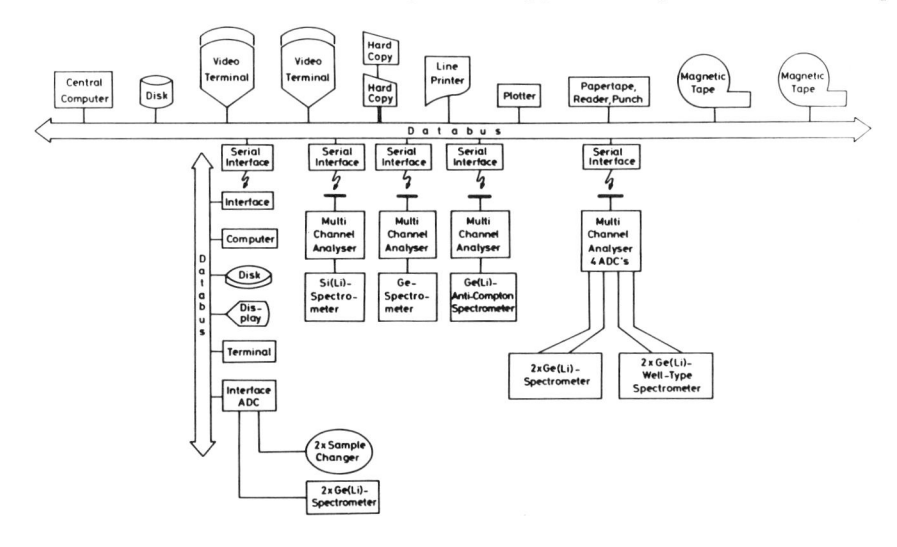

Fig.3:   Block-diagram of an automated multi-purpose measuring system.

spectrometer. All basic measuring systems are "on-line" connected via serial interfaces with a central computer and its periphericals, as e.g. disks, magnetic tapes, terminals, line-printer, plotter etc. The central computer can provide the control of data collection and storage, as well as analyzing all data. The software for the control of data acquisition usually is device-specific. On the other hand, the data should be transferred and stored in a standard-format. This makes it possible to develop the analyzing programs independently of the special detection system.

4. HINTS TO THE RANGE OF APPLICATIONS

The applicability of high-resolution gamma-ray spectroscopy in the different pure and applied fields of sciences is so manifold demonstrated that it is nearly impossible to give a complete review of the literature dealing with this subject. Therefore, finally without being entitled to a completeness, it should be only referred to some comprehensive treatises. So, counting and high resolution spectroscopy involving nuclear activation and radioisotopic methods are analytical tools in various fields of technology, medicine, geo- and cosmochemistry and even archaeology and they are described in (10,22-36). The spectroscopy is also a basic tool for radiation protection, nuclear medicine, nuclear fuel cycle, waste handling etc. (37-51). Nevertheless, the analysis of radionuclides by high-resolution gamma-spectroscopy is a well-known and wide-spread analytical method and the fields of science and industrial application dealing with gamma-rays in any way cannot be imagined without this method.

REFERENCES

(1) H.P.Blok, C.J.de Lange and J.W.Schotman, Nucl.Instr.Meth. <u>128</u> (1975) 545-556.

(2) I.de Lotto and A.Ghirardi, Nucl.Instr.Meth. <u>143</u> (1977) 617-620.

(3) M.Hillman, Nucl.Instr.Meth. <u>135</u> (1976) 363-368.

(4) M.Kitamura, Y.Takeda, K.Kawase and K.Sugiyama, Nucl.Instr.Meth. <u>136</u> (1976) 363-367.

(5) G.W.Phillips and K.W.Marlow, Nucl.Instr.Meth. <u>137</u> (1976) 525-536.

(6) N.Sasamoto, K.Koyama and S. Tanaka, Nucl.Instr.Meth. <u>125</u> (1975) 507-523.

(7) P.C.Stevenson, Processing of Counting Data, NAS-NS 3109, Washington DC (1966).

(8)   R.K.Webster, The Application of Computers in Radiochemistry, Chapter 4, in D.I.Coomber (Ed.), Radiochemical Methods in Analysis, Plenum Press, New York (1975).

(9)   B.S.Carpenter, M.D.D'Agostino and H.P.Yule (Eds.), Computer in Activation Analysis and Gamma-Ray Spectroscopy. Proceedings of the American Nuclear Society Topical Conference at Mayaguez, Puerto Rico, April 30-May 4, 1978. Published by Technical Information Center/U.S.Department of Energy (1979) CONF-780421.

(10) T.Braun, E.Bujdosó, W.S.Lyon and R.G.W.Hancock (Eds.), Proceedings of the 1981 International Conference on Modern Trends in Activation Analysis, J.Radioanal.Chem. 69-72 (1982).

(11) CAMAC, Instrumentation and Interface Standards, IEEE Document No. SHO6437. Distributed by Wiley Interscience, New York (1979).

(12) CAMAC, Tutorial Issue, IEEE Trans.Nucl.Sci., NS 20, 2 (1973).

(13) Proceedings of the "Workshop on Activation Analysis with Short-Lived Nuclides" Vienna, February 4-8, 1980. J.Radioanal.Chem. 61 (1981).

(14) G.P.Westphal, Nucl.Instr.Meth. 136 (1976) 271-283.

(15) G.P.Westphal, Nucl.Instr.Meth. 138 (1976) 467-470.

(16) G.P.Westphal, Nucl.Instr.Meth. 146 (1977) 605-606.

(17) G.P.Westphal, Nucl.Instr.Meth. 163 (1979) 189-196.

(18) G.P.Westphal, J.Radioanal.Chem. 70, No.1-2 (1982) 387-410.

(19) O.Brandstädter, F.Girsig, F.Grass and R.Klenk, Nucl.Instr.Meth. 104 (1972) 45-53.

(20) F.Grass, J.Radioanal.Chem. 70, No.1-2 (1982) 411-425.

(21) G.P.Westphal, Nucl.Instr.Meth. 113 (1973) 77-80.

(22) Proceedings of the 1976 International Conference on Modern Trends in Activation Analysis. J.Radioanal.Chem. 37-39 (1977)

(23) G.R.Choppin and J.Rydberg, Nuclear Chemistry, Theory and Applications, Pergamon Press, Oxford (1980).

(24) D.De Soete, R.Gijbels and J.Hoste, Neutron Activation Analysis, Wiley-Interscience, London (1972).

(25) J.Hoste, J.Op de Beeck, R.Gijbels, F.Adams, P.van den Winkel and D.De Soete, Activation Analysis, Butterworths, CRC London (1971).

(26) R.H.Kogan, Gamma Spectrometry of Natural Environments and Formations: Theory of the Method and Application to Geology and Geophysics, (ISBN-0-7065-1059-3, Pub.by IPST) Intl.Schol.Bk Serv. (1971).

(27) J.Krugers, Instrumentation in Applied Nuclear Chemistry, Plenum Press, New York (1973).

(28) J.M.A.Lenihan and S.J.Thomson, Advances in Activation Analysis, Academic Press, New York (1969).

(29) G.Pfrepper, W.Goerner and S.Niese, Spurenelementbestimmung durch Neutronenaktivierung, Akad.Verlagsgesellschaft, Leipzig (1981).

(30) H.W.Thümel, Isotopenpraxis 11, 1-12 (Heft 1), 41-49 (Heft 2), 87-98 (Heft 3), 117-125 (Heft 4), 172-180 (Heft 5) (1975).

(31) K.Heydorn, Neutron Activation Analysis for Clinical Trace Element Research, Vol.I and II, CRC Press (1984).

(32) Nuclear Activation Techniques in the Life Sciences, IAEA, Vienna (1967).

(33) Nuclear Activation Techniques in the Life Sciences 1972, IAEA, Vienna (1972).

(34) P.Brätter and P.Schramel (Eds.), Trace Element Analytical Chemistry in Medicine and Biology, Vol.1: Proceedings of the First International Workshop, Neuherberg, FRG, April 1980, Walter de Gruyter, Berlin, New York (1980).

(35) P.Brätter and P.Schramel (Eds.), Trace Element Analytical Chemistry in Medicine and Biology, Vol.2: Proceedings of the Second International Workshop, Neuherberg, FRG, April 1982, Walter de Gruyter, Berlin, New York (1983).

(36) A.O.Brunfelt and E.Steinnes (Eds.) Activation Analysis in Geochemistry and Cosmochemistry, Universitetsforlaget Oslo-Bergen-Tromsö (1971).

(37) F.Attix and W.Roesch, Radiation Dosimetry, 2nd.Ed., Vol.1 (1968) and Vol.2 (1967) Academic Press, New York.

(38) J.W.Boag, Ionization Chambers, in F.H.Attix and W.C.Roesch (Eds.) Radiation Dosimetry, Vol.II, Academic Press, New York (1966).

(39) J.F.Boland, Nuclear Reactor Instrumentation (In-Core), Gordon and Breach, New York (1970).

(40) J.R.Cameron, N.Suntharalingam and G.N.Kennedy, Thermoluminescent Dosimetry, University of Wisconsin Press, Madison (1968).

(41) J.M.Harrer and J.G.Beckerley, Nuclear Power Reactor Instrumentation Systems Handbook, Vol.I, CHS.2&3, TID-25952-PI (1973).

(42) L.J.Herbst (Ed.), Electronics for Nuclear Particle Analysis, Oxford University Press, London (1970).

(43) International Atomic Energy Agency, Personnel Dosimetry Systems for External Radiation Exposures, Technical Report No.109, IAEA, Vienna (1970).

(44) H.Kiefer and R.Maushart, Radiation Protection Measurement, Pergamon Press, Oxford (1972).

(45) E.Kowalski, Nuclear Electronics, Springer-Verlag, New York (1970).

(46) J.B.Marion and F.C.Young, Nuclear Reaction Analysis, North-Holland, Amsterdam (1968).

(47) J.Millman, Microelectronics, Mc-Graw-Hill Book Co., New York (1979).

(48) P.W.Nicholson, Nuclear Electronics, John Wiley & Sons, New York (1974).

(49) G.D. O'Kelley, Detection and Measurement of Nuclear Radiation, NAS-NS 3105, Washington D.C. (1962)

(50) A.W.Rogers, Techniques of Autoradiography, Elsevier, Amsterdam (1973).

(51) J.Sharpe, Nuclear Radiation Detectors, 2nd Edition, Methuen and Co., Ltd., London (1964).

Multi-Element-Analysis by Means of Alpha Particle Spectrometry

M. Helmbold
HTA-Projekt, Kernforschungsanlage Jülich, GmbH

SUMMARY

During burnup of fissile materials many transmutations occur with the result that discharged fuel has a multi-element composition although fresh fuel has only some constituents. The actinide and fission product isotopic composition of spent fuel is of great interest. For the determination of fission products gamma and beta spectrometry is well suited whereas for the determination of actinide isotopes mass and alpha particle spectrometry are applied.
This alpha particle spectroscopy of spent nuclear fuel has been carried out using different techniques for the preparation of the alpha samples by deposition from aqueous solutions of spent fuel, annealing of fuel particles in an oven and evaporating of fuel material by laser beam, the latter one allowing reliable multi-element-analysis. The very thin samples require mostly low level measurements with high resolution silicon surface barrier detectors. The measured spectra must be evaluated with sophisticated computer codes because of accuracy requirements. In this paper 1) the preparation, measurement and evaluation techniques are discussed in detail and 2) it is shown that the methods can be well applied for the analysis not only of nuclear fuel but also for samples with natural radioactivity.

INTRODUCTION

Solid state detector alpha particle spectrometry has now reached a high level of maturity and has found its own growing user community. This is due to a number of facts as there are improved techniques of source preparation, superior surface barrier detectors, high-tech electronics and sophisticated evaluation programme systems running on small to big size computers.

Combining all these features reliable multi-element-analysis by means of alpha particle spectrometry was done in the frame High Temperature Reactor (HTR) fuel cycle research at Kernforschungsanlage Jülich GmbH (KFA). The stringent accuracy requirements for fuel cycle work gave rise to many improvements in alpha particle spectrometry. In this paper methods and techniques as they are developed in KFA to date are presented in detail together with some fuel cycle relevant results as illustration of the general performance.

This improved alpha particle spectrometry can be applied to the investigation of nearly every alpha active material and it is proposed to use it specially for multi-element-analysis of samples with natural radioactivity.

It is in the scope of this paper to mention also work other than that from the KFA thus allowing a broader survey of the state of alpha particle spectrometry. But to shorten the comments the reader will be referred to literature if appropriate. For the special KFA alpha particle spectrometric work relevant to the HTR fuel cycle see Helmbold (1).

1. Sample Preparation

1.1 Laser Micro Boring

The irradiated fuel investigated by means of alpha spectrometry at the KFA has always been a special type developed for the German pebble-bed HTR. Typically it consists of some thousand fuel kernels, surrounded by pyrolytic carbon coatings. The particles are compressed with graphite to one spherical fuel element of 6 cm diameter. The fuel itself is mainly of the mixed oxide type $(Th,U)O_2$ or UO_2 and typical fuel kernels have diameters of about 500 μm.

The laser micro boring sample preparation technique was developed by Overhoff (2) and Max (3) for the determination of fission product concentrations and their local distributions inside the coated particle fuel and for the production of well-defined defects within the coatings. A short but comprehensive description of the laser system, a discussion of the influence of the self-channelling effect upon bore hole geometry, the experience with boring and sampling and some applications are given by Allelein et al (4).

This laser micro boring was first used for alpha sample preparation by Helmbold et al (5) and has been explained comprehensively by Helmbold (1): A solid state YAG-Nd^{3+} doped pulsed laser, with a wave length of 1.06 μm was used. The laser system operates in TEM_{00}-mode and is Q-switched. The pulse length can be restricted to about 30ns. Pulses are Gaussian in space and time. Pulse energy can be varied from $10^{-7} - 3 \times 10^{-3}$ J, which leads to maximum power densities of about 10^{11} W cm^{-2}.

Samples for laser micro-boring are prepared in the following way. A fuel particle is encapsulated in epoxy resin which is allowed to harden. One surface is then ground until the hemispherical plane of the fuel particle is exposed. This surface is uniformly polished and the sample mounted on a positioning table in a vacuum chamber. The laser pulses are focussed in the surface of the polished coated particle. Any point of the surface can be reached with a precision of ± 1 μm. With bore hole diameters of about 10 μm the determination of radial isotope distributions is possible in both the particle kernel and the surrounding multilayer coatings. Before striking the specimen surface the laser beam penetrates a glass plate. The distance between the specimen surface and glass plate is 1 cm. This plate acts as a collecting device for the evaporated material, and on completion of deposition it is analyzed for the amount of alpha, beta, or gamma-emitters. The surface of the specimen is illuminated by a He-Ne laser so that the surface can be observed during sample preparation with the aid of a television camera. A borehole acts as a light conducting channel for the subsequent incoming light pulses. This leads to the formation of holes with depths up to 250 μm. The cone shape of the hole leads to a narrow cone shaped beam of evaporated material guaranteeing that at least 95% of the totally evaporated material is collected on the glass plate within a well defined diameter (6). Depending on the volume of the bore hole 5 to 30 holes are needed to collect enough evaporated material on one plate for an alpha-acitvity measurement.

The alpha spectrum form such a sample is shown in figure 1 and demonstrate the effectiveness of this sample preparation technique for solid fuel materials to which it is restricted. This laser bored sample was prepared from an irradiated mixed oxide kernel. By comparing the ^{137}Cs activity of the whole kernel with the activity of the sample the mass of the sample was calculated to be 0.45 μg, using a kernel density of 9.5 g cm^{-3}.

Taking into account the cone shape of the bore holes and the distance between the surface of the specimen and the collecting plate, the area of the sample on the plates is calculated as 4.9 mm^2 leading to a thickness of 9 μg cm^{-2}.

For such a thin sample the variance δ_2 of the alpha energy-straggling distribution within the samples can be given as follows (7)

$$\delta_2 = \frac{4 \pi e^4}{(4\pi\epsilon_o)^2} \quad 4 \quad ZN \quad \frac{\Delta X}{2}$$

where
- e = charge of the electron
- ε_o = dielectric constant of the vacuum
- ZN = charge density within the sample
- ΔX = thickness of the sample

A thickness of 9 μg cm^{-2} corresponds to ΔX equal 0.015 μm and therefore the FWHM of the energy distribution will be about 3 keV. The electronic energy loss of the alpha particles within the samples can be calculated using the BETHE formula

$$\frac{\Delta E}{\Delta X} = \frac{4\,e^4}{8\varepsilon_o^2\,E\,m_e/m_\alpha}\,ZN\,\ln\left(\frac{4\,m_e/m_\alpha}{I}\right)$$

where: m_e = mass of the electron, m_α = mass of the alpha particle. $E = E(X)$ energy of the alpha particle, I = mean ionization energy of the electrons in the sample. Using $E = 5.49$ MeV, ΔE will be about 1.5 keV for the emitted alphas. The FWHM values for energy straggling distribution and electronic energy loss indicate that the resolution will be mainly limited by the detector performance.

1.2 Other Preparation Techniques

é Silva (8) also succeeded in measuring the actinide isotope release from intact and defective fuel particles during vacuum annealing. The well-developed annealing techniques also permit extremely thin alpha samples to be prepared. Extremely good alpha spectra are shown in (8). The samples were obtained by annealing an irradiated particle which was specially prepared by laser-micro-boring. A small hole was bored from the outside through all the coatings just into kernel. Annealing this "defect" particle for 37 hours at 1600 °C and collecting the evaporated material on a substratum yielded a sample sufficient for high accuracy alpha spectroscopy.

W. van der Eijk et al (9) have given a comprehensive review how to prepare thin, homogeneous and chemically and mechanically stable samples for all kind of nuclear spectroscopy. In this paper (9) an extensive bibliography is included and reference to source preparation procedures are given for each element.

^{241}Am sources are widely used as calibrated samples in alpha particle spectrometry. It was found very useful to check by independent means the specifications (as given by the supplier). A simple method for preparing thin ^{241}Am sources is given by U. Shreter and R. Kalish (10).

2. Measuring Devices

2.1 Surface Barrier Detectors for Alpha Particle Spectrometry

Since more than twenty years semiconductor surface barrier detector are used for charged particle spectroscopy. Nowadays their operation is well understood and they can be produced in excellent qualities. An introduction manual (11) for example helps much to choose the type well suited for the specific applications. The performance of such an detector might be checked according to the IEEE Test Procedures for Semiconductor Radiation Detectors (12) if needed.

Using proper electronics with the latest signal processing methods accurate alpha samples energy resolution of about 10 keV (FWHM) at 5.5 MeV alpha particle energy can be reached with superior surface barrier detectors (13).

Depending on the special measuring situation cooling and thermally screening of the surface barrier detector will be helpful to reach the desired energy resolution (14).

Alpha-recoil atoms ejected directly from the alpha sample might be tranferred to the detector surface (15). Injection and some back transportation by subjected alpha-recoil will occur.

So the detector will deliver alpha samples after some time even without measuring an alpha sample. Therefore background measurements before and after low activity measurements are needed. Specially measuring alpha-decay-chain members gives accuracy problems because of the varying detection probabilities for each isotope in a surface barrier detector source face-to-face set up. W. Westmeier (16) has given an algorithm to overcome that difficulty.

2.2 Vacuum Chamber

High precision alpha particle spectrometry must be done under vacuum conditions to avoid energy loss and straggling in air both affecting seriously the energy resolution. The vacuum must be better than one mPa to avoid microplasma breakdown. When applying the bias to the detector the noise can be observed easily using a noise meter or an oscilloscope. A sudden very large increase in noise is an indication of complete microplasma breakdown .

The vacuum chambers as offered commercially might not fit to the special measuring purposes. So some modifications are needed. The better way is however to construct a vacuum chamber according to your own wishes. Some useful design details are listed:

- short connection from the outside to the surface barrier detector
- detector surface over sample surface
- exact and variable positioning of the sample

- the chamber must be large enough to allow easy changing of
  detector and sample
- the free diameter of the connection to the vacuum pump must be
  large enough to reach a good vacuum
- clean the chamber carefully before installing the detector
- use only oil free vacuum pumps (e.g. a turbomolecular pump)
  to avoid contaminations
- avoid mechanical vibrations during measurement.

2.3 Preamplifier, Amplifier, Pulser, Multichannel Analyzer

The performance of the chain from preamplifier to the multichannel analyzer should be checked carefully according to the related IEEE test procedures (17). A high precision pulser is extremely useful first during the checks and later on during the measurements. This holds specially for long term measurements as often needed for alpha spectrometry of low activity samples.
Today multichannel analysers offer great possibilities in data handling and spectrum analysis. But high accuracy multi-element analysis normally includes the use of a computer. Therefore the multichannel analyzer should have a connection (bus or via magnetic tape device) to the computer.

3. Evaluation of the Alpha Sepctra by use of SAMPO

For the analysis of the measured alpha spectra the use of the computer program SAMPO (18) in connection with the alpha energy table as given by W. Westmeier and R.A. Esterlund (19) was most useful. The program SAMPO is nowaday available also as SAMPO 80 (20) for minicomputers and even vor microcomputers (21).
The alpha spectrum as shown in figure 1 was measured for five days under vacuum with a 100 mm^2 silicon surface barrier detector and a distance of 12 mm between the sample surface and detector. The FWHM of the ^{238}Pu 5.499 MeV line in the spectrum shown was 15.5 keV, which corresponds to the detector specification (15 keV at 5.59 MeV).
The measured alpha spectrum has been analysed using a fitting program which is based on SAMPO.
The parameters of the energy-dependent line shape (Gaussian with exponential tail) were determined in a fit of the intense lines of the spectrum. In fitting the complex structure the line shapes were fixed. The number of lines fitted to one complex structure and their rough positions were chosen by interactively minimizing χ^2 and the shape of the corresponding residual spectrum section. As an example the computer deconvolution of the peak at 5.15 MeV into ^{239}Pu and ^{240}Pu lines is

shown in figure 1 (b).

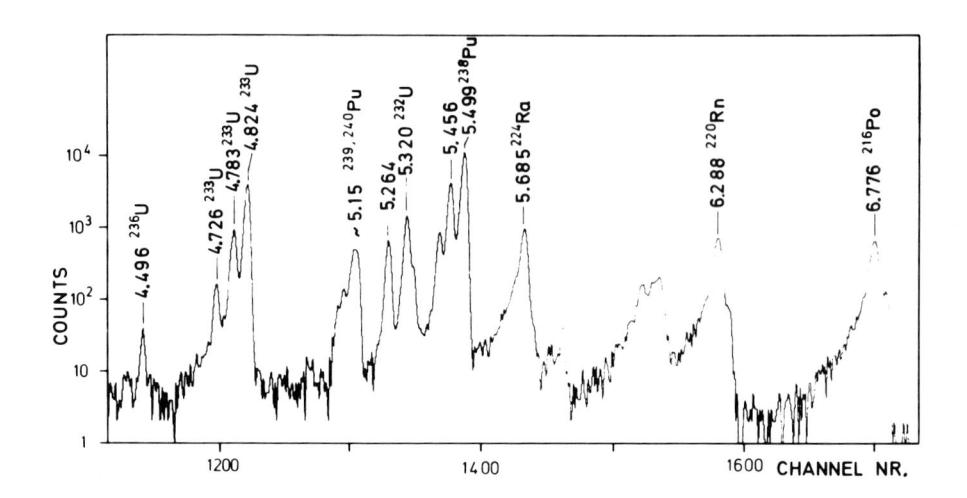

Fig. 1. (a) Alpha spectrum of a laser bored sample. The line energies are given in MeV.

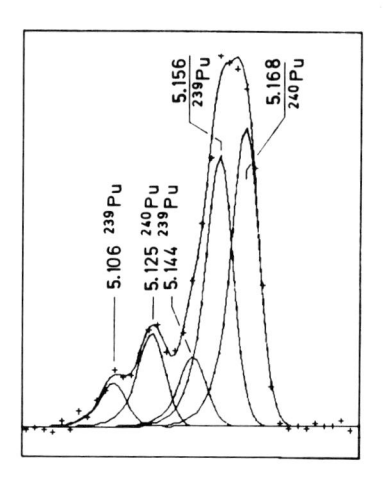

Fig. 1. (b) Decomposition of the complex group at 5.15 MeV of the spectrum shown in fig. 1 (a).

We have indentified 24 lines of 13 actinide isotopes in this spectrum using energies and intensities given in (19). Identification of the ^{232}Th, ^{235}U and ^{238}U lines was not possible because of the very low specific activities of these isotopes. The use of SAMPO is not without problems. To illustrate the above for five isotopes, the number of counts in the main lines obtained from a computer deconvolution of a measured spectrum together with their abundance and errors are given in table 1.

Table 1, Values calculated by deconvolution

Isotope	Energy (MeV)	Abundances (%)	Counts	Error	Error (%)
^{233}U	4.824	84.4	10974	206	1.9
^{234}U	4.776	72.5	1521	393	25.9
^{238}Pu	5.499	71.6	452903	904	0.2
^{239}Pu	5.155	73.3	1173	57	4.9
^{240}Pu	5.168	73.4	1818	93	5.1

The error in the 4.776 MeV line of ^{234}U is very large because of the influence of the stronger 4.782 line of ^{233}U (13.2% abundance). The energy difference between these two lines is only 6 keV which is half the energy resolution of the detector and at the limit of the deconvolution program. Also, if the slope of the background under the peaks to be identified is too steep, the numerics of SAMPO breaks down and further deconvolution of such peaks is impossible.

It is a great advantage that the quality of alpha spectra is little affected by beta- or gamma-emitters present in the sample. Therefore, measurements can be made without further chemical separation on samples prepared by laser micro-boring, whatever the gamma activity level. To give an example of the performance of our alpha spectrometric method the alpha activity ratio ^{238}Pu/ (^{239}Pu + ^{240}Pu) was determined in HEU mixed oxide fuel with 10.9% fima (fissions per initial metal atoms). The resulting total error for this ratio was 1.6% using the areas of all lines of the three isotopes. In the literature (22) for the determination of the same ratio the error is given as 1 to 3%, but the samples referred to had a ^{238}Pu/ (^{239}Pu + ^{240}Pu) ratio of the order of 1 to 4, whereas the present sample has a ^{238}Pu / (^{239}Pu + ^{240}Pu) ratio value of 152 and contains all alpha-, beta- and gamma-emitters present in irradiated fuel.

Measuring of the actinide isotope inventory of a whole kernel (or fuel element) by our alpha spectrometric method has the following error components:

The measurement of the ^{137}Cs activity of the whole kernel and of the sample (both with 1 to 5% error depending on the gamma-spectrometric procedures), the measurement of the detection angle to the surface barrier detector with a calibrated alpha-source (approximately 1% error) and the evaluation-dependent error. This altogether normally yields an error of 7% for ^{238}Pu and 11% ^{239}Pu content.

It is also possible to analyse the measured spectra automatically using the code ALFUN (23). Background free spectra can be computed by the use of GAMBAK (24) which facilitates the automatic spectra analysis. A comparison of the effectiveness of SAMPO and ALFUN is given in (25).

4. Applications

4.1 Spent Fuel Isotopic Composition

Comprehensive theoretical work in the field of HTR fuel cycle analysis has been carried out by the KFA. Predictions have been made about spent fuel isotopic composition and the related alpha activity of the nuclear waste using the burnup computer code ORIGEN (26). Many of those predicitons could be checked experimentally by our alpha spectrometric approaches. Because some of the actinide isotopes show no or only little alpha activity, mass spectroscopy (27) was also applied for these investigations. As an example of the calculational results and the measured actinide isotope inventory in spent fuel, some actinide isotopic ratios are given in table 2 for a fuel particle of the irradiation experiment DR-K4. This fuel reached a burn-up of 10.9% fima during 742 days of irradiation in the DRAGON reactor. Because of the deviations between calculated and measured ratios a special HTR neutron cross section library was generated and with those corrected effective cross sections it was possible to calculate all the ratios in good agreement with the experimental figures.

Table 2, Actinide isotopic ratios of an irradiated fuel kernel. Comparison of experimental values, as delivered by alpha- and mass-spectroscopy, and calculated values by use of ORIGEN.

	Experimental values	Calculated values	Deviation in %	Calculated values*	Deviation in %
$^{233}U/^{235}U$	0.524	0.53	+ 1.15	0.524	.0
$^{234}U/^{245}U$	0.0776	0.106	+ 36.6	0.0776	.0
$^{236}U/^{235}U$	0.478	0.569	+ 19.	0.478	.0
$^{238}U/^{235}U$	0.217	0.227	+ 4.6	0.217	.0
$^{234}U/^{233}U$	0.161	0.20	+ 25.	0.148	.0
$^{238}Pu/^{233}U$	0.025	0.07	+ 180.	0.024	- .4
$^{239}Pu/^{233}U$	0.0187	0.0255	+ 36.4	0.0185	- 1.07
$^{240}Pu/^{233}U$	7.2E-03	8.06E-03	+ 11.9	6.96E-03	- 3.3

*by use of fitted effective neutron cross sections

4.2 Proposal for Multi-element Analysis of Natural Radioactive Samples

Natural alpha active material is normally bulky. Preparing alpha samples by cutting yields thick samples. Nevertheless alpha spectroscopy can be well applied as shown by B. Sansoni (28).

But high resolution computer deconvolution programs will break down in the analysis of those spectra for the main reason that the straggling distributions cannot be calculated for extremely large energy losses (29) as they occur in thick samples.

So it is proposed to use the laser micro boring sample preparation technique for samples with many alpha active isotopes. The alpha activity of natural samples prepared this way will be sufficient for accuracy measurements. For example the specific alpha acitivity of the ^{239}Pu isotope in the fuel sample mentioned earlier was only about 0.008 counts per second. Natural samples may contain about 100 µg ^{238}U per gram (28). The alpha sample as prepared by the laser micro-boring method would show a specific ^{238}U alpha activity of about 0.006 counts per second. Those natural samples than could be measured and evaluated as shown in this paper allowing reliable multi-element analysis by means of alpha particle spectrometry.

REFERENCES

1) M.Helmbold, "Alpha Particle Spectroscopy - A Usefuel Tool for the Investigation of Spent Nuclear Fuel from High Temperature Gas-Cooled Reactors", Seminar on Alpha Particle Spectrometry and Low Level Measurement, Harwell, England May 10-13, 1983
2) Th. Overhoff, in Jül - 1421 (1977) 67-84
3) A. Max, Jül - 1496 (1978)
4) H.J. Allelein, R. Hecker, A. Max, Th. Overhoff and D. Stöver J. Appl. Phys. 50 (1979) 6162-6167
5) M.Helmbold, H.J. Allelein and H.R. Koch, NIM 169 (1980) 235-238
6) B.Hagmann, Jül - 1708 (1981)
7) C.Tschalär and H.D.Maccabee, Phys. Rev. B 1 (1970) 2863-2869
8) A.T. éSilva, Jül - 1833 (1983)
9) W. van der Eijk, W.Oldenhof and W.Zehner, NIM 112 (1973) 434-451
10) U.Shreter and R. Kalish, NIM 166 (1979) 117-120
11) ORTEC, Inc. "Silicon Surface Barrier Radiation Detectors Instruction Manual"
12) ANSI/IEEE Std 300-1969 "IEEE Test Procedure for Semiconductor Radiation Detectors" The Institute of Electrical and Electronical Engineers, Inc. 1976
13) K. Kandiah, "Instrumental factors which affect the observed spectrum", Seminar on Alpha Particle Spectrometry and Low Level Measurement, Harwell, England May 10-13, 1983

14) F.Calligaris, P.Ciuti, I.Gabrielli and R.Giacomich, NIM 112 (1973) 591-595

15) T.Hashimoto, K.Taniguchi, H.Sugiyama and T.Sotobayashi, Journal of Radioanalytical Chemistry 52 (1979) 133-142

16) W.Westmeier, NIM 163 (1979) 593-595

17) ANSI/IEEE Std 301-1976 "IEEE Standard Test Procedures for Amplifiers and Preamplifiers for Semiconductor Radiation Detectors for Ionizing Radiation", The Institute of Electrical and Electronical Engineers, Inc. 1976

18) J.T.Routti and S.G.Prussin, NIM 72 (1969) 125-142

19) W.Westmeier and R.A.Esterlund, "Alpha Energy Table" (1979) unpublished

20) M.Koskelo, P.A.Aarnio and J.T.Routti, NIM 190 (1981) 89-99

21) P.A.Aarnio, J.Routti, J.V.Sandberg and M.J.Winberg, NIM 219 (1984) 173-175

22) W.Beyrich and G.Spannagel, "Performance and Results of the AS-76 Interlaboratory Experiment on Alpha Sepctrometric Determination of the PU-238 Isotope",1st Annual Symposium on Safeguards and Nuclear Material Management, Brussels, April 25-27, 1979

23) W.Wätzig and W.Westmeier, NIM 153 (1978) 517-542

24) W.Westmeier, NIM 180 (1981) 205-210

25) M.Helmbold and B.Hagmann, in EUR 6629 EN (1979) 141-155

26) M.J.Bell, ORNL - 4628 (1973)

27) M.Stöppler, Jül - 633 - Ca (1969)

28) B.Sansoni, "Areas of Higher Natural Radioactivity at the Fichtelgebirge", Second Special Symposium on Natural Radiation Environment, Bombay, India January 19-23, 1981

29) C.Tschalär, NIM 64 (1968) 237-243

Review on the Monostandard Method in Multielement Neutron Activation Analysis (x)

A. Alian (xx), B. Sansoni

Central Department for Chemical Analysis
Kernforschungsanlage Jülich GmbH
D-5170 Jülich, Federal Republic of Germany

Abstract

The rising needs for multi-element analysis of a large number of samples has initiated many trials for developing simple and automated non-destructive neutron activation analysis techniques. One such approach is the monostandard or single comparator method.

The monostandard method of activation analysis, discovered first by GIRARDI, GUZZI, PAULY in 1965, is an interesting approach for calibration and calculation of element contents from the gamma-ray intensities induced after activation of the sample with (a) the commonly used whole reactor neutron spectrum, (b) more selectively by pure thermal neutrons far away from reactor core and (c) by pure epithermal neutrons under cadmium cover in the reactor. It is an intermediate between the absolute and the relative method within the large variety of instrumental methods for determining elements and isotopes, resp.

The various procedures and formalisms are described, compared and discussed. The advantages, disadvantages and limitations of each technique are critically evaluated.

(x) 5th Communication on Improvements of Instrumental Analysis for Service Analysis. 4th Communication: A. Alian, R.G. Djingova, B. Kröner, B. Sansoni, Fresenius Z. Anal. Chem., in print; also Report Jül-1822 (January 1983), ISSN 0366-0885

(xx) Guest Scientist from Egypt, present address: Professor of Inorganic and Nuclear Chemistry, Faculty of Science, Garyounis University, Benghazi, Libya

1 Introduction

Neutron activation analysis is based on the measurement of intensities of gamma-rays with definite and characteristic energies between 0.1 and 2.7 MeV. As the analytical signal they are emitted from radioisotopes produced from inactive isotopes of the elements to be determined by neutron activation. The intensity, measured as peak area in counts per second, is only a relative signal and neutron activation analysis, therefore, is a relative method as most other instrumental chemical analysis methods. Fortunately, the gamma-ray peak area is linearly proportional to the radionuclide content produced during neutron activation and therefore to the inactive element to be determined.

In order to calculate the element content to be determined from the relative signal measured via the calibration function, two different approaches and a third one in-between are available.

1) The <u>absolute</u> method calculates the element content with help of the fundamental neutron activation equation from nuclear constants of the nuclides involved and by additional physical parameters of the neutron activation process. The element content to be determined is linearly proportional to the effective neutron cross section $\tilde{\sigma}$, the fractional abundance of the isotope activated f, the absolut intensity of the measured gamma line b, the detector efficiency E and to the saturation factor $S = 1 - e^{-\lambda t_1}$, and the decay factor $D = e^{-\lambda t_2}$, where t_1 is the irradiation time, λ the decay constant and t_2 the decay time (see equation (1) below). Unfortunately, the experimental parameters of the neutron activation process are not well known for each special case and are exhibiting reasonable variations. By this reason, the absolute method in most practical cases cannot be applied. On the other hand it would have the big advantage, that each element, which produces a gamma-ray peak, could be determined quantitatively, without using standards or reference materials.

2) The <u>relative</u> method relates the measured peak area for the sample to be analysed with a corresponding peak area of a standard with exactly known and similar element content and the same matrix. By dividing the two activation equations for the sample and the standard, all parameters of the neutron activation, which may be uncertain, are eliminated. The standard must be irradiated exactly in the same position and under the same experimental conditions as the sample. Furthermore the standard has to be the same matrix composition and almost the same element content. In

addition, only such elements can be determined in the sample, for which certified element contents in the standard are available. Lack of appropriate standards is one of the main disadvantages of the relative method. If standards are available, very often only a smaller number of element contents are certified with the necessary accuracy.

3) The monostandard method is inbetween the absolute and the relative method. A monostandard (single comparator), this means a standard for only one element, is irradiated at the same time and under exactly the same irradiation conditions as the sample. It is used as a monitor of the real neutron flux during neutron irradiation at the given irradiation site in the reactor. From the measured photopeak areas of the gamma rays emitted by isotopes induced from the monostandard and the elements to be determined and the actual parameters of the activation process, using the fundamental activation equation, element contents can be calculated. As monostandard, alloys of gold or cobalt are used. There have been used also a few other elements such as iron, either in a pure form or as components of multielemental standards.

The monostandard method has the advantage of the absolute method, that for all measured gamma peak areas the corresponding element contents can be calculated independent of the availability of appropriate element standards with certified contents for each of the elements to be determined.

In the following, the different approaches to the monostandard method are discussed in detail.

2 Whole reactor neutron activation

2.1 Girardis Formalism (12)

This first critical evaluation of the monostandard method was published in 1965 (12). The weight m of an irradiated element is related to the photopeak counting rate A of the radio-isotope measured by the relation

$$m = \frac{A}{\phi S D} \cdot \frac{M}{\bar{\sigma} f b N E}, \qquad (1)$$

where

M = atomic weight of the irradiated element
f = isotopic abundance of the target nuclide
$\bar{\sigma}$ = effective activation cross section for the neutron energy spectrum used
N = Avogadro's number
E = efficiency of the detector for the γ-ray measured
b = absolute γ-intensity in decay scheme
ϕ = neutron flux
S = saturation factor; $1 - e^{-\lambda t_1}$, where λ is decay constant and t_1 irradiation time
D = decay factor; $e^{-\lambda t_2}$, where t_2 is decay time.

When the neutron flux is measured by irradiating a known weight of an element (neutron flux monitor) and measuring the induced radioactivity by γ-ray spectrometry, the same relation holds:

$$\phi = \frac{A^*}{S^* D^* m^*} \cdot \frac{M^*}{\bar{\sigma}^* f^* b^* N E^*}, \qquad (2)$$

where the asterisk refers to the neutron flux monitor (cobalt). By substituting ϕ taken from Equation 2 in Equation 1,

$$m = k \cdot \frac{A}{A^*} \cdot \frac{S^* D^*}{S D} \cdot m^*, \qquad (3)$$

where

$$k = \frac{M \bar{\sigma}^* f^* b^* E^*}{M^* \bar{\sigma} f b E}. \qquad (4)$$

The Girardi's factor k can be determined either from Equation 4 (absolute method) or experimentally from Equation 3 by irradiating a known amount of element and of the monostandard (or single comparator) and measuring their photopeak counting rates and saturation and decay factors.

Once the k-factor has been determined for a given element, the unknown sample can be irradiated with a known weight of the monostandard and equation 3 used, if it is assumed that k is constant. The accuracy of the results depends on the accuracy of the determination of the quantities appearing in Equation 3, and the constancy of the k values.

Girardi et al (12) studied the effect of variation of reactor neutron spectrum on the accuracy of the results in terms of the ratio of the effective cross section of any element to be determined to that of the monostandard. The authors showed that the error thereby encountered can be estimated experimentally by measuring the cadmium radio (CdR) of the flux monitor.

$$\frac{m_{found}}{m} = \frac{1}{CdR}\left(\frac{I_o \, \sigma_o^*}{I_o^* \, \sigma_o} - 1\right) + 1 \quad , \quad (5)$$

where I_o is the resonance integral and σ_o is the "2200 m/sec cross section".

Girardi's method can be successfully used in case of steady well thermalized reactors with periodic control of the constancy of the irradiation and measuring conditions. The CdR gives also the fraction of the epithermal neutrons in the spectrum ($\phi_{epi}/(\phi_{th} + \phi_{epi})$).

Thus

$$\frac{\phi_{epi}}{\phi_{th} + \phi_{epi}} = \frac{\sigma_o^*}{I_o^*(CdR - 1) + \sigma_o^*} \quad ;$$

$$\bar{\sigma} = \sigma^o \frac{\phi_{th}}{\phi_{th} + \phi_{epi}} + I_o \frac{\phi_{epi}}{\phi_{th} + \phi_{epi}} \quad .$$

2.2 De Corte's Formalism (11)

De Corte et al could extend the monostandard method to a triple comparator method using ^{60}Co, ^{114m}In and ^{198}Au as comparators.

In the standard irradiation position one can estimate the k_S factors in terms of resonance integral I_o and thermal cross section σ_{th} ($\sigma_{th} \approx g\sigma_o$), where g is the Westcott's factor representing deviation from 1/v thermal cross sections:

$$k_s = \frac{A_{sp}}{A_{sp}^*} = \frac{M^* f \, b \, E}{M \, f^* \, b^* E^*} \cdot \frac{\sigma_{th}\left(\frac{\phi_{th}}{\phi_{epi}}\right)_s + I_o}{\sigma_{th}^*\left(\frac{\phi_{th}}{\phi_{epi}}\right)_s + I_o^*} \cdot (6)$$

For the k_x value at any other irradiation place (index x), the same relation holds

$$k_x = \frac{M^* f b E}{M f^* b^* E^*} \cdot \frac{\sigma_{th}\left(\frac{\phi_{th}}{\phi_{epi}}\right)_x + I_0}{\sigma_{th}^*\left(\frac{\phi_{th}}{\phi_{epi}}\right)_x + I_0^*} \quad (7)$$

The factors k_s and k_x are thus related as follows:

$$k_x = k_s \cdot \frac{\left[\left(\frac{\phi_{th}}{\phi_{epi}}\right)_x + \frac{I_0}{\sigma_{th}}\right]\left[\left(\frac{\phi_{th}}{\phi_{epi}}\right)_s + \frac{I_0^*}{\sigma_{th}^*}\right]}{\left[\left(\frac{\phi_{th}}{\phi_{epi}}\right)_x + \frac{I_0^*}{\sigma_{th}^*}\right]\left[\left(\frac{\phi_{th}}{\phi_{epi}}\right)_s + \frac{I_0}{\sigma_{th}}\right]} \quad (8)$$

From Equation 8, it is clear that this formalism relies on two important parameters: ϕ_{th}/ϕ_{epi} and I_0/σ_{th} ratios. I_0/σ_{th} can be considered a nuclear constant for a given isotope and can be taken from the literature or should be determined. ϕ_{th}/ϕ_{epi} should be determined in the standard and the other reactor irradiation sites. Also the k_x and K_s (or k_{anal} and k_{ref}) values must be determined experimetally by estimating A_{sp} and A^*_{sp}:

$$A_{sp} = \frac{A}{SDm}$$

De Corte et al. suggest that ϕ_{th}/ϕ_{epi} be determined by the triple comparator method from the A_{sp} values measured for the three comparators:

$$A_{sp} = \frac{NfbE}{M}\left(\sigma_{th}\phi_{th} + I_0\phi_{epi}\right) = C\left(\sigma_{th}\phi_{th} + I_0\phi_{epi}\right) \quad (9)$$

When irradiating and counting three comparators, for which all parameters of the K-factor are known, a system of three equations with two unknowns (ϕ_{th} and ϕ_{epi}) arises, thus enabling the determination of ϕ_{th}, ϕ_{epi} and their ratio. The inaccuracies introduced by weighing and absolute counting can be eliminated by making use of an element with two isotopes (1 and 2) for the determination of ϕ_{th}/ϕ_{epi} ratios as suggested by Maenhaut et al (3):

$$\frac{\phi_{th}}{\phi_{epi}} = \frac{\frac{f_1}{f_2} \cdot \frac{b_1}{b_2} \cdot \frac{E_1}{E_2} \cdot \frac{\sigma_{th,1}}{\sigma_{th,2}}\left(\frac{I_0}{\sigma_{th}}\right)_1 - \frac{A_{sp,1}}{A_{sp,2}}\left(\frac{I_0}{\sigma_{th}}\right)_2}{\frac{A_{sp,1}}{A_{sp,2}} - \frac{f_1}{f_2} \cdot \frac{b_1}{b_2} \cdot \frac{E_1}{E_2} \cdot \frac{\sigma_{th,1}}{\sigma_{th,2}}} \quad (10)$$

or

$$\frac{\phi_{th}}{\phi_{epi}} = \frac{\left(\frac{A_{sp,1}}{A_{sp,2}}\right)_{ref}\left(\frac{I_0}{\sigma_{th}}\right)_1\left[\left(\frac{\phi_{th}}{\phi_{epi}}\right)+\left(\frac{I_0}{\sigma_{th}}\right)_2\right] - \left(\frac{A_{sp,1}}{A_{sp,2}}\right)_{anal}\left(\frac{I_0}{\sigma_{th}}\right)_2\left[\left(\frac{\phi_{th}}{\phi_{epi}}\right)_{ref}+\left(\frac{I_0}{\sigma_{th}}\right)_1\right]}{\left(\frac{A_{sp,1}}{A_{sp,2}}\right)_{anal}\left[\left(\frac{\phi_{th}}{\phi_{epi}}\right)_{ref}+\left(\frac{I_0}{\sigma_{th}}\right)_1\right] - \left(\frac{A_{sp,1}}{A_{sp,2}}\right)\left[\left(\frac{\phi_{th}}{\phi_{epi}}\right)_{ref}+\left(\frac{I_0}{\sigma_{th}}\right)_2\right]} \quad (11)$$

2.3 Kim's Formalism (13, 14)

All the above formalisms are based on the experimentally-determined or converted k factors. Kim et al in 1971 formulated the monostandard method on the basis of calculating the k-factor, taking all the nuclear data from the literature. Experimental work is needed only for the determination of the parameters ϕ_{th}/ϕ_{epi} and E^*/E, which describe the experimental conditions that are considered stable. Kim used the more accurate Westcott's convention (20):

$$m = m^* \cdot \frac{A}{A^*} \cdot \frac{S^*}{S} \cdot \frac{D^*}{D} \cdot \frac{f^*M}{fM^*} \cdot \frac{b^*}{b} \cdot \frac{E^*}{E} \cdot \frac{\bar{\sigma}^*}{\bar{\sigma}}, \quad (12)$$

where

$$\bar{\sigma} = \sigma_0\left(g + r\sqrt{\frac{T}{T_0}} \cdot S_0\right);$$

where $r\sqrt{\frac{T}{T_0}}$ -relative density of the epithermal component of the neutron spectrum (here used instead of ϕ_{epi}/ϕ_{th}), g, s_0-factors depending on the departure of the cross-section from the 1/v law:

$$S_0 = (2/\sqrt{\pi\sigma_0})\int_{\mu kT}^{\infty}\left(\sigma(E) - g\sigma_0\sqrt{E_0/E}\right)dE/E = \left(\frac{2}{\sqrt{\pi}}\right)\left(\frac{I_0'}{\sigma_0}\right). \quad (13)$$

$r\sqrt{\frac{T}{T_0}}$ can be determined from a CdR measurement in the irradiation site:

$$r\sqrt{\frac{T}{T_0}} = g\bigg/\left[(CdR-1)s_0 + 4CdR\sqrt{E_0/\pi E_c}\right], \quad (14)$$

where E_0 is the neutron energy at 293.6 °K (0.0253 ev) and E_c the Cd cut-off energy.

2.4 Simonits's Formalism (18)

Simonits et al suggested a new approach in which they introduced a generalized k_0 factor, combining the simplicity of the "almost absolute" methods with nearly the same accuracy attained in the relative methods. Replacing the Høgdahl's convention used by De Corte (Equation 6) by the more accurate Stoughton-Halperin one (19):

$$k_0 = \frac{M^* f\, b\, \sigma_M}{M f^* b^* \sigma_M^*} = \frac{A_{sp}}{A_{sp}^*} \cdot \frac{E^*}{E} \cdot \frac{\frac{\phi_{th}}{\phi_{epi}} + \frac{I_0'^*}{\sigma_M^*}}{\frac{\phi_{th}}{\phi_{epi}} + \frac{I_0'}{\sigma_M}} \quad (15)$$

where $\phi_M = n_M v_0$, conventional thermal flux for Maxwellian neutrons, with $n_M = \int_0^\infty n_m(v)\,dv$, the total Maxwellian neutron density,

$\sigma_M = g\sigma_0$, neutron cross-section, where Westcott's g-factor corrects for the non-1/v cross section, for use with ϕ_M,

$I_0' = \int_{\mu kT}^\infty \sigma(E)\,dE/E$, infinitely-dilute resonance integral, where the integration is taken from μkT, a cutoff point between the Maxwellian thermal and the 1/E epithermal flux,

μ = 5 for most reactors,

$E_T = kT$, the most probable neutron energy at temperature T. At room temperature $T = T_0$, $E_0 = 0.0253$ eV and $\mu kT = 0.165$ eV.

k_0 can thus be calculated (first term) or determined experimentally (second term)

For any practical "analysis" case, k_{anal} can be calculated simply from:

$$k_{anal} = k_0 \cdot \frac{\left(\frac{\phi_M}{\phi_{epi}}\right)_{anal} + \frac{I_0'}{\sigma_M}}{\left(\frac{\phi_M}{\phi_{epi}}\right)_{anal} + \frac{I_0'^*}{\sigma_M^*}} \cdot \frac{E}{E^*} \cdot \quad (16)$$

From Equation 16 it is clear that k_0 is independent of the reactor spectrum and of the detector characteristics.

3 Epithermal (epicadmium) neutron activation

3.1 Direct Formalism (1, 4, 5, 6, 7)

By selective irradiation in the epithermal neutron energy range (e.g. under cadmium cover), the reaction rate per atom of the (n,γ) reaction will be

$$R = \phi_{epi} I_0 \qquad (17)$$

and the weight of the element can accordingly be directly obtained from the relation

$$m = \frac{m^* A M I_0^* f^* b^* E^* S^* D^*}{A^* M^* I_0 f b E S D} = m^* \frac{A}{A^*} \cdot \alpha \cdot \frac{S^* D^* E^*}{S D E}, \qquad (18)$$

where α is constant that can be either calculated:

$$\alpha = \frac{M I_0^* f^* b^*}{M I_0 f b}, \qquad (19)$$

or determined experimentally:

$$\alpha = \frac{m A^* S D E}{m^* A S^* D^* E^*},$$

thus testing the analytical irradiation sites with the ideal conditions. However, this formalism can be best tested by estimation of the I_0 values for the elements to be determined using an equation similar to Equation 18, through irradiation of known weights of these elements and the monostandard under a Cd-cover:

$$I_0 = \frac{m^* A M I_0^* f^* b^* E^* S^* D^*}{m A^* M^* f b E S D}. \qquad (20)$$

Equation 20 gives an easy way for the determination of I_0 values in the given irradiation sites. Therefore these values are experimental (analytical) values; they incorporate all the values of nuclear constants and the detector efficiency for the elements to be determined and the monostandard.

The I_0 values evaluated using relation 20 are the ones given by

$$I_0 = \int_{E_{Cd}}^{\infty} \sigma(E) \frac{dE}{E}. \qquad (21)$$

Their values depend on the cadmium cut-off energy E_{Cd}, hence on the shape and thickness of cadmium cover.

Therefore similar cadmium covers (cylinder, box, etc) should be used in the subsequent work with the monostandard method. The deviation of the neutron energy spectrum from a 1/E distribution may be associated by up to 40 percent change of these I_o values (9).

Alian and Sansoni (6, 7) showed, however, that this will not cause any serious error, if Equation 18 and 20 are used for the determination of I_o values and subsequent monostandard evaluation of the concentration of the elements under study. These I_o values contain also the 1/v tail of the thermal neutron component. They are related to the absolute value of the resonance integral I_o by the equation

$$I_o = I_o' + 2g\sigma_o \sqrt{\frac{E_o}{E_{Cd}}} \quad . \tag{22}$$

3.2 Indirect Formalisms (8, 17, 16)

Bereznai et al in 1977 published a paper in which they proposed to convert the De Corte k-factors in K_{epi}-ones, a step that seems easy, thus (8)

$$K_{epi} = K_{ref} \cdot \frac{\left[\left(\frac{\phi_{th}}{\phi_{epi}}\right)_{ref} + \left(\frac{I_o}{\sigma_{th}}\right)^*\right]\left(\frac{I_o}{\sigma_{th}}\right)}{\left[\left(\frac{\phi_{th}}{\phi_{epi}}\right)_{ref} + \left(\frac{I_o}{\sigma_{th}}\right)\right]\left(\frac{I_o}{\sigma_{th}}\right)} \tag{23}$$

This equation is simply obtained from Equation 8 by neglecting ϕ_{th}/ϕ_{epi}, when irradiating under Cd-cover, which absorbs most of the thermal neutrons. The authors investigated the effect of errors on I_o/σ_{th} values as well as on $(\phi_{th}/\phi_{epi})_{ref}$, when k_e is estimated using Equation 23. According to their formalism, the general rules of error propagation bring about a relative error of K_e given by:

$$\Delta k_e \% = \left(\left[\Delta k_{ref}\%\right]^2 + \left[C_\sigma \cdot \Delta\left(\frac{I_o}{\sigma_{th}}\right)\%\right]^2 + \left[C_\sigma \Delta\left(\frac{I_o}{\sigma_{th}}\right)^*\%\right]^2 + \left[C_\phi \cdot \Delta\left(\frac{\phi_{th}}{\phi_{epi}}\right)_{ref}\%\right]^2\right)^{\frac{1}{2}}, \tag{24}$$

where C_σ and C_ϕ are the error coefficients. The error on K_{epi} arises mainly from inaccurate I_o/σ_{th} ratios of the elements to be determined being ± 10 % and to a lesser extend is due to erros on I_o/σ_o of the monostandard being ± 5 % and on k_{ref} being ± 3 %, thus also showing the advantages of the direct formalism.

Another indirect formalism based on the k_o factor was published recently (16, 17). Here the weight of an element in a sample co-irradiated under Cd-cover together with a monostandard is estimated by the following equations:

$$m = \frac{A\, F_{Cd}^*\, E^*\, Q_o^*(\beta)\, Q_o\, S^*\, D^*}{A^*\, k_{e,p}\, F_{Cd}\, E\, Q_o^*\, Q_o(\beta)\, S\, D}$$

$$= \frac{A\, Q_o^*(\beta)\, F_{Cd}^*\, E^*\, S^*\, D^*}{A^*\, k_o\, Q_o(\beta)\, F_{Cd}\, E\, S\, D} \quad , \qquad (25)$$

where $Q_o = I_o/\sigma_o$ and $Q_o(\beta) = I_o(\beta)/\sigma_o$
and F_{Cd} is the cadmium epithermal transmission factor (mostly = 1 or slightly less). Neglecting the F_{Cd} factor which will not introduce any error for most elements and replacing the k_o by its value (relation 15), Equation 25 will be exactly the same as Equation 18 (direct formalism). However, the authors suggest to determine $k_{e,o}$ experimentally from the relation (similar to direct formalism):

$$k_{e,o} = \frac{A\, F_{Cd}^*\, E^*\, I_o^*(\beta)\, I_o\, S^*\, D^*}{A^*\, F_{Cd}\, E\, I_o^*\, I(\beta)\, S\, D} \quad , \qquad (26)$$

or calculated theoretically from (indirect formalism)

$$k_{e,o} = k_o\, \frac{I_o\, \sigma_o^*}{\sigma_o\, I_o^*} \quad , \qquad (27)$$

and $I_o(\beta)$ is calculated through an estimation of $Q_o(\beta)$ from the following relation, which requires an estimation of the β factor (10):

$$Q_o(\beta) = \frac{I_o(\beta)}{\sigma_o} = \frac{(1\,ev)^\beta}{\sigma_o} \int_{E_{Cd}}^{\infty} \frac{\sigma(E)\, dE}{E^{1+\beta}} \quad .$$

4 Thermal neutron activation analysis (2)

Monostandard thermal neutron activation analysis in a pure thermal neutron site can be achieved through a direct formalism (2), where the weight of the element is given by

$$m = m^* \cdot \frac{AM\sigma_0^* g^* f^* b^* E^* S^* D^*}{A^* M^* \sigma_{th} f b E S D}, \qquad (28)$$

or

$$m = m^* \cdot \frac{A E^* S^* D^*}{A^* k_0 E S D}. \qquad (29)$$

The values of σ_{th} of the elements to be analysed can be experimentally determined by a relation similar to relation (28)

$$\sigma_{th} = \frac{\sigma_0^* g^* m^* AM f^* b^* E^* S^* D^*}{m A^* M^* f b E S D}. \qquad (30)$$

5 Conclusion and determinable elements

It can be concluded that, for use of the monostandard with whole reactor neutron irradiation, the generalized k_0 method is most convenient and elaborate, in particular, for computer assisted analysis. The new available compound constants k_0, Q_0, etc. as data library would save much efforts of the analyses. On the other hand, for epithermal (epicadmium) neutron irradiation, which is used less frequently, a direct formalism (relation 18), that does not incorporate thermal neutron cross section, is preferred than the other two indirect formalisms (relations 23 and 25) both incorporating thermal neutron cross section.

The elements that can be determined by the above four methods can be classified into 3 groups according to the halflife periods of the suitably produced radioisotopes. These are in the <u>minutes</u> range for the 9 elements Al, Cl, Mg, I, Pd, Rh, S, Ti and V; in the <u>hours</u> range for the 11 elements Dy, Er, Ga, Gd, Ge, K, Mn, Na, Pb, Pr and W; in the <u>days</u> range for the 36 elements As, Au, Ba, Br, Ca, Cd, Ce, Cr, Fe, Hg, In, Ir, La, Lu, Mo, Os, Pt, Rb, Ru, Sb, Sc, Se, Sm, Sn, Sr, Ta, Tb, Te, Th, Tm, U, Y, Yb, Zn and Zr and in the <u>years</u> range for the 3 elements Co, Cs and Eu. Analysis of all elements of the first group and some of the second group requires a rapid transfer system. The rest of elements can be analyzed after cooling periods ranging from few days up to one month.

The elements that can be determined by each technique are presented and discussed in a following paper by A. Alian, R.G. Djingova and B. Sansoni (3) for geological materials.

References

1) A. Alian, H.-J. Born, J.I. Kim, International Conference on Modern Trends in Activation Analysis, Saclay 2-6 October 1972; J. Radional. Chem. $\underline{37}$, 1085 (1973)

2) A. Alian, R.G. Djingova, B. Kröner, B. Sansoni, Berichte der Kernforschungsanlage Jülich, Nr. 1822, (January 1983) ISSN 0366 - 0885; Fresenius Z. Analyt. Chem., in print

3) A. Alian, R.G. Djingova, B. Sansoni, in: B. Sansoni (Edit.), Instrumental Multi-Element Analysis Verlag Chemie, Weinheim, to be published in 1985

4) A. Alian, B. Sansoni, Seminar on Activation Analysis Jülich, November 1975, Abstracts p. 17

5) A. Alian, B. Sansoni, Proceedings of a Symposium on Trace Elements in Drinking Water, Agriculture and Human Life, Cairo, Jan. 1977, p. 340; see also A. Alian and B. Sansoni, Colloquium of the Deutsche Forschungsgemeinschaft on the Geochemistry of Environmental Trace Constituents, Main, March 1979

6) A. Alian and B. Sansoni, J. Radioanal. Chem. $\underline{59}$, 511 (1980)

7) A. Alian and B. Sansoni, Spezielle Berichte der Kernforschungsanlage Jülich, Nr. 75 (April 1980)

8) T. Bereznai, D. Bodizs, G. Keomley, J. Radioanal. Chem., $\underline{36}$, 509 (1977); See also T. Bereznai, Fresenius Z. Anal. Chem., $\underline{302}$, 353 (1980); see references therein.

9) J.W. Connolloy, A. Rose, T. Wall, AAEC/TM191, 1963

10) F. De Corte, K. Sordo-El Hommami, L. Moens, A. Simonits, A. De Wispelaere, J. Hoste, J. Radioanal. Chem. $\underline{62}$, 209 (1981)

11) F. De Corte, A. Speecke and J. Hoste: J. Radioanal. Chem. **3**, 205 (1969)

12) F. Girardi, G. Guzzi and J. Pauly, Anal. Chem. **37**, 1085 (1965)

13) J. Kim, H.-J. Born, J. Radioanal. Chem. **15**, 535, (1973); see also **63**, 121 (1981); see references therein.

14) J. Kim and H. Staerk, in: Activation Analysis in Geochemistry and Cosmochemistry (A.I. Brunfelt and E. Steinnes, Eds.) Universitets forlaget. Oslo, pp. 397 (1971)

15) W. Maenhaut, F. Adams and J. Hoste, J. Radioanal. Chem **9**, 325 (1971)

16) T. El Nimr, F. De Corte, L. Moens, A. Simonits, J. Hoste, J. Radioanal. Chem. **67**, 421 (1981).

17) A. Simonits, L. Moens, F. De Corte, A. De Wispelaere, A. Elekt, J. Hoste, J. Radioanal. Chem. **60**, 461 (1980)

18) A. Simonits, F. De Corte, J. Hoste: J. Radioanal. Chem. **24**, 31 (1975). See also F. Moens, F. De Corte, J. Hoste, IBID **88**, 319 (1977)

19) R.W. Stoughton, J. Halperin, Nucl. Sci. Eng. **6**, 100 (1959)

20) C.H. Westcott, AECL - 1106, 1960

Further references on monostandard neutron activation analysis:

21) Shu-de Tu, K.H. Lieser, Multielement analysis of Chinese Biological Standard Reference Materials by monostandard instrumental neutron activation analysis, J. Radioanal. Chem., **81** (1984), 345

22) Lin Xilai, F. De Corte, L. Moens, A. Simonits, J. Hoste, Computer-assisted reactor NAA of geological and other reference materials, using the K_0-Standardization method: Evaluation of the accuracy, J. Radioanal. Chem., **81**, (1984), 333

23) L. Moens, Doct. Thesis, University of Gent, Belgium, 1981

24) A. Simonits, S. Jovanovic, F. De Corte, L. Moens, J. Hoste. A method for experimental determination of effective resonance energies related to (n, γ) reactions, J. Radioanal. Chem., **82** (1984) 169

Comparison of Different Neutron Activation Analysis Methods
for Multielement Analysis of Geological Material

Atif Alian (a), Rumiana G. Djingova (b), Bruno Sansoni

Zentralabteilung für Chemische Analysen
Kernforschungsanlage Jülich GmbH
D-5170 Jülich-1

1. INTRODUCTION

The Chemical Analysis Service at the Nuclear Research Establishment Jülich GmbH has a large spectrum of methods for element and especially trace element analysis available. Among these methods, neutron activation analysis (NAA) is only one method among others. It is used in cases, where it is superior over other methods. Main advantages are its multi-element character, optimal sensitivity for a definite number of important elements, reliability with respect to high accuracy and good precision, acceptable interferences and freedom from contamination by trace elements after irradiation. It is, however, expensive, needs trained personnel and has a low speed because of time consuming reactor irradiation, cooling and measuring. Improvements can be obtained with respect to lower detection limits, selectivity and freedom of interferences by using additional radiochemical steps. They need, however, highly trained radiochemists, and they are too slow and expensive.

By this reason, in Chemical Analysis Service, there is an evident trend to pure instrumental neutron activation analysis (INAA). In case of automation in addition to the advantages mentioned above, large sample numbers can be handled and sometimes the human error is reduced and precision improved. For multielement analysis of large sample series INAA is the only choice.

Two different calibration methods are available. The multielement standard method (relative method) uses multielement standards with certified contents for each element to be determined and with similar matrix compo-

---

(a) Guest scientist from Egypt, present adress: Garyounis University, Faculty of Science, Benghazi, Libya

(b) Guest scientist under IAEA-Fellowship from Bulgaria, University of Sofia, Faculty of Chemistry, Sofia

sition. The monostandard method (comparator or k_o-method) needs only one single standard element for monitoring the integral neutron flux during irradiation. From this, other elements can be calculated from their corresponding gamma-ray peaks using the fundamental neutron activation equation. This needs a detailed knowledge of the whole reactor and epithermal neutron spectrum and its variation with time. Even small contributions of epithermal neutrons may remarkably change the neutron cross section and resonance integrals, resulting in wrong calculations of the element content.

Irradiation with pure thermal neutrons from a thermal column of a reactor, therefore, has definite advantages. It eliminates the influence of variations in neutron energy on neutron cross sections and epithermal resonance integrals. The general equations are becoming more simple and normal nuclear data from literature can be used. Effective cross sections, experimentally to be determined, can be avoided. The theoretical concepts and practical applications have been outlined in (2, 3, 4).

It is the main scope of this investigation, with respect to daily routine analysis under realistic conditions to compare different methods of neutron activation analysis, including whole reactor (RNAA), epithermal (ENAA) and thermal (TNAA) neutron activation as well as calibration according to the multielement (rel)- and the monostandard (mono) method. As examples, geological and pedological samples from a DFG Research Project and standard granite JG-1 have been used. After description of the experimental parameters, the number of elements, which can be determined by each method, are given, followed by detection limits, precision and accuracy.

2. EXPERIMENTAL

2.1 Equipment

For RNAA and ENAA, the Merlin-Reactor FRJ-1 (5 MW; 5.10^{13} n.cm^{-2}.sec^{-1}; range of epithermal neutron flux component 3 - 7 %) has been used. 8 K multichannel analyser (Nuclear Data ND 4420) supplied with a Pertec magnetic tape drive; 33 cm^3 Ge(Li)-detector (Canberra 7249; 2,4 KeV resolution, peak/compton ratio 18: 1 for the 1332 KeV gamma line of ^{60}Co) or 25 cm^3 cylindrical coaxially driftet Ge(Li) detector (6TCW-170-20; 1,0 KeV; 38:1); linear amplifier (Tennelec TC 205). Printer (Centronics 30 b) and X-Y recorder (Hewlett Packard 700B); desk calculator (Hewlett Packard 9821 A).

Cadmium capsules (3 cm long, 1 cm outer Ø 1 mm wall thickness) and Cd-boxes (cylindrical, 8 cm long, 2 - 5 mm Ø, 1 mm wall thickness). Aluminium irradiation cans (different sizes). Quartz ampules (suprasil; 3 - 4 cm long, 0,5 cm Ø).

For TNAA, the thermal column of the 25 MW Research Reactor FRJ-2 (DIDO) has been available in the irradiation position BE 43. The thermal neutron flux was about $9 \cdot 10^{12} \, n \cdot cm^{-2} \cdot s^{-1}$, the fraction of epithermal neutrons 0.03 % of the total flux.

2.2 Standards and Reference Materials

As certified standards, mainly Japanese standard granite JG-1 (granodiorite type) and coal fly ash NBS 1633 and orchard leaves NBS 1571 have been used.

Reference rock and soil samples have been available from the Deutsche Forschungsgemeinschaft, Coordinated Research Project Geochemistry of Environmental Trace Substances: Andesit KA-3 (Santorin Greece), Granit KA-1 (Albtal, Schwarzwald), Kristallgranit 1 (porphyrisch Regensburger Wald), Dazit KA-2 (N. Kameni, Santorin, Greece), Humus pseudogley (Neuhausen-Filder) together with Japanese standard granite JG-1. The elemental compositions has been given in (2). - As monostandard, an iron wire of 5-10 mg weight has been taken in cases, where corresponding multi-element standards containing Fe have not been available.

2.3 Procedures

About 100 to 200 mg powdered samples were sealed in clean quartz ampoules (suprasil). As a reference, Japanese standard granite JG-1 was used. The ampoules were wrapped in thin aluminium foil and subsequently placed in aluminium irradiation cans. For epithermal NAA, samples and standards in the quartz ampoules were enclosed in the cadmium box before being placed in the irradiation can.

For RNAA and ENAA, irradiation time was 3 to 8 houres at $5 \cdot 10^{13} \, n \cdot cm^{2} \cdot sec^{-1}$, 4 to 7 days cooling, 1 to 8 hours measuring time (2). For TNAA, the samples were irradiated for 10 hours at a thermal neutron flux of $9 \cdot 10^{12} \, n \cdot cm^{2} \cdot sec^{-1}$ (epithermal contribution 0.03 %) and allowed to cool for 5 to 15 days. Further experimental data details are given in (2,3).

3. RESULTS

3.1 Qualitative multielement analysis

The results are given in Table 1 for neutron activation analysis using whole reactor neutrons according to the multielement standard (relative) method with epithermal neutron activation under cadmium cover and multi-element as well as monostandard method. For the soil sample, the epithermal monostandard method has been calculated in addition also with experimental effective resonance integrals $\tilde{I}$.

As a result, from Ba, Br, Ca, Ce, Co, Cr, Cs, Eu, Fe, Hf, La, Lu, Mo, Sb, Sc, Sm, Ta, Tb, Th, U, Yb in andesite 13 to 17 elements, in granite 12 to 17 and in crystal granite 13 to 16 elements could be determined. With whole reactor activation and relative method in the same order 13, 12, 14 elements, with epithermal neutron activation and relative method 16, 14,

Table 1 Elements Determined by Different Instrumental Neutron Activation Analysis Methods in DFG Rock and Soil Reference Material
R: Whole reactor neutron activation analysis;
E: Epithermal neutron activation analysis.
r: Relative method; m: monostandard method;
m, Ĩ: monostandard method using effective resonance integrals Ĩ;
+: can be determined, -: cannot be determined.

	Andesit KA-3	Dazit KA-2	Granit KA-1	Granit JG-1	Kristall-granit I	DFG Soil SR
	Rr Er Em	Rr Er Em	Rr Er Em	Rr Er Em	Rr Er Em	Rr Er Em Em, Ĩ
Ba	+ + +	+ + +	+ + +	+	+ + +	+ + + +
Br	- - -	- + -	- - +	-	- + -	- + - -
Ca	+ - -	- + -	- + -	+	- + -	- + - -
Ce	+ + +	+ + +	+ + +	+	+ + +	+ + + +
Co	+ + +	+ + +	- + +	+	- + +	+ + + +
Cr	+ - -	- - -	+ - -	+	+ - -	+ - - -
Cs	- + +	+ + +	+ + -	+	+ + +	+ + + +
Eu	+ - -	+ - -	+ - -	-	+ - -	+ - - -
Fe	+ + +	+ + +	+ + +	+	+ + +	+ + + +
Hf	+ - +	+ - +	+ - +	+	+ - +	+ - + -
La	+ + +	+ + +	+ + +	+	+ + +	+ + + +
Lu	+ - -	+ - -	+ - -	+	+ - -	+ - - -
Mo	- - -	- - -	- - -	-	- - -	- - - -
Rb	+ + +	+ + +	+ + +	+	+ + +	+ + + +
Sb	- + +	- + +	- + +	+	- + +	+ + + +
Sc	+ + +	+ + +	+ + +	+	+ + +	+ + + +
Sm	+ + +	+ + +	+ + +	+	+ + +	+ + + +
Sr	- + +	- + +	- + +	+	- + +	- + + -
Ta	- + +	- + +	- + +	+	+ + -	+ + + +
Tb	- + -	- + -	- + -	+	- + -	- + - -
Th	+ + +	+ + +	- + +	+	+ + +	+ + + +
U	- + +	- + +	- + +	+	- + +	+ + + +
Yb	+ - +	- + -	- + -	-	- + -	- - - -
Sum:	13 17 16	13 18 16	12 17 14	17	14 16 13	18 15 14 13

Table 2: Multielement Neutron Activation Analysis
Contents in ppm (µg/g), except Ca and Fe in %
[1] x̄ from 8 parallel determinations ± standard deviation
[2] x̄ from 4 parallel determinations ± standard deviation

Andesit KA-3 (DFG)

Element	RNAA, relative[1] without Cd-Cover	ENAA, relative[2] with Cd-Cover	ENAA, monostandard[2] with Cd-Cover
Ba	214 ± 77	185 ± 21.6	188 ± 22
Br	-	-	1.0 0
Ca (%)	-	6 ± 0.7	-
Ce	29.9 ± 2	45.7 ± 9.0	48 ± 11.4
Co	37 ± 12	25.3 ± 3.4	24 ± 3.5
Cr	112 ± 39	-	-
Cs	6.23 ± 1.0	1.1 ± 0.16	1.28 ± 0.25
Eu	1.1 ± 0.13	-	-
Fe (%)	5.8 ± 0.6	5.4 ± 0.5	-
Hf	3 ± 0.95	-	2.65 ± 0.34
La	24.4 ± 3.8	14.8 ± 1	14.0 ± 1
Lu	0.5 ± 0.11	-	-
Rb	65 ± 0.71	42 ± 4	40 ± 3.4
Sb	-	0.46 ± 0.08	0.45 ± 0.09
Sc	31 ± 1.4	28.0 ± 2	30.5 ± 2.4
Sm	4.7 ± 0.2	5.1 ± 0.37	4.7 ± 0.3
Sr	-	261.5 ± 3.5	260 ± 0
Ta	-	0.8 ± 0.16	0.4 ± 0.06
Tb	-	0.43 ± 0.05	-
Th	7.9 ± 0.7	7.0 ± 0.43	7.23 ± 0.46
U	-	1.9 ± 0.15	1.93 ± 0.13
Yb	-	1.35	-

Granit KA-1 (DFG)

Element	RNAA, relative[1] without Cd-Cover	ENAA, relative[2] with Cd-Cover	ENAA, monostandard[2] with Cd-Cover
Ba	727 ± 137	897 ± 115	886 ± 107
Br	-	-	0.8 ± 0.2
Ca (%)	-	3.47 ± 0.32	-
Ce	86 ± 9	78 ± 10	78 ± 12
Co	-	7.8 ± 3.9	7.0 ± 3.6
Cr	50 ± 7.6	-	-
Cs	11.1 ± 2.0	11.75 ± 1.2	-
Eu	1.6 ± 0.14	-	-
Fe (%)	2.06 ± 0.41	2.48 ± 0.6	2.1 ± 0.24
Hf	11.0 ± 2.3	-	-
La	78 ± 6.6	51 ± 5	47 ± 5.0
Lu	0.6 ± 0.16	-	-
Rb	106 ± 22.5	228.5 ± 31	220 ± 29.4
Sb	-	0.48 ± 0.165	0.54 ± 0.2
Sc	11.5 ± 0.68	8.2 ± 0.9	8.9 ± 1
Sm	9.4 ± 0.7	9.2 ± 1.1	6.7 ± 0.6
Sr	-	380 ± 80	373 ± 78
Ta	-	2.8 ± 0.8	1.4 ± 0.4
Tb	-	0.4 ± 0	-
Th	-	32 ± 3.5	33 ± 3.6
U	-	9.85 ± 1.04	13 ± 1.6
Yb	-	1.1	-

Kristallgranit 1 (DFG)

Element	RNAA, relative[1] without Cd-Cover	ENAA, relative[2] with Cd-Cover	ENAA, monostandard[2] with Cd-Cover
Ba	1024 ± 121	930 ± 155	903 ± 173
Ca (%)	-	3.6 ± 1.16	-
Ce	97 ± 9	111 ± 21	102 ± 25.5
Co	-	7.10 ± 0.70	6.5 ± 0.5
Cs	15.4 ± 2.5	10.6 ± 1.8	10.2 ± 1.7
Eu	1.3 ± 0.15	-	-
Fe (%)	2.84 ± 0.32	2.98 ± 0.44	2.67 ± 0.4
Hf	6.8 ± 1.8	-	-
La	86.2 ± 11	65.5 ± 10.4	59.7 ± 8.5
Lu	0.4 ± 0.11	-	-
Rb	247 ± 17	286.8 ± 56.7	257 ± 47
Sb	0.8 ± 0	0.67 ± 0.17	0.6 ± 0.1
Sc	7.5 ± 0.44	9.45 ± 1.17	9.38 ± 1.15
Sm	10.2 ± 1.5	11.63 ± 2.0	10.47 ± 2.34
Sr	-	216.0 ± 50	217 ± 57
Ta	1.9 ± 0.3	3 ± 0.3	-
Tb	-	0.6 ± 0.08	-
Th	21 ± 3	23.5 ± 4.0	23.3 ± 4
U	-	3.5 ± 0.6	3.4 ± 0.6

Element	Andesit KA-3 (DFG) Mono x̄	±s	±V_r	Rel x̄	±s	±V_r	Granit KA-1 (DFG) Mono x̄	±s	±V_r	Rel x̄	±s	±V_r	Kristallgranit 1 (DFG) Mono x̄	±s	±V_r	Rel x̄	±s	±V_r
As	-	-	-	-	-	-	-	-	-	-	-	-	-	-	-	-	-	-
Ba	193	6	3.1	190	32	17	829	56	7	830	21	3	900	27	3	942	81	9
Br	-	-	-	-	-	-	-	-	-	-	-	-	-	-	-	-	-	-
Ca (%)	6.9	0.7	10	7.4	0.8	11	2.2	0.1	6	1.9	0.8	42	2.2	0.2	9	1.9	0.8	42
Ce	22.8	2.8	12	-	-	-	108	2	2	-	-	-	144	2	1	-	-	-
Co	39.3	5.1	13	32	4	13	8.1	1.2	15	8.6	1.0	12	8.2	0.6	7	7.8	0.9	12
Cr	129	13	10	150	30	20	58.6	4.9	8	53.1	1.7	3	16.3	2.4	15	14.9	3.1	21
Cs	1.0	0.7	70	-	-	-	11.3	0.4	4	-	-	-	10.8	0.6	6	-	-	-
Eu	3.3	0.9	27	-	-	-	1.3	0.2	15	-	-	-	1.8	0.1	6	-	-	-
Fe (%)	-	-	-	7.3	0.5	7	-	-	-	2.4	0.2	8	-	-	-	3.2	0.2	6
Hf	2.8	0.4	14	-	-	-	6.9	1.5	22	-	-	-	8.6	2.1	24	-	-	-
K	-	-	-	-	-	-	-	-	-	-	-	-	-	-	-	-	-	-
La	16.8	0.2	1.2	17.2	0.6	3	60.1	1.9	3	52.0	1.7	3	54.3	4.1	8	60.2	5.8	10
Lu	2.2	0.5	23	-	-	-	0.9	0.2	22	-	-	-	0.7	0.3	43	-	-	-
Na (%)	-	-	-	2.56	0.03	1	-	-	-	2.50	0.03	1	-	-	-	2.48	0.02	1
Rb	-	-	-	-	-	-	-	-	-	-	-	-	-	-	-	-	-	-
Sb	0.39	0.09	23	0.40	0.10	25	0.61	0.07	11	0.69	0.06	9	0.68	0.11	16	0.62	0.08	13
Sc	39.8	1.2	2	36.1	0.9	2	9.1	0.4	4	8.7	0.2	2	7.8	1.3	17	9.9	1.2	12
Sm	4.6	0.5	11	-	-	-	8.1	0.2	3	-	-	-	11.2	1.2	11	-	-	-
Th	6.1	1.6	26	-	-	-	30.3	8.2	27	-	-	-	17.7	4.8	27	-	-	-

Table 3: Results of Monostandard and Relative Thermal Neutron Activation Analysis: Comparison of Element Contents x̄ and Precisions V_r.
Origin of the samples see chapter 2.2. x̄: element content (ppm, µg/g), mean of 4 separate determinations; s: standard deviation (ppm, µg/g); V_r: relative coefficient of variation (rel.-%); all element contents in ppm (µg/g), except Ca, Fe and Na in %.

12 elements; with thermal neutron activation and relative method 9 and with monostandard 14 elements (Ba, Ca, Ce, Co, Cr, Cs, Eu, Hf, La, Lu, Sb, Se, Sm, Tm) for all three samples.

3.2 Quantitative multielement analysis

The element contents determined by individual methods are given in Table 2 for whole reactor and epithermal neutron activation, in Table 3 for thermal neutron activation with the thermal column.

3.3 Detection Limits

Detection limits for the individual methods have been calculated approximately from the corresponding peak areas of the gamma spectra. The following Table 4 gives an overview.

3.4 Precision

Precisions are given in Table 2 and 3 as standard deviation s and/or as relative variation coefficient V_r in rel.-% for 4 to 8 independent single determinations each. The experimental conditions have been a day-by-day multielement analysis routine without any optimization for one single element. Therefore, they are by far not optimal, but they are realistic. For whole reactor and epithermal neutron activation according to Figure 1 and Figure 2 the overall precision for 248 determinations had a median of ± 14 %.

Element	NAA with whole reactor neutrons		NAA with epithermal neutrons		NAA with epithermal neutrons, for soil	
	µg	µg/g	µg	µg/g	µg	µg/g
Ba	1.0	10	2.0	20	3.0	30
Br	0.02	0.2	0.05	0.5	0.2	02
Ca	50.0	500	100	1000	>100	>1000
Ce	0.005	0.05	0.1	1	0.06	0.6
Co	0.002	0.02	0.02	0.2	0.02	0.2
Cr	0.005	0.05	0.2	2	–	–
Cs	0.008	0.08	0.015	0.15	0.03	0.3
Eu	0.0002	0.002	0.002	0.02	–	–
Fe	1.5	15	30.0	300	30	200
Hf	0.01	0.1	0.1	1	0.03	0.3
La	0.001	0.01	0.02	0.2	0.03	0.2
Lu	0.0005	0.005	0.02	0.2	–	–
Mo	0.04	0.4	0.08	0.8	0.06	0.6
Rb	0.01	0.1	0.04	0.4	–	–
Sb	0.02	0.2	0.04	0.4	0.06	0.6
Sc	0.001	0.01	0.04	0.4	0.01	0.1
Sm	0.001	0.01	0.05	0.5	0.02	0.2
Sr	4.0	40	7.0	70	–	–
Ta	0.005	0.05	0.01	0.1	0.01	0.1
Tb	0.005	0.05	0.01	0.1	0.01	0.1
Th	0.0008	0.008	0.003	0.03	0.005	0.05
U	0.02	0.2	0.04	0.4	–	–
Yb	0.0002	0.002	0.005	0.05	–	–

Table 4 : Detection Limits of Elements Determined by Neutron Activation Analysis with Whole Reactor Neutron Spectrum and with Epithermal Neutrons.
Irradiation time 6 hours; cooling time 8 days; counting time 1 hour. Total neutron flux $5 \cdot 10^{13} n \cdot cm^{-2} \cdot sec^{-1}$; epithermal to thermal neutron flux ratio 5%

It was similar for all four investigated methods and between $\pm 14,0$ and $\pm 15,0$ rel-%. In thermal neutron activation, according to Figure 3, the median of precision for elements, which could be determined according to the monostandard as well as relative method was $\tilde{x} = \pm 8$ rel-% ($\bar{x} = \pm 10,2$) for the monostandard method and $\tilde{x} = 9,5$ rel-% ($\bar{x} = \pm 13,4$) for the relative method. The slightly better precision is mainly due to the more simple conditions for thermal neutron activation.

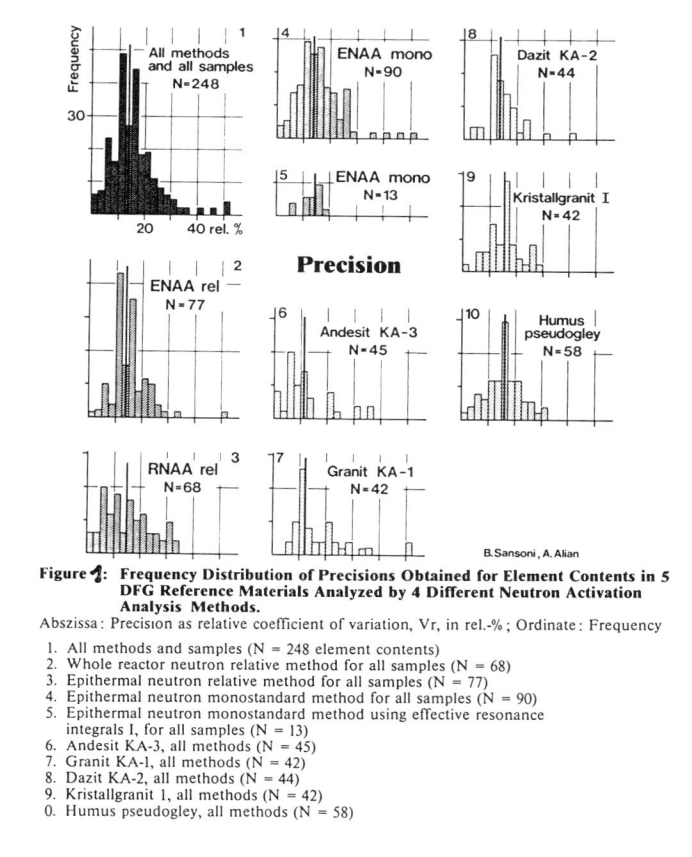

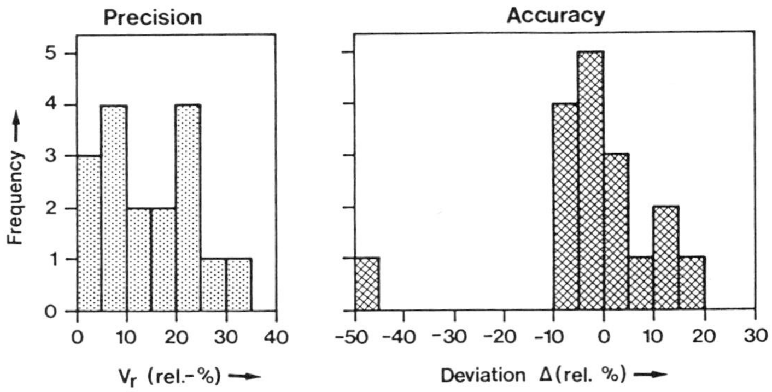

Figure 2: Comparison of Frequency Distributions of Precision and Accuracy of Neutron Activation Analysis of Standard Granite JG-1.
Precision as relative coefficient of variation Vr (rel.-%), accuracy as relative difference between the element content determined and the certificated value (rel.-%).

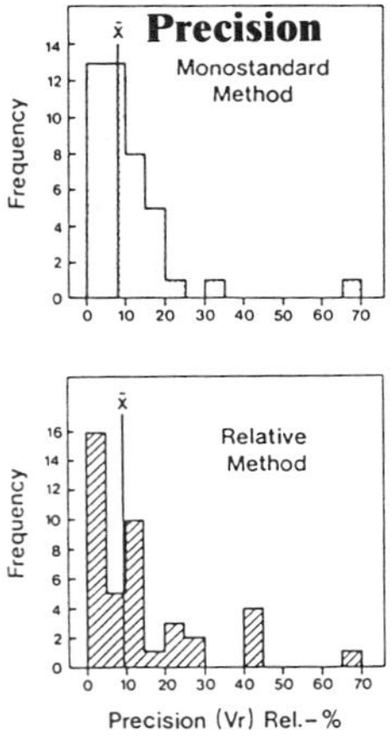

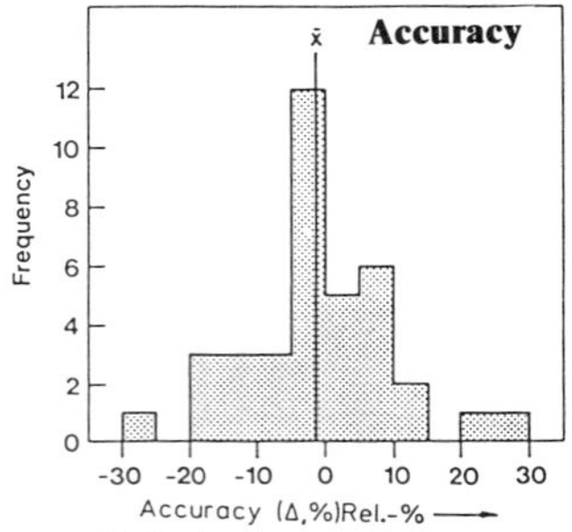

Fig. 4. Frequency distribution for accuracies of the monostandard method in thermal neutron activation analysis. Accuracies (Δ, %) in rel.-% from Table. Median $\bar{x}$: −1.4 rel.-%; Range: −29 to +29 rel.-%

Fig. 3. Frequency distribution of precision (V_r) of element contents: Comparison of monostandard and relative methods in thermal neutron activation analysis. Precisions from Table only for elements, which could be analysed according to both methods.

B. Sansoni, A. Alian

Figure 1: Frequency Distribution of Precisions Obtained for Element Contents in 5 DFG Reference Materials Analyzed by 4 Different Neutron Activation Analysis Methods.

Abszissa: Precision as relative coefficient of variation, Vr, in rel.-%; Ordinate: Frequency

1. All methods and samples (N = 248 element contents)
2. Whole reactor neutron relative method for all samples (N = 68)
3. Epithermal neutron relative method for all samples (N = 77)
4. Epithermal neutron monostandard method for all samples (N = 90)
5. Epithermal neutron monostandard method using effective resonance integrals I, for all samples (N = 13)
6. Andesit KA-3, all methods (N = 45)
7. Granit KA-1, all methods (N = 42)
8. Dazit KA-2, all methods (N = 44)
9. Kristallgranit 1, all methods (N = 42)
0. Humus pseudogley, all methods (N = 58)

3.4 Accuracy

Accuracies can be obtained only from analysis of the certified standards (granite JG-1; NBS coal). In case of epithermal neutron activation analysis according to the monostandard method, accuracy as well as precision are given in Table 5, whereas Figure 2 shows the distribution of accuracy. It was between -10 and +20 %, neglecting -47 % as an outlier for traces of Ta. The arithmetic mean is as good as -1,7 rel-%. For ca. 90 % of all elements analyzed, accuracy was within the range of precision.

For thermal neutron activation, Table 6 gives accuracies for granite JG-1, coal fly ash and orchard leaves for comparison, Figure 4 shows the corresponding distribution. Similar as with the other two activation methods, accuracy range is about -29 to +29 %, with a median of $\tilde{x} = -1,4$ rel-%.

3.6 Comparison of different NAA methods

The results of neutron activation analysis based on whole reactor, epithermal and thermal neutron activation are compared for crystal granite 1 in Table 7. For several elements, agreement is good, for a few others not.

This shows, that in important cases more than only one NAA method should be used.

3.7 Intercomparison

Intercomparison of the element determinations by epithermal NAA according to relative and monostandard method with other conventional analytical methods by participants of the DFG Research Project (as grey block) is shown in Figure 5 for the elements cobalt and antimony in all five sample materials. Since these reference materials had no officially certified element contents, it cannot be decided, which values are the better ones. Because of the relatively low concentrations of Co and Sb, NAA in several cases might be more reliable. Drastically wrong results, however are obtained in case of the soil samples with the relative epithermal method, whereas the epithermal monostandard method without (M) and with (M,$\bar{I}$) use of effective resonance integrals gives good agreement. Especially in case of soil samples it has been shown (2), that application of these experimentally determined effective resonance integrals ($\bar{I}$) instead of the literature values, has great advantages.

Element	Granite JG-1				Coal Fly Ash NBS 1633				Orchard Leaves NBS 1571				
	Element Content			Accuracy	Element Content			Accuracy	Element Content				Accuracy
	Experimental		Certified		Experimental		Certified		Experimental		Certified		
	$\bar{x}_{exp.}$	±s	$\bar{x}_{theor.}$ ±s	Δ, %	$\bar{x}_{exp.}$	±s	$\bar{x}_{theor.}$ ±s	Δ, %	$\bar{x}_{exp.}$	±s	$\bar{x}_{theor.}$	±s	Δ, %
As	–	–	– –	–	143	6.3	145 15	–1.4	9.1	0.7	10	2	–9
Ba	456	14	450 –	+1.3	0.12	0.01	(0.15) –	–20	44.8	4.6	46.3	–	–3.2
Br	–	–	– –	–	–	–	– –	–	9.9	0.9	9.14	–	+8.3
Ca (%)	1.54	0.05	1.56 –	–1	1.22	0.8	1.11 0.01	+11	2.1	0.3	2.9	0.03	–27
Ce	–	–	42 –	–	182	3	(180) –	+1.1	–	–	–	–	–
Co	3.6	0.4	4.0 –	–10	45.3	1.2	(46) –	–1.5	0.12	0.03	0.17	–	–29
Cr	54.6	7.1	50 –	+9.2	193	6	196 6	–1.5	–	–	–	–	–
Cs	10.2	0.7	10.3 –	–1	8.8	0.7	(11) –	–20	–	–	–	–	–
Eu	–	–	– –	–	4.2	0.2	(4) –	+5	–	–	–	–	–
Fe (%)	1.85	0.04	1.54 –	+20	–	–	9.4 0.1	–	369	12	300	20	+23
Hf	3.0	0.4	3.5 –	–14	6.5	0.3	(7.6) –	–14	–	–	–	–	–
K	–	–	– –	–	–	–	– –	–	1.46	0.16	1.47	0.03	–0.7
La	26.1	2.7	23.8 5.5	+9.7	85.3	2.2	– –	–	1.23	0.12	1.24	–	–3.2
Lu	–	–	– –	–	1.1	0.3	– –	–	–	–	–	–	–
Na (%)	–	–	– –	–	–	–	0.17 0.01	–	–	–	–	–	–
Rb	–	–	180 8	–	–	–	– –	–	–	–	12	–	–
Sb	0.56	0.06	0.57 0.1	–2	6.9	0.1	(7) –	–	2.4	0.2	2.9	0.3	–17
Sc	7.0	0.2	7.2 0.7	–2.8	44.7	2.3	(40) –	+11	0.009	0.001	0.007	–	+29
Sm	4.7	0.1	4.8 1.1	–2.1	2.0	0.5	– –	–	0.11	0.02	0.13	–	–15
Sr	–	–	180 28	–	–	–	– –	–	–	–	–	–	–
Th	14.9	0.6	14 2.7	+6.4	25.4	0.6	24.7 0.3	+2.8	–	–	–	–	–
U	–	–	3.5 0.5	–	9.3	1.3	10.2 0.1	–8.8	–	–	–	–	–
Yb	–	–	– –	–	22.1	3.4	– –	–	–	–	–	–	–
Zn	–	–	– –	–	–	–	– –	–	25.8	2.5	25	2	+3.2
Hg	–	–	– –	–	–	–	– –	–	0.17	0.05	0.155	0.015	+9

Table 5: Accuracy of the Monostandard Thermal Neutron Activation Analysis: Experimental Element Contents, $x_{exp.}$, Compared with Certified Values of Standards, $x_{theor.}$

Accuracy Δ, % = $(x_{exp.} – x_{theor.}) \cdot 100/x_{theor.}$. $\bar{x}_{exp.}$ is the mean element content calculated from 4 independent determinations in different samples; $\bar{x}_{theor.}$: Certified element content of the standard. Experimental conditions are the same as in Table 2 and Chapter 3. Element contents in ppm (µg/g), except Ca, Fe, Na in %.

Element	Thermal Neutron Activation				Whole Reactor Neutron Activation		Epithermal Neutron Activation			
	Mono (TNAA, mono)		Rel (TNAA, rel)		Rel (RNAA, rel)		Mono (ENAA, mono)		Rel (ENAA, rel)	
	$\bar{x}_{exp.}$	±s	$\bar{x}_{exp.}$	±s	$\bar{x}_{exp.}$	±s	$\bar{x}_{exp.}$	±s	$\bar{x}_{exp.}$	±s
Ba	900	27	942	81	1024	121	903	173	930	155
Ce	144	2.0	–	–	97	9	102	26	111	21
Co	8.2	0.6	7.8	0.9	–	–	6.5	0.5	7.1	0.7
Cr	16.3	2.4	14.9	3.1	–	–	–	–	–	–
Cs	10.8	0.6	–	–	15.4	2.5	10.2	1.7	10.6	1.8
Eu	1.7	0.12	–	–	1.3	0.2	–	–	–	–
Hf	8.6	2.1	–	–	6.8	1.8	–	–	–	–
La	54.3	4.1	60.2	5.8	86.2	11	59.7	8.5	65.6	10.4
Lu	0.7	0.3	–	–	0.4	0.1	–	–	–	–
Sb	0.68	0.11	0.62	0.08	0.8	0.1	0.6	0.1	0.67	0.17
Sc	7.8	1.3	9.9	1.2	7.5	0.4	9.4	1.2	9.5	1.2
Sm	11.2	1.2	–	–	10.2	1.5	10.5	2.4	11.6	2.0
Th	17.7	4.8	–	–	21	3	23.3	4	23.5	4.0
U	4.1	0.8	–	–	–	–	3.4	0.6	3.5	0.6
Yb	6.1	1.3	–	–	–	–	–	–	–	–
Fe (%)	–	–	3.19	0.23	2.84	0.32	2.67	0.4	2.9	0.4
Ca (%)	2.2	0.2	1.90	0.8	–	–	–	–	3.6	1.2
Na (%)	–	–	2.48	0.02	–	–	–	–	–	–

Table 6: Comparison of Different Neutron Activation Analysis Methods for Crystal Granite 1

Activation by thermal (T), epithermal (E) and whole reactor (R) neutron spectra according to monostandard (mono) and relative (rel) method. Experimental conditions for TNAA see chapter 3, for RNAA and ENAA (5). Element content and standard deviation in ppm (µg/g), except for Fe, Ca and Na in %.

Element	Thermal Neutron Activation				Whole Reactor Neutron Activation		Epithermal Neutron Activation			
	Mono (TNAA, mono)		Rel (TNAA, rel)		Rel (RNAA, rel)		Mono (ENAA, mono)		Rel (ENAA, rel)	
	$\bar{x}_{exp.}$	±s	$\bar{x}_{exp.}$	±s	$\bar{x}_{exp.}$	±s	$\bar{x}_{exp.}$	±s	$\bar{x}_{exp.}$	±s
Ba	900	27	942	81	1024	121	903	173	930	155
Ce	144	2.0	–	–	97	9	102	26	111	21
Co	8.2	0.6	7.8	0.9	–	–	6.5	0.5	7.1	0.7
Cr	16.3	2.4	14.9	3.1	–	–	–	–	–	–
Cs	10.8	0.6	–	–	15.4	2.5	10.2	1.7	10.6	1.8
Eu	1.7	0.12	–	–	1.3	0.2	–	–	–	–
Hf	8.6	2.1	–	–	6.8	1.8	–	–	–	–
La	54.3	4.1	60.2	5.8	86.2	11	59.7	8.5	65.6	10.4
Lu	0.7	0.3	–	–	0.4	0.1	–	–	–	–
Sb	0.68	0.11	0.62	0.08	0.8	0.1	0.6	0.1	0.67	0.17
Sc	7.8	1.3	9.9	1.2	7.5	0.4	9.4	1.2	9.5	1.2
Sm	11.2	1.2	–	–	10.2	1.5	10.5	2.4	11.6	2.0
Th	17.7	4.8	–	–	21	3	23.3	4	23.5	4.0
U	4.1	0.8	–	–	–	–	3.4	0.6	3.5	0.6
Yb	6.1	1.3	–	–	–	–	–	–	–	–
Fe (%)	–	–	3.19	0.23	2.84	0.32	2.67	0.4	2.9	0.4
Ca (%)	2.2	0.2	1.90	0.8	–	–	–	–	3.6	1.2
Na (%)	–	–	2.48	0.02	–	–	–	–	–	–

Table 7: Comparison of Different Neutron Activation Analysis Methodes for Crystal Granite 1

Activation by thermal (T), epithermal (E) and whole reactor (R) neutron spectra according to monostandard (mono) and relative (rel) method. Experimental conditions for TNAA see chapter 2, for RNAA and ENAA (2) Element content and standard deviation in ppm (µg/g), except for Fe, Ca and Na in %.

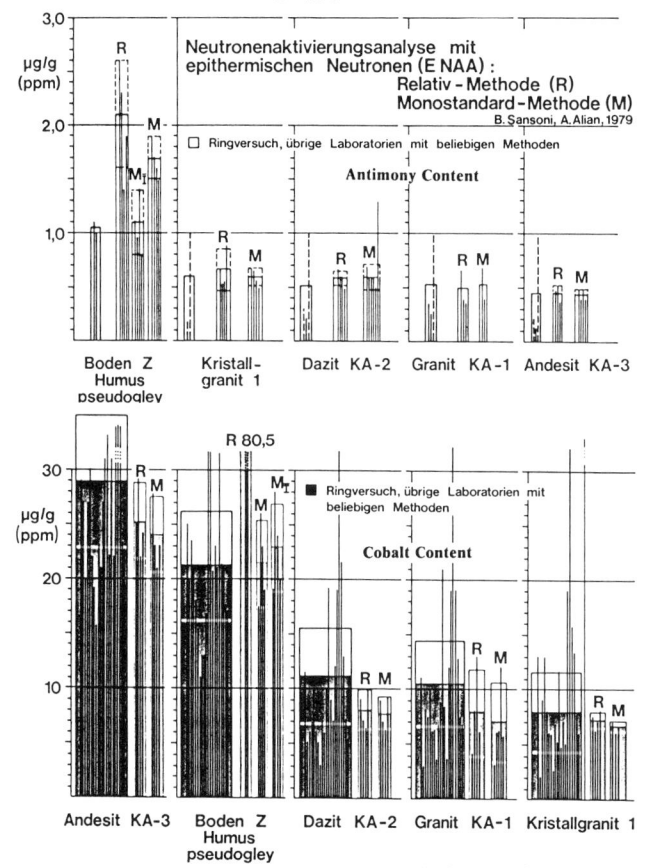

Figure 5: Intercomparison of the Neutron Activation Analysis Methods Investigated and Methods from other Laboratories.
Schwerpunktprogramm "Geochemie umweltrelevanter Spurenstoffe" DFG

4. DISCUSSION

This investigation shows for the sample types investigated, that (1) epithermal neutron activation gives the largest number of elements determinable, but may have serious difficulties in calculations and its theroretical background. (2) This is much more simple with pure thermal neutron activation, which here seems to be applied for first time to monostandard method (Lit. (3)). The precisions obtained have been slightly better for the thermal neutron activation. The results show clearly, that if possible, more than only one NAA method should be used. Monostandard method enlarges the number of elements to be determined practically to all elements, which show resonably peaks in the gamma spectrum, regardless, if there is a certified standard available or not.

5. LITERATURE

More details and literature are given in:

1) A. Alian, B. Sansoni, Instrumental neutron activation analysis of geological and pedological samples. Further investigation of epithermal neutron activation analysis using monostandard method, J. Radioanal. Chem. 59 (1980) 511-543

2) A. Alian, B. Sansoni, Comparison of different methods for activation analysis of geological and pedological samples: reactor and epithermal neutron activation, relative and monostandard method, Spezieller Bericht der Kernforschungsanlage Jülich (ISSN 0343-7639), JÜL-Spez-75, April 1980

3) A. Alian, R.G. Djingova, B. Kröner, B. Sansoni, The monostandard method in thermal neutron activation analysis, Bericht der Kernforschungsanlage Jülich, (ISSN 0366-0885) JÜL-1822, January 1983; 24 S.

4) A. Alian, B. Sansoni, Review on the mono-standard method in multielement neutron activation analysis, in: Instrumentelle Multi-Element-Analyse (B. Sansoni, Hrsg.), Verlag Chemie, Weinheim, etc., 1985, in print

SINGLE COMPARATOR METHOD IN 14 MEV NEUTRON ACTIVATION ANALYSIS

Jerzy Janczyszyn

Institute of Physics and Nuclear Techniques
Academy of Mining and Metallurgy, Cracov, Poland

The known idea of the single comparator method with k_o-factors [1,2] is discussed here in relation to the 14 MeV neutron activation analysis. The theoretical background of the reactor version of the method is adapted by introducing the following assumptions:
1. The cross section curves $\sigma(E)$ can be approximated, between 14 and 15 MeV, by linear function (Fig.1.),
2. The neutron spectrum function of the D-T source - $\Phi(E)$ is narrow and symmetrical with respect to E_o - the most probable energy of the source neutrons.
3. Contribution from secondary, low energy, neutrons to the reaction rate is negligible.

Then, the activation reaction rate $R = \int_0^\infty \sigma(E)\Phi(E)\,dE$ can be expressed by:

$$R = \Phi[\sigma_o + \Delta E_o (\frac{\partial \sigma}{\partial E})_o]\ ,\quad \Phi = \int_0^\infty \Phi(E)\,dE\ ,\quad \Delta E_o = E_o - 14.5\,\text{MeV} \qquad (1)$$

and the ratio of specific count rates - A_{sp} of the determinand and the comparator element (denoted by asterisk) is:

$$k = \frac{A_{sp}}{A_{sp}^*} = \frac{M^* \theta \gamma \varepsilon_p [\sigma_o + \Delta E_o (\frac{\partial \sigma}{\partial E})_o]}{M \theta^* \gamma^* \varepsilon_p^* [\sigma_o^* + \Delta E_o (\frac{\partial \sigma}{\partial E})_o^*]}\ ,\quad A_{sp} = \frac{A}{SDCm} \qquad (2)$$

After introducing:

$$k_o = \frac{M^* \theta \gamma \sigma_o}{M \theta^* \gamma^* \sigma_o^*}\ ,\quad Q = \frac{1}{\sigma_o}(\frac{\partial \sigma}{\partial E})_o \qquad (3)$$

equation (2) becomes:

$$k = k_o \frac{1 + \Delta E_o Q}{1 + \Delta E_o Q^*}(\frac{\varepsilon_p}{\varepsilon_p^*}) \qquad (2a)$$

Meanings of the used above symbols are:

σ_o — cross section for 14.5 MeV, the reference energy,

$(\frac{\partial \sigma}{\partial E})_o$ — slope of the excitation function, for 14.5 MeV,

ε_p — full energy peak efficiency, A — measured count rate,

S,D,C — saturation, decay and counting factors.

The value of ΔE_o, characteristic for a particular neutron source (D-T type) and irradiation geometry, can be calculated from eqs. (4):

$$\Delta E_o = -\frac{\alpha - 1}{\alpha Q_2 - Q_1}\ ,\quad \alpha = \frac{A_{sp,1} k_{o2} \varepsilon_{p2}}{A_{sp,2} k_{o1} \varepsilon_{p1}} \qquad (4)$$

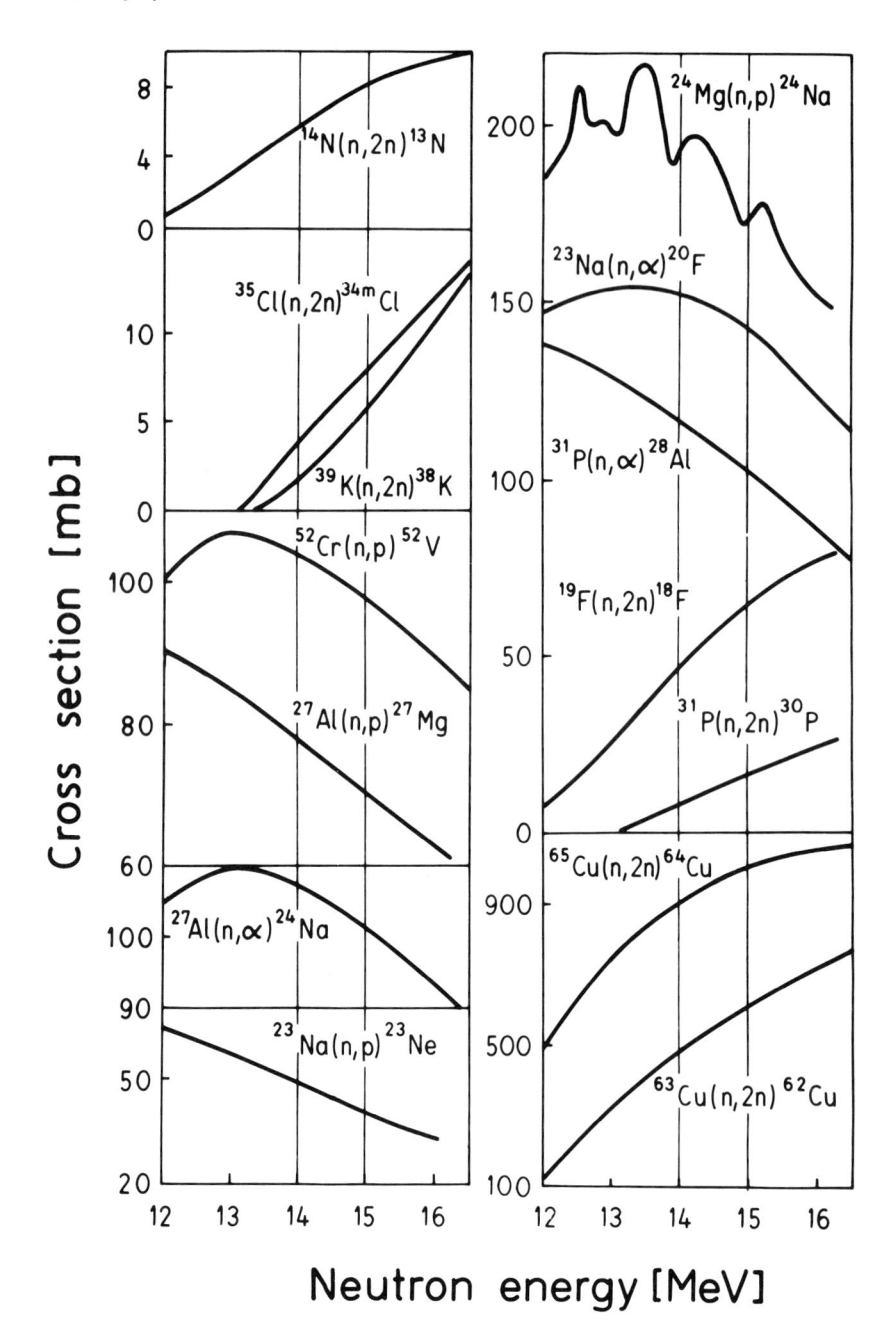

Fig.1. Excitation functions of the most frequently used activation reactions.

where 1 and 2 denote the two reactions used for the experimental determination of ΔE_0. E_0 can be also determined using the method proposed by Lewis and Zieba[3] and modified by Csikai[4].

The equations (2,2a) can be applied for the determination of Q. If two neutron sources 1 and 2 are used for simultaneous irradiation of an element, for which Q is to be determined, and the comparator element, the value of Q can be computed from:

$$Q = -\frac{\beta - \gamma}{\beta \Delta E_{02} - \gamma \Delta E_{01}}, \quad \beta = \frac{A_{sp,1} A^*_{sp,2}}{A_{sp,2} A^*_{sp,1}}, \quad \gamma = \frac{1 + \Delta E_{02} Q^*}{1 + \Delta E_{01} Q^*} \quad (5)$$

The nuclear data necessary for the calculation of k_0-factors and specific activities[5-7] were critically revieved. A small part of the data is shown in table 1 together with the calculated relative errors and values of Q.

Table 1. Examples of the cross section data.

Reaction	σ [mb]			Q [MeV^{-1}]	σ_0 error [%]
	14.1 MeV	14.5 MeV	14.9 MeV		
^{14}N (n,2n) ^{13}N	6.1	7.0 ± 0.5	8.2	0.375	7
^{19}F (n,2n) ^{18}F	47	55 ± 4	60	0.295	7
^{23}Na (n,p) ^{23}Ne	48.0	44.3 ± 4	40.8	-0.203	9
^{24}Mg (n,p) ^{24}Na	196.5	190.7 ± 1.1	171.8	-0.162	0.6
^{27}Al (n,α) ^{24}Na	121.9	116.0 ± 0.7	111.3	-0.114	0.6
	122.0	115.8 ± 2.4	113.3	-0.110	2.1
^{27}Al (n,p) ^{27}Mg	77.0	74.0 ± 2.4	71.1	-0.100	3.2
^{28}Si (n,p) ^{28}Al	273	260 ± 25	248	-0.120	10
^{52}Cr (n,p) ^{52}V	90	94 ± 10	96	0.080	11
^{56}Fe (n,p) ^{56}Mn	112	103 ± 6	103	-0.109	6
^{63}Cu (n,2n) ^{62}Cu	480	522 ± 20	585	0.251	3.8
^{65}Cu (n,2n) ^{64}Cu	922.7	967.5 ± 20	1003	0.104	2.1

Error of analysis is discussed on the basis of the equation for the determined amount of an element:

$$m = \frac{m^*(SDC)^* A}{k_0 \, SDC \, A^*} \cdot \frac{1 + \Delta E_0 Q^*}{1 + \Delta E_0 Q} \left(\frac{\varepsilon^*_p}{\varepsilon_p}\right) \quad (6)$$

In table 2 error propagation coefficients and values of absolute and relative partial errors are shown. Contributions of $m^*, S, D, C, S^*, D^*, C^*$ errors are neglected.

Table 2. Components of the total error of analysis.

Source of error - x	Coefficient $-\frac{1}{m}(\frac{\partial m}{\partial x})$	Value of error	
		absolute - s_x	relative - δ_x [%]
k_o	$1/k_o$	-	< 20
A	$1/A$	-	< 3
A^*	$1/A^*$	-	< 1
$\varepsilon_p^*/\varepsilon_p$	$\varepsilon_p/\varepsilon_p^*$	-	1 - 2
ΔE_o	$\frac{Q^* - Q}{(1+\Delta E_o Q^*)(1+\Delta E_o Q)}$	~ 0.1 MeV	0 - 7
Q	$\Delta E_o / (1+\Delta E_o Q)$	0.1 - 0.2	0 - 12
Q^*	$\Delta E_o / (1+\Delta E_o Q^*)$	~ 0.01	0 - 0.5

The most important source of error is the k_o-factor and in it the cross section. The improving state of cross section data and the possibility of precise experimental determination of k_o-factors lead to the conclusion that the method can be suitable for multi--elemental analyses, especially with the use of intense neutron generators. At present, taking into consideration the accuracy of nuclear data, high saturation activity, half-lifes and γ-line energies, Al and Fe can be proposed for comparator elements. When introducing the method, the influence of secondary neutrons should be avoided, if possible, by proper construction of the tritium target holder.

References

1. A.Simonits, F.DeCorte, J.Hoste, J.Radioanal.Chem. 24(1975) 31.
2. A.Simonits, L.Moens, F.DeCorte, A.DeWispelaere, A.Elek, J.Hoste, ibid. 60(1980) 461.
3. V.E.Lewis, K.J.Zieba, Nucl.Instr.Meth. 174(1980) 19.
4. J.Csikai, in "Nuclear Data for Science and Technology", K.H.Böckhoff(ed.), ECSC, EEC, EAEC, Brussels and Luxenbourg, 1983, p.414.
5. Handbook on Nuclear Activation Cross Sections, IAEA, Techn. Rept. Ser. 156, 1974.
6. C.M.Lederer, V.S.Shirley, E.Brown, J.M.Dairiki, R.E.Doebler, "Table of Isotopes", Wiley-Interscience, New York, 1978.
7. Cross Sections of Threshold Neutron Induced Reactions. Handbook. (in Russian), Energizdat, Moscow, 1982.

SIMULTANEOUS DETERMINATION OF THE HALOGENS Cl, Br, I IN BIOLOGICAL MATERIALS WITH EPITHERMAL NEUTRON ACTIVATION; RESULTS AND COMPARISON WITH CONVENTIONAL METHODS.

Armin Wyttenbach, Leonhand Tobler, Verena Furrer

Swiss Federal Institute for Reactor Research,
CH-5303 Wuerenlingen, Switzerland

1 INTRODUCTION

Epithermal activation is a very sensitive method for the determination of iodine. Because the halflife of ^{128}I is only 25 min, irradiation and measuring times can be kept short. This is a prerequisite for dealing with many samples per day and for keeping the unit cost low. Since the same measurement allows the determination of bromine (via ^{80}Br, τ = 17.5 min) and chlorine (via ^{38}Cl, τ = 37 min), valuable additional information can be obtained for the same cost.

2 PRINCIPLES OF OPERATION

Samples (100 - 200 mg) are weighed in polyethylene containers and irradiated in a container made of boron nitride (320 mg B/cm^2), using the pneumatic transfer system to the reactor SAPHIR (thermal flux 1.3 · 10^{13} n/cm^{2}s). The BN-container stops most of the thermal neutrons and thus accentuates the activation of nuclides with high resonance integrals (Table 1). Reducing ^{56}Mn and ^{38}Cl, which are abundant in the thermal activation of most biological matrices, is a prerequisite to the determination of I by instrumental neutron activation analysis. B is a more appropriate neutron absorber in this work than Cd, because it does not give rise to high levels of shortlived activities which would hinder manipulation of the samples just after activation. A minor disadvantage of B is the copious amount of α-particles generated during irradiation by ^{10}B(n,α)^{7}Li; this leads to a substantial heating of the samples. Irradiation time has therefore been limited to 2 min.

After irradiation, the samples are counted on a 45 cm^3 Ge(Li)-detector. Counting starts after 7 min and lasts for 20 min. A shorter decay time is precluded by the reaction ^{23}Na(n,p)^{23}Ne. The ^{23}Ne (τ = 37 s, 440 keV) interferes with the detection of ^{128}I.

The samples are irradiated singly and every fifth sample is accompanied by a standard. A sample is irradiated and cooled while the preceeding one is being counted, leading to a maximum throughput of 20 samples in 8 hours.

Table 1   Nuclear parameters

Nuclear reaction	$\tau_{1/2}$ (min)	Resonance integral (barn)	R_B (a)	Measured γ-line (keV)	Sensitivity (b)
$^{37}Cl(n,\gamma)^{38}Cl$	37.2	0.3	320	1642.4	3
$^{79}Br(n,\gamma)^{80}Br$	17.4 c)	92	12	616.3	1100
$^{127}I(n,\gamma)^{128}I$	25.0	150	5	442.9	4900
$^{55}Mn(n,\gamma)^{56}Mn$	155	14	130	846.8	890

(a) R_B is the activity ratio between irradiation without and with the B-absorber, under otherwise identical conditions.
(b) Sensitivity is given as counts/μg for the irradiation and measuring conditions defined above.
(c) The actual halflife is greater than 17.4 min because of the concurrent reaction $^{79}Br(n,\gamma)^{80m}Br$ (4.4 h) → ^{80}Br. A halflife of 20 min has been found appropriate during the first hour after irradiation.

3 RESULTS

The sensitivity is given in Table 1. For the halogens, it varies roughly reciprocally with their content in most samples thus leading to a well balanced γ-spectra.

The experimental reproducibility was evaluated by the analysis of replicates (usually n > 10), and is mainly determined by counting statistics. Table 2 gives typical values for natural systems. In systems with increased levels of halogens such as those arising from testing medicaments or food-additives on animals, the reproducibility falls rapidly to 4 - 5 %. This level is dictated by the other parameters, especially the stability of the neutron flux.

The detection limit is about 40 ng for I, 150 ng for Br and 40 µg for Cl. It is somewhat influenced by the ^{56}Mn activity of the samples. The calculated detection limits are given in Table 2. If detection limits are a problem, the samples can alternatively be counted in a Ge(Li)-well-detector. This lowers the limit by a factor of approximately 2.

Table 2   Natural concentration, experimental reproducibility and detection limit of the halogens in biological materials.

Matrix	Concentration (ppm)				Reproducib. (%)			Detection Limit (ppm)		
	Mn	Cl	Br	I	Cl	Br	I	Cl	Br	I
Citrus leaves	23	414	8.2	1.88	10	5	5	100	0.61	0.14
Faeces (rat)	137	580	3.6	0.26	a	a	a	140	0.87	0.15
Blood (rat, b)	< 1	3100	11.5	0.05	6	3	17	60	0.15	0.03
Milkpowder (c)	< 2	7200	11.8	1.12	5	6	4	290	1.25	0.30

(a) Faeces have not been homogenized, so the reproducibility could not be determined.
(b) Because of their small I-concentration, natural blood samples were measured in the well detector.
(c) Concentration in milk is 8 times less than in milkpowder.

4   COMPARISON WITH OTHER METHODS

The accuracy of the determination of I by the present method was estimated by a comparison with results obtained by a microspectrophotometric method (catalysis of Ce^{+4}/As^{+3}-reaction), on 12 spiked milk samples covering a range between 0.1 ppm (natural level) and 2.2 ppm. Correlation between the two sets of measurements showed that

(a) the regression goes through the origin (intercept 3 ± 3 ppb), i.e. neither method has a significant blank or bias

(b) the slope of the line has a value of 0.999 ± 0.017, i.e. neither method has a calibration problem

(c) the residual standard deviation of the regression is 5.2 % corresponding to a standard deviation for each method of 3.7 %.

Whereas this comparison showed perfect agreement between activation analysis and spectrophotometry, determinations made with an ionspecific electrode showed both bias and incorrect results for reasons that were not investigated.

5 HALFTIME OF IODINE EXCRETION WITH MILK

Part of the samples for the comparison of methods where obtained by administration of a single dose of KI together with the fodder to a cow and by sampling the milk at intervalls of 12 hours up to a total of 4 days. The resulting concentration c(t) of I in the milk is very well described by

$$c(t) = a + b \exp(-\lambda/t) \qquad (eq.\ 1)$$

where a is the concentration before the administration of the KI, b and λ are constants and t is the time after the administration. λ was found to be $0.61\ d^{-1}$, meaning that the part of the iodine which is excreted with the milk has a halftime of excretion of 27 hours.

6 ACKNOWLEDGMENTS

The samples for the intercomparison of methods where prepared by Dr. Hunziker and Zimmerli from the BAG, Berne and the microspectrophotometric determinations were carried out by Dr. Lauber, University of Berne. Their collaboration is greatly appreciated.

THE RATIO METHOD AS THE BASIS FOR A PREDICTION-OPTIMIZATION INAA-PROGRAM

R.Gwozdz

Institute of Petrology University of Copenhagen
Øster Voldgade 10 DK-1350 Copenhagen K

The choice of correct irradiation, decay and counting conditions has fundamental effect upon the quality of final results and on effectiveness of the measuring and irradiation facilities used. Already 20 years ago a program was described for optimization of the irradiation and decay times in activation analysis /1/.

The optimization method was based on a purely mathematical approach to solving of the maximization problem. The drawback of such approach is mainly connected with impossibility to account for several very important 'variables' of administrative, financial or purely technical character. An example of that 'variable' type can be a 5-days working week, cost of irradiation and measurement, a periodic working scheme of the irradiation center, time limitations for irradiation due to the kind of container, or sample involved.

V.P.Guinn /2/ described a prediction method based on calculation of the approximate pulse-height gamma-spectrum for a sample with composition roughly known, and with subsequent calculation of the lower limit of detection for any activable elements. Replicate calculations for different experimental conditions allow to chose optimum parameters for any real situation.

In a way similar to that of the ETZR method, described by Korthoven and de Bruin /3/ , the present method applies the concept of two lines ratio of two different isotopes. For the line i of element x, the number of counts is expressed as

$$A_{ix} = R_i * FE * W * K * \exp(-\lambda * t_d)$$

where
- R_i - count ratio of one microgram of the element x to the 1099.3 keV line of Fe (10000 microgram).
- FE - counts of 10000 micrograms of Fe in normalized counting, decay and irradiation conditions.
- W - content in micrograms of element x in a sample.
- K - normalization coefficient accounting for difference in counting geometry and in the irradiation time.

The background area B_i under the peak A_{ix} can be described as a linear combination of interaction products of all the gamma lines emitted by the irradiated sample, with material of the detector, shielding and the sample itself.

$$B_i = \Sigma_j C_{ij} * A_j$$

where A_j - peak j of other elements present in the sample.
C_{ij} - the fraction of background continuum produced by the element j in position under peak A_{ix}.

Coefficients C_{ij} are previously determined by means of irradiation and measurement of pure single element samples. The library of R_i values for 180 lines of 40 isotopes, C_{ij} coefficients for the main 20 gamma-emitters and detector resolution calibration are used as the constant part of input data for calculation. The varying input consists of energy calibration, irradiation, decay and measurement time, sample weight, concentration of 20 main gamma-emitters, and coefficient characterizing geometrical conditions.

The following values are calculated by the program:

Cps - total count per second in the spectrum.
B - background area = 2.548 * FWHM * B_i/gain
N - the total width of the peak expressed in channels
Peak - predicted peak area in counts (A_i)
Sig-% - 100 * (Sqrt(2 * B + Peak)/Peak)
LD - detection limit according to Rogers /4/.
LD = 2.71 + 2.326 * Sqrt(N*B)

Fraction in % - list of the first five elements contributing most to the background.

For 45 gamma-lines there are predicted values of detection limits for 11 various measurement times and about 40 decay times - all obtained in a single run of the program.

During the last 4 years there were collected more than 60 measurements of international geological standards, GSP,BCR,AGV,Soil-5,and SL. Sample weight, irradiation and measurement conditions varied in a wide range. Measurements were performed on the same coaxial Ge(Li)-detector with 12% relative efficiency. Irradiation times were from 1 to 6 hour in DR-3 Reactor(Risø National Laboratory), in a position with thermal neutron flux of $3*10**13$ and epithermal part less than 0.25% .Measurement times were from 25 to 900 minutes, decay times from 8 to 110 days. The detector - sample distance varied from 4 to 30 cm. A comparison of experimental results with numerical experiments demonstrated clearly

- an agreement better than 30% between experimental and calculated standard deviation and detection limit for all the compared spectra.
- that in all the calculated spectra only four elements were responsible for more than 90% of background continuum. For short decay time these elements were Na,La,Sc,Fe, for medium and long decay time - Sc,Fe,Co,Eu.
- that for every isotope there exists a characteristic decay time interval - when both the calculated standard deviation and detection limits assume a minimum value.

The execution cost of the program is about 5% of irradiation cost for one sample, and its applications can save considerable expenses, and improve the quality of results due to possibility of adequate selection of the experimental conditions prior to experiment.

The support of the Danish Natural Science Research Council is gratefully acknowledged.

References

/1/ Isenhour,T.L., Morrison,G.H, Anal.Chem. 36 (1964) 1089-1092

/2/ Guinn,V.P. J.Radioanal.Chem. 15 (1973) 473-475

/3/ Korthoven,P.J.M.,DeBruin,M.J., J.Radioanal.Chem. 35 (1977) 127-137

/4/ Rogers,V.C., Anal.Chem. 42 (1970) 807-808

DETECTION LIMITS IN MULTI ELEMENT NEUTRON ACTIVATION ANALYSIS - SOME
OBSERVATIONS ON SIGNAL-TO-NOISE RATIO AND COUNTING STATISTICS

M. Sankar Das and S. Yegnasubramanian

Analytical Chemistry Division, Bhabha Atomic Research Centre,
Trombay, Bombay 400 085, India

1. INTRODUCTION

It is not uncommon to see analytical data reported by the activation technique with the uncertainties in the measurements expressed in concentration units obtained from counting statistics. The reported precision of the data does not give information whether these values are obtained based on well defined signals or from measurements near the detection limits[1]. The apparent high precision could be obtained from long counting periods, or, from high gross signals over equally large spectral background. Such signals may be considered as those which are approaching the detection limit of the technique. In view of the above, it may be worthwhile reviewing the quality of the analytical data by INAA in relation to the basic definitions of limit of detection, LOD, normally applied to analytical techniques.

2. DEFINITIONS

2.1 General

According to the IUPAC definition[1], 'the limit of detection, expressed as a concentration, c_L, (or quantity, q_L) is derived from the smallest measure, x_L, that can be detected with reasonable certainty for a given analytical procedure'. There are several other definitions, for example, by ACS[2]. Many of them offer wide-ranging terminologies - lower limit of detection[3], critical level[4], limit of determination[5], or limit of quantitation[2], limit of guarantee of purity[6] etc. As expected, these definitions are general.

2.2 Conditional Detection Limits

Most techniques adopt the concept of a 'conditional detection limit' in practice. Any number stated for a LOD does not carry adequate meaning unless the experimental conditions (and the parameters) are completely defined. The definitions by Smales[7] and Guinn[8] for neutron activation analysis are well defined conditional LODs. These values are interference free LODs and are for single nuclides and that too, in favourable circumstances for single gamma rays, but do not suffice for multielement instrumental neutron activation analysis methods.

The experimental measurements involving gamma ray spectrometry always measures the spectral background (matrix effect) associated with a given signal and this background is not invariant with time. Hence one encounters situations of varying spectral background from sample-to-sample and for the same sample, for spectra measured at different elapsed times. Mathematical procedures for optimising the decay and counting-times and derivations of LODs are of limited use when it involves measurement of a nuclide in a multicomponent complex spectrum of unknown samples.

2.3 Derivation of Detection Limits

Several authors have described the methods of deriving LODs, both for the case of paired observations and for cases with 'well known' blanks. Parsons[5] arrived at a method, first by experimentally obtaining a limit of determination and extrapolating to the appropriately defined detection limit for many atomic and molecular techniques. Currie[4] had derived the functions for calculating LODs and limit of quantitation (LOQ) and applied them to radioactivity and activation analysis. Long and Winefordner[9] had observed that the IUPAC definition is valid only if the major sources of error are in the blank; he had also recommended a propagation of errors approach for the determination.

3. SIGNAL-TO-NOISE RATIO (S/N) AND RELATIVE STANDARD DEVIATION (RSD)

3.1 Since the measurement process of the gross signal itself is statistical, estimates of the statistical error associated with the net signal is possible after taking into account the laws of propagation of errors. However, in the instrumental approach, the errors associated with the counting statistics alone do not reflect the quality of the signal involved in the measurements, as evidenced in our experience in the analysis of several samples. It is advantageous to consider the quality of the signal through the signal-to-noise ratio, defined as

(S/N) = Net peak counts/spectral background.

For example, Table 1 gives the results of our analysis of copper foundry samples where it is seen that for the same magnitude of error due to counting statistics (obtained after propagation through spectral background and standard) for cobalt and rhenium, namely 5 %, the S/N for cobalt is 7, whereas for rhenium, it is only 0.2, which speaks volumes about the confidence attributable to the rhenium value, as against the cobalt value.

3.2 Significance of S/N and RSD

To understand the significance of S/N and the RSD of the measurements, the RSD of net counts, NPPc (Gross, GPPc - Background, BLC) is plotted at two magnitude levels of counting, one at 10^3 and the other 10^4 counts, as a function of S/N in Figure 1. The improvement of precision in the high quality signals is worth the time spent in counting, but not on the poor signals; worse, any apparent improvement represents false security of large numbers; rhenium in the above example.

Table 1: INAA of Copper Foundry Samples

Element	Concentration (ppm)	E (keV)	Per cent error due to counting statistics	S/N
Co	55	1332	5	7
As	2700	559	1	18
Se	28000	264	1	9
Te	1500	364.5	3	1
Sb	60	602.7	5	3
Re	200	155	5	0.2
Au	0.6	412	3	0.4

5.3 (S/N) of Standard and Sample

For a standard - sample combination, the compounded variance is$^{(10)}$,

$$s^2(x_1/x_2) = (x_1/x_2)^2 \cdot \left[(s_1^2/x_1^2) + (s_2^2/x_2^2)\right]$$

where x_1, s_1, x_2, s_2 are the measured NPPc's and the standard deviations of sample and standard respectively. If expressed in units of percent RSD,

$$\text{Net RSD (\%)} = \left(\frac{s_1^2}{x_1^2} + \frac{s_2^2}{x_2^2}\right)^{1/2} \cdot 100 = \left[(RSD_{x1})^2 + (RSD_{x2})^2\right]^{1/2} \cdot 100$$

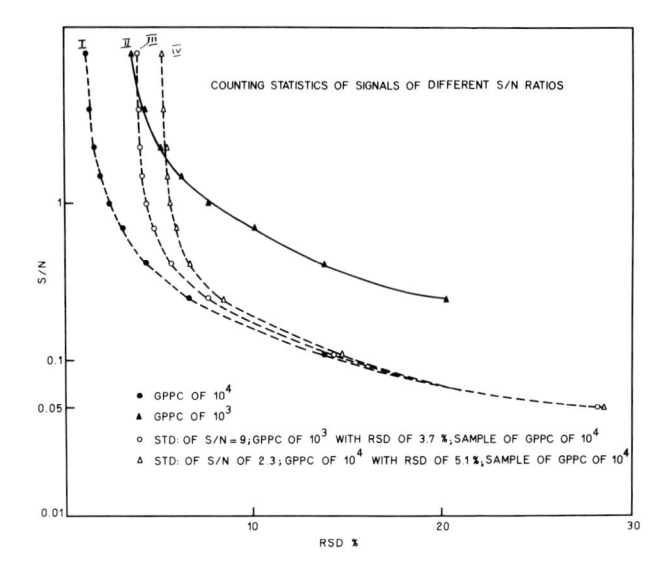

Fig. 1

Typical values computed from the data used to generate curves I. and II of Figure 1 are plotted in curves III and IV of the same figure, which reiterates the fact that, unless the quality of the signals of the comparator and standard are good and comparable, the result is not improved. Some practical examples from our laboratory in the analysis of the animal muscle (H-4) sample are given in Table 2. The above instances are meant to demonstrate that the results obtained by comparing a sample and a multielement reference material as standard, should not have the same analytical acceptability, even if they carry the same errors due to counting statistics, unless their S/N also are comparable.

Table 2: Propagation of Errors and (S/N) - Animal Muscle, H-4

Element	Standard	Concentration ppm	$\frac{S}{N}$ Std.	$\frac{S}{N}$ H-4	Counting errors		Concentration ppm	Net RSD%
					H-4 RSD%	Std. RSD%		
Br	Kale	25.6	0.4	0.1	30.7	5.9	5.3	31.3
Cr	Kale	0.33	0.48	0.08	16	3.6	0.08	17
Cu	Bovine Liver	193	2.7	0.95	3.1	1.1	4.35	3.3
Fe	Bovine Liver	270	2.45	1.19	5.4	1.8	47.3	5.7
Rb	Bovine Liver	18.3	2.86	9.8	1.4	1.9	16.4	2.4
Se	Kale	0.12	0.15	0.18	5.6	6.9	0.43	8.9

4. SITUATION NEAR DETECTION LIMITS IN INAA

4.1 Case i. A signal equal to '3s' of background

According to Kaiser[6], 'the difference between the analytical measure 'x' and the mean blank value $\bar{x}_{bl}$ must be greater than a definite multiple 'k' of the standard deviation, s_{b1}, of the blanks' and 'if at least 20 blank analyses are carried out and 'k' is taken as 3, then the risk of erroneously eliminating a measure as not sound is at most a few per cent'. According to the IUPAC definition[1], 'the minimum concentration or quantity detectable is, therefore, the concentration or quantity corresponding to $c_L = (x_L - \bar{x}_{bl})/s$; $q_L = (x_L - \bar{x}_{bl})/s$, where 's' is assumed to be constant for low values of 'c' and 'q' (when counting statistics are involved, as in x-ray spectroscopy, s_{bl} is often estimated directly from a single measurement of s_b', a value of 3 for 'k' is strongly recommended)'.

In paired observations (as counting experiments) it is useful to have a high magnitude for the background since it gives a better confidence in the measurement of the background; however, acceptance of '3s' of the background to define the detection limits can lead to the risk of acceptance of impure signals. An example is given in Table 3.

Table 3: Purity of Signals at Detection Limits for Different Magnitudes of Background Limit of Detection = 3s Background

Background counts	s_B counts	Per cent counting statistics	$3s_B$	Derived GPPc, counts	NPPc counts	S/N	RSD % $s = (GPPC+BLC)^{1/2}$
100	10	10	30	130	30	0.3	50
900	30	3.3	90	990	90	0.1	48
10000	100	1	300	10300	300	0.03	48
90000	300	0.33	900	90900	900	0.01	47

One finds from Table 3, that the RSD does not change but the S/N becomes poorer by more than an order, an instance of false security of large numbers arising from the definition of detection limits.

4.2 Case ii. A signal equal to '10s' background

According to the ACS recommendations[2], 'the numerical significance of the apparent analyte concentration increases as the analyte signal increases above the LOD. As a minimum criterion, the region of quantitation should be clearly above the limit of detection' and 'the limit of quantitation (LOQ) is located above the measured average blank, s_b, by the following definition,

$$s_t - s_b > k_q \sigma$$

It is recommended that the minimum value be $k_q = 10$'. Here s_t is the gross signal and has the same significance as 's' in case i. Hence if the LOQ is defined as the signal equivalent of '10s' background, data as given in Table 4 have to be considered.

Table 4: Purity of Signals at Limit of Quantitation for Different Magnitudes of Background; Limit of Quantitation = '10s' Background

Background counts, BLC	Per cent error counting statistics of BLC	GPPc	NPPc '10s'BLC	S/N	RSD (%)
50	14	120	70	1.4*	20
100	10	200	100	1	17
900	3.3	1200	300	0.3	15
10000	1	11000	1000	0.1	14
90000	0.33	93000	3000	0.03	

*A good signal with poor counting statistics.

The behaviour is similar to case 4.1 except that the per cent RSD is numerically smaller.

4.3 Case iii. A signal equal to the background (s^2)

If the limit of quantitation is modified to represent the signal equivalent of the background itself, the pertinent data are given in Table 5.

Table 5: Measurements at Net Signal Equal to Background

Background	GPPc	NPPc	S/N	RSD, %
50	100	50	1	25
100	200	100	1	17
900	1800	900	1	6
10000	20000	10000	1	0.6

Under this condition, for the same S/N, the RSD goes down by a factor of nearly 30; hence it recommends itself.

5. COMMENTS / OBSERVATIONS

5.1 Definition of limit of detection or limit of quantitation in INAA, even if conditional by stating the flux, duration of irradiation etc., will not reflect the quality of the signal, unless the S/N also is taken as a parameter.

5.2 The definition of limit of quantitation may be modified as the concentration equivalent of the background itself so that a S/N of unity is assured particularly for signals measured with RSDs poorer than 20 %; this value of 20 % is assumed to be the maximum permissible coefficient of variation for concentrations determined below 1 ppm [11].

5.3 When results are reported in an unknown sample using a certified reference material as standard, unless the measurement parameters of S/N and RSD are within comparable limits for the sample-reference material combination, the results for all the elements should not be accepted with the same confidence. Further reporting the S/N value along with the final result will help the agencies for screening the data before deriving a recommended value for the analyte.

REFERENCES

1. Spectrochim. Acta B 33B (1978) 242
2. Anal. chem 52 (1980) 2242
3. Altshuler, B. and Pasternack, B., Health Physics, 9 (1963) 293

4. Currie, L.A., Anal. Chem. 40 (1968) 586
5. Parsons, M.L., J. Chem. Education, 46 (1969) 290
6. Kaiser, H. "Two Papers on the Limit of Detection of a Complete Analytical Procedure", Adam Hilger Ltd., London, 1968
7. Smales, A.A. 'Neutron Activation Analysis' in Trace Analysis, 518-546, Joe, J.H. and Koch, H.J. (Eds.) John Wiley, New York, 1957
8. Guinn, V.P., Report GA-8171, 1966
9. Gary L. Long and Winefordner, J.D., Anal. Chem. 55 (1983) 713A
10. Svehla, G., "The Application of Mathematical Statistics in Analytical Chemistry, Mass Spectrometry, Ion Selective Electrodes, p. 70; Comprehensive Analytical Chemistry, Wilson and Wilson (Eds.) Elsevier, 1981
11. Pure and Appl. Chem., 51 (1979) 1183

MINIMIERUNG DER BLINDWERTE BEI DER INSTRUMENTELLEN MULTI-ELEMENT-BESTIMMUNG IN BIOLOGISCHEN PROBEN

Franz Lux, Tamas Bereznai und Susanne Knobl

Institut für Radiochemie der Technischen Universität München, D-8046 Garching

SUMMARY

In our wound healing experiments we have tried to minimize the blank values that are caused by different sources of contamination during surgery, sampling and the activation analysis procedure.

The following topics have been investigated: The abrasions from scalpels of stainless steel, titanium and quartz; the type of surgery; the homogenisation of the samples before irradiation by use of a ball mill; the surface contaminations of the quartz ampules that pass into the digestion solution of the irradiated samples. The appropriate measures to be taken in order to reduce the blank values are described.

All these measures have been taken in a series of analyses of muscle controls of rabbits. The achieved low level of contamination during the whole procedure is especially indicated by the obtained very small value of the chromium content: 1.2 ng/g. Lowest value in the literature: 5 - 6 ng/g.

In preceding investigations we have found iron enrichments in the wound area. In order to distinguish between enrichments caused by bleeding during wounding and enrichments due to increased iron requirement by the healing processes, the erythrocytes of the animals were labelled with ^{55}Fe, and the ^{55}Fe content of one part of each sample was determined by means of a liquid scintillation counter. In order to avoid quenching effects, a procedure has been worked out for preparation of colourless, transparent counting samples.

1 Einleitung

Ein wichtiger Vorteil der Aktivierungsanalyse ist bekanntlich, daß Verunreinigungen, die *nach* der Bestrahlung in die Probe gelangen, das Analysenergebnis nicht beeinflussen. In dem *vor* der Bestrahlung durchzuführenden Teil eines Versuchsprogramms (einschl. des Abschnittes vor der Probennahme, z. B. während der Tierexperimente) können jedoch durch die Art der Handhabung oder durch sonstige Einflüsse störende Blindwertniveaus in gleicher Weise wie bei anderen Spurenelementbestimmungsverfahren verursacht werden. Im folgenden wird berichtet, welche Erfahrungen wir bei unseren Wundheilungsuntersuchungen (1, 2, 3, 4) hinsichtlich der von verschiedenen Kontaminationsquellen bedingten Blindwerte und deren weitgehender Verringerung gemacht haben. Diese Ergebnisse sind von allgemeiner Bedeutung für die Aktivierungsanalyse biologischen Materials.

2 Skalpell-Abrieb

Die Tierexperimente bei den Wundheilungsuntersuchungen werden wie folgt durchgeführt: An geeigneter Stelle eines Kaninchens wird eine Schnittwunde gesetzt und das Tier nach einer vorgegebenen Heilungsdauer getötet. Anschließend werden Proben aus dem Wundbereich und Kontrollproben entnommen. Bei Verwendung der üblichen Edelstahlskalpelle für Wundsetzung und Probennahme ist eine nicht zu vernachlässigende Beeinflussung der Analysenergebnisse durch Abrieb zu erwarten, da zu bestimmende Spurenelemente wie Eisen, Chrom und Kobalt auch Hauptbestandteile von V4A-Stahl sind. Im Hinblick auf die im Wundbereich stets gefundenen Eisenanreicherungen und die daraus gezogenen Folgerungen (1, 2, 3, 4) war die Möglichkeit einer Beeinflussung dieser Resultate durch den Eisenabrieb vom Skalpell speziell zu prüfen. Auch die Bestimmung der äußerst niedrigen Chromgehalte von Gewebe könnte durch Abrieb - V4A-Stahl enthält 19,5 % Chrom - verfälscht werden. In dem Zusammenhang ist zu erwähnen, daß die in der Literatur angegebenen Werte für den Chromgehalt von menschlichem Muskel (5) um vier Größenordnungen differieren.

Wegen der offenkundigen Kontaminationsprobleme bei Verwendung von Edelstahlskalpellen benutzt man Skalpelle aus Titan (6) bzw. Quarz (7). In diesen beiden Fällen kann ein Abrieb der Hauptbestandteile nur bei Titanmessern stören, und zwar wegen der Bildung des Calciumbezugsnuklids ^{47}Sc durch $^{47}Ti(n,p)^{47}Sc$. Es muß allerdings auch der Abrieb von Oberflächenverunreinigungen in Betracht gezogen werden.

Zur Klärung der Abriebsproblematik wurden Skalpelle aus V4A-Stahl, Titan und Quarz zusammen mit Multi-Element-Standards im Reaktor bestrahlt, mit den bestrahlten Messern nacheinander an mehreren Fleischproben Schnitte vorgenommen und aus den γ-Spektren dieser Fleischproben in üblicher Weise die von den Skalpellen abgeriebenen Elementmassen ermittelt.

In einem ersten Versuch wurde geprüft, wie weit sich die durch Skalpellober-
flächenverunreinigungen verursachte Probenkontamination reduzieren läßt, wenn man
die Messer vor ihrer Benutzung mit einer heißen Mischung aus conc. Salpetersäure,
conc. Perchlorsäure und conc. Schwefelsäure (3/2/1) behandelt. Dafür wurden Titan-
skalpelle verwendet, da bei diesen das Problem der Oberflächenverunreinigungen besonders offenkundig ist, und zwar wegen der Weichheit des Materials: Die Messer
müssen vor jeder Verwendung neu geschliffen werden; Abrieb von den Schleifwerkzeugen
bleibt in der Außenschicht des Skalpells gut haften. Die Ergebnisse dieses Experiments sind in Tab. 1 zusammengestellt.

Tab. 1: Abrieb von Titanskalpellen

Reinigung nach Schleifen nur mit bidest. H_2O oder mit heißer Mischung von conc.
HNO_3/conc. $HClO_4$/conc. H_2SO_4 = 3/2/1, dann Abspülen mit bidest. H_2O

Element	Reinigung mit		ng/200 mg Kaninchenmuskel (Frischgewicht)
	bidest. H_2O	Säuremischung	
	ng pro Schnittprobe 1 *		
Ti(^{47}Sc)	12000	4410	
Ca(^{47}Sc)	22000	7790	5000 - 14000
Sc	0,028	0,01	
Cr	16	0,55	0,24
Fe	500	8,1	600 - 1800
Co	1,1	0,09	0,8 - 2
Zn	200	8,3	1400 - 3800
Se		< 0,11	8 - 84
Cs		< 0,01	0,6 - 2,8

* vgl. Tab. 2

Tab. 2 enthält die Resultate einer anschließend mit Skalpellen aus allen drei genannten Materialien durchgeführten Versuchsreihe, wobei die Titan- und Quarzmesser vor der Bestrahlung mit der beschriebenen Säuremischung gereinigt worden waren. Die Edelstahlskapelle wurden nicht vorbehandelt, da bei diesen, wie erwähnt, hauptsächlich der Abrieb ihrer Hauptbestandteile stört.

Tab. 2: Abrieb von Skalpellen
Mit neutronenbestrahlten Skalpellen wurden nacheinander in 3 Fleischproben jeweils 6 Schnitte vorgenommen.

Skalpell	Element	Gehalt des Skalpells % oder ng/g	ng pro Fleischprobe Nr. 1	Nr. 2	Nr. 3	Für den Bulk-Abrieb in Probe 1 berechnete Elementmasse [a] ng	ng/200 mg Kaninchenmuskel (Frischgewicht)
V4A-Stahl	Cr	19,5 %	390	90	9	1091	0,24
	Fe	69,7 %	3900	1000	240	-	600 - 1800
	Co	0,027 %	1,2	0,6	0,6	1,5	0,8 - 2
Titan	Ti(^{47}Sc)	99,9 %	4410 [b]	2358 [b]	1089 [b]	-	
	Ca(^{47}Sc)	-	7790 [c]	4164 [c]	1924 [c]		5000 - 14000
	Sc	1190	0,01	0,0045	0,0025	0,0052	
	Cr	550	0,55	0,3	< 0,06	0,0024	0,24
	Fe	170000	8,1	4,1		0,75	600 - 1800
	Co	25	0,09	0,09	0,08	0,00011	0,8 - 2
	Zn	-	8,3	7,5	6,0	-	1400 - 3800
	Se	< 71	< 0,11	< 0,09	< 0,08	< 0,00031	8 - 84
	Cs	< 26	< 0,01	< 0,01	< 0,01	< 0,00011	0,6 - 2,8
Quarz	Ca(^{47}Sc)	-	< 54	< 63	< 51		5000 - 14000
	Sc	0,01	0,001	0,001	0,001		
	Cr	0,88	0,25	< 0,05	< 0,004		0,24
	Fe	-	< 2,4	< 2,4	< 2,5		600 - 1800
	Co	0,15	0,03	0,02	0,03		0,8 - 2
	Zn	-	2,0	2,3	2,2		1400 - 3800
	Se	1,2	0,19	< 0,05	< 0,04		8 - 84
	Cs	-	< 0,004	< 0,004	< 0,004		0,6 - 2,8

Die Spurenelementgehalte des Titans (Spalte 3) wurden aktivierungsanalytisch bestimmt.

a) Die in den Proben Nr. 1 gefundenen Eisen- bzw. Titanmassen wurden als Maß für die in diese Proben gelangten Skalpell-Bulkabriebe genommen. Aus diesen Bulkabriebs-Werten und den Bulkgehalten (Spalte 3) wurde von jedem Element die Masse berechnet, die als Anteil des Bulkabriebs in die Probe 1 gelangte. Wenn die in Probe Nr. 1 gefundene Masse eines Elements größer ist als die in Spalte 7 angegebene berechnete Masse, ist dies ein Hinweis auf eine Oberflächenkontamination des Skalpells durch das entsprechende Element.

b) Die Titanbestimmung erfolgte unter Bezug auf ^{47}Ti(n,p)^{47}Sc. Eine mögliche Verfälschung der Analysenergebnisse durch ^{46}Ca(n,γ)^{47}Ca $\xrightarrow{\beta^-}$ ^{47}Sc war zumindest in merklichem Maße nicht gegeben, da in den γ-Spektren des Skalpells und der Fleischproben keine γ-Linien des ^{47}Ca nachweisbar waren.

c) Masse des Calciums, die der gemessenen Aktivität von ^{47}Sc auf Grund von ^{46}Ca(n,γ)^{47}Ca $\xrightarrow{\beta^-}$ ^{47}Sc entsprechen würde.

Aus den Daten in Tab. 2 ergibt sich erwartungsgemäß, daß nur bei Verwendung von Quarzskalpellen die Probenkontaminationen durch den Abrieb akzeptabel sind. Die Kontamination durch Chrom war jedoch noch zu hoch. Man sieht aus dem Vergleich der abgeriebenen mit der pro Messer gefundenen Chrommasse (0,69 ng, Quarzmesser = 0,7 g), daß Chromoberflächenverunreinigungen abgerieben wurden. Die genannte Reinigung mit dem Salpeter-, Perchlor-, Schwefelsäuregemisch ist demnach nicht ausreichend (vgl. auch Anmerkung a) in Tab. 2, und zwar in Bezug auf Titanskalpell). Die Quarzskalpelle werden daher jetzt folgendermaßen behandelt: Abätzen mit conc. Flußsäure, Abspülen mit bidestilliertem Wasser, Abextrahieren mit conc. Salpetersäure, nochmaliges Abätzen mit conc. Flußsäure und Abspülen mit bidestilliertem Wasser.

Abschließend ist zu diesen Untersuchungen noch zu bemerken, daß die gefundenen Abriebe sicher Maximalwerte sind, da infolge der Strahlenschäden der Oberfläche von den bestrahlten Skalpellen mehr abgerieben wird als von unbestrahlten.

3 Modifizierung der Tierversuche *

Die Modellwunden müssen so plaziert werden, daß ihre möglichst ungestörte Heilung gewährleistet ist, wobei die Verhaltensweise des Tieres mitberücksichtigt werden muß. Vom medizinischen Standpunkt interessiert sehr die Heilung der Hautwunde. Um die beiden genannten Bedingungen zu erfüllen, wurden anfangs am Rücken von Kaninchen Schnitte gesetzt, und zwar außer in die Haut auch in den Hautmuskel, um zusätzlich zur Heilung der Haut auch die des Muskels untersuchen zu können. Es ergaben sich aber folgende Nachteile:

1. Der Bereich des Hautschnittes ist Kontaminationen aus der Umgebung, z.B. durch Berührungen mit dem Stallgitter ausgesetzt.

2. Ein Hautschnitt heilt durch Neubildung von Haut. In einem Muskelschnitt entsteht kein neues Gewebe, sondern er wird durch Bildung von Bindegewebe repariert. Haut und Bindegewebe enthalten jeweils bis zu 35 % Kollagen, d. h. in beiden Fällen ist die Heilung mit einer Kollagenbildung verknüpft. Es besteht die Hypothese, daß die festgestellten Eisenanreicherungen im Wundgebiet durch den bekannten Eisenbedarf zur Kollagensynthese hervorgerufen sein könnten (1, 2). In dem Zusammenhang interessiert außer der Ermittlung der Eisenanreicherung auch, wie groß die Menge des Kollagens ist, die im Wundspalt während der vorgegebenen Heilungsdauer gebildet wurde, um Aussagen hinsichtlich einer Korrelation zwischen Eisenanreicherung und Kollagenbildungsrate machen zu können.

---

* Die Tierversuche erfolgten in Zusammenarbeit mit dem Institut für Experimentelle Chirurgie der Technischen Universität München (Direktor Prof. Dr. G. Blümel) und wurden von Priv.-Doz. Dr. habil. W. Erhardt und Dr. H. Rechl ausgeführt.

Die Hautwundenproben können nicht frei von beträchtlichen Anteilen an angrenzender Haut erhalten werden. Die aus dieser Quelle stammende Kollagenmenge ist so groß, daß man aus dem Analysenresultat (Gesamtkollagengehalt der Probe) den Anteil des im Wundspalt neu gebildeten Kollagens kaum ermitteln kann; m. a. W.: die Feststellung der neu gebildeten Kollagenmenge wird durch einen zu hohen Blindwert beeinträchtigt.

Bei Proben aus dem Hautmuskelwundbereich besteht dieses Problem nicht, da die an den Proben anhängenden Muskelreste nur einen Kollagengehalt von 1 - 2 % haben. Der Hautmuskel ist aber nur etwa 3 mm dick, so daß man oft Proben von weniger als 50 mg zur Verfügung hat, eine für die Aktivierungsanalyse und biochemische Analyse zu geringe Menge.

Es wurde daher dazu übergegangen, als Modellwunden etwa 1 cm tiefe und 3 cm lange Schnitte in die relativ dicke Oberschenkelmuskulatur von Kaninchen zu setzen. Damit wird folgendes erreicht:

1. Es wird nur der Hautschnitt vernäht und dadurch eine Kontamination der allein interessierenden Muskelwunde durch die Nadel vermieden. Wird straff vernäht, so ist auch das erwünschte Zusammenfügen der Flanken des Muskelschnitts gewährleistet. Durch die Hautschnittvernähung ist die Muskelwunde vor Kontaminationen aus der Umgebung während der Heilungsdauer geschützt.

2. Der Muskelschnittbereich ist groß genug, um aus ihm eine ausreichend große Probenmasse mit einem nur geringen Anteil an Gewebe zu gewinnen, das nicht unmittelbar zum Wundbereich gehört. Diese kleinen Fremdgewebeanteile sind außerdem Muskel mit dem schon erwähnten geringen Kollagengehalt von 1 - 2 %. Infolge dieser beiden Fakten ist der Blindwert bezüglich der Bestimmung des neugebildeten Kollagens klein.

4 Homogenisierung der Proben

Die entnommenen Gewebeproben werden mit einem sog. Dismembrator (8) homogenisiert, um für die Aktivierungsanalyse und die biochemische Kollagenbestimmung Proben einsetzen zu können, die die gleiche Verteilung von Spurenelementen und Gewebebestandteilen haben. Das PTFE-Homogenisierungsgefäß und die PTFE-Mahlkugeln werden zur Vermeidung bzw. Minimierung einer Probenkontamination mit conc. Flußsäure und anschließend mit bidestilliertem Wasser gewaschen. Es wird bei Raumtemperatur gearbeitet; für die vorgeschlagene Homogenisierung von biologischem Material in tiefgefrorenem Zustand (8) sind PTFE-Kugeln zu leicht.

5 Quarzampullen

Die bestrahlten Proben werden mit einem Gemisch aus conc. Salpetersäure, conc. Schwefelsäure und Wasserstoffperoxid aus den Ampullen, die vorher außen gereinigt wurden, unter Zerstörung der organischen Matrix herausgelöst, um jeweils gleiche Meßgeometrie für Probe und Standard einstellen und den ^{32}P abtrennen zu können (9). Durch diese Probenauflösung werden gleichzeitig diejenigen Blindwertanteile beseitigt, die von den auch in hochreinem Quarz enthaltenen Verunreinigungen stammen und die bei einer Messung der Proben in den Bestrahlungsampullen, die wir früher auch durchgeführt haben, sehr stören können. Oberflächenkontaminationen der Quarzampullen gelangen jedoch in die Aufschlußlösung. Um diese Kontaminationen möglichst gering zu halten, wurden zwei Reinigungsverfahren für Quarzampullen geprüft:
- Abextrahieren der Quarzrohre vor der Ampullenherstellung und der bestrahlten Ampullen mit conc. Salpetersäure
- Abätzen mit conc. Flußsäure, und zwar der Quarzrohre vor der Ampullenherstellung, der Innenseiten der Ampullen nach halbseitigem Zuschmelzen der Quarzrohre und der bestrahlten Ampullen.

Die Ergebnisse der beiden Reinigungsverfahren sind in Tab. 3 dargestellt. Es ist ersichtlich, daß nur die Flußsäureätzung den gewünschten Erfolg bringt, wobei es mit diesem Verfahren auch gelingt, einen akzeptablen Blindwert für die Bestimmung des besonders geringen Chromgehalts von Muskel zu erreichen.

Tab. 3: Reduzierung der Oberflächenkontamination von Quarzbestrahlungsampullen

Element	Bei Reinigung * der Ampullen durch wurden nach Behandlung ** der bestrahlten Ampullen gefunden				Elementmasse in 200 mg Kaninchenmuskel (Frischgewicht)
	Abextraktion mit conc. HNO_3		Abätzen mit conc. HF		
	in der Ampulle ng	in der Aufschlußlösung ng	in der Ampulle ng	in der Aufschlußlösung ng	ng
Ca	< 1000	< 300	< 100	< 80	5000 - 14000
Cr	4,4	3,5	0,5	0,04	0,24
Fe	360	< 40	< 3	< 1,3	600 - 1800
Co	0,24	0,07	0,07	0,02	0,8 - 2
Zn	2,1	< 5	8	8	1400 - 3800
Se	< 0,5	< 0,5	< 0,1	< 0,02	8 - 84
Rb	< 20	< 2	< 0,09	< 0,05	560 - 1080
Cs	0,1	< 0,08	< 0,008	< 0,004	0,6 - 2,8

* Vor der Bestrahlung innen und außen, nach der Bestrahlung außen.
** Die bestrahlten Leerampullen wurden außen gereinigt, geöffnet und dann mit heißer Aufschlußlösung (s. Text) behandelt.

6 Ergebnisse der Analysen von Kaninchen-Muskelproben

Tab. 4 enthält die Ergebnisse von Kaninchenmuskel-Analysen, bei denen alle beschriebenen Maßnahmen zur Blindwertminimierung getroffen wurden. Als Indikator für das erreichte tiefe Niveau der Kontaminationen während des gesamten Verfahrens kann der gefundene niedrige Chromgehalt von 1,2 ng/g angesehen werden. Der kleinste bisher in der Literatur angegebene Wert für den Chromgehalt von Kaninchenmuskel ist 5 - 6 ng/g (10).

Tab. 4: Analysen von Kaninchenmuskel-Kontrollproben

$\Phi_{th} = 2 \cdot 10^{13}$ n·cm^{-2}·s^{-1}, t_b = 100 h, t_m = 12 h

Element	E_γ keV	Gehaltsrelation bez. auf Frischgewicht	2 d	5 d	8 d	10 d	Heilungsdauer 2 d		8 d	
							Erythrozytenmarkierung mit ^{55}Fe			
							linker Oberschenkel	rechter Oberschenkel	linker Oberschenkel	rechter Oberschenkel
Na(^{24}Na)	1368	µg/g	340	268	310	370	315	325	245	477
Ca(^{47}Sc)	159	µg/g	44	26	< 33	-	< 123	52	36	61
(^{47}Ca)	1297	µg/g	< 40	33	38	41	< 97	58	< 34	71
Fe(^{59}Fe)	1099	µg/g	4,8	3,3	3,7	3,1	4,9	8,9	3,5	8,6
	1292	µg/g	3,4	3,6	3,7	2,9	4,9	8,7	3,4	7,8
Co(^{60}Co)	1173	ng/g	4,2	5	7	5,2	8	10	8	8
	1332	ng/g	5,8	6	7	5,1	8	10	8	9
Zn(^{65}Zn)	1115	µg/g	7,5	7,2	7,6	6,9	11	19	7,5	17
Se(^{75}Se)	136	ng/g	50	120	99	-	280	400	210	370
	265	ng/g	70	120	120	42	300	420	240	370
Br(^{82}Br)	777	µg/g	-	-	-	-	0,94	1,3	0,8	1,1
Rb(^{86}Rb)	1077	µg/g	2,8	3,3	3,8	2,8	4,6	5,4	4,4	5,3
Cs(^{134}Cs)	795	ng/g	3	11	8	7	10	14	10	13
Cr(^{51}Cr)	320		1,2 ng/g; $\Phi_{th} = 7,5 \cdot 10^{13}$ n·cm^{-2}·s^{-1}, t_b = 260 h, t_m = 337 h								

7 Erythrozytenmarkierung mit ^{55}Fe

Auf die während der ersten 2 - 3 Wochen einer Heilung stets beobachtete Eisenanreicherung im Wundgewebe wurde bereits hingewiesen. Der Hypothese, daß diese Anreicherung mit dem Eisenbedarf zur Kollagensynthese zusammenhängt (1, 2), steht das Argument entgegen, daß Eisenanreicherungen auch als Folge des Blutens bei der Wundsetzung auftreten können (sog. Hämatomeisen). Um hinsichtlich der zweiten Möglichkeit zu einer eindeutigen Aussage zu gelangen, wurden auf Vorschlag von Prof. W. Forth, Institut für Pharmakologie der Universität München, die Erythrozyten der Versuchstiere mit ^{55}Fe markiert. Die Erythrozyten von Kaninchen haben eine Lebensdauer von ca. 50 Tagen (11), die Heilungsversuche dauern maximal 14 Tage. Infolgedessen gehört alles in den Wundbereichsproben gefundene ^{55}Fe zum Bluteisen, und es gilt:

$$\frac{\text{Masse Bluteisen in der Wundbereichsprobe}}{\text{Masse Gesamteisen in der Wundbereichsprobe}} = \frac{\text{Spezifische }^{55}\text{Fe-Aktivität des Bluteisens}}{\text{Spezifische }^{55}\text{Fe Aktivität des Eisens in der Wundbereichsprobe}}$$

Damit kann eindeutig entschieden werden, ob die Eisenanreicherungen im Wundgebiet nur Hämatomeisen sind oder ob sie eine andere Ursache haben.

Abb. 1 zeigt schematisch die im Institut von Prof. Forth in Zusammenarbeit mit Dr. S.G. Schäfer vorgenommene Durchführung der Markierung.

Spenderkaninchen (1)   Empfängerkaninchen (2)

1. Anämisch machen
2. Injektion von ^{55}Fe, das in Plasma gelöst ist
3. Blutentnahme nach 3 d

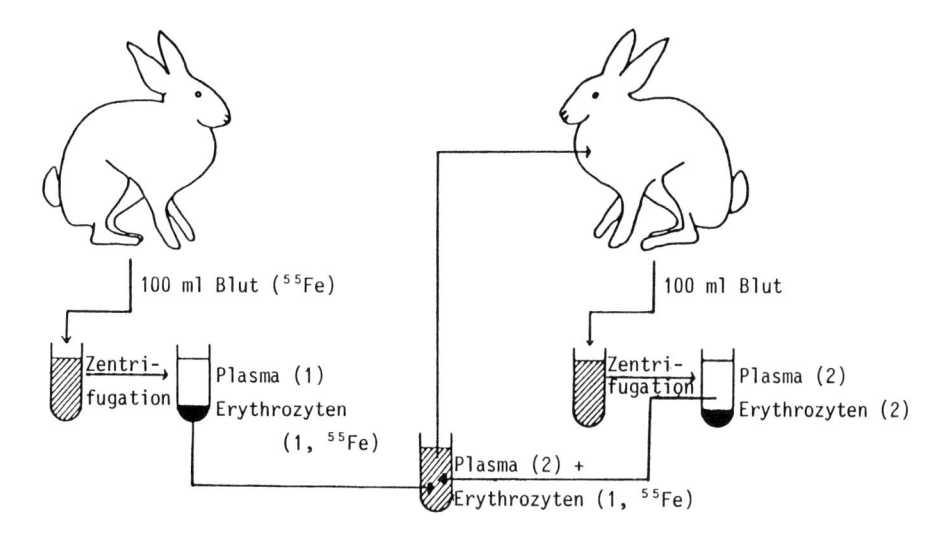

Abb. 1: Erythrozytenmarkierung mit ^{55}Fe

Die Messung des ^{55}Fe erfolgte mit Hilfe eines Flüssigszintillators (^{55}Fe $\xrightarrow{\varepsilon}$ ^{55}Mn, Messung von K_α(Mn) = 5,9 keV). Dabei spielen bekanntlich Quencheffekte, die zu Zählverlusten führen, eine große Rolle. Einwandfreie Ergebnisse wurden durch Messung von Proben erhalten, deren Herstellung in nachfolgend beschriebener Weise erfolgte: Die Gewebeprobe wird nach Zusatz von 1 mg Eisenträger mit rauchender Salpetersäure aufgeschlossen, die Lösung bei 90 °C zur Trockene eingedampft, der Rückstand zuerst auf 250 °C erhitzt, um die Ammonsalze abzurauchen, und schließlich bei 700 °C verglüht. Der Glührückstand wird in 0,5 ml conc. Salzsäure gelöst, bei 90 °C zur Trockene eingedampft und dann in 0,2 ml 1 M Salpetersäure gelöst. Es werden 1,8 ml 0,1 M Phosphorsäure zugesetzt, wobei sich der farblose Eisen-Phosphato-Komplex bildet. Durch Einrühren dieser Lösung in 18 ml eines Flüssigszintillators, der wäßrige Phase aufnimmt, erhält man ein farbloses, klares Meßpräparat.

Eine erste Versuchsreihe mit erythrozytenmarkierten Tieren verlief operationstechnisch noch nicht optimal. Die Schnitte klafften zu stark auseinander, so daß die Bedingungen einer sog. sekundären Heilung und nicht die der erwünschten primären Heilung vorlagen. Außerdem hatten sich in den klaffenden Wunden relativ große Hämatome gebildet. Bemerkenswert ist immerhin, daß der Bluteisenanteil der Hämatome nur 60 % betrug. Hinsichtlich anderer Spurenelemente ergaben die Versuche (s. Tab. 5) einmal eine Bestätigung unserer früheren Ergebnisse (1, 2), daß im Wundbereich zunächst eine sehr starke Calciumanreicherung erfolgt, die mit der Heilungsdauer abnimmt. Außerdem wurden Anreicherungen von Natrium und Brom gefunden.

Tab. 5: Anreicherung von Natrium, Calcium und Brom im Wundbereich

$$\text{Anreicherungsfaktor} = \frac{\text{Gehalt der Wundbereichsprobe}}{\text{Gehalt der Kontrollprobe}}$$

	2 d	2 d		8 d	8 d		10 d
		^{55}Fe-Erythrozytenmarkierung linker Oberschenkel	rechter Oberschenkel		^{55}Fe-Erythrozytenmarkierung linker Oberschenkel	rechter Oberschenkel	
Na	5,3	4,8	4,9	3,8	2,6	1,7	-
Ca	33	23	32	7,9	11	7,3	2,9
Br	-	5,5	4	-	3	2	-

Die Untersuchungen wurden vom Bundesministerium für Forschung und Technologie und vom Fonds der Chemischen Industrie unterstützt.

LITERATUR

(1) F. Lux, D. Božanić, G. Blümel and W. Erhardt, Mikrochim. Acta 1978 I, 35.
(2) F. Lux, D. Božanić, G. Blümel, W. Erhardt, J. Radioanal. Chem. 58 (1980) 289.
(3) T. Bereznai, S. Knobl, F. Lux, A. Ghermai, G. Blümel, Vortrag auf der Tagung "Kern-, Radio-, Strahlenchemie - Grundlagen und Anwendungen -", Karlsruhe, 20. bis 24. September 1982.
(4) T. Bereznai, S. Knobl, F. Lux, G. Blümel, H. Rechl, Poster auf dem "9th International Symposium on Microchemical Techniques", Amsterdam, August 28 - September 2, 1983.
(5) G. V. Iyengar, W. E. Kollmer H. J. M. Bowen, The Elemental Composition of Human Tissues and Body Fluids. Verlag Chemie, Weinheim - New York 1978.
(6) R. Parr, International Atomic Energy Agency, Wien, Privatmitteilung.
(7) G. V. Iyengar, K. Kasperek, J. Radioanal. Chem. 39 (1977) 301.
(8) G. V. Iyengar, Radiochem. Radioanal. Lett. 24 (1976) 35.
(9) S. Knobl und F. Lux, Beitrag "Reduction of the background in the γ-spectra of neutron activated biological samples by separation of ^{32}P" in diesem Band.
(10) J. Hofmann, Dissertation, Universität Köln 1981.
(11) The Biology of the Laboratory Rabbit (Steven H. Weisbroth, Ronald E. Flatt, Alan L. Kraus, Eds), Academic Press, New York - San Francisco - London 1974, S. 65.

REDUCTION OF THE BACKGROUND IN THE γ-SPECTRA OF NEUTRON ACTIVATED BIOLOGICAL SAMPLES
BY SEPARATION OF ^{32}P

Susanne Knobl and Franz Lux

Institut für Radiochemie der Technischen Universität München, D-8046 Garching

1 Introduction

·The relatively high activity of ^{32}P in neutron irradiated biological samples
causes a considerable increase of the background count rate in the low-energy region
(below about 400 keV) of the respective γ-spectra due to bremsstrahlung.

Owing to this measurement and evaluation of the γ-spectra are restricted:
- The high bremsstrahlung count rate contributes to a great extent to the total
  count rate of the spectrum. Therefore, in order to avoid dead time effects, a
  large distance between sample and detector has to be chosen. As a consequence,
  the counting efficiency is low.
- The critical levels L_C(1, 2) for elements with reference nuclides with E_γ < 400 keV
  (^{47}Sc(Ca), ^{51}Cr, ^{75}Se) are too high.
- As a consequence of the given contents of calcium, chromium and selenium of
  biological material and of the activation properties of these elements the net
  peak areas (N) of ^{47}Sc, ^{51}Cr and ^{75}Se in the γ-spectrum of an irradiated
  biological sample are small. The small N and the considerably high background,
  produced by ^{32}P in the region of these peaks, cause especially low N/B-ratios
  (B = background) which result in falsified N-values and a high relative standard
  deviation in the sense of counting statistics.

2 Modification of a chromatography method for the separation of ^{32}P

The procedure described by others (3) for separating ^{32}P by chromatography of a
1 M HNO$_3$ solution over acid alumina should be tested for a solution resulting from
mineralization of irradiated biological material. The digestion mixture which we have
used to date consists of 2 ml of conc. HNO$_3$/conc. HClO$_4$/conc. H$_2$SO$_4$ = 3/2/1 and is
diluted after mineralization with 0.5 M HNO$_3$ to 10 ml. The essential requirement for
the validity of the method is that all interesting reference nuclides (^{47}Sc(Ca),
^{51}Cr, ^{59}Fe, ^{60}Co, ^{65}Zn, ^{75}Se, ^{86}Rb, ^{134}Cs) are obtained free of losses in the eluate
allowing them to be simultaneously counted.

An initial chromatography experiment with irradiated (NH$_4$)$_3$PO$_4$ in the digestion
mixture described above resulted in a decontamination factor of 1000 for ^{32}P. In
further investigations an irradiated synthetic multi element standard consisting of

all interesting elements (see above) was treated with the digestion mixture and after dilution the chromatography was carried out. The following results were obtained:

- ^{75}Se was completely adsorbed on the column, which is not astonishing, because in the mixture only SeO_3^{2-} can occur.
- 60 to 90 % of the ^{51}Cr were found in the eluate. This fact might be a consequence of the well-known partial oxidation of chromium by $HClO_4$ to $CrO_4^{2-}/Cr_2O_7^{2-}$.
- The other reference nuclides of interest, which all form cations in the given solution, were eluted to 100 %.

In order to avoid the losses of ^{75}Se and ^{51}Cr the procedure was modified as follows:

- A digestion of the irradiated biological sample in a mixture of conc. HNO_3/conc. H_2SO_4 = 3/1, followed by an addition of H_2O_2, provides the exclusive existence of Cr^{3+}.
- As described in (4, 5), SeO_3^{2-} was transformed into the uncharged piazselenol at p_H 1 - 2 (6) by addition of an equimolar quantity of 1,2-diaminobenzene in ethanolic solution:

$$\text{o-C}_6\text{H}_4(\text{NH}_3^+)(\text{NH}_2) + H_2SeO_3 \longrightarrow \text{piazselenol-Se} + 2H_2O + H_3O^+$$

The elution with a mixture of 0.5 M or 1 M HNO_3/ethanol = 5/2 yielded 100 % of the selenium in the eluate.

However, by application of the modified procedure described some of the scandium remained on the column. In model experiments with ^{46}Sc it was found that aqueous 0.5 M or 1 M HNO_3 elutes the scandium to 100 %, but in order to obtain the same result with an eluant containing ethanol, the nitric acid concentration had to be increased to 2 M HNO_3/ethanol = 5/2. This slight modification of the eluant did not affect the elution of the piazselenol.

A possible reason for the observed behaviour of scandium during the chromatography could be that $[Sc(H_2O)_6]^{3+}$ dominates in solution of lower acid concentration (7, 8), whereas $[ScNO_3]^{2+}$ and $[Sc(NO_3)_2]^+$ are formed with increasing HNO_3 concentration (9). We suppose that due to some characteristics of alumina (10, 11, 12) the hexaquo complex is adsorbed to some extent on the column, whereas the nitrato complexes are not retained.

3 Definitive procedure for the separation of ^{32}P

- Wet digestion: To about 100 mg of irradiated biological material (dry weight basis) are added 0.5 ml of a multi element carrier solution (1 mg of each element/ml), 30 mg H_2SeO_3, 2 ml conc. HNO_3/conc. H_2SO_4 = 3/1. The mixture is heated under

reflux for 3 hours. 0.5 ml H_2O_2 are added and the mixture is heated again for half an hour. The transparent solution is transferred into a polyethylene counting vessel and diluted with 0.5 M HNO_3 to 10 ml.
- γ-spectrum of the solution (P).
- Addition of NaOH (solid) until p_H 1 - 2 is reached.
- 27 mg of 1,2-diaminobenzene are dissolved in 4 ml ethanol and added to the sample solution.
- Chromatography: The 14 ml (aqueous acid/ethanol = 10/4) are pipetted on the column (4.6 g acid alumina (Merck, activity I) in 1 M HNO_3, frit P1, ∅ = 1 cm) and eluted. The column is washed with a mixture of 2 M HNO_3/ethanol = 5/2 until an eluate volume of 20 ml is reached.
- γ-spectrum of the eluate (P').

4 Assessment of the method

The total count rate (0 - 2000 keV) of P' is only 1/8 of that of P (see Fig.1).

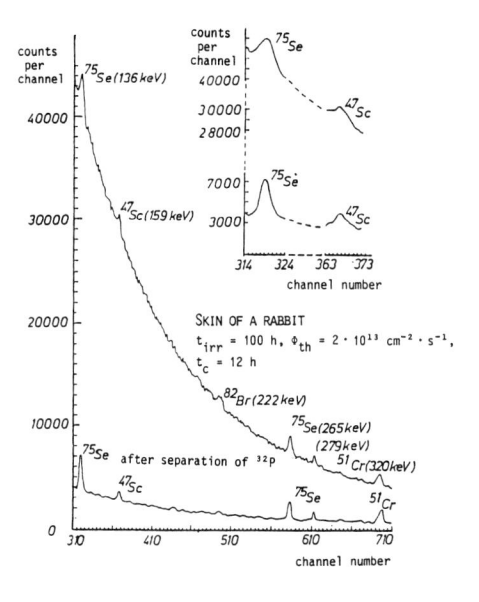

Figure 1. γ-spectrum of an neutron irratiated sample of the skin of a rabbit before and after separation of ^{32}P.

The effects of the separation of ^{32}P on the increase of N/B and on the decrease of the detection limits are shown in Tables 1 and 2.

Table 1: Effect of the separation of ^{32}P on the increase of N/B
380 mg thigh muscle of a rabbit (fresh weight), $\Phi_{th} = 2 \cdot 10^{13}$ cm$^{-2} \cdot$s^{-1},
$t_{irr} = 100$ h, $t_c = 24$ h
N_{10}, B_{10}: counting of P at a distance of 10 cm from the detector
N_{10}', B_{10}': counting of P' at a distance of 10 cm from the detector
N_0', B_0': counting of P' on the detector

Nuclide	E_γ keV	N_{10}	B_{10}	N_{10}'	B_{10}'	N_0'	B_0'	N_{10}/B_{10}	N_{10}'/B_{10}'	N_0'/B_0'
^{47}Sc	159	-	1267926	1508	119452	12672	911150	-	0.01	0.01
^{51}Cr	320	-	174946	-	34319	-	154526	-	-	-
^{75}Se	136	27908	1217078	29119	211873	292360	1662708	0.02	0.14	0.18
	265	17980	225274	17413	50961	157879	357762	0.08	0.34	0.44

Table 2: Effect of the separation of ^{32}P on the decrease of the detection limits
380 mg thigh muscle of a rabbit (fresh weight), $\Phi_{th} = 2 \cdot 10^{13}$ cm$^{-2} \cdot$s^{-1}, $t_{irr} = 100$ h, $t_c = 24$ h.
m_C = mass of the element corresponding to L_C (critical level)
B_{10}, $m_{C,10}$: counting of P at a distance of 10 cm from the detector, dead time 2 %
B_{10}', $m_{C,10}'$: counting of P' at a distance of 10 cm from the detector, dead time 1 %
B_0', $m_{C,0}'$: counting of P' on the detector, dead time 2 %.

Element	E_γ keV	B_{10}	$m_{C,10}$ µg	B_{10}'	$m_{C,10}'$ µg	B_0'	$m_{C,0}'$ µg	$m_{C,10}'/m_{C,10}$	$m_{C,0}'/m_{C,10}$
Ca(^{47}Sc)	159	1267926	18	119452	7.8	911150	3.4	0.44	0.19
(^{47}Ca)	1297	1230	16	170	12	-	-	0.75	-
Cr(^{51}Cr)	320	174946	0.01	34319	0.008	154526	0.001	0.66	0.11
Fe(^{59}Fe)	1099	2453	0.23	1482	0.25	12172	0.074	1.08	0.32
	1292	830	0.25	203	0.19	708	0.042	0.78	0.17
Co(^{60}Co)	1173	1554	0.0004	690	0.0004	1352	0.00006	0.95	0.14
	1332	716	0.0004	369	0.0003	1328	0.00006	0.71	0.15
Zn(^{65}Zn)	1115	4388	0.023	2609	0.022	19870	0.006	0.95	0.26
Se(^{75}Se)	136	1217078	0.0086	211873	0.0034	1662708	0.0015	0.40	0.17
	265	225274	0.0071	50961	0.0054	357762	0.0015	0.76	0.21
Rb(^{86}Rb)	1077	5393	0.021	2588	0.02	22525	0.007	0.94	0.32
Cs(^{134}Cs)	795	10261	0.0007	5447	0.0007	31815	0.0001	0.96	0.20

The accuracy of the results using this procedure is shown by the data in Table 3.

Table 3: Analysis of IAEA Animal Muscle H-4, n = 7

Element	E_γ keV	Unit	P mean ± s	P' mean ± s	IAEA value ± s
Na(^{24}Na)	1368	mg/g	2.05 ± 0.09	*)	2.06 ± 0.12
K (^{42}K)	1525	mg/g	16 ± 0.4	*)	15.8 ± 0.6
Ca(^{47}Sc)	159	µg/g	197 ± 18	185 ± 14	188 ± 25
(^{47}Ca)	1297	µg/g	193 ± 19	195 ± 21	
Cr(^{51}Cr)	320	ng/g	30 (n = 1)	95 (n = 1)	80
Fe(^{59}Fe)	1099	µg/g	47 ± 3.5	49 ± 3.4	49 ± 2
	1292	µg/g	47 ± 3.6	48 ± 4.8	
Co(^{60}Co)	1173	ng/g	5.9 ± 0.84	4.6 ± 0.5	8
	1332	ng/g	3.8 ± 1.3	4.7 ± 1.6	
Zn(^{65}Zn)	1115	µg/g	85.8 ± 8.4	85.9 ± 7.3	86 ± 4
Se(^{75}Se)	136	µg/g	0.30 ± 0.05	0.33 ± 0.02	0.28 ± 0.04
	265	µg/g	0.33 ± 0.03	0.34 ± 0.03	
Br(^{82}Br)	777	µg/g	4.9 ± 0.8	*)	4.1 ± 0.6
Rb(^{86}Rb)	1077	µg/g	19.6 ± 2.1	20 ± 1.9	19 ± 1
Cs(^{134}Cs)	795	µg/g	0.14 ± 0.01	0.14 ± 0.02	0.12 ± 0.02

*) Cooling time > 10 $t_{1/2}$ → reference nuclides ^{24}Na, ^{42}K, ^{82}Br not detectable in P'

5 Some ideas concerning the mechanism of the separation of ^{32}P by chromatography over acid alumina

The mechanism of the separation of ^{32}P by chromatography over acid alumina is not yet understood completely. G.M. Schwab (10) has reported a sequence of anions that are adsorbed with decreasing strength: $OH^- > PO_4^{3-} > F^- > [Fe(CN)_6]^{4-}/CrO_4^{2-} > SO_4^{2-} > [Fe(CN)_6]^{3-}/Cr_2O_7^{2-} > Cl^- > NO_3^- > MnO_4^- > ClO_4^- > S^{2-}$. In agreement with this sequence is, of course, that phosphate is retained very well, and also the fact that some of the chromium is adsorbed, if the digestion mixture conc. HNO_3/conc. $HClO_4$/conc. H_2SO_4 is used (partial formation of chromate/dichromate, no displacement of chromate by sulfate).

Alumina can behave both as electron donor and as electron acceptor (11), indicated by anion and cation radicals being formed by compounds adsorbed on the surface of this material (12). From these properties of alumina, however, no conclusion on the mechanism of the separation of ^{32}P, i.e. $H_2^{32}PO_4^-$ can be drawn.

Acknowledgement
This work was supported by the Bundesministerium für Forschung und Technologie and the Fonds der Chemischen Industrie.

REFERENCES

(1) L.A. Currie, Anal. Chem. 40 (1968) 586.
(2) V.C. Rogers, Anal. Chem. 42 (1970) 807.
(3) F. Girardi, R. Pietra, E. Sabbioni, J. Radioanal. Chem. 5 (1970) 141.
(4) O. Hinsberg, Chem. Ber. 22 (1889) 862.
(5) O. Hinsberg, Chem. Ber. 22 (1889) 2895.
(6) U. Shimoishi, J. Chromatogr. 136 (1977) 85.
(7) Gmelin Handbuch der Anorganischen Chemie, Seltenerdelemente, Teil C2, Sc, Y, La und Lanthanide, 8. Auflage, Springer-Verlag, Berlin-Heidelberg-New York 1974, S. 228.
(8) L.N. Kommissarova, Russ. J. Inorg. Chem. 25 (1980) 75.
(9) A.P. Samodelov, Sov. Radiochem. 6 (1964) 558.
(10) G.-M. Schwab, G. Dattler, Angew. Chem. 33 (1937) 691.
(11) J. Steigman, Int. J. Appl. Radiat. Isot. 33 (1982) 829.
(12) J. Bodrikov, K.C. Khulbe, R.S. Mann, J. Catal. 43 (1976) 339.

A SIMPLE DEAD TIME CORRECTION FOR ROUTINE INSTRUMENTAL NEUTRON ACTIVATION ANALYSIS OF SHORT-LIVED ISOTOPES

A Faanhof

Nuclear Development Corporation of South Africa (Pty) Ltd, Private Bag X256, Pretoria, South Africa.

1. INTRODUCTION

Accurate and relatively simple procedures to correct for residual dead time losses in gamma-ray-spectrometry can be obtained from literature (1, 2). These require the determination of the total dead time fraction at the beginning of the counting period (D_0), as well as a mathematical decay function for the sample matrix. The application of these procedures in routine analysis is limited due to the following factors:
- Most of the modern multichannel analyzers calculate the dead time fraction at regular intervals during the counting period by using the real and clock time values. Only the averaged dead time fraction over the counting period becomes available at the end of the measurement, and the value of D_0 must be obtained by extrapolation.
- The decay function must be determined empirically. The large variation in elemental composition between batches of samples and even amongst samples in one particular batch, requires that this has to be done at least for each batch and even for every sample if significant matrix variations are to be expected.

To obtain the necessary data will obviously be rather time consuming, which is not acceptable in routine analysis and might require modifications to existing hardware that are beyond the capabilities of some laboratories. Consequently, a simple correction factor, obtainable from data provided by any standard spectrometer system, had to be developed.

2. LIST OF SYMBOLS

f'_τ = fractional loss in the observed signal due to dead time
t_m = preset counting period
t'_m = automatically extended counting period
λ = decay constant of the isotope of interest (= $\ln 2 / t_{\frac{1}{2}}$)
λ' = decay constant of the sample matrix over the counting period (= $\ln 2 / t'_{\frac{1}{2}}$)
D_0 = dead time at the beginning of the counting period
$(f'_\tau)'$ = reduced value of f'_τ according to the limitation set for λ' towards λ
D_{f_τ} = deviation from the true fractional loss {= $(f'_\tau)' / f'_\tau$ }

3. MATHEMATICAL FORMULATION

3.1. General

According to reference (1) and solving the equation for $D_0 \leq 0.5$, one comes to;

$$f'_\tau = \frac{\lambda}{1 - e^{-\lambda t_m}} \int_0^{t'_m} \frac{e^{-\lambda t} dt}{1 + \frac{D_0}{1 - D_0} e^{-\lambda' t}} \rightarrow \text{for } D_0 \leq 0.5 \rightarrow f'_\tau = \frac{\lambda}{1 - e^{-\lambda t_m}} \sum_{n=0}^{\infty} \frac{(\frac{D_0}{D_0 - 1})^n}{\lambda + n\lambda'} \{1 - e^{-(\lambda + n\lambda') t'_m}\}$$

(1) Das, H A et al. Report ECN 131, 1983.
(2) Woittiez, J W R et al. J Radioanal. Chem. 53 1979. 191.

For the automatically extended counting time one comes to the following formulation;

$$t_m = \int_0^{t'_m} \frac{dt}{1 + \frac{D_0}{1-D_0}e^{-\lambda' t}} \rightarrow\rightarrow t'_m = \frac{1}{\lambda'}\ln\left\{\frac{e^{\lambda' t_m} - D_0}{1 - D_0}\right\} \quad \text{or} \quad D_0 = \frac{1 - e^{-\lambda'(t'_m - t_m)}}{1 - e^{-\lambda' t'_m}}$$

3.2. Assumptions

Three simplifications to the general formulae can be made;

FOR $\lambda' \ll \lambda$

$$(f'_\tau)' = \frac{t_m(1 - e^{-\lambda t'_m})}{t'_m(1 - e^{-\lambda t_m})}$$

$$(D_0)' = \frac{t'_m - t_m}{t'_m}$$

FOR $\lambda' = \lambda$

$$(f'_\tau)' = \frac{\lambda(t'_m - t_m)}{e^{\lambda(t'_m - t_m)} - 1}$$

$$(D_0)' = \frac{1 - e^{-\lambda(t'_m - t_m)}}{1 - e^{-\lambda t'_m}}$$

FOR $\lambda' \gg \lambda$

$$(f'_\tau)' = \frac{1 - e^{-\lambda t'_m} - \lambda(t'_m - t_m)}{1 - e^{-\lambda t_m}}$$

$$(D_0)' = \frac{1 - e^{-\lambda'(t'_m - t_m)}}{1 - e^{-\lambda' t'_m}}$$

From this it can be seen that the first and second set of equations are independent of λ' and D_0 and are a function of only t_m, t'_m and λ. In the last set of equations D'_0 is still dependent of λ', however the assumption made for this case can be easily overcome on application of longer decay periods and normally has no practical value.

4. PRACTICAL IMPLICATIONS

4.1. Choise of formulae

For normal routine analysis of short-lived isotopes we may expect that $\lambda' > \lambda$ for most samples and $\lambda' \approx \bar{\lambda}$ for well chosen multi-element standards. In order to obtain the deviation from the "true" correction factor by the indicated simplifications, we define

$$D_{f_\tau} = (f'_\tau)'/f'_\tau$$

4.2. Graphical representation

Figures 1 and 2 show typical examples of the variation of D_{f_τ} with λ'/λ as a function of $t_m/t_{\frac{1}{2}}$ for the assumptions made for $\lambda' \gg \lambda$ and $\lambda' = \lambda$ respectively. For the same assumptions graphs are made (figures 3 and 4) to obtain the variation of $t'_{\frac{1}{2}}/t_m$ with

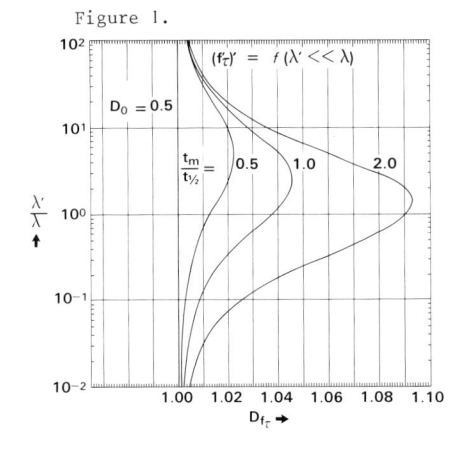

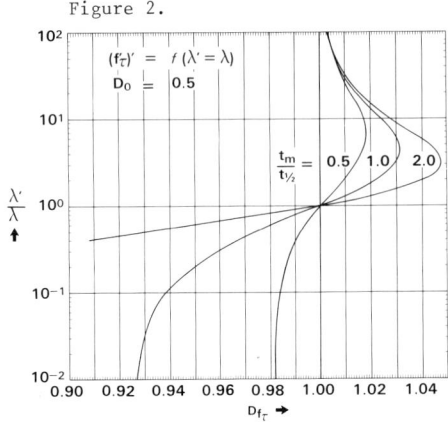

Figure 1.

Figure 2.

Dead time correction 171

$t_m/t_{\frac{1}{2}}$ as a function of D_0 to fulfill $Df_\tau \leq 2\%$. From this it will de easy to make a rough estimation on the applicability of our formulation in routine multi-element analysis.

4.3. Example on some biological materials

As an example we will work out the possibility of simultaneous determination of Se and Ag in biological materials. Table 1 shows the limitation towards $t'_{\frac{1}{2}}$ for standards and samples, whereas table 2 shows some experimental data on several biological matices.

Figure 3.

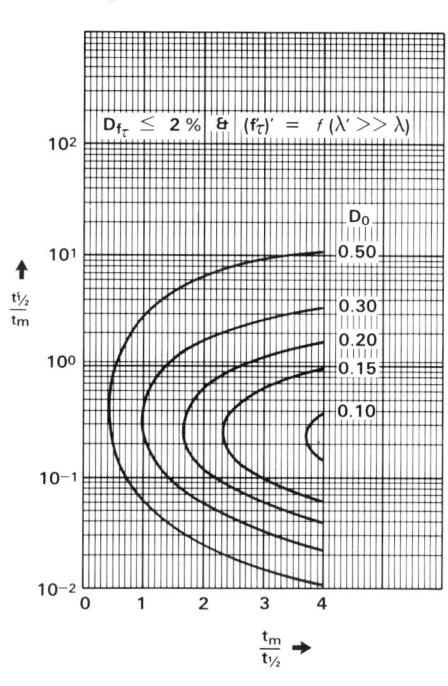

Figure 4.

Table 1.

EXAMPLE ON THE DETERMINATION OF Se AND Ag IN SOME BIOLOGICAL MATERIALS

PRESET COUNTING TIME: 30 s.   ALL OTHER VALUES IN SECONDS

	$t_{\frac{1}{2}}$	$t_m/t_{\frac{1}{2}}$	$t_{\frac{1}{2}}$ measured
Se STANDARD	17.5	1.71	17.33
Ag STANDARD	24.6	1.22	25.01
Se-Ag STANDARD	≈ 21.0	≈ 1.43	20.90

CALCULATED $t'_{\frac{1}{2}}$ FOR STANDARDS TO FULFILL $D_{f_\tau} \leq 2\%$

D_0	Se	Ag	Se-Ag
0.5	$13.2 \leq t'_{\frac{1}{2}} \leq 22.5$	$15.3 \leq t'_{\frac{1}{2}} \leq 37.3$	$14.1 \leq t'_{\frac{1}{2}} \leq 30.0$

CALCULATED $t'_{\frac{1}{2}}$ FOR SAMPLES TO FULFILL $D_{f_\tau} \leq 2\%$

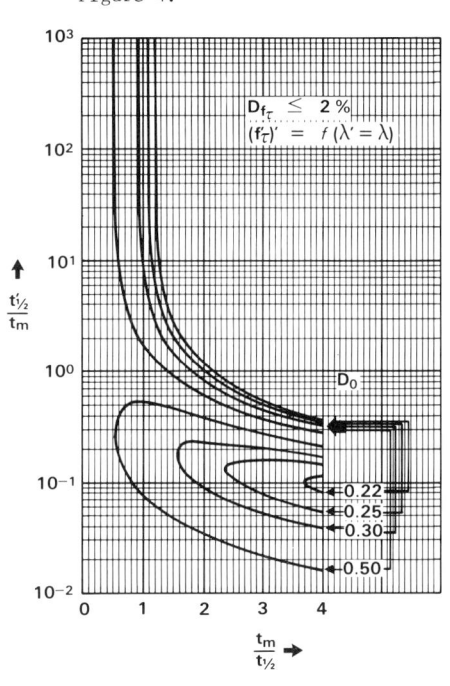

D_0	Se	Ag
0.5	> 175	> 114
0.3	> 42	> 24
0.2	> 0	> 0

Table 2.

EXPERIMENTAL DATA* ON SOME BIOLOGICAL MATRICES

MATRIX	N	AVERAGE WEIGHT (mg)	$t_{\frac{1}{2}}$ MEASURED (s)	D_0 MEASURED	D_{f_τ} CALCULATED Se	D_{f_τ} CALCULATED Ag
Se STANDARD	5	0.200	17.33	0.493 8	1.000 7	–
Ag STANDARD	5	0.200	25.01	0.487 1	–	0.999 2
Se-Ag STANDARD	5	0.1/0.1	20.90	0.491 0	0.987 1	1.007 2
ANIMAL BLOOD	116	46.4	171	0.172 7	1.002 8	1.002 0
ANIMAL MUSCLE	15	95.2	> 1E4	0.029 5	1.000 0	1.000 0
ANIMAL LIVER	21	97.6	105	0.033 9	1.000 6	1.000 4
ANIMAL KIDNEY	21	79.1	> 1E4	0.038 2	1.000 0	1.000 0
ANIMAL HEART	2	94.7	> 1E4	0.036 0	1.000 0	1.000 0
LUCERN	3	138.6	165	0.209 0	1.003 6	1.002 6
FOOD CUBES (DOG)	3	277.6	> 1E4	0.228 0	1.000 1	1.000 1
FOOD CUBES (RABBIT)	1	258.4	> 1E4	0.365 0	1.000 2	1.000 2
ENAMEL (HUMAN)	4	250.4	> 1E4	0.055 1	1.000 0	1.000 0
DENTINE (HUMAN)	4	296.4	> 1E4	0.047 6	1.000 0	1.000 0

*SAFARI – 1 REACTOR OPERATING AT 5 MW
IRRADIATION POSITION : $\Phi_{th} = 4.2\ E12\ cm^{-2}.s^{-1}$., cadmium ratio : 66.6
IRRADIATION TIME : 30 s.
DECAY TIME : 30 s.
COUNTING TIME : 30 s.
Ge-Li CRYSTAL : 90 cm³ flat, 10 % relative efficiency

ANALYTICAL INVESTIGATIONS USING THE 14 MEV NEUTRON ACTIVATION FACILITY KORONA

R. Pepelnik, B. Anders*, E. Bössow, H.-U. Fanger

Institut für Physik, GKSS Research Center Geesthacht,
*I. Institut für Experimentalphysik, University of Hamburg,
Federal Republic of Germany

At the GKSS Research Center Geesthacht an intense 14 MeV neutron generator is in operation since 1981. The main components of the facility, named KORONA, are a sealed neutron tube and an integrated fast pneumatic rabbit system [1]. Solid as well as liquid samples, in polyethylene containers with volumes of 0.55 cm^3, can be activated in the center of the cylindrical neutron generator target. Due to the high neutron flux of more than 3×10^{10} n/cm^2 s, the irradiation of natural samples, containing abundant elements such as oxygen or sodium, induces high initial activities of short-lived isotopes with more than 10^6 Bq, typically. At rates of 800000 cps, counting losses of 99 % occur (see Fig. 1). Therefore, the γ-ray spectroscopy system [2] has been improved considerably by a novel method for real-time correction of counting losses [3, 4].

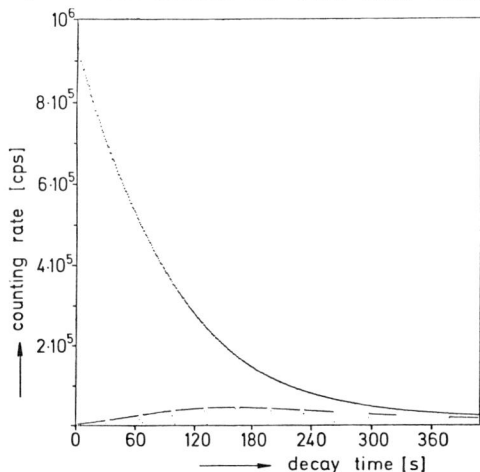

Fig. 1: True integral (above) and acceptance counting rate (lower curve) during a measurement of ^{139m}Ce with a half-life of 56.5 s.

The neutron flux variations within the dimensions of the sample container were determined to be ± 5 % [5]. The neutron energy distribution was investigated by experimental [6] and theoretical [7] methods, yielding a median at 14.7 MeV with a FWHM of 600 keV. The contribution of thermal neutrons to the total neutron flux was estimated to be less than 5×10^{-3}.

To enhance the accuracy of the activation analysis at KORONA, various neutron-induced reaction cross sections were measured either with better precision or for the first time [6, 8]. These values, together with other reported nuclear data, were used to recalculate the analytical sensitivities of 78 elements using the activation facility KORONA. In this sensitivity study a neutron flux of 3×10^{10} n/cm^2 s and a total analysis time of 2000 s are used, assuming that 100 counts per peak (registered with a 17 % Ge(Li) detector) are sufficient for the analysis of an element, neglecting any matrix effects. To improve the counting statistics for short-lived radioisotopes, up to 30 repetitions of activation and measurement are admitted. The calculated detection limits are illustrated in Fig. 2.

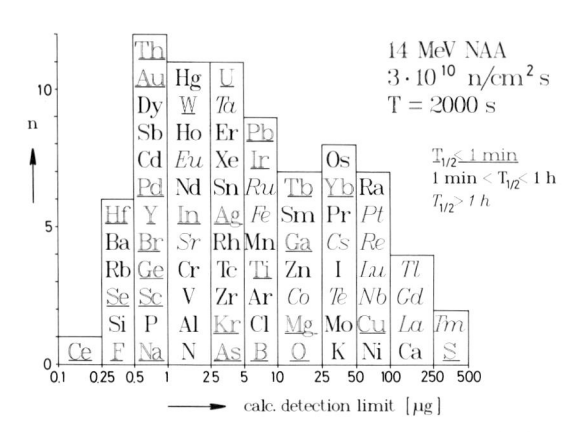

Fig. 2: Calculated elemental sensitivities using KORONA

The main application of KORONA is in the field of analytical investigations for environmental research. Several types of samples have been analyzed such as minerals, sediment, suspended matter [9], soil, air dust [2], wood and fish material.

The 14 MeV neutron activation analysis (NAA) is suited to determine all major components in natural samples, except carbon. The detection of trace elements by reaction products with half-lives similar to those of the major components is often impeded. For example, the low-energy γ-ray lines produced by elements like Se and As with high analytical sensitivity (see Fig. 2) are often masked by a high Compton background produced by ^{16}N and ^{28}Al, e.g. But with enlarged measurement time (~ 60 h) As can be analyzed by its long-lived reaction products down to concentrations in the ppm-region.

The analysis of soil (IAEA) and sediment (IAEA) resulted in the detection of 21 and 19 elements, respectively, determined by 14 MeV NAA, 20 and 22 elements by TXRF and 31 and 40 elements by thermal NAA. The concentrations determined are given in paper No. 72, this conference.
The samples were analyzed first using 10 cycles with activation and measurement times of 15 s. Afterwards they were irradiated for 1000 s, followed by two measurements of 20 min and 80 h.

A serious problem often arising in 14 MeV NAA are interferences by different reaction channels [2]. As shown in Tab. 1, there are numerous cases where identical radioisotopes are produced via (n,p), (n,α) and (n,γ) reactions from three different elements. Thus, usually the contributions by interfering reactions have to be considered or less sensitive reactions have to be used to get clear-cut results, which leads to a reduction of accuracy, in general.

Table 1: Interferences of 14 MeV neutron induced reactions

IAEA-Sediment SD-N-1/2

Reaction	Ab. [%]	σ [mb]	$T_{1/2}$	E_γ [keV]	I_γ [%]	Ab σ A [mb]	Cum. Conc. [mg/g]	Conc. [mg/g]
$^{16}O(n,p)^{16}N$	100	39	7.1 s	6129	69	2.43	564	564 ± 17
$^{19}F(n,\alpha)^{16}N$	100	21				1.11	1240	0 ± 0
$^{19}F(n,p)^{19}O$	100	19	27 s	197	96	1.00	0	0 ± 0
$^{23}Na(n,\alpha)^{20}F$	100	125	11 s	1634	100	5.43	10.5	10.5 ± 1.0
$^{19}F(n,\gamma)^{20}F$	100	0.05				0.003	21700	0 ± 0
$^{24}Mg(n,p)^{24}Na$	79	180	15 h	1369	100	5.85	36.2	8.2 ± 0.9
$^{27}Al(n,\alpha)^{24}Na$	100	114				4.23	50.1	38.8 ± 1.8
$^{23}Na(n,\gamma)^{24}Na$	100	2.9				0.13	1680	10.5 ± 1.0
$^{27}Al(n,p)^{27}Mg$	100	69	9.5 min	844	72	2.56	47.3	38.8 ± 1.8
$^{30}Si(n,\alpha)^{27}Mg$	3.1	70				0.077	1560	285 ± 11
$^{26}Mg(n,\gamma)^{27}Mg$	11	0.4				0.002	66600	8.2 ± 0.9
$^{28}Si(n,p)^{28}Al$	92	265	2.2 min	1779	100	8.70	285	285 ± 11
$^{31}P(n,\alpha)^{28}Al$	100	118				3.81	652	0 ± 0
$^{27}Al(n,\gamma)^{28}Al$	100	1.6				0.06	42700	38.8 ± 1.8
$^{29}Si(n,p)^{29}Al$	4.7	131	6.6 min	1273	91	0.218	290	290 ± 30

In conclusion, the advantages of the 14 MeV NAA method may be summarized as follows: Due to the high neutron flux of KORONA good analytical sensitivities are achieved for a wide region of elements, especially for light elements. The short half-lives of the majority of the reaction products, allow to detect many elements within short activation and measurement times. No chemical sample preparation is necessary, thus contaminations or losses are avoided. Hitherto, during the 220 h operation of the neutron generator, no serious problems arised.

REFERENCES

[1] H.-U. Fanger, R. Pepelnik, W. Michaelis, J. Radioanal. Chem. 61 (1981) 147

[2] B.M. Bahal, R. Pepelnik, GKSS 83/E/13

[3] G.P. Westphal, J. Radioanal. Chem. 70 (1982) 387

[4] R. Pepelnik, G.P. Westphal, B. Anders, submitted to Nucl. Instr. and Meth.

[5] B. Anders, E. Bössow, GKSS 83/E/23

[6] B. Anders, R. Pepelnik, H.-U. Fanger, GKSS 83/E/29

[7] B.M. Bahal, H.-U. Fanger, Nucl. Instr. and Meth. 211 (1983) 469

[8] W. Michaelis, Proc. IAEA Consultants' Meeting on Nucl. Data for Bore-hole and Bulk-media Assay Using Nuclear Techniques 14. - 18. Nov., 1983, Krakow, Poland

[9] R. Pepelnik, H.-U. Fanger, W. Michaelis, B. Anders, J. Radioanal. Chem. 72 (1982) 393

INDICATOR ACTIVATION WITH A 14 MeV NEUTRON GENERATOR

Vadim Cercasov

Institute of Physics, Hohenheim University, D-7000 Stuttgart 70

The modern trend in marker techniques for biological and environmental studies consists in the use of inactive markers and the analysis of the collected samples by means of different analytical methods. An important place among these methods is taken by the activation analysis. The combination of inactive tracer technique with activation analysis is called indicator activation method /1/.

Since the number of reactor activation facilities decreases permanently, we applied for that purpose, as an alternative, a 14 MeV neutron generator. This is possible owing to the great flexibility in the choice of the markers and of the marker quantities used in biological applications.

The 14 MeV neutron activation equipment available at the University of Hohenheim has already been described /2/. The formerly used Philips neutron generator is now replaced by a Kaman A-711 (neutron output $> 10^{11}$ $n \cdot s^{-1}$). The 14 MeV neutron flux in the irradiation position of the rabbit system reaches approximately 10^9 $n \cdot s^{-1} \cdot cm^{-2}$. Due to this change the calculated determination limits in /2/ were improved by a factor of about three.

In order to investigate the matter flow in the gastro-intestinal tract of animals (llamas, rock hyraxes and sheep) the food is marked with rare-earth elements (Ce, Sm) and recently with Cr. Samples of matter out of the forestomach (of llamas) and samples of feces are taken after various time intervals. With tracer contents above 0.5 mg a rapid analysis of these samples by neutron activation is possible with our facility /3/. The nuclear reactions and data, that are to be considered in this case, are compiled in Table 1.

For the determination of Sm the slow neutron reaction $^{154}Sm(n,\gamma)^{155}Sm$ is particularly useful. To increase the number of slow neutrons the samples are irradiated inside a polyethylene moderator block surrounding the neutron tube and the transfer tube of the rabbit system. By that means the thermal neutron flux in the irradiation position exceeds 10^7 $n \cdot s^{-1} \cdot cm^{-2}$.

Table 1: Nuclear data

Element	Nuclear reaction	Isotopic abundance %	Cross section $10^{-24} cm^2$	Half-life	Gamma-energy, MeV (Abundance in %)
Ce	$^{140}Ce(n,2n)^{139m}Ce$	88.5	2.5'	56 s	0.754 (93)
Cr	$^{52}Cr(n,p)^{52}V$	83.8	0.11'	3.76 min	1.434 (100)
Sm	$^{144}Sm(n,2n)^{143m}Sm$	3.1	0.7'	66 s	0.754 (100)
	$^{152}Sm(n,\gamma)^{153}Sm$	26.7	206"	46.8 h	0.103 (28)
	$^{154}Sm(n,2n)^{153}Sm$	22.8	2.0'		
	$^{154}Sm(n,\gamma)^{155}Sm$	22.8	5.5"	22.4 min	0.104 (73)
Si	$^{28}Si(n,p)^{28}Al$	92.2	0.2'		1.779 (100)
Al	$^{27}Al(n,\gamma)^{28}Al$	100	0.23"	2.2 min	1.268 (pair peak)
P	$^{31}P(n,\alpha)^{28}Al$	100	0.12'		0.757 (pair peak)

') for 14 MeV neutrons   ") for thermal neutrons

The production of ^{143m}Sm gives rise to an interference in the simultaneous determination of Ce. This can however be corrected using a separate Sm standard.

Another limiting factor for the sensitivity of this method results from the matrix activity, particularly of ^{28}Al, produced by different reactions from Si, Al and P. Besides a high spectrum background, the activity of ^{28}Al causes a further interference with ^{139m}Ce due to its double escape peak. The latter contribution can be calculated with the aid of the full-energy peak of ^{28}Al.

The method was checked by comparison with reactor neutron activation analysis and atomic absorption spectrometry. A good agreement was found.

REFERENCES

/1/ H. GLUBRECHT, W. KÜHN: Indikatoraktivierungsmethode (Indicator Activation Method), Verlag Karl Thiemig, München, 1976

/2/ V. CERCASOV: J. Radioanal. Chem. 52 (1979) 399-410

/3/ V. CERCASOV, R. HELLER: J. Radioanal. Chem. 60 (1980) 453-459

OBERFLÄCHENPROBLEME BEI AKTIVIERUNGSANALYSEN MIT GELADENEN TEILCHEN

H. Weniger, K. Bethge, G. Wolf

Institut für Kernphysik, J.W. Goethe-Universität,
Frankfurt am Main, Deutschland

1. Aktivierungsanalyse mit geladenen Teilchen

Die Aktivierungsanalyse mit geladenen Teilchen (insbesondere ^{3}He-Ionen) eignet sich sehr gut zum zerstörungsfreien Nachweis leichter Elemente in einer schweren Matrix. Untersucht wurden bereits Proben aus Al, Si, Ti, Ge, Zr, Nb, Ag, Au, Pb und Stahl. Die Nachweisgrenzen des Verfahrens sind neben weiteren Parametern in Tabelle 1 aufgeführt.

Tabelle 1: Reaktionen und Nachweisgrenzen der Aktivierungsanalyse mit geladenen Teilchen

Nachweis-Isotop	Reaktion	Halbwertz. Endprodukt min	Nachweisgrenze in Si ppm	in Pb
^{10}B	^{10}B(d,n)^{11}C	20,3	0.030	0,050
^{12}C	^{12}C(^{3}He,α)^{11}C	20,3	0,008	0,014
^{14}N	^{14}N(^{3}He,α)^{13}N	9,96	0,060	0,100
^{16}O	^{16}O(^{3}He,p)^{18}F	109,7	0,002	0,004

Eine Analyse kann ohne chemischen Aufschluß der Probe durchgeführt werden. Dies gilt auch für den Fall, daß die Matrix aktiviert wird, wenn die Halbwertzeit dieser Aktivität von den anderen Aktivitäten deutlich getrennt ist. Abbildung 1 zeigt die Abklingkurven bestrahlter Silizium-Proben. Die Silizium-Matrix wird zwar angeregt, diese Aktivität kann jedoch von den weiteren Komponenten gut getrennt werden. Die Reichweite der anregenden Ionen beträgt etwa 100 µm. Somit können Oberflächenverunreinigungen einen großen Einfluß auf das Analysenergebnis haben. Eine Abtragung der Oberfläche ist in den meisten Fällen unumgänglich.

a) $^{28}Si(^{3}He,p)^{30}P$
b) $^{12}C(^{3}He,\alpha)^{11}C$
c) $^{16}O(^{3}He,p)^{18}F$

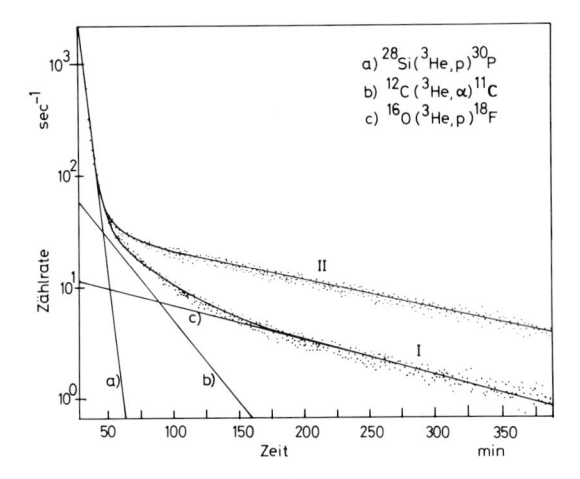

Abb. 1: Gegenüberstellung der β^{+}-Aktivitäten nach Bestrahlung von zonengezogenen (Kurve I) und tiegelgezogenen (Kurve II) Silizium-Einkristallen.

2. Oberflächenuntersuchung mit der Kernreaktionsanalyse

Konzentrationsprofile der Nachweis-Isotope können mit der Kernreaktionsanalyse gemessen werden /1/. Ein Beispiel für das Ergebnis einer Messung zeigt Abb. 2.

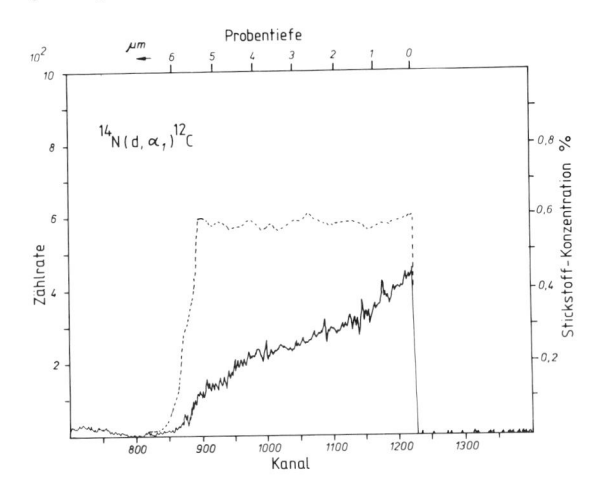

Abb. 2: Ausschnitt aus dem Teilchenspektrum einer Kernreaktionsanalyse bei Bestrahlung von nitriertem Stahl mit Deuteronen von 1,3 MeV.

Abb. 3 zeigt einen Ausschnitt aus dem Teilchenspektrum einer Kernreaktionsanalyse an einer Silizium-Probe. Hinzugefügt ist gestrichelt

der entsprechende Bereich des Spektrums nach Sputtern der Probe
(Abtrag 6 µm). Dabei zeigt es sich, daß der Kohlenstoffgehalt in
der Tiefe von 6 µm noch bei etwa 200 ppm liegt.

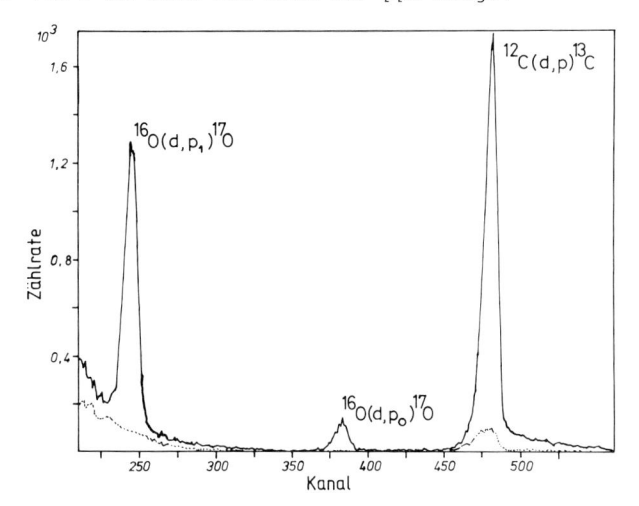

Abb. 3: Gegenüberstellung von Spektrenausschnitten von Kernreaktions-
analysen vor und nach einem Sputtervorgang.

Das Konzentrationsprofil von O und C kann auch durch sukzessives
chemisches Ätzen in Verbindung mit der Restaktivität einer aktivie-
renden Bestrahlung gemessen werden. An den Silizium-Proben ergeben
sich dabei die in Tabelle 2 genannten Werte.

Tabelle 2: Konzentrationsprofil von O und C in Silizium-Einkristallen (nicht poliert) gemessen durch Aktivierungsanalyse nach
chemischem Abtrag von Oberflächenschichten

Ätzabtrag	Konzentration in abgetragener Schicht	
µm	O	C
	ppm	
–	1530	635
2	140	800
4	1,5	200
6	1,4	200
8	1,4	190
12	0,05	4,4
16	0,05	0,4

Sputtern hat gegenüber dem chemischen Ätzen Vorteile, z.B. kann das Verfahren in-situ in der Vakuum-Bestrahlungskammer angewendet werden, so daß kein Kontakt mehr mit atmossphärischen Verunreinigungen möglich ist.

3. Bestimmung von Sputterraten mit Hilfe der Kernreaktionsanalyse.

Sputterraten können mit der Kernreaktionsanalyse bestimmt werden, wobei die Tiefeninformation des Verfahrens ausgenutzt wird. Die Methode enthält folgende Schritte:
a) Implantation eines Tiefenmarkers (z.B. ^{14}N)
b) Messung der Reichweiteverteilung mit der Kernreaktionsanalyse
c) Sputtern mit einem niederenergetischen Ionenstrahl (z.B. Argon-Ionen von 15 keV)
d) erneute Bestimmung der Tiefenverteilung des implantierten Markers.

Aus dem verringerten Energieverlust der Reaktionsteilchen in Schritt d) gegenüber b) läßt sich die Sputterrate errechnen. In Abb. 4 ist die Sputterrate von Silizium bei Beschuß mit Argon-Ionen in Abhängigkeit vom Einschußwinkel dargestellt.

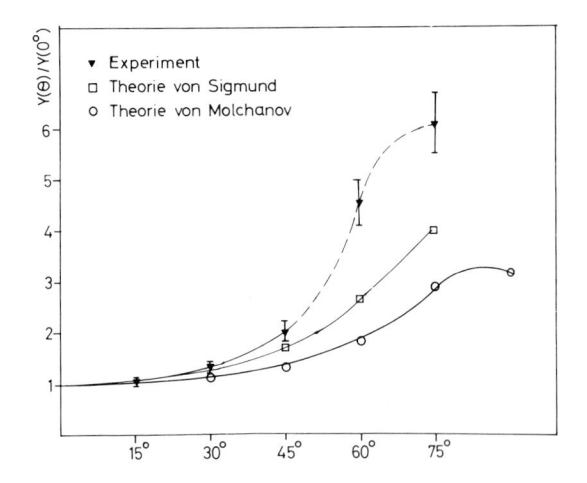

Abb. 4: Sputterrate von Silizium bei Bestrahlung mit Argon-Ionen von 15 keV in Abhängigkeit von Einschußwinkel; Gegenüberstellung mit theoretischen Kurven /2, 3/

4. Literatur

/1/ G. Amsel et al.; Nucl. Instr. Meth. 92 (1971) 481
/2/ P. Sigmund; Phys. Rev. 184 (1969) 383
/3/ V.A. Molchanov, Sov. Phys. Dokl. 6 (1961) 137

SOME REMARKS ON VOLATILITY LOSSES OF SEVERAL SAMPLE COMPONENTS DURING ACTIVATION WITH HIGH ENERGY BREMSSTRAHLUNG

C. Segebade
Bundesanstalt für Materialprüfung Berlin

Particularly in instrumental radiometric analysis methods the analyst has to be aware of partial decomposition of the sample material, especially organic matter. In particular this is true in radioactivation analysis techniques using any kind of incident activating radiation. Hereby, caused both by radiation energy and heat, several elements are volatilised, depending upon the physicochemical properties of the molecule which they are part of, and its chemical environment, i.e. the composition of the sample matrix. In activation analysis with thermal pile neutrons - as the most frequently applied activation method - one usually has circumvented volatility losses by sealing the samples in high purity silica containments[1] or by cooling during activation[2,3]. This method cannot be used in charged particle activation analysis, but in photon activation as well. In this method, however, the use of quartz vessels is entailed with some difficulty due to their dimensions. In the authors'laboratory, for instance, samples to be irradiated simultaneously usually have been packed closely adjacent to each other so as to optimally exploit the narrow activating photon beam close to the bremsstrahlung converter. In the case of quartz vials used for volatilisation prevention, because of the large total sample volume, this method would not be possible if many samples are to be irradiated simultaneously. Another way of prevention of component losses is the chemical "trapping". This has been applied particularly in the case of mercury analysis. Mercury is most problematic since nearly all Hg - components are easily volatilised and only three are fairly heat- and radiation-resistant, namely HgS, HgSe and HgTe. Takeuchi et al.[4,5] proposed to mix samples, organic ones in particular, with several sulfur-bearing compounds prior to exposure to reactor neutrons. Thereby, mercury eventually volatilised during activation is caught and transfered to HgS which has a better chance to stand heat and radiation undestroyed.

In photon activation analysis, LiHS has been used successfully to trap mercury in organic matrix[6]. In table 1 the volatility behavior of mercury in different matrices during bremsstrahlung exposure with and without addition of LiHS is indicated.

Losses of other elements (Se, Br, I and Pb) during activation with various particles were also noted in the authors'laboratory and/or other workers[7,8,9]. Therefore, the volatility behavior of these elements under bremsstrahlung attack with and without LiHS addition was investigated; it was anticipated that the lithium salts or sulphides, respectively, of the concerned elements would show more stability than the components in their original state. Soil, air-dust and plant material was studied. The results in terms of relative losses during five hours exposure to 30 MeV bremsstrahlung (mean electron beam current: 150 microamperes) are compiled in table 2. It is shown that mercury and lead is retained in the sample by sulphide quite efficiently, whereas selenium and bromine in air-dust suffer from significant losses during irradiation and cannot be trapped by lithium sulphide addition. Iodine was obviously not volatilised at all under the named conditions.

The method is easy to perform and the samples do not require significantly more irradiation volume than without sulphide imprenation. Moreover, LiHS does not produce any cumbersome radiation background through photon activation and does not cause any activating beam attenuation during bremsstrahlung exposure (as would be the case in neutron activation due to the huge thermal neutron absorption cross section of lithium). One drawback should be mentioned: Any additive to the sample prior to activation bears the danger of contamination; therefore it is recommendable to run blanks for contamination monitoring during analysis.

Table 1: Volatility behaviour of mercury during bremsstrahlung exposure, values given in relative % residual concentration (initial = 100)

Matrix	Exposure time, hours				
	1	2	3	4	5
Soil (1)	72	65	34	27	22
dto. (2)	101	100	96	101	102
Plant (1)	------------less than 20------------				
dto. (2)	104	95	99	94	94
Hg-Nitrate (1)	85	68	51	45	40
dto. (2)	100	94	97	95	102

(1) original material, (2) LiHS added

Table 2: Losses of Se, Br, I, Hg and Pb after 5 hours bremsstrahlung exposure given in relative % (initial content = 100) without/with LiHS added

Matrix	retained, rel. %				
	Se	Br	I	Hg	Pb
Soil	90/82	100/98	101/97	22/102	101/100
Air-dust	(a)	92/101	100/100	(a)	82/101
Plant	(a)	98/100	100/102	0/94	104/103

References

[1] Bate, L. C., Radiochem. Radioanal. Letters 6 (1971), 139

[2] Brune, D., Anal. Chim. Acta 44 (1969), 15

[3] Brune, D., Landström, O., Radiochim. Acta 5 (1966), 228, see also Brune, D., Jirlow, K., ibidem 8 (1967), 161

[4] Takeuchi, T., Shinogi, M., Mori, I., Journ. Radioanal. Chem. 53 (1979), 81

[5] idem, Trans. Amer. Nuclear Soc. 32 (1979), 189

[6] Raghi-Atri, F., Segebade, C., Angew. Botanik 53 (1979), 175

[7] Chattopadhyay, A., Proceed. Internat. Conf. on Measurement, Detection and Control of Environmental Pollutants, IAEA Wien/Österreich, March 15 - 19, 1976, 383

[8] Williams, D. R., Hislop, J. S., Journ. Radioanal. Chem. 39 (1977), 359

[9] Whitehead, D. C., Journ. Soil Science 24 (1973), 260

MASSENSPEKTROMETRIE ZUR MULTI-ELEMENT-ANALYSE IN FESTKÖRPERN.
MÖGLICHKEITEN DURCH OPTIMALE WAHL DER IONISIERUNGSMETHODE.

H.E. Beske

Zentralabteilung für Chemische Analysen der Kernforschungsanlage
Jülich GmbH, Postfach 19 13, D-5170 Jülich

EINLEITUNG

Die Massenspektrometrie ist ursprünglich eine physikalische Methode
zur Bestimmung von Atommassen. Mit ihr wurden die Isotope entdeckt, die
natürlichen stabilen Isotopenverhältnisse gemessen und die Massen der
Nuklide als wohl genaueste Naturkonstante bestimmt.

Nach Verbesserung der Meßtechnik läßt sich die Massenspektrometrie
in idealer Weise in der Analytik einsetzen. Dies ist vor allem durch
4 Gründe gegeben:

- Die Atommasse erlaubt die eindeutige Zuordnung zum Isotop eines
  Elementes.
- Jede Atommasse kann bestimmt und damit jedes Element nachgewiesen
  werden.
- Die Identifizierung der Elemente ist mit einer räumlichen Trennung
  verbunden.
- Es sind nachweisstarke Detektoren verfügbar.

Mit anderen Worten, die Massenspektrometrie benutzt eine unveränderli-
che, ständig vorhandene Eigenschaft der Atome - die Masse - zur Identifizierung. Sie ist damit eine element- und sogar isotopenspezifische
Trennmethode mit im Extremfall Einzelionennachweis.

Wegen der geringen Masse der Atome ($10^{-24} - 10^{-22}$ g) lassen sich zu
ihrer Bestimmung mechanische Methoden nicht mehr einsetzen. Es werden

dafür elektrische und vor allem magnetische Felder verwendet. Auf die Beschreibung dieser Methoden wird hier nicht eingegangen. Anzumerken ist aber, daß die elektrischen und magnetischen Felder Kräfte nur auf elektrisch geladene Teilchen, also auf Ionen ausüben. Die Probenatome müssen durch Wegnahme oder Anlagerung von Elektronen ionisiert sein, bevor sie mit dieser Methode nach Massen getrennt werden können.

Dieser Vortrag befaßt sich mit einer Auswahl aus der Vielfalt der Ionisierungsmethoden und zwar mit denen, die sich direkt zur Ionisation fester Proben einsetzen lassen. Diese Methoden müssen nämlich einen weiteren Analysenschritt, die Atomisierung der festen Proben vor oder gleichzeitig mit der Ionisation, ermöglichen.

Diese beiden Analysenschritte - Atomisierung und Ionisation - bilden eine Schlüsselstellung für den analytischen Einsatz der Massenspektrometrie. Sie liefern aber gleichzeitig die Möglichkeit der Anpassung des Analysenverfahrens an das Analysenproblem.

IONISIERUNGSMETHODEN FÜR FESTKÖRPER

Im folgenden werden Ionisierungsmethoden beschrieben, die eine direkte Ionisation der festen Probe ohne chemische Probenvorbereitung - also ohne Lösen, Abtrennen der Matrix, Anreichern oder ähnliches - ermöglichen. In der Abb. 1 sind sechs der leistungsfähigsten Ionisierungsmethoden, ihre Arbeitsweisen und ihre Eigenschaften schematisch dargestellt.

Die Hochfrequenz-Funken-Entladung (a) benutzt Probenpaare in Stiftform, die im Vakuum in kleinem Abstand gegenüberstehen. Durch hochfrequente Wechselspannung wird zwischen den Elektroden ein Funkenplasma erzeugt. Beide Probenelektroden werden abgebaut. Das Probenmaterial gelangt vorwiegend atomar in das Plasma und wird dann in Wechselwirkung mit Elektronen ionisiert.

Die Gleichspannungs-Bogenentladung (b) verwendet gleiche Probenformen und Anordnungen. Die Bogenentladung wird jedoch durch einen Gleichspannungsimpuls gezündet und mit elektronischen Maßnahmen für etwa 100 µs aufrecht erhalten. Der Probenverbrauch erfolgt hier wegen der unipolaren Entladung nur an einer der Elektroden.

Beide Ionisierungsmethoden werden in der Funken-Massenspektrometrie oder englisch "Spark Source Mass Spectrometry" eingesetzt.

Abb. 1: Ionisierungsmethoden für Festkörper

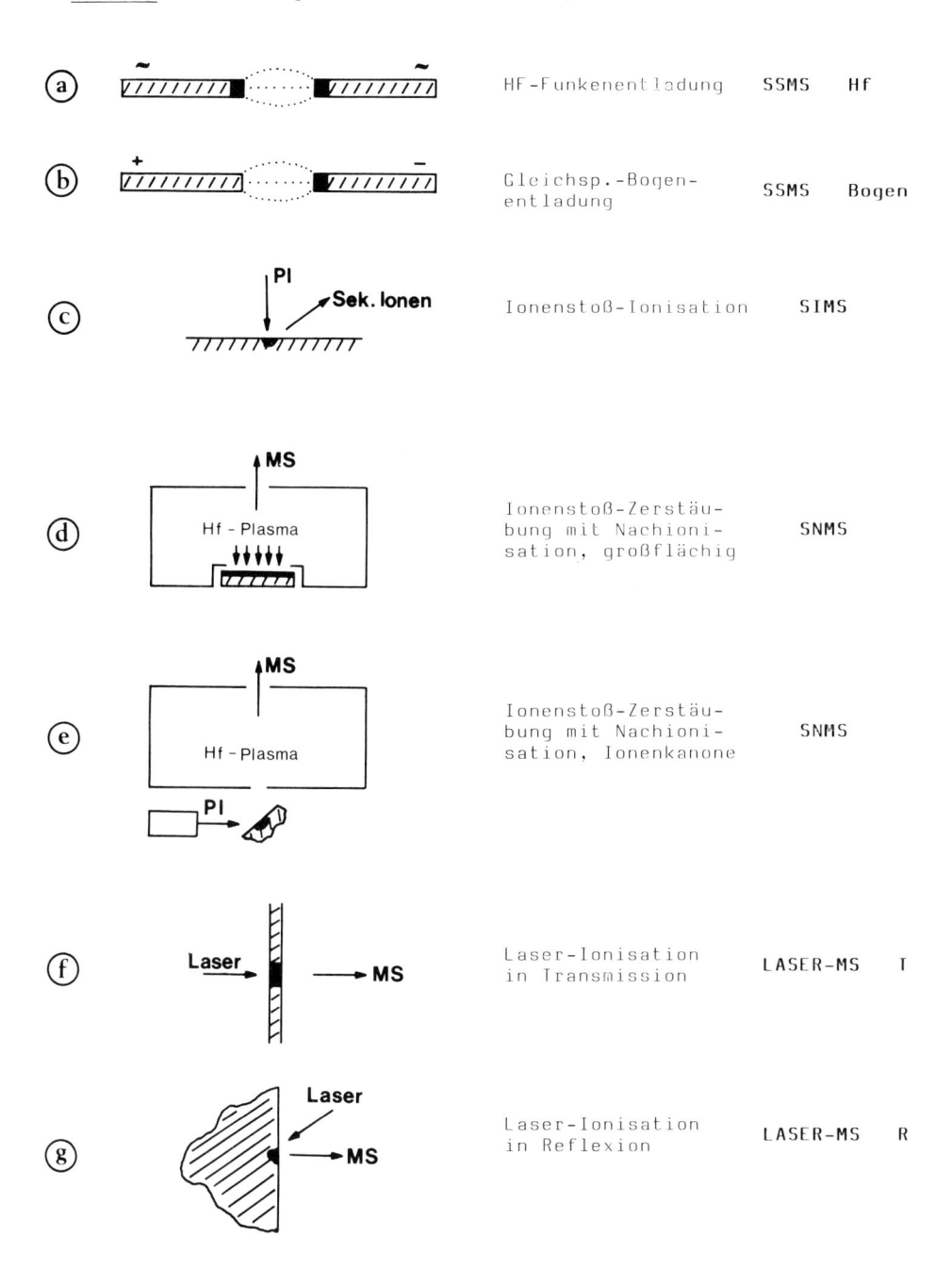

(a) HF-Funkenentladung — SSMS Hf

(b) Gleichsp.-Bogen-entladung — SSMS Bogen

(c) Ionenstoß-Ionisation — SIMS

(d) Ionenstoß-Zerstäubung mit Nachionisation, großflächig — SNMS

(e) Ionenstoß-Zerstäubung mit Nachionisation, Ionenkanone — SNMS

(f) Laser-Ionisation in Transmission — LASER-MS T

(g) Laser-Ionisation in Reflexion — LASER-MS R

Der Ionenbeschuß ⓒ führt ebenfalls zur Zerstäubung (Sputtering) der Festkörperoberfläche. Es werden vorwiegend neutrale Atome emittiert, ein Teil ist aber positiv oder negativ geladen - also direkt ionisiert. Durch geeignete Wahl der Beschußbedingungen kann der Anteil der Ionen selektiv gesteigert werden. Der Primärionenstrahl kann als Sonde zur lokalen Analyse benutzt werden. Der Einsatz dieser Ionisierungsmethode erfolgt in der Sekundärionen-Massenspektrometrie, SIMS.

Eine Variante der Ionenbeschuß-Ionenquelle benutzt den Ionenbeschuß eines Hochfrequenzplasmas ⓓ zur Zerstäubung der Probenoberfläche und die Elektronen des Plasmas zur Nachionisation. Die Zerstäubung kann auch durch eine Ionenkanone ⓔ erfolgen. Dabei werden auch nur die neutral emittierten Atome verwendet, die in dem nachgeschalteten Hf-Plasma nachionisiert werden. Diese Ionenquelle findet Anwendung in der Sekundären Neutralteilchen Massenspektrometrie, SNMS.

Die neueste Form der Festkörperionisation ist durch den Laser-Beschuß ⓕ + ⓖ gegeben. Die hohe Leistungsdichte eines Laserspots erlaubt die entmischungsfreie Verdampfung eines kleinen Probenvolumens. Dabei ist die übertragene Energie so groß, daß hohe Ionisierungsausbeuten durch Bildung eines Plasmas erreicht werden. Eingesetzt werden heute vor allem Nd-YAG Laser mit einer Leistungsdichte > 10^9 W/cm^2. Zwei Varianten sind im Einsatz. Der Laser im Durchschuß. In diesem Fall ist die Probe eine dünne Folie. Im anderen Fall wird der Laser in Reflexion betrieben.

Alle sechs Ionisierungsmethoden liefern positive atomare Ionen in mehreren Ionisierungsstufen, aber auch negative atomare Ionen und sogar mehratomige Ionen, sogenannte Clusterionen. Von Bedeutung ist aber, daß auch nicht ionisierte - neutrale - Atome emittiert werden und vor allem, daß das Verhältnis aller dieser Spezies element- und matrixabhängig ist. Alle diese Ionenanteile lassen sich im Massenspektrometer messen, die Neutralteilchen dagegen nicht. Hier liegt die Begründung für die Notwendigkeit der Eichung in der Massenspektrometrie.

EIGENSCHAFTEN DER FESTKÖRPER-IONISIERUNGSMETHODEN

Es sollen nun die für den analytischen Einsatz wichtigen Eigenschaften dieser Ionenquellen zusammengestellt werden. Vier Eigenschaftsbereiche werden behandelt:

- Eigenschaften, die zur Wahl des MS-Systems führen,
- Eigenschaften, die die Probe betreffen,
- Nachweiseigenschaften.
- Richtigkeiten.

In Tabelle 1 sind diese Eigenschaften zusammengestellt. Im folgenden werden diese für die drei Ionenquellengruppen diskutiert und die besonderen Leistungen hervorgehoben. Alle diese Angaben sind als Richtwerte zu betrachten.

Ionenquellen mit Hf- und Bogen-Entladungen haben zwar unterschiedliche Energiebreiten ΔE, sie benötigen aber beide doppelfokussierende ionenoptische Trennsysteme. Das ist wichtig, da hierdurch ein hoher Aufwand und Preis verursacht wird.

In beiden Fällen werden Stiftelektrodenpaare verwendet. Diese Elektroden müssen aus elektrisch leitendem Probenmaterial bestehen oder durch Preßzusätze leitend gemacht werden. Verwendet werden dafür Graphit oder Reinstmetallpulver. Die Hf-Ionenquelle hat bei dieser Präparation für Nichtleiter gewisse Vorteile. Der Probenverbrauch liegt für die Nachweisgrenze in beiden Fällen bei etwa 10 mg. Es lassen sich Isotope und damit Elemente nachweisen, keine Moleküle oder Phasen.

Die Nachweisgrenze für alle Elemente liegt bei 0,1 ppm at bei Fotoplattenregistrierung. Bei elektrischer Registrierung etwa Faktor 10 niedriger. Die Nachweisgrenzen der Elemente unterscheiden sich nur gering. Natürlich bringt die Aufteilung auf viele Isotope einen Verlust an Nachweisstärke gegenüber einem einisotopigen Element. Hier kann ein Faktor 5 gegeben sein. Der Rest bis zu einem Faktor 10 liegt in der Ionisierungswahrscheinlichkeit oder in der unterschiedlichen Verteilung auf Ionen mehrerer Ladungsstufen. Beim Zünden der Entladung werden Krater einer mittleren Tiefe von 0,5 - 1 µm erzeugt. Damit ist die Informationstiefe gegeben. Eine laterale Auflösung gibt es nicht, da der Funke oder Bogen über die ganze Stirnfläche der Elektroden wandert. Die erreichbare Richtigkeit ohne Eichung bei Hf liegt bei Faktor 3, beim Bogen unter Verwendung aller Ladungszustände besser Faktor 2. Bei Eichung über Standards ohne chemische Probenvorbereitung, also auch ohne Isotopenverdünnungsanalytik, bei 10 - 20 %.

Ionenbeschuß-Ionenquellen liefern Ionen einer Energiebreite von etwa 20 eV und können darum mit einfachen Quadrupol-Systemen auskommen. Bei geforderter Hochauflösung, z. B. bei Ionensonden, werden aber doppelfokussierende MS-Systeme benötigt. Bei SIMS werden ebene Analysenflächen als Probenform benötigt. Eine Fläche von 1 mm^2 reicht für eine vollständige Analyse aus. Bei der beschriebenen Anordnung bei SNMS sind Flächen von 0,5 bis 1 cm^2 erforderlich. Auch hier werden elektrisch leitende Proben benötigt, um elektrische Aufladungen zu vermeiden. Es gibt eine Reihe von Maßnahmen, um dies zu erreichen: Bedampfen mit Gold, Beschuß mit negativen Primärionen oder Ladungskompensation aus einer Elektronenquelle.

Tab. 1   Eigenschaften der Festkörper-Ionisierungsmethoden

Ionisierungs-methode	Ionenoptik		Probe				Nachweiseigenschaften					Richtigkeit	
	Energiebreite ΔE	MS - System	Probenform	El. Leiter	El. Nichtleiter	Probenverbrauch bei NWG	Nachw.-Möglichk. I; E; M; P	Nachweis-grenzen	Unterschied NWG	Informations-tiefe	Laterale Auflösung	ohne Eichung	mit Eichung
	keV					g	I;E;M;P	ppmat	Faktor	µm	µm	Faktor	%
SSMS Hf	1	DF	Stift-paar	x	Preßzusatz	10 mg	I + E	0,1	10	1	Keine	3	10 - 20
SSMS Bogen	0,1	DF	Stift-paar	x	Preßzusatz	10 mg	I + E	0,1	10	1	Keine	< 2	10-20
SIMS	0,02	DF QP	Flächen 1 mm²	x	bedampfen	1 - 10 ng	I + E (M + P)	0,1 bis 100	10³	nm	2 µm	100	F 2
SNMS	0,02	QP	Flächen 0,5 cm²	x	bedampfen	1 - 10 ng	I + E (M)	1	10	nm	Keine	10	20
LASER-MS T	1	Laufzeit	Folien 1 - 3 µm	bel.		10 pg	I + E (M)	1	10	1	1µm	100	F 2
LASER-MS R	1	Laufzeit DF	bel.	bel.		10 pg 20 mg	I + E (M)	0,1	10	1 - 5	10-20µm	3	20

I = Isotope   E = Elemente   M = Moleküle   P = Phasen

Der Probenverbrauch liegt bei einigen ng. Bei beiden Ionenquellen lassen sich Isotope und Elemente, in bestimmten Fällen auch Moleküle über charakteristische Molekülbruchstücke bestimmen. Auch Informationen über Phasen sind mit Einschränkungen zu erhalten. Die Nachweisgrenzen sind bei SIMS stark element- und matrixabhängig. Bei SNMS sind keine so starken Elementabhängigkeiten zu beobachten. Die Zerstäubungsausbeuten liegen nahe beieinander und die Ionisierungsausbeuten auch. Damit ist ein Vorteil gegenüber SIMS gegeben. Die Informationstiefe ist bei beiden Ionenquellen gleich. Sie liegt bei 1 - 2 Atomlagen. Wegen der geringen Primärionenenergie ist eine bessere Tiefenauflösung bei SNMS gegeben. Die laterale Auflösung ist nur bei SIMS durch den Betrieb als Ionenmikrosonde gegeben. Bei SNMS in der Anordnung (d) gibt es keine laterale Auflösung. In der Variante (e) mit separater Primärionenquelle kann ebenfalls eine laterale Auflösung bei Verlust an Empfindlichkeit erhalten werden. Die erreichbaren Richtigkeiten sind bei SIMS ohne Eichung sehr schlecht, bis Faktor 100, mit Eichung über LTE Faktor 2. Bei SNMS wird bereits ohne Eichung Faktor 10 erreicht und mit Eichung 10 %.

Für die Laser-Ionenquellen ist es gegenwärtig noch schwer, allgemeine Aussagen zu machen, weil der verwendete Lasertyp, vor allem die Laserenergie, aber auch andere Betriebsparameter noch nicht optimiert sind. Wegen der hohen Energiebreite werden Flugzeit-Massenspektrometersysteme, aber auch doppelfokussierende Systeme eingesetzt. Ein großer Vorteil bei Laser-Ionenquellen ist die direkte Verwendbarkeit von elektrisch nichtleitenden Proben. Leitfähigkeit wird also nicht gefordert. Bei der Laser-MS in Transmission ist die Folienform, vor allem die geringe Dicke, ein stark begrenzender Faktor. Praktisch sind Probleme der Werkstoffwissenschaften nicht zu bearbeiten. Diese Einschränkung ist bei der Laser-MS in Reflexion nicht gegeben.

Der Probenverbrauch bei Verwendung von Flugzeitsystemen liegt in Transmission und Reflexion bei etwa 10 pg. In dieser Betriebsweise wird nur ein Laserschuß für die Analyse verwendet. Beim Einsatz doppelfokussierender Systeme wird mit der Fotoplatte über viele Laserschüsse integriert. Wegen der geringen Transmission werden etwa 10 - 20 mg an Probematerial benötigt. Entsprechend liegen die Nachweisgrenzen. Bei Einzelschuß etwa 1 ppm at. Bei Integration über viele Schüsse bei 0,1 ppm. Die elementabhängigen Unterschiede in der Nachweisgrenze sind gering wie bei den Entladungs-Ionenquellen.

Die Informationstiefe ist in Transmission durch die Foliendicke und in Reflexion durch die Kratertiefe gegeben. Die laterale Auflösung in Transmission liegt bei 1 µm und in Reflexion bei 10 - 20 µm.

Die erreichbare Richtigkeit ohne Eichung im Einzelschuß liegt bei Faktor 10 - 100, bei Integration über viele Laserschüsse wie bei den Entladungs-Ionenquellen bei Faktor 3. Mit Eichung lassen sich Faktor 2 bzw. 10 - 20 % erreichen.

ZUSAMMENFASSUNG

Aus der Darstellung dieser Eigenschaften lassen sich zusammenfassend die wichtigsten analytischen Einsatzgebiete dieser 6 Festkörper-Ionenquellen ableiten.

Tabelle 2: Analytische Einsatzgebiete der Festkörper-Massenspektrometrie in Abhängigkeit von der Ionisierungsmethode

	Analysenziel	Einsatzgebiete
SSMS Hf	Multielement-bulk Analyse (fast alle Elemente)	Geologische Proben
SSMS Bogen	Empfindlicher Spurennachweis	Reinheitskontrolle
SIMS	Oberflächenanalyse (alle Elemente)	Halbleiter (Tiefenprofile)
SNMS	Tiefenprofile Elementverteilungen Mikrobereichsanalyse	Diffusionsprofile Isotopieeffekte
Laser-MS T	Mikrobereichsanalyse	Biologische und medizinische Probleme
Laser-MS R	Multielement-bulk Analyse (fast alle Elemente)	Elektrisch nichtleitende Proben

SSMS-Hf und -Bogen sind Methoden für die Multielement-bulk- Analyse zur Reinheitskontrolle von Reinststoffen. Es sind die nachweisstärksten Ionenquellen, wenn das gesamte Elementspektrum gefordert ist. Die Hf-Ionenquelle findet außerdem ein großes Einsatzgebiet in der Multielementanalyse geologischer Proben. Beide sind besonders geeignet für die Multielement-Isotopenverdünnungsanalyse.

Die Ionenbeschuß-Ionenquelle für SIMS und SNMS liefern die Voraussetzung für die Oberflächenanalyse. Ihre Informationstiefe beträgt nur 1 - 2 Atomlagen. Beide sind darum geeignet für die Bestimmung von Konzentrationstiefenprofilen senkrecht zur Oberfläche, SNMS mit höherer

Tiefenauflösung als SIMS. Der Nachweis von Elementverteilungen mit hoher lateraler Auflösung ist mit SIMS empfindlicher möglich als mit SNMS. Die quantitative Analyse dünner Schichten ist dagegen mit SNMS eher möglich als mit SIMS.

Laser-ionenquellen werden mit Vorteil an elektrisch nichtleitenden Proben zur bulk-Analyse eingesetzt. Die Nachweisstärke und die Richtigkeit ist vergleichbar mit der der Entladungs-Ionenquellen. In Transmission werden Lokalanalysen, vor allem an biologisch-medizinischen Dünnschnitten möglich. In Reflexion steht die empfindliche bulk-Analyse im Vordergrund.

Diese Übersicht sollte zeigen, wo die Stärken der sechs Festkörper-Ionenquellen liegen, um durch geeignete Wahl der Ionenquelle ein optimales Analysenergebnis zu erreichen.

HIGH PRECISION SPARK SOURCE MASS SPECTROMETRY BY MULTIELEMENT ISOTOPE DILUTION

Klaus Peter Jochum

(Max-Planck-Institut für Chemie, Saarstraße 23, D-6500 Mainz, F.R. Germany)

SUMMARY

Isotope dilution-spark source mass spectrometry (ID-SSMS) is the most precise and accurate SSMS technique. It yields absolute concentrations directly and does not depend on comparisons with standard samples. ID-SSMS has been applied to the quantitative determination of 25 trace and minor elements in terrestrial and extraterrestrial samples. The technique yields an analytical uncertainty as low as $\pm$ 1% to $\pm$ 3%.

INTRODUCTION

Spark source mass spectrometry (SSMS) is an analytical technique that is capable of determining nearly all elements with detection limits varying normally between 0.01 ppm and 0.1 ppm /1/. The application of isotope dilution to SSMS (ID-SSMS) has considerably increased the reliability of the results. This technique is restricted to elements having two or more naturally occuring or long-lived isotopes. For each element to be determined, a known amount of a "spike" (that is an element whose isotopic composition is different from the natural composition) is mixed with the sample. The unknown concentration of the element in the sample is computed from the measured isotopic abundances of the mixture.

ID-SSMS was pioneered by Leipziger /2/ in 1965. Since that time ID-SSMS analyses were used for the determination of trace elements in metal samples /2 - 5/ and geological samples /6,7/.

In our laboratory ID-SSMS has been used for geochemical and cosmochemical investigations /8 - 10/. The purpose of this paper is to outline the ID-SSMS technique and to illustrate the analytical capability of this method by showing analyses of terrestrial and meteoritical samples.

ANALYTICAL TECHNIQUE

ID-SSMS has been applied for the quantitative determination of the minor and trace elements K, Ni, Cu, Ga, Ge, Sr, Zr, Mo, Pd, Sn, Sb, Te, Ba, Nd, Sm, Eu, Gd, Dy, Yb, Hf, W, Re, Pt, Pb and U in terrestrial rocks, stony and iron meteorites and minerals. Many other, mainly mono-isotopic, elements are analyzed using the isotope dilution data as internal standard elements.

The enriched isotopes are obtained from the Oak Ridge National Laboratory. About 10 mg are dissolved in suprapure reagents /11,12/. Calibration is performed with certified standards by SSMS and thermal ionization mass spectrometry.

Two different techniques of sample preparation are used for ID-SSMS: (a) sample dissolution method, (b) spiked graphite method.

(a) Sample dissolution method

This method is performed for isotope dilution analyses of metal and silicate samples /13/. The metal as well as the silicate samples are dissolved in a Teflon bomb together with spikes of isotopically enriched solutions using a mixture of HNO_3 + HCl and HF + HNO_3 + HCl, respectively (about 10 h at 150°C). The dissolved samples are dried on a hot plate in a clean box. A few drops o nitric acid are added repeatedly to the nearly dry samples in order to convert them mainly to nitrates or oxides which produce simple mass spectra. The dried residues are finally powdered and mixed with ultrapure graphite (1:1 weight ratio).

(b) Spiked graphite method

In order to simplify the sample preparation, the spiked graphite method has been used for the analyses of powdered rocks and minerals. This technique has been evolved by Jaworski and Morrison /6/ and further developed by Knab and Hintenberger /11/ and Jochum et al. /12/. In this method the samples is mechanically mixed with spiked graphite (that is a graphite containing up to 20 isotopically enriched elements) in an agate mill (1:2 to 2:1 weight ratio). The advantage of this technique is that undissolved samples can be directly analyzed. Contamination of the sample introduced by acids used in the dissolution procedure can be avoided. If a set of similar samples is to be analyzed, the tedious work of sample handling can be reduced by making a large batch of spiked graphite.

Mass spectrometric measurements

The spark source mass spectrometers Type AEI-MS 7 and AEI-MS 702 R are used for this work /14/. Two techniques are utilized for ion detection:
(a) photoplate detection with complete elemental coverage and high resolution,
(b) electrical detection on a restricted number of elements.

(a) Photoplate detection

About five Ilford Q2 photoplates with 15 exposures each, ranging from 0.001-2000 nC are taken, so that for each element about 20 measurements can be performed. Line densities are recorded automatically by a Steinheil densitometer. The photoplate evaluation is based on the transparency curve by Franzen et al. /15/.

(b) Electrical detection

The measurements are performed by magnetic or electrical peak switching. An on-line computer controls the magnetic field and processes the data. On each mass, ion currents are registered by a multiplier until a total charge of 1 nC has been collected at a monitor. About 10 blocks with 5 integrations each are performed on each isotope, taking the elements in rotation.

Interferences

Low level, unresolvable interferences can cause a relative high error in the analytical result. In these cases the doubly charged — instead of the normally evaluated single charged — spectrum is used where the interferences are negligibly small.

RESULTS

PRECISION

The mass spectrometric precision can be determined by measuring the abundance ratio of unspiked to spiked isotopes. It depends mainly on the magnitude of the measured isotope ratio and ranges from about ± 0.6% to ± 3% for photoplate and from ± 0.5% to ± 1% for electrical detection (Fig. 1).

There is no significant difference in precision of the analyses of fine pulverized samples done by the spiked graphite method and by the dissolution method, indicating that good mechanical mixtures are attained between spiked graphite and sample. The dissolution method, however, has improved the precision of the analyses of inhomogeneous metal samples, like iron meteorites /13/.

In order to measure the reproducibility of ID-SSMS, some standard samples have been analyzed at different times using spikes from different batches. The overall observed relative standard deviation of the mean is about ± 2%.

Multielement isotope dilution has also improved the precision of those elements, e.g. the mono-isotopic elements, which cannot determined directly by this method. Many isotope dilution data of different concentration and mass ranges can be used as internal standard elements. This technique yields a precision between ± 3% and ± 10%.

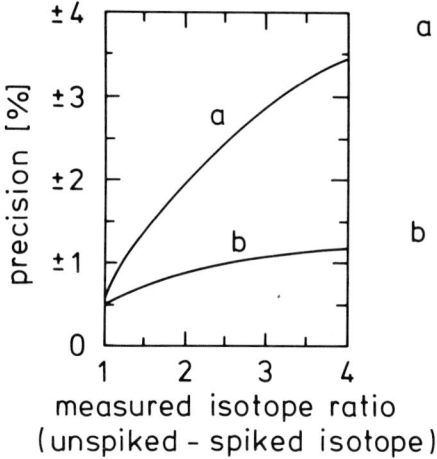

Fig. 1: Mass spectrometric precision for photoplate (a) and electrical (b) detection versus the magnitude of the measured ratio of unspiked to spiked isotope.

ACCURACY

The isotope dilution data are calculated using the equation formulated by Hintenberger /16/. Table 1 and Fig. 2 show the ID-SSMS results for some well-analyzed standard samples together with comparative data from the literature. Fig. 2 also shows the results of the elements which are calibrated by relative sensitivity factors (RSF-SSMS) using different ID-SSMS data as internal standard elements. There is a very satisfactory agreement. Nearly all elements have deviations smaller than ± 10%, and about 75% of the ID-SSMS data deviate less than ± 5%.

These results indicate that ID-SSMS is the most precise and accurate SSMS technique. ID-SSMS is capable to analyze simultaneously many elements with a precision of about ± 2%. Even greater precision of less than ± 1% can be achieved if the ratio of unspiked to spiked isotope is suitably choosen (~1) and/or electrical detection is used. These data indicate an improvement in precision by about a factor of ten compared to the usual procedure using the matrix element as internal standard element and calibration with standard samples. This results mainly from the proximity of the spiked and unspiked isotopic lines. This is expecially important for photoplate detection, because variations in the photoplate sensitivity and exposure errors are minimized by having spiked and unspiked lines readable within the same exposure and mass range. The accuracy is also considerably increased because ID-SSMS yields absolute concentrations directly without resorting to standard samples.

Tab. 1: Comparison of ID-SSMS analyses together with literature values
(Concentrations in ppm by weight)

Sample	USGS BCR-1 geological standard rock		Orgueil stony meteorite	
	ID-SSMS	Lit./17/	ID-SSMS /9/	Lit./18/
Cu	19.9	18.4	121	112
Sr	327	330	7.92	7.91
Zr	187	190	3.87	3.69
Sn	2.33	2.6	1.75	1.68
Ba	661	675	2.31	2.27
Nd	28.9	29	0.499	0.462
Sm	6.55	6.6	0.172	0.142
Eu	1.93	1.94	0.054	0.0543
Gd	6.52	6.6	0.191	0.196
Dy	6.38	6.3	0.233	0.242
Yb	3.36	3.36	0.164	0.166
Hf	4.99	4.7	0.124	0.119
Pt			0.931	0.953
Pb	17.2	17.6	2.77	2.43
U	1.70	1.74		

Sample	Campo del Cielo iron meteorite		NBS-1161 steel standard	
	ID-SSMS/8/	Lit./19/	ID-SSMS/13/	Certificate
Cu	110	109	3270	3400
Ga	92	90		
Ge	420	410		
Mo	7.3	7.8	3130	3000
Pd	5.4	3.7		
W			119	120
Pt	9.0	9.7		

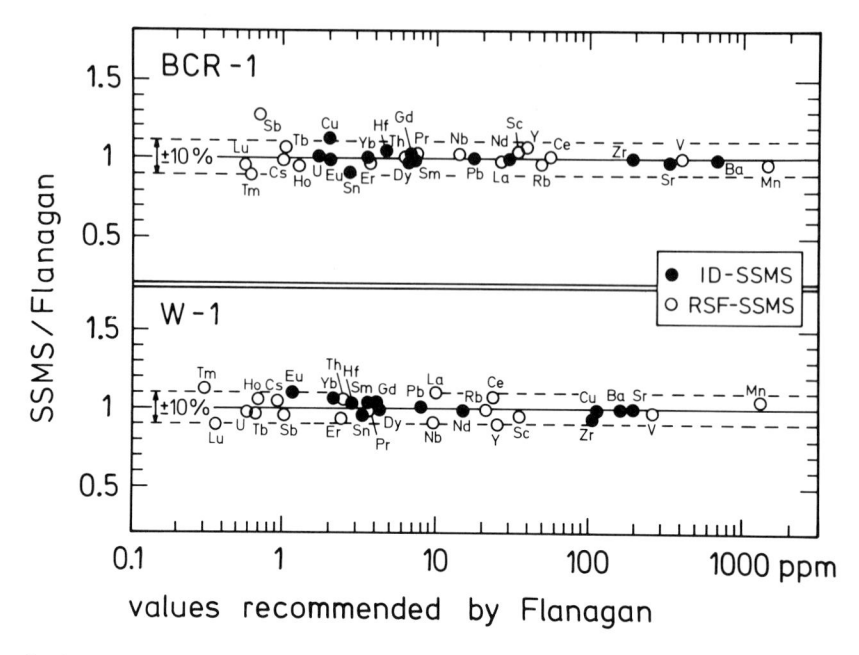

Fig. 2: Comparison of SSMS analyses of the standard rocks BCR-1 and W-1 with the values recommended by Flanagan /17/.

REFERENCES

1. H.E. Beske, R. Gijbels, A. Hurrle, K.P. Jochum, Fresenius Z. Anal. Chem. 309 (1981), 329-341.
2. D.F. Leipziger, Anal. Chem. 37 (1965), 171-172.
3. R. Alvarez, P.J. Paulsen, D.E. Kelleher, Anal. Chem. 41 (1969), 955-958.
4. P.J. Paulsen, R. Alvarez, C.W. Mueller, J. Appl. Spectrosc. 30 (1976), 42-46.
5. M.A. Haney and J.F. Gallagher, Anal. Chem. 47 (1975), 62-65.
6. J.F. Jaworski, G.H. Morrison, Anal. Chem. 47 (1975), 1173-1175.
7. J. van Puymbroeck, R. Gijbels, Fresenius Z. Anal. Chem. 309 (1981), 312-315.
8. K.P. Jochum, M. Seufert, F. Begemann, Z. Naturforsch. 35a (1980), 57-63.
9. H.-J. Knab, Geochim. Cosmochim. Acta 45 (1981), 1563-1572.
10. K.P. Jochum, A.W. Hofmann, E. Ito, H.M. Seufert, W.M. White, Nature 306 (1983), 431-436.
11. H.-J. Knab, H. Hintenberger, Anal. Chem. 52 (1980), 390-394.
12. K.P. Jochum, M. Seufert, H.-J. Knab, Fresenius Z. Anal. Chem. 309 (1981), 285-290.
13. K.P. Jochum, M. Seufert, S. Best, Fresenius Z. Anal. Chem. 309 (1981), 308-311.
14. K.P. Jochum, M. Seufert, Geol. Rundschau 69 (1980), 997-1012.
15. J. Franzen, K.H. Maurer, K.D. Schuy, Z. Naturforsch. 21a (1966), 38-62.
16. H. Hintenberger, in "Electromagnetically enriched isotopes and mass spectrometry" (Ed. M.L. Smith), Butterworths Sci. Publ. London (1956), 177-189.
17. F.J. Flanagan, Geochim. Cosmochim. Acta 37 (1973), 1189-1200.
18. E. Anders, M. Ebihara, Geochim. Cosmochim. Acta 46 (1982), 2363-2380.
19. V.F. Buchwald, Handbook of Iron Meteorites, University of California Press, Berkeley 1975.

LASER MASS SPECTROSCOPY, A USEFUL INSTRUMENT FOR THE MULTI-ELEMENT ANALYSIS OF SOLIDS

by

A.W. Witmer
Philips Research Laboratories,
P.O. Box 80.000, 5600 JA Eindhoven
The Netherlands

ABSTRACT

The Author describes briefly the analytical activities of his laboratory. Laser mass spectroscopy is presented, mainly as it is applied there. This presentation touches the following points: theory, literature, experimental set-up, some examples, characteristics, disadvantages and advantages.

INTRODUCTION

Since this is a general symposium on multi-element analysis I think it useful to start my paper with a brief description of how we apply multi-element analysis in the whole of our analytical activities and secondly why we chose amongst others laser mass spectroscopy for this purpose. I will then enlarge upon its special features and how we practise this method.

    In the Analytical Group of the Philips Research Laboratories in Eindhoven to which I belong, various methods of multielement analysis are practised such as: Emission Spectroscopy, X-ray Spectrometry, Neutron Activation Analysis, Mass Spectroscopy.
In our laboratories an important application of multi-element analysis is survey analysis by which in a sample the presence is checked of all detectable elements. The concentrations of the elements found are determined and limits of detection are calculated for the remaining detectable elements. Often this investigation produces enough information. Sometimes it is followed by a more accurate and/or more sensitive, dedicated determination on one or a few elements. In the

latter case the preinformation originating from the survey analysis is more or less indispensable. The most wanted characteristics of a survey analysis are: speed, wide element range, high sensitivity, high precision, simple quantization.

Having all this obtained you mostly have to make concessions as to accuracy. A quick, universal, sensitive and accurate analysis is the old dream of every analyst and that is probably what it will stay.

I mentioned already the methods of multi-element analysis, practised in our analytical group. They are all very useful in the way we apply them, which is, of course, for a great deal specific for our work. Still it must be said that not all of them are equally suitable for survey analysis.

X-ray Spectrometry has a wide element range and a rapid qualitative survey is possible. A quantitative analysis, however, requires extensive measures as to chemical analysis or software or both and the limits of detection are often unfavourable. You can determine components but hardly traces. This applies to both wavelength and energy dispersive instruments.

Radiochemical analysis is very sensitive; quantization is done by standard calculations which yield reasonably accurate results. But there are gaps in the range of detectable elements and the method is mostly time consuming. I refer here also to the paper of Mr. Verheijke.

In our laboratories survey analysis started in the late forties with emission spectroscopy. A d.c. arc method was set up by Dr. Addink, probably well remembered by the seniors in the audience (1). The main feature of his method was a large decrease of matrix effects owing to high dilution of the sample with graphite powder and its total evaporation. This allowed a rapid semiquantitative survey analysis with general sensitivity factors for the element lines thus excluding the daily need for standard samples. Essentially the method is still in use, but it is executed now with all modern aids of automation and computerization. The method is quick and has for many purposes acceptable though not optimum sensitivity and accuracy; but the element range has gaps.

The ICP plasma source using nebulization of solutions offers higher sensitivity and accuracy. A disadvantage for survey analysis is the necessity to bring solid samples into solution with the risk that an unknown element is lost by evaporation or precipitation.

An ICP for direct analysis of solids misses the advantage of easy quantization. Moreover, the specimen in the source should be conducting. By the way, the same objections apply when the ICP source is coupled with a mass spectrometer.

Other spectral sources, <u>arc, spark and the glow discharge lamp</u> also lack easy quantization and ask for a conducting specimen in the source.

When in 1958 spark source mass spectrometers became commercially available we placed an MS7 from the firm AEI in our spectrochemical department (2). We found <u>mass spectroscopy</u> can solve many problems as to survey analysis much better than emission spectroscopy owing to its wider element range and better limits of detection. The accuracy is comparable but the analysis time is longer. I also refer here to the papers of Dr. Beske and Dr. Jochum, the latter showing high accuracies.

A few years ago we replaced the spark source of our mass spectrometer by a laser source. This presented a considerable simplification of the sample preparation, a greater possibility of microanalysis and often a better accuracy. The main application is still survey analysis. In our laboratory laser mass spectroscopy has acquired a special position in this field which cannot be taken over by the two other favorite methods of survey analysis: emission spectroscopy and neutron activation analysis.

So I have attained my actual theme "Laser Mass Spectroscopy". The material I can present on this subject is limited in 2 ways: <u>Firstly</u> by the way of application in our laboratory, mainly survey analysis. <u>Secondly</u> by the quantity of experience collected in a few years, mainly based on practical and little on theoretical work.

The development of modern lasers has allowed an efficient evaporation and ionization of sample material. Coupling with a mass spectrometer offers a sensitive method of bulk analysis and even microanalysis since a laser beam may cover a small spot.

Ready discusses in his book "Effects of High-Power Laser Radiation" the impact of such radiation on solids (3). He states that the radiation originating from Q-switched lasers with flux densities in excess of 10^9 W/s is absorbed by the surface of the specimen at the beginning of the laser pulse. This causes explosive evaporation of material with vapour velocities in the order of 100 km/s. At the same time starts the ejection of so called blow-off material, a mixture of various particles which produces a plasma. This plasma impedes further ejection of sample material by recoil, moreover the plasma is opaque and absorbs the laser radiation. At the end of the laser pulse the plasma has reached such a high temperature (it is estimated 10^5K) that it heats up the sample surface by irradiation which causes a renewed material evaporation. Most probably the evaporation phenomenon is even more complicated. We did no special research on those processes and limited ourselves to the empirical determination of the optimum power density per sample material, on which I shall come back.

The application of a laser as an ion source in mass spectroscopy is not new. Already in 1963 Richard Honig from RCA reported about his experiences with a ruby laser (4). Yet, it is only in the recent years that we saw a larger application of laser ion sources. A general view of the literature shows mainly two types of application: First the coupling of a laser ion source with a double focusing mass spectrometer, often as a replacement of or an addition to some electrical source (5-7). We belong to this group (8). The detection is mostly both electric and photographic. Microanalysis is possible but the first application is bulk analysis. The relative detection limits are good owing to the integrating detector. Second the coupling of a laser ion source with a time-of-flight mass spectrometer with electrical detection (9,10). These instruments are preferably used as scanning microprobes. They are mostly provided with a microscope and X-Y sample manipulators for this purpose. The mass resolution is limited; the relative detection limits might be better. Sometimes the laser beam is trasmitted by the specimen, sometimes it is reflected. Only the latter version permits bulk analysis. Also organic analysis is possible.

EXPERIMENTAL

Now I shall briefly describe our experimental set-up:
Ionization is performed with a Q-switched Nd-YAG-laser, System 2000, from the firm J.K. Lasers. Its main features are:
wavelength   1060 nm
output energy   10 mJ in TEM 00 mode
pulse duration   15 ns
repetition rate   variable from single shot up to 50 Hz.
This laser source is coupled to a double focusing mass spectrometer of the Mattauch-Herzog type (2). It was originally an AEI MS 702, but both the electronic and the vacuum system have been modernized. Energy focusing is obtained from the electrostatic analyzer, mass discrimination from the magnetic analyzer. The ions produced in the source are accelerated into the analytical system by a voltage of 20 kV. The source chamber can be vented and pumped separately from the analy-

zers. Both electrical and photographic detection are possible, in the latter case simultaneously in a wide mass range. The photoplates are prepumped in a magazine before introduction into the analyzer.

Fig. 1 shows schematically the focussing optics of the laser. The beam from the Nd-YAG laser is passed into the source chamber with the aid of mirrors. A He-Ne laser is used for alignment of the sample. The erosion process can be followed by a TV camera and monitor. The laser beam enters the ion source chamber through a bakeable window and is focused on the specimen by a lens. The lens holder is of ceramic material. It should be insulating to prevent breakdown of the acceleration voltage. The specimen makes equal angles with the laser beam and the optical axis of the spectrometer. Since both optical axis and laser beam are in fixed positions, the only way to obtain optimum focusing is moving the sample holder. This may be observed on the TV monitor. The decisive check is done with the ion current meter which should indicate maximum value.

The distance from the sample to the entrance slit is 17 mm, to the lense 20 mm. The lense is protected against deposition of sputtered sample material by thin glass disks. Carbon is notorious in this respect. The sample preparation is very simple, one of the main advantages of the laser ion source. Any piece of material, conductor or insulator, with a diameter smaller than 10 mm can directly be analysed. The ion yield depends amongst others on the transparency for the laser radiation used. Until now the wavelength of this radiation has been 1060 nm. Other wavelengths are possible using a frequency multiplier but this implies high energy losses.

Small sample pieces are fixed in pure Indium. Powders are pressed in an Indium cup. The greatest concern is the avoiding of contamination, a risk that grows with increasing sensitivity of a method.

The linear part of the density curve of the photographic emulsion used is limited and therefore it is necessary to make a number of exposures of the same specimen with increasing intensities. Low intensity exposures may be made with single shots. The energy of a shot may be regulated with the aid of filters or with the flash lamp. High intensity exposures are made at 20-50 Hz repetition rate. It is clear that the material consumption depends on both energy and number of shots.

The highest energy gives the highest sample consumption but not the highest ion yield, expressed in nC, as is shown on the following table. There the maximum yield is at 2 mJ.

Amount of sample vaporized and charge collected for a sample of NBS 444 steel as function of the laser-beam energy per shot. The crater diameter was roughly 30 μm.

Energy of laser beam per shot (mJ)	Crater depth for single shot (μm)	Amount of material vaporized (g)	Charge collected (nC)
10	10	1×10^{-8}	1×10^{-3}
2	3	3×10^{-9}	2×10^{-3}
0.5	1	1×10^{-9}	1×10^{-3}

The optimum energy should be selected for each type of sample. High sample homogeneity is required due to the low material consumption. The small dimensions of the craters allow micro, thin layer, and even profile analysis.

The number of multiply charged ions decreases when the energy per shot is lowered. The production of clusters is much smaller with the laser than with the radio frequency spark.

The evaluation of the spectra is done in the standard way, that is: Line position, here indicating isotope mass, is used for qualitative analysis. The intensity of the line, deduced from its blackening, is used for quantitative analysis. The evaluation is performed with the aid of an automatic microphotometer, connected with a computer.

A thin layer analysis is obtained by superficial scanning with the laser beam.

I have various examples of laser mass spectrometric analyses (8), compared with the results obtained by other methods which show good agreements. Since my time is limited I shall give one example: the analysis of a geological standard AGV1. Fig. 2 shows the results obtained directly from the intensity ratios, i.e. without standard samples and without sensitivity factors, plotted against the reported values on the abcissa. The accuracy is mainly within a factor 3. The deviations appear to be systematic rather than random. Fig. 3 shows the results obtained after calibration with a reference standard. Then the accuracy is mostly within 25%. The improvement is mainly due to correction of the concentration of the reference element: Silicon.

Now I have come to the summary of the characteristics of the method, to begin with bulk analysis:
- element range: Li to U
- concentration range: from ppb to 100%
- accuracy without reference standards: up to a factor 2
- accuracy with reference standards: up to 10%
- relative standard deviation: up to 10%
- absolute detection limit: 10^{-12} g
- relative detection limit: 10 ppb
- material consumed: 100 µg

Here are specific characteristics for microanalysis:
- spot diameter: 20 µm minimum
- spot depth: 0.1-10 µm
- absolute detection limit: 10^{-12} g
- relative detection limit: 0.1%
- reproducibility per shot: 10%
- thin film and profile analysis

CONCLUSION

I wil end my paper with a recapitulation of advantages and disadvantages of Laser Mass Spectroscopy and I begin with the disadvantages:

Disadvantages Laser Mass Spectroscopy
- Transparency of material affects ion yield
- Laser optics in source chamber vulnerable to contamination
- Long analysis time
- Accuracy is reasonable, not superb (compare spark source)

These disadvantages are no impediment to analysis, moreover, there are many advantages:

Advantages Laser Mass Spectroscopy
- Simple sample preparation
- Possibility to analyze a large variety of samples: bulk, incl. crystals, powders, liquids; thin layers incl. profiles; microsamples
- High degree of ionization-high sensitivity
- Low degree of vaporization-low memory effects
- Simple spectra
- Wide element range
- Low detection limits

- Possibility of semiquantitative analysis without standards
- Rapid simultaneous multi-element analysis with reasonable precision and accuracy

I repeat:

   Laser Mass Spectroscopy covers a special domain between neutron activation analysis and emission spectroscopy where it is often irreplaceable.

   It certainly deserves a larger part in multi-element analysis than it has acquired until now.

REFERENCES

(1) N.W.H. Addink, D.C. Arc Analysis. MacMillan, London (1971).

(2) A.J. Ahearn, Trace Analysis by Mass Spectrometry. Academic Press, New York (1972).

(3) J.F. Ready, Effects of High-Power Laser Radiation. Academic Press, New York (1971).

(4) R.E. Honig, Appl. Phys. Letters 3,8 (1963).

(5) Y.A. Bykovskii c.s., Prib. Tech. Eksp. 2, 163 (1977).

(6) R.A. Bingham and P.L. Salter, Anal. Chem. 48, 1735 (1976).

(7) R.J. Conzemius and H.J. Svec, Anal. Chem. 50, 1854 (1978).

(8) J.A.J. Jansen and A.W. Witmer, Spectrochim. Acta 37B, 483 (1982).

(9) T. Dingle and B. Griffith, J. Phys. E. Scient. Instrum. 14, 513 (1981).

(10) H.J. Heinen c.s., Int. J. Mass. Spectrom. Ion Phys. 47, 19 (1983).

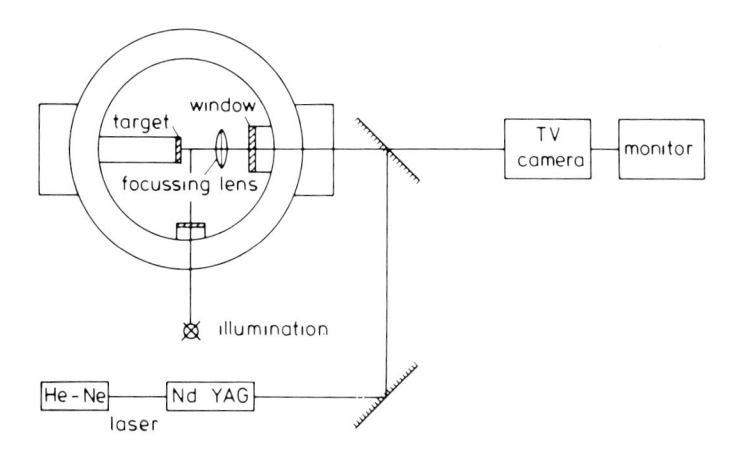

FIG. 1

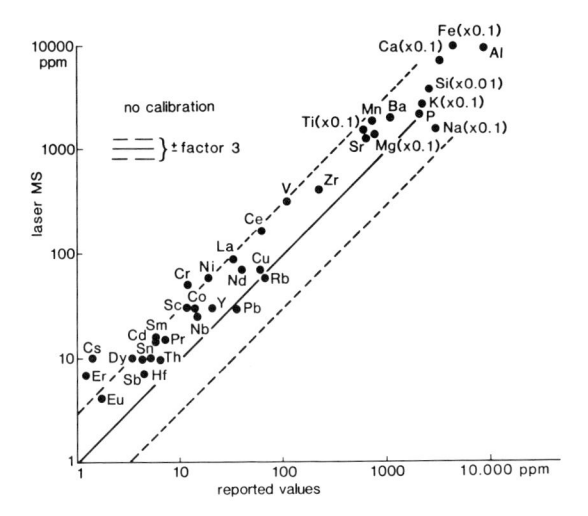

FIG. 2

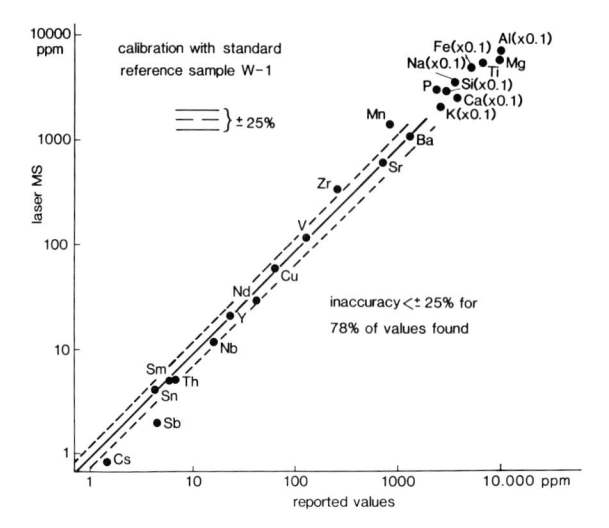

FIG. 3

IN SITU MULTI ELEMENT TRACE ANALYSIS WITH SIMS

M. Grasserbauer

Institute for Analytical Chemistry, Technical University Vienna,
A-1060 Wien, Getreidemarkt 9.

SUMMARY

SIMS in its most modern instrumental configuration combining sophisticated micro and surface analysis with high standard mass spectrometry enables a significant progress in multi element trace analysis of materials. The major areas where the unique features of this technique provide advantages over other methods are: ultra trace element characterization, trace analysis in microdomains of materials, micro and surface distribution analysis. The potential and limitations of SIMS for multi element trace characterization of technical materials like refractory metals, zinc ores, cast alloys and thin film metallization structures are described.

1. INTRODUCTION

The development and increasing technical use of pure and ultra pure materials - like semiconductors, sinter metals, ceramics - demands an extensive characterization of the trace elements which largely influence

---

Symbols used:

PI = primary ion, E_0 = excitation energy, i_B = beam intensity, d_A = diameter of analyzed area.

or even determine the properties of such materials [1]. Also for the study of origin and prospection or use of raw materials (ores) extensive knowledge of the trace elements is necessary [2].

These trends in technology require the further development of methods and procedures for trace analysis, particularly multi element characterization at ultra trace levels (ng/g-range), micro and surface distribution analysis. These analytical requirements - high detection power and spatial resolution - can only be fulfilled by analytical techniques using focussed particle beams like PIXES, LAMMS, CPAA and SIMS. Among these techniques SIMS offers the largest capabilities for many technical analytical problems due to the recent progress in instrumentation which combines the features of sophisticated micro and surface analysis with high standard mass spectrometry [3].

2. PRINCIPLES AND FEATURES OF SIMS

2.1 Signal Generation [4]

Due to the bombardment of the material's surface with primary ions (Ar^+, O_2^+, O^-, Cs^+) with an energy of several keV secondary particles are emitted from the sample surface with a typical information depth of 3 - 5 atomic layers. Ionization of a fraction of these particles during emission can be described in a semiquantitative manner by the local thermal equilibrium model [5] if reactive sputtering with oxygen is used. Basically this means that relative secondary intensities can be predicted from atomic properties like ionization energy and electron affinity and material functions (plasma temperature, electron concentration in the plasma) which can be determined by measurement of 1 or 2 internal standards (e.g. matrix elements).

During the sputter process primary ions are implanted into the sample surface leading to a zone of changed composition and structure ("altered layer"). Structural effects are mainly amorphization in semiconductors and generation of defects in metals [6].

The depth of this zone is about the double mean projected range (R_p) of the primary ions. In Si R_p is about 25 nm for particles with an energy of 1 keV per nucleon (for 60° beam incidence). This leads to a thickness of the altered layer for typical analytical conditions (primary ion energy ca. 5 keV, angle of ion incidence 60°) of 10 nm for O_2^+, 20 nm for O^-, 8 nm for Ar^+ and 2 nm for Cs^+. Within this thin surface layer the signal does not reflect the actual elemental distribution but a mixture of the influence of primary ion implantation and sputtering with composition (Fig. 1).

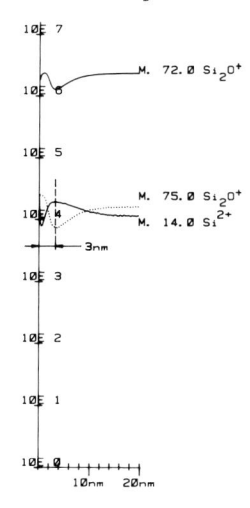

Fig. 1: Secondary ion signals at surface of Si (depth profile, PI = O_2^+, E_0 = 5.5 keV): Stable secondary ion emission is reached at a depth of 10 nm corresponding to 2 R_p

After sputtering of a surface layer thicker than 2 R_p an equilibrium between implantation and erosion is achieved. Then the secondary ion signal exhibits the following relation to material parameters and experimental conditions (for positive secondary ions):

$$I_{S(A)} = I_p \cdot S \cdot \alpha_A \cdot c_A \cdot i_{S(A)} \cdot \eta$$

$I_{S(A)}$ = secondary ion intensity of measured isotope of element A [cps]
I_p = primary ion intensity [ions/s]
S = sputter yield [atoms per primary ions]
α_A = (positive or negative) ionization yield of sputtered atoms
$i_{S(A)}$ = isotopic abundance
η = efficiency of secondary ion measurement (counts/ion) (ion extraction, mass spectrometer, detection)

The magnitudes of these variables are for oxygen bombardment:
I_p = 1 - 5 µA = ca. 10^{13} ions/s; I_2 = 0.1; S = 0.5 - 10; α_A = 0.1 - 10^{-5}; $i_{S(A)}$ = 0.01 - 1

From this relation the (theoretical) detection power of SIMS can be calculated. For most elements the detection limits are in the ng/g-range.

2.2 Quantification

For materials analysis oxygen is mostly used as a primary ion due to its high yield for positive ions and the reduction of the chemical matrix effect. Under oxygen bombardment the altered layer is enriched with oxygen (concentration in the percent range) and the sample surface transformed into an oxide-like state. The influence of different chemical bonding of an element on the secondary ion signal is reduced typically to less than a factor of 2.

For trace elements the major influence on the signal yield originates from the matrix (S and α_A). Therefore the prime approach to quantification is the use of external standards with the same matrix as the sample (relative sensitivity factor method).

Such standards should be homogeneous on the nm-scale. This condition can however often not be fulfilled and has to be considered in the characterization of analytical accuracy.

Often materials are available which contain the same matrix as the sample but a much higher trace content. These materials can be analyzed with several reference methods (like XRF, AAS, NAA) and used as standards. Homogeneity has to be checked with SIMS. Since linear relations between c_A and $I_{S(A)}$ extending over more than 3 decades are obtained for trace elements in a specific matrix extrapolations from concentrations in the 100 µg/g range as present in the standard to ng/g levels can be made with high accuracy.

If the trace element is not dissolved in the lattice of a matrix but precipitated (forming a different phase) then the chemical bonding of the trace element influences accuracy. Analytical errors of typically a factor of 2 are then encountered. In such cases accuracy is usually also effected by the inhomogeneity of the sample. A measure for this effect in bulk analysis can be obtained by repeated measurements on different areas of a sample and statistical evaluations of the data.

Considerable attention is presently given to the development of techniques for the preparation of standards specially suited for SIMS, e.g. by ion implantation, diffusion, co-sputtering of materials, controlled doping of materials.

Often the production of external standards for a particular matrix is not possible or too costly (especially for multielement trace analysis). On the other hand for many problems in materials research semiquantitative trace results are sufficient. In such cases calculation of the trace element content using the LTE-model [5] can be

applied. The one-parameter version of Morgan and Werner [7] (QUASIMS)
yields results which exhibit a typical accuracy of a factor 2 - 5. The
most critical point in the application of the LTE-calculation procedure
is the selection of the optimal plasma temperature [8]. If possible it
is useful to select several element combinations (which are either
known or can be analyzed with another technique) as internal standards
to optimize the temperature parameter.

In trace analysis of heterogeneous materials (e.g. metals consisting
of 2 or more phases, thin film structures of different materials) the
dependance of sputter and ionization yield on chemical composition has
to be taken into account. The approaches described above can be used to
quantify results in each phase or layer separately. Special problems
arise at interfaces or areas where the major components change gradually.
For bulk analysis of heterogeneous materials the distribution of the
trace element between the different phases must be known.

3. APPLICATIONS OF SIMS FOR MULTI ELEMENT TRACE ANALYSIS OF MATERIALS

3.1 Ultra Trace Bulk Analysis

The basic advantage of SIMS for ultra trace bulk analysis is that the
method is contamination free for most elements (exceptions: H, C, O, N
from residual gas in sample chamber - $p = 10^{-8}$ mbar, Cr, Ni, Fe, Ta from
metallic parts in the instrument). The multielement capability is also
valuable for systematic studies of the influence of trace elements on
material properties.

Reliable ultra trace analysis depends on a thorough evaluation of the
complex spectrum for possible interferences by other ions (molecules,
clusters, multiply charged atomic ions). That can be done by recording
of mass spectra with high mass resolution ($\frac{M}{\Delta M} = 1000$) or calculations
of possible atomic combinations. Interferences can be eliminated by use
of high mass resolution or energy filtering (separation of atomic and
molecular ions on the basis of their different energy distributions).
In either case a loss of intensity and detection power is encountered
which is in the range between 10 and 100.

Stringent requirements concerning detection power for the analytical
techniques are posed in the characterization of pure sinter metals,
like tantalum or molybdenum used in high temperature processes [9,10].
Recrystallization of these materials during heating is strongly influenced by trace elements. Controlled doping with trace amounts of

elements which prevent or reduce recrystallization (e.g. Y for Ta; K, Si, O for Mo) significantly improves life time and material properties.

In order to study the influence of these trace elements systematic analysis of materials of different properties is necessary. This demands for ultra trace and multielement capability.

Fig. 2 shows as an example important parts of the mass spectrum measured in a sintered tantalum wire [11]. The trace elements Na, K, Fe, Mn, Co, Y, Zr, Mo, Nb and W can be analyzed practically without interferences. Since neither energy filtering nor high mass resolution is necessary maximal detection power is achieved. It is for most of these elements in the low ng/g-range (Table 1).

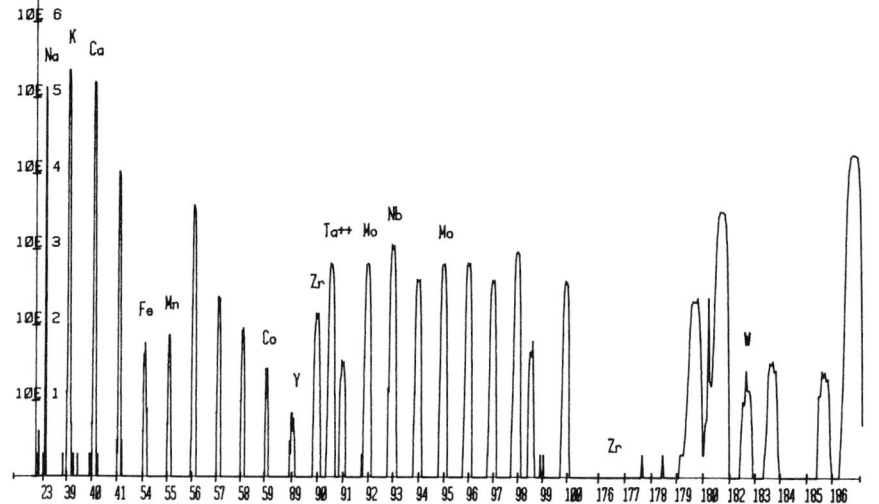

Fig. 2: Mass spectrum of sintered tantalum wire for quantitative determination of trace elements (PI = O_2^+, E_0 = 5.5 keV, d_A = 60 µm, i_B = 500 nA, thickness of wire = 120 µm)

For Y quantification can be performed with the use of a doped Ta-powder (40 µg/g Y, reference method: XRF). Accuracy of an individual result is estimated to be better than 25 % rel. The inhomogeneous distribution of Y however (precipitations at grain boundaries) leads to variations in the results of more than a factor 5 when different areas of the sample are analyzed. For the other elements semiquantitative concentration levels were calculated with the QUASIMS procedure. The results for trace element analysis can be correlated with the microstructure of Ta-wires after heat treatment.

Table 1: Multi Element Bulk Trace Analysis of Ta-Wire: SIMS-Results and Detection Limits, Mass Spectrum Fig. 2 (I = Isotopic abundance, $Si_{(rel)}$ = Relative secondary ion yield)

Element	Mass	I	$(Si)_{rel}$	Concentration [µg/g]	Detection limit [ng/g]
Na	23	1.00	35	5	0.1
K	39	0.931	110	7	0.03
Ca	40	0.970	24	20	0.2
Fe	54	0.058	1.2	3	100
Mn	55	1.00	1.2	0.2	3
Co	59	1.00	0.75	0.2	3
Y	89	1.00	3.2	0.009	3
Zr	90	0.515	5.3	0.3	5
Nb	93	1.00	3.5	2	3
Mo	92	0.158	2.6	10	30
	95	0.157		10	30
Hf	178	0.27	3.5	0.005	15
W	182	0.264	3.0	0.5	10

3.2 Trace Analysis of Microdomains

Multi element trace analysis of microdomains is of special importance for mineralogy and geochemistry. Trace contents of individual mineral phases may yield information about genesis or for prospection.

Multi element analysis with high detection power can be performed in phases of at least ca. 10 µm size [12]. Smaller inclusions are still difficult to analyze and the analytical figures of merit not well established. For mineralogical analysis a lateral resolution of 10 µm is usually sufficient, however.

More than in metals in such substances extensive interferences in the mass spectra are encountered. Strong energy filtering (up to 100 eV) is necessary which reduces the detection power significantly. Fig. 3 shows as an example the mass spectrum of a ZnS-phase in the energy range of secondary ions between 95 and 225 eV (diameter of analyzed area 8 µm). Even under such problematic conditions practical detection limits in the low µg/g-range can be obtained (Table 2). Quantification can be performed by a combination of analysis of external standards (homogeneous ZnS-crystals with ICP-OES, AAS and EPMA) and LTE-correction [12,13].

Accuracy for the elements calculated with QUASIMS is about a factor 2. Elements for which external standards exist (like Ge) can be determined with an accuracy of about 20 to 50 % rel. in microdomains of ZnS ore samples.

Table 2: Multi Element Trace Analysis of Microdomains of ZnS: SIMS-Results and Detection Limits, Mass Spectrum in Fig. 3

Element	Concentration [ppma]	Detection limit [µg/g]
Mn	35	15
Fe	1.2 %	100
Ge	1000	1
As	60	15
Cd	450	5
Tl	25	0.5
Pb	1300	5

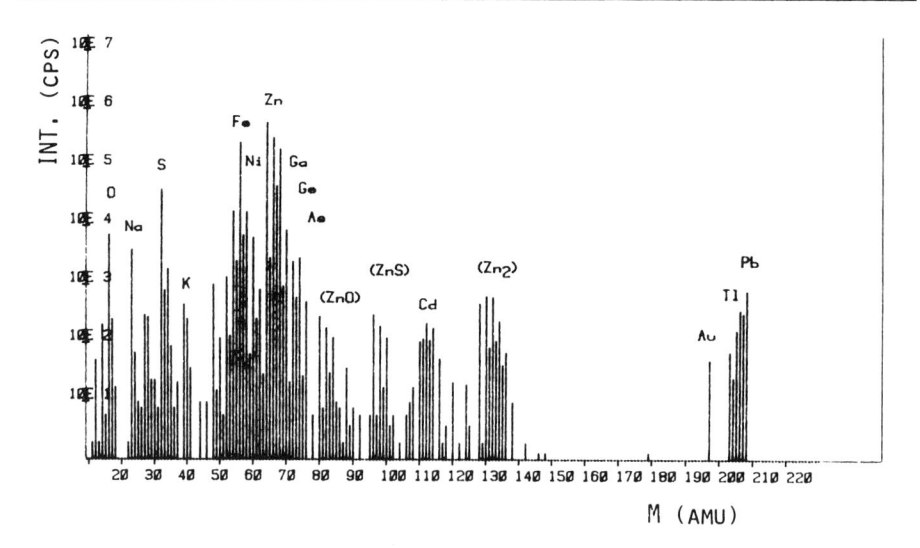

Fig. 3: Mass spectrum of ZnS-phase in mineral deposits (PI = O^-, E_0 = 10 keV, i_B = 240 nA, d_A = 8 µm, energy acceptance 90 - 225 eV)

3.3 Micro Distribution Analysis

Micro distribution analysis is performed on polished cross sections of materials - either by qualitative ion imaging or quantitative step scanning technique with a finely focussed ion beam.

3.3.1 Ion Microscopy

Ion microscopy [14] yields (primarily) qualitative information about the distribution of trace elements between different phases and allows to study trace element precipitations.

Ion images of sensitive elements (high secondary ion yield) can be obtained when their local concentration is at least 100 to 1000 µg/g. For elements with a low secondary ion yield a local enrichment up to the percent range is necessary.

Ion microscopy can be used advantageously to study the influence of trace elements on the microstructure of a material. In such investigations it is important to know whether the trace elements are dissolved in the matrix or other phases or precipitated at grain boundaries. Furthermore the homogeneity of precipitations of trace elements is of interest.

Fig. 4 shows as an example the distribution of trace elements added to an AlSi 22.5 cast alloy to achieve a specific microstructure (eutectic, lamellar, spherulitic) [15]. It can be seen that all trace elements are locally enriched (precipitated).

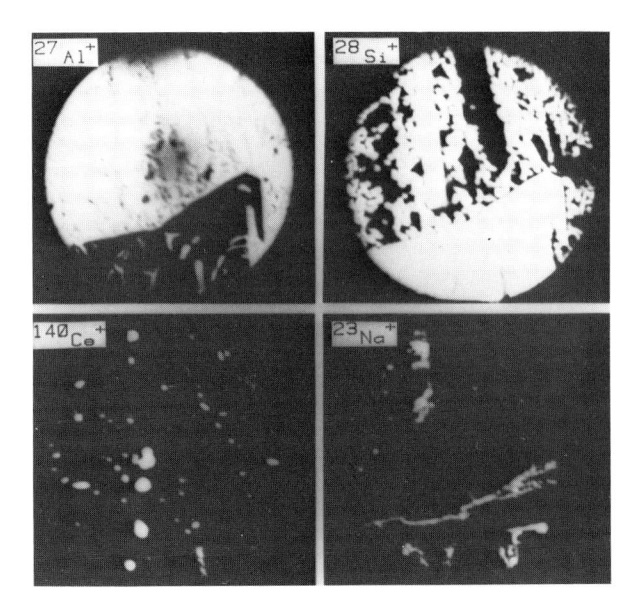

Fig. 4 : Micro distribution analysis of trace elements (1 wt %) in AlSi 22.5 cast alloy by ion microscopy

The size of small inclusions cannot be determined from ion micrographs since the image resolution of the mass spectrometer is about

0.5 µm. Even smaller precipitations yield image points of this dimension. It can be estimated that precipitations consisting of a sensitive element must have at least a size of 10 to 100 nm to be registered as ion micrographs (yielding ca. 10^4 counts per image point).

3.3.2 Qualitative and Quantitative Elemental Profiles

In order to obtain elemental profiles the step scanning technique is used. Simultaneous registration of at least 10 different masses or elements is possible. Precleaning of the surface with the ion beam is usually necessary. The lateral resolution is ca. 2 µm but for most problems only ca. 10 µm are used since a higher detection power is obtained.

Multi element distribution analysis is of particular interest for geochemical studies since information can be obtained about ore formation processes and mining strategy.

Fig. 5 shows as an example the distribution of Ge, Fe, Pb, Tl, As and Cd in a zonar grown ZnS mineral. A strong correlation of the trace elements Fe, Pb, Tl and As can be observed. Quantification of these profiles is possible with relative sensitivity factors.

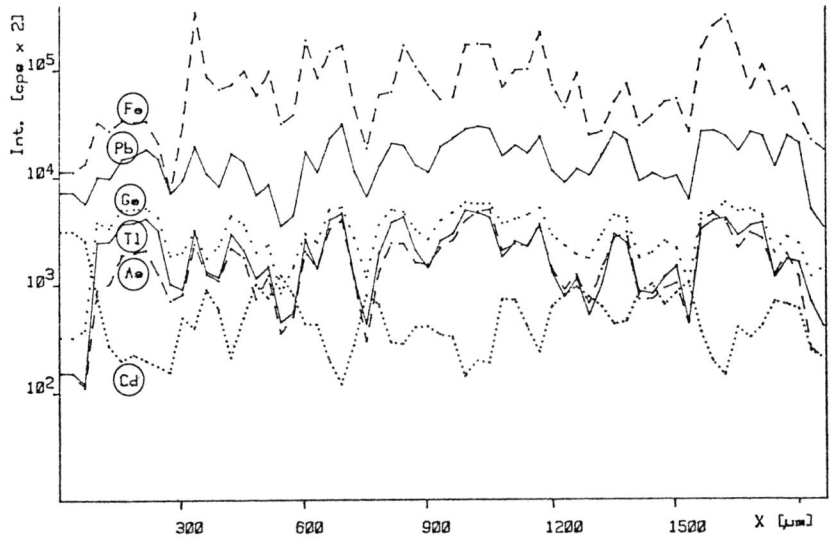

Fig. 5: Lateral Distribution Step Scan Profile of Zonar Grown ZnS [12] (PI = O^-, E_0 = 10 keV, i_B = 20 nA, raster 50 x 50 µm, d_A = 25 µm)

$$c_A = \frac{c_B}{RSF_{A/B}} \cdot \frac{I_{S(A)}}{I_{S(B)}} \qquad RSF_{A/B} = \text{rel. sensitivity factor}$$

With Zn as reference element the relative sensitivity factors for a ZnS matrix calculated with the LTE procedure are:

$RSF_{Fe/Zn}$ = 15, $RSF_{Cd/Zn}$ = 0.45, $RSF_{Ge/Zn}$ = 2, $RSF_{Pb/Zn}$ = 0.4, $RSF_{Tl/Zn}$ = 3.5, $RSF_{As/Zn}$ = 0.15.

Concentration calculated for the yellow, dense zones of ZnS exhibiting higher Pb, Tl, As and Cd content are:

Fe = 0.12 %, Ge = 0.2 %, As = 0.4 %, Cd = 0.04 %, Tl = 0.07 %, Pb = 1.4 %.

Analytical accuracy of quantitative distribution analysis is determined by the LTE model and is in the order a factor 2 - 5.

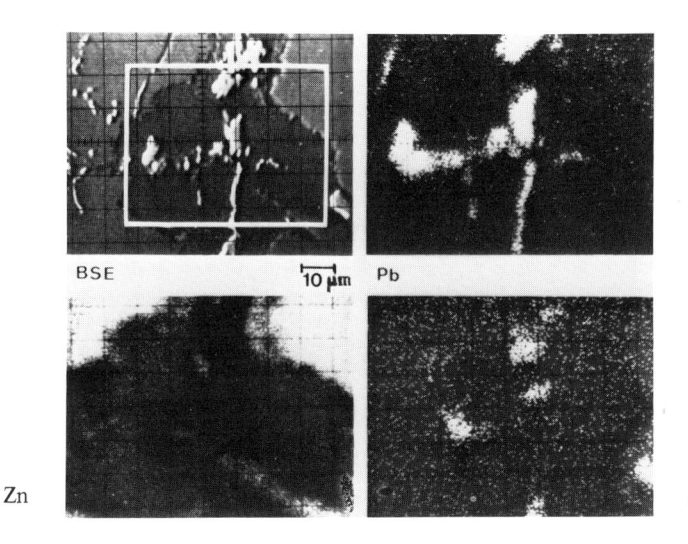

Fig. 6: EPMA-micrographs of $Pb_4As_2S_7$-precipitations in zinc ore

Multi element distribution analysis and evaluation of elemental correlations enables to study the incorporation of trace elements into mineral phases even in cases when these precipitations are too small to be analyzed individually. Fig. 6 shows as an example the EPMA-micrographs of a $Pb_4As_2S_7$-phase in a zinc ore sample. The geochemical question of interest was if this phase contained Tl. Distribution analysis of Pb, As and Tl with SIMS across an area of the ZnS which contained these precipitations and mathematical evaluation yielded the result of a clear correlation of Tl with Pb (Fig. 7). This proved that the $Pb_4As_2S_7$-phase is Tl-bearing with concentrations up to 1000 µg/g.

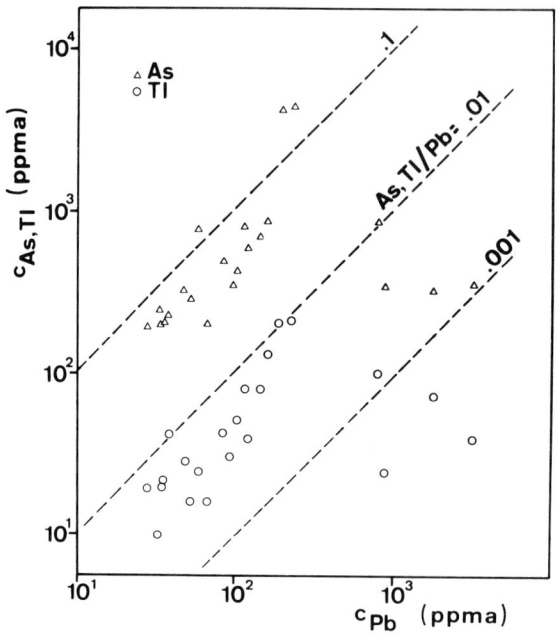

Fig. 7: Correlation of Tl and As with Pb obtained from SIMS step scan profiles across precipitations in ZnS (d_A = 8 µm)

3.4 Surface Distribution Analysis

Surface distribution analysis of trace elements uses the depth profiling technique. These profiles can be measured for at least 10 masses or elements simultaneously and quantified using relative sensitivity factors or LTE-procedures for the intensity scale and sputter rates or crater depth measurements (profilometry, interference microscopy) for the time scale [16 - 18].

Multi element surface distribution analysis is particularly important for technical thin film metal structures - e.g. for metallization films used for electronic devices.

Fig. 8 shows as an example of such a multielement analysis the mass spectrum measured in a Cr-metallization layer of only ca. 50 nm thickness. Qualitative and semiquantitative evaluation can be performed with the LTE-procedure. Trace elements in the low µg/g-level can be characterized even in such thin layers (Table 3).

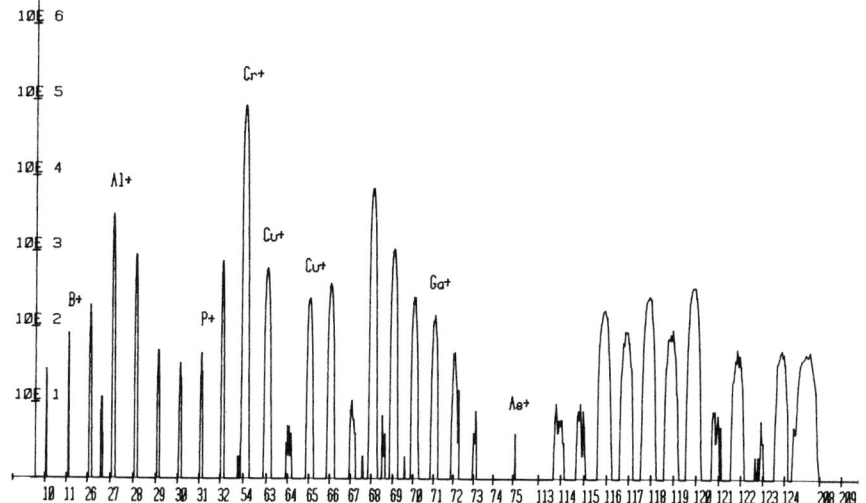

Fig. 8: Multi element surface trace analysis: Mass spectrum of Cr-layer of 50 nm thickness on Cu (PI = O_2^+, E_0 = 5.5 keV, d_A = 150 µA, raster 500 x 500 µm, i_B = 100 nA, energy filtering 50 eV)

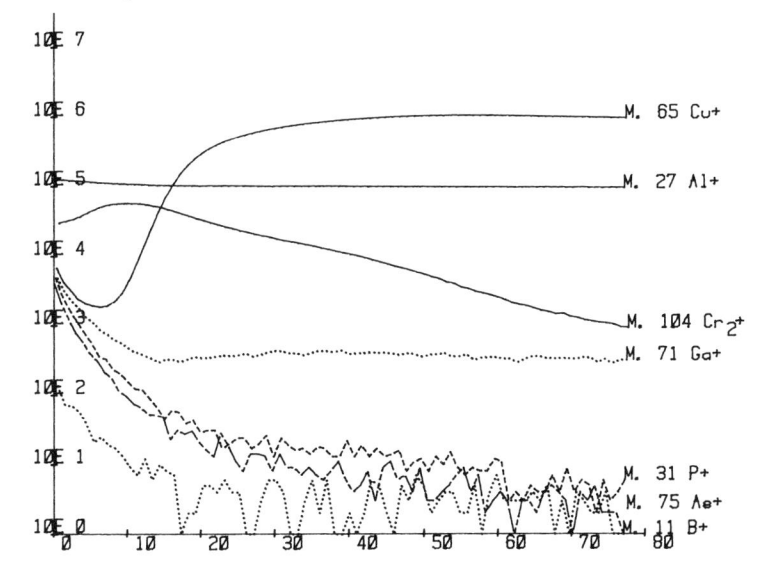

Fig. 9: Depth profiles of Cr, Cu and B over metallization structure: 50 nm Cr-layer on Cu (PI = O_2^+, E_0 = 5.5 keV, d_A = 150 µA, raster 500 x 500 µm, i_B = 100 nA, energy filtering 50 eV)

Table 3: Multi Element Surface Trace Analysis: SIMS-Results and
Detection Limits for 50 nm Cr-Layer on Cu, Depth Profile Fig.9

Element	Concentration [µg/g]	Detection limit [µg/g]
Cu	369	0.15
B	12	0.1
Al	228	0.005
P	103	1
Ga	16	0.02
As	19	0.15

Distribution information is obtained by depth profiling. Fig. 9 shows that the most important III-V-elements are enriched at the surface of the Cr-layer.

4. COMPREHENSIVE EVALUATION

SIMS based on the most modern instrumental techniques can be used successfully for the following multi element trace analytical problems:

i) Ultra Trace Bulk Analysis: practical detection limits in the low ng/g range can be obtained for many elements. Analytical accuracy depends on the availability of standards and is in the range from ca. 25 % rel. to a factor of 5 (when physical calculation procedures are applied). The multielement potential of SIMS is advantageous for many problems: SIMS supplements more routinely used techniques like OES, ETAAS, NAA, SSMS.

ii) Multi Element Trace Analysis of Microdomains: at a lateral resolution of ca. 10 µm detection limits are typically a few µg/g for most elements. SIMS is the major technique available for this purpose and is supplemented by EPMA, LAMMS and PIXES.

iii) Micro Distribution Analysis:
    Qualitative distribution information can be obtained by ion microscopy with a lateral resolution of 0.5 µm. Elemental distribution analysis of trace elements with a lateral resolution of ca. 2 µm is possible for elements with a very high ionization yield. Elements having a lower ionization yield can often be analyzed by measurement of high intensity molecular ions formed by chemical surface reaction during analysis. In such cases

detection limits in the low µg/g-range can be obtained for high spatial resolution.

SIMS is one of the major techniques for micro distribution analysis of trace elements and is supplemented by LAMMS, PIXES, AES and EPMA.

iv) Surface Distribution Analysis:
Qualitative and quantitative distribution of trace elements can be obtained with a depth resolution of some nm (depending on structure and morphology of the surface). Detection limits are in the ng/g to the µg/g-range. SIMS is the major technique for surface distribution analysis of trace elements and is often combined with AES, XPS (for distribution studies of major elements) and other sensitive techniques like RBS, CPAA, NAA, PIXES.

ACKNOWLEDGEMENT

Support of this work by the Austrian Scientific Research Council (Project No. 3603 and 4508) and the University Jubilee Fund of the City of Vienna is gratefully acknowledged. The author thanks his coworkers, Dr. G. Stingeder, Dipl.-Ing. P. Wilhartitz, then Prof. M. Schroll (Geotechnical Institute, Vienna), Prof. Chabicovsky (Technical University Vienna), Doz. H. Ortner (Metallwerk Plansee) and Dr. I. Cerny (Bleiberger Bergwerks Union) for their cooperation.

REFERENCES

[1] Grasserbauer M., Zacherl M.K. (Eds.): Progress in Materials Analysis, Vol. 1, Mikrochim. Acta, Suppl. 10, Springer Wien 1983

[2] Schroll E.: Analytische Geochemie, Bd. 1 und 2, Enke, Stuttgart 1975 und 1976

[3] Grasserbauer M., Stingeder G., Pimminger M.: Z.Analyt.Chemie 315, 575 (1983)

[4] Heinrich K.F.J., Newbury D.E.: NBS SP 427

[5] Andersen C.A., Hinthorne J.R.: Analyt.Chemistry 45, 1421 (1973)

[6] Dearnaley G.: IAEA SMR 15/44, p. 167

[7] Werner H.M., Morgan A.E.: J.Appl.Phys. $\underline{47}$, 1232 (1976)

[8] Pimminger M.: Thesis, Technical University Vienna, 1983

[9] Miller G.K.: Tantalum and Niobium, Metallurgy of Rarer Metals - 6, Butterworth. 1959

[10] Ortner H.M.: Techniques for Bulk Chemical Analysis, Part 1 - Trace Analysis of Refractory Metals, in: Analysis of High Temperature Materials (O. van der Biest, ed.), London: Appl.Science Publ. 1983

[11] Grasserbauer M., Ortner H.M., Wilhartitz P., Pimminger M., Leuprecht R.: Z.Analyt.Chemie $\underline{317}$, 539 (1984)

[12] Pimminger M., Grasserbauer M., Schroll E., Cerny I.: Z.Analyt. Chem. $\underline{316}$, 293 (1983)

[13] Pimminger M., Grasserbauer M., Schroll E., Cerny I.: Tschermak's Min.Petr.Mitt., in press

[14] Morrison G.H., Slozdian G.: Analyt.Chemistry $\underline{47}$, 932 (1975)

[15] Paesold D.: Thesis, Technical University Vienna, 1982

[16] Zinner E.: J.Electrochem.Soc. $\underline{130}$, 199 C (1983)

[17] Stingeder G., Grasserbauer M., Guerrero E., Pötzl H., Tielert R.: Z.Analyt.Chemie $\underline{314}$, 304 (1983)

[18] Pimminger M., Grasserbauer M.: Mikrochim.Acta Suppl. 10, 61 (1983)

THE POTENTIAL OF ICP SOURCE MASS SPECTROMETRY

Alan L. Gray

Department of Chemistry, University of Surrey,
Guildford, Surrey, GU2 5XH, England

SUMMARY

The concept of sampling analyte ions from a plasma and introducing them to a mass spectrometer, instead of using it as an emission source, is a little over 12 years old, and has now reached the stage of practical application, two commercial instruments recently becoming available.

Over 90% of the elements of the periodic table are accessible to the method and detection limits for most elements lie between 0.01 and 1 ng. ml^{-1} in aqueous solution. Dynamic range extends from below 1 to 10^7 ng. ml^{-1} with good matrix tolerance in solutions of up to 1% solids content. Since samples remain at atmospheric pressure and analysis times of 1 minute are usual, sample throughput is high, typically a fresh sample every two minutes.

The simplicity of the predominantly singly ionized spectra make interpretation straightforward and the response from analyte oxide and molecular peaks is very small. Good resolution and abundance sensitivity is available from the large quadrupole mass analyzers used and rapid isotope ratio measurements may be made at concentrations of about 1 µg.ml^{-1} with precisions of 0.5% or better.

The origin and development of the technique are described and the current performance available illustrated with examples from the authors own work on the original research instrument.

1. INTRODUCTION

In the continual search for lower detection limits in elemental analysis the analyst has repeatedly been forced to contemplate more complex, costly and time consuming techniques to achieve his aim. Major steps forward were taken with the successive introduction of AAS, AFS and ICP-AES which generally offer considerable operational convenience, and in the last case multi-element capability with relatively few matrix problems. However the continued further development of the material sciences, in both the raw material and refined product stages, the increasingly sophisticated requirement for elemental analysis in geological exploration and the growing understanding of the role of trace elements in the life sciences have identified many areas where concentrations below the reach of these widely used methods are now important.

Hitherto at ppb levels and below it has been necessary to use either neutron activation analysis or solids mass spectrometry. Both these techniques offer very low detection limits but they are generally complex, require skilled interpretation and are usually capable of only a low rate of sample throughput, at most a few samples per day. These characteristics impose their own restraints on the planning of an analytical programme and in many cases preclude the use of sample suites large enough to provide good statistical or survey data.

At the time when the exciting new technique of ICP-AES was approaching commercial exploitation it seemed appropriate to consider what new methods might be developed in the future to yield even lower detection limits without sacrificing convenience. It was evident that mass spectrometry, already such a powerful technique in organic chemistry, might offer an attractive solution. The limitations of the method for inorganic analysis had however always lain in the difficulties of sample introduction, dissociation and ionization, since the samples usually consisted of strongly bound, often refractory, inorganic molecules in solid or solution form. It was realised that the atmospheric pressure electrical plasmas then being extensively investigated for solution analysis contained many analyte ions and possessed properties that could make them very attractive ion sources if only the ions could be extracted into a mass spectrometer.

The first steps were therefore taken towards the realisation of a new hybrid technique, ICP Source Mass Spectrometry which now, after an interval of 12 years is attracting increasing attention by its unique combination of the convenience and speed of the ICP source with the flexibility and sensitivity of mass spectrometry.

2. DEVELOPMENT HISTORY

The foundations of ICPSMS were laid over 20 years ago by T.M. Sugden and his group at Cambridge who were then developing methods for the diagnosis of combustion processes in atmospheric pressure flames by mass spectrometry. By 1964 Sugden was able to describe, in a contribution to Mass Spectrometry, edited by Reed (1), most of the essential features of the equipment used to extract ions from a plasma. He included the use of large apertures to induce continuum flow, mass analysis by a quadrupole analyzer and the nebulization of salt solutions in the ppb-ppm concentration range into the flame, producing ions in equilibria described by the Saha equation. The only step Sugden omitted was the concept of using the method for the analysis of solutions rather than for the flame processes.

This suggestion came later from Alkemade in his Redwood lecture on flame spectroscopy (2) but by that time the feasibility of the concept was already being explored by the author and his colleagues (3), using an electrical plasma coupled to a diagnostic system similar to that described by Sugden. At that time valuable assistance was given by Sugden's co-workers A.N. Hayhurst and P.F. Knewstubb. It initially seemed possible (4) that any of the plasmas being examined as emission sources could be used and the work was started with a small DC plasma which was available in the laboratory.

The early work demonstrated the feasibility of the technique and that strong ion signals could be obtained with low background levels so that detection limits in aqueous solution in the $ng.ml^{-1}$ region were possible. Very simple spectra were obtained from the quadrupole mass analyzer. Publication of this work (5) attracted attention in two other centres that were to start parallel programmes and make important contributions.

Although encouraging, these results also revealed the limitations of the plasma chosen (6). Gas temperature, about 3,500 K, was too low to provide adequate sample volatilization and dissociation, and the ionization temperature was also too low to provide adequate ionization of elements with ionization energies above 8 ev. However the promise of the technique was apparent if a more suitable plasma could be used. Of the alternatives the MIP was rejected because of its low gas temperature and the ICP was considered the most suitable. Support for further development was received in the U.K. from the Institute of Geological Sciences. Both the other interested centres also started work soon after the feasibility was demonstrated, in Toronto with an MIP and at the Ames Laboratory in the U.S.A. with an ICP.

The author was invited to join the work at Ames for short periods in 1978 and 1979 and following the exploratory work on the interface to an ICP by Houk and his colleagues analyte ions were first extracted from an ICP in 1979 (7). Development then proceeded rapidly on slightly different lines in all three centres. The ICP at that time was only used in the U.K. and U,S.A., but problems were experienced at both places in using large enough apertures to achieve continuum flow from the higher plasma gas temperature of 7000 - 8000 K. Quite apart from the greater pumping capacity needed to handle the gas inflow from the apertures above 0.2 mm diameter, an intende discharge was found to occur in the aperture (8). The early work with an ICP was thus limited to apertures too small to prevent the formation of a boundary layer. Passage of the extracted ions through this layer caused delay in the lower temperature region during which molecular recombination occured and solids recondensed so that the anticipated advantages of the ICP were not realised.

During this time important advances had been made in atmospheric pressure sampling and molecular beam formation in which the Toronto group had considerable experience (9). They applied this to their work with the MIP (10) taking the important step of using an intermediate high pressure stage at about 1 torr between a large sampling aperture of 0.4 mm diameter and the later vacuum stages, thus greatly reducing the pumping requirement. This technique was applied with the ICP in the U.K. soon after this and it was found that the pinch discharge was largely quenched at this pressure and the performance expected from an ICP achieved (11). From then on progress was rapid. The Toronto group, Sciex, also changed to an ICP and became the first manufacturer to exhibit a commercial instrument.

3. SYSTEM OUTLINE

The following discussion of system characteristics and performance is based entirely on the author's own work at the University of Surrey. The commercial instruments are similar in principle and performance but considerably more sophisticated in control philosophy.

The system is described in detail elsewhere (12). The plasma source is a standard type used for AES with the torch remounted horizontally. A conventional pneumatic nebulizer without desolvation is used for solution samples. Other types of sample introduction developed for ICP-AES are equally applicable.

A gas sample is extracted from the centre of the plasma tail flame, usually at a point between 5 and 10 mm from the load coil, by the primary aperture, of between 0.5 and 0.8 mm diameter, drilled in a shallow water cooled cone. The core of the jet of gas formed by the supersonic expansion is extracted by a sharp skimmer cone about 10 mm behind the aperture into the high vacuum stages, the bulk of the entering gas being removed by a mechanical pump. No potentials are used in the first stage but an ion lens in the second and third stages focusses the ions into the mass analyser, a large quadrupole, the VG Type 12-12S.

Although the plasma flame plays continuously on the cool, conducting sampling cone, plasma operation is quite uncritical. A wide range of power levels from 1-2.5 kW and associated gas flows may be chosen by the operator to suit particular samples. For aqueous solutions operation at 1.2 kW and coolant gas flow of 12 l/min is usual but for organic solvents higher power and flow are needed. Auxiliary and injector flow may be varied over a wide range although the former is not usually used after start up.

Ions are detected by a channel electron multiplier, operated either as a pulse counter or a mean current detector. Single ion monitoring or mass scanning may be employed synchronised with a digital memory scan for the storage of signal response. When scanning the channel address corresponds to mass number and a spectrum may thus be produced on a V.D.U. In normal use the selected mass range is scanned once per second, a total of 60 scans being stored in a 1 minute run. For short duration signals, such as those from a thermal evaporation device or from flow injection of the sample, scan times of as little as 30 ms may be used. For good precision in isotope ratio measurement scan times of 0.1 sec are usually used.

4. PERFORMANCE CHARACTERISTICS

Most species are almost fully ionized in the plasma, the degree of ionization falling to about 50% at ionization energies of 10 ev and to 1% by about 14 ev. Fully ionized mono isotopic elements give signals of between 2 and 5 x 10^5 counts/sec, per $\mu g.ml^{-1}$, the precise signal depending on setting up conditions. The response appears about 10 seconds after the solution is introduced to the nebulizer and decays to about 10^{-3} of the full value in a few minutes after its removal. Random background levels are low, typically 20 counts/sec or less and detection limits for most elements below 1 $ng.ml^{-1}$ are thus obtained. Values obtained in the single ion mode for a selection of elements are shown in Table I.

TABLE I

SOLUTION DETECTION LIMITS - $ng.ml^{-1}$ (2σ)

Li	B	Be	Co	Zn	Ge	As	Se	Ag	Cd
0.06	0.2	0.02	0.03	0.1	0.02	0.02	0.5	0.02	0.04

Sn	Te	La	Ce	W	Au	Hg	Pb	Th	U
0.04	0.06	0.03	0.04	0.07	0.04	0.01	0.03	0.02	0.02

Measured on 10 sec. integrations in single ion mode. σ determined on minimum of 10 integrations of 10 sec. each on blank.

Only four elements are inaccessible to the technique as positive ions, He, F, Ne and Ar. Major background peaks arise from the plasma gas argon and any impurities in it. When aqueous or organic solutions are introduced by nebulizer major peaks from the elements H, C, O and acid components such as N or Cl also occur. These and their minor isotopes can cause some interference with analyte peaks, principally at m/z 28 by N_2(Si), 32 by O_2(S), 40 by Ar(Ca) and 80 by Ar_2(Se). Above m/z 80 the spectrum is normally free of background peaks although some supplies of argon have been found to contain traces of krypton and xenon.

The spectrum is normally very simple, as may be seen in Fig. 1. This shows the response to a multi-element calibration sample containing 26 elements, each at 1 $\mu g.ml^{-1}$, in 4% HCl solution. The background peaks from water and plasma gas have been removed by spectrum stripping the response to a 4% HCl blank solution in the data system. All 26 elements may be seen although Ca is only visible as the 4% abundant ^{44}Ca isotope. The main isotope coincides with ^{40}Ar which saturates the counting system so no additional response from Ca remains after stripping.

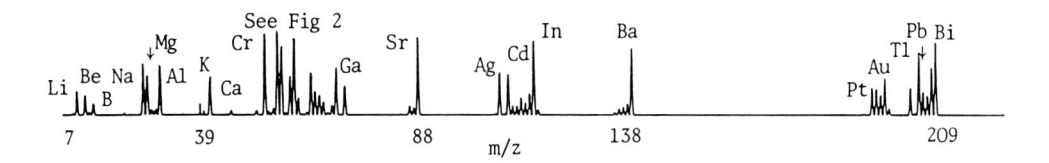

Fig. 1. Response to 1 µg.ml^{-1} of 26 elements in 4% HCl, from Li to Bi (^{m}/z 7-209). Peak rate ^{88}Sr 3.3 x 10^5 counts/sec. See also Fig.2.

Only one significant peak is obtained for each isotope and each is well isolated. This may be appreciated more easily from Fig. 2 where part of this spectrum around the transition elements is expanded. Selection and area integration of the desired peak is very simple.

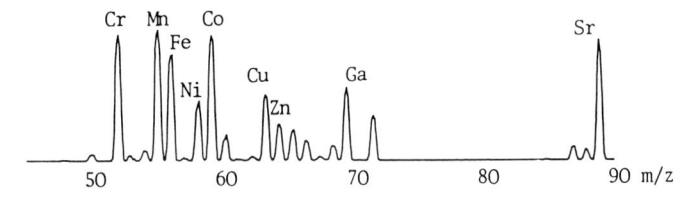

Fig.2. Expanded portion of Fig.1 showing transition elements.

Peaks from doubly charged ions or from oxide and hydroxides are very small. Fig.3 shows a spectrum of two refractory metals, Th and U, and Ba, the element with the lowest second ionization energy. Oxide and doubly ionized peaks are about 1% of the main peak.

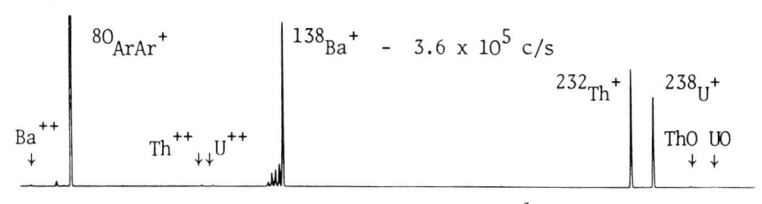

Fig.3. Spectrum of Ba, Th and U at 1 µg.ml^{-1} in 1% HNO$_3$.

It has been found that the proportion of Ba^{++} ions can be varied by minor changes in plasma parameters from about 0.5% to 90% of the total Ba ions. It is usually preferable to avoid doubly charged ions and this has been done for these illustrations.

Similar plasma conditions are used in ICP emission and mass spectrometry and similar freedom from volatilization interference is found. There is no equivalent however in mass spectrometry to atom lines in emission spectrometry which may be used when the ionization equilibria are disturbed by ionization suppressants in the sample. Significant suppression occurs above about 1000 $\mu g.ml^{-1}$ of Na, about 40% being found at 4000 $\mu g.ml^{-1}$ Na. In these conditions the use of standard additions for calibration is preferable to matrix matching.

Solids condensation around the aperture was a problem in early work, but it is found that with the plasma conditions now used and aperture of 0.75 mm diameter, solutions with solids contents of 1% or more may be analysed. Fig.4 (a) and (b) show spectra of the major constituents in 1% Na solution and the seawater standard NASS-1, both run without dilution. NASS-1 contains about 1.4% Na and preliminary results suggest that trace elements down to about 1 $ng.ml^{-1}$ can be detected. The major peaks are non linear here due to counting system saturation but Na, Mg, S, Cl, K and Ca can all be identified.

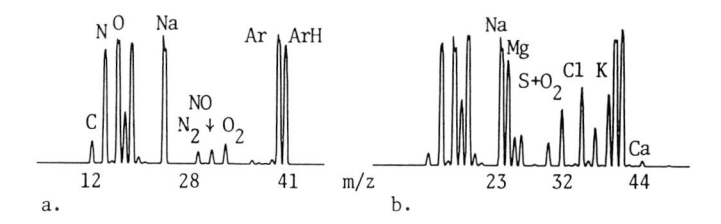

Fig.4. Spectra of (a) 1% Na in 1% HNO_3 (b) Sea water standard NASS-1. Both run undiluted. Peak level 9 x 10^6 counts/sec.

The useful response may be extended to at least 1% if integrated current is measured instead of counting pulses.

A typical high solids matrix is that of blood. A group of blood samples were ashed in HNO_3/H_2O_2 and then dissolved in 5% HNO_3 to an equivalent dilution of 1:1 at which the sodium concentration was approximately 1600 $\mu g.ml^{-1}$. The calibration curve for Cd obtained by single ion monitoring of $^{114}Cd^+$ with 25 second integrations at each addition is shown in Fig.5 after background subtraction. Determination down to at least 0.1 $ng.ml^{-1}$ is possible. The level in the original diluted blood sample was found to be 0.67 $ng.ml^{-1}$.

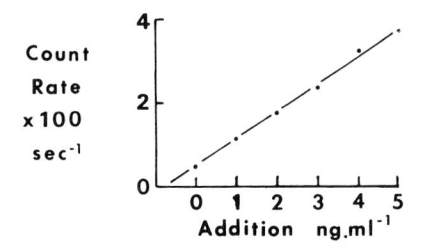

Fig.5. Calibration plot for Cd in blood at 1:1 dilution in 5% HNO_3.

An attractive feature of the technique is the ability to determine isotope ratios to moderate precision with reasonably short integration times. Table 2 shows results obtained on a natural platinum solution at 1.5 $\mu g.ml^{-1}$. Twelve integrations each of 2 minutes were accumulated, each consisting of 1200 scans of 0.1 sec each.

TABLE 2

Isotope Ratios of Natural Platinum

Ratio	$\frac{192}{195}$	$\frac{194}{195}$	$\frac{196}{195}$	$\frac{198}{195}$
True value	0.2308×10^{-2}	0.9734	0.7485	0.2133
Measured value	0.2300×10^{-2}	0.9695	0.7457	0.2155
Error %	-0.35	-0.40	-0.37	+1.03
RSD %	1.3	0.34	0.39	0.51
Counting RSD %	0.56	0.12	0.13	0.20

5. CONCLUSIONS

True applications work with this technique is only as yet being done by Date at The British Geological Survey[13] with an instrument based on that at Surrey. However as commercial deliveries commence information should rapidly accumulate on advantages and disadvantages. This necessarily brief survey of the history of the method and its overall characteristics must leave many questions unanswered. It is hoped however that it provides some impression of the power of the technique to extend multielement analysis by laboratory instrument to detection limits one to two orders of magnitude lower than hitherto available on a rapid, routine basis.

6. ACKNOWLEDGEMENTS

The development of ICPSMS has been supported by The British Geological Survey and the R & D Programme on Uranium Exploration of The European Communities. This paper is published by permission of The Director, British Geological Survey (NERC).

REFERENCES

1. Sugden, T.M., Mass Spectrometry, ed. R.I. Reed, 347-358, (1965), Academic Press.

2. Alkemade, C.Th.J., Proc. Soc. Anal. Chem., 10, (1973), 130-143.

3. Gray, A.L., Proc. Soc. Anal. Chem., 11, (1974), 182-3.

4. Brit. Patent, 1,371,104, (1971).

5. Gray, A.L., Analyst, 100, (1975), 289-299.

6. Gray, A.L. Dynamic Mass Spectrometry, 4, (1976), Heyden, 153-162.

7. Houk, R.S., Fassel, V.A., Flesch, G.D., Svec, H.J., Gray, A.L., and Taylor, C.E., Anal. Chem., 52, (1980), 2283-2289.

8. Houk, R.S., Fassel, V.A., and Svec, H.J., Dynamic Mass Spectrometry, 6, (1981), Heyden, 234-251.

9. Stearns, C.A., Kohl, F.J., Fryburg, G.C., and Miller, R.A., Nat. Bur. Stand. Spec. Publ., 561, 1, (1979), 303-355.

10. Douglas, D.J., and French, J.B., Anal. Chem., 53, (1981), 37-41.

11. Date, A.R., and Gray, A.L. Spectrochimica Acta., 38B, (1983), 29-37.

12. Gray, A.L., and Date, A.R., Analyst, 108, (1983), 1033-1050.

13. Date, A.R., and Gray, A.L., 1984 Winter Conference on Plasma Spectrochemistry, San Diego. Submitted to Spectrochimica Acta.

ADVANCES IN X-RAY FLUORESCENCE ANALYSIS OF SOFT AND ULTRASOFT X-RAYS

Dr. Tomoya Arai

Rigaku Industrial Corporation, Osaka, Japan

The spectroscopic analysis of heavy and medium elements by fluorescent X-rays has become a powerful instrument in many laboratories.

Further more instrumental improvements in the analysis of low-atomic number elements have been made so that the method can be applied more widely and new applications have been studied.

In the electron and proton exitation techniques for light element detection, the measured intensity of characteristic X-rays is fairly high, but it is strongly influenced by the surface condition of the sample owing to the low penetration of irradiating electrons and protons.

The effects of chemical combination on soft and ultrasoft X-ray emission spectra with long wavelength have been investigated by many scientists using high resolution mass spectrometer.

Since the analyzing depth of the X-ray exitation method is much greater than that of electron and proton excitation techniques, the X-ray excitation method is more suitable for the analysis of the concentrations of low atomic number elements.

The fluorescent intensity of soft and ultrasoft X-rays emitted from low atomic number elements has generally been somewhat low, owing to their low excitation efficiency, small fluorescence yield and low penetration into materials.

Using a conventional sealed-off-X-ray tube and a gas flow-proportional counter, a total reflection mirror was used as a high intensity monochromator, instead of the Bragg reflection from a single crystal, soap multi-layer or gratings which greatly reduces the reflecting X-ray intensity.

For the total reflection of soft and ultrasoft X-rays, the reflected angle becomes larger than that of short wavelength X-rays and the intensity is very high.

The purpose of this presentation is to discuss the analytical applications of carbon and boron analysis by the method of a total reflection monochromatization and oxygen analysis using a regular X-ray fluorescent spectrometer.

In Experimentation

Rigaku S/Max, equipped with a Rh-target and end-window X-Ray tube with a thin beryllium window was employed for this study.

The fluorescent X-rays of Oxygen was monochromatized by a flat TAP analyzing crystal under vacuum conditions.

A flat multilayer soap crystal and a total reflection mirror were employed. The monochromatized fluorescent X-rays were detected by a gas flow proportional counter equipped with a specially prepared thin polyester film window covered with a thin layer of aluminium metal.

This polyester film is effective for the detection of Oxygen and Nitrogen because of its low absorption characteristics.

Figure no. 1 shows the relationship between the K, L, M and N series of elements in the soft X-ray region. When $O-K\alpha$ and $N-K\alpha$ X-rays are detected, the overlapping interferences of the weaker L, M and N series had to be investigated before measurement.

Also, higher order X-rays of short wavelength radiation interfere to measuring X-rays. 2nd order X-rays of $NaK\alpha$ and 7th order X-rays of $CaK\alpha$ interfere with $O-K\alpha$. 2nd order X-rays of $Co-L\alpha$ interfere with $N-K\alpha$.

In the case of total reflection monochromatization for carbon detection, M series X-rays of molybdenum and niobium and N series X-rays of tungsten and tantalum give rise to overlapping deviations.

For the boron detections, L series X-rays of chlorine and sulfur, M series X-rays of zirconium N series of rare earth elements give rise to plus deviations.

The absorption relationship of $O-K\alpha$, $C-K\alpha$, and $B-K\alpha$ X-rays are shown in figure no. 2.

Oxygen compounds containing C, Ca, Ti, V, Zr or Mo, exhibit decreased intensities because of strong absorption of $O\ K\alpha$ X-rays.

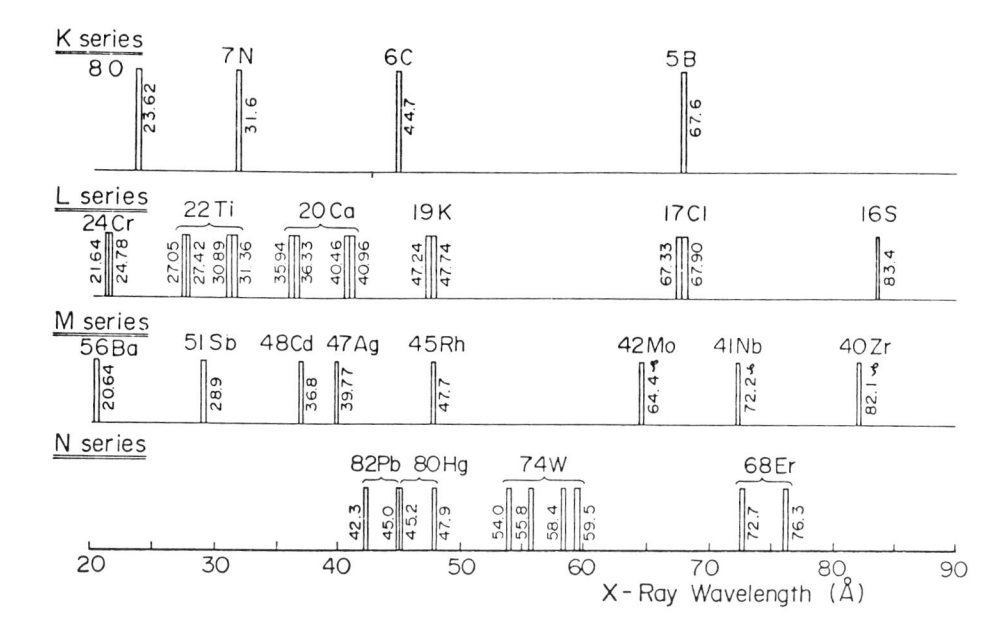

Fig. no. 1: Emission Spectra in Soft and Ultrasoft X-rays

Oxygen compounds with B, Mg, Al, Si, Cr, Mn, Fe or Ba, exhibit relatively high intensities because of low absorption of O-Kα X-rays.

As N-Kα X-rays are located at high absorption region of Carbon, Calcium, Potassium and Palladium, lower intensity of N-Kα X-rays can be found.

In the comparison of monochromators for X-ray Fluorescent spectrometry, reflecting intensity, resolution to interfering X-rays and special problems which are large thermal expansion, the emission of fluorescent X-rays from monochromator, etc., have been investigated.

Figure no. 3 shows the comparison of measurable X-rays with some remarks.

TAP crystal has to be used for the shorter wave-length X-rays of O-Kα because of high resolution and relatively high intensity. It is important that measured intensity of B-Kα, C-Kα and N-Kα X-ray is very high when using a total reflection mirror.

Recently, a sythetic multilayer which is a stacking layer composed of amorphous silicon or carbon layer and thin heavy metal film like tungsten has been studied for soft X-ray monochromatization, its reflec-

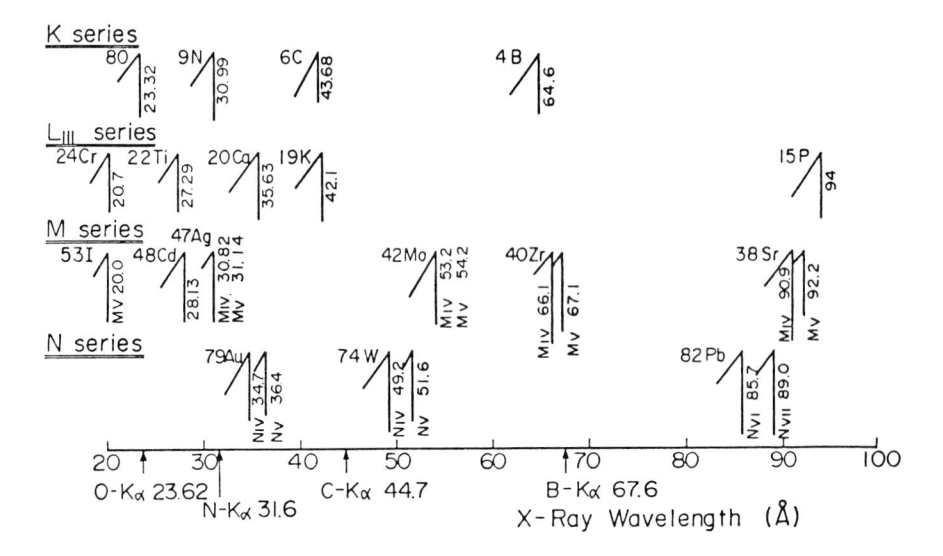

Fig. no. 2: Adsorption Edges in Soft and Ultrasoft X-rays

tivity is relatively high and the resolution is also better.

When comparing the analytical results using a total reflection mirror and this synthetic multilayer for the analysis of carbon in steel, the analytical result of the total reflection mirror is superior to that of the synthetic multilayer.

X-Rays	C-Kα 44.7	N-Kα 31.6	O-Kα 23.62	F-Kα 18.32	Na-Kα 11.91	Note
Emitter	Graphite	Si_3N_4	SiO_2	NaF	NaCl	
Analyzer						
Glass mirror	7731 c/s at 8.0°	330 c/s at 7.0°	610 c/s at 6.0°	4855 c/s at 5.0°	2403 c/s at 4.0°	Width of X-ray beam / Mirror material
TAP 2d=25.763Å	—	—	1552 at 132.7°	9830 at 90.6°	79259 at 54.9°	Reflecting surface
LSD 2d=100.7Å	947 at 52.6°	24.1 at 34.0°	294 at 31.0°	3074 at 20.9°	12026 at 13.5°	Number of layers / Life

S/MAX X-Ray conditions : OEG-75 Rh-Target 40KV 70mA
       Soller slit : 0.52°

Fig. no. 3: Comparison of Reflected Peak Intensity of Various Monochromators

Analysis of soft and ultrasoft X-rays

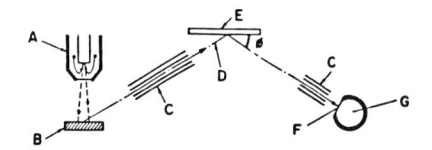

Components	Factors
A X-Ray Tube	i) Primary X-rays — white and characteristic ii) X-Ray take off angle — 90° iii) Thin Be window iv) Short distance between target and sample

B Analyzing Sample	i) Low fluorescent yield
	ii) Surface analysis
	  • Low excitation efficiency by short wave length X-rays
	  • Large absorption for measuring X-rays

C Collimator	i) Soller slits
D X-Ray Path	i) Absorption of residual air
E Reflecting Mirror	i) Mirror material and its surface condition ii) Total reflecting angle (ϕ)
F Window of Detector and Filter	i) Filtering monochromatization
G Detector	i) Gas flow proportional counter

Fig. no. 4: Monochromatization and Related Factors for Soft X-rays

Figure no. 4 shows all of the related factors that contribute to the measurement of soft X-rays for the demonstration of the principle of total reflection monochromatization.

It was decided in principle, to use a regular sealed-off X-ray tube for convenience in routine laboratory applications. In order to compensate for the intensity reduction which arises from the low excitation efficiency, a high efficiency monochromator was adopted which consisted of a total reflection mirror and an appropriate filter. For the measurement of C-Kα X-rays the combination of an LiF mirror which has high reflectivity for C-Kα X-rays and low reflectivity for interfacing X-rays of O-Kα and Fe-Lα X-rays and a polypropylene film of one or five microns is used. In the case of the detection of B-Kα X-rays the combination of a molybdenum-mirror and boron filter on the thin polypropylene film was adopted.

Since Carbon determination in steel and iron has been made in production control laboratories, practical procedures in sampling, X-ray mea-

1) Routine check of instrument
   - PHA condition
   - X-ray intensity/background intensity
   - X-ray intensity check from X-ray intensity of standard material

2) Measurement (Standard operation)
   - Measurement time -------40—80 sec
   - Repeat   --------2 times

3) Sample surface
   - Fresh surface
   - No wiping with cloth wetted by organic solvent
   - Carbon Contamination during x-ray measurement
       ----- negligible small

4) Standard sample
   - Primary standard sample : JSS, NBS, SAB, BAM
   - Secondary standard sample : Self prepared samples with homogeneity test and closs check on the basis of chemical analysis
   - Unknown sample to be analyzed : Casting procedure should be as same as secondary standard sample

Fig. no. 5: Operation of "Simultix" for Carbon Analysis

suring and data calculating process have been investigated in order to obtain better analytical results.

Figure no. 5 shows the summarization of practical key points at each stage.

Figure no. 6 shows the relation between chemical content and measured X-ray intensity of carbon in cast iron.

In this case, measured X-ray intensity is strongly influenced by sampling conditions, a rapid quenching method must be applied in order to make a solid sample homogeneous. The analyzing surface was prepared on a grinding wheel which had a 80 Mesh Al_2O_3 grit for finishing.

The matrix effects influence the measured X-ray intensity which consist of the absorption of silicon for C-Kα X-rays and the overlapping deviation of N series X-rays of Molybdenum.

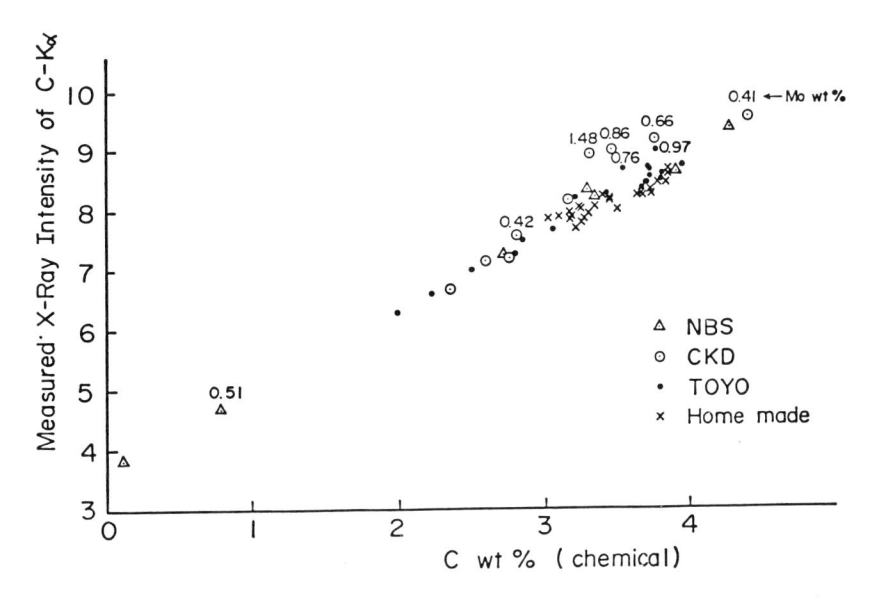

Fig. no. 6: Relation between Measured X-ray Intensity of Carbon and Carbon Content in Cast-Iron

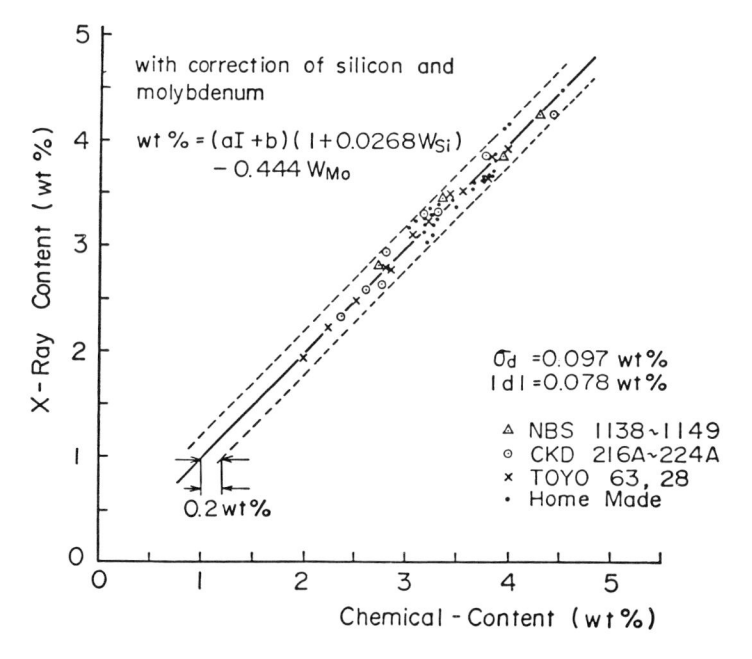

Fig. no. 7: Carbon Determination in Cast-Iron by X-ray Analysis

For the carbon determination, data processing equation was derived which is shown in Figure no. 7. The analytical accuracy is about 0.1 wt %.

In order to exhibit the analytical meaning of carbon determination in castiron which should be a good homogeneous sample, the relationship among X-ray precision, grinding errors and analytical accuracies of many different elements is shown in Figure no. 8.

Figure no. 9 shows the relationship between the chemical analysis of carbon in steel and measured X-ray content which was observed at the process control of a certain castiron manufacturer.

A great difference between chemically determined and X-ray content of carbon was found when there were pinholes and cracks on the analyzing surface of a cast sample.

The analytical accuracy was about 0.01 wt % at the concentration of 0.1 to 0.5 wt % carbon.

In the direct analysis of raw coal powders, which are, of course, typical natural resources, complex analytical problems will be described which are a result of mixture of small grained constituents consisting of organic and mineral matter.

Figure no. 10 shows the experimental results of the measured X-ray intensity of C-Kα X-rays against carbon content which is determined by

Analyzing elements	C	Si	Mn	P	S	Cu	Ni	Cr	Mo	Ti	Mg
Content range	1.9 ~4.3	0.1 ~3.5	0.09 ~1.8	0.01 ~1.0	0.007 ~0.2	0.01 ~1.2	0.02 ~2.5	0.02 ~24	0.005 ~1.5	0.005 ~0.7	0.01 ~0.2
Content	3.8	2.4	0.3	0.045	0.016	0.46	0.026	0.55	0.002	0.035	0.04
X-ray precision	0.011	0.003	0.0005	0.0002	0.0001	0.0006	0.0004	0.0006	0.0009	0.0001	0.0007
Grinding errors	0.024	0.010	0.0006	0.0003	0.0003	0.0006	0.0005	0.0010	0.0001	0.0002	0.0008
Analytical accuracies	0.100	0.083	0.0220	0.0140	0.0050	0.0210	0.0150	0.0290	0.0120	0.0170	0.0060

(wt %)

Simultix, 50 KV 50mA 40 sec F.T.

Fig. no. 8: Relation among X-ray Precisions, Grinding Errors and Analytical Accuracies in the Analysis of Cast Irons

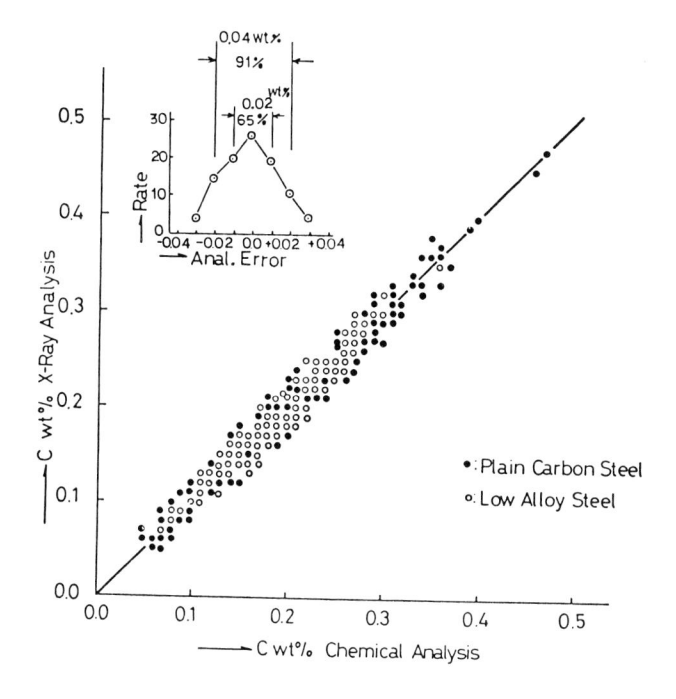

Fig. no. 9: Relationship between X-ray Analyzed and Chemically Determined Concentration of Carbon Steel in case of Routine Analysis for Production Control

standard combustion method. The black points indicate pure material of anthracene, methacrylite, cellulose and sugar.

Using the least-squares method, the experimental equation for the carbon determination in raw coal was derived and its analytical accuracy was about 1.5 wt %.

As is shown in Figure no. 11, it was found that there is an adsorption effect on sulfur in organic matter for C-Kα X-ray.

Since the boron content can have a marked influence on a number of the material properties, the quantitative analysis of this component is quite important for the production process control. Owing to the volatility of boron oxide, the rapid analysis of melted glass (see Figure no. 13) during the melting process is very important instead of regular chemical analysis.

It is expected that this analytical technique will be useful in glass industry.

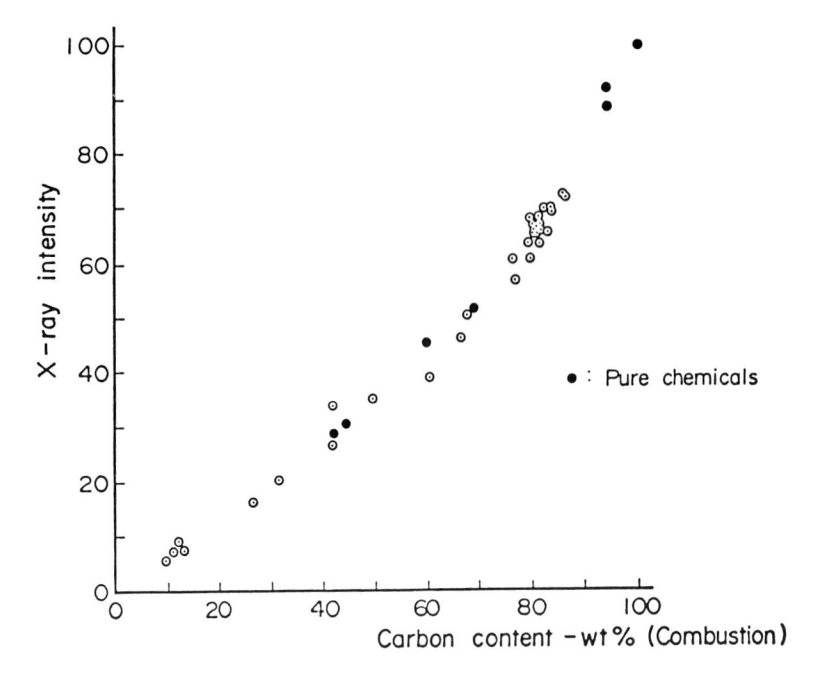

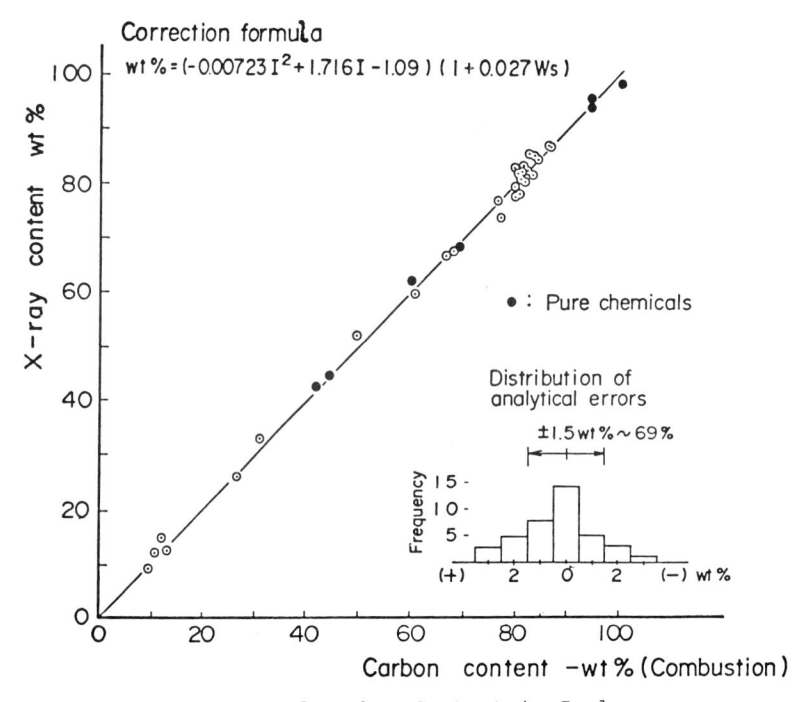

Fig. no. 10: Relation between Measured Intensity of C-Kα and Carbon Content in Coal

Fig. no. 11: Determination of Carbon Content in Coal

Correction formula
$$wt\% = (-0.00723 I^2 + 1.716 I - 1.09)(1 + 0.027 W_s)$$

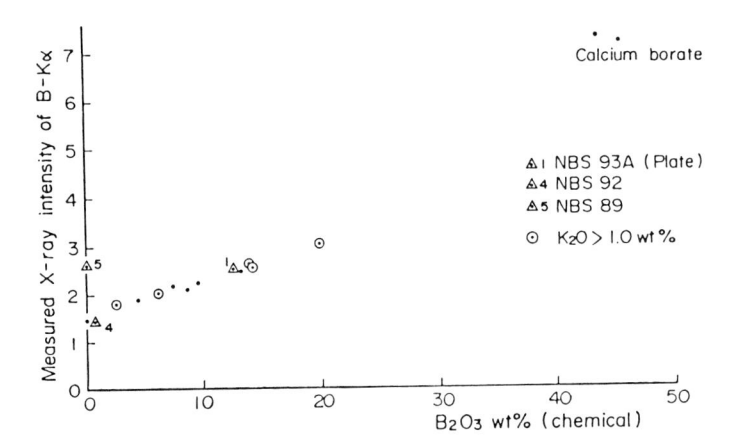

Fig. no. 12: Measured X-ray Intensity of B-Kα against B_2O_3 Content in Glass Powder Samples

This is the relation of measured X-ray intensity of B-Kα versus boron oxide content as determined chemically, using the combination of a molybdenum mirror and one micron polypropylene film covered with boron filter.

These points are the results for calcium borate which is the raw material for boron silicate glass.

The increase of measured X-ray intensity of NBS 89 is found which is arising from the overlapping interference of N series of lead.

These are the analytical results for boron oxide in glass: for boron content between 1 and 20 percent, the analytical accuracy varies between 0.5 and 1 percent.

Figure no. 14 is a table of analytical results for samples of glass showing the chemically determined contents of boron oxide, measured X-ray intensities and analytical accuracies.

The boron filter can eliminate C-Kα and K-Lα X-rays, however, a small influence from the L series X-rays of potassium was found which is shown in the calculating equation for the determination of boron oxide.

Also because of the interference of the N series X-rays of lead, X-ray intensity was found when there was no boron present, as is observed in lead-barium glass of NBS 89.

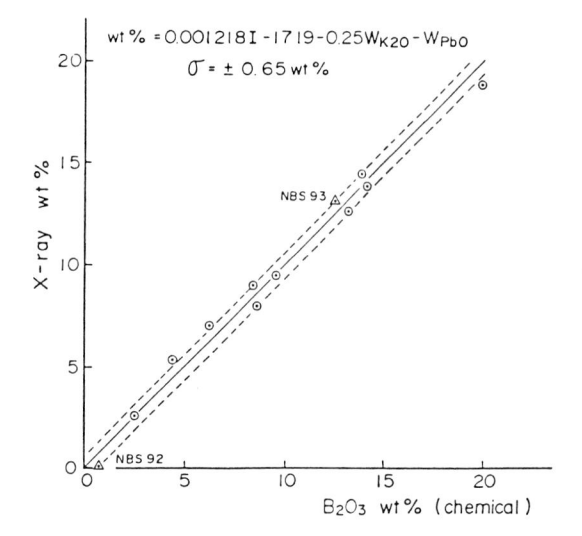

Fig. no. 13: Relation between X-ray and Chemical Content of Boron Oxide in Glass

From this result, the correction coefficient for lead oxide was calculated.

Figure no. 15 shows the relations between O-Kα X-rays and Oxygen content regarding chemical compounds.

Sample No.	Kinds of glass	B_2O_3 (chemical)	X-Ray Intensity (I)	B_2O_3 (X-Ray)	Δ (chemi-X-ray)	K_2O	PbO
R-501	High boron borosilicate	13.2 wt%	245.7 c/s	12.65 wt%	-0.55 wt%	0.30 wt%	wt%
S-2	E glass for fiber	8.6	207.3	8.01	-0.59	0.20	
3	for optical	19.9	302.4	18.92	-0.98	2.8	
4	for thermos	13.9	262.3	14.47	+0.57	1.1	
5	for microwave oven	14.2	255.4	13.88	-0.32	0.1	
6	for watch cover	2.5	178.4	2.58	+0.08	7.7	
R-1	Soda-Lime for sheet	0.0	145.1	0.26	+0.26	0.84	
NBS-92	Low boron borosilicate	0.7	143.6	0.16	-0.54	(0.55)	
93A	High boron borosilicate	12.56	249.7	13.21	+0.65	0.01	
89	Lead-Barium	0.0	257.7	—	—	8.40	17.50
NP-20	Borosilicate	4.38	186.6	5.33	+0.95	0.80	
21		8.39	216.1	9.03	+0.64	0.35	
22		6.21	201.2	7.04	+0.83	(1.08)	
23		9.54	219.7	9.48	-0.06	0.33	
30	Calcium-Borate	43.6	729.2		$\sigma = \pm 0.65$		
31		45.5	719.7				

Calculating Formula wt% = 0.1218 I - 17.19 - 0.25 W_{K_2O} - 0.69 W_{PbO}

Fig. no. 14: X-ray Analysis Content of B_2O_3 in Glass

If the analyzing samples contain those elements having high absorption coefficients to O-Kα X-rays, the O-Kα X-rays from them exhibit lower intensity such as Titanium Oxide, and Coal.

The intensity of samples containing low absorption elements such as SiO_2 Al_2O_3 MnO_2 etc., to O-Kα X-rays become higher.

The 2 Theta values of the intensity peak maximum changed around one degree and the deviation range of the measured intensity of various samples at peak maximum of SiO_2 were 10 %.

As coal is a mixture of organic and mineral matter the oxygen in coal is contained as components of organic matter. Since carbon has large absorption coefficient to O-Kα X-rays, X-ray emission from the coal particles shows lower intensity.

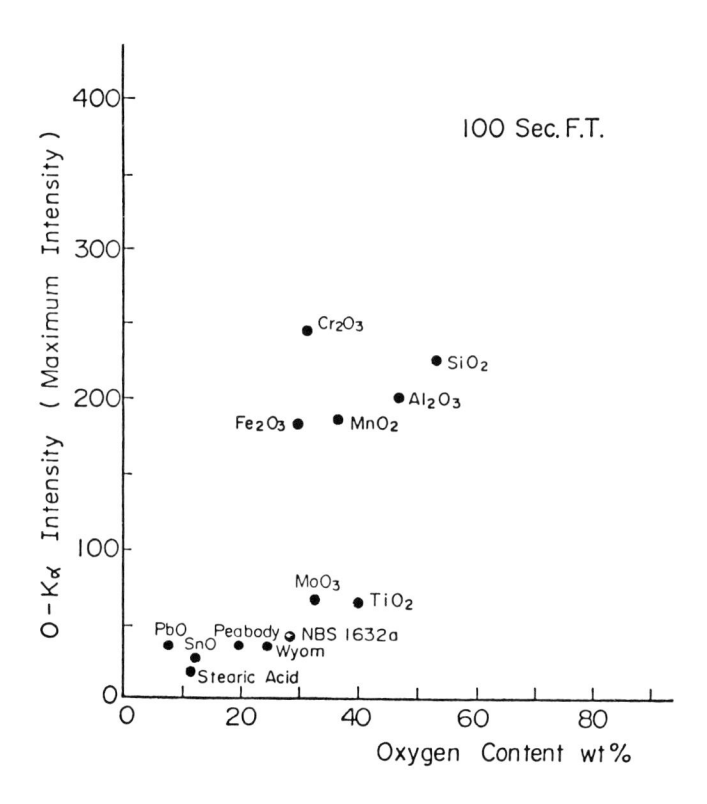

Fig. no. 15: Measured Intensity of O-Kα X-rays from Chemical Oxygen Compounds against Oxygen Content

O-Kα X-rays from mineral matter particles show higher intensity owing to the low-absorption to O-Kα X-rays.

As measured O-Kα intensity is a sum of a low and high intensity component mineral matter and organic material, the complex matrix effects for O-Kα in coal powder samples should be expected, as can be seen in Figure no. 16.

Using the regression technique, shown in Figure no. 17, with a correction term consisting of the ratio of oxygen and carbon, analytical results of oxygen in coal was obtained.

Analytical accuracy was 1.7 wt % and it was 8 times larger than that of measured standard deviations of same kind of samples.

In the case of no correction of the ratio of oxygen and carbon content, the analytical accuracy was 2.2 wt %.

For the improvement of analytical accuracy, the matrix effects of O-Kα and chemical content should be studied more.

The data presented in Figure no. 18 are the results from an aluminium metal plate covered by very thin material which is a complex chemical compound containing chromium and oxygen for the binding of aluminium metal and paints. Thickness of the material is 50 ∼ 300 Å.

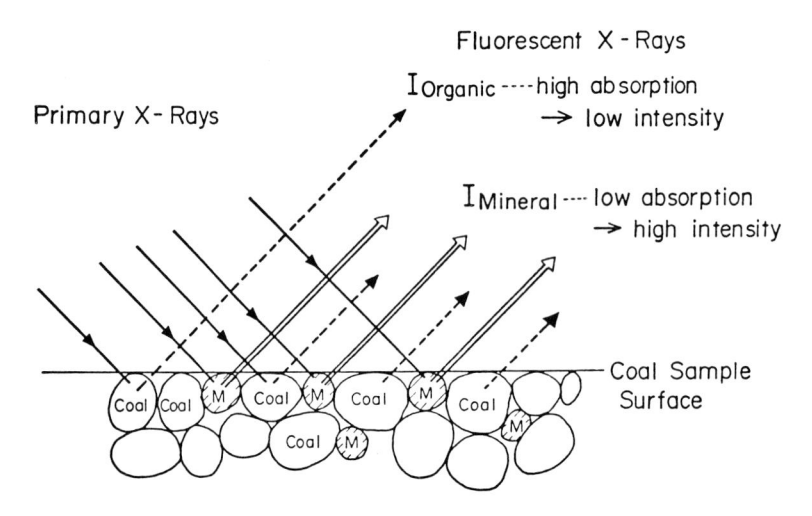

Fig. no. 16: Schematic Model of Analyzing Surface of Coal Powder Sample

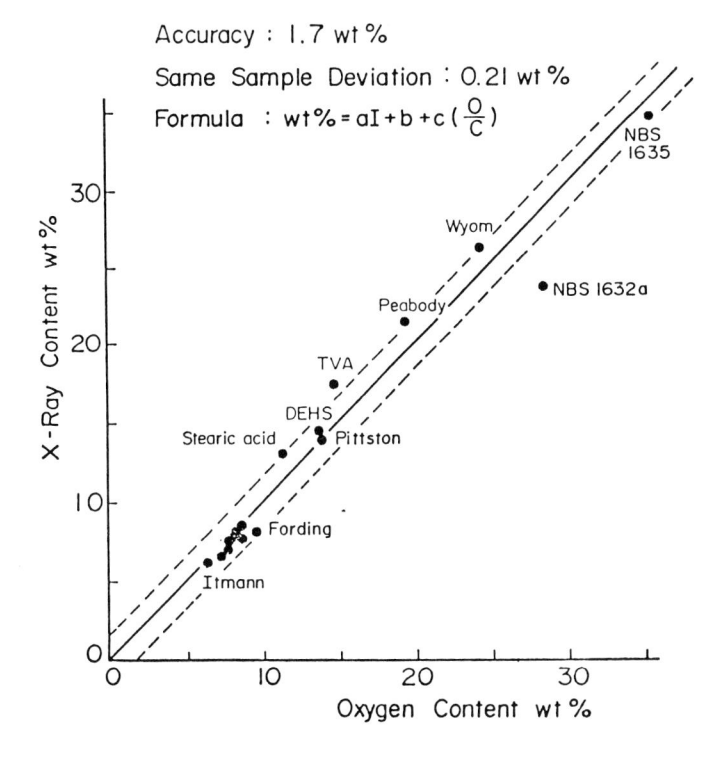

Fig. no. 17: Oxygen Determination in Coal by X-ray Fluorescence Analysis

This very thin thickness was measured using 23,62Å X-rays of O-Kα and 21.6Å X-rays of Cr-Lα.

On the basis of simple calculations, estimated thickness of saturated intensity of O-Kα X-rays in Chromium Oxide is approximately 2.2 μm.

Figure no. 19 shows the experimental results of Cr-Lα X-rays which is located at high absorption region of oxygen.

Therefore, the intensity of Cr-Lα X-rays are about 20 times lower than that of O-Kα.

The estimated thickness of the saturated intensity of Cr-Lα X-rays in Chromium Oxide is approximately 7400Å.

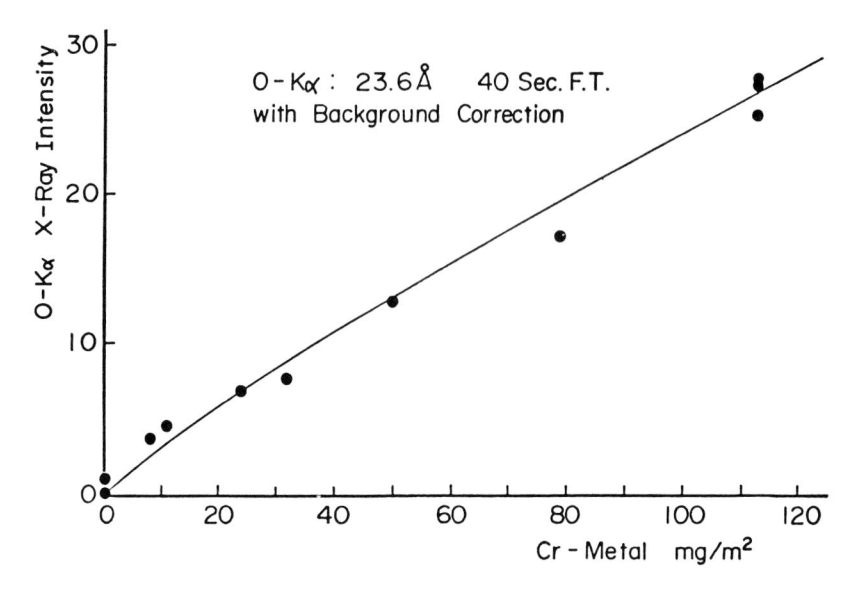

Fig. no. 18: O-Kα Intensity Measurement of Chromate Treatment Film on Aluminium Metal Plate

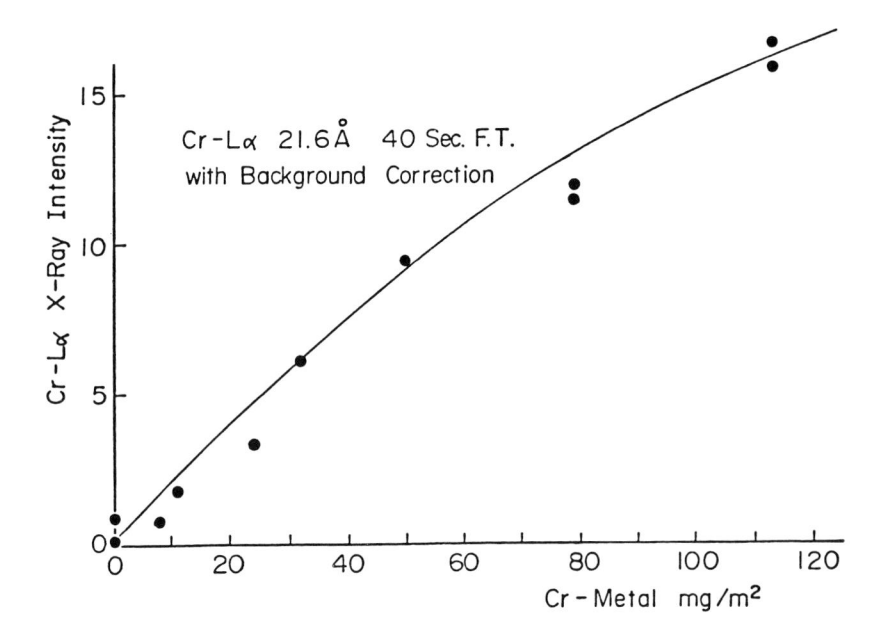

Fig. no. 19: Cr-Lα Intensity Measurement of Chromate Treatment Film on Aluminium Metal Plate

Relation between O-Kα and Cr-Lα X-rays is shown in Figure no. 20. Since the good linear relation is shown, it appears that there is no change in the ratio of chromium and oxygen elements.

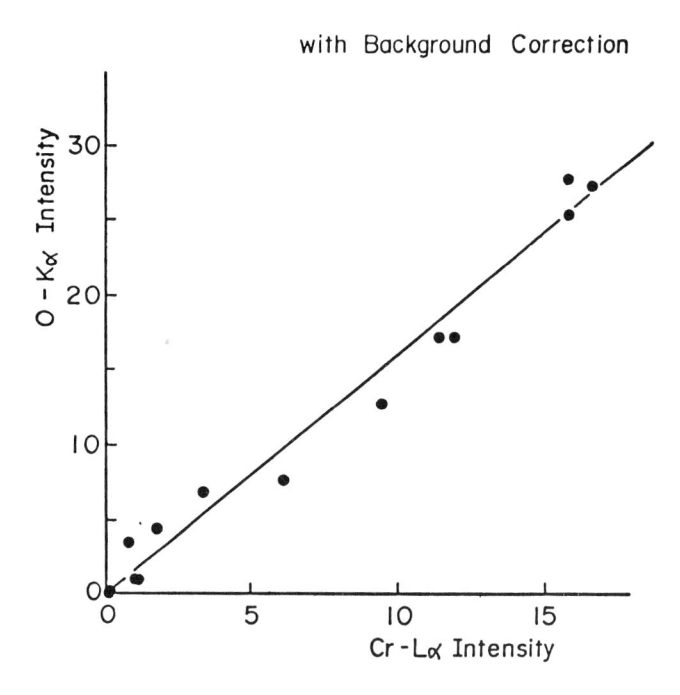

with Background Correction

Fig. no. 20: Relation between O-Kα and Cr-Lα X-rays of Chromate Treatment Film on Aluminium Metal Plate

The applications of X-ray fluorescence analysis using a total reflection monochromator for the detection of boron and carbon X-rays and a TAP-crystal for oxygen X-rays have been shown on the basis of a commercially available instrument. As an additional possibility of this method, the very thin thickness measurement was performed.

DIE RÖNTGENBOX - RFA IM RASTERELEKTRONENMIKROSKOP

Richard Eckert, Forschungszentrum SEL, 7 Stuttgart 40

Zur Materialanalyse ist das Rasterelektronenmikroskop (REM) üblich mit einem energiedispersiven Spektrometer (EDS) ausgestattet. Die Kombination REM + EDS ermöglicht Materialanalysen hoher Ortsauflösung (1 µm) aber begrenzter Nachweisempfindlichkeit (500 ppm). Ein Zusatz zur Röntgenfluoreszenz (RFA) würde
- die Nachweisgrenze senken auf 3 bis 30 ppm
- Tiefeninformation bis 1 mm Materialdicke liefern
- eine Metallisierung der Probe vermeiden
- selbst Pulver einer REM-Analyse zugänglich machen.

Anders als bei der ersten RFA-Konstruktion für ein REM /1/ ist hier
- der RFA-Zusatz als komplette Box auf einem Probenteller montiert
- ein robuster Aufbau mit Massivanode.

Ein Wechsel von den üblichen Untersuchungen zur RFA im REM beschränkt sich damit auf einen Probenwechsel mit Umschalten auf höheren Strom.

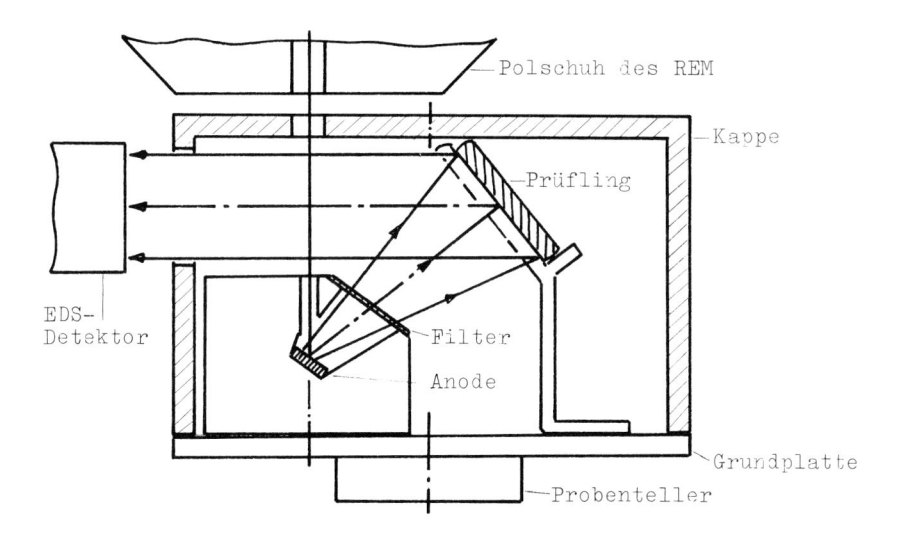

Bild 1: RFA-Zusatz "Röntgenbox" im REM

Wie aus Bild 1 zu erkennen, lassen sich Anode und Filter als Einheit stecken oder auch - in einer anderen Ausführung - unabhängig auswechseln. In der Röntgenbox /2/ dient die Grundplatte als Träger von Anode, Filter und Prüfling, die Kappe als Abschirmung des Prüflings vor Streuelektronen sowie vor gestreuter Röntgenstrahlung.

Der meßbare Elementebereich umfaßt praktisch alle mit EDS erfaßbaren Stoffe. Jedoch wird im REM mit seinem beschränkten Elektronenstrom ($I_{max} \approx 1$ µA) die Zählrate für Elemente mit $Z \leq 10$ zu gering. Eine Verbesserung in der Nachweisgrenze von EDS + RFA gegenüber EDS ist damit gegeben für Elemente ab einschließlich Natrium (Z = 11).

Für Analysenzeiten von je 20 min bei einer Röntgenzählrate von 2000 Imp/s ergeben sich folgende Nachweisgrenzen in ppm, vgl. /3/:

AlSi-Legierung, c_{mdl} in ppm:

	Ti	Mn	Fe	Zn	Sr
Mo-Anode mit 25 µm Al-Filter	26	19	13	10	7

NBS-Glas 612, c_{mdl} in ppm:

	Na	Ca	Mn	Ni	Sr	Bi	U
Al-Anode mit 7 µm Al-Filter	300	25	7	7	-	-	-
Mo-Anode mit 50 µm Mo-Filter	-	250	5	5	3	3	5

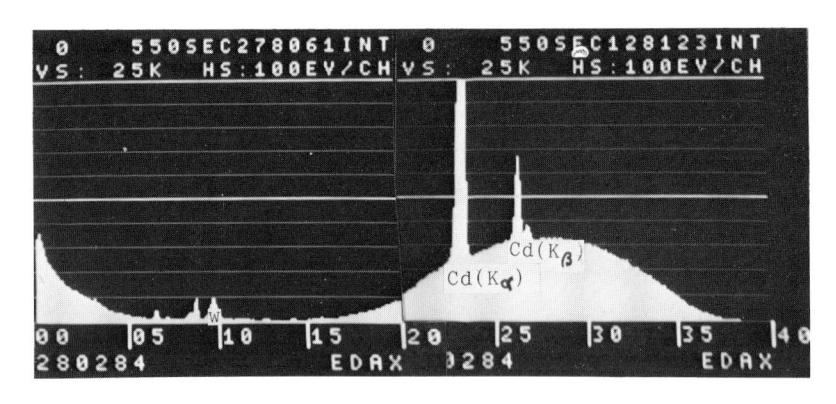

Bild 2: Kunststoffgranulat mit 1000 ppm Cadmium, W-Anode mit 30 µm W-Filter, c_{mdl} = 3 ppm für Analysenzeit 20 min, Zählrate 2000 Imp/s.

Literatur
/1/ Middleman LM, Geller JD. (1976). Trace element analysis using X-ray excitation with an energy dispersive spectrometer on a scanning electron microscope. Scanning Electron Microscopy 1976/I: 171-178, 762.
/2/ PLANO-AGAR-catalog Nr. 4 (1983). Röntgenbox.Friedrichsplatz9,Marb.
/3/ Eckert R. (1983). XRF in the SEM with a massive anode. Scanning Electron Microscopy 1983/IV: 1535-1545.

ENERGY DISPERSIVE X-RAY FLUORESCENCE ANALYSIS WITH MULTIPLE TOTAL REFLECTION -
AN IMPROVEMENT OF DETECTION LIMITS

Knut Freitag

Rich. Seifert & Co., Bogenstrasse 41   D-2070 Ahrensburg

INTRODUCTION

   The fundamental Advantages of Total Reflection X-Ray Fluorescence Analysis
(TXRF) is its capability to detect minute quantities of elements in the pg
range contrary to - at best - nanograms for XRF (1).   This increase of
detection power is based on a suggestion of Yoneda and Horiuchi (2) and
further investigations of Wobrauschek and Aiginger (3).
In principle, it is achieved by altering the geometry arrangement of the
main components of the energy dispersive X-ray fluorescence analysis, i.e.
X-ray tube, sample, and Si (Li) detector along with the use of polished
quartz slices as sample supports.   The specimen forms a thin film on the
surface of the quartz carrier, and is irridiated in grazing incidence by a
narrow collimated X-ray beam.   At glancing angles below a few minutes of
arc, total reflection occurs on the surface.
The scattered radiation from the sample support is thus virtually eliminated,
thereby drastically reducing the background of the fluorescence spectra.

1. DEVELOPMENT OF THE INSTRUMENT

Substantial development work had to be done in establishing the TXRF method as a competitive tool for routine and trace element analysis. Prototypes of the instrument were built at the GKSS Forschungszentrum Geesthacht (4, 5, 6). The actual production model is a slightly modified version of the construction described elsewhere (5, 6) with multiple reflection of the exciting beam (fig. 1).

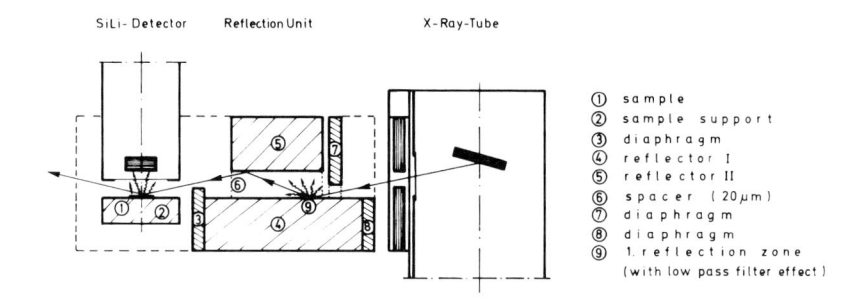

Fig. 1: Schematic design of TXRF spectrometer

① sample
② sample support
③ diaphragm
④ reflector I
⑤ reflector II
⑥ spacer (20 μm)
⑦ diaphragm
⑧ diaphragm
⑨ 1. reflection zone (with low pass filter effect)

The utilization of simple X-ray optics (7, 8) instead of the direct excitation of the specimen facilitates the observation of small tolerances for manufacturing and adjustment. At higher anode voltages it leads additionally to further reduction of the background by the low pass filtering effect from reflection of the primary radiation.
The shaping of the exciting spectrum along with the resulting improvement of the detection performance is shown in fig. 2.

Resulting from a close cooperation between the GKSS Forschungszentrum Geesthacht and Rich. Seifert & Co., the presently available instrument meets - in addition to its outstanding detection power - the demands of the user, e.g. automatic sample changing for 35 specimens, radiation safety, and a mechanical stability which eliminates any readjustment.

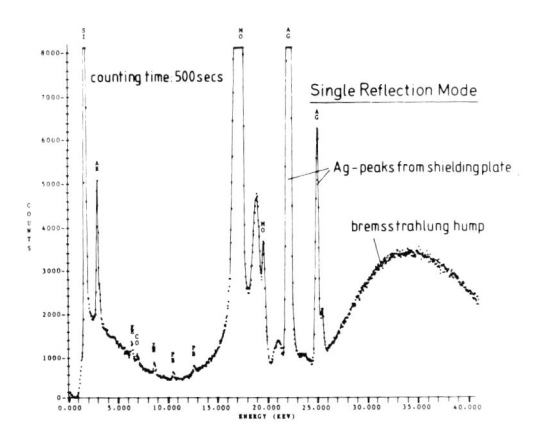

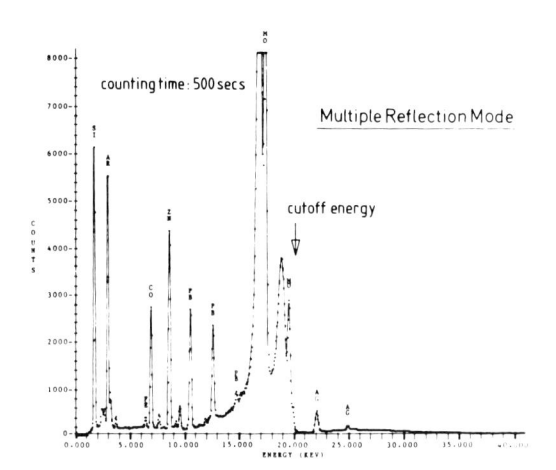

Fig. 2: Demonstration of Background Suppression by
        Multiple Reflection

        Excitation: Mo tube at 60 kV; filter 60 μm Mo;

        Sample: 100 μl of a 10 ppb Co-Zn-Pb standard solution,
                    i.e. 1 ng each

2. PERFORMANCE DATA OF THE INSTRUMENT

To some extent, the performance of an instrumental method can be characterized by its detection limits, its reproducibility, and its dynamic ranges. Convincing information with respect to accuracy - probably the most important property - can be obtained in the framework of intercomparison tests, and is even then hard to quantify. Therefore, we refer to methodical comparisons of the accuracy which are published elsewhere (9).

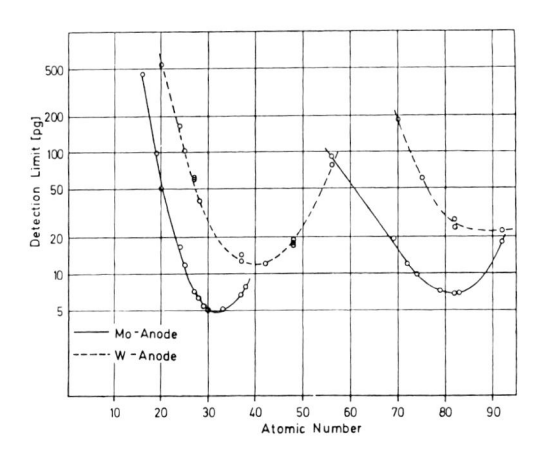

Fig. 3 : Curves of Detection Limits

The present detection limits (3 sigma above background, counting time 1000 secs) are shown in fig. 3 for the molybdenum anode and for excitation with the filtered Bremsstrahlungsspektrum of a tungsten tube.
It is necessary to note that those values resulted from measurements of diluted aqueous solutions (fig. 4 and 5) which are virtually free from any matrix and contain only metal salts at concentrations as low as 20 ppb (ng/ml) with non-overlapping peaks. Detection limits of e.g. 10 pg correspond with minimum concentration of 0.2 ppb for a 50 microliter sample.

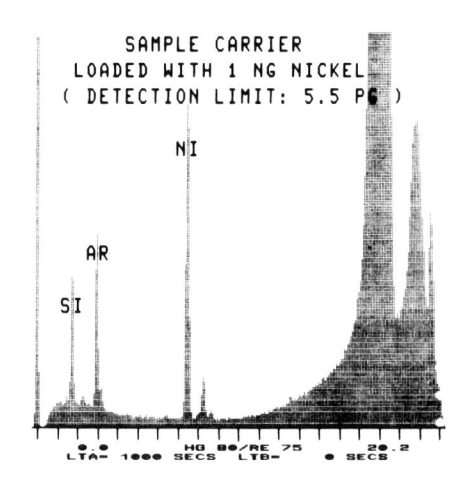

Fig. 4: Mo excitation of 1 ng Ni

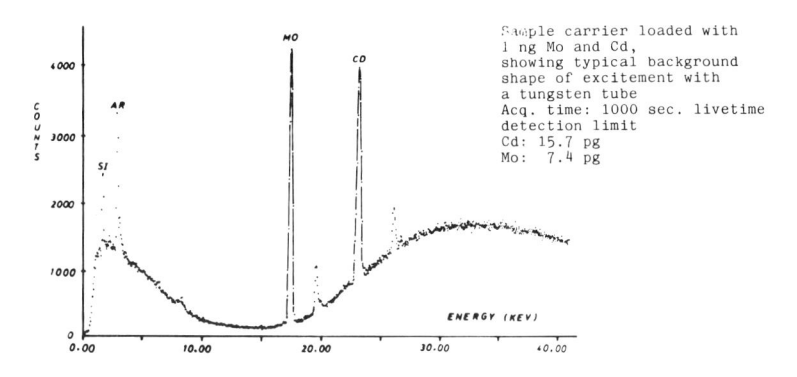

Sample carrier loaded with
1 ng Mo and Cd,
showing typical background
shape of excitement with
a tungsten tube
Acq. time: 1000 sec. livetime
detection limit
Cd: 15.7 pg
Mo: 7.4 pg

Fig. 5: W excitation of 1 ng Mo and Cd

Higher matrix contents - specifically of substances with higher atomic numbers - deteriorate the detection performance. As an example, the influence of NaCl on the detection limit of lead is shown in fig. 6.

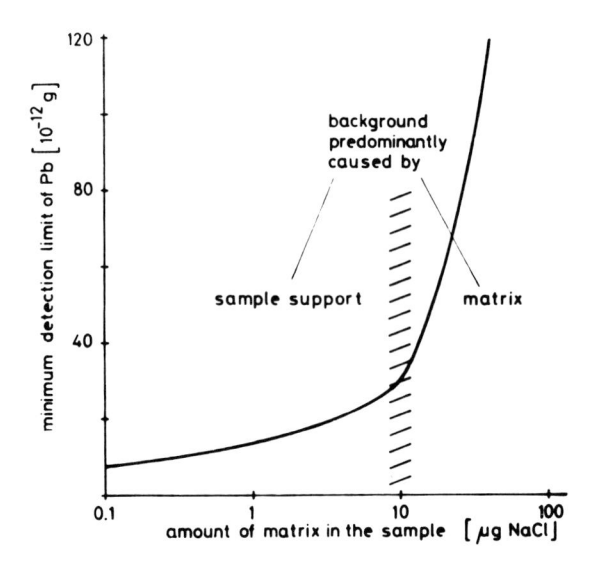

Fig. 6: Influence of matrix on the detection limites in case
of Pb in aqueous NaCl solution

Therefore, matrix removal techniques are often necessary to contribute to a high basic sensitivity. Such a procedure, specifically adapted to TXRF, was developed for seawater samples (10, 11).

Due to the exclusive application of a thin film technique combined with internal standardization (see below) which favours the accuracy at the cost of precision, only a reproducibility of, at best, 1% (usually 2 to 5%) can be obtained with TXRF up to date. The comparitively poor precision, however, which - similar to Neutron Activation Analysis - does normally not substantially deviate from the accuracy, has turned out to be fully adequate for trace and microanalysis. TXRF can be of use even for the determination of major constituents, particulary in case of missing standards.

A linear dynamic range of four orders of magnitude is demonstrated in fig. 7, showing the measuring of Pb concentration from 2 to 20,000 ppb against an internal Co standard of 200 ppb.

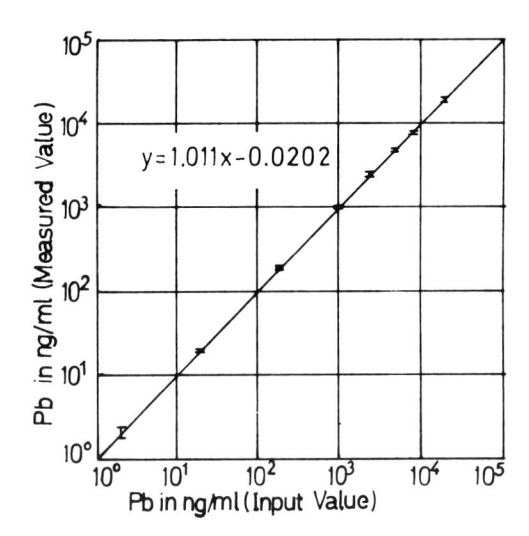

Fig. 7: Linear dynamic range

3. CALIBRATION AND SPECTRUM PROCESSING SOFTWARE

It is the outstanding sensitivity of TXRF that allows the unrestricted application of the thin film technique with international standardization, because minute specimens are one prerequisite for applying this technique successfully. The films have a diameter of only some millimeters and are a few micrometers thick. The small diameter restricts the influence of an irregular distribution in the primary radiation field and, thereby, the effect of an inhomogeneous metal distribution in the specimen. The low weight of the specimen films excludes any self-absorption of the fluorescence radiation down to the atomic number of sulfur (A = 16). On these conditions, the following relations can be of use for quantitative analysis:

$$C_i = \frac{P_i}{P_k} \cdot \frac{E_k}{E_i} \cdot C_k$$

C_i = required concentration of the element i

C_k = known concentration of the internal standard element k

$P_{i,k}$ = peak area of the corresponding fluorescence radiation

$E_{i,k}$ = firm (i.e. independent from sample matrix, intensity of the exciting radiation, or sample volume) calibration factor of the element i as compared with the standard element k.

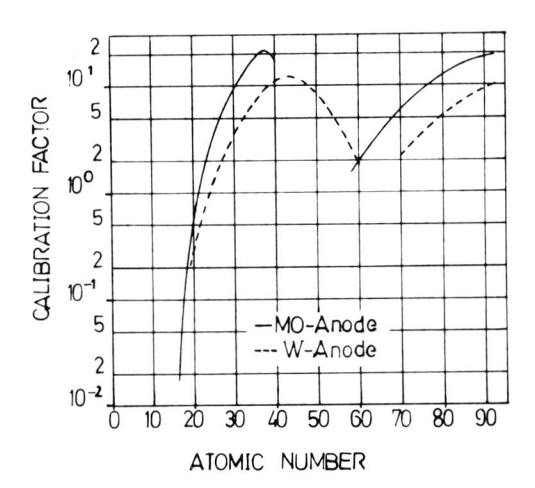

Fig. 8: Calibration factors of an instrument

The calibration curve (fig. 8) is determined by repeated measurements of various multi-element standard solutions. Once carefully established - normally at the installation phase - the calibration should never be altered or adapted except, of course, if hardware is varied or a sensitive part of the software is changed. It maintains its validity for all matrixes, for the entire concentration range, and for any element choosen as internal standard.

The spectrum processing software plays an important role in view of the capacity, the ease of operation, and the reliability of the method.

Because of the complexity of X-ray spectra from real samples, deconvolution of overlapping peaks and automatic background subtraction is mandatory, e.g. for the determination of arsenic and lead in the presence of selenium and bromine. Past experience has shown that adequate programs are commercially available.

4. PREPARATION OF THE MEASURING SAMPLE

Apart from special cases the sample is submitted for measurement in dissolved state. Concentrated solutions are diluted to ppm concentration levels. The solvent must be easily evaporable. Solid samples are preferably processed by digestion in pressurized PTFE vessels with nitric acid. In case of a silicate matrix small amounts of fluoric acid are added. Furthermore, open decomposition techniques or suspending or fine ground materials were used.

In any case, a standard solution of a metal salt must be added for quantitative analysis. Spiking with an internal standard at an early stage of the sample preparation procedure results additionally in reducing some sources of error. The solvent serves to mix the sample with the internal standard and to transport and to dose the otherwise unweighable amounts remaining on the quartz carrier after its evaporation.

The sometimes imminent formation of larger crystals on the sample carrier which would separate the metals in the specimen and may, thereby, cause objectionable inhomogenities, should be prevented by adding an appropriate homogenizing agent, e.g. cellulose acetate dissolved in HNO_3.

Samples are pipetted onto the quartz carrier in volumes between 1 and 50 µl, and subsequently allowed to dry. For aqueous solutions, the sample carrier should be made water-repellant (e.g. by a silicone solution) in order to hold a pipetted volume of up to 100 µl in the center of the quartz surface. For organic solvents, a particularly developed receptacle for up to 300 µl is in use.

Element	TXRF concentration in % + s.d	reference values in % + s.d.*)
Ca	4.14 ± 0.27	
Ti	0.20 ± 0.02	
Fe	2.47 ± 0.08	
Cu	1.80 ± 0.06	
Sr	0.044± 0.003	
Y	1.51 ± 0.03	1.38 ± 0.25
Zr	0.18 ± 0.05	
La	8.09 ± 0.21	7.80 ± 0.71
Ce	16.47 ± 0.24	16.22 ± 0.58
Pr	1.62 ± 0.21	1.62 ± 0.44
Nd	7.38 ± 0.14	6.48 ± 0.33
Sm	1.00 ± 0.04	1.02 ± 0.23
Eu	n.d.	0.022± 0.010
Gd	0.75 ± 0.03	0.82 ± 0.28
Tb	n.d.	0.070
Dy	0.33 ± 0.05	0.34 ± 0.06
Ho	n.d.	0.045
Er	0.13 ± 0.01	0.097
Tm	n.d.	0.016
Yb	n.d.	0.040
Lu	n.d.	0.005
Pb	0.282± 0.004	
Th	5.83 ± 0.04	5.21 ± 0.35
U	0.107± 0.003	

*) values without standard daviation are not certified

Table 1: Monazite sample NIM 66/69

Element	TXRF	NBS
Ag	1.07± 0.17	0.89 ± 0.09
As	13.3 ± 1.1	13.4 ± 1.9
Ba	5.4 ± 1.0	—
Ca	1317 ± 130	1500 ± 200
Cd	3.67± 0.35	3.5 ± 0.4
Cu	65.3 ± 5.0	63.0 ± 3.5
Fe	202 ± 12	195 ± 34
K	7930 ± 1600	9690 ± 50
Mn	16.3 ± 1.3	17.5 ± 1.2
Mo	0.21± 0.05	(≤ 0.2)
Ni	1.17± 0.20	1.03 ± 0.19
Pb	0.5 ± 0.2	0.48 ± 0.04
Rb	4.5 ± 0.5	4.45 ± 0.09
S	7050 ± 1400	(7600)
Se	2.1 ± 0.2	2.1 ± 0.5
Sr	9.8 ± 0.8	10.36 ± 0.56
Ti	10.3 ± 3.0	—
V	2.1 ± 0.4	2.3 ± 0.1
Y	0.4 ± 0.1	—
Zn	863 ± 40	852 ± 14

( ) non-certified values

Table 2: NBS Oyster tissue 1566 (element concentration in ng/g)

5. APPLICATIONS

Examples of application are given for three different analytes (tables 1 and 2, fig. 9). Those examples demonstrate the capabilities of the method in view of concentration and detectable elements. A large number of analyses was carried out up to date, specifically in the environmental and biological fields as well as for mineralogical samples. For a more detailed description, especially for sample preparation procedures, we refer to (12) of these proceedings.

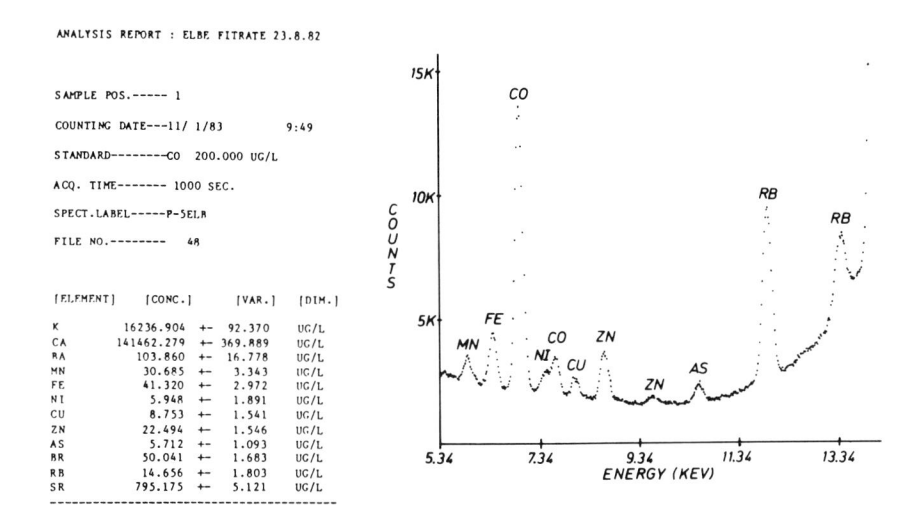

```
ANALYSIS REPORT : ELBE FITRATE 23.8.82

SAMPLE POS.----- 1

COUNTING DATE---11/ 1/83      9:49

STANDARD--------CO 200.000 UG/L

ACQ. TIME------- 1000 SEC.

SPECT.LABEL-----P-5ELB

FILE NO.--------    48


[ELEMENT]    [CONC.]        [VAR.]    [DIM.]
K            16236.904   +-  92.370   UG/L
CA          141462.279   +- 369.889   UG/L
BA             103.860   +-  16.778   UG/L
MN              30.685   +-   3.343   UG/L
FE              41.320   +-   2.972   UG/L
NI               5.948   +-   1.891   UG/L
CU               8.753   +-   1.541   UG/L
ZN              22.494   +-   1.546   UG/L
AS               5.712   +-   1.093   UG/L
BR              50.041   +-   1.683   UG/L
RB              14.656   +-   1.803   UG/L
SR             795.175   +-   5.121   UG/L
```

Fig. 9: Filtrate sample of Elbe water, direct measurement

6. CONCLUSIONS

In spite of its limitations, e.g. the insensibility to light elements or - even more severe - the deterioration of detection power by means of infavourable matrixes the TXRF has proven to be competitive with respect to other, better known analytical techniques, especially in intercomparison tests.
In addition to its multi-element features down to the picogram level, particularly its universal calibration function has turned out to be a great help in the analytical practice.

REFERENCES

(1) J. V. Gilfrich, L.S. Birks, Anal. Chem., 1984, 56, 77-79
(2) Y. Yoneda, T. Horiuchi, Rev. Scien. Instrl., Vol. 2, No. 7 (1971) 1069-70
(3) P. Wobrauschek, H. Aiginger, Anal. Chem., 1975, 47, 852-955
(4) J. Knoth, H. Schwenke, Fresenius Z., Anal. Chem., 1978, 291, 200-204
(5) J. Knoth, H. Schwenke, Fresenius Z., Anal. Chem., 1980, 301, 7-9
(6) H. Schwenke, J. Knoth, Nucl. Meth., 1982, 193, 239-243
(7) German Patent DE 29 11 596 C 3
(8) US Patent 4,426,717
(9) W. Michaelis et. al. "Intercomparison of the Multielement Analytical Methods TXRF, NAA and ICP with Regard to Trace Element Determinations in Environmental Samples" in these proceedings
(10) Knöchel A. and Prange, A.: Microchim. Acta (II), 1980, 395-408
(11) Prange, A. Knöchel, A., and Michaelis, W.: submitted to Anal. Chim. Acta
(12) Michaelis W. et. al.: "Recent Applications of TXRF in Multielement Analysis", in these proceedings

APPLICATIONS OF TOTAL REFLECTION X-RAY FLUORESCENCE
IN MULTI-ELEMENT ANALYSIS

W. Michaelis, A. Prange, J. Knoth

Institut für Physik, GKSS Forschungszentrum Geesthacht
D-2054 Geesthacht, Germany

SUMMARY

Although Total Reflection X-Ray Fluorescence Analysis (TXRF) became available for practical applications and routine measurements only few years ago, the number of programmes that make use of this method is increasing rapidly. The scope of work is widespread over environmental research and monitoring, mineralogy, mineral exploration, oceanography, biology, medicine and biochemistry. The present paper gives a brief survey on these applications and summarizes some of them which are typical for quite different matrices.

1 INTRODUCTION

Total Reflection X-Ray Fluorescence Analysis (TXRF) is a rather new trace analytical method. Nevertheless, a lot of papers have already appeared during the past few years which obviously demonstrates the capabilities of TXRF. Part of these publications is devoted to the principle of the method (1 - 9) or procedures for sample preparation and trace-element pre-enrichment (10 - 13). The present state of TXRF is summarized in Refs. (14, 15), and intercomparisons with neutron activation analysis and inductively coupled plasma optical emission spectroscopy are given in another contribution to the Jülich conference (16).

In the present paper an attempt is made to review the applications of TXRF which meanwhile are found to be widespread over various disciplines in research and monitoring programmes. Above all, the method has been used in studies of transport phenomena in tidal rivers (17 - 20), pollution studies of estuaries and the Wadden Sea (20 - 22), sediment analysis (19, 22 - 24), air-dust analysis and monitoring (25 - 31), mineralogical investigations (32, 33), mineral exploration (34), sea-water analysis (12, 13, 34 - 36), biology (37 - 39), medicine (40, 41) and biochemistry (42). Accordingly, numerous matrices have been handled. The motivation to utilize TXRF for tackling the respective analytical problems has, in general, arisen from one of the characteristic features associated with the method or combinations of them: the high sample throughput achievable, the low sample masses required, the simple calibration procedures and the high detection power. Most of the results have so far been obtained by using instruments manufactured by Rich. Seifert & Co. (43) under licence of the GKSS Research Centre, or previous prototypes of them. In the following sections some applications of TXRF which are typical for quite different matrices are briefly summarized.

2 HEAVY METAL TRANSPORT IN TIDAL RIVERS

The knowledge of the heavy metal transport phenomena in tidal rivers and the ability to make up a balance are of great ecological importance since they are a prerequisite for predicting pollution trends, for assessing the self-purification power and for determining the net discharge into the sea. Unfortunately, the investigation of the transport processes is considerably complicated by the pronounced spatial heterogeneities and temporal variabilities that are typical for tidal rivers. Amount and sign of the flow and thus also the water level and the river width vary periodically with the tides. Settling of suspended particulates and resuspension change with the flow and influence the heavy metal load since remarkable parts of the trace elements are taken up by the particulate matter. Flocculation, sorption, desorption and disintegration processes control the interaction between the suspended material and the liquid phase. In addition, heavy metals may be remobilized from the sediment. To tackle this rather complex problem, a methodology has been developed which combines trace analytical methods with hydrographic field measurements and mathematical simulation models (17, 18).

Within this concept, a detailed knowledge of the heavy-metal concentrations in the water, the suspended load and the sediment is required. A great number of samples have to be collected, and high precision and accuracy must be ensured. TXRF has proved to be an efficient, economic and powerful tool in these research programmes. Details of the sample preparation may be found in the literature cited (6, 10, 16).

As an example, Fig. 1 presents some selected data from an extensive study performed in 1982 on the Lower Elbe River (17 - 20). The specific load of the suspended particulate matter and the metal content of the

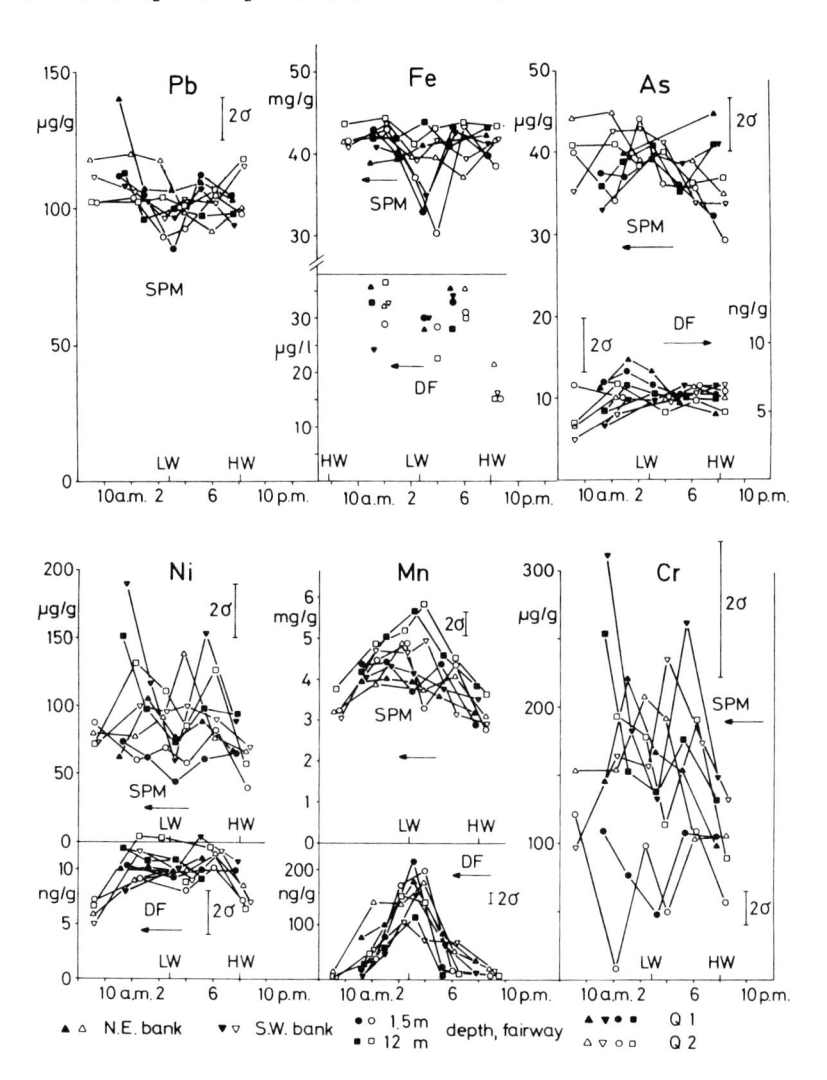

Fig. 1. Specific load of the suspended particulate matter (SPM) and metal content of the dissolved fraction (DF) as a function of tidal phase for some of a total of 35 elements determined in the Lower Elbe River. LW and HW denote low and high water, respectively. Sampling on two cross sections Q1 and Q2 1.2 km apart on August 23rd, 1982.

dissolved fraction are plotted vs. time (tidal phase) with the sampling location on the river cross section as a parameter. All data given for the particulate matter refer to the dry substance. The observed pattern is the outcome of the complex interplay of changing flow, diverse influxes up- and downstream, particle size distribution, composition of the suspended matter and the various processes mentioned above. Two groups of elements may be distinguished. Elements like Fe, As, Zn, Pb as well as, with some reservation, Ti and V show a rather uniform behaviour with regard to tidal phase and sampling site and, therefore, are directly accessible to transport calculations. Other elements such as Cr, Mn and Ni exhibit a time dependence of the load and, in part, strong spatial heterogeneities. In these cases further work is required, in particular with respect to the formulation of the boundary conditions for the mathematical models and a better understanding of the processes that lead to the observed pattern. A comprehensive discussion of the complete data is given in Ref. (20).

As a result of the transport calculations, Fig. 2 displays some examples of the time integrals vs. tidal phase for the river cross section-averaged transport of suspended particulate matter and some trace elements (44). It is shown that during the tidal period considered in this figure there was a net transport upstream. The dissimilar courses of the integrals are a consequence of the different percentages of the total heavy metal content that are fixed to the particulate matter.

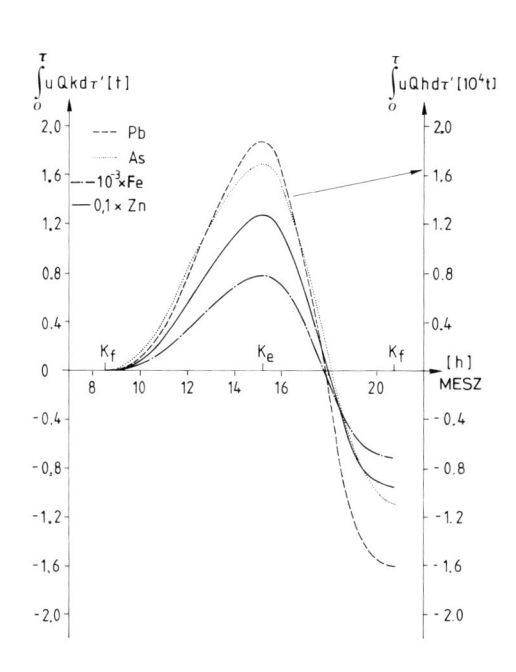

Fig. 2 Time integral of the cross section-averaged transport of suspended particulate matter (right-hand scale) and some elements (left-hand scale) vs. tidal phase τ. u = averaged flow velocity, Q = river cross section, k = total trace element content, h = suspended matter content, K_f, K_e = high and low tide phase of slack water.

3 SEDIMENT ANALYSIS

The chronology of heavy metal pollution of surface waters and the geochemical background may be assessed by trace element analysis of sediment cores in combination with dating of the strata. Sediment profiles are obtained by means of a gravity corer handled from board a ship (19, 24). In tidal rivers, a practical problem lies in the difficulty to find undisturbed profiles since besides bioturbation extreme upper water, heavy tidal waves or dredging work may destroy the regular stratification from the sedimentation rates. Therefore, a number of cores are taken and sufficiently undisturbed profiles are selected by X-ray screening (24).

Applications of TXRF to the analysis of sediment cores from the Elbe River have been described in Refs. (19, 24). Details on the sample preparation procedures are given in Ref. (19). Chronology is derived in these studies from low-level counting of ^{137}Cs released in nuclear weapon tests or of ^{210}Pb which is a member of the uranium decay series (45). An example for the depth dependence of some heavy metal concentrations in the < 63 µm fraction, of the percentage of the grain size < 63 µm and of the ^{137}Cs activity is shown in Fig. 3. The region of increased concentrations of Cu, Zn, As, Cd and Pb extends approximately over twice the depth with appreciable ^{137}Cs activities. This observation suggests that pollution started long before the numerous nuclear weapon tests in the atmosphere during the fifties.

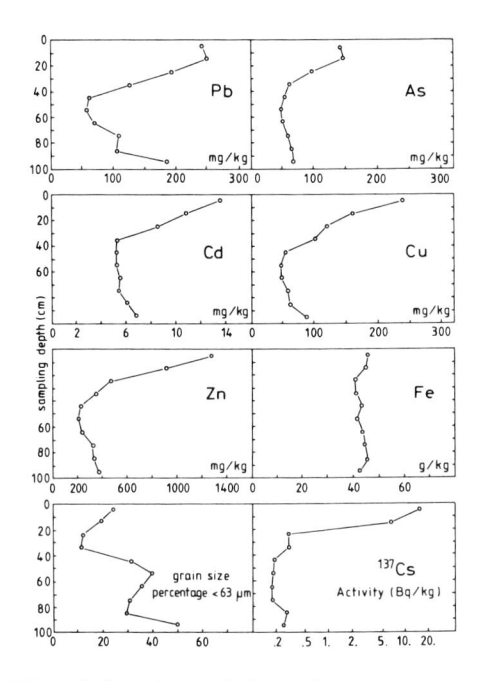

Fig. 3 Depth profiles of some metals, the percentage of grain size < 63 µm and the ^{137}Cs activity in a sediment core from the Elbe River.

Sediment analyses with TXRF performed in the framework of pollution studies of the North Friesian Wadden Sea have been reported in Ref. (22). Other authors utilized TXRF for grain size corrections of trace metal contents of sediments (23).

4 AIR-DUST ANALYSIS AND MONITORING

Sampling and analysis of airborne particulate matter are important measures in air quality monitoring programmes. While the total mass of the sample is a criterion of only limited value, multi-element analyses of the collected samples represent an important step towards air pollution assessment since the composition of the particulates to a high degree determines the ecological impact.

TXRF has been applied in two large-scale air-dust monitoring programmes. The first one has dealt with the deposition of settling air dust at numerous sampling sites throughout the area of Hamburg (26, 27) and will briefly be outlined in the following. In the course of the other programme samples of suspended air dust have been taken over a period of about one year at four urban sites in different cities and, for comparison, in one rural area (29, 30). General aspects of air-dust analysis may be found in Refs. (16, 25, 28, 29, 31). For applications in large-scale studies with high sample throughput, quite simple and fast sample preparation procedures have proved a success (29). They prepare a homogeneous fine-grained suspension and dispense with a digestion.

The aim of the Hamburg programme was to elaborate the deposition of settling air dust on the basis of a statistically significant number of samples and to investigate possible correlations between the total deposition and the heavy metal input into the ground. Using the Bergerhoff technique, more than 600 samples were collected at 90 locations over a

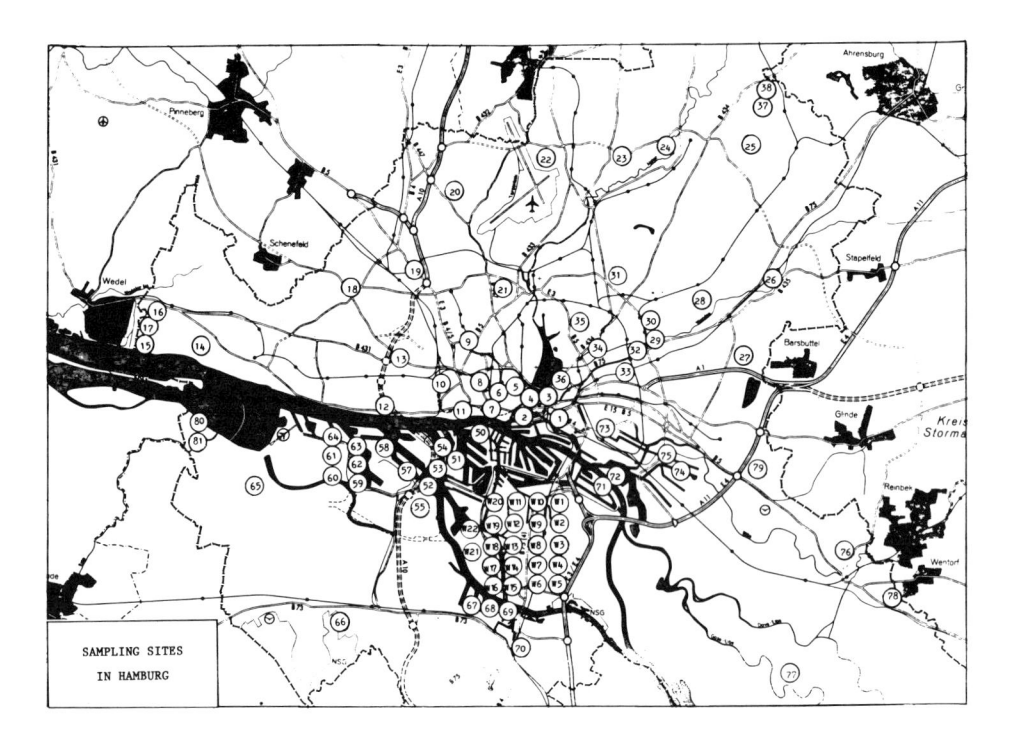

Fig. 4 Sampling sites for settling air dust in the area of Hamburg.

period of seven months (December, 1978 - June, 1979). The position of the sampling sites is shown in Fig. 4.

According to the guide-lines valid in the FRG since 1974 ('Technische Anleitung Luft'), the long-term average of the daily deposition of dust in Bergerhoff devices should not exceed 350 mg/m^2. The corresponding value for Pb is 0.25 mg/m^2. In Table 1 a few selected results of the monitoring programme for the total deposition and some trace metals are

Table 1. Deposition of settling air dust and heavy metals in the area of Hamburg. Mean values for a 7 months period from December, 1978 to June, 1979

	Deposition in mg/m^2 per day						
	Total dust	Cr	Fe	Cu	Zn	As	Pb
Mean value from monitoring sites* with low metal load**	104	0.01	2.3	0.07	0.17	0.005	0.08
Mean value from monitoring sites with medium metal load***	222	0.03	7.5	0.34	0.58	0.019	0.22
Monitoring sites with high metal load							
5	2555	0.15	91.0	1.0	2.5	0.06	2.62
10	145	0.03	7.0	0.5	3.1	0.02	0.20
19	990	0.07	47.3	0.4	1.4	0.04	2.12
50	344	0.11	47.2	3.2	4.4	0.10	0.55
51	286	0.07	31.2	1.3	4.8	0.03	0.45
57	330	0.08	38.0	0.3	26.9	0.02	0.14
69	804	0.15	12.1	0.3	5.1	0.01	0.30
71	149	0.06	7.0	4.3	1.8	0.78	3.7
72	229	0.02	7.6	1.2	0.6	0.14	0.70
W1	388	0.14	43.5	4.1	4.7	0.25	1.64
W10	200	0.02	6.8	1.4	1.2	0.10	0.53

* The positions of the sampling sites are given in Fig. 4.
** Monitoring sites with low load: 14, 15, 17, 22, 23, 24, 25, 26, 27, 28, 31, 37, 60, 65, 66, 76, 77, 78, 80, W16.
*** Monitoring sites with medium load: 1, 2, 3, 4, 6, 7, 8, 9, 11, 12, 13, 16, 18, 20, 21, 29, 30, 32, 33, 34, 35, 36, 38, 52, 53, 54, 55, 58, 61, 62, 63, 64, 67, 68, 70, 73, 74, 75, 79, 81, W2, W3, W4, W5, W6, W7, W8, W9, W11, W12, W13, W14, W15, W17, W18, W19, W21, W22.

listed. In order to arrive at some classification in the various urban areas, three classes were introduced depending on whether or not two individual thresholds for one or more of the elements Cr, Cu, Zn, As and Pb were exceeded. As shown in the table, the data measured for the dust deposition are beyond the guide value in a few cases. However, much more attention should be directed to the fact that the total dust precipitation does not reflect the ecological impact by toxic heavy metals. This is more clearly represented in Fig. 5 in which the daily deposition of dust averaged over one month is compared with the associated deposition of lead. For instance, the results for the sampling site 71 are characterized by rather low values for the total amount of dust and extreme data for lead, while the measurements at the monitoring location 73 yielded a high deposition of dust, but low values for lead. On the other hand, the classification used for the heavy metal load represents quite well the pollution states in the various urban areas (Fig. 4) and allows, in some cases, the identification of the heavy metal sources.

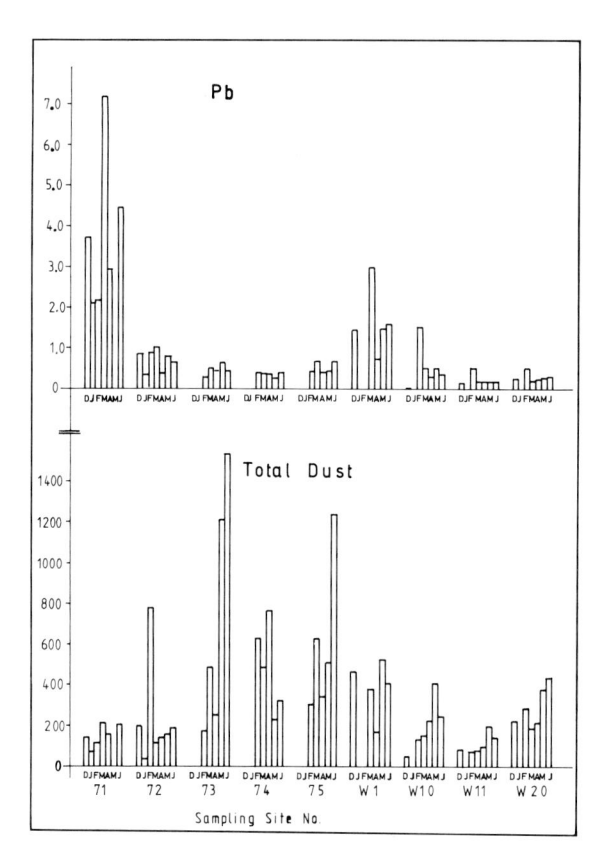

Fig. 5 Monthly average of the daily deposition of Pb and settling dust in mg/m^2 per day at some sampling sites for the period December, 1978 to June, 1979.

5 RELATIONS BETWEEN CHEMICAL AND OPTICAL PROPERTIES OF COLUMBITES AND IXIOLITES

Columbites are mixed crystals of Niobite (Fe,Mn)(Nb,Ta)$_2$O$_6$ and Tantalite (Fe,Mn)(Ta,Nb)$_2$O$_6$. The content of Mn may exceed that of Fe. A small portion of (Nb,Ta) might be replaced by W. The elements Ti, (Y,Ce), Mg and U have also been found, but are considered as impurities.

Ixiolites cannot be distinguished from Columbites through their chemical composition. However, the dimensions a_o, b_o, c_o of their elementary cell differ in b_o.

In Ref. (33) single crystals of 13 Columbites and 2 Ixiolites have been investigated with respect to their chemical and optical properties. The samples had been collected all over the world. Although a wavelength-dispersive X-ray fluorescence spectrometer was available, TXRF has been chosen for element analysis mainly for the following reasons: (i) there was a lack of standards, (ii) difficulties arose with smelting sample pellets with lithiumtetraborate in Pt crucibles, (iii) some of the crystals were quite small. Samples were prepared by digestion with HNO_3 and HF at $160°C$ in a teflon bomb.

Table 2. Chemical composition of 13 Columbites and 2 Ixiolites in wt-%

Sample No.	FeO	MnO	Nb_2O_5	Ta_2O_5	TiO_2
1	3.72	10.05	9.23	76.98	
2	12.95	4.57	67.77	13.95	0.74
3	2.73	12.93	39.07	44.53	0.56
4	5.58	10.20	43.89	39.70	0.64
5	0.47	13.38	5.48	80.68	
6	13.37	3.01	38.94	42.44	
7	11.94	4.62	52.40	30.13	0.91
8	1.59	14.44	28.37	55.59	
9	13.58	2.98	46.36	33.20	2.08
10	4.84	10.65	41.42	42.85	0.23
11	14.76	4.24	78.81	1.02	0.45
12	4.15	11.22	73.02	8.96	2.16
13	1.09	14.90	28.57	54.86	0.59
X_1	11.65	1.51	5.62	80.21	1.02
X_2	11.68	1.65	3.07	83.13	0.46

The results of the analyses are summarized in Table 2. Only the most important constituents have been listed. Two optical properties of the crystals were measured. The main refractive index n_γ was determined via the Brewster angles at 337 nm using a method described in Ref. (46). The relative reflectance $\bar{R}$ was measured by means of a mini-photometer (47).

Figs. 6 and 7 show the resulting correlations between the Ta_2O_5 contents c and the refractive indices n_γ and the reflectance $\bar{R}$, respectively. Both optical parameters are well correlated to the chemical composition. The corresponding relations and the pertinent correlation coefficients are:

$$n_\gamma = -0.00183 \cdot c_{Ta_2O_5} + 2.62 \quad ; \quad r_{n_\gamma} = -0.97$$

and

$$\bar{R} = -0.0288 \cdot c_{Ta_2O_5} + 17.15 \quad ; \quad r_{\bar{R}} = -0.99.$$

In Ref. (32) similar investigations of complex titano-niobo-tantalates (betafites) are described.

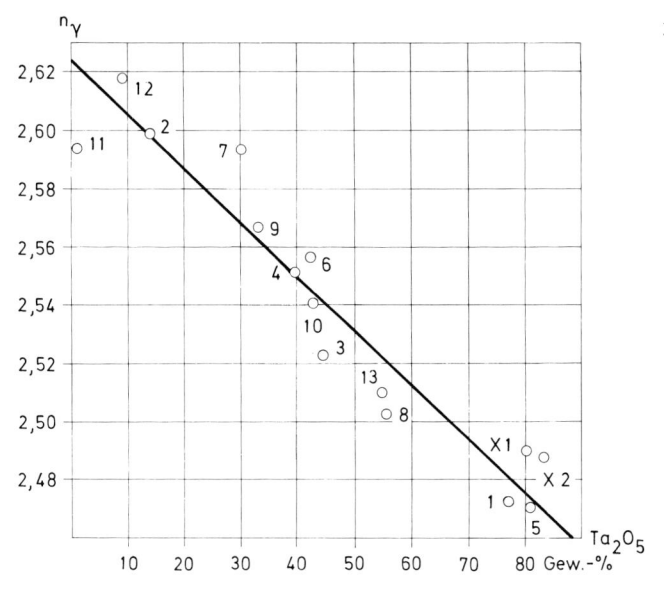

Fig. 6 Dependence of the main refractive index n_γ at 337 nm on the Ta_2O_5 content for 13 Columbites (1 – 13) and two Ixiolites (X1, X2).

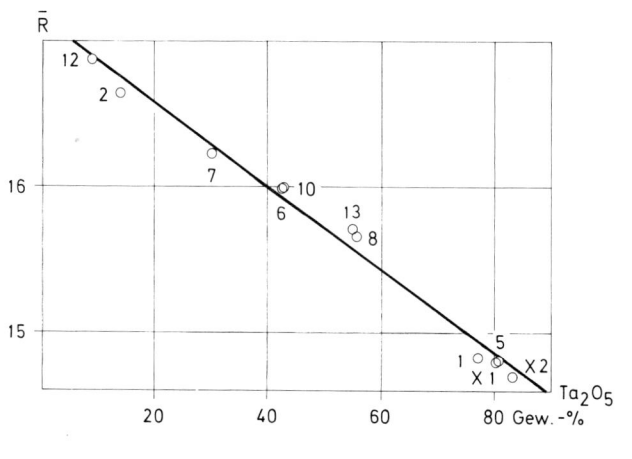

Fig. 7 Relation between the reflectance $\bar{R}$ and the Ta_2O_5 content of Columbites and Ixiolites (cf. Fig. 6).

6 DISTRIBUTION OF TRACE METALS IN SEA-WATER

In open ocean waters a vertical distribution of dissolved heavy metal traces is to be expected which is controlled or dominated by involvement in one or more of the biogeochemical cycles operative in the oceans. Nutrient components like phosphate and silicate as well as nutrient-type trace elements like Cd, Zn or Ni undergo numerous internal cycles within the ocean prior to final removal in the sediments (48). This internal cycle involves upwelling of nutrient-rich deep water to the surface euphotic zone, the fixation or removal of the nutrient elements in association with primary productivity, the resultant vertical particulate transport to depth, and the regeneration of the elements back to the dissolved form. Reactive trace metals can be enclosed in the cycle in different ways: either actively as micronutrients and/or skeleton material, or passively by adsorption on particle surfaces and co-precipitation with diverse inorganic or organic particulate matter, respectively.

A completely different situation is perceptible from the vertical concentration profiles of elements like Mn, Co or Pb (49 - 51). Their profiles indicate that these elements are usually scavenged rather than regenerated at depth.

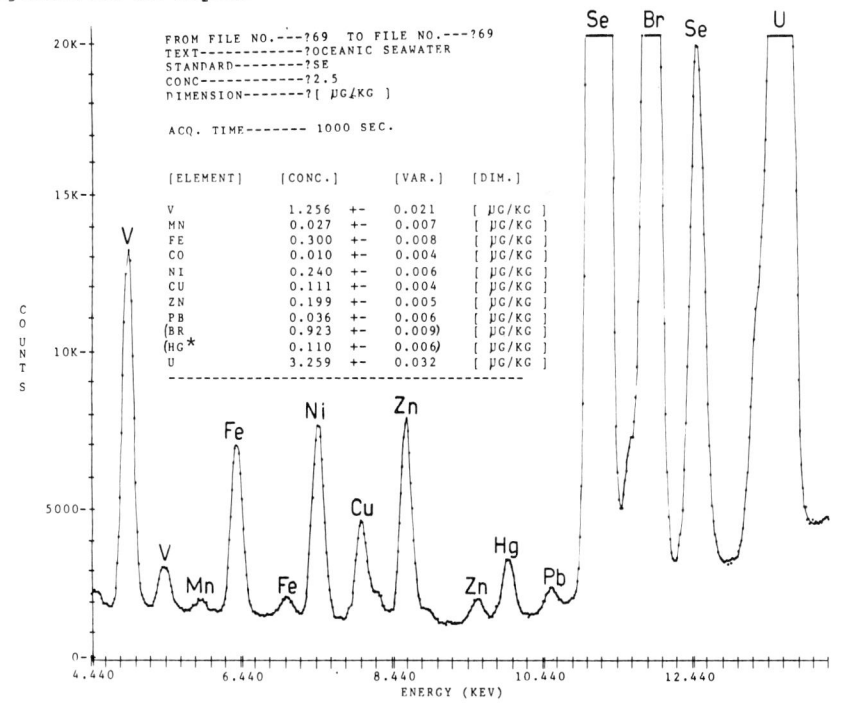

Fig. 8 Region of interest in the TXRF spectrum of an oceanic sea-water sample including spectrum evaluation. The asterisk labels contamination from sampling and storage.

Dissolved heavy metal concentrations in sea-water are normally present in the nmol/kg (μg/kg - ng/kg) region and their determination makes high demands on the trace analytical procedures.

Based on TXRF a trace analytical procedure has been developed which allows multielement determinations of dissolved heavy metal traces in sea-water down to the nmol/kg concentration region by combination of TXRF with simple sample preparation steps. The necessary separation of the element traces from the salt matrix was accomplished through complexation with sodiumdibenzyldithiocarbamate and subsequent extraction of the metal complexes using a reverse-phase technique. The following elements can be determined: V, Mn, Fe, Co, Ni, Cu, Zn, Se, (Hg), Pb, U as well as Mo and Cd. In the course of two research campaigns to the Baltic Sea and to the Middle Pacific the performance of the procedure has exemplarily been demonstrated for trace element distributions in connection with other oceanographic parameters. On this occasion, the use of TXRF was successfully tested aboard research vessels. Details of systematic investigations of the trace analytical procedure itself as well as comprehensive determinations of trace-metal concentration distributions in different areas of the sea are described elsewhere (12, 13, 36).

In this summary, vertical distributions of the elements Mn and Ni from the Middle Pacific are outlined because they represent different types of vertical trace element concentration profiles and demonstrate the capability of TXRF in ultra-trace analysis. Fig. 8 shows a typical TXRF spectrum (region of interest) of an oceanic sea-water sample including evaluation data. Fig. 9 shows the concentration profiles of Mn and Ni as well as of temperature, salinity, phosphate and silicate.

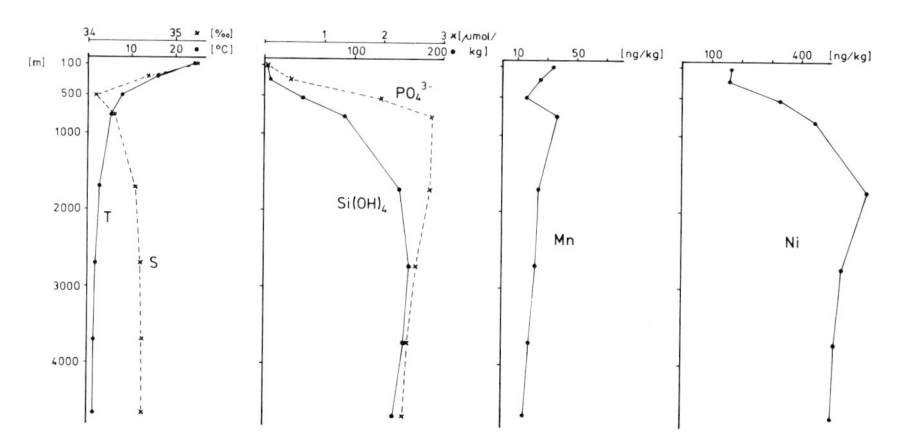

Fig. 9 Vertical distributions of temperature, salinity, nutrients, Mn and Ni in the Middle Pacific

It turns out that the Mn distribution differs distinctly from the vertical profile of the nutrient-type element Ni. While Ni shows evidence of surface depletion to values about 2.7 nmol/kg (160 ng/kg), deep water maxima of about 10.6 nmol/kg (620 ng/kg) and a slight decrease to the bottom water, dissolved Mn shows evidence of a minor rise from a minimum with 0.3 nmol/kg (15 ng/kg) at 500 m depth to values of about 0.7 nmol/kg (40 ng/kg) at the surface. There is a steep gradient of the Mn profile through the upper pycnocline, it exhibits a slight maximum at about 750 m corresponding with the well-known oxygen minimum at this depth and decreases gradually below this depth. These results agree well with those obtained by other authors (48, 49, 52)

The significant correlation of the Ni concentrations with the nutrient components phosphate and silicate is obvious from the linear regression coefficients r. The resulting relations are:

$$[Ni][nmol/kg] = (2.10 \pm 0.53) + (2.47 \pm 0.25) \cdot [PO_4^{3-}][\mu mol/kg] \quad \text{(eq. 1)}$$
$$r = 0.94$$

$$[Ni][nmol/kg] = (2.97 \pm 0.32) + (0.04 \pm 0.003) \cdot [Si(OH)_4][\mu mol/kg] \quad \text{(eq. 2)}$$
$$r = 0.97$$

Due to this dual behaviour of Ni, a multiple regression can be performed with Ni being a function of both phosphate and silicate:

$$[Ni] = a + b [PO_4^{3-}] + c [Si(OH)_4] \quad \text{(eq. 3)}$$

Table 3 gives the resulting correlation parameters in comparison with those reported in the literature (48, 53). The results show good agreement. From the slope values b and c in equation (3) it could be verified that the role of phosphate in relation to Ni dominates over silicate in the biogeochemical cycle. That means that similar to phosphate Ni is incorporated in organic material (outer covering and soft parts) which can be mineralized more easily, and less in skeleton material which needs silicate for its growth.

Table 3. Correlation parameters for Ni as a function of phosphate and silicate

Author		a	b	c	r
Bruland	[48]	2.74	0.95	0.033	0.995
Sclater et al.	[53]	3.5	1.07	0.033	0.93
Prange	[12]	2.36	1.04	0.027	0.99

7 APPLICATIONS IN BIOLOGY

For the purpose of the ecological assessment of polluted surface waters, trace element determinations in organisms, along with analyses of water, suspended particulate matter and sediment, play an important role as a diagnostic tool. Algae are of particular interest in this respect since they are able to accumulate heavy metals with remarkable concentrations. This enrichment facilitates the analysis, and, as an additional advantage, the analytical results reflect the pollution integrated over a longer period of time. However, this kind of biomonitoring remains fragmentary with respect to local and seasonal variations if resident algae grown in the open water are utilized.

During the past few years, these drawbacks have been eliminated by exposing laboratory-cultured unpolluted algae to the polluted water in specially constructed capsules (37 - 39). This biomonitoring method is independent of the naturally occurring populations, the physiological state is well-defined, the load before exposure is known, the accumulation time can be chosen to be constant, and there are no local and only limited seasonal restrictions for the exposure experiments.

The authors have used atomic absorption spectrometry for Cd, Mn and Pb, and TXRF for multi-element analyses which cover the elements As, Ba,

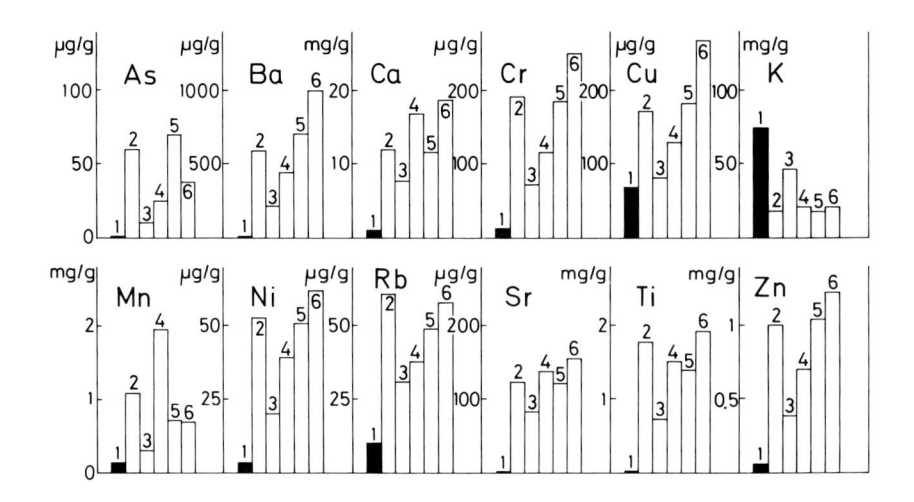

Fig. 10 Accumulation of some elements in *fritschiella tuberosa* after 24 h exposure in the Elbe River together with the load of the unpolluted algae, all referred to the dry substance. Data from 1979. 1: before exposure, 2: April 28, 3: May 8, 4: May 22, 5: October 24, 6: November 7.

Br, Ca, Cr, Cu, K, Mn, Ni, Pb, Rb, Sr, Ti and Zn. The algae were freeze-dried and digested with HNO_3 and $HClO_4$. Seasonal variations and longitudinal profiles of filtrated water and the test algae were investigated in the Elbe River over a period of several years. It was found that the accumulation time at which a steady state with the environment is attained amounts to at most 4 to 6 h. Correlations were studied between the trace metal contents and significant water quality parameters. The method was proved using the alga *fritschiella tuberosa*. Comparisons were made with *cladophora fracta* which was collected from the river.

As an example for the results obtained, Fig. 10 shows some of the data from multi-element analyses before and after a 24 h exposure at different seasons in 1979. All elements investigated except K are enriched. A possible explanation for the behaviour of K may be that this element occupies the available binding sites and is then displaced by the other elements (37).

Other applications in biology deal with the analysis of micro-roots (54) and growth rings (16) from normal and diseased trees. Particularly in the first case, the minute sample masses that are required for TXRF are of great advantage.

8 MICROANALYSIS OF METALS AT THE HYDROGENASE ENZYME FROM THE BACTERIUM *nocardia opaca*

Enzymes control the metabolism in biological systems. The enlightment of their structure is therefore of great importance. One means to approach to this object is the determination of functional metals in these proteins. The extraction of enzymes, especially of those which are produced by bacteria, requires laborious procedures for their isolation and purification which normally yield only small amounts. Moreover, the finally available quantities should be subjected to various distinct investigations apart from element analysis. Therefore, samples as small as approximately 50 µg diluted with 50 mM potassium phosphate buffer to about 50 µl must be sufficient for metal analysis. TXRF is obviously well prepared for this task because of its sensitivity combined with multi-element capability and standardless calibration. Therefore, this technique has been used to investigate the hydrogenase isolated from the bacterium *nocardia opaca* (42). Hydrogenase enzymes are particularly interesting because of their ability to utilize and to evolve gaseous hydrogen, a reversible reaction which can be expressed by the equation

$$H_2 \rightleftarrows 2H^+ + 2e^-.$$

An extraordinary property of the *nocardia*-hydrogenase is its tetramer structure: the enzyme is composed of 4 non-identical subunits with molecular weights of 64000, 56000, 31000, and 27000, respectively (55). Under certain conditions the tetramer dissociates into two subunit dimers (Fig. 11).

NATIVE ENZYME

64 000   56 000

31 000   27 000

+NiCl$_2$ ↕ -NiCl$_2$

LARGE DIMER   SMALL DIMER

64 000   56 000

31 000   27 000

CONSTITUENTS   CONSTITUENTS
1 FMN   2 Ni
1 [2Fe-2S]$^{2+}$ 1 [4Fe-4S]$^{3+}$/[3Fe-xS]$^{3+}$
2 [4Fe-4S]$^{2+}$

REACTIVITY   REACTIVITY
Diaphorase activity   Hydrogenase activity

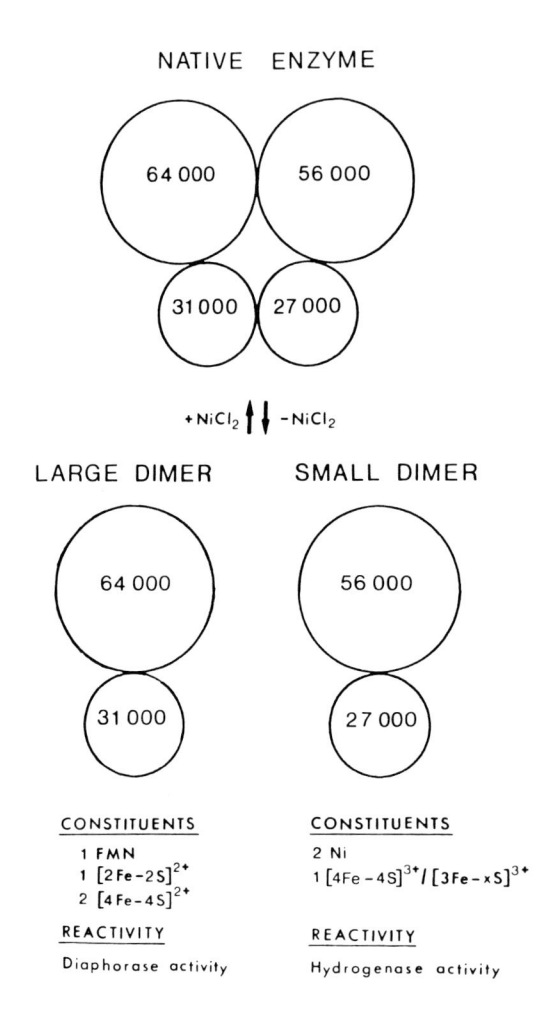

Fig. 11 Structure of hydrogenase from *nocardia opaca*.

In order to find out in which subunit dimer metals are definitely localized, which of the detected components are involved in which reaction, and thus to better understand the complexity of the enzyme structure and the mechanism of the hydrogenase-catalyzed reaction, the metal content of the native hydrogenase as well as of both separated enzyme dimers was determined. The analyses were carried out at about 100 enzyme samples, many of them to control or to exclude metal contamination, which might occur in the procedures needed for isolation, purification and identification of the enzyme and its dimers. The result is given in Table 4.

It has been found that the hydrogenase contains 14 atoms of iron and 4 atoms of nickel and that cobalt, zinc, copper, manganese and selenium - metals which have been detected in other enzymes - are no functional components of hydrogenase. Furthermore, it was possible to elucidate the distribution of iron and nickel among the two subunit dimers into which hydrogenase can be separated. 10 of the 14 iron atoms were localized in the larger dimer (molecular weight of the subunits 64000, 31000) and 4 in the smaller dimer (MW of the subunits 56000, 27000). Nickel was essentially detected in the smaller dimer. Because only this dimer showed hydrogenase activity, whereas the larger dimer was completely inactive, the results indicate that nickel plays a specific role in the enzyme reaction with H_2. The total nickel content of 2 atoms referred to both dimers is in discrepancy with the value of 4 analyzed for the whole enzyme. As it has been observed that the native hydrogenase only remains intact if nickel ions are present in the enzyme solution but that it dissociates into its subunit dimers if the nickel is removed (55), it can be concluded that of the 4 nickel atoms detected in hydrogenase, 2 are tightly bound and directly involved in enzyme catalysis and 2 are only loosely bound and may have a function in subunit binding.

Table 4. Metal content (atoms/molecule) in native hydrogenase and subunit preparations

Metal	Native hydrogenase (MW 178 000)	Large subunit dimer (MW 95 000)	Small subunit dimer (MW 83 000)
Manganese	< 0.2	< 0.2	< 0.2
Iron	13.6	9.6	3.9
Nickel	3.8	0.18	1.8
Cobalt	< 0.2	< 0.2	< 0.2
Copper	0.3	0.37	0.63
Zinc	0.14	0.23	0.36
Selenium	< 0.05	< 0.05	< 0.05
Rubidium	0.15	0.16	0.16
Lead	< 0.05	< 0.06	< 0.13

9 CONCLUSIONS

In summary, the diversity of the applications presented in this paper has demonstrated the meanwhile wide-spread use of Total Reflection X-Ray Fluorescence Analysis (TXRF). That implies that TXRF is neither a method of limited applicability and thus confined to few special analytical tasks nor is it difficult to use. On the contrary, it proves to be a versatile tool for tackling various analytical problems. TXRF is a genuine multi-element technique thereby accomplishing many real demands in analytical chemistry. Analyses of specimens with quite different matrices and concentration ranges have elucidated the characteristic features of the method: the inherent calibration function with the associated possibility of internal standardization, the high detection power and the minute sample masses required.

10 REFERENCES

(1) Y. Yoneda, T. Horiuchi, Rev. Sci. Instr. 42 (1971) 1069.

(2) H. Aiginger, P. Wobrauschek, Nucl. Instr. and Meth. 114 (1974) 157.

(3) P. Wobrauschek, H. Aiginger, Anal. Chem. 47 (1975) 852.

(4) J. Knoth, H. Schwenke, Fresenius Z. Anal. Chem. 291 (1978) 200.

(5) P. Wobrauschek, H. Aiginger, X-Ray Spectrom. 8 (1979) 57.

(6) J. Knoth, H. Schwenke, Fresenius Z. Anal. Chem. 301 (1980) 7.

(7) H. Schwenke, J. Knoth, W. Michaelis, Proc. 4th Int. Conf. on Nuclear Methods in Environmental and Energy Research, April 14 - 17, 1980, Columbia, MO, USA, CONF-800433, p. 313.

(8) H. Schwenke, J. Knoth, Nucl. Instr. and Meth. 193 (1982) 239.

(9) K. Freitag, this conference.

(10) J. Knoth, H. Schwenke, Fresenius Z. Anal. Chem. 294 (1979) 273.

(11) A. Knöchel, A. Prange, Fresenius Z. Anal Chem. 306 (1981) 252.

(12) A. Prange, Dissertation, Universität Hamburg, 1983.

(13) A. Prange, A. Knöchel, W. Michaelis, Anal. Chim. Acta (submitted for publication).

(14) J. Knoth, H. Schneider, H. Schwenke, K. Freitag, Anal. Chem. (submitted for publication)

(15) W. Michaelis, J. Knoth, A. Prange, H. Schwenke, Advances in X-Ray Analysis, 33rd Annual Denver X-Ray Conference, July 30 - August 3, 1984, Denver, CO, USA.

(16) W. Michaelis, H.-U. Fanger, R. Niedergesäß, H. Schwenke, this conference.

(17) W. Michaelis, GKSS 83/E/39.

(18) W. Michaelis, Proc. Int. Conf. on Heavy Metals in the Environment, September 6 - 9, 1983, Heidelberg, Germany, Vol. II, p. 972.

(19) B. Anders, W. Junge, J. Knoth, W. Michaelis, R. Pepelnik, H. Schwenke, Proc. 5th Int. Conf. on Nuclear Methods in Environmental and Energy Research, April 2 - 6, 1984, Mayaguez, Puerto Rico.

(20) W. Michaelis, H. Böddeker, K. Kramer, R. Niedergesäß, R. Racky, C. Schnier, K. Weiler in W. Michaelis, H.-D. Knauth (Hrsg.): Das Bilanzierungsexperiment 1982 (Bilex '82) auf der Unterelbe - Experimentelle und theoretische Hilfsmittel, Ergebnisse und ihre Bewertung , GKSS (to be published).

(21) GKSS Research Centre, Gewässeranalytische Untersuchungen auf der Unterweser im Herbst 1979, Teil 2: Hochauflösende hydrographische Messungen und Spurenanalytik, GKSS 80/E/27.

(22) GKSS Research Centre, Schadstoffuntersuchungen an ausgewählten Standorten mit Muschelvorkommen im nordfriesischen Wattenmeer.

(23) B. Schneider, K. Weiler, Environ. Techn. Lett. (submitted for publication).

(24) H.-U. Fanger, W. Michaelis, Tagung der Arbeitsgemeinschaft der Großforschungseinrichtungen (AGF) "Strahlung und Radionuklide in der Umwelt", 8. - 9. November 1984, Bonn-Bad Godesberg.

(25) K. Freitag, J. Knoth, H. Schwenke, GKSS 79/E/9.

(26) B. Bartels, Diplomarbeit, Universität Hamburg, 1981.

(27) B. Bartels, K. Freitag, R. Rath, Forum Städte-Hygiene 32 (1981) 220.

(28) P. Ketelsen, Dissertation, Universität Hamburg, 1982.

(29) W. Michaelis, H. Böddeker, J. Knoth, H. Schwenke, VIth World Congress on Air Quality, May 16 - 20, 1983, Paris, France, Vol. I, p. 391; GKSS 83/E/33.

(30) AFR-Berichte, Elementanalyse von Schwebstäuben, KFK-AFR 006.

(31) P. Ketelsen, A. Knöchel, Fresenius Z. Anal. Chem. 317 (1984) 333.

(32) W. Junge, J. Knoth, R. Rath,
Neues Jahrbuch Miner. Abh. 147 (1983) 169.

(33) U. Hentschke, W. Junge, R. Rath (to be published).

(34) A. Knöchel, R. Makhraban, A. Prange, R. Marten, W. Michaelis,
J. Knoth, H. Schneider, GKSS 84/E/32.

(35) A. Knöchel, A. Prange, Microchim. Acta (II) (1980) 395.

(36) A. Prange, K. Kremling, Marine Chemistry (submitted for publication).

(37) W. Ahlf, Dissertation, Universität Hamburg, 1982.

(38) W. Ahlf, A. Weber, Environ. Techn. Lett. 2 (1981) 317.

(39) B. Schmidt, Dissertation, Universität Hamburg (to be published).

(40) J. Knoth, H. Schwenke, R. Marten, J. Glauer,
J. Clin. Chem. Clin. Biochem. 15 (1977) 557.

(41) B. Hein, Diplomarbeit, Universität Hamburg, 1981.

(42) K. Schneider, R. Cammack, H.G. Schlegel, Eur. J. Biochem. (in press).

(43) Rich. Seifert & Co., Röntgenwerk, D-2070 Ahrensburg: Extra II.

(44) A. Müller, B. Kunze in W. Michaelis, H.-D. Knauth (Hrsg.):
Das Bilanzierungsexperiment 1982 (Bilex '82) auf der Unterelbe
- Experimentelle und theoretische Hilfsmittel, Ergebnisse und
ihre Bewertung, GKSS (to be published).

(45) L.D. Krishnaswami, J.M. Martin, M. Meybeck,
Earth Planet. Sci. Lett. 11 (1971) 407.

(46) K. Enke, R. Rath, Neues Jahrbuch Miner. Abh. 141 (1981) 1.

(47) K. Enke, W. Tolksdorf, Rev. Sci. Instr. 49 (11) (1978) 18.

(48) K.W. Bruland, Earth Planet. Sci. Lett. 47 (1980) 176.

(49) W.M. Landing, K.W. Bruland, Earth Planet. Sci. Lett. 49 (1980) 45.

(50) G.A. Knauer, J.H. Martin, R.M. Gordon, Nature 297 (1982) 49.

(51) B. Schaule, Dissertation, Universität Heidelberg, 1979.

(52) M.L. Bender, G.P. Klinkhammer, D.W. Spencer,
 Deep Sea Res. 24 (1977) 799.

(53) F.R. Slater, E. Boyle, J.M. Edmond,
 Earth Planet. Sci. Lett. 31 (1976) 119.

(54) GKSS Research Centre (unpublished data).

(55) K. Schneider, H.G. Schlegel, K. Jochim,
 Eur. J. Biochem. 138 (1984) 533.

PIXE ANALYSIS - PHYSICAL BASIS AND EXAMPLES OF APPLICATIONS

B. Gonsior and M. Roth
Institut für Experimentalphysik
Ruhr-Universität Bochum, Postfach 10 21 48
D-4630 Bochum 1

SUMMARY

The analytical use of particle induced X-ray emission (PIXE) spectroscopy has become an important tool in trace element analysis in different fields, especially in those cases, where only small amounts of sample material are available.

To this end ion beams of particle accelerators with beam diameters about 1 mm or narrow beams with micrometer dimensions have been applied. The availability of semiconductor detectors with high energy resolution allowed for measuring the induced characteristic X-rays.

As far as the field of applications is concerned many different regions of research took advantage of the PIXE method (1,2). In this fast growing field we confine ourselves here onto specific aspects. Due to the role of trace elements in metabolic processes the investigation of biological materials plays a decisive role in bioengineering research to give one example. A general survey as well as the spatial distribution of trace elements in the microstructure might be of interest.

1. INTRODUCTION

From the analytical point of view one can investigate matter under two main different aspects: mean elemental composition of a piece of material as a whole and micro structural properties of matter where the local property cannot be extrapolated from the one in bulk material.

To meet these requirements during the last ten years a broad development of analytical techniques based on particle induced X-ray emission (PIXE) took place. This was caused by the availability of particle accelerators and of semiconductor detectors with high energy resolution. Ion beams from accelerators are used to induce characteristic X-rays.

The principal aim of this paper is to describe the physical basis, facts of importance using PIXE for quantitative analytical purposes in a wide range of concentrations and some main topics of application.

2. THE PHYSICAL BASIS OF PIXE

The production of a vacancy in an inner shell, using ions like protons, deuterons, α-particles or heavier ions and the subsequent relaxation of the atomic shell by emission of characteristic X-rays is the physical basis of trace element analysis by PIXE. The ionisation of the inner shell is considered to arise by the electromagnetic interaction between the bound electron and the incident particle. The registration of characteristic X-rays is the basis for the analyzing process.

The Moseley law gives a relationship between the wavelength of the emitted radiation and the atomic number of the respective element. This radiation is therefore characteristic in energy regarding the emitting atom. There are other kinds of transitions of less importance for analytical purposes; so, the ratio of vacancies filled by emission of X-ray quanta to the total number of vacancies in a shell, called fluorescence yield ω, is < 1.

According to these considerations the cross section for production of X-rays σ_x can be described as the product of the cross section for producing vacancies in a certain inner shell σ_I and the fluorescence yield ω. Therefore we have

$$\sigma_x = \omega \cdot \sigma_I \quad .$$

The production cross section σ_x is related to the actual number of X-ray quanta detected and the X-ray emitting target atoms under consideration. To find this number of atoms is the analytical aim. Actually this number of atoms can be given as a mass per cm^2 crossed by the beam. This is the so called mass layer of the target. One considers the area of the sample, which is transversed by the particle beam. The mass layer is given in units of $g \cdot cm^{-2}$. On the basis of the area under irradition we can directly speak of the mass of a substance given in g. In this way we get a measure for the amount of a certain substance under investigation in relation to the target specimen as a whole. We arrive at concentrations.

3. PRINCIPLES OF THE USE OF PIXE FOR TRACE ELEMENT ANALYSIS

The spectrum of the characteristic X-ray contains the lines of different transitions corresponding to the amount and kind of elements present in the sample. In this way identification of trace elements is possible. The most popular detectors for X-ray spectroscopy are semiconductor detectors. The energy resolution is sufficient to resolve lines of neighbouring elements rendering possible the simultaneous determination of these elements. K- as well as L-radiation can be used for analysis according to whether K- oder L-shell ionization results in a higher yield in the detector. There might arise problems in certain cases. This is possible e.g. when one tries to determine a trace element in presence of large amounts of neighbouring elements. There are also problems associated with interferences where the K-line of one element falls close to the L-line of another element.

Applicability and sensitivity of the method depend on the ratio of the characteristic X-ray yield to the amount of background radiation present in the energy region of the X-ray line. Therefore an estimation of sensitivities achievable requires a consideration of background contributions. The background in PIXE is dominated by bremsstrahlung created by the stepwise slowing down of energetic secondary electrons arising through scattering of the ion beam in the target. A continuous spectrum is resulting.

In case of PIXE analysis sensitivities in the ppm region can be achieved for all elements. Small amounts of material are sufficient. Experiments have been done with amounts of less than 1 mg. The detection limits obtained as far as the absolute amount of trace elements is concerned depend on the sample area irradiated. Using a beam of 5 mm^2, a sample thickness of 10 $\mu g/cm^2$ and a sensitivity of 10^{-5} quantities of

$5 \cdot 10^{-12}$ g can be detected. Collimating the impinging beam to a diameter of 10 μm and a beam current of 1 nA quantities of 10^{-16} g become detectable.

There are various kinds of definitions of the detection limit. In a simple way the detection limit can be defined as the ratio of the X-ray peak to the background radiation yield both taken in the energy region of the double full width half maximum (FWHM) of the peak. This ratio should be at least equal to one.

Then a certain detection limit can be stated as a minimum detectable concentration $C_{min} = T^{trace}/T^{matrix}$ where T is the respective mass layer given in g/cm².

Important for the achievable power of detection is the material of the substratum chosen. In cases where only deposits on surfaces are to be analyzed the substratum foil should be taken as thin as possible in order to keep the background due to bremsstrahlung small.

4. EXPERIMENTAL ARRANGEMENTS

In principle a system for trace element analysis by emitted characteristic X-rays is composed of a suitable irradiation source, a containment for a sample target and an X-ray detector. To perform PIXE analysis one must have an ion accelerator at one's disposal. The chamber containing the sample target is connected to the beam line from the accelerator and is generally under vacuum as well as the beam line system. This restricts the kind of samples to be investigated since the sample in question has to be stable under vacuum conditions and must not be destroyed by overheating during particle irradiation. Close to this irradiation chamber the X-ray detector is mounted, separated from the vacuum system by a thin beryllium window.

The beam spot at the target position amounts to a few mm² in most cases of application. Regarding the peculiar applicational possibilities using a highly focused microbeam with a diameter of some μm review is given below. To get absolute concentrations the beam current must be measured; this is done by counting the ions of the beam, e.g. protons, crossing the target, i.e. measuring the integrated beam charge.

To overcome the difficulty of overheating or evaporation in the case of organic samples, with their poor thermal conductivity, PIXE analysis was used in combination with the external beam technique (3). The accelerated particle beam leaves the vacuum system through a thin vacuum-tight exit foil and impinges on the sample which is located about 10 mm downstream. Heating-up of the sample is efficiently reduced by air cooling.

A powerful system for the analysis of microstructures is the nuclear microprobe (4). In certain cases it is desirable to investigate trace element distributions in small areas. Therefore in a number of laboratories beams of accelerated protons with spot sizes as small as 1 µm have been produced. This can be achieved by using a series of quadrupole magnets to focus the particle beam. The electron microprobe is already a standard tool in many laboratories. There is a strong limitation in power of detection, however, being a factor of more than 10^2 smaller than in the case of the proton microprobe. This is due to the high bremsstrahlung background from the primary electrons. On the other hand spot sizes of 0.1 µm can be gained with an electron microprobe which only in future developments can be achieved with proton microprobes.

5. TARGET PREPARATION

The preparation of sample targets is one of the main problems of induced X-ray emission analysis. Most simple is the investigation of the so-called "thin" targets, i.e. such targets where the energy loss of the impinging particles while crossing the target is small in comparison with the energy of the particles. In many cases one can use filters, e.g. nuclepore filters as substratum, thin in the described kind, for aerosol or solution samples or the like. Also thin carbon foils are used in those cases. They are strong but unfortunately often show too high an amount of impurities.

Biological and medical specimens can be irradiated directly if prepared as microscopic cuts. The same can be done in the case of samples in the material sciences or in geology.

There are other samples where by special preparation it becomes possible to raise the detection limit drastically, i.e. by orders of magnitude, applying preconcentration procedures. In this case a larger amount of material is a prerequisite.

6. EXAMPLES OF APPLICATION

Soon after the introduction of PIXE into the field of applications many different regions of research took advantage of this method. Therefore today it seems to be difficult to give a complete survey on this fast growing field. Our purpose here is to confine ourselves to some special examples.

Before going into some details we prefer to give an outline of applicational work. In many applications a general survey of the multi-element content is desirable. On the other hand one might be interested especially in certain elements e.g. the toxic heavy metals in the field of environmental research. In such cases to get highest power of detection one can take advantage of using selected experimental conditions, e.g. prefered projectile energies or preconcentration techniques. In addition there are special applications where interest arises regarding aspects of spatial distribution of trace elements in the sample. This might be of concern regarding distribution profiles with specific consideration for surface sensitivity as well as regarding lateral distributions. Let us first present one example, where preconcentration techniques are applied to investigate trace metals in water.

6.1 TRACE METALS IN WATER SAMPLES

Water is one of the main parts of our environment and of large importance in the food chain of men. It contains essential trace elements as well as toxic elements due to the industrial activities. The list of essential trace elements meanwhile amounts to not less than 60 elements whereas most of them also seem to be toxic if they are present in too large an amount. It turns out that most of the trace element concentrations have to be maintained in certain limits, i.e. above a lower one, in order to establish the role as an essential trace element and below an upper one, not to run the risk of their toxic properties. It seems to be obvious that the knowledge of the trace element concentration of water is highly important.

Analyzing trace metals in water in some cases is performed using preconcentration techniques already mentioned. Then the limits of detection are improved appreciably. In most cases also matrix interferences can be eliminated at the same time. The sample targets for PIXE prepared in this way have to be extremely homogeneous. If targets are quite sensitive with respect to overheating they are investigated by use of an external proton beam (5,6).

In cooperation with Jackwerth et al. a preconcentration technique has been developed based on collector precipitation with chelating material (7,8). Professor Jackwerth is presenting this method at the Symposium. Therefore in this paper we can refrain from details and concentrate ourselves on some results shown in Tab. 1. Given are the results for different trace metals. In general high recovery could be achieved, whereas detection limits of 1 ppb or lower have been observed. The standard deviation, using this method, amounted to (5 - 20) % depending

Tab. 1   Analysis of drinking water.
Recovery and limit of detection

Trace Metal	Recovery [%]	Limit of Detection $[\mu g \cdot l^{-1}]$ (3 σ, N = 20)
Bi	$\geq$ 95	1.3
Cd	$\geq$ 95	3.5
Co	$\geq$ 95	0.2
Cu	$\geq$ 95	0.2
Fe	$\geq$ 95	0.8
Hg	$\geq$ 95	1.5
In	$\geq$ 95	4.0
Mn	80	0.2
Ni	$\geq$ 95	0.2
Pb	$\geq$ 95	2.0
Pd	$\geq$ 95	3.5
Tl	70	1.5
Zn	$\geq$ 95	0.6

on the trace metal under consideration.

In Tab. 2 a comparison between PIXE and AAS is shown, investigating two samples from the water pipe.

Tab. 2   Analysis of drinking water. Comparison of different methods

Sample no	Method	Concentration $[\mu g \cdot l^{-1}]$				
		Zn	Fe	Ni	Cu	Mn
1	PIXE	208	30	5.8	5.4	3.6
	AAS	198	28	5.3	5.5	3.4
2	PIXE	580	43	5.0	6.7	3.7
	AAS	615	50	5.7	6.8	3.6

6.2 TRACE ELEMENTS IN BIOLOGICAL MATERIALS

Due to the role of trace elements in the living organism, especially in metabolic processes, there has been a drastic increase in investigating biological materials.

In recent years we have used the Bochum proton microprobe (9,10) in combination with PIXE for trace element analysis of biological samples. We investigated the element concentrations of different sample regions by irradiating single spots. Beyond that we have installed a deflection unit at our microprobe, in order to gain complete information on the element distributions by scanning the beam over a contiguous sample area. This procedure requires an efficient data-acquisition system, as it is necessary to process every X-ray event with respect to the actual beam position. One of our projects using the Bochum proton microprobe consists in investigating rabbit liver samples. The sample was taken

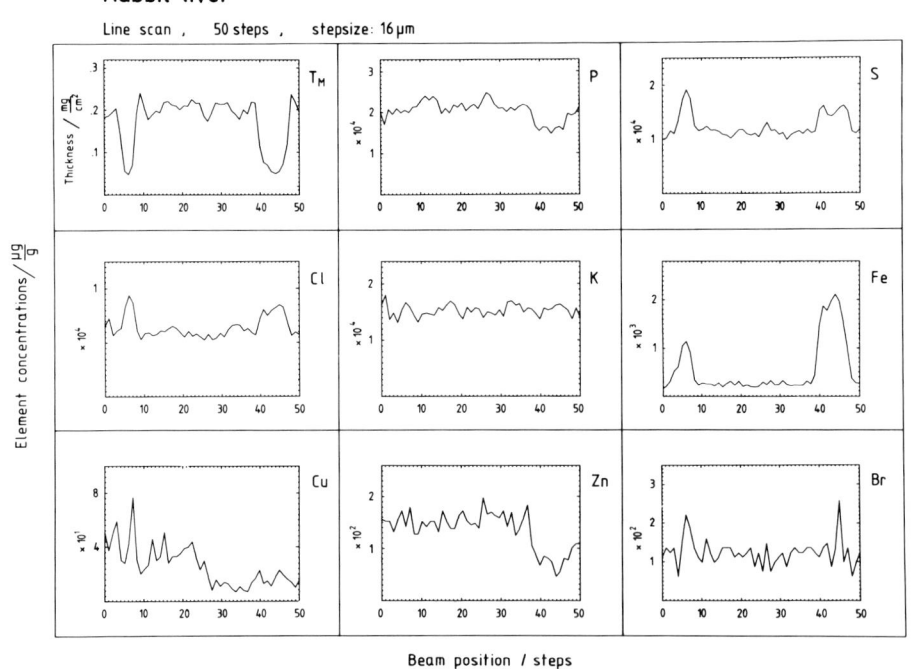

Fig. 1: Results of the line-scan measurements of a rabbit liver sample. For Different elements the concentration along the scan is given. In the upper left drawing the target thickness is depicted, indicating the two veins as holes.

after anaesthesia of the rabbit using a bromine containing narcotic gas. We have been interested in the distribution of the bromine between two veins. The determined concentration profiles as well as the target thickness are depicted in fig. 1. These measurements were performed in cooperation with Dr. E. Rokita of the Jagellonian University of Krakow. The scientific problem consists in the pathways of metabolization of trace substances, e.g. bromium.

Special care must be observed in respect to the sample preparation. To avoid the migration of ions i.e. in order to get a realistic picture of spatial distributions the samples must be immediately quench-frozen at -190°C before cryosectioning them.

CONCLUSIONS

In view of the work which has been done so far in respect to the development of the technique as well as the application is concerned PIXE stand well up to the claims which should be made for a reliable trace element analytical method. Trace element abundances can be obtained for all elements with an atomic number $Z > 12$ simultaneously in one analysis. A disadvantage results from the fact that interferences between K and L X-rays from light and heavy elements, respectively, might limit the number of detectable elements, at least near the detection limit of 10^{-6} to 10^{-7} g/g. Very small amounts of a sample can be analyzed whereby the analyzing time generally is of the order of magnitude of only 10 minutes. Using a proton microprobe with the present lower limit of resolution structures of 1 μm diameter can be investigated. This means a minimum detectable trace element amount of about 10^{-16} g.

The field is still open for developments just to mention the technique of the microprobe where a spatial resolution of 0.1 μm seems to be achievable and the possibility of preconcentration where in some cases already a lowering of the detection limit by more than three orders of magnitude could be achieved.

Taking for granted an interdisciplinary cooperation as a necessary basis PIXE is a method of choice promising interesting successes in many areas of application.

The work was supported in part by the Minister für Wissenschaft und Forschung des Landes Nordrhein-Westfalen, and by the Bundesminister für Forschung und Technologie.

REFERENCES

(1)   S.A.E. Johansson and T.B. Johansson, Nucl. Instr. Meth. 137 (1976) 473-516

(2)   B. Gonsior and M. Roth, Talanta, 30 (1983) 385-400

(3)   G.G. Seaman and K.C. Shane, Nucl. Instr. Meth. 126 (1975) 473-474

(4)   J.A. Cookson, Nucl. Instr. Meth. 165 (1979) 477-508

(5)   B. Raith, H.R. Wilde, M. Roth, A. Stratmann and B. Gonsior, Nucl. Instr. Meth. 168 (1980) 251-257

(6)   B. Raith, A. Stratmann, H.R. Wilde, B. Gonsior, S. Brüggerhoff and E. Jackwerth, Nucl. Instr. Meth. 181 (1981) 199-204

(7)   S. Brüggerhoff und E. Jackwerth, B. Raith, A. Stratmann und B. Gonsior, Fresenius Z. Anal. Chem. 311 (1982) 252-258

(8)   S. Brüggerhoff, E. Jackwerth, B. Raith, S. Divoux und B. Gonsior, Fresenius Z. Anal. Chem. 316 (1983) 221-226

(9)   W. Bischof, M. Höfert, B. Raith, H.R. Wilde and B. Gonsior: Trace Element Analysis of Biological Samples by Means of a Proton Microprobe. In: Analytical Chemistry in Medicine and Biology, Vol. 2, P. Brätter, P. Schramel, (Ed.), Walter de Gruyter & Co., Berlin, New York, 1983, p. 1053-1067

(10)   M. Höfert, W. Bischof, A. Stratmann, B. Raith and B. Gonsior, Nucl. Instr. Meth. 231 (1984), in press

Röntgenfluoreszenzanalyse mit Hilfe der Synchrotronstrahlung (SYRFA)

P. Ketelsen, A. Knöchel, W. Petersen und G. Tolkiehn
Universität Hamburg
Institut für Anorganische und Angewandte Chemie
Martin-Luther-King-Platz 6
D-2000 Hamburg 13

Die Benutzung der Synchrotronstrahlung als Anregungsquelle für Röntgenfluoreszenzanalysen (SYRFA) bietet gegenüber konventionellen Anregungsquellen eine Reihe von Vorteilen, die Multielementanalysen im Spurenbereich gestatten.

- Die hohe Photonenflußdichte und die starke Kollimation erlauben es, die bestrahlte Fläche und damit die bestrahlte Masse der Probe klein zu halten.
- Der hohe Polarisationsgrad verbessert das Peak/Untergrund-verhältnis. Es wird bei einem Polarisationsgrad von 90% um den Faktor 10 verbessert, die Polarisation der Synchrotronstrahlung ist i.a. größer.
- Das bis ca. 35 keV weiße Spektrum ermöglicht gegenüber der monoenergetischen Röhrenanregung eine gleichmäßigere Anregung aller Elemente, wobei noch bis $Z \leq 60$ die K-Fluoreszenzlinien zur Bestimmung herangezogen werden können.
- Die Berechenbarkeit der Anregungsquelle gestattet es, die Methode zu einem Absolutverfahren auszubauen, das ohne zusätzliche Standards auskommt. In der Praxis arbeitet man aber üblicherweise mit einem weitgehend frei wählbaren inneren Standard. Es wurde kritisch überprüft, daß dieser sich bei der Meßprobenpräparation gleichartig wie die Elementspuren verhält.

Zur Charakterisierung der Möglichkeiten und besonderen Vorteile, die das neue Analysenprinzip bietet, wurde die SYRFA hinsichtlich ihrer Nachweisstärke und Eichfunktionen untersucht. Um die Vorteile als Multielementmethode voll auszunutzen, lag der Schwerpunkt der Entwicklung auf der durchleuchtenden RFA mit weißer Anregung. Abb. 1 zeigt die hierzu entwickelte Meßanordnung.

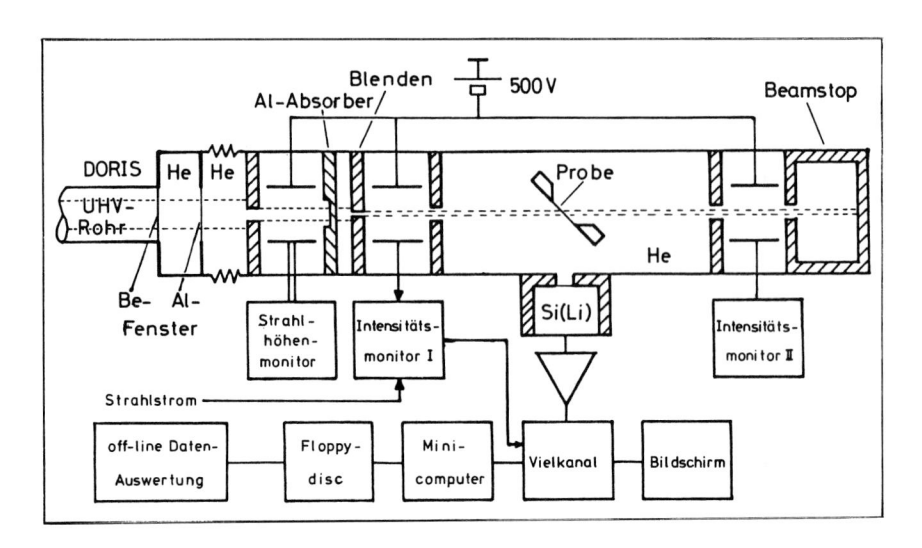

Abb. 1 : Prinzipaufbau der Meßanordnung

Um den Streuuntergrund zu reduzieren und Absorptionsprobleme
zu vermeiden, sollten die Proben möglichst dünn (Flächendichte
$d \leq 1$ mg/cm^2) sein. Speziell hierfür entwickelte Probenpräparationstechniken gestatten es, verschiedenartige Probengüter
zu analysieren[1]. Wässrige und organische Lösungen sind durch
Einbettung in eine Polymermatrix und anschließendes Eintrocknen
im Probenträger zu dünnen hydrophilen bzw. lipophilen Filmen
($d \sim 1$ mg/cm^2) zu verarbeiten. Stark verdünnte Lösungen werden
einer chemischen Anreicherung unterworfen und in eine organische
Lösung der komplexierten Elemente überführt, die dann als lipophile Filmprobe präpariert wird. Dünne Proben mit einer Flächendichte von $d \leq 2$ mg/cm^2 können direkt vermessen werden, wenn eine
homogene Verteilung vorliegt. Mögliche Anwendungen ergeben sich
so für Anreicherungsverfahren mit Hilfe von Ionenaustauscherpapieren oder beaufschlagte Filterproben (z.B. Luftstaubproben).
Die Bestimmung des physikalischen Nachweisvermögens ohne Berücksichtigung der Probenvorbereitung ergab relative Werte im Bereich
von 0.05 µg/g bis 1 µg/g in Abhängigkeit von der Ordnungszahl
und den Anregungsbedingungen. Aufgrund der gleichmäßigeren Anregung durch die weiße Primärstrahlung ist die Abhängigkeit von
der Ordnungszahl weniger stark ausgeprägt als bei RFA-Methoden

mit monoenergetischer Anregung (wie z.B. bei der röhren- oder radionuklidangeregten RFA). Abb. 2 zeigt einen Vergleich des experimentell bestimmten Nachweisvermögens mit den entsprechenden, aus den Charakteristika der Synchrotronstrahlung berechneten Werten.[2,3]

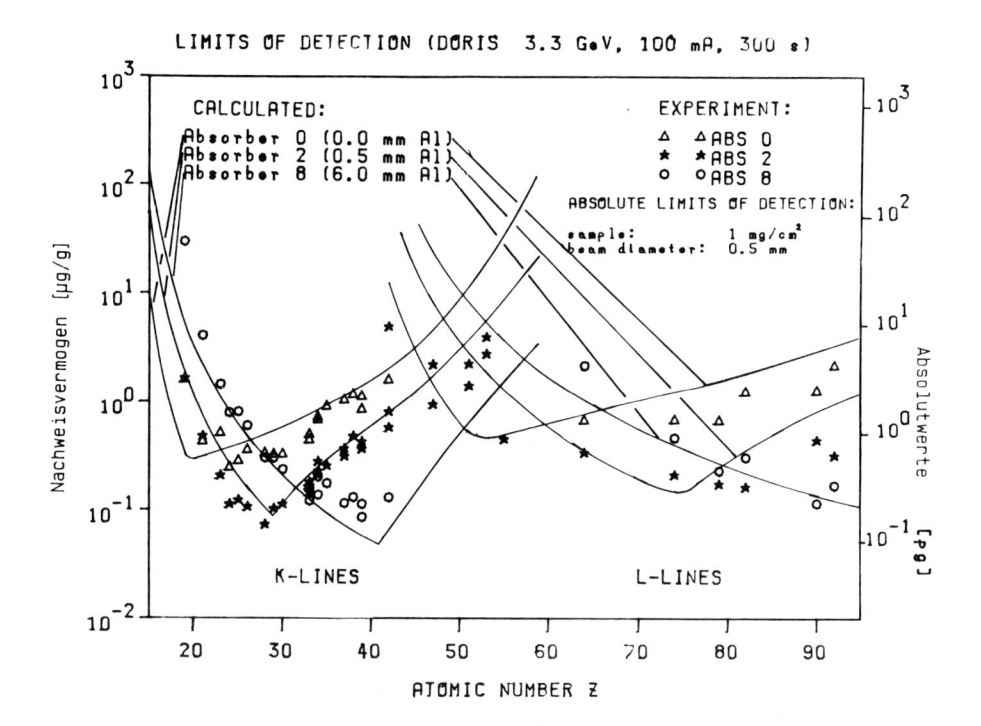

Abb. 2 : Theoretische und experimentelle Werte für das Nachweisvermögen

Gegenüber den bisher bekannten Verfahren wird ein mindestens um den Faktor 10 besseres relatives Nachweisvermögen erreicht. Dies ist vor allem auf die wesentlich geringere Matrixempfindlichkeit im Vergleich mit anderen Methoden zurückzuführen. Aufgrund der hohen Intensität und der starken Kollimation der Synchrotronstrahlung zeigt die Methode ein besonders gutes absolutes Nachweisvermögen. Die Werte, die jedoch immer nur für eine definierte Probenmatrix und einen definierten Strahl angegeben werden können, betragen für eine bestrahlte kreisförmige Fläche von 0.5 mm Durchmesser und eine Probe mit einer Flächendichte von 1mg/cm^2 (Kohlenstoffmatrix) 0.1-0.4 pg.

Das Nachweisvermögen läßt sich sowohl für die Elemente mit K-Linien als auch für die mit L-Linien für bestimmte Z-Bereiche durch Filterung der Primärstrahlung noch weiter verbessern, wobei das Optimum zu höheren Ordnungszahlen verschoben wird. Während bei der Anregung mit der ungefilterten weißen Primärstrahlung das Optimum für die K-Linien emittierenden Elemente im Bereich von Z=20 liegt, kann es durch Einsatz von stärkeren Absorbern im Primärstrahl zu Z=40-50 verschoben werden. Bei einer Probe mit der Dicke d~1mg/cm^2 wird dann noch immer mit einem Anregungsstrahl von 2 mm Durchmesser eine ausreichende Zählrate erreicht.

Zur Auswertung der erhaltenen Spektren wurde ein spezielles Auswerteprogramm entwickelt, welches es gestattet, halbautomatisch und interaktiv am Bildschirm die Spektren auszuwerten, wobei das Ergebnis, die Fitqualität, sofort überprüft werden kann. Das Programm fittet die Fluoreszenzlinien über Gaußfunktionen mit einer definierten Halbwertsbreite. Die als feste Parameter in das Programm eingehenden Linienverhältnisse eines Elements wurden experimentell in Abhängigkeit von den Anregungsbedingungen bestimmt. Mit diesem Programm werden anschließend über den mitgemessenen inneren Standard die Massenbestimmungen durchgeführt werden.[1]

Zur Bestimmung der Reproduzierbarkeiten und der Richtigkeit wurde die Methode durch vergleichende Analysen von Referenzmaterialien überprüft [1]. Zur Anwendung kamen Proben des National Bureau of Standard (NBS) sowie im Rahmen einer Ringanalyse untersuchte Proben. Die erzielten Übereinstimmungen liegen im Bereich der Fehlergrenzen. Die Reproduzierbarkeiten sind abhängig von der Konzentration, liegen aber größtenteils im Bereich von 10%.

Die Tabelle 1 zeigt als Beispiel die Analysenwerte für die NBS-Probe pine needles (NBS SRM 1575). Die Proben wurden zur Homogenisierung einem Druckaufschluß (HNO$_3$) unterworfen und ein Aliquot davon wurde als Polyvinylpyrrolidonfilm präpariert.

Tabelle 1 : Vergleichsanalyse der NBS-Probe SRM 1575

	SYRFA [µg/g]			NBS-Werte [µg/g]		
Mn	631.9	±	6.1	675.	±	15
Fe	176.0	±	8.7	200	±	10
Cu	3.49	±	0.40	3.0	±	0.3
Zn	61.2	±	1.4	63.	±	6.0
As	0.9	±	0.14	0.21	±	0.04
Rb	12.0	±	1.23	11.7	±	0.1
Sr	4.86	±	0.30	4.8	±	0.2
Zr	0.65	±	0.07	-		-
Pb	13.3	±	3.3	10.8	±	0.5

Untersuchungen auf systematische Fehler ergaben bei den hydrophilen Filmproben inhomogene Verteilungen bei den Alkali- sowie den Erdalkalielementen und den Halogenen. Bei diesen Elementen kann das Analysenergebnis verfälscht werden, da immer nur ein sehr kleiner Teil der Probe ($\emptyset \leq 2$ mm) ausgeleuchtet wird. Bei den lipophilen Filmen wurden keine Inhomogenitäten festgestellt. Elementverflüchtigungen aufgrund der Strahlenbelastung waren bei keiner Probenform festzustellen. Die bei den Multielementproben möglichen Störungen von Interferenzen wurden an Hand der beiden für die energiedispersive RFA typischen Beispiele Ba/Ti und As/Pb näher untersucht. Bei Barium und Titan können die Probleme durch Auswertung der bei weißer Anregung ebenfalls emittierten K-Linien von Barium aufgelöst werden. Hierbei muß jedoch eine Verschlechterung des relativen Nachweisvermögens in Kauf genommen werden.

Besonders attraktive Anwendungsmöglichkeiten der SYRFA ergeben sich aufgrund der hohen Intensitätsdichte für ortsauflösende Analysen. Hierzu reicht es aus, den Strahl mit Hilfe von Blenden stark auszublenden. Bis zu einem Blendendurchmesser von ~ 10 µm ist noch genügend Intensität vorhanden, um bei Probendicken von 10-50 µm (Kohlenstoffmatrix) ausreichende Zählraten zu erzielen. Gegenüber anderen Mikromethoden wie PIXE, Massenspektroskopie (z.B. LAMMA) bietet die Röntgenfluoreszenzanalyse mit Hilfe der Synchrotronstrahlung folgende Vorteile :

- Einfache Ausblendung des Strahls ohne weitere Kollimierung

- Bewegung der Probe im Strahl
- Besseres Nachweisvermögen
- Möglichkeit der Absoluteichung
- Analyse von dicken Proben
- Keine Verflüchtigungsprobleme (Messung unter He-Normaldruck)
- Geringere Strahlenbelastung

Für die ortsabhängige Analytik wurde inzwischen eine spezielle Apparatur zur Abrasterung aufgebaut, die es gestattet, verschiedene Proben in Schritten von 0.5 µm in X- und Y-Richtung abzuscannen. Erste Messungen zur Auflösungsbestimmung zeigt das folgende Bild einer linearen Abrasterung eines Kupfergitters mit einer Maschenweite von ca. 50 µm:

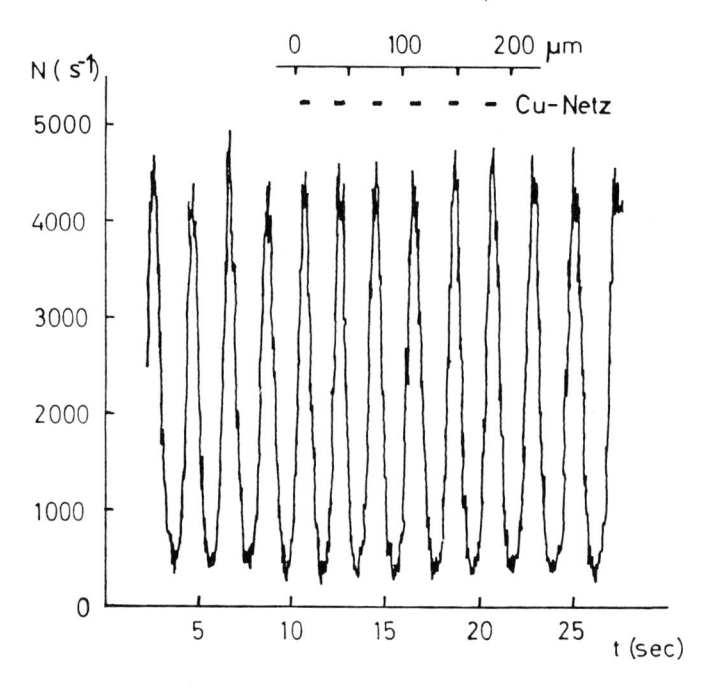

Abb. 3 : Abrasterung eines Gitters

Die aus diesen Messungen resultierende Auflösung liegt bei ca. 20 µm. Durch verbesserte geometrische Verhältnisse sollten jedoch Auflösungen im Bereich von 10 µm erreichbar sein. Erste Abschätzungen über das Nachweisvermögen ergaben bei 10 µm Auflösung und 100 sec. Meßzeit pro Pixel ein Nachweisvermögen im 10^{-15} g-Bereich.

Erste praktische Ergebnisse zeigt die Abrasterung eines Dünnschnitts einer Kohlenprobe. Abbildung 4 zeigt den Ausschnitt einer linearen Abrasterung, wobei die erhaltenen Spektren hintereinander gelegt wurden.

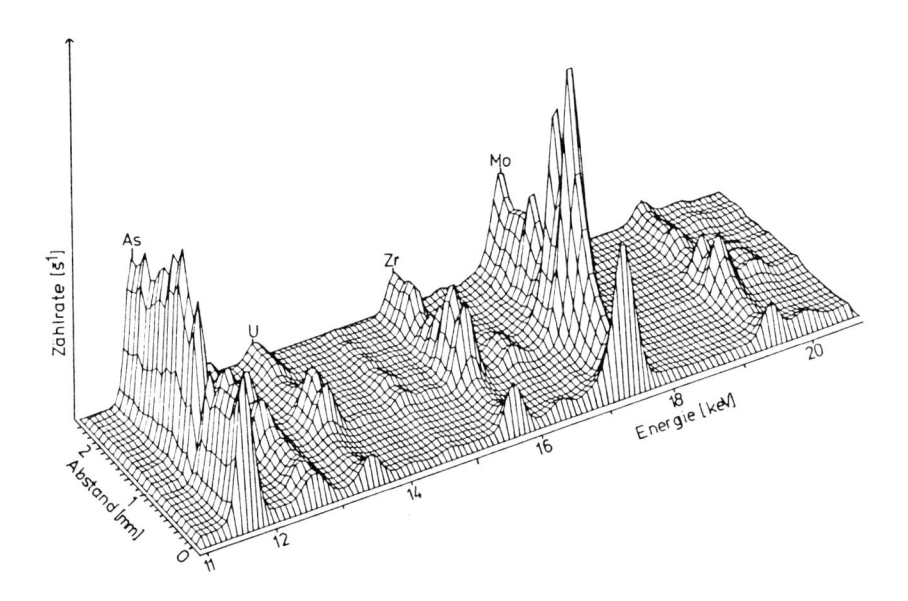

Abb. 4 : Abrasterung einer Kohleprobe

Aus den Messungen sind die Elementverteilungen deutlich zu erkennen, wobei einige Elemente miteinander korrelieren.
An der Fortentwicklung dieser Analysentechnik wird gearbeitet.
Zusammenfassend ergeben sich folgende Vorteile bei der Verwendung der Synchrotronstrahlung als Anregungsquelle für die Röntgenfluoreszenzanalyse :
- Geringere Abhängigkeit des Nachweisvermögens von der Ordnungszahl
- Besseres relatives Nachweisvermögen aufgrund des hohen linearen Polarisationsgrades
- Sehr gutes absolutes Nachweisvermögen aufgrund der hohen Strahlungsdichte
- Selektive Anregung mit durchstimmbarem Monochromator in komplizierten und sonst nicht lösbaren Elementgemischen

Nachteilig sind die Standortfixierung und die meist geringe Verfügbarkeit der Strahlungsquelle, die die praktische Anwendung auf sehr spezielle Analysenprobleme, die mit anderen Methoden nicht oder nur mit erheblich größerem Aufwand gelöst werden können, beschränken.

Ein praktisches Anwendungsbeispiel sind z.B. beaufschlagte Luftstaubfilter, die direkt und ohne weitere chemische Präparationsschritte und den damit verbundenen Kontaminationsrisiken analysiert werden können. Für die meisten hierbei relevanten Elemente wird dabei auch ein ausreichendes Nachweisvermögen erzielt.

Die Abbildung 5 zeigt einen entsprechenden Vergleich der Nachweisgrenzen zwischen der konventionellen RFA (durchleuchtete Proben und Anregung mit der Mo-Röhre), der Röntgenfluoreszenzanalyse mit totalreflektierendem Probenträger (TRFA)[4] und der SYRFA.

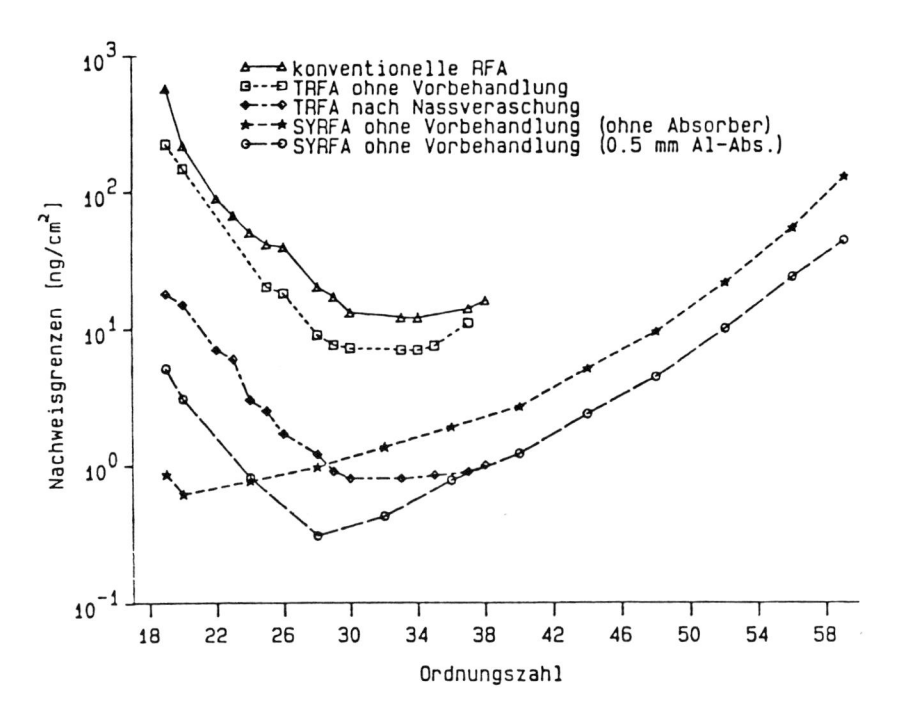

Abb. 5 : Vergleich des Nachweisvermögens bei der Analyse von Luftstaubfiltern (Flächendichte 5 mg/cm^2) und der Verwendung verschiedener RFA-Methoden)

Die Abbildung zeigt deutlich, daß mit Hilfe der SYRFA und
Direktmessung der Proben gegenüber der TRFA, bei der
normalerweise ein Aufschlußschritt vorangeht, ein vergleich-
bares bzw. sogar ein besseres Nachweisvermögen erzielt wird und
auch schwere Elemente noch über K-Linien erfaßt werden können.

Literatur

1. W. Petersen
   Dissertation, Universität Hamburg 1984
2. W. Petersen und G. Tolkiehn
   Nucl. Instr. Meth., 215 (1983) 515-520
3. A. Knöchel, W. Petersen und G. Tolkiehn
   Nucl. Instr. Meth., 208 (1983) 659-663
4. P. Ketelsen und A. Knöchel
   Z. Anal. Chem., 317 (1984) 333-342

HIGH-RESOLUTION ICP SPECTROSCOPY USING A COMPUTER-CONTROLLED ECHELLE SPECTROMETER WITH PREDISPERSER IN PARALLEL SLIT ARRANGEMENT

P.W.J.M. BOUMANS AND J.J.A.M. VRAKKING

Philips Research Laboratories
P.O.Box 80.000, 5600 JA Eindhoven,
The Netherlands

ABSTRACT

This paper discusses the application of high-resolution spectroscopy to inductively coupled plasma (ICP) atomic emission spectrometry (AES) with particular reference to work in the authors' laboratory involving the exploration of a new type of echelle spectrometer. The paper discusses this work in the perspective of earlier studies by LAQUA et al. on the application of high resolving power to spectrochemical analysis and in the light of recent results of measurements of physical line widths in ICPs. The main body of the paper summarizes the authors' work recently published in Spectrochimica Acta, Part B, dealing with
(a) the instrumental characteristics of the echelle spectrometer and the measurement of physical widths of lines emitted by an argon ICP,
(b) the analytical optimization of the instrument, that is, the slit width optimization with a view to both signal-to-background ratio (SBR) and relative standard deviation $(RSD)_B$ of the background signal, so as to minimize the detection limit by minimizing the quotient $(RSD)_B/SBR$, and
(c) line wing interference as a major cause of background enhancement in line-rich spectra.
The latter topic is considered in this context, because the work emphasizes the benefits of high-resolution spectroscopy for samples that emit line-rich spectra. These benefits manifest in particular in (ultra-) trace analysis and consist of improved detection power, precision, and accuracy, the last one because a substantial part of unpredictable background structures is shifted out of the spectral windows of the analysis lines, which reduces much of the uncertainty associated with the measurement of both the line peaks and the background. It is shown that the echelle spectrometer explored in this work provides, under analytically usable conditions, a practical resolving power that reasonably well approaches the ideal.

1. INTRODUCTION

The past decade has seen the maturing and popularization of inductively coupled plasmas (ICP) as powerful tools for simultaneous or sequential multielement analysis of liquids and dissolved solids by atomic emission spectroscopy (AES). ICP-AES is capable of covering major, minor and trace constituents in a wide variety of samples. However, a persistent challenge remains (ultra) trace analysis of samples having constituents that emit complex, line-rich spectra. Most transition elements, the rare earths, the platinum metals, uranium and thorium belong in this category. The presence of such elements may considerably worsen the detection limits, compared to pure water, and endanger the accuracy of the analysis. The reasons are enhancements of the background and overlapping of analysis lines by lines of the concomitants. Consequently, as the analyte concentration decreases, it becomes increasingly more difficult to detect analysis signals in the structured background and to correctly isolate net analyte signals from the composite signals in the "spectral windows" viewed.

One of the targets of present-day emission spectroscopic research therefore is to solve these problems. Initial steps have been the establishment of libraries of prominent analysis lines for universally acceptable ICP compromise operating conditions [1-3]. Subsequent steps involved the compilation of libraries of interfering lines in the form of tables [4-7], atlases of spectral scans [8,9], or software packages [3,10], all intended to facilitate the rational and flexible selection of analysis lines compatible with the composition of the sample under consideration. Also, polychromators are provided with software and data that permit correction for spectral interferences on the basis of the signals obtained for the various lines incorporated [11-14]. The complexity and extensiveness of the problem of spectral interferences entail, however, that the phase of collecting and storage of data relevant to ICPs and the design of appropriate computer software for fast and efficient retrieval and interpretation of such data are by far from completion yet.

The increased application of ICPs to diverse analytical problems has also increased the awareness that not only detailed knowledge of interferences is indispensable and useful, but that also further approaches should be made to circumvent these interferences. Evidently, the chemical separation of analytes and interferents is one way to achieve this, but generally this approach is applicable only to routine samples for which the laborious development of standard procedures repays this investment. Therefore other methods should be also available, in particular to deal with trace analysis of non-routine samples.

An obvious alternative to the separation of the elements is the separation of the spectral signals by the use of high-resolution spectroscopy. This alternative is not uncommon to the elder generation of emission spectroscopists, who were well aware of the necessity that the analysis of the spectra of molybdenum, tungsten, rare earths, yttrium, zirconium and uranium, for instance, required the use of such instruments as a 3.4 m Ebert spectrograph, equipped with a 600, 1200 or 1800 lines/mm grating to reach a reciprocal linear dispersion of 0.5 to 0.16 nm/mm, corresponding to a theoretical resolving power of 72,000 to 216,000. By searching the older literature on spectrographic analysis it would not be difficult to expand the article "ICP: dc arc in a new jacket?" [15], which focussed attention in particular to a basic problem of emission spectroscopy: spectral interferences. Studies of this problem led more or less a catacomb existence during the years that the development and successes of atomic absorption spectroscopy absorbed a great deal of the attention of atomic spectroscopists and analytical chemists. Inevitably it came into the limelight again with the renaissance of emission spectroscopy that resulted from the development of new radiation sources, such as ICPs. The successful combination of ICPs with polychromators and monochromators of small to medium resolving power could not conceal the fact that trace analysis of samples that emit line-rich spectra demands instruments of high resolving power, in the same way as this has been always recognized in dc arc spectrography. Consequently, there is renewed interest in high-resolution spectroscopy, the more so since technical advancements have led to more powerful optical components and refined electronic and mechanical devices for instrument control. Advanced systems of the type such as the one described in this paper permit the precise and accurate acquisition, storage and processing of detailed data on ICP spectra at a level of resolution that approaches the physical resolution of the spectrum. Such advanced facilities enable one to explore new possibilities and to apply, test and extend earlier developed theories with greater ease and under conditions that could not be easily achieved in a classical set-up.

The first and most fundamental approach to the use of high resolution in analytical emission spectroscopy was made by LAQUA and his associates [16], who in their initial work applied a 3.4 m Ebert spectrograph with a 300 lines/mm grating, combined with an external order sorter. Those authors not merely claimed the benefits of high resolution, but also clearly recognized a basic drawback: the sacrificed background radiant flux per unit of wavelength interval. The noise associated with the available flux sets a limit to the practical resolution that can be actually benefited of. In dc arc spectrography LAQUA et al. circumvented this problem by using multiple exposures and installing in the spectrograph a specially made, 50 cm long cylindrical lens of quartz, located close to the focal plane so as to produce a 10-fold demagnified image of the slit on the photographic plate, with a consequently 10-fold increase in the irradiance [17]. Approaches of this type however fundamentally important are incompatible with common practice.

LAQUA et al. extended the above study [16] to a complete description of the dependence of the power of detection in emission spectrographic analysis on the characteristics of the spectral apparatus and the radiation detector [18-21]. This description covers both photographic and photoelectric radiation detectors and includes properties such as theoretical resolving power, the resultant slit width and the geometrical conductance of the apparatus as parameters. *) The most recent paper also presents a tabular survey of the characteristics of typical grating spectrometers and an echelle spectrometer with crossed dispersion, the theoretical resolving power of these instruments at 425.4 nm varying from about 70,000 to 400,000. The work of LAQUA et al. [16-21] represents a systematic and fundamental approach and covers many details of interest for whatever emission spectroscopic work. However, its main purpose was to compare fundamentally the detection powers achievable with photographic and photoelectric radiation measurement in the case of unlimited sample consumption and for spectral background without structure due to lines or bands. The basic conclusion than is that the photographic emulsion can be distinctly better than a photoelectric detector as long as the necessary measuring time is not too long.

Although the answers to a number of questions related to the effect of spectral resolution on the detection power are implicit in the above studies, their aim and scope had to leave many other questions unanswered. Pointing out such problems and providing solutions is one of the targets of studies at present made in this laboratory.

Essentially this work is a systematic study of the benefits and penalties of high resolution spectroscopy in conjunction with the ICP and with emphasis on samples whose major constituents emit line-rich spectra. We thus involve the problem of spectral interferences and in addition will cover both detection power and accuracy, the latter in connection with the accuracy with which gross line and background signals can be defined and measured in line-rich spectra.

---

*) Results of this fundamental analysis are also reflected in the I.U.P.A.C. proposals "Nomenclature, symbols and their usage in spectrochemical analysis — IX: Instrumentation for the spectral dispersion and isolation of optical radiation", Draft 3, prepared by L.R.P. BUTLER and K. LAQUA (1983). The terminology used in the present paper links up with these proposals as far as we feel this useful.

A special feature of this work is the use of spectral apparatus that is new in analytical spectroscopy, i.e. a 1.5 echelle monochromator with predisperser, the latter having its slits parallel to those of the echelle monochromator. An instrument of this type has been described by PRESTON [22] and FREEMAN et al. [23]. The version used in this work was custom-built by SOPRA*).

An echelle monochromator with predisperser in parallel slit arrangement should be well distinguished from the more common type of echelle monochromator incorporating a prism whose direction of dispersion is crossed with that of the echelle so that the orders are separated in a direction perpendicular to that of the dispersion within the orders. The latter arrangement is well known in the literature [24-44] and is used in some types of commercial instruments [30-32, 36, 37, 40]. The two-dimensional display of the spectrum sets limits to the height of the entrance slit and may thus severely limit the (background) radiant flux reaching the detector, in contrast to an arrangement with parallel dispersion.

Both types of instrument exploit the intrinisic property of an echelle that it can be used at a high angle of incidence and therefore in high orders [24], yielding a theoretical resolving power

$$R_0 = n N_r \tag{1}$$

where n is the spectral order and N_r the total number of lines of the echelle. For the present instrument n varies from 10 to 30 and N_r is 65096, so that R_0 is in a range between 600,000 and 2,000,000 approximately.

The novelty of applying an instrument of this type in ICP-AES [45] induced us to study the optical and electronic-mechanical characteristics in detail with the dual purpose of (a) providing a sound basis for the analytical optimization and (b) developing a method for the measurement of physical line widths. We started this work in 1981 and the results obtained hitherto will be published shortly in a series of three articles dealing respectively with
(1) the instrumental characteristics and the measurement of physical line widths [46],
(2) the analytical optimization [47], and
(3) line wing interference in line-rich spectra [48].

The present article provides a brief survey of this work, which centres about the application of this new type of echelle spectrometer in ICP spectroscopy. Exploiting the full advantage of high resolving power in general requires insight in all the factors involved in the optimization and optimum use of the complete ICP system, source plus spectrometer. Providing such integrated insight is therefore the prime purpose of our work, which consequently deals not only with items specific for high-resolution spectroscopy, but also involves the investigation of relationships that are of general interest in ICP spectroscopy, irrespective of resolution.

2. INSTRUMENTATION

2.1. Principle of the spectrometer [46]

Figure 1 is a schematic diagram of the spectrometer, which essentially consists of a 1.5 m Ebert-Fastie monochromator, equipped with an echelle, and a 0.4 m Seya-Namioka monochromator. The monochromators are arranged with their slit parallel.

The echelle equation can be written in this form:

$$n \lambda = 2d \sin \alpha \cos \epsilon \tag{1}$$

where λ is the wavelength, d the grating constant ($= 10^6/316$ nm), α the echelle angle and ϵ the (constant) semi-angle of the monochromator. Eqn 1 can be represented by a set of straight lines with slopes $n/2d \cos \epsilon$ (Fig. 2).

---

*) SOPRA, 68, rue Pierre Joigneaux, Bois-Colombes, France.
A modified version based on the same principle has become commercially available as standard equipment.

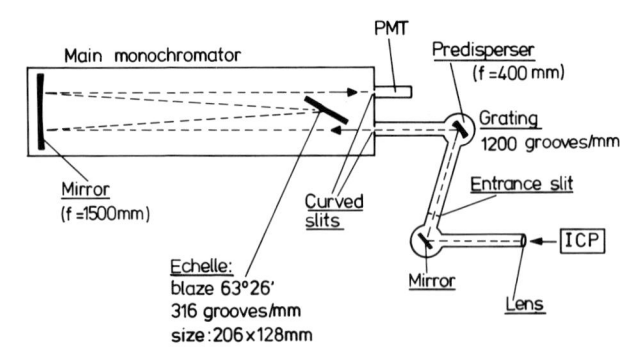

Fig. 1. Schematic diagram of the echelle monochromator with predisperser in parallel slit arrangement.

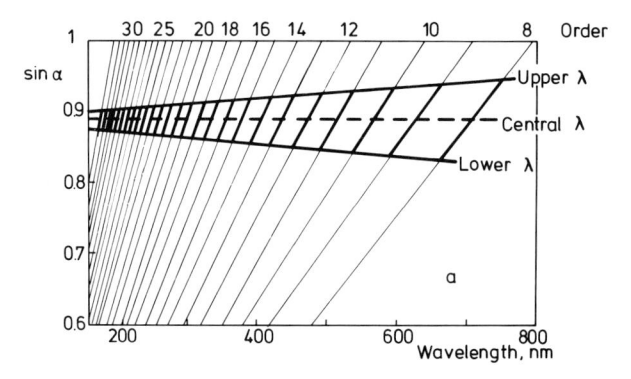

Fig. 2. Plots of the sine of the echelle angle α versus wavelength for orders 8-30 (eqn 1). The broken horizontal line represents the sine of an echelle angle close to the blaze angle: its intersections with the plots of eqn. 1 define the "central" wavelengths in the orders. The continuous lines labelled "upper λ" and "lower λ" define the upper and lower wavelengths in the orders (eqns 2b and 2c).

To obtain maximum radiation throughput the echelle should be always used at an angle close to its blaze angle. The wavelengths for which the echelle angle equals the blaze angle are found as the intersections of the straight lines sine $\alpha = f(\lambda)$ and a horizontal line $\sin \alpha = \sin 63°26' = 0.8930$. The best conditions are actually achieved at an angle for which $\sin \alpha = 0.8830$, so that

$$\lambda_{central}(nm) = 5580/n \qquad (2a)$$

defines the so-called "central wavelengths" for the various orders, which are represented in Fig. 2 by the intersections of the broken horizontal line with the plots of eqn 1.

The upper and lower wavelengths in the various orders are then defined by the conditions (cf. Fig. 2):

$$\lambda_{lower}(nm) = 5580/(n + 0.5) \qquad (2b)$$

$$\lambda_{upper}(nm) = 5580/(n - 0.5) \qquad (2c)$$

Consequently, during a wavelength scan the echelle angle follows a saw-tooth pattern.

2.2. Additional features of the spectrometer [46]

Basically the following characteristics of the spectrometer should be noted:
- completely computer-controlled wavelength adjustment of both the main monochromator and the pre-disperser, involving stepping motors, optical encoders and microprocessors;
- mechanical resolution: 1 motor step corresponds to 0.01/n (nm), where n is the spectral order, (n = 29 for λ = 190 nm and n = 10 for λ = 550 nm);
- reciprocal linear dispersion, theoretical resolving power (R_0), and spectral slit (= theoretical spectral bandwidth) for 60 μm entrance and exit slits: Fig. 3;
- detection and read-out: EMI 9789 QA photomultiplier followed by a photon counting system (dynamic range: 6 orders of magnitude);
- Apple II computer with an extensive software package in Pascal, developed in this laboratory;
- automatic wavelength calibration using 6 neodymium lines emitted from the ICP.

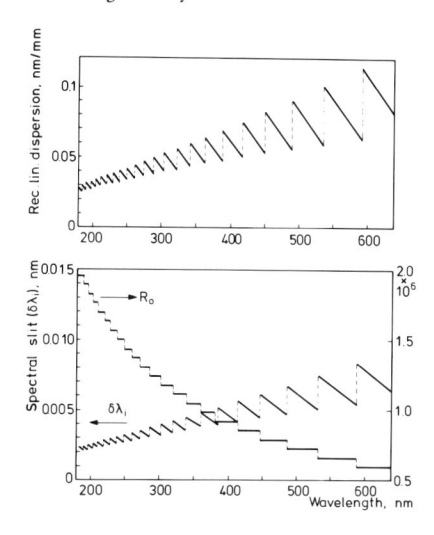

Fig. 3. Upper diagram: reciprocal linear dispersion as a function of wavelength. Lower diagram: theoretical spectral slit ($\Delta\lambda_i$) for 60 μm slit widths and theoretical resolving power (R_0) as functions of wavelength.

2.3. Inductively coupled plasma

The ICP used is the 50 MHz Philips ICP detailed in Refs. [49,50]. ICP compromise conditions as stated in [47,50] are commonly used in this work.

3. SLIT WIDTH OPTIMIZATION — TRADE-OFF BETWEEN PRACTICAL RESOLVING POWER AND BACKGROUND RADIANT FLUX: SIGNAL-TO-BACKGROUND RATIO VERSUS RELATIVE STANDARD DEVIATION OF BACKGROUND SIGNAL

3.1. Statement of the problem [47]

The slit widths dictate the practical resolving power (hereafter often referred to as "resolution") and practical spectral bandwidth. It has turned out that, for the present instrument, it serves no useful purpose to employ unequal slit widths, whence we shall in the following only refer to "the" slit width, which can be varied over a practical range from 15 to about 240 μm. Accordingly, the resolution can be varied over a wide range, since the spectral bandwidth is approximately proportional to the slit width. By using narrow slits one may easily demonstrate an extreme level of resolution such that the ICP spectra are completely physically resolved. A more important question is which level of resolution can be used under practical analytical conditions, in particular in trace analysis. To answer this question one must consider that the slit width not only dictates the resolution, but also governs the radiant flux that reaches the detector and therefore determines the contribution from shot noise to the total noise in the measured signal. This problem has been first recognized by LAQUA et al. [16], who elaborated a generalized theory that describes the dependence of the detection limits on the spectral apparatus and the detector [16,18-21]. Although the present approach basically follows similar lines as in the work of LAQUA et al., it is chiefly empirical and specifically refers to the echelle spectrometer; it closely links up with an earlier investigation of the effects of shot noise and flicker noise on the detection power achievable with a 50 MHz ICP and and a classical monochromator [50]. The latter work involved an assessment for pure aqueous solutions and a matrix that emits a line-rich spectrum, typically a mixture of nickel and cobalt nitrates. This approach is continued in the present work in which we essentially connect high-resolution spectroscopy with spectral interferences and do not confine ourselves to an idealized continuous background spectrum, as has been done in the fundamental studies of LAQUA et al. [16, 18-21]. Evidently, also in the present approach the first step in the optimization is that for pure aqueous solutions and ideal continuous background.

In summary, a realistic assessment of the benefits of high resolution for the detection power does not only consider possible gains in signal-to-background ratio (SBR) resulting from increased resolution, but essentially takes into account the trade-off between maximizing the SBR and minimizing the relative standard deviation (RSD) of the background signal, in other words, it implies the minimization of the ratio of the RSD of the background, on the one hand, and the SBR, on the other; this ratio is proportional to the detection limit.

3.2. Minimization of the detection limit: pure aqueous solutions

We formulate the detection limit as [50]

$$c_L = 0.01 \cdot k \, (RSD)_B \cdot c_0/SBR \qquad (3)$$

where c_L is the detection limit, $(RSD)_B$ the relative standard deviation of the background signal (x_B) in percent, and c_0 the analyte concentration at which the signal-to-background ratio (SBR) is measured. The value of the constant k is irrelevant and is taken to be 2 in this work.

The background RSD can be formulated as

$$(RSD)_B = [a^2 + b/x_B + c/x_B^2]^{1/2} \qquad (4)$$

where a, b and c are coefficients associated with the source flicker noise, the shot noise and the detector noise, respectively [50]. For the present system c = 0, while b = 1 if x_B is expressed in counts. The flicker noise coefficient has been experimentally determined to be about 0.005 if the RSD is expressed as a fraction or about 0.5% if the RSD is expressed in percent; in the latter case b must be put equal to 10000 instead of 1. Since the background RSD is a function of the background signal and thus depends on the absolute number of counts recorded, it is a function of the integration time. In the present work all numerical values of the background RSD and thus detection limits always refer to a 10 s integration time.

The dependence of the SBR, the background RSD and the detection limit on slit width is thus completely defined by the dependences of the net line and background signals on slit width. For the present system the two functions behave rather similarly over the practical range of slit width so that the SBR changes little. Therefore it is the background signal and the associated RSD which govern the picture. As a typical example Fig. 4 shows the SBR, the background RSD and the detection limit for Mn II 257.610 nm in dependence on slit width. This behaviour is found at all wavelengths. A detailed analysis leads to straightforward conclusions as to the optimum slit widths at different wavelengths and also shows to what extent the intrinsic detection power of the ICP can be fully exploited, that is, under which conditions the background RSD is dictated by source flicker noise only. For pure aqueous solutions this ideal cannot be achieved at all wavelengths. However, for samples whose major constituents emit line-rich spectra substantial background enhancements may occur (section 4), which, on the one hand, decrease the SBRs, but on the other hand, also decrease the shot noise. This complicates slit width optimization in that the situation has to be judged for each analysis line and sample type separately. In practice, however, a rather simple rule can be followed by considering eqn 4 for the background RSD: chose the minimum slit width such that the background signal (for a 10 s integration time) is at least 10000 counts: the background RSD then closely approaches 1%, while the SBR has the best value compatible with an acceptable noise level.

If this rule is followed for line-rich spectra, a slit width of 60 μm can be often used over virtually the entire wavelength range. This slit width provides for practical spectral bandwidths that generally are only a factor of 1 to 3 larger than the physical widths of spectral lines emitted by the ICP (section 5), which implies that the ideal of using a spectral bandwidth equal to the physical line width can be rather closely approached under practical analysis conditions.

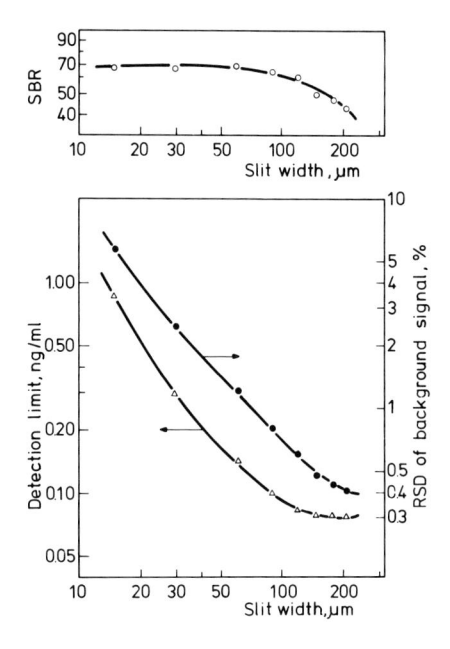

Fig. 4. Double logarithmic plots of SBR, $(RSD)_B$ and c_L vs slit width for Mn II 257.610 nm. Mn concentration for SBR plot: 400 ng/ml.

3.3. Detection limits: numerical values for pure aqueous solutions [47]

Table 1 shows two sets of detection limits obtained in this work (columns II and III) and compares these results with values reported in the literature (columns IV to VI), all results being normalized to a "2σ basis", i.e. k = 2 in eqn 1, to ease comparison.

The results in column II are for ICP compromise conditions [49,50] and for optimum slit width, that is 200 μm for λ < 250 nm and 60 μm for λ > 250 nm. The results in column III are for the same conditions except for the ICP power, which was 1.15 instead of 1.4 kW. Actually the lower power level yields the better detection limits. The benefits of lower power are marginal, however, for wavelengths below about 260 nm. This is so because the well-known gain in SBR achieved for pure aqueous solutions when the power is reduced [53,54] is virtually balanced by a larger background RSD due to enhanced shot noise associated with the decrease in background radiant flux. At higher wavelengths the background radiant flux decreases too, but it remains at a level high enough to let the flicker noise dominate the background RSD, also at the lower ICP power; consequently the gain in SBR resulting from the power reduction translates into improved detection limits.

We now compare the results in column III with the literature values. For wavelengths < 250 nm our results are seen to be a factor of 3 to 5 better than the values of WINGE et al. [1]; for higher wavelengths this difference amounts to a factor of 10 to 20. We explain the differences as follows. First, the results of WINGE et al. are based on SBR measurements and the assumption of a constant background RSD of 1%. If we then consider that the background RSD in this work varies systematically from 0.5 at high wavelengths to 1.5% in the low UV, we can conclude that the SBRs in this work are generally an order of magnitude larger than those of WINGE et al. On the basis of previous results we estimate that this order of magnitude is composed of a factor of 3 to 5 due to a difference between the ICPs (as pointed out previously [50]) and a factor of 3 to 2 due to the high resolution. On the whole, we may conclude that the present set-up thus yields excellent detection limits, a result which is partly due to the high resolution, partly due to the ICP.

If we compare our results with those of FERNANDO (column V) for an echelle spectrometer with crossed dispersion and an argon ICP approximately optimized for each element separately [51] we find that the detection limits attained in the present approach are usually substantially lower. We attribute the difference in part to a difference between the ICPs and for another part to differences in the resolution and noise characteristics of the two spectrometers.

Column VI finally lists results of FLOYD et al. [52] for a classical computer-controlled monochromator and an argon ICP similar to that used by WINGE et al. [1], but operated with an ultrasonic rather than a pneumatic nebulizer. It is well known [56,57] that the amount of sample that can be injected into an ICP using an ultrasonic nebulizer exceeds that for a pneumatic nebulizer by a factor of 5 to 10. Considering this point and the above comparison between our results and those of WINGE et al. we can understand that the results obtained in this work with pneumatic nebulization are of the same magnitude as those reported by FLOYD et al. for ultrasonic nebulization.

Table 1. Detection limits (ng/ml) obtained in this work using a 50 MHz argon ICP with pneumatic nebulization (PN) and three sets of results reported in the literature for 27 MHz ICPs with either pneumatic or ultrasonic nebulization (USN) and monochromators different from that used in the present work. All results have been normalized to $k = 2$ (eqn 3).

I	II	III	IV	V	VI
Spectral line (nm)	This work	This work	WINGE	FERNANDO	FLOYD
	a) ≠	b) ≠	c) ¤	d) ‡	e) ≠
	PN	PN	PN	PN	USN
As I 193.696	7	7	35	88	—
As I 197.197	23	21	50	—	7
Mo II 202.030	2	1.7	5	11 *	4
Ge I 209.426	11	9	27	—	14 *
Pb I 220.353	7	7	28	36	14
In II 230.606	13	11	42	90	14
Co II 238.892	1	0.9	4	5	1.4 *
Rh II 249.077	7	6	38	14 *	7 *
Mn II 257.610	0.08	0.06	0.9	1.3	0.07
Mg II 279.553	0.01	0.005	0.1	0.05	0.03
Ga I 294.364	4	1.6	31	12	4
V II 309.311	0.33	0.16	3.3	13 *	0.14 *
Ti II 334.941	0.2	0.07	2.5	0.4	0.3
Y II 371.030	0.14	0.06	2.3	0.12	0.05
Nd II 401.225	2.4	0.9	33	—	2 *
Sm II 442.434	4	1.4	36	—	8 *
Ba II 493.409	0.25	0.16	1.5	—	0.3 *

a) 1.5 m echelle monochromator with predisperser (grating), parallel slit arrangement; echelle 316 grooves/mm; ICP compromise conditions as stated in [47,50].

b) Same as a), but with 1.15 kW ICP power instead of 1.4 kW.

c) WINGE et al. [1]; 1-m grating monochromator, grating 1180 grooves/mm.

d) FERNANDO [51]; 0.75 m echelle monochromator with crossed dispersion (prism order sorter); echelle 79 grooves/mm.

e) FLOYD et al. [52]; 0.5 m grating double monochromator, grating 1200 grooves/mm.

≠) Variation of background RSD with wavelength taken into account.

¤) Constant background RSD of 1% assumed.

‡) For consistency this column lists 2 times the standard deviation of the blank in ng/ml as reported by FERNANDO [52], rather than 2/3 times the rounded detection limit reported by that author.

*) Different line used.

4. HIGH-RESOLUTION SPECTROSCOPY APPLIED TO LINE-RICH SPECTRA

4.1. Case studies of typical situations [47]

It is generally recognized that a substantial part of the detection power may be lost if a sample contains major elements that emit line-rich spectra. The main reason is the partial or complete overlap of analysis lines by lines of the concomitants, where the term "overlap" refers to the interaction of spectral images located within a few spectral bandwidths from each other. However, line overlap in this sense is not the only reason for the deterioration of the detection power, as will be clarified below.

For a case study we used a mixture of nickel and cobalt nitrates containing 1 mg/ml of each nickel and cobalt and investigated for 18 analysis lines the background enhancements, signal-to-background ratios, background RSDs and detection limits at various spectral bandwidths. Three typical cases of spectral interference were distinguished:

(1) *Line wing interference*, which manifests as additional background that optically behaves as continuous background, but in its dependence on the concentration of the relevant (matrix) element behaves as a line spectrum, that is, its intensity is proportional to the concentration of the pertinent element.

In the case of line wing interference background enhancements expressed as background-to-background ratios (solution with matrix with respect to pure solvent) are independent of the spectral resolution, while SBRs improve in the same way with increasing resolution as for the pure solvent. As a result of the higher background level, the optimum values of the slit width are smaller than for the pure solvent. On the whole, the maximum gain in detection power over the practical range of slit width is the same as for pure aqueous solutions, that is a factor of 2 to 3, usually not more than 2, because the flicker noise tends to be somewhat increased by the presence of the matrix.

(2) *Complete line coincidence*, that is the situation with an interfering line is located at exactly the same wavelength as the analysis line, or at a wavelength distance from it smaller than the mean physical width of the two lines involved.

The SBR of the analysis line with respect to the gross intensity of the interfering line changes little with resolution and, obviously, the benefits from high-resolution spectroscopy can only be marginal.

(3) *Partial line overlap*, conveniently defined as a situation in which analysis line and interfering line are physically completely resolved, but are observed as lines that partially overlap each other as a result of insufficient instrumental resolving power.

Evidently, it is in this case that the greatest benefits can be expected from high-resolution spectroscopy. Increasing the spectral resolution by reducing the slit width enables one to completely separate the lines and thus to "shift" the interfering line out of the spectral window of the analysis line. This may lead to a substantial increase in SBR and a concomitant improvement in detection limit, provided that the effect of increased SBR is not balanced or outweighed by an increase in the background RSD. However, the often substantial background enhancements in line-rich spectra resulting from the line wings permit the use of relatively narrow slits without reducing the background intensity to such a level that excess shot noise would dominate the background RSD.

Finally, we have reasons to believe that the profits of high resolution for trace analysis of materials with line-rich spectra are only partly reflected in the detection limits themselves; more important is the expectation that in line-rich spectra accurate determinations will still be possible down to concentration levels where poorer resolution fails to yield reliable results. This point is briefly discussed in Ref. [47], but is now subject of a quantitative investigation.

We add here the following examples as illustrations of the points made above. For a detailed discussion of these and related examples the reader is referred to Refs. [47] and [48].

(a) Fe II 239.562 nm in the Ni-Co spectrum with interference from Co 239.552 nm: Fig. 5. The Fe line is physically completely separated from the Co line. The decrease in slit width from 210 to 60 μm implies a change in practical spectral bandwidth from 0.011 to 0.004 nm. The associated increase in SBR is nearly a factor of 3 while the detection limit improves by a factor of over 2. More important is the fact that both the peak and the background are clearly defined at the 0.004 nm spectral bandwidth, whereas the measurement of peak and background is cumbersome at the 0.011 nm bandwidth. The situation at a poorer resolution level, typically medium resolution (0.015 – 0.02 nm spectral bandwidth), could not be simulated with the present apparatus, but it will be clear that the profits of high-resolution spectroscopy with respect to that level of resolution will be greater than what we could reveal here.

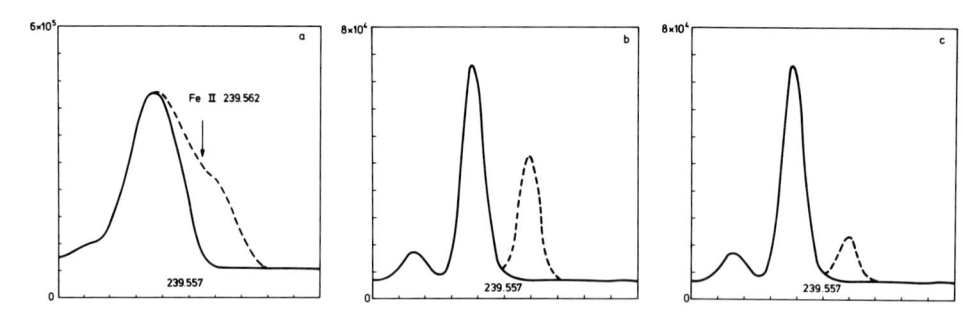

Fig. 5. Scans over a 0.05 nm spectral window centred about 239.557 nm. Continuous curve: Ni-Co matrix. Broken curve: contribution from Fe in spiked Ni-Co matrix. Each frame shows Fe 239.562 nm (right) and Co 239.552 nm (left). The arrow indicates the position of the Fe line.
(a) Slit width = 210 µm, Fe concn. = 1.8 µg/ml, SBR = 2;
(b) Slit width = 60 µm, Fe concn. = 1.8 µg/ml, SBR = 5.8;
(c) Slit width = 60 µm, Fe concn. = 0.6 µg/ml, SBR = 2;
Ordinate: intensity (counts/s), abscissa: wavelength ± 0.025 nm about central λ.

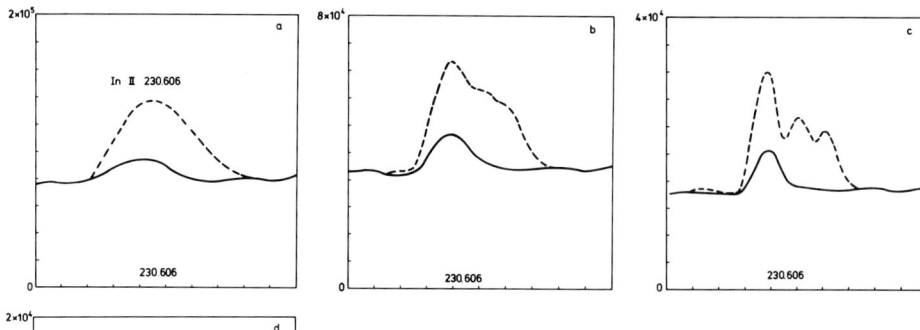

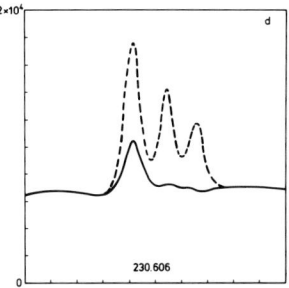

Fig. 6. Scans over a 0.05 nm spectral window centred about 230.606 nm, that is the wavelength normally listed for the prominent line In II 230.606 nm. Upper scan: Ni-Co matrix spiked with In. Lower scan: Ni-Co matrix (1 mg/ml of each Ni and Co). All scans are for 10 µg/ml In. Consecutive frames are for slit widths of 210, 135, 90 and 60 µm.

Ordinate: intensity (counts/s), abscissa: wavelength ± 0.025 nm about central λ.

(b) In II 230.606 nm in the Ni-Co spectrum: Fig. 6. This prominent line actually is a triplet (230.601 − 230.606 − 230.612 nm), of which one component coincides with a Co line. The increase in resolution (from 0.011 to 0.004 nm spectral bandwidth) in frames a to d shows the gradual resolution of the In line into its three components. The high-resolution level permits the use of the central component, which does not experience interference from Co, has a better SBR and a clearly defined background.

(c) Nd II 396.312 nm with interference from La II 396.304 nm: Figs. 7 and 8. The spectral bandwidth of 0.002 nm permits an unambiguous definition of the gross line and background readings for the Nd line to at least 0.3 µg/ml Nd. In contrast, Fig. 8 shows the situation for a spectral bandwidth of 0.018 nm: although the Nd line distinctly contributes to the intensity at its own wavelength position (marked by a broken line), it does not show resolved from the La line at even 2 µg/ml, while a correct intensity

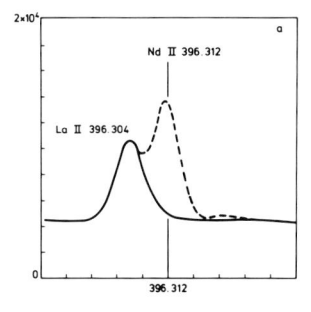

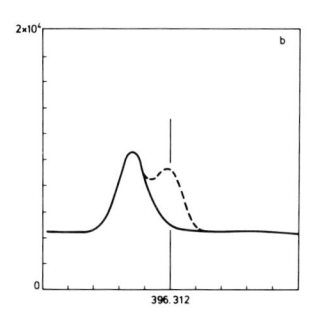

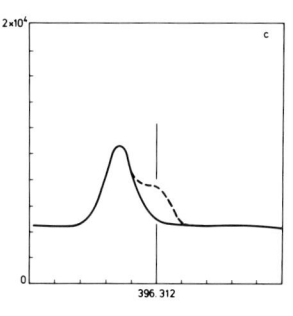

Fig. 7. Scans over a 0.05 nm window about the Nd line at 396.312 nm with interference from La 396.304 nm. Continuous curves: scans for 100 µg/ml La solution. Broken curves: La solution spiked with Nd. The vertical line indicates the position of the Nd line. All scans are for a practical spectral bandwidth of 0.0023 nm (15 µm slit width). The Nd concentrations are as follows: (a) 1 µg/ml, (b) 0.5 µg/ml and (c) 0.3 µg/ml.

Ordinate: intensity (counts/s), abscissa: wavelength ± 0.025 nm about central λ.

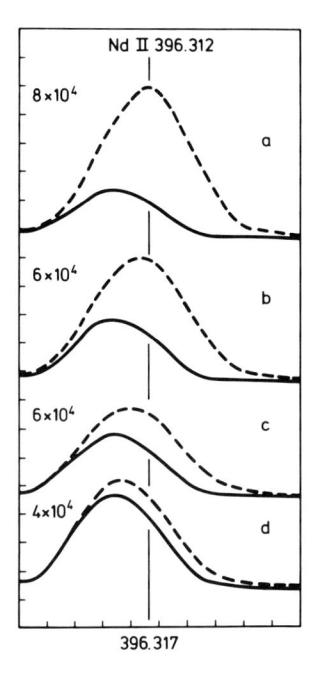

Fig. 8. Scans for the same Nd line as shown in Fig. 7, but for a practical spectral bandwidth of 0.018 nm. As in Fig. 7, the La concentration is 100 µg/ml, while the Nd concentrations are as follows: (a) 2 µg/ml, (b) 1 µg/ml, (c) 0.5 µg/ml, (d) 0.2 µg/ml.

Ordinate: intensity (counts/s), abscissa wavelength ± 0.025 nm about central λ. The scans have been fitted into one frame. Owing to the normalization of each scan separately the ordinate scales are different. The relevant labels indicate the number of counts/s per scale division.

determination at concentrations below 1 µg/ml will require measurements to be made on the sloping shoulder of the La line profile. This makes high demands upon the accuracy and reproducibility of the wavelength setting of the monochromator and requires an accurate correspondence between wavelength setting and measured signals. This precludes, for example, the acquisition of meaningful results from scans registered with a simple chart recorder coupled to a classical scanning monochromator.

In the case that a computer-controlled slew-scan monochromator is used and the measurement is based on a "peak find" routine, then the wavelength at which the measurement is made will drift to the peak of the La line as the Nd concentration decreases. Such a drift results in severe curvature of log intensity vs log concentration calibration curves.

Preliminary tests further indicate that in cases of line overlap, such as illustrated in Figs. 5 to 8, a measurement at the wavelength position of maximum SBR rather than at the true peak wavelength of the analysis line will yield both the best detection limit and the highest accuracy if the latter is limited by the precision with which the wavelength can be alternately adjusted to line and background positions during a series of measurements on different analysis or reference samples. This point is being examined more closely.

(d) Nd II 396.312 nm with interference from neighbouring Ce lines: Fig. 9. This figure contrasts spectral scans for 0.2 and 2 µg/ml Nd in the presence of 100 µg/ml Ce taken at three levels of the practical spectral bandwidth, 0.018, 0.0058 and 0.0023 nm. The right-hand frames include the values of the SBR, which show the progression expected for this increase in resolution. If we take here a background RSD of 1%, we find the detection limits to be 33, 15 and 11 µg/ml respectively. An interesting point now is to establish quantitatively whether there will be a difference in the accuracy with which the gross line and background signals (and thus the net line signal) can be determined at the three levels of resolution for a concentration of, say, 10 times the detection limit, thus for 330, 150 and 110 µg/ml Nd.

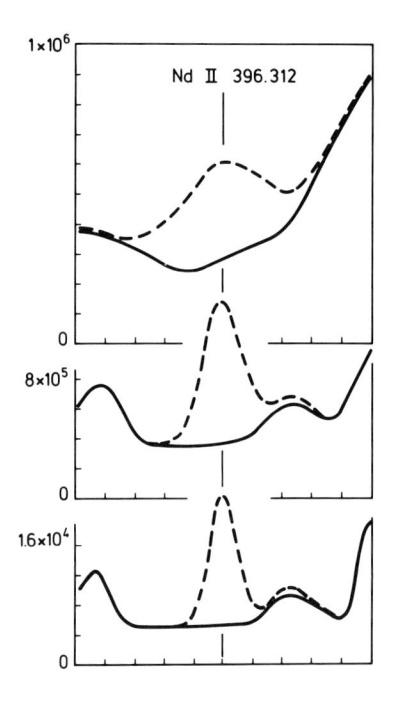

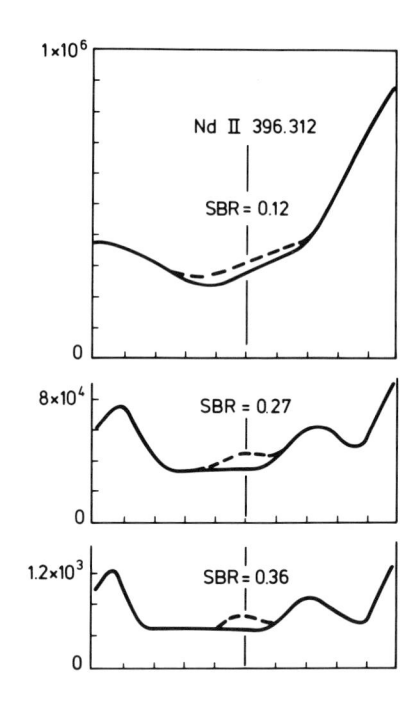

Fig. 9. Left: scans over a 0.05 nm window about the Nd line at 396.312 nm with interference from not further identified Ce lines. Continuous curves: scans for 100 µg/ml Ce solution. Broken curves: scans for Ce solution spiked with Nd. The vertical line indicates the position of the Nd line. All scans are for a 2 µg/ml Nd concentration. The practical spectral bandwidth was as follows: (a) 0.018 nm (215 µm slit), (b) 0.0058 nm (60 µm slit), (c) 0.0023 nm (15 µm slit).

Right: scans similar to those shown in the left-hand frames but for a 0.2 µg/ml Nd concentration. The vertical line indicates the wavelength position of the Nd line; the SBRs given in the figure refer to this wavelength.

Ordinate: intensity (counts/sec), abscissa: wavelength ± 0.025 nm about central λ.

We made some estimates assuming a wavelength positioning error of plus or minus one step, the latter being taken equal to about 1/10 of the "full width" of the experimental line profile. These estimates indicate a substantial advantage for a bandwidths of 0.0058 or 0.0023 nm compared to 0.018 nm, the main reason being the flat background level obtained at the higher resolution levels. For the case of partial line overlap we thus have some indications that, at a given ratio of the concentration present to the detection limit, high resolution tends to yield better accuracy than medium resolution, while we have already proved that better detection limits can be attained. However, in the present stage of this work, we consider the above statement about improved accuracy as a thesis rather than an experimental finding. In general, there are many arguments indicating that the removal of substantial amounts of unwanted radiation (especially that from interfering lines) from the spectral window of an analysis line must be beneficial for the achievable accuracy, but only a quantitative experimental treatment of the problem will be able to reveal the true advantages in concrete situations. Therefore we shall investigate this problem along with the related question of the choice of the best wavelength position for the measurement, that is, the choice between the wavelength position of maximum SBR or that of the peak of the analysis line.

4.2. Closer inspection of line wing interference

It is well known from the work of LARSON and FASSEL [57] that line wings may appreciably contribute to the background up to large distances from the line peaks. Those authors limited their investigation to lines of Ca, Mg and Sr. These elements do not emit line-rich spectra, but a few very strong lines, which may cause severe stray light problems [58, 59] and give rise to line wing interference [57]. Although it is very likely that an appreciable part of the background enhancements caused by major elements emitting line-rich spectra is also due to line wings, this has never been proved quantitatively. Experimental investigations were chiefly limited to strong lines of the alkaline earths [57,60], while a general discussion of line broadening effects relevant to the ICP [60] made reference only to wing interference from Ga I 250.070 – Ga I 250.017 nm and the effect of the wing of Fe 249.699 nm on B I 249.678 nm. The scarcity of quantitative data on line wing interference in line-rich spectra and the fact that better knowledge of line wing interference is of paramount importance for analytical interference in general induced us to study the line wing behaviour more elaborately in the scope of our work on high-resolution ICP spectroscopy [48]. The following results are mentioned here.

Figure 10 shows the behaviour of line wings for 7 spectral lines as plots of SBR versus wavelength distance to peak, where the background refers to the pure solvent blank. The ordinate has been normalized in that all plots refer to concentrations of the elements yielding an SBR of 6000 in the line peak (see figure caption). This is a concentration level of a few 100,000 times the detection limit. Evidently, the wings then still essentially contribute to the background up to a distance as large as 0.5 nm from the line peaks. The curves are seen to run closely together, except for the In line, which was also found to be exceptionally broad. The small scatter between the wing functions induced us to describe the wing behaviour by the average empirical function represented by the broken line in Fig. 10. The mathematical expression for this curve was used for the description of the superposition of line wings in the Ni-Co spectrum. Thus it could be quantitatively proved that the background enhancement by a factor of 22 found at 230.6 nm in the Ni-Co spectrum was due to the superposition of the wings of Ni and Co lines, a set of 16 strong lines being responsible for 95% of the enhancement. It was also proved quantitatively that the background contributions from Ni, Co and solvent blank are additive at all wavelengths to within the limits sets by multiplicative interferences (i.e. to within − 5%). The fit of the Mg II line at 279.553 nm in the picture further indicates that the notority of the Mg wing interference is not due to a peculiarity of the wing, but to the exceptionally high intensity and consequently low detection limit of this Mg line. Probably the same will apply to the strong Ca and Sr ionic lines.

On the whole, line wings contribute essentially to the background in line-rich spectra. This background behaves optically as a (quasi-) continuum but as a line spectrum as to its dependence on the concentration of the relevant element and the ICP operating conditions. This has various consequences for ICP optimization in that for samples yielding line-rich spectra other rules should be followed than for samples that emit simple spectra. This point is exemplified and further discussed in Ref. [48].

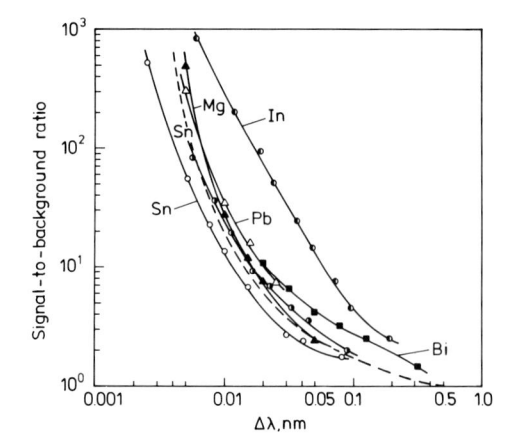

Figure 10. Double logarithmic plots of signal-to-background ratio vs distance to peak for various lines emitted by the ICP. The results are for the lines and element concentrations (c_n) listed below. These concentrations are normalized values chosen such that SBR = 6000 in the line peak. The concentrations (c) at which the measurements were actually made are given in parentheses.

(1) ○ Sn I   235.484 nm; c_n = 2.5   mg/ml (c = 500 μg/ml);
(2) ◐ Sn I   242.949 nm; c_n = 2.4   mg/ml (c = 500 μg/ml);
(3) ▲ Mg II 279.553 nm; c_n = 0.006 mg/ml (c =   5 μg/ml);
(4) △ Pb II 283.306 nm; c_n = 10   mg/ml (c = 50 μg/ml):
(5) ■ Bi I   223.061 nm; c_n = 1.0   mg/ml (c = 1000 μg/ml);
(6) ◑ In I   271.026 nm; c_n = 35   mg/ml (c = 1000 μg/ml);

5. REQUIRED AND ACHIEVABLE PRACTICAL RESOLVING POWER [46]

What resolving power is needed for the complete physical resolution of the spectra encountered in emission spectroscopy has been answered in the early work of LAQUA et al. [16]. The condition is that the practical spectral bandwidth is of the same magnitude as the physical width of the lines in the spectrum, a condition that can be also formulated in terms of the practical resolving power ($\lambda/\Delta\lambda$). The necessary data are the physical widths of the lines. LAQUA et al. [16] determined such data for a fairly large number of lines emitted from various types of d.c. arc and a soft spark. Mean values are included in Fig. 11, which further brings results of line width measurements for argon ICPs taken from recent literature [46,61-67]. Figure 11 is a diagram in which physical line widths reported by various authors have been plotted in dependence on wavelength. The bases of the data are discussed in Ref. [46]. The picture includes some lines with hyperfine structure (HFS) [16,65].

The bulk of the data follows an increase with wavelength in agreement with the proportionality of the Doppler width with wavelength. The scatter is primarily caused by the dependence of the Doppler width on the atomic mass, but must be also partly due different contributions from Lorentz broadening, HFS and experimental errors.

By considering the points in Fig. 11, one can derive that ideally practical spectral bandwidths ranging from about 0.0015 to 0.005 nm are required depending on the wavelength and the (atomic mass of) the element, which translates into practical resolving powers in a range between 100,000 and 200,000, as derived by LAQUA et al. [16] for d.c. arcs and soft sparks. The same order of magnitude applies to argon ICPs, since obviously the line widths of ICP lines do not differ essentially from those emitted by arcs and soft sparks. Thus in this respect nothing really new can be added to the existing knowledge.

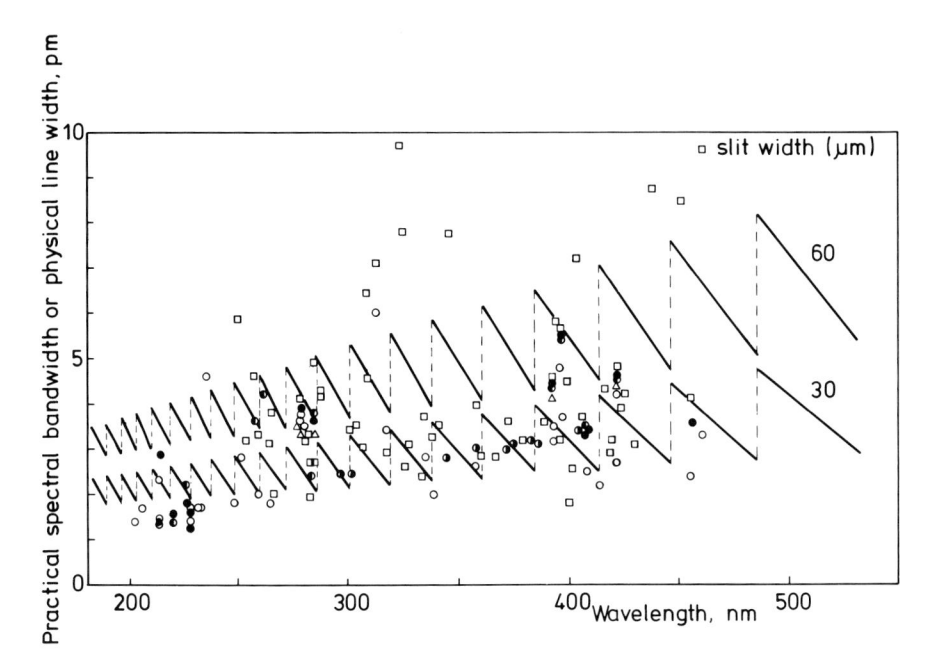

Fig. 11. Matching of the practical spectral bandwidth attainable with the echelle monochromator and physical line widths reported in the literature. The figure shows the saw-tooth patterns of the bandwidth for 30 and 60 μm slit widths and data on line widths from LAQUA et al. [16]: ◻, HASEGAWA et al. [63-65]: ○, KAWAGUCHI et al. [61]: ●, BATAL and MERMET [66]: △, FAIRES et al. [62]: ◐, and this work: ◉. The data refer to argon ICPs except for those of LAQUA et al., which are mean values for four types of d.c. arc and a soft spark.

A different question is to what extent present-day apparatus does meet the above condition and yield a practical resolving power of 100,000 to 200,000 under real analysis conditions. The question has been considered by McLAREN and MERMET [68] for a conventional grating monochromator and by the present authors [46,47] for the echelle monochromator with predisperser. From the latter work we derive that for line-rich spectra — thus where high resolution is really needed — a 60 μm slit width is compatible with the requirements to be made upon the background radiant flux. The practical spectral bandwidth for this slit width is shown by the upper saw-tooth pattern in Fig. 11. This implies that the ratio of the practical spectral bandwidth and physical line widths will generally be in a range between 1 and 3. The minimum distance between two lines of equal intensity at which they will appear resolved depends on the precise situation, but it can be derived from Fig. 11 that this distance for the 60 μm slit widths will usually lie between 5 and 10 pm (0.05 and 0.1 Å). It is also shown in Ref. [46] that the mechanical resolution perfectly fits the effective line widths obtained with 60 μm slits, that is, the effective line width (expressed in terms of full width at half maximum) is covered in 6 to 10 steps over the wavelength range between 190 and 500 nm.

Figure 11 also contains the saw-tooth pattern for a 30 μm slit width, that is the slit width that would be ideally required. It meets two objections for normal practice. First, the background radiant flux is too low in the ultraviolet so that shot noise will predominate and the effect of the higher practical resolution (better SBR) is offset by the effect of the poorer background RSD. Second, for wavelengths above 400 nm the mechanical resolution becomes marginal in combination with the effective line width associated with a 30 μm slit width. However, the 30 μm slit width can be used for physical measurements, in particular line width measurements, as has been shown in Ref. [46]. Then an even better resolution than shown in Fig. 11 can be achieved at this slit width when the slit height is reduced so as to eliminate the effect of aberrations and to make the practical bandwidth virtually equal to the theoretical resultant slit.

We should further point out two unique properties of an echelle instrument compared to apparatus equipped with a common grating:
(i) the increase of spectral bandwidth with wavelength at constant slit width links up favourably with the overall trend of physical line widths with wavelength as dictated by the Doppler effect (Fig. 11);
(ii) the radiation throughput is high at all wavelengths because the echelle is used only over a small angle about the blaze angle (Fig. 2): this provides for a broad wavelength coverage, which in the present instrument is limited at the upper end by the spectral response characteristic of the photomultiplier tube rather than the optics.

6. CONCLUSIONS

It has been shown that modern instruments such as the present computer-controlled echelle spectrometer with predisperser can be used under practical analysis conditions to exploit the benefits of high resolving power in ICP spectroscopy.

Although high resolution spectroscopy brings an advantage in detection power for pure aqueous solutions and generally for samples that emit relatively simple spectra, its true benefits primarily manifest themselves for sample types that emit complex, line-rich spectra. These benefits are the following.

(1) Improved detection limits, partly because signal-to-background ratios are generally somewhat increased, but chiefly because more lines free from spectral interference become available.

(2) Higher precision in ultra-trace analysis, inherent in the improved detection power.

(3) Far better accuracy because a substantial part of unpredictable background structures is shifted out of the spectral windows of the analysis lines; this reduces much of the uncertainty associated with the measurement of both the line peaks and the background.

On the whole, high-resolution spectroscopy brings us close to the ideal of being able to cast a look under the spectral line, while the complete computer control of a high-resolution spectroscopic instrument provides for full quantitative control over spectral positions that follow up each other by less then 1 pm in wavelength. This feature in particular permits the computer manipulation of spectral scans [47] in both systematic studies and common analytical practice.

REFERENCES

1. R.K. Winge, V.J. Peterson, and V.A. Fassel, *Appl. Spectrosc.* **33**, 206 (1979).

2. P.W.J.M. Boumans and M. Bosveld, *Spectrochim. Acta* **34B**, 59 (1979).

3. P.W.J.M. Boumans, *Spectrochim. Acta* **38B**, 747 (1983).

4. M.L. Parsons, A. Foster, and D. Anderson, *An Atlas of Spectral Interferences in ICP spectroscopy*. Plenum, New York (1980).

5. P.W.J.M. Boumans, *Line Coincidence Tables for Inductively Coupled Plasma Atomic Emission Spectrometry*. Pergamon Press, Oxford (1980); second revised edition (1984).

6. A.R. Forster, T.A. Anderson, and M.L. Parsons, *Appl. Spectrosc.* **36**, 499 (1982).

7. T.A. Anderson, A.R. Foster, and M.L. Parsons, *Appl. Spectrosc.* **36**, 504 (1982).

8. R.K. Winge, V.A. Fassel, V.J. Peterson, and M.A. Floyd, *Appl. Spectrosc.* **36**, 210 (1982).

9. R.K. Winge, V.A. Fassel, V.J. Peterson, and M.A. Floyd, *Atlas of Spectral Information for ICP-AES* (1982?).

10. P.W.J.M Boumans, in: *"Recent Advances in Analytical Spectrometry"* (Edited by K. Fuwa), p. 61. Pergamon Press, Oxford (1982).

11. R.I. Botto, *Developments in Atomic Plasma Spectrochemical Analysis* (Edited by R.M. Barnes), p. 141. Heyden, London/Philadelphia (1981).

12. R.I. Botto, *Anal. Chem.* **54,** 1654 (1982).

13. R.I. Botto, *Spectrochim. Acta* **38B,** 129 (1983).

14. R.I. Botto, *Spectrochim. Acta* **39B,** 95 (1984).

15. P.W.J.M. Boumans, *Proc. 21th Coll. Spectr. Int. and 8th Int. Conf. Atomic Spectr., Cambridge 1979, Keynote Lectures,* p. 49. Heyden, London (1979); *Spectrochim. Acta* **35B,** 57 (1980).

16. K. Laqua, W.-D. Hagenah, and H. Waechter, *Fresenius Z. Anal. Chem.* **225,** 142 (1967).

17. K. Laqua, Institut für Spektrochemie, Dortmund, FRG, personal communication (1965).

18. U. Haisch, *Spectrochim. Acta* **25B,** 597 (1970).

19. U. Haisch, K. Laqua, and W.-D. Hagenah, *Spectrochim. Acta* **26B,** 651 (1971).

20. U. Haisch, *Fresenius Z. Anal. Chem.* **259,** 1 (1972).

21. U. Haisch, K. Laqua, W.-D. Hagenah, and H. Waechter, *Fresenius Z. Anal. Chem.* **316,** 157 (1983).

22. R.C. Preston, *Ph.D. Thesis,* University of London (1972).

23. G.H.C. Freeman, M. Outred, and L.R. Morris, *Spectrochim. Acta* **35B,** 687 (1980).

24. G.R. Harrison, *J. Opt. Soc. Amer.* **39,** 522 (1949).

25. G.R. Harrison, J.E. Archer, and J. Camus, *J. Opt. Soc. Amer.* **42,** 522 (1952).

26. G.R. Harrison, S.P. Davis, and H.J. Robertson, *J. Opt. Soc. Amer.* **42,** 522 (1952).

27. N.A. Finkelstein, *J. Opt. Soc. Amer.* **43,** 90 (1953).

28. D. Richardson, *Spectrochim. Acta* **6,** 61 (1953).

29. E.G. Loewen, *The Echelle Story.* Bausch & Lomb, Rochester, New York (1971).

30. W.G. Eliott, *Amer. Lab.* **2,** (3), 67 (1970).

31. W.G. Eliott, U.S. Patent 3,658,423, April 25 (1972).

32. W.G. Eliott, U.S. Patent 3,658,424, April 25 (1972).

33. A. Danielsson and P. Lindblom, *Phys. Scripta* **5,** 227 (1972).

34. M.S. Cresser, P.N. Keliher, and C.C. Wohlers, *Anal. Chem.* **45,** 111 (1973).

35. A. Danielsson and P. Lindblom, *Appl. Spectrosc.* **30,** 151 (1976).

36. P.N. Keliher and C.C. Wohlers, *Anal. Chem.* **48,** 333A (1976).

37. P.N. Keliher, *Res. Developm.* **27,** (6), 26 (1976).

38. D.J. Schroeder, *Appl. Optics* **6,** 1976 (1967).

39. R. Hoekstra, T.M. Kamperman, C.W. Wells, and W. Werner, *Appl. Optics* **17,** 604 (1978).

40. A.T. Zander and P.N. Keliher, *Appl. Spectrosc.* **33,** 499 (1979).

41. D.J. Schroeder and R.L. Hilliard, *Appl. Optics* **19**, 2833 (1980).

42. D.L. Anderson, A.R. Foster, and M.L. Parsons, *Anal. Chem.* **53**, 770 (1981).

43. S. Engman and P. Lindblom, *Appl. Optics* **21**, 4356 (1982).

44. S. Engman and P. Lindblom, *Abstracts 23rd Coll. Spectrosc. Int., Amsterdam 1983. Spectrochim. Acta* **38B**, Supplement, Abstracts p. 392 (1983).

45. P.W.J.M. Boumans, *Invited Lecture, 9th Int. Conf. Atom. Spectrosc. and 22nd Coll. Spectrosc. Int., Tokyo 1981.*

46. P.W.J.M. Boumans and J.J.A.M. Vrakking, *Spectrochim. Acta* **39B**, Nos 10/11 (1984).

47. P.W.J.M. Boumans and J.J.A.M. Vrakking, *Spectrochim. Acta* **39B**, Nos 10/11 (1984).

48. P.W.J.M. Boumans and J.J.A.M. Vrakking, *Spectrochim. Acta* **39B**, Nos 10/11 (1984).

49. P.W.J.M. Boumans and M.Ch. Lux-Steiner, *Spectrochim. Acta* **36B**, 97 (1981).

50. P.W.J.M. Boumans, R.J. McKenna, and M. Bosveld, *Spectrochim. Acta* **36B**, 1031 (1981).

51. L.A. Fernando, *Spectrochim. Acta* **37B**, 859 (1982).

52. M.A. Floyd, V.A. Fassel, R.K. Winge, J.M. Katzenberger, and A.P. D'Silva, *Anal. Chem.* **52**, 431 (1980).

53. P.W.J.M. Boumans and F.J. de Boer, *Spectrochim. Acta* **30B**, 309 (1975).

54. P.W.J.M. Boumans and F.J. de Boer, *Spectrochim. Acta* **32B**, 365 (1977).

55. P.W.J.M. Boumans and F.J. de Boer, *Spectrochim. Acta* **31B**, 355 (1976).

56. K.W. Olson, W.J. Haas, Jr., and V.A. Fassel, *Anal. Chem.* **49**, 632 (1977).

57. G.F. Larson and V.A. Fassel, *Appl. Spectrosc.* **33**, 592 (1979).

58. G.F. Larson, V.A. Fassel, R.K. Winge, and R.N. Kniseley, *Appl. Spectrosc.* **30**, 384 (1976).

59. V.A. Fassel, J.M. Katzenberger, and R.K. Winge, *Appl. Spectrosc.* **33**, 1 (1979).

60. J.M. Mermet and C. Trassy, *Spectrochim. Acta* **36B**, 269 (1981).

61. H. Kawaguchi, Y. Oshio, and A. Mizuike, *Spectrochim. Acta* **37B**, 809 (1982).

62. L.M. Faires, B.A. Palmer, and J.W. Brault, *1984 Winter Conference on Plasma Spectrochemistry, San Diego*, Abstract No. 50; *Spectrochim. Acta* **40B**, Nos 1/2 (1985).

63. H. Haraguchi, T. Hasegawa, and K. Fuwa, *Abstracts 23rd Coll. Spectr. Int., Amsterdam 1983, Spectrochim. Acta* **38B**, Supplement, Abstracts, p. 135 (1983).

64. T. Hasegawa, H. Haraguchi, and K. Fuwa, *Chem. Lett.* 397 (1983).

65. T. Hasegawa and H. Haraguchi, *1984 Winter Conference on Plasma Spectrochemistry, San Diego; Spectrochim. Acta* **40B**, Nos 1/2 (1985).

66. A. Batal and J.M. Mermet, *Spectrochim. Acta* **36B**, 993 (1981).

67. J.A.C. Broekaert, F. Leis and K. Laqua, *Spectrochim. Acta* **34B**, 73 (1979).

68. J.W. McLaren and J.M. Mermet, *Spectrochim. Acta* **39B**, Nos 10/11 (1984).

LOW-CONSUMPTION ICP EMISSION SPECTROMETRY

Leo de Galan

Laboratorium voor Analytische Chemie, Technische Hogeschool,
Jaffalaan 9, 2628 BX Delft, The Netherlands

SUMMARY

For economical reasons the development of ICP-torches that require less RF-power and argon gas is evidently of interest, but only if their analytical performance matches that of conventional ICP-torches. The three approaches reported in the literature (miniaturized, high-efficiency and externally cooled torches) will be discussed and it is concluded that a torch running on 500 W and 1 l Ar/min is entirely feasible. Analytical data for aqueous solutions are already satisfactory, but more study on interferences and organic solvents is required. A novel approach designated as a radiatively cooled torch, will be introduced.

INTRODUCTION

In retrospect it is clear that Inductively Coupled Plasma Atomic Emission Spectrometry only gained acceptance as an analytical technique, when the original high-power torches (5-15 kW) were replaced by medium-power facilities. Indeed, all torches presently on the market for the argon ICP require about 1 kW for aqueous solutions to at most 2 kW for organic solvents. The originally also very high consumption of argon has decreased to a much lesser extent and now varies between 10 and 20 l/min. Manufacturers of ICP equipment have been surprisingly reluctant to appreciate the user's interest in more economical ICP-torches. One reason may be that the first reports on low-argon ICP-torches maintained the RF-power at the 1-2 kW level. Practical and theoretical concepts developed in the last few years have shown, however, that simultaneous reduction of the argon flow (below 5 l/min) and the RF-power (to 0.5 kW) is possible. The latter result indicates that the current bulky power generators can be replaced by modules, that fit into a 19" rack. This would allow the design of a true table-top instrument, which in the author's opinion would open up an entirely new market for the ICP. Ultimately, however, the introduction of a new generation of ICP instruments is not decided by convenience or economy, but by analytical performance. These aspects will be considered in the present article. Full details and references can be found in the recent review by Hieftje (1).

APPROACHES TO LOW-CONSUMPTION TORCHES

If the aim is to reduce the consumption of argon gas, it seems logical to diminish the size of the ICP torch. Indeed, early attempts by Hieftje and Savage (2) and Allemand et al. (3) were based on torches with half the diameter (9-13 mm) of conventional torches (20-25 mm). Although it was demonstrated that a stable plasma could be generated at 500 W and 7 l/min argon, the analytical performance was less than desired. Our personal experience agrees with these observations. The tentative explanation is that the electromagnetic field of the RF-coil requires a minimum depth of penetration for the plasma gas to transfer the power efficiently. This so-called skin-depth (a term adopted from solid conductors) dictates a minimum diameter for the plasma. Because the skin depth is inversely proportional to the square root of the RF-frequency, miniaturized torches might become profitable again when RF-generators are used that operate at 100 MHz rather than 27 MHz. The argon flow rates and RF-powers reported for miniaturized torches are roughly 5-10 l/min and 0.5-1 kW, respectively.

To appreciate the other two approaches to low consumption torches, it is instructive to consider a simplified power balance for the ICP-torch (4).

$$P_{torch} = P_{sample} + P_{plasma} + P_{conv} + P_{cond} \qquad (1)$$

where P_{torch} is the power dissipated in the torch, which for a well-tuned generator amounts to roughly 75% of the generator power. The four terms on the right hand side of Eq. (1) denote, respectively:

P_{sample} = the power needed to volatilize and atomize the sample solution, about 25 W for aqueous solutions and 150 W for organic solvents

P_{plasma} = the power needed to maintain the argon plasma, about 60 W for 1 l/min argon gas.

P_{conv} = the power drained away by excess argon, again 60 W for every 1/min.

P_{cond} = the power lost by conduction through the torch wall and to the environment.

In conventional torches a large excess of 'cooling' argon is used to protect the silica tube against the hot plasma. Wasteful torch designs require some 20 l/min, but with a more careful design of the torch geometry (5) a reduction to 10 l/min is feasible. These so-called high-efficiency torches have received new interest by the work of Hieftje and co-workers (6). Through careful optimization of all torch dimensions they were able to maintain a plasma with an argon flow of 5 l/min and a power as low as 125 W (for aqueous solutions). However, acceptable analytical performance required higher values, especially for the power. Indeed, a more recent study by Montaser et al. (7) showed that this torch can operate successfully at 720 W and 4 l/min of argon. Nevertheless, *optimum* analytical performance is only obtained at 920 W and 11 l/min of argon, conditions that come very close to those in current commercial torch designs. It is not surprising then that the analytical performance of these torches comes close to that of conventional ICP in terms of detection power and freedom from interference. It should be remarked, however, that real-life tests have been limited to a few samples and to aqueous solutions.

It thus appears that the two modifications of the *conventional* ICP still require an argon flow of 5-10 l/min. Obviously, a much more dramatic reduction in argon consumption results when the excess of cooling argon is abandoned completely. In that case the outer tube of the ICP torch can no longer be shielded effectively and external cooling is necessary to protect the silica tube from overheating. In terms of Eq. (1) it means that convective power dissipation (P_{conv}) is reduced to zero, but that the conductive power transfer through the tube (P_{cond}) must increase. Because silica is a fairly good heat conductor, the critical step is the transfer of heat from the hot tube to the environment.

A very efficient transfer is realized when the torch is equipped with a water-cooling jacket. In fact, water is such an efficient cooling agent that the outer torch tube is maintained virtually at room temperature. As a result, the water drains away a substantial amount of power and, hence, water-cooled ICP torches still require 1 kW or more RF-power (4,8). The analytical performance seems to vary with the amount of plasma argon used. Thus, Kawaguchi et al. report excellent detection limits for a water-cooled torch using a total argon flow of 5 l/min (9). Somewhat less favourable detection limits and interference resistance are reported by Ripson et al. for a design that uses only 1 l/min of total argon (10).

The latter authors obtain much better results when water is replaced by air as a forced external coolant medium. While the argon flow was maintained at 1 l/min, the RF-power was reduced to 500 W. In fact, the limited cooling capacity of air prohibited the use of higher power. Already, the temperature of the outer silica tube increased to 1000 K (4). Although the analytical results obtained for aqueous solutions are comparable to those of a conventional ICP, the results for organic solvents were decidedly inferior. Obviously, the efficiency of air-cooling must be enhanced to allow the higher power levels needed for organic solvents.

It should be realized that the use of either internal cooling by argon or external cooling by water or air stems from the need to keep the silica outer tube at a sufficiently low temperature, well below its softening point. If, instead of silica, another ceramic material is used that can withstand a higher temperature, no forced cooling would be needed and the argon flow could still be kept very low. Indeed, the power transfered from the plasma to the torch and conducted through the tube, would then be dissipated through heat radiation, according to

$$P_{rad} = A\varepsilon\sigma T^4 \tag{2}$$

where σ is the Stefan-Boltzmann constant and A, ε and T are the area, the emissivity and the temperature of the tube, respectively. For reasonable values of A and ε, temperatures up to 2000 K would be sufficient to release about 800 W.

The feasibility of this approach has recently been realized in our laboratory. Of the ceramics tested so far (Al_2O_3, ZrO, SiO_2 and BN) only boronnitride proved to be suitable. To maintain the possibility of spectroscopic observation the opaque boronnitride was provided with an extension of silica. This segmented torch was mounted inside the ordinary three-turn coil of a Plasmatherm RF-generator. A stable plasma could be ignited and maintained at a power of 600 W and an argon flow of 1 l/min. The temperature of the boronnitride segment inside the RF-coil was estimated to be 1600 K. However, it was observed that in the presence of sample solution the liberated oxygen slowly oxidized the boronnitride to cause volatilization and consecutive deposition of boric acid. Clearly, other high-temperature ceramics must be tested to study this interesting design further.

ANALYTICAL PERFORMANCE OF TORCHES RUN ON 1 L/MIN ARGON

In this final section we shall report on some analytical characteristics of ICP torches designed to run on a total argon flow of 1 l/min. Three designs will be discussed (Fig. 1):

(i)   a two-tube torch with an outside diameter of 16 mm and equipped with a water cooling jacket
(ii) a two-tube torch with an outside diameter of 18 mm to which a forced air-flow is blown
(iii) a three-tube, radiatively cooled torch with a segmented outer tube (18 mm) with a central part of boron nitride

The results reported for the externally cooled torches have been obtained with a special two-turn flat plated copper coil (11). Recently, we have operated the air-cooled torch also with the standard three-turn tubular coil that has been used with the segmented boron nitride torch.

All results refer to the standard operating conditions listed in Fig. 1 that are the results of an optimization study using the signal to background ratio of a representative spectral line as the criterion. Clearly, all torches operate on a total argon flow of 1 l/min or less. A major part is used as plasma gas, leaving only 0.2 l/min for the introduction of the aqueous solution. This has been achieved with a narrow bore (0.1 mm) Babington nebulizer made from either stainless steel or Kel-F. All torches can be easily ignited and have been operated satisfactory for many hours. No special precautions need to be taken for sample interchange and the range of the operating conditions is suffi-

Fig. 1 Schematic drawing and standard operating conditions of three ICP torches designed for a total argon flow of 1 l/min

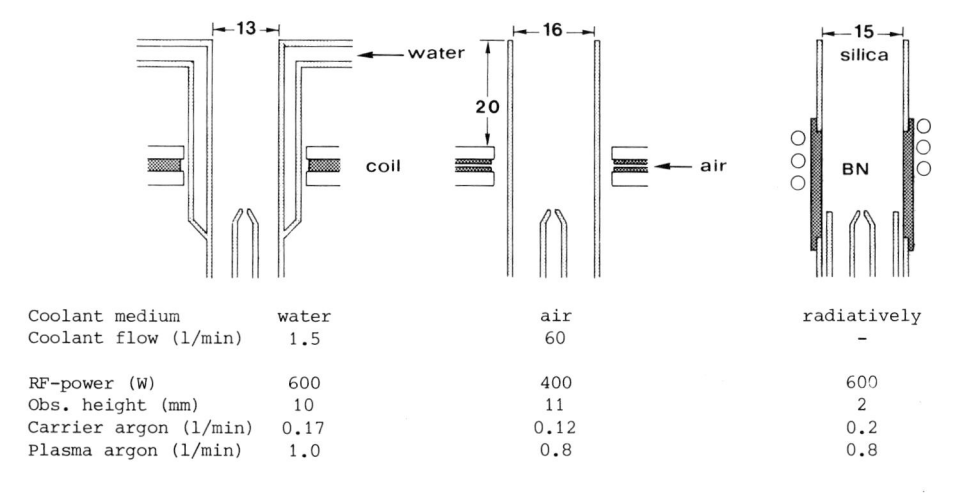

Coolant medium	water	air	radiatively
Coolant flow (l/min)	1.5	60	-
RF-power (W)	600	400	600
Obs. height (mm)	10	11	2
Carrier argon (l/min)	0.17	0.12	0.2
Plasma argon (l/min)	1.0	0.8	0.8

ciently wide to allow convenient and unattended operation. However, to prevent air entrainment all torches require an extended tube and so the plasma must be observed through the tube. Visual end-on observation reveals the usual ring-shaped discharge with a dark central hole present even if no carrier gas is introduced.

Table 1. Detection limits (in µg/l) for ICP-torches operating on 1 l/min shown in Fig. 1

Element and wavelength (nm)		water cooled	air cooled	radiatively cooled	conventional (ref. 12)
Al	I	396.2	70	40	50	28
B	I	249.8	35	10	-	4.8
Ca	II	393.4	-	0.13	0.1	0.19
Co	II	238.9	35	44	10	6.0
Cr	II	267.7	-	10	8	7.1
Cu	I	324.8	17	5	5	5.4
Fe	II	238.2	55	11	15	6
Mg	II	279.6	1	0.3	0.4	0.15
Mn	II	257.6	-	2	2	1.4
Na	I	589.0	24	7	5	29
Ni	II	221.6	-	26	6	10
Pb	II	220.4	-	230	100	42
V	II	309.3	24	8	-	5
Zn	I	213.9	-	12	9	1.8

Detection limits of a few representative elements are listed in Table 1. Whereas the water-cooled torch is inferior to the conventional ICP, the values for the other two torches come very close to the reference data of Winge et al. (12). It is recalled that these data refer to a single set of standard operating conditions and thus demonstrate the capability for simultaneous multi-element analysis.

The *dynamic range* (4-5 decades), the *precision* (1-2%) and the *long-term stability* (5-10% over 6 hours) are equal to the experience gained in operating a conventional ICP. Similarly, all torches allow the introduction of highly salted solutions (up to 3% NaCl). The introduction of organic solvents (methanol, xylene, MIBK) is possible, but only if the carrier gas flow is reduced and the power increased relative to the standard operating conditions. Even then, however, the power levels are too low to obtain adequate signals and detection power.

Matrix interferences have been studied in some more detail for the two externally cooled torches than for the radiatively cooled torch. Typical volatilization interferences studied for the anions nitrate and phosphate are insignificant up to the maximum concentration level of 10 g/l considered. As is usual in the ICP the most severe interference arises from the presence of easily ionized elements, such as sodium. From the representative examples shown in Fig. 2 it is seen that the

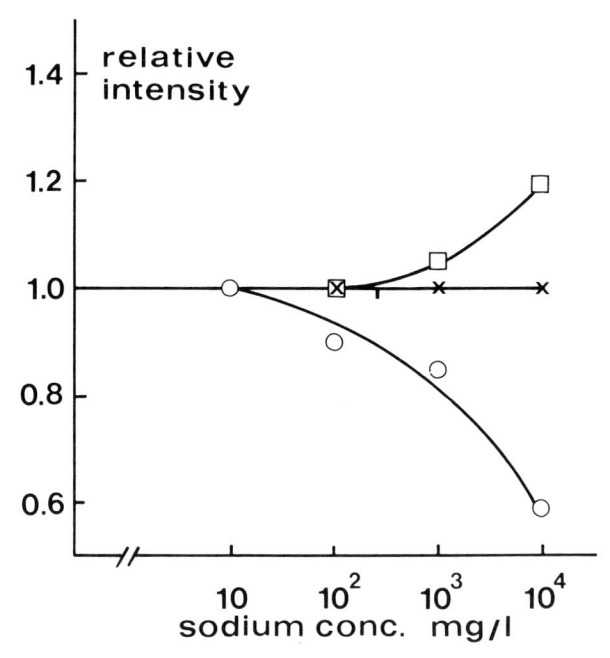

Fig. 2 Interfering effects of sodium in torches shown
in Fig. 1
○ Mn II   257.6 nm - water-cooled torch
□ Mn II   257.6 nm - air-cooled torch
✗ Ca II   393.3 nm - radiatively cooled torch

water-cooled torch is more subject to this type of interference. In the air-cooled and the radiatively cooled torch the line intensity remains stable up to a sodium concentration of 1 g/l and is only slightly affected at 10 g/l. This behaviour agrees closely to that of a conventional ICP.

CONCLUSION

The present discussion has shown that low-cost torches for ICP with satisfactory analytical performance can be realized provided that we break away from the conventional design and explore novel concepts. The common feature of the air-cooled and the radiatively cooled torch is the high temperature of the outer tube of the torch. It seems that this feature constitutes the means to obtain a high temperature plasma at low power. Indeed, with aqueous solutions good analytical properties have been obtained at 500 W and 1 l/min total argon. Probably, higher power is needed for organic solvents and this may be feasible with a combination of air-cooling and a high-temperature ceramic. The road towards a cheaper but equivalent ICP instrument is wide open.

Acknowledgement

The loan of an RF-generator from Plasmaproducts Inc. is gratefully acknowledged. This research is supported by the Stichting Technische Wetenschappen.

REFERENCES

1. G.M. Hieftje, Spectrochim. Acta 38B (1983) 1465-1481.

2. R.N. Savage and G.M. Hieftje, Anal. Chem. 51 (1979) 408-413; 52 (1980) 1267-1272.

3. C.D. Allemand, R.M. Barnes and C.C. Wohlers, Anal. Chem. 51 (1979) 2392-2394.

4. P.A.M. Ripson and L. de Galan, Spectrochim. Acta 38B (1983) 707-726.

5. C.D. Allemand and R.M. Barnes, Appl. Spectrosc. 31 (1977) 434-443.

6. R. Rezaaiyaan, G.M. Hieftje, H. Anderson, H. Kaiser and B. Meddings, Appl. Spectrosc. 36 (1982) 627-631.

7. A. Montaser, G.R. Huse, R.A. Wax, S.K. Chan, D.W. Golightly, J.S. Kane and A.F. Dorrzapf, Anal. Chem. 56 (1984) 283-288.

8. H. Kawaguchi, T. Ito, S. Rubi and A. Mizuike, Anal. Chem. 52 (1980) 2440-2442.

9. H. Kawaguchi, T. Tanaka, S. Miura, J. Xu and A. Mizuike, Spectrochim. Acta 38B (1983) 1319-1327.

10. P.A.M. Ripson, E.B.M. Jansen and L. de Galan, Anal. Chem. (in press)

11. P.A.M. Ripson, L. de Galan and J.W. de Ruiter, Spectrochim. Acta 37B (1982) 733-738.

12. R.K. Winge, V.J. Petterson and V.A. Fassel, Appl. Spectrosc. 33 (1979) 206-219.

SPECTRAL INTERFERENCES AND MATRIX EFFECTS IN OPTICAL EMISSION
SPECTROSCOPY

J.A.C. BROEKAERT

Institut für Spektrochemie und Angewandte Spektroskopie, Postfach 778, D-4600 Dortmund 1, Bundesrepublik Deutschland.

SUMMARY: - A discussion on the significance of spectral interferrences and matrix effects as sources for systematic errors in optical emission spectroscopy is presented. With special reference to working with plasma sources (ICP, MIP, a.o.), methods for preventing such errors are treated. They include both instrumental techniques which allow it to acquire the data for correction methods and suitable calibration techniques. Special attention is paid to background correction techniques including the use of a multichannel detector and techniques for the case of transient signals. The possibilities for eliminating matrix effects by optimization of the operating parameters or by calibrating with standard addition are exemplary discussed.

I. INTRODUCTION

Spectral interferences in optical emission spectroscopy are of different origin:
- Line interferences. Due to the complexity of atomic term diagrams the elemental spectra are line-rich, e.g. over 100 000 atomic lines are listed in the MIT-tables [1]. The latter have been made for arc and spark sources only and they are even for that purpose far from complete. Work with new spectroscopic radiation

sources, among them the inductively coupled plasma (ICP) and the microwave induced plasma (MIP) discharges indeed revealed many new lines. Due to the variety of sample matrices, it is clear that tracing of interferences of matrix lines and analyte lines will be an important task for the analyst using emission spectroscopic methods.
- Band interferences. Band spectra originate from radicals and molecular species in the radiation source. Indeed, despite the fact that in most sources temperatures are high, the atomization of the sample is incomplete and a series of metal-oxides and other species (CN, NH,...) remain. Further, the use of molecular working gases or simply the surrounding atmosphere and the solvents used contribute here.
- Contributions arising from the spectrometer. Stray light aris-ing from matrix lines and the overlap of different orders may further complicate the optical emission spectrum.

From this variety of sources for spectral interferences, one may conclude that the use of matrix-matched blank samples for estimating their contribution to blank and background intensities will require much skill and knowledge about the sample nature. Especially in the case of real samples, such attempts often will fail. They must be replaced by instrumental techniques which allow the examination of the spectral environment of the analytical line for the sample itself and the estimation of its true net line intensity accordingly. It is the aim of this paper to evaluate a number of such techniques and to show their possibilities from applications in plasma optical emission spectroscopy.

The above discussed estimation of the true net line intensities is of prime importance for trace element determinations. Indeed, when performing a determination at a concentration level of e.g. ten times the detection limit (defined on a 3 s base, as proposed by Kaiser and Specker [2]) and accepting relative standard deviations of 0.01 for the intensity measurements, the net intensities are only 40% of the background intensities. Consequently, an error of 10% in the estimation of the true background intensity would produce an analytical error of about 25%. As in ICP-OES e.g. matrix composition may affect the background intensities considerably, while net line intensities nearly remain unchanged, background correction by back-estimation at the analytical sample itself is indispensible as shown in [3].

Matrix effects, defined as the influence of the matrix composition

on the net line intensities of the analytical lines, as in any
physical method of analysis relate to the different stages of the
signal generation.
- The aerosol generation step.
In thermal vaporization, a selective volatilization of species
occurs and consequently, the aerosol generation is matrix-dependent.
This applies much less to pneumatic nebulization of liquids e.g.
- Plasma geometry. As the analyte number densities may considerably influence the temperature distribution in the plasma, they also may change the volume and the intensity distributions in the radiation source.
- Excitation and vaporization. Easily ionizable elements (e.g.
alkalines) may considerably influence the electron pressure in the
plasma and therefore, as known, may change the intensity ratios for
atom and ion lines. Apart from their influence on the ionization
equilibrium, their presence may influence the energy exchange processes in the radiation source. As many processes contribute to
analyte excitation, e.g. collisions with fast electrons, collisions
with excited working gas atoms or ions and photon excitation, the
influence of the matrix composition on analyte excitation is complex.

It will be shown at the example of results of ICP and MIP-OES
that by optimizing the working conditions, part of the matrix
effects arising from aerosol generation (e.g. nebulization effects)
may be eliminated. However, residual matrix effects can often only
be overcome by matrix-matching or calibration by standard addition,
together with instrumental techniques which enable the acquisition
of true net line intensities.

2. ACQUISITION OF DATA FOR BACKGROUND AND SPECTRAL INTERFERENCE
   CORRECTIONS.

When using a photographic emulsion for radiation measurements, the
contributions of background and interfering lines to be substracted
can directly be obtained from one recorded spectrogram. The information is available at high wavelength resolution and as the information at all wavelengths is gathered simultaneously, fluctuations of
the radiation source influencing both the analyte, the interferent
and the background intensities are eliminated to a large extent.
This enables true corrections only limited by the graininess of the
photographic emulsion. However, due to limitations in precision and
especially to laborous and time-consuming processing, emulsion

calibration a.s.o., the use of the technique nowadays is practically confined to research, documentation or semi-quantitative survey analysis.

With photoelectric techniques, the high parallel input and wavelength resolution obtained in photographic work is difficult to be realized. At the present state-of-the-art, spectral apparatus using one photomultiplier behind an exit slit for radiation measurements may be operated in an alternating stepping and measurement mode. However, this makes multielement determinations slow or at a reduction of the measurement time per single step it decreases the analytical precision. By using wavelength modulation, this limitation can partially be circumvented. Dual-channel instruments with flexible wavelength selection in each channel are of special interest for background correction in the case of transient signals. Multichannel detectors such as photodiode arrays or a vidicon e.g. enable the recording of spectra within a restricted wavelength range with simultaneous signal acquisiton at all wavelengths.

2.1 WAVELENGTH MODULATION

The most common systems base on:
- displacement of the entrance slit,
- the use of a quartz refractor plate e.g. in front of the exit slit as proposed by NORDMEYER [4].

Both methods may easily be controlled by a computer. A wavelength shift of 0.05 nm wich is almost sufficient for finding a wavelength representative of the true spectral background, may be realized by each of the two methods.

A computer-controlled displacement of the entrance slit may easily be incorporated in a simultaneous spectrometer and is standard, nowadays. The method enables only a rather slow wavelength modulation. However, in the case of radiation sources with a constant radiation output such as the ICP with pneumastic nebulization, the technique is suitable for background correction. This is, e.g. shown for the analysis of oils by ICP-OES [5]. As here, the background continuum intensities are highly sensitive to the type of oil e.g., the method was highly effective for eliminating matrix effects.

The quartz refractor plate technique is especially versatile for sequential spectrometers. The instrument then may slew from one analytical wavelength to another by turning the grating and perform dynamic background correction with the aid of a quartz plate rotation. At a reciprocal linear dispersion of 0.1-0.5 nm/mm which is common for plasma spectrometers, a 1-2 mm thick quartz plate enables the required spectral shift at a deflection below 15° where reflection losses still are negligible. The technique was used in ICP-OES for the analysis of minerals [3]. It enabled the correction for background intensity variations caused by changes of the concentrations of Ca, Mg and Na between 0 and 500 µg/mL.

The quartz refractor plate technique is also of use for transient signals having a duration of 1-3 s. The latter are encountered in plasma optical emission spectroscopy, e.g. in the case of injection techniques (see e.g. Ref [5]), when evaporating dry solution residues from a graphite furnace and subsequently transporting the aerosol into an ICP (e.g. Ref. [7]) or into an MIP (e.g. Ref. [8]), or when introducing samples directly into the ICP (e.g. Ref. [9,10,11]). However, the time resolution is limited to a certain degree by the drive system, but still much more so by counting statistics. In the case of advanced stepping motors, start-stop frequencies of 500-700 Hz, but also reset times of a few ms must be respected. Counting statistics limitations were evident e.g. in the case of ICP-OES. In the case of a 3 kW argon/nitrogen ICP, the photon flux at the exit slit of a 1 m Czerny-Turner monochromator for the spectral background at 250 nm was found to be 80 000 photons/s [12]. In the case of a photomultiplier with high sensitivity and low dark current (e.g. the EMI 9789 QB), the photoelectrons ($1.3 \cdot 10^4$ e/s) then produce an error of $(1.3 \cdot 10^4)^{-1/2}$ or 0.9% at a 1 s integration time. This shows that when the integration time per step is decreased below 55 ms e.g. the relative standard deviation for the background intensities depasses 4%. Consequently, the maximum operation frequency that can be obtained with the quartz refractor plate techniques is only 1-2 Hz. Thus, the system has lack of time resolution or measurement precision. The sequential measurement of line and background intensities further limits the acquisition of the true background intensities. Therefore, the technique has the advantage that all measurements are made with the same optical set-up and photomultiplier thus avoiding the calibration of channels.

2.2 DUAL-CHANNEL OPERATION

In sequential instuments, dual-channel operation may be realized by reflecting a part of the sorting beam into a second exit slit. A flexible choice of the wavelength shift between both channels may be realized e.g. with the aid of a quartz refractor plate or by a displacement of one of the exit slits. The methods will necessitate a calibration of the two channels and problems may arise from a non-uniform illumination of the beam-splitter. Therefore, it allows a simultaneous acquisition of line and background intensities with high time resolving power. From the background intensity values for the ICP stated before, one may conclude that the measurement frequency of this technique might be higher than 10 Hz. Thus, it is of potential interest for most plasma OES techniques with transient signals.

2.3 MULITCHANNEL DETECTION WITH PARALLEL INPUT

These detectors enable simultaneous measurements at all wavelengths within a small spectral range as required for accurate and flexible background correction. The capabilities and limitations of the following detectors have been regularly reviewed (see e.g. Ref. [13]):
- silicon photodiode arrays,
- vidicon and SIT-vidicon systems,
- microchannel plates.

Silicon photodiode arrays have been used in several investigations for radiation measurements in plasma emission spectroscopy (see e.g. Ref. [13,14]). Their interest for analytical work, however, is hampered up to now, by detector noise limitations. The latter might be overcome in detection systems containing microchannel plates, as recently introduced commercially.

The featurs of a SIT vidicon for radiation measurements in a sequential ICP-spectrometer e.g. have been studied by FURUTA et al. [15], who reported detector noise limitations. It was attempted to eliminate them by cooling the SIT-vidicon down to -20° C. A series of experiments was performed with an OMA 2 system and a high-quality photomultiplier (EMI 9789 QB) to the same monochromator. A 1.3 kW argon ICP was used as radiation source.

Table 1 Detection limits in ICP-OES using a photomultiplier and a cooled SIT [12].

1 m Czerny-Turner monochromator; grating: constant 1/1800 mm, 90x90 mm² entrance slit width: 17 µm, exit slit width: 25 µm; integration time: 5 s, G.M.K. nebulizer.

Element / line (nm)	Detection limits (µg/mL) SIT	Photomultiplier
Cr 425.4	0.6	0.6
Pb 405.7	1.4	0.7
Ni 352.4	0.35	0.2
Cu 324.7	0.1	0.02
Mg 279.6	0.01	0.003
Tl 276.7	2	0.8
Fe 260.0	0.06	0.03
Mn 257.6	0.02	0.004
Co 238.9	0.2	0.06
Cd 226.5	0.4	0.03
Zn 213.8	0.5	0.08

Detection limits (Table 1) in the case of the cooled SIT were found to be 3-5 times higher compared to those obtained with the photomultiplier. This applies especially to analytical wavelengths below 250 nm. Both, counting statistics limitations as well as low signal-to-noise ratios in the case of background intensity measurements are responsible for this. The low performance of the SIT vidicon and the low background radiances of plasma sources in this wavelength range act together here. The linear dynamic range is 3.5 decades in the case of the cooled SIT and 4 decades of intensity in the case of the photomultiplier. However, by peak-area measurements in the case of the cooled SIT differences could be eliminated. The "blooming effect" caused by cross-talk between neighbouring channels limits the wavelength resolution obtained. Indeed, with the monochromator used, the SIT channel width (25 µm) could enable a resolution of 5 pm. However, the practical resolution found was only 9 pm. Due to its limited dimensions and pixel density, the SIT vidicon has only a limited wavelength coverage and is only of interest for sequential multielement determinations. Apart from these limitations, the use of a SIT in sequential systems enables random access to analytical wavelengths. An effective correction for spectral interferences and background changes is possible, as information at all wavelengths within the spectral window is gathered simultaneously. Reference lines and analytical lines, when they are spectrally close enough, can be measured simultaneously and both can be corrected for the true background at the same time.

3. SOME POSSIBILITIES OF OPTIMIZATION TOWARDS LOWEST MATRIX EFFECTS IN PLASMA OES

In plasma optical emission spectroscopy using continuous nebulization of liquid samples, a considerable portion of matrix effects are nebulization effects, as already known from early ICP work (see e.g. Ref. [14]). It has been described that they can be minimized by selecting a proper aerosol carrier gas pressure [18,19]. Provided that an accurate background correction is applied, matrix effects of Ca, Na and Mg at concentrations of up to 500 µg/mL in ICP-OES still are negligible [3]. At a concentration of 2500 µg/mL Na (as $Na_2B_4O_7$), matrix effects after a careful optimization of the aerosol carrier gas flow were 4% [18].

In the case of electrothermal evaporation coupled to plasma optical emission spectroscopy, matrix effects are much higher. This is known to be due to the selective volatilization of species and related anion effects [7], as well as by transport effects [20]. The strategy for improving the analytical accuracy should include:
- a selection of the proper analytical signal (peak-maximum or peak-area measurements).
- the acquisition of the true net line intensities as discussed above
- calibration by standard addition.

The way described leads to accurate results in ICP-OES coupled to evaporation from a graphite furnace. The latter is shown by multi-element trace determinations in biological matrices [7]. Indeed, trace element concentrations of cadmium, manganese and zinc could be accurately determined in NBS 1571 and 1577. Systematic errors in the determination of lead e.g. were found to be due to an increase of the spectral background intensity during analyte evaporation [21]. In the case of MIP-OES coupled to evaporation from a graphite furnace, a total de/composition of the matrix by thermal treatment may be necessary, so as to keep the MIP discharge stable. This was found to be the case for trace determinations in serum samples [8]. After this precaution, here also, calibration by standard addition leads to accurate analysis results.

4. CONCLUSION

Systematic errors in emission spectroscopy which are caused by interfering lines or changes in background intensities can be efficiently eliminated by applying a number of instrumental techniques which are available now. As discussed with the aid of results in plasma optical emission spectroscopy, this applies to both methods with steady and those with transient signals. Matrix effects in optical emission spectroscopy have been shown to be of complex origin. They can be minimized by selecting proper working conditions and in some cases, calibration by standard addition is useful.

REFERENCES

[1] G.R. Harrison, "M.I.T Wavelength Tables", J. Wiley & Sons, New York (1939).

[2] H. Kaiser and H. Specker, Z. Anal. Chem. 149 (1956) 46-66.

[3] J.A.C. Broekaert, F. Leis and K. Laqua, Spectrochim. Acta 34B (1979) 73-84.

[4] N. Nordmeyer, Z. Anal. Chem. 225 (1967) 247-252.

[5] J.A.C. Broekaert, F. Leis and K. Laqua, Talanta 28 (1981) 745-752.

[6] A. Aziz, J.A.C. Broekaert and F. Leis, Spectrochim. Acta 36B (1981) 251-260.

[7] A. Aziz, J.A.C. Broekaert and F. Leis, Spectrochim. Acta 37B (1982) 369-379.

[8] A. Aziz, J.A.C. Broekaert and F. Leis, Spectrochim. Acta 37B (1982) 381-389.

[9] D. Sommer and K. Ohls, Fresenius' Z. Anal. Chem. 304 (1980) 97-103.

[10] E.D. Salin and G. Horlick, Anal. Chem. 51 (1979) 2284-2286.

[11] G.F. Kirkbright and S.J. Walton, Analyst 107 (1981) 276-281.

[12] J.A.C. Broekaert and F. Leis, unpublished work.

[13] Y. Talmi, Anal. Chem. 47 (1975) 658A-670A.

[14] H.R. Betty and G. Horlick, Applied Spectrosc. 32 (1978) 31-37.

[15] H. Bubert, Spectrochim. Acta 37B (1982) 533-548.

[16] N. Furuta, C.W. McLeod, H. Haraguchi and K. Fuwa, Applied Spectrosc. 34 (1980) 211-216.

[17] S. Greenfield, H. MsD. McGeachin and P.B. Smith, Anal. Chim. Acta 84 (1976) 67-78.

[18] J.A.C. Broekaert, F. Leis and K. Laqua, Spectrochim. Acta 34B (1979) 167-175.

[19] J.A.C. Broekaert and F. Leis, Anal. Chim. Acta 109 (1979) 73-83.

[20] D.L. Millard, H.C. Shan and G.F. Kirkbright, 105 (1980) 502-508.

[21] J.A.C. Broekaert, F. Leis and K. Laqua, Paper presented at the 9th Intern. Symposium on Microchemical Techniques, Amsterdam (1983).

SAMPLE INTRUDUCTION, SIGNAL GENERATION AND NOISE
CHARACTERISTICS FOR ARGON INDUCTIVELY-COUPLED
PLASMA OPTICAL EMISSION SPECTROSCOPY

G.F. Kirkbright

Department of Instrumentation and Analytical Science,
UMIST, Manchester, England

SUMMARY

Different methods of sample introduction and signal generation for optical emission spectroscopy using inductively-coupled plasma (ICP) systems are considered. These include graphite rod electrothermal and direct graphite rod sample introduction of small liquid and solid samples and the use of several types of hydride generation systems for those elements forming volatile hydrides; the determination of the halogens using gas generator cells with both the ICP and microwave induced plasma is discussed. The potential of noise spectrum analysis in plasma sources to allow optimisation of nebuliser parameters and spectrometer operation is also discussed.

INTRODUCTION

It has previously been demonstrated (1,2) that the high-frequency argon inductively-coupled plasma (ICP) provides an effective excitation source for the simultaneous multi-channel determination of metals and metalloids over a wide concentration range by pneumatic nebulisation of solution samples. Optical emission spectrometry with the ICP source offers detection limits in the parts per 10^9 (ppb) range for many elements, linear dynamic concentration ranges of typically five orders of magnitude and freedom from many of the chemical interferences encountered with other sources. The most commonly used technique for the introduction of sample solutions into the ICP is based on the injection of a liquid aerosol generated by a pneumatic nebuliser. In earlier papers from our group we described the development of a graphite rod electrothermal vaporisation device for sample introduction (3,4). In this device the microlitre volumes of liquid samples were placed on a graphite rod contained within a cylindrical glass manifold, remote from the plasma (see Fig.1). The graphite rod was heated in an argon stream by a low-voltage, high-current power supply fitted with a programmer allowing control of power and time during desolvation, ashing and vaporisation cycles. The vaporised analyte was then carried into the plasma using the argon injector gas stream and the transient emission signal at the atomic line of the analyte element was detected.

In recent publications (5,6) preliminary studies of the analytical performance of an instrumental device in which microlitre volumes of liquid samples were applied to a graphite rod and direct axial insertion of the rod into a continuously operating low power (less than 1.5kW) inductively-coupled argon plasma have been described. With this technique some volatile elements were determined by optical emission spectrometry at the sub-nanogram level with adequate precision and high powers of detection in small sample volumes (5 microlitres). The use of an argon-0.1% Freon 23 injector gas has been shown to permit efficient volatilisation of refractory carbide-forming elements with this system. This early work has indicated that the direct graphite rod sample technique is a potentially powerful method for the determination of trace elements and may be extended to simultaneous multi-element analysis. A microprocessor-controlled graphite rod direct sample insertion device has now been developed. The insertion of the rod into the plasma is effected by using a stepper motor assembly. The software design enables the plasma to be used for the desolvation and ashing of sample solutions as well as for their vaporisation. A schematic

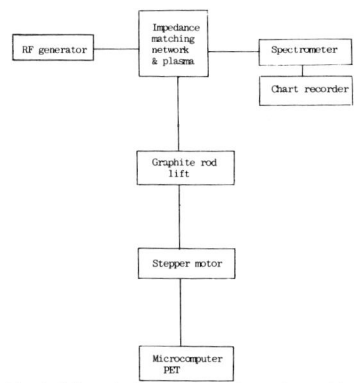

Fig. 1. Schematic diagram of experimental assembly.

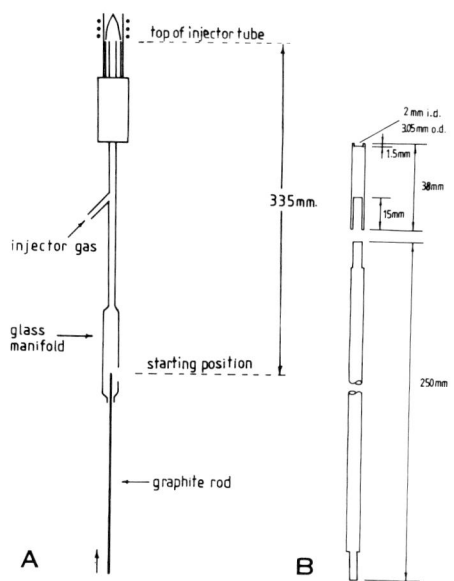

Fig. 2. A, Direct sample introduction system; graphite rod and detachable graphite cup.

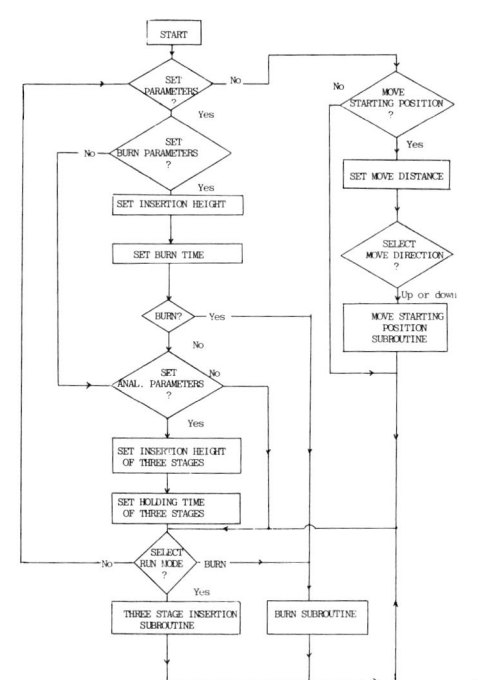

Fig. 3. Main BASIC program flow chart.

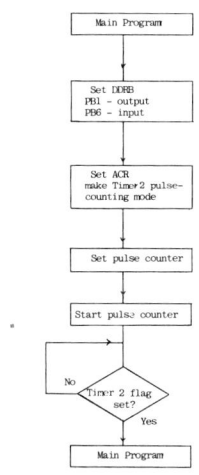

Fig. 4. Pulse counting subroutine (Machine code) flow chart. DDRB, data direction register "B"; ACR, auxiliary control register.

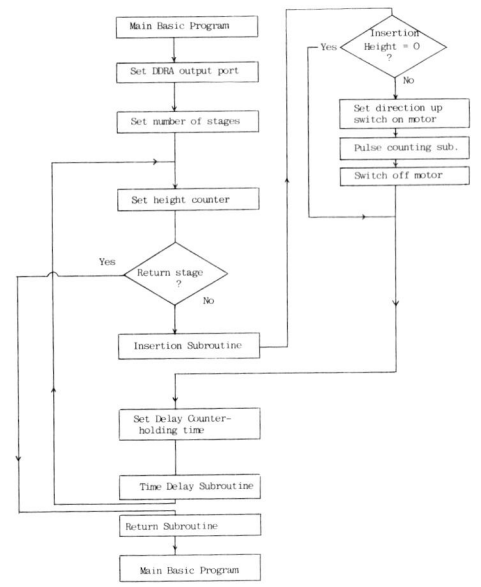

FIG. 5. Three-stage (drying, ashing, and atomization) insertion subroutine (Machine code) flow chart. DDRA, data direction register "A."

TABLE 1 Detection limits obtained for elements of environmental interest, using direct rod sample insertion device into ICP.

Element	Line, nm, I/II		Optimised operating conditions			Detection limits, µg/ml	
			Viewing Height, mm	Carrier gas flow-rate, dm^3/min.	Applied power, kW	Single element conditions	Compromise conditions+
As	193.69	I	27	0.72	1.0	1	2
As	228.81	I	27	0.72	1.0	0.6	0.6
Cd	228.80	I	27	0.61	1.0	0.03	0.06
Cd	226.50	II	23	0.84	1.0	0.04	0.07
Cr	357.87	I	37	0.91	0.8	0.003	0.003
Hg	253.65	I	23	0.81	0.8	0.2	0.1
Ni	341.47	I	37	0.92	0.8	0.006	0.02
Pb	280.20	I	26	0.78	1.0	0.2	0.3
Sb	231.15	I	23	0.92	1.0	0.4	0.3
Se	196.02	I	22	0.70	0.8	1	3
Zn	213.85	I	22	0.68	1.0	0.02	0.3

+Compromise conditions:· viewing height 27 mm; carrier gas flow-rate 0.8 dm^3/min; applied power, 1 kW.

diagram of the sample experimental assembly is shown in Fig.1. Automation of the graphite rod insertion device in the manner described improves the precision attainable by this technique; the system is suitable for trace element determinations in small liquid samples or directly with small solid samples. Figure 2 shows the manner in which the direct sample introduction system is operated and Figure 3 and 4 illustrate the main BASIC programme flow chart and the pulse counting sub-routine (machine code) flow chart (7).

A commercial 27.12MHz argon inductively-coupled plasma source (Plasma-Therm Inc., Kresson, NJ) of nominal maximum forward operating power od 1.5kW was interfaced to a single channel 1-m plane grating spectrometer (SPEX Industries Inc., Metuchen, NJ). A demountable silica torch was modified so that the injector tube had no constriction at its tip and was 3.7mm i.d.; the injector tube was lengthened so that its lower end extended through an aperture cut into the base of the torch box and was connected with a bulb pipet-shaped glass manifold (see Fig.2a) on which there was a port to allow the delivery of sample solution to the top of the graphite rod. The injector tube was fitted with an injector gas inlet below the brass base of the demountable torch. The sample insertion system consisted of a microcomputer (PET 2001-16NBS, Commodore Business Machines, Inc.), a stepper motor-controlled graphite rod lead screw lift, and a graphite cup (38mm in length) attached to a graphite rod (250mm in length) (Fig.2b), which was push-fitted onto the top of a glass tube. This glass tube was fixed into the side of the graphite rod lift. The stepper motor (23 PM-C 401 Miniangle Stepper Motor, 200 steps/rev, Shinkon Communication Industry Co.Ltd.) was controlled by the microcomputer.

The output of the photomultiplier (EMI 6256B) was directly taken to a fast response chart recorder (series 3000, Oxford Instrument Co.Ltd.) for signal registration. One of the features of the PET is that it can be programmed in both BASIC and Machine code. The programme used to implement the drying, ashing, and atomization stages of sample analysis by the technique used, consists of a main BASIC programme and several Machine code subroutines. The main BASIC programme sets the operation parameters of the graphite rod lift (insertion height and burn time for burn operation; insertion height of the rod and the hold time at the drying, ashing and atomization stages of sample analysis, respectively), and also selects the operation modes of the graphite rod lift, including running the sample, preburning of the graphite cup, and changing the start position of the graphite rod. The preburning stage is used to eliminate contamination or memory effects in the graphite cup.

A stepper motor will move an angular increment for each digital pulse it receives and is thus easily controlled by a microcomputer. With a PET there are two methods of producing output pulses in the free-running mode whose frequency is not affected by monitor interrupt routines. Both methods utilise the 6552 VIA (versatile interface adapter) timers. One method utilises timer 2 and shift register. A repeated pattern of eight bits can be shifted out onto the CB2 line of peripheral B control line at a particular frequency that is controlled by timer 2. This free-running output on the CB2 line is a useful source of pulses and is in full synchronization with the PET timing. Another method utilises the timer 1 free-running mode to produce a square wave on peripheral B port 7 (PB7). A stepper motor requires controlled acceleration and deceleration especially when required to move a load having significant inertia as is the case for the graphite rod driver. A microprocessor can effect control of the stepper motor acceleration and deceleration by generating a variable pulse rate through software. However, the acceleration and deceleration are not smooth because they can be generated only by digital acceleration/deceleration tables that have been stored in memory, thereby giving rise to step-structured ramps. Such step-structured ramps may cause the stepper motor to stall when a high step rate is used; the maximum usable step rate was thus <7000 steps/s.

In order to obtain reliable high-speed operation, the stepper motor drive board, Digicard 053 (PKS Designs Ltd.) was used to output pulses. The Digicard 053 has a built-in wide range oscillator with separate control of fast and slow speeds. When the fast control is shorted to ground the motor will accelerate to the preset speed and when it is isolated from ground the motor will decelerate. By correct choice of the components on the board a maximum usable step rate of approximately 12,000 steps/s is possible. The output of pulses is counted by VIA timer 2. This is accomplished by first loading a predetermined number of pulses into T2. Writing into T2 high order counter 8t2 C-H) clears the interrupt flag and allows the counter to decrement each time a pulse is applied to PB6. The interrupt flag will be set when T2 reaches zero. Fig.4 shows the counting subroutine flow chart.

The three-stage insertion subroutine is shown in Fig.5. The programme consists of data (insertion height and holding time) setting, insertion subroutine, return subroutine, and time delay subroutine. After running the three stages of analysis for an applied sample, the graphite rod automatically returns to its original starting position.

The application of the microprocessor-controlled direct graphite rod sample introduction system to the determination of elements of environmental interest by ICP-OES has been investigated. Detection limits for aqueous solutions of these elements under simplex optimised conditions for each and under compromise conditions are shown in Table 1. While the system performs well for relatively volatile elements using argon alone in the injector channel we have again confirmed that for those elements which form thermally stable oxides and carbides the use of an active atmosphere, in this case 1% Freon 23 in the injector carrier gas assists volatilisation of analyte species such as boron, molybdenum, tungsten, titanium.

The analytical utility of the direct sample introduction system depends on two main parameters:

(1) Reproducible positioning of the graphite furnace along the axis of the plasma for drying, ashing and vaporisation to provide the desired precision.

(2) A rapid heating rate to facilitate the efficient removal of the sample from the furnace and to produce a relatively high transient and localised concentration of the analyte thus resulting in high sensitivity.

Figure 6 illustrates four furnace configurations that have been evaluated. These furnaces were manufactured within the laboratory using a jewellers lathe (Emco, Austria) from RW-O graphite rod and ES40 glassy carbon rod (Ringsdorff GmbH, FRG). The same programme was used to drive each of the four furnaces into the plasma, hence each furnace experienced the same environment for the desolvation and vaporisation stages.

The recorder traces in Fig.7 show a successive increase in the heating rates from electrode type A to type D and also indicate that the time taken to achieve the equilibrium temperature (T_E) decreases in the same order. Although the contributions from radiation and induction to the heating of the furnace have not been determined, the trend in the heating rates shown in Fig.2 is empirically consistent with the geometries of the furnace and the materials of their construction. The cooling curves obtained appear to approximate to Newton's law of cooling; thus:

$$\text{rate of heat loss} \propto (\theta - \theta_o)$$

where θ and θ_o are the temperatures of the body and its surroundings respectively. The constants of proportionality take in account the conditions of the surface k, and the surface area, A; then:

$$\text{rate of heat loss} = kA(\theta - \theta_o).$$

Hence the rate of fall of temperature is:

$$= \frac{kA(\theta - \theta_o)}{mc}$$

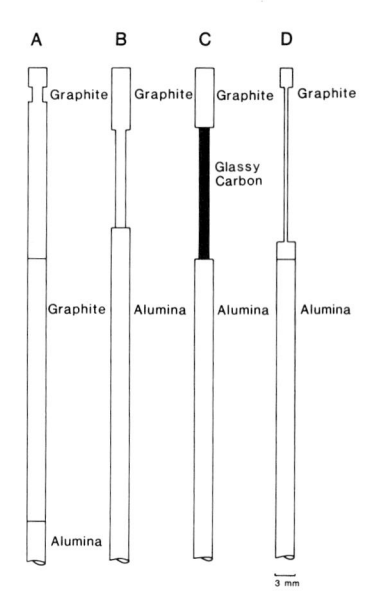

Figure 6 Rod Configurations employed

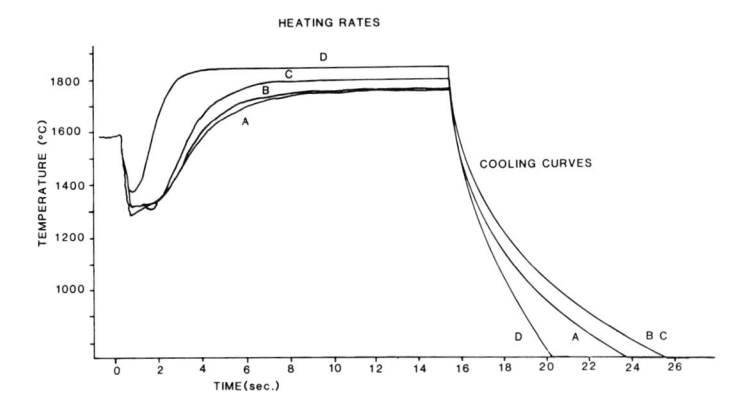

Figure 7 Heating and Cooling curves for Rod designs employed.

where m is the mass of the heated body and c is its specific heat capacity. It follows that the rate of temperature fall of a body is proportional to the ratio of the surface area to volume and inversely proportional to any one linear dimension. This relationship is observed in the cooling curves shown in Fig.7 (8).

The most important factor in the practical evaluation of the various furnace configurations is analytical performance. The peak shapes and relative sensitivities for nickel, chromium, manganese and lead are shown in Figs.8a-8d respectively. The significant improvement in sensitivity for the two less volatile elements (Ni and Cr) reflects, to some extent, the difference in heating rates between furnace configurations. The observed differences in

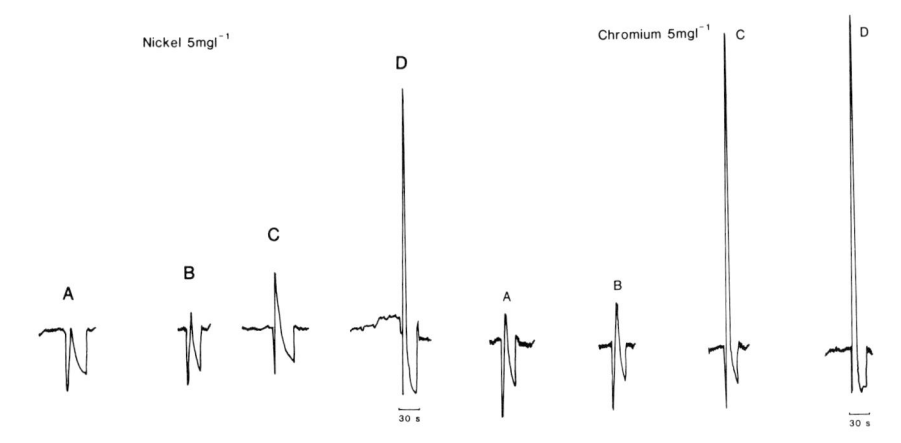

Figure 8a Responses for Ni with different rod configurations.

Figure 8b Responses for Cr with different rod configurations.

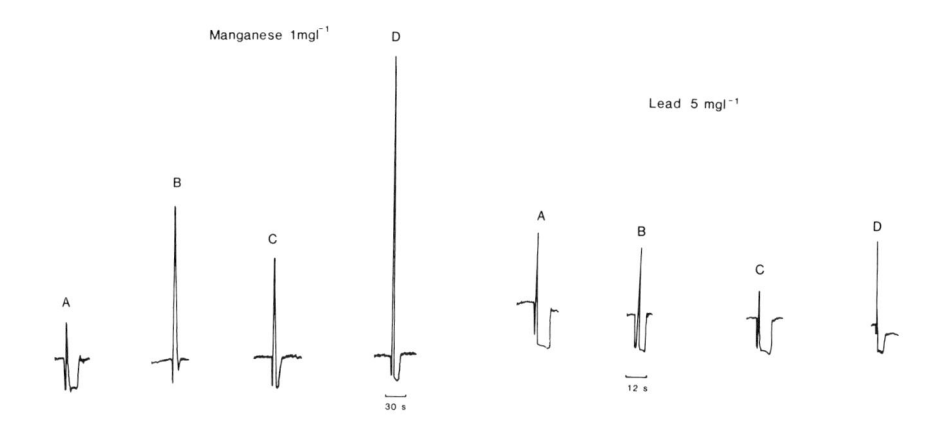

Figure 8c Responses for Mn with different rod configurations.

Figure 8d Responses for Pb with different rod configurations.

sensitivity and response time for chromium for furnaces C and D are only small compared to nickel and using the same two furnace configurations. This observation cannot be explained on the basis of the boiling points of the elements or the compounds from which the standard solutions were prepared. It would seem evident that a more complex mechanism of vaporisation operates and that this does not entirely depend on volatility alone. the nature and condition of the internal surface and the physical nature of the residue which remains after drying may influence the rate and efficiency of the vaporisation process.

The response characteristics for manganese (Fig.8c) exhibit no clear trend between furnaces A, B and C with respect to heating rates and equilibrium temperatures; the significant signal improvement for furnace D is consistent with the results obtained for the other three elements

Figure 9

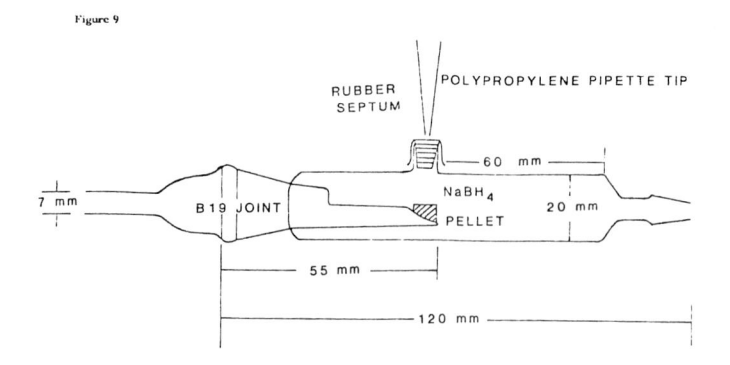

RUBBER SEPTUM
POLYPROPYLENE PIPETTE TIP
7 mm
B19 JOINT
NaBH$_4$ PELLET
60 mm
20 mm
55 mm
120 mm

Figure 10A

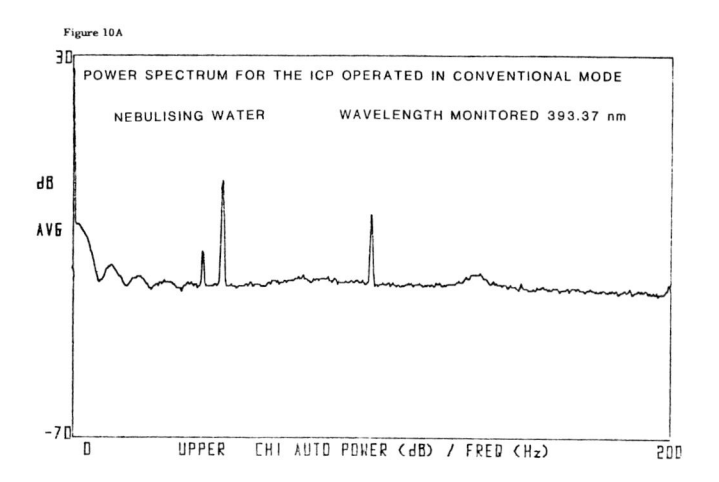

POWER SPECTRUM FOR THE ICP OPERATED IN CONVENTIONAL MODE
NEBULISING WATER   WAVELENGTH MONITORED 393.37 nm

dB AVG

UPPER   CHI AUTO POWER (dB) / FREQ (Hz)

Figure 10B

POWER SPECTRUM FOR THE ICP OPERATED IN CONVENTIONAL MODE
NEBULISING 100 ppm Ca   WAVELENGTH MONITORED 393.37 nm

dB AVG

UPPER   CHI AUTO POWER (dB) / FREQ (Hz)

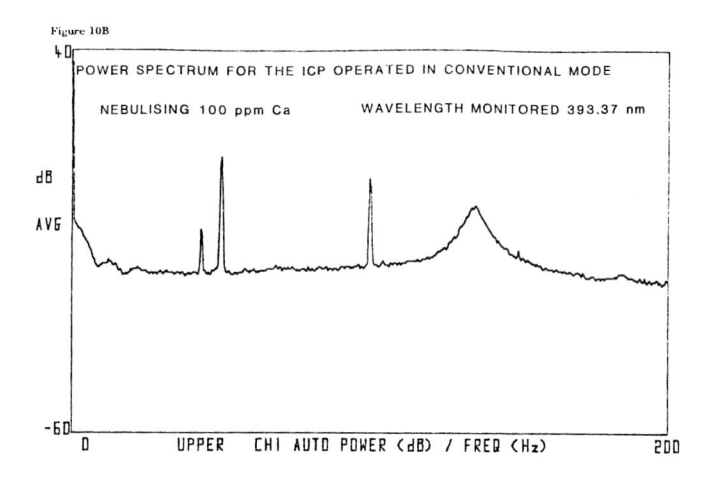

using this furnace. The lead sensitivity shown in Fig.8d, for furnaces, A,B and C, appears to be inversely related to the observed heating rates. As expected, furnace D gives a more rapid signal response and approximately the same sensitivity.

Hydride Generation

In a separate study a hydride generation system which permit the use of either an inductively-coupled plasma or a microwave induced plasma systems for optical emission spectrometry for the elements arsenic, selenium, tellurium, tin, bismuth and germanium in continuous or discrete sampling mode have been developed. In a mini-hydride system small liquid samples (5-10 microlitres) (Figure 9) are introduced directly onto a solid pellet of sodium borohydride; by this technique the volatile hydrides of these elements are generated rapidly with minimum dilution in the carrier gas via which they are transported to the plasma. The technique has been shown to be useful for a number of applications.

Noise spectrum analysis for the ICP and microwave induced plasma has been conducted with a system which is designed to allow optimisation of nebuliser parameters and to permit effective simplex optimisation of all of the variable operating parameters for the plasma systems to minimise noise at the frequencies which govern the attainable signal to noise and therefore detection limits obtainable. Figure 10 shows typical power spectra obtained for the ICP for pneumatic nebulisation.

References

1. Fassel, V.A. and Kniseley, R.N., Anal.Chem, 46, 1110A and 1155A (1974.
2. Greenfield, S., McGeachin, H.McD. and Smith, P.B., Talanta, 23, 1 (1976).
3. Gunn, A.M., Millard, D.L. and Kirkbright, G.F., Analyst, 103, 1066 (1978).
4. Millard, D.L., Shan, H.C. and Kirkbright, G.F., Analyst, 105, 502 (1980).
5. Kirkbright, G.F. and Walton, S.J., Analyst, 107, 276 (1982).
6. Kirkbright, G.F. and Zhang Li-Xing, Analyst, 107, 617 (1982).
7. Zhang Li-Xing, Kirkbright, G.F., Cope, M.J. and Watson, J.M., Applied Spectroscopy, 37, 250 (1983).
8. Barnett, N.W., Cope, M.J., Kirkbright, G.F. and Taobi, A.A., Spectrochimica Acta, 39B, 343 (1984).
9. Barnett, N.W., Chen, L.S. and Kirkbright, G.F., Analyst, (1984), In press.

A CONTRIBUTION TO THE DIRECT ANALYSIS OF SOLID SAMPLES BY SPARK
EROSION COMBINED TO ICP-OES

J.A.C. BROEKAERT, F. LEIS and K. LAQUA

Institut für Spektrochemie und Angewandte Spektroskopie, Postfach
778 D-4600 Dortmund 1, Bundesrepublik Deutschland

1. INTRODUCTION

When performing routine analysis of solids with plasma optical
emission spectroscopy, the samples are normally brought in solution.
This step, however, is time-consuming and introduces risks of contamination. Further, the analyte is diluted and the power of detection with respect to the solid sample decreases. With the aid of
a spark discharge, an aerosol may be produced directly from the
solid sample. The aerosol may be transported into a plasma radiation source (ICP, DCP, MIP) where the analyte material is excited.
Already in 1976, HUMAN et al. [1] reported the use of spark erosion
coupled to ICP-OES for the direct analysis of solids. Up to now,
however, the power of detection of such methods is inferior to that
of ICP-OES of the dissolved samples.

It is the aim of this work to investigate the factors limiting
the power of detection and to find a way to improve it. It will be
shown at the example of analyses of compact metallic samples and of
non-conducting powders, in how far the advantages of analyzing
liquid samples by ICP-OES may be realized in the case of spark erosion coupled to ICP-OES.

2. EXPERIMENTAL

2.1 INSTRUMENTATION

Spark erosion was performed with the aid of a medium voltage
spark at low repetition rate (up to 50 s^{-1}) and a high repetition
rate spark (Monoalternance, Philips, PV 8530; up to 400 s^{-1}). The
analyte material was transported into a high power argon/nitrogen
ICP and measurements were performed with the aid of a 1 m Paschen-

Runge spectrometer and a 1 m Czerny-Turner monochromator. Further instrumental details may be found in Ref. [2].

2.2 OPTIMIZATION

The electrical spark parameters, the argon carrier-gas flow and the operation parameters of the ICP were optimized towards highest power of detection and operation stability. It was found that the ablation and thus, also the power of detection increased with the condensor voltage. However, also the particle diameter increased with the condensor voltage. From 1.5 kV on, particles became so large that they were no longer completely evaporated in the ICP and made the discharge unstable. A good compromise between highest power of detection and optimal operation stability was obtained at the following working conditions: condensor voltage: 1 kV, spark repetition rate: $25s^{-1}$, Argon carrier-gas flow: 1.7 L/min and ICP operation power: 3 kW.

2.3 POWER OF DETECTION

Detection limits found in the case of aluminium alloys were a factor of 10-50 higher than those obtained with spark optical emission spectroscopy or ICP-OES subsequent to sample dissolution. The main cause for this lay in the low amount of analyte material introduced into the ICP (ca. 5-10 µg/min) and not in the particle size, which was found to be below 1 µm.

In order to improve the power of detection, one should find a way to increase the sample introduction rate without increasing the mean particle size. This could be realized by using a high repetition rate spark at medium voltage. Indeed, by varying the spark repetition rate from 50 s^{-1} to 400 s^{-1} the sample introduction rate for aluminium samples increased from 25 to 70 µg/min and the mean particle size practically remains unchanged. At a repetition rate of 400 s^{-1} the detection limits of spark erosion coupled to ICP-OES were found to be as low as those of spark optical emission spectroscopy or ICP-OES of the dissolved samples (Table 1).
The aerosol obtained by spark erosion may be transported over considerable distances, as it was reported for arc erosion earlier [5]. Indeed, at a distance of 5-6 m, intensity losses were found to be ≤ 50%.

Table 1: Detection limits (3 s) for aluminium alloys.

Element/line (nm)	$c_L(\mu g/g)$ ICP-OES acid dissolution (20 g/L) [3]	spark optical emission spectroscopy [4]	spark erosion coupled to ICP-OES (400 s^{-1})
Cu I 324.8	2	1	0.5
Fe II 259.9	0.3	3°	1
Mg II 279.6	0.05	1°	0.2
Mn II 293.3	-	-	10
Si I 288.3	-	-	2

(°) other lines used

2.4 MATRIX EFFECTS

Very different types of aluminium alloys could be analyzed by using the same straight calibration curves. This applied e.g. to the determination of Mg in Al, AlMn and AlMgSi alloys where c_{Si} varies between 1000 and 10 000 µg/g. It was also found to be true for the determination of Mg, Cu, Fe and Si in Al, AlMn, AlSiCuNi, AlSiCu and AlMgSi alloys. Only for alloys where the silicon concentration depasses the eutectic concentration of Si in Al, matrix effects were found.

2.5 ANALYSIS OF ELECTRICALLY NON-CONDUCTING POWDERS

Materials such as Al_2O_3 and $CaCO_3$ can be mixed with copper powder (1:4) and the mixture can be briquetted into pellets, as it is known from glow discharge work [6]. The pellets may be directly analyzed by spark erosion coupled to ICP-OES. Detection limits obtained were of the same order of magnitude as those found in ICP-OES subsequent to sample dissolution. The method thus is of potential interest for substances which are difficult to be dissolved (e.g. Al_2O_3).

Spark erosion in the case of non-conducting powders was found to depend considerably on the sample base. In the case of Al_2O_3 e.g. the ablation rate was high and a homogeneous burning spot was obtained, not however, in the case of $CaCO_3$. Consequently, the analytical precision also varies from one sample base to another. This could not be corrected by using a reference signal. In the case of Al_2O_3 relative standard deviations of the estimate [7] at a concentration level of 5 mg/g were found to be 0.05. This shows that the method may be used for quantitative determinations.

3. CONCLUSIONS

Spark erosion is a suitable technique for the direct analysis of solids with ICP-OES. This applies both to compact metallic samples and to non-conducting powder samples. Routine applications are possible as the memory effects were low. As transport losses are also low, the method may be used for remote sampling. The detection limits in the case of a high repetition rate spark are of the same order of magnitude as those obtained with spark emission spectroscopy or with ICP-OES subsequent to sample dissolution. The analytical precision for a number of applications is sufficient. The matrix effects are comparable to those found in ICP-OES of the dissolved samples. They are at any case lower than those encountered in spark emission spectroscopy.

REFERENCES

[1] H.C.G. Human, R.H. Scott, A.R. Oakes and C.D. West, Analyst 101 (1976) 265-271.

[2] A. Aziz, J.A.C. Broekaert, F. Leis and K. Laqua, Spectrochim. Acta B, in press.

[3] A. Aziz, J.A,.C. Broekaert and F. Leis, Spectrochim. Acta 37B (1982) 369-379.

[4] K. Slickers, "Die Automatische Emissionsspektalanalyse", Brühlsche Universitätsdruckerei, Lahn-Gießen (1977).

[5] K. Ohls and D. Sommer, Fresenius' Z. Anal. Chem. 296 (1979) 241-246.

[6] S. El Alfy, K. Laqua and H. Maßmann, Fresenius' Z. Anal. Chem. 263 (1973) 1-14.

[7] V.V. Nalimov, "The Application of Mathematical Statistics to Chemical Analysis", p.189. Pergamon Press, Oxford (1963).

THREE-FILAMENT AND HOLLOW-CYLINDER MICROWAVE INDUCED PLASMAS AS EXCITATION SOURCES FOR EMISSION SPECTROMETRIC TRACE ANALYSIS OF SOLUTIONS AND GASEOUS SAMPLES

D. Kollotzek, P. Tschöpel and G. Tölg

Max-Planck-Institut for Metal Research, Institute of Materials Science, Laboratory for High-Purity Materials, Katharinenstr. 17, D-7070 Schwäbisch Gmünd, F.R.G.

The capability of a microwave induced argon plasma as an excitation source for atomic emission spectrometry in the extreme trace analysis of elements can be increased if special plasma configurations are used. Two new Ar-MIPs and their analytical performance are described: A stable 3-filament Ar-MIP for the determination of volatile elements or compounds and a hollow-cylinder Ar-MIP for nebulized solutions. The detection limits for the elements Be, Ca, Cd, Co, Cr, Cu, Ga, Hg, Mg, Mn, Sr, Tl and Zn in aqueous solutions were found to be in the range of 1 to 50 ng/ml.

For the analysis of samples with high contents of organic compounds an oxygen-argon mixed-gas MIP consisting of two concentric quartz tubes was used. This plasma tolerates methanol-water mixtures with a methanol content of up to 90 % and can therefore be used as an on-line element-specific detector in HPLC. The capability of the HPLC-MIP system is illustrated by way of the separation of some mercury species as 2-mercapto-ethanol complexes and by investigations of immobile mercury compounds in highland peat bog soils. Detection limits for organically bound Hg are in the nanogram range.

1. Introduction

Direct instrumental multi-element determination procedures as well as multi-stage combined procedures via solutions can show extremly high systematic errors (1-3), which with decreasing concentration of the analytes can cause the analytical results to be incorrect by orders of magnitude. This is clearly demonstrated by interlaboratory comparative control studies (4,5).
The best way of obtaining accurate results leads via multistage-combined-procedures, in which the trace elements in the sample are determined - after dissolution of the sample, separation and preconcentration - in an isolated form. The determination can then be performed by calibration with aqueous solutions.
For AES solution techniques electrically generated plasmas, such as the direct current plasma (DCP), the inductively coupled plasma (ICP), the capacitively coupled microwave plasma (CMP) and the microwave induced plasma (MIP) are useful excitation sources (6).
In a recent paper (7) we described some new forms of the argon MIP. These novel

sources and their analytical performance shall be described and demonstrated by some special applications.

2. The filament Ar-MIP (F-MIP)

An argon MIP burns as a filament located exactly along the axis of the quartz capillary (i.d. 2.5 to 4 mm) at a low gas flow of 50 to 100 ml/min and a microwave power of 70 to 150 W at atmospheric pressure. Already small changes in the plasma parameters (gas flow, power, temperature, aerosol load etc.) cause irreproducible variations of the shape and the spatial structure of the plasma.
Since the sample aerosol bypasses the plasma, the filament is pushed towards the wall of the capillary and the analyte atoms are therefore insufficiently excited. Accordingly different zones appear along the cross section, showing different excitation conditions. Therefore the optimum power of detection of the F-MIP will be reached only if a definite zone of the cross section of the capillary is observed. This zone is very small and difficult to define. Simultaneous multielement determination is feasible only under compromise conditions. Also the long term stability of the F-MIP is poor.
These difficulties are circumvented with other discharge types. The use of the F-MIP is only useful if the other forms are unstable, as happens to be so, e.g., if high microwave power (< 140 W) is used or oxygene is added to the plasma gas. Such conditions may be stringent if the excitation potential of the analysis line or the atomization energy of the analyte is high or if organic solvents instead of water are used.

3. The 3-filament-Ar-MIP (3-F-MIP)

An essential improvement in excitation should result from a more efficient introduction of the analyte into the plasma or from heating the sample aerosol outside the filament up to higher temperatures with aid of a plasma of suitable shape. In capillaries with an inner diameter of 2.2 to 4.1 mm, stable plasmas consisting of several uniform filaments with equal distances from each other and all attached to the wall can be maintained. The optimum observation zone is situated at the axis of the tube. Here the background is very low and the SBR is very good.
A multi-filament MIP can be realized, if the capillary is fixed by a chuck of brass, screwed in the centre hole of the back of the TM_{010} resonator. The number of the filaments is dictated by the number of jaws. The jaws also fix the position of the filaments.
Three filaments have proved to be an optimum compromise of maximum excitation and stability of the plasma, the latter of which is nearly independent of gasflow and power within wide ranges.

To compare the 3-F-MIP and the conventional F-MIP, a routine procedure for the determination of mercury /8/ was used. Here the 3-F-MIP is superior to the F-MIP. When the same monochromator is used, we found for the detection limits: 5×10^{-12} g (3-F-MIP), 10^{-10} g (F-MIP); for the linear range: four and two orders of magnitude respectively.

In principle the 3-F-MIP can be used in connection with all sampling devices with which only small amounts of water or organic solvents are introduced into the plasma. Especially electrothermal vaporization techniques and gas chromatographic techniques should be mentioned here. Gaseous samples can be analysed very sensitively and metal vapours produced by a laser microprobe (9) can be introduced into the 3-F-MIP without problems.

Since mg amounts of water or organic solvents interfere severely, the analysis of solutions with a nebulizer is impossible. For that purpose one has to use the hollow cylinder-MIP.

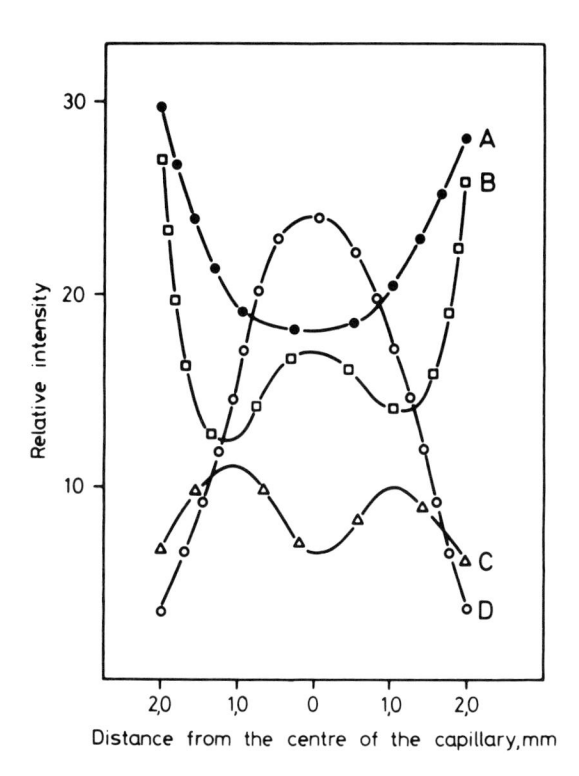

Fig. 1. Radial net line intensity profiles of some elements in a HC-MIP.
A) Cd (228,80 nm); B) Tl (377,57 nm); C) Co (240,73 nm); D) Ca (422,67 nm).
Inner diameter of the capillary: 4,0 mm; Uptake rate: 0.35 ml/min; Gas flow: 500 ml/min; Microwave power: 110 W.

4. The hollow-cylinder MIP (HC-MIP)

Bollo-Kamara and Codding (10) succeeded in giving the MIP an ICP-like configuration by using a special discharge device consisting of two concentric quartz tubes with a central carrier gas flow and a second plasma support gas with a spiral trajectory.

A similar plasma configuration with good spatial and temporal stability can be obtained using a single, common quartz capillary (7).

Decisive is a separate mounting (7) for the capillary, independent of the TM_{010} resonator, thus allowing exact adjustment of the tube along the axis of the cavity. Another condition is the presence of at least small amounts of water vapour in the plasma gas.

The HC-MIP burns in a thin layer close to the wall of the quartz tube with a tunnel in the centre, through which the plasma gas and the sample aerosol flow. In this configuration the sample aerosol is completely envelopped by the plasma and is optimally excited.

The thickness of the plasma layer increases with power and decreasing gas flow. Axially the plasma tends only less than 2 mm beyond the resonator, irrespective of the microwave power and the gas flow. The long term stability is high and memory effects due to condensation processes do not occur.

Radial profiles of the signals, the background signals and the SBR showed, that the intensity of the background always is high in the plasma layer and low in the tunnel region, the shapes of the curves, showing the net line signals, differ from element to element (Fig. 1). The intensity maxima of the analysis signals of elements with low excitation potential lie in the centre of the capillary, those of elements with high excitation potentials have different distances to the plasma. But for the majority of the investigated elements, the maximum of the SBR is situated in the centre and simultaneous multielement determination is possible under optimum conditions.

It is only for a few elements (e.g., Mo, P, B) that the excitation conditions in the centre are inadequate. Then useful signals are only found in the plasma layer and the detection limits are poor owing to the high background.

The detection limits (3s-criterion) for the elements Be, Ca, Cd, Co, Cr, Cu, Ga, Hg, Mg, Mn, Sr, Tl and Zn were found to be in the range of 1 to 50 ng/ml. In table 1 the detection limits of some elements obtained with the HC-MIP are compared to those reported for a F-MIP (11). The detection limits of the HC-MIP are lower by a factor of 5 to 7 than those of the F-MIP.

Because of the relative low uptake rate of the nebulizer system the use of an injection method instead of continuously sucking of the solution is favourable. Already with an injection volume of 20 µl the full signal height is obtained. Therefore the detection limits mentioned above correspond with detectable amounts of 0.02 to 2 ng.

The linear range of the calibration functions of the HC-MIP (3 to 4 orders of

Table 1. Detection limits of F-MIP (11) and HC-MIP sustained in TM_{010}-cavity

Element	Wavelength [nm]	Detection limit: HC-MIP [μg/ml]	F-MIP[1] [μg/ml]
Co I	240.73	0.05	0.36
Cr I	425.43	-	0.21
	357.86	0.01	-
Cu I	324.75	0.003	0.015
Ga I	417.20	0.003	0.015
Mg I	285.21	0.002	0.008
Mg II	279.55	0.002	0.009
Mn I	403.08	0.015	0.08
Mn II	257.61	0.006	0.05
Sr II	407.77	0.002	0.008

[1] For reason of a more convenient comparison the detection limits given by Boumans (11) were recalculated from 2 s to 3 s.

magnitude) surpass the reported ranges of an F-MIP (12) for about one order of magnitude.

Hydrochloric acid and nitric acid do not affect the stability of the HC-MIP but in high concentrations cause a depression of emission signals up to 50 %. As an example for the application of the HC-MIP in extreme trace analysis Cd and Cu were determined in some natural water samples, e.g. drinking water, glacial ice and snow. After precipitation exchange in thin sulphide layers (13) copper and cadmium were determined by the injection method (20 μl). Detection limits of 0.05 and 0.03 ng/ml for Cu and Cd could be obtained. The relative standard deviations were 5 to 8 %.

Taking into account all these points, it is obvious, that the HC-MIP as well as the 3-F-MIP, inspite of their lowered load limits relative to microwave power, aerosol loading and matrix concentration, offer great advantages in comparison with the conventional F-MIP.

5. The MIP as an on-line detector in HPLC

Due to its extremely low tolerance of organic compounds the conventional MIP is incompatible to high performance liquid chromatography (HPLC). On the other hand the combination of HPLC and the MIP as an element specific detector promises to be a powerful analytical tool because of its ability to perform speciation studies even of trace elements.

5.1 The oxygen-argon-MIP (O_2-Ar-MIP)

The operation of a mixed-gas O_2-Ar-MIP with oxygen contents of the plasma gas between 3 and 10 % can be performed in a special discharge tube. It consists of two concentric quartz capillaries of different length, the inner and outer diameters of which are 3.8 and 6.0 or 1.6 and 3.0 mm respectively. This construction enables a separate supply with oxygen and argon gas, the latter of which is loaded with organic aerosol. The oxygen flows along the inner surface of the plasma capillary, thus avoiding carbon deposits. The O_2-Ar-MIP is operated in a TM_{010} cavity (14) and tolerates solution aerosols containing up to 90 % methanol. It burns in the shape of a filament or a hollow-cylinder dependent on operational parameters and the mounting of the discharge tube (15). The pros and cons of the different plasma forms were already discussed above. The O_2-Ar-MIP in the filament form shows wider ranges of load capacity with regard to variability of gas flows and loading with organic solvents. The toroidal O_2-Ar-MIP stands out for easier adjusting and better temporal and spatial stability of the optimum observation zone. The operational conditions, e.g., microwave power, concentration of oxygen in the plasma gas and argon pressure at the nebulizer are closely associated with each other. The concentration of the organic solvent affects essentially the oxygen flow. The oxygen flow on the other hand determines the optimum microwave power. Moreover, with increasing oxygen content of the plasma gas, the intensities of net line signals decrease. So always a compromise has to be made between stability of the plasma and emission intensity.

As organic solvent methanol was investigated intensively because it often can be used as mobile phase in HPLC. While ethanol shows quite the same characteristics, other organic compounds, e.g., aceton, chloroform and carbon tetrachloride severely disturb even the O_2-Ar-MIP.

5.2 Interfacing MIP with HPLC

For coupling the O_2-Ar-MIP with HPLC the pneumatic nebulizer is directly connected with the outlet of the UV-vis detector cell of the HPLC. Thus, the eluent, a water-methanol mixture of changing composition, is nebulized after passing the UV-vis detector. The aerosol is transported by the argon plasma gas to the discharge tube, where the oxygen is admixed. The plasma is observed end-on, so that with the aid of an intermediate diaphragm the optimum observation zone can be selected. It was found to be next to the plasma filament. The optimum argon flow is determined by the nebulizer characteristic and influences the residence time of the analyte in the plasma. The necessary oxygen content depends on the methanol concentration of the eluent, e.g., for a methanol content of 80 % an oxygen flow of at least 2 l/h is necessary.

5.3 Separation and determination of organic mercury compounds with MIP-HPLC

To demonstrate the applicability of the mixed-gas oxygen-argon-MIP as detection system for HPLC and to compare it with an UV-vis detector, some Hg-compounds were separated as 2-mercapto-ethanol complexes by reversed-phase-chromatography and detected by both the UV-vis and the MIP detector (16). A mixture of Hg-species, containing 300 ng Hg of each Hg^{2+}, CH_3Hg^{2+}, $C_2H_5Hg^+$ and phenyl-Hg was separated by gradient elution (Fig. 2). The chromatograms demonstrate that inspite of the

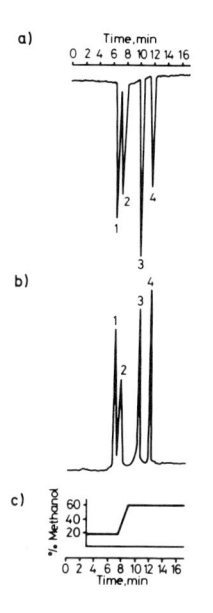

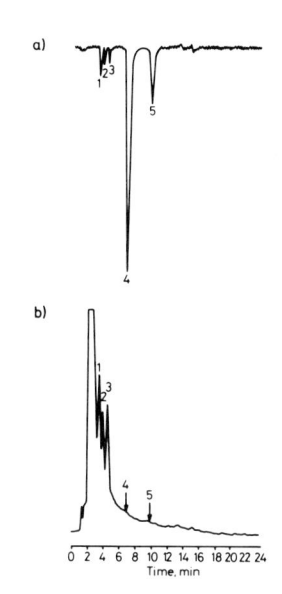

Fig. 2

Fig. 3

Fig. 2. UV-vis and MIP detection of different Hg-species after separation by gradient elution RP-HPLC as 2-mercapto-ethanol complexes.
Peaks: 1) Hg^{2+}; 2) CH_3Hg^+; 3) $C_2H_5Hg^+$, 4) Phenyl-Hg^+; 300 ng Hg in each case.
Column: RP 6,5 μm, 250 mm x 4,6 mm i.d.;
(a) MIP detection: Hg(I), 253,6 nm. (b) UV-vis detection: Absorption at 250 nm, reference wavelength 430 nm. (c) Course of gradient elution.
Fig. 3. Chromatographic separation of different Hg-species after their extraction with ethanol from highland peat bog soil.
Column: RP 6,5 μm, 250 mm x 4.6 mm i.d.;
Mobile phase: Water/methanol (40/60, v/v), flow rate 1 ml/min.
(a) MIP-detection: Hg(I), 253.6 nm. (b) UV-vis detection: Absorption at 250 nm, reference wavelength 430 nm.

unavoidable additional dead volumes (e.g., spray chamber) no significant peakbroadening occurs in the MIP detection system compared to the UV-vis detector.
The different heights of peaks correlating to equal amounts of 300 ng of mercury in different types of bonding show that obviously the thermal energy of the MIP is not sufficient for the complete dissociation and atomization of all mercury compounds. This major disadvantage exists in the UV-vis detection, too. Because of this varying sensitivity of the MIP-Detector it is impossible to state an universally valid detection limit. They can only be given for single compounds. As an example, the detection limit (3 s, n = 10) of mercury in Hg-dihizonate solutions was found to be about 75 ng. The linear range of the calibration curve reaches up to 1.2 µg.
Pollution or toxicity studies on heavy trace metals always have to take into consideration that mobility and bioavailability of the different metal species are strongly dependent on its varying valencies, forms of bonding and complex states. Such investigations, therefore, need analytical methods that allow the separation of the dieffrent metal species and their sensitive and elementspecific detection. The present MIP-HPLC system seems to offer these capabilities as shall be demonstrated by its use for the separation and detection of immobile mercury compounds in highland peat bog soils. For this purpose, peat samples were extracted with ethanol. Samples of this solution separated by HPLC and detected by both the UV-vis and the MIP detector resulted in chromatograms (Fig. 3), which show the superiority of the element specific MIP. The Hg-species No. 4 and 5 are detected only by the MIP and do not appear in the UV-vis chromatogram. Additionally, the MIP-chromatogram ensures that the compounds 1 to 3 contain mercury, a fact that with the UV-vis detector could only be verified by the retention times of corresponding standards.
Though the detection power and the relative sensitivities for various compounds for the O_2-Ar-MIP are not as good as for an analogous ICP detection system (17), the MIP detector might become a usefull supplementation to conventional HPLC detection systems due to its element-specific detection capability and its lower costs in comparison to the ICP.

Acknowledgement-This research was supported by the Arbeitsgemeinschaft Industrieller Forschungsvereinigungen e.V., from funds of the Federal Minister of Economics of the FRG.

REFERENCES

(1) K. Gretzinger, L. Kotz, P. Tschöpel, G. Tölg, Talanta 29, 1011 (1982)

(2) P. Tschöpel, Pure & Appl. Chem. 54, 913 (1982)

(3) P. Tschöpel, G. Tölg, J. Trace Microprobe Techn. 1, 1 (1982)

(4) G. Tölg, Fresenius Z. Anal. Chem. 294, 1 (1979)

(5) G. Tölg, Erzmetall 28, 390 (1975)

(6) P. Tschöpel, Plasma Excitation in Spectrochemical Analysis in: G. Svehla (Ed.), Wilson & Wilson's Comprehensive Analytical Chemistry, Vol. IX, pp 173, Elsevier, Amsterdam, Oxford, New York (1979)

(7) D. Kollotzek, P. Tschöpel, G. Tölg, Spectrochimica Acta 73B, 91 (1982)

(8) G. Kaiser, D. Götz, P. Schoch, G. Tölg, Talanta 22, 889 (1975)

(9) J. Bitzenauer, Diplomarbeit, University of Stuttgart, FRG (1982)

(10) A. Bollo-Kamara, E.G. Codding, Spectrochim. Acta 36B, 973 (1981)

(11) C.J.M. Beenakker, B. Bosman, P.W.J.M. Boumans, Spectrochim. Acta 33B, 373 (1978)

(12) H. Kuwaguchi, M. Hasegawa, A. Mizuike, Spectrochim. Acta 27B, 205 (1972)

(13) A. Disam, P. Tschöpel, G. Tölg, Fresenius Z. Anal. Chem. 295, 97 (1979)

(14) J.P.J. van Dalen, P.A. de Lezenne Coulander, L. de Galan, Spectrochim. Acta 33B, 545 (1978)

(15) D. Kollotzek, Doctoral Thesis, University of Stuttgart, FRG (1982)

(16) D. Oechsle, Doctoral Thesis, University of Stuttgart, FRG (1982)

(17) C.H. Gast, J.C. Kraak, H. Poppe, F.J.M.J. Maessen, J. Chromat. 185, 549 (1979)

ERFAHRUNGEN MIT DER LICHTLEITERKOPPLUNG BEI EINEM
ICP-SPEKTROMETER

A. Golloch[1], H.-M. Kuß[1], R. Rütjes[1], K.H. Schmitz[2]

Universität -GH- Duisburg[1]
FB 6 - Instrumentelle Analytik, D-4100 Duisburg
Mannesmannröhren-Werke AG, D-4100 Duisburg[2]

Gegenüber der herkömmlichen Ankopplung eines Plasmas an einen Polychromator bietet die Lichtübertragung mittels Lichtleiter einige Vorteile. Bei einer Länge des Lichtleiterkabels bis zu 10 m ist die räumliche Anordnung der Anregungseinheit zum Polychromator flexibel. Die Stabilität der "beliebig gekrümmten" optischen Achse ist gewährleistet durch die feste Positionierung des einen Lichtleiterendes direkt vor dem Eintrittsspalt des Polychromators und die in den drei Raumrichtungen variable Anordnung des anderen Endes in der unmittelbaren Nähe des Plasmas.
Das frei bewegliche Ende des Lichtleiters ist auf einem in den drei Raumrichtungen variabel und reproduzierbar einzustellenden Kreuztisch angebracht. Mit dieser Anordnung kann die Intensitätsverteilung eines Emissionssignals über den vertikalen Querschnitt des Plasmas ermittelt werden.
Bei der seitlichen Verschiebung s (Abb. 1) wurde das Maximum der Intensitätsverteilung in der Mitte des Plasmas gefunden vorausgesetzt, das Plasma ist symmetrisch. Bei der Höhenvariation h macht sich der Einfluß des Temperaturgradienten im Plasma erheblich bemerkbar, so daß sich die Intensität des Emissionslichtes sehr stark ändert. Eine Änderung in der dritten Raumrichtung a ergab ein um so größeres Signal, je näher das Lichtleiterende an das Plasma herangeführt wurde.
Ist die seitliche Verschiebung optimiert und der Abstand möglichst klein, so beschränkt sich die weitere Optimierung der Lichtleiterstellung auf die Variation der Höhe h.
Neben der geometrischen Stellung des Lichtleiterendes zu einem induktiv gekoppelten Plasma hängen die Empfindlichkeit und die BEC-Werte noch wesentlich von der Plasmaleistung ab. In der Abbildung 2a-e sind die Empfindlichkeiten und die BEC-Werte für ausgewählte Elemente graphisch dargestellt.
Die Leistungsfähigkeit dieses Systems wurde an Standard-Stahlproben ermittelt (Abb. 3).

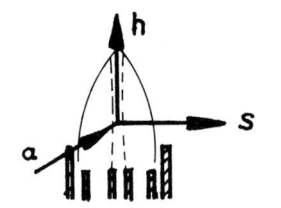

Abb.1 Plasmaquerschnitt

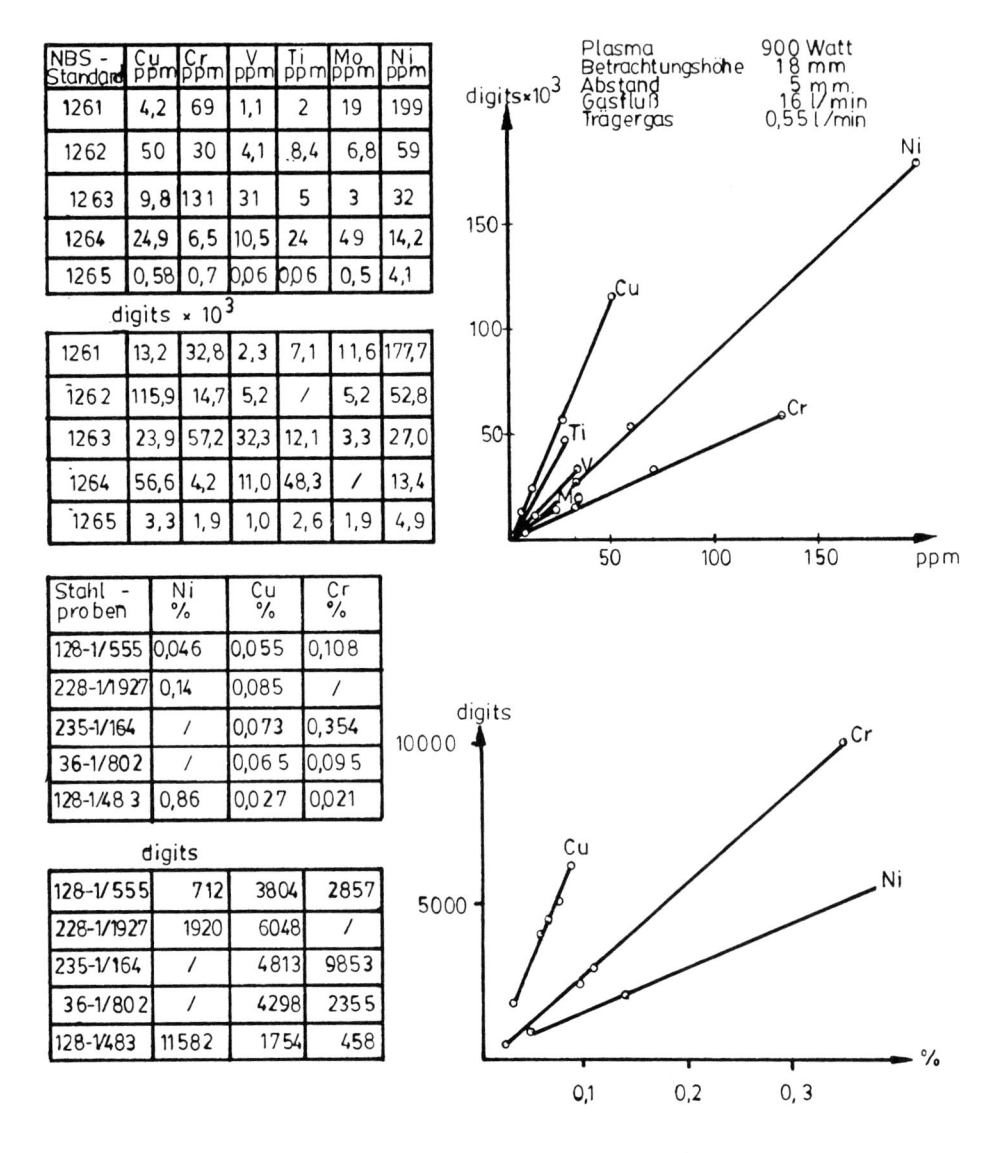

NBS-Standard	Cu ppm	Cr ppm	V ppm	Ti ppm	Mo ppm	Ni ppm
1261	4,2	69	1,1	2	19	199
1262	50	30	4,1	8,4	6,8	59
1263	9,8	131	31	5	3	32
1264	24,9	6,5	10,5	24	49	14,2
1265	0,58	0,7	0,06	0,06	0,5	4,1

digits × 10^3

1261	13,2	32,8	2,3	7,1	11,6	177,7
1262	115,9	14,7	5,2	/	5,2	52,8
1263	23,9	57,2	32,3	12,1	3,3	27,0
1264	56,6	4,2	11,0	48,3	/	13,4
1265	3,3	1,9	1,0	2,6	1,9	4,9

Stahl-proben	Ni %	Cu %	Cr %
128-1/555	0,046	0,055	0,108
228-1/1927	0,14	0,085	/
235-1/164	/	0,073	0,354
36-1/802	/	0,065	0,095
128-1/483	0,86	0,027	0,021

digits

128-1/555	712	3804	2857
228-1/1927	1920	6048	/
235-1/164	/	4813	9853
36-1/802	/	4298	2355
128-1/483	11582	1754	458

Plasma 900 Watt
Betrachtungshöhe 18 mm
Abstand 5 mm
Gasfluß 16 l/min
Trägergas 0,55 l/min

Abb. 2 Analysen- und Meßdaten von Stahl-Standardproben

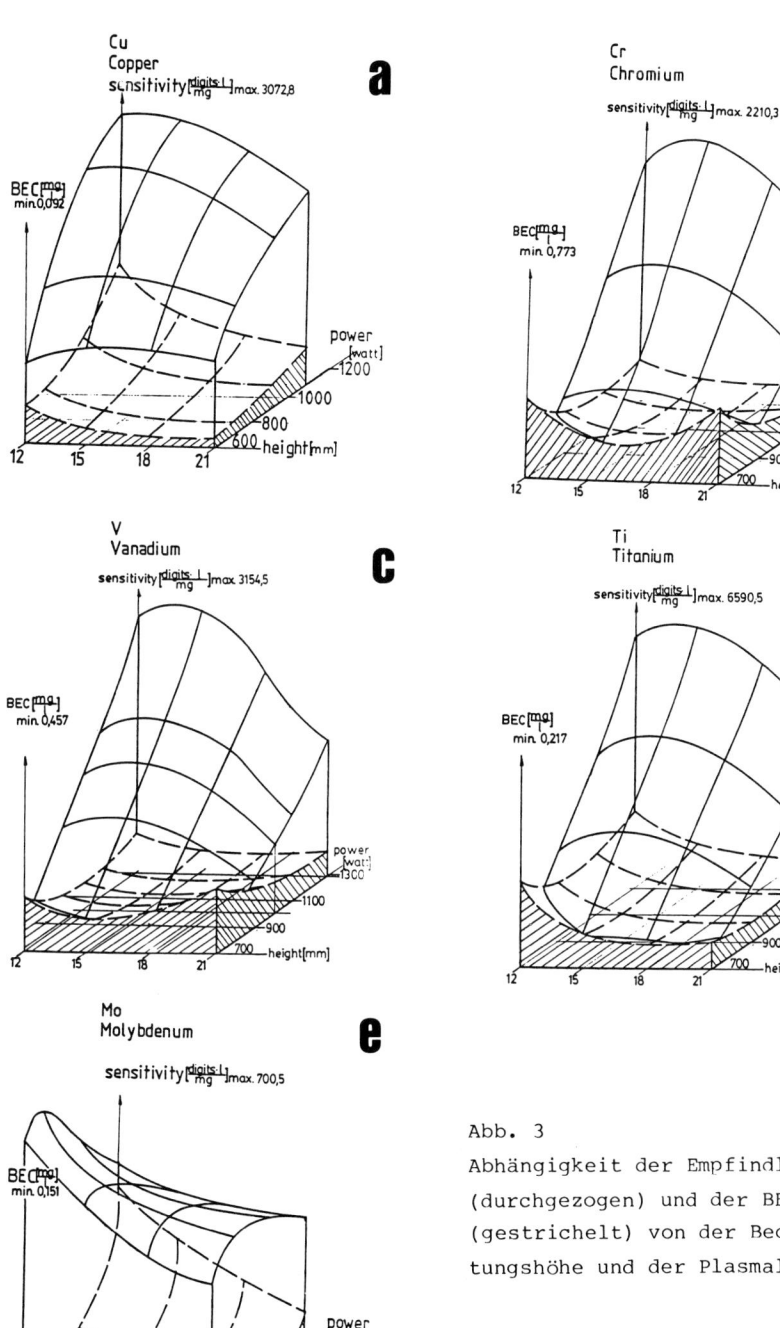

Abb. 3
Abhängigkeit der Empfindlichkeit (durchgezogen) und der BEC-Werte (gestrichelt) von der Beobachtungshöhe und der Plasmaleistung

Development of an Optimization Criterion for ICP Atomic Emission Spectrometers

Henning Friege and Peter Werner

Staatliches Amt für Wasser- und Abfallwirtschaft Düsseldorf, D-4000 Düsseldorf 11

1. Introduction

Determination of major and trace elements in water monitoring programs have typically been performed with single-element methods, e.g. conventional molecular or atomic absorption spectrometry. The use of ICP atomic emission spectrometers as a multi-element method offers a number of attractive advantages to environmental monitoring (1). On the other hand, detection limits in ICP-AES are not as low as in furnace AAS. For many heavy metals, indigenous concentrations in natural waters are seldom above some ppb. Therefore, some authors have used preconcentration steps (2-4) risking undesirable contamination or loss of analyte species and adding length to the procedure. For these reasons, there have been several attempts to reduce detection limits by optimizing the working conditions of ICP spectrometers (5-9). Optimization has to be aligned to a certain criterion which should reach a maximum or minimum value. The criterion normally used is the signal-to-background ratio (SBR). From preliminary experiments with SBR, we learned that optimization yielded a set of working conditions leading to bad precision.

2. Theoretical

We studied several optimization criteria with respect to the detection limit and to analytical precision. It is very well known that the power coupled into plasma, the Ar flow rates, and the height of observation affect the analytical performance of an ICP (7,9). When a peristaltic pump is used, the sample flow rate should be of importance, too. All these parameters are clearly interrelated (7,8,10). Thus, a true optimum cannot be achieved by varying one factor keeping the others constant (univariate method). We therefore decided to use the variable step-size simplex; this is a multivariate technique where all parameters are changed in one step (11). Because of the multi-element character of ICP-AES optimum working conditions form a compromise. All experiments were performed using an Fe II emission line at 238.20 nm which is located in the region of prominent lines, because working conditions do not differ considerably with wavelength provided that the region < 350 nm is chosen (9).

Usually, optimization of an ICP-AES is performed using signal-to-background ratio (SBR) for a given concentration c_q as a criterion,

$$SBR(c_q) = \frac{I_q}{I_o},$$

I_q being the gross intensity, i.e. $I_q = I_n + I_o$. Assuming a linear analytical function and defining the detection limit as a concentration equivalent to twice the standard deviation (SD) of the background signal I_o one may derive the detection limit as a function of the relative standard deviation of the background, $RSD(I_o)$, and of SBR.

Criterion A: $\quad c_{lim} = \dfrac{0.02 \cdot c}{SBR(c_q)} q - \cdot RSD(I_o)$

If the sensitivity reflected by SBR is maximized (6-9), the influence of $RSD(I_o)$ on the detection limit will be neglected and precision will not be taken into consideration. Optimization of SBR is performed using criterion B.

Criterion B: $\quad c_{lim} = \dfrac{0.02 \cdot c}{SBR(c_q)} q -$

Criterion B will be sufficient if $RSD(I_o)$ scatters about 1% (12). On the other hand, one may argue that $RSD(I_o)$ is dominated by shot-noise leading to $SD(I) \sim \sqrt{I}$ (13). Additionally, analyte flicker noise is not reflected in $RSD(I_o)$. The dominant contributions to flicker noise derive from the nebulizer. For background intensity, the flicker term seems to be of lower magnitude than for net line intensity (13). Thus, one may also include the relative standard deviation of the gross signal, $RSD(I_q)$, in an optimization criterion (criterion C) accepting the loss of a direct

Criterion C: $\quad c^x = \dfrac{0.02 \cdot c}{SBR(c_q)} q - \cdot RSD(I_q)$

relation of the optimization criterion to the detection limit. An empirical combination of criterion B and criterion C is used in the criterion D,

Criterion D: $\quad c^* = \dfrac{0.01 \cdot c}{SBR(c_q)} q - \cdot RSD(I_q) + RSD(I_o)$

with the aim to take advantage of the possibly different dependences of $RSD(I_q)$ and $RSD(I_o)$ from noise characteristics. (13-15)

3. Experimental

The ICP-spectrometer used was an Instrumentation Laboratory IL 100 equipped with a double monochromator (330 mmm/165 mm); spectral resolution in second order is 0.02 nm. For optimization experiments, four parameters were varied in the simplex process:
1) Power level, adjustable in seven steps between 1.07 and 1.64 kW.
2) Vertical observation height (mm above load coil).
3) Pressure of carrier Ar (kp/cm^2).
4) Sample flow rate, adjustable by computer-controlled peristaltic pump (ml/min).
The IL 100 is a sequential spectrometer. Therefore, the integration time was optimized for each element in question after the simplex procedure. Each measurement cycle consisted of 30 s fast pump delay and a settling period of 30 s followed by ten consecutive readings (2 s) of a 2 ppm Fe standard or blank solution (1% HNO_3).

4. Results and Discussion

The results of the optimization experiments are summarized in Table 1 (working conditions) and Table 2 (detection limits and long-time precision).

Table 1: Results of simplex optimization procedures (Fe II, 238.20 nm)

Criterion	A	B	C	D
Energy (kW)	1.29	1.04	1.29	1.29
Ar pressure (kp/cm^2)	2.1	1.9-2.0	2.4	2.0-2.2
Flow of analyte (ml/min)	1.5-1.6	1.9	1.4	1.2-1.4
Observation height (mm)	18-19	20	17	16-18

Table 2: Detection limits (ppb) and long-time precision over 3 h (%-RSD) for several elements yielded from criterion B and criterion D.

Int. time	Detection limits (ppb)								Long-time precision (%-RSD) (integration time)	
	1s	3s	5s	10s	1s	3s	5s	10s		
Fe 238.20	4.4	3.2	2.6	1.4	3.4	2.2	1.6	1.4	0.77%	0.54% (3s)
Mn 257.61	2.6	1.4	1.0	1.0	1.0	1.0	1.0	0.5	0.63%	0.54% (1s)
Cr 284.33	6.4	6.2	3.8	2.4	3.2	2.8	2.4	2.2	0.80%	0.61% (5s)
Zn 213.86	3.8	2.8	1.4	1.4	2.0	1.6	1.0	1.0	0.74%	0.49% (3s)
Cu 324.75	6.6	1.0	1.2	1.0	7.0	4.2	2.0	1.6	1.07%	0.73% (5s)
Ni 231.60	29.6	24.0	9.4	7.0	8.6	6.4	3.6	1.6	0.78%	0.53% (10s)
Cd 226.50	15.8	8.2	6.4	3.4	2.4	1.0	1.0	0.8		
Pb 220.35	71.0	21.2	21.0	14.0	34.6	16.2	12.0	7.6		
	Criterion B				Criterion D				Crit. B	Crit. D

It is demonstrated quite clearly that criterion D yields best results with respect to sensitivity and precision whereas criterion B leads to bad long time precision. The same is true for criterion A. This effect is even more pronounced in routine analysis possibly due to matrix or nebulization effects which may be overcome best with the working conditions derived from criterion D.

Optimum values for Ar pressure p_{Ar} and for the analyte flow $\dot{V}$ are different for all criteria, but the ratio $\dot{V}/p_{Ar}$ is nearly identical for criterion C and criterion D. So, one may suggest an optimum analyte flow rate for a fixed carrier Ar flow leading to maximal precision. As to be seen in Fig. 1 there is a clear minimum for $RSD(I_q)$ at $\dot{V} \approx 1.2$ ml/min, whereas SBR increases at higher analyte flow rate. As outlined above, the integration time was varied after optimization of the other working conditions. In contrast to earlier suggestions (16) we found significantly lower detection limits with integration time increasing (cf. Table 2) due to the influence of shot-noise represented best in $RSD(I_o)$ (14,15). Especially in the case

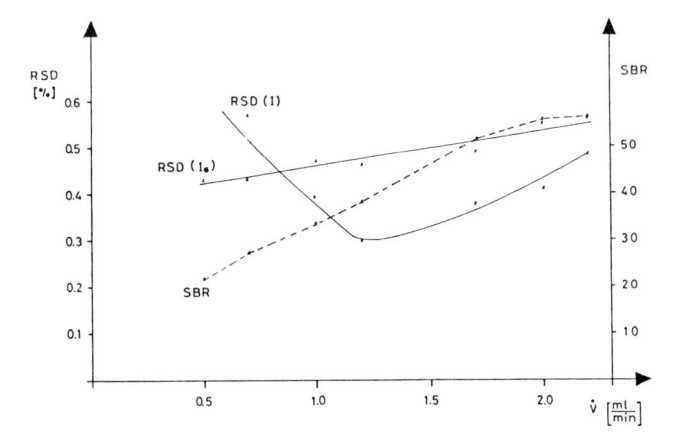

Fig. 1: SBR and RSD's as a function of analyte flow rate $\dot{V}$

of sequential ICP-AES, one may obtain a good compromise between low detection limits and fast analysis procedure.

Summing up, we may state that $RSD(I_o)$ as well as $RSD(I_q)$ have to be included in an optimization procedure leading to low detection limits and excellent analytical precision. The usual maximization of SBR does not lead to good analytical performance. Detection limits may be lowered considerably by longer integration times obviously due to the time-dependent shot-noise.

REFERENCES

(1) H. Friege, Umwelt 1983, 291-294.
(2) T. Kempf and W. Sonneborn, Vom Wasser 83 (1981), 83-94.
(3) M.Thompson, M.H.Ramsey and B. Pahlavanpour, Analyst 107 (1982), 1330-1334.
(4) R.K. Winge V.A. Fassel, R.N. Kniseley, E. de Kalb and W.J. Haas, Spectrochim. Acta 32B (1977), 327-345.
(5) J.R. Garbarino and H.E. Taylor, Appl. Spectrosc. 33 (1979), 220-226.
(6) S. Greenfield and D.Th. Burns, Anal. Chim. Acta 113 (1980), 205-211.
(7) L. Ebdon, M.R. Cave and D.J. Mowthorpe, Anal. Chim. Acta 115 (1980), 179-185.
(8) S.P. Terblanche, K. Visser and P.B. Zeeman, Spectrochim. Acta 36B (1981), 293-297.
(9) S.S. Bergman and J.W. McLaren, Appl. Spectrosc. 32 (1978), 372-377.
(10) S. Imai, ICP Information Newsl. 8 (1982), 85-91.
(11) St.N. Deming and St.L. Morgan, Anal. Chem. 45 (1973), 278A-286A.
(12) P.W.J.M. Boumans: "Line Coincidence Tables for Inductively Coupled Plasma Atomic Emission Spectrometry", Pergamon Press, Oxford (1980).
(13) P.W.J.M. Boumans, R.J. McKenna and M. Bosveld, Spectrochim. Acta 36B (1981), 1031-1058.
(14) S. Greenfield: "Developments in Atomic Plasma Spectrochemistry" (Edited by R.M. Barnes), Heyden and Sons, London (1981).
(15) P. Werner and H. Friege, Spectrochim. Acta, submitted for publication.
(16) R.M. Belchamber and G. Horlick, Spectrochim. Acta 38B (1982), 71-82.

SIMULTANE ODER SEQUENTIELLE MESSUNGEN MIT DER ICP-ATOMEMISSIONS-SPEKTROMETRIE?
Erfahrungen am System ARL-3580

G. DREWS, Römisch-Germanisches Zentralmuseum, Mainz

Seit Ende 1981 betreibt das Römisch-Germanische Zentralmuseum Mainz (RGZM) gemeinsam mit der Johannes Gutenberg-Universität Mainz im Inst. f. Geowissenschaften das Spektrometer-System ARL-3580. Seine Haupteinsatzgebiete sind:
1) Elementanalyse archäologischer Proben
2) Elementanalyse geologischer Proben
3) Elementanalyse von Umweltproben
4) Allgemeine anorganisch-chemische Spurenanalyse

Entscheidend für die Wahl der ICP-Atomemissionsspektrometrie waren:
1) Niedrige Nachweisgrenzen und großer dynamischer Meßbereich
2) Hohe Flexibilität für die Analyse unterschiedlichster Materialien
3) Angeblich sehr geringe Matrixabhängigkeit und leichte Eichbarkeit des Meßverfahrens.

Das das Gerät vornehmlich zur Analyse von üblicherweise nur in sehr geringen Probemengen vorliegenden archäologischen Proben bestimmt war, hatte die simultane On-Peak-Meßmethode Präferenz vor der theoretisch besseren Methode der sequentiellen Peak-Abtastung. Wegen des geplanten weiten Einsatz-Spektrums sollte dennoch größtmögliche Flexibilität erreicht werden. Vor dem Hintergrund dieser Forderungen entstand das simultan und sequentiell On-Peak-messende Spektrometersystem ARL 3580. Hierbei handelt es sich im Prinzip um ein System auf der Basis des ARL 34000, das um ein On-Peak-messendes Sequenzspektrometer erweitert wurde. Beide Gerätekomponenten können über eine gemeinsame Software entweder getrennt oder gemeinsam benutzt werden. Beide Gerätekomponenten sind auch als völlig eigenständige Anlagen als 3520 (Sequenzgerät) und 3560 (Simultangerät) erhältlich.

Wie das Schema der Primäroptik in Abb. 1 zeigt, werden im Fall des Kombinationstyps 3580 beide Spektrometer von derselben Lichtquelle gleichzeitig ausgeleuchtet. Dabei hat das über die Kondensorlinse von der Lichtquelle abgegriffene Strahlenbündel durch den Spalt eines geschlitzten Hohlspiegels direkten Zutritt zu dem Simultanspektrometer, während es erst nach zweifacher Umlenkung in das Sequenzspektrometer eintritt (s. Abb. 2). Im Falle des 3520 als Einzelgerät entfallen diese Spiegel.

Zur Abschätzung der Leistungsfähigkeit des Sequenzspektrometers 3520 im Vergleich mit einem Simultangerät wurde ein kommerziell erhältlicher Mehrelement-Standard unter Verwendung identischer Analysenlinien in gleicher Ordnung sowohl mit dem 3520 des Applikations-Labors der ARL in Ecublens und dem Simultangerät des 3580 des Röm.-Germ. Zentral-Museums vermessen. Die Auswahl der Analysenlinien richtete sich nach den im Simultangerät installierten Linien (s. Tab. 1). Beide Geräte sind mit dem sog. Internationalen Stativ mit fest justierter Beobachtungshöhe ausgerüstet. Das ARL-Gerät verwendet einen Meinhard-Zerstäuber, das RGZM-Gerät einen Babbington-GMK-Zerstäuber.

Als Maß für die Leistungsfähigkeit wurde das Peak/Background-Verhältnis herangezogen. Wie Tabelle 2 zeigt, sind – von einigen extremen Ausnahmen abgesehen – hierin beide Gerätetypen nahezu gleichwertig; das im Schnitt leicht bessere Abschneiden des RGZM-Gerätes kann durch den effektiveren Babbington-Zerstäuber erklärt werden (s.a. Tab. 3).

Gravierende Unterschiede ergeben sich jedoch, betrachtet man die Ergebnisse, die bei der Messung mit dem Simultan- und Sequenzteil des 3580 – unter Zwischenschaltung der Umlenkspiegel – erhalten wurden (s. Tab. 3). Mit zunehmend kürzerer Wellenlänge treten erhebliche Empfindlichkeitsverluste auf. Verluste mit dem Faktor 30 lassen sich bei dem langen Strahlengang innerhalb der Luft-Spektrometer nicht mit dem kurzen zusätzlichen Weg zwischen den Umlenkspiegeln erklären. Ihre Ursache muß in den Umlenkspiegeln liegen.

Als Konsequenz muß daraus abgeleitet werden, daß bei Kombinationssystemen jegliche vermeidbaren optischen Bauteile tunlichst vermieden werden sollten. Ist dies – wie bei dem vorgestellten System – nicht vermeidbar, sollte in der Zusammenstellung des analytischen Programms des Simultangerätes dieser Aspekt berücksichtigt werden.

Im Gegensatz zu der Empfindlichkeit sind bei dem System 3580 sowohl Genauigkeit als auch erzielbare Richtigkeit der Analyse für beide Komponenten bei entsprechend sorgfältiger Programmgestaltung identisch. Bei konstanten äußeren Bedingungen sind Meßergebnisse mit rel. Standardabweichungen von ± 0,5% zu erreichen.

Tabelle 1
VERGLEICHS-STANDARD UND VERWENDETE LINIEN STANDARD:MERCK 15474

Element	Linie	(Å)Ordnung	Konzentration im Standard (ppm)
B	2496,8	2	15
Al	3961,5	1	100
Cr	2677,2	2	25
Mn	2576,1	2	5
Fe	2599,4	2	15
Co	2388,9	3	20
Ni	3414,4	2	50
Cu	3247,5	2	20
Zn	2138,6	2	20
Sr	4077,7	1	1
Cd	2288,0	2	20
Tl	1908,6	3	400
Pb	2203,5	3	200
Bi	2230,6	2	200

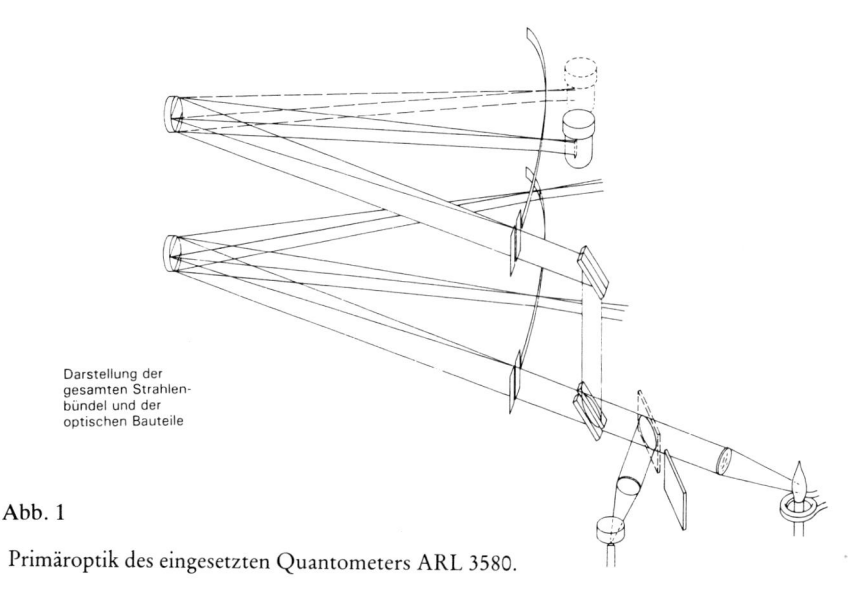

Darstellung der gesamten Strahlenbündel und der optischen Bauteile

Abb. 1
Primäroptik des eingesetzten Quantometers ARL 3580.

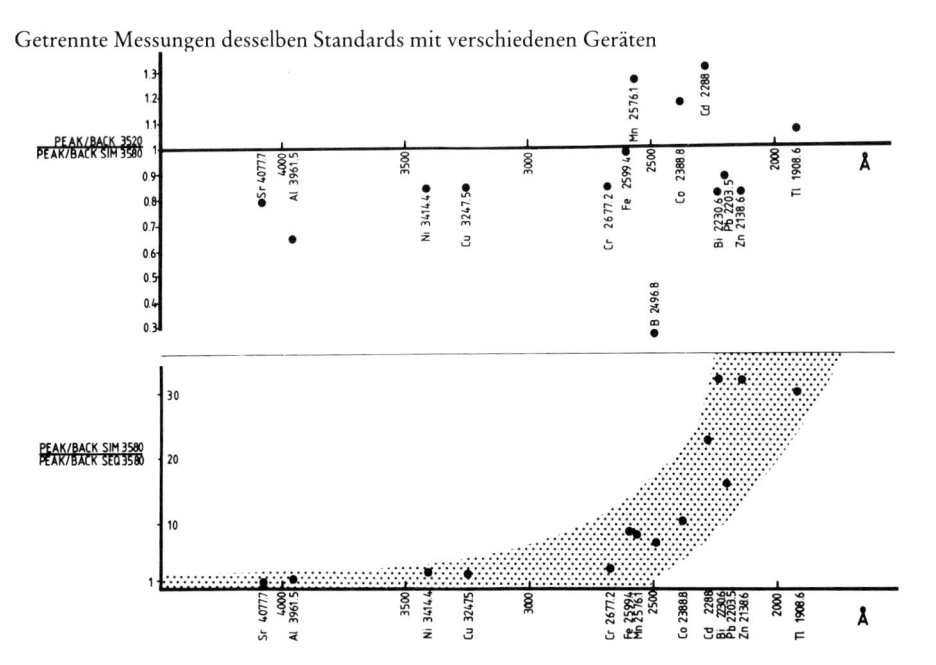

Tabelle 2
PEAK/BACKGROUND MERCK 15474 Einzelgeräte-Messungen

Sequenzgerät 3520 (Applikationslabor ARL Ecublens)			Simultangerät eines 3580 (RGZM Mainz)
Element	A	A:B	B
B	12,69	0,27:1	46,35
Al	30,42	0,65:1	46,85
Cr	77,95	0,85:1	91,41
Mn	124.70	1,26:1	98,91
Fe	68,24	0,98:1	69,75
Co	82,68	1,17:1	70,84
Ni	18,91	0,84:1	22,46
Cu	55,07	0,84:1	65,63
Zn	141,52	0,82:1	172,40
Sr	36,76	0,79:1	46,44
Cd	178,33	1,31:1	136,00
Tl	379,20	1,06:1	358,67
Pb	125,21	0,88:1	141,78
Bi	82,57	0,82:1	101,18

ARL-3520 Zerstäubersystem: Meinhard-Glas Lösungsverbrauch: ca 2-2,5ml/min
Integrationszeit: 5 sec

ARL-3580 Zerstäubersystem: Babbington-GMK Lösungsverbrauch: 1,75 ml/min
Integrationszeit: 10 sec

Getrennte Messungen desselben Standards mit verschiedenen Geräten

Tabelle 3
PEAK/BACKGROUND MERCK 15474 Kombinationsgerät ARL-3580 als

	Simultangerät		Sequenzgerät
Element	A	A:B	B
B	46,35	7,10:1	6,53
Al	46,85	1,37:1	34,18
Cr	91,41	2,96:1	30,93
Mn	98,91	8,23:1	12,02
Fe	69,75	9,76:1	7,15
Co	70,84	11,41:1	6,21
Ni	22,46	2,44:1	9,22
Cu	65,63	2,12:1	31,02
Zn	172,40	32,10:1	5,37
Sr	46,44	0,95:1	48,90
Cd	136,00	22,91:1	5,94
Tl	358,67	30,24:1	11,86
Pb	141,78	16,13:1	8,79
Bi	101,18	32,74:1	3,09

Zerstäubersystem: Babbington GMK
Lösungsverbrauch: 1,75 ml/min
Integrationszeit: 10 sec

Gleichzeitige Messung mit Simulta und Sequenzspektrometer aus einer Lösung.

APPLICATION OF FORWARD SCATTERING TO SIMULTANEOUS MULTIELEMENT DETERMINATION

H. Debus*, S. Ganz, W. Hanle, G. Hermann, A. Scharmann

I. Physikalisches Institut der Justus-Liebig-Universität Giessen, FRG

MESSPRINZIP UND VERSUCHSAUFBAU

Die Vorwärtsstreuung von Licht an freien Atomen im Magnetfeld ist eine Methode zur Spurenanalyse, die sich durch hohe Empfindlichkeit und niedrige Nachweisgrenzen auszeichnet /1,2/. Hervorzuheben ist darüber hinaus der simultane Nachweis vieler Elemente in einem Meßvorgang.

Der prinzipielle Aufbau der Meßapparatur besteht aus wenigen Komponenten (Abb. 1).

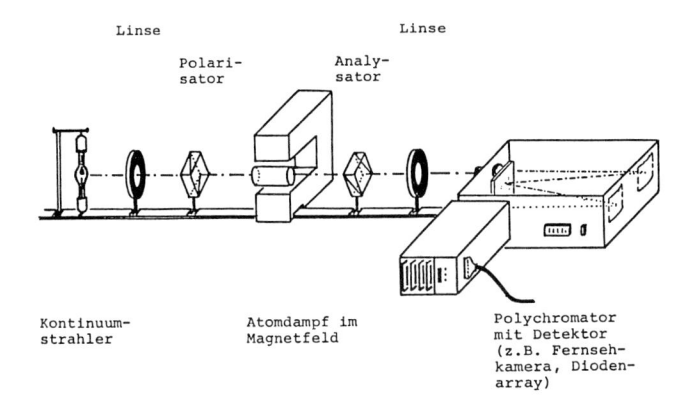

Abb. 1: Schematische Darstellung der Apparatur

Die Meßzelle - ein Graphitrohrofen - befindet sich zwischen gekreuzten Polarisatoren in einem transversalen Magnetfeld, wobei Magnetfeldrichtung und Polarisationsebenen einen Winkel von jeweils 45° einschließen. Als Lichtquelle wird für alle Messungen ein einziger Kontinuumstrahler verwendet. Das in Vorwärtsrichtung gestreute Licht wird mit Hilfe eines Spektrumanalysators detektiert.

---

*Part of thesis (D 26)

Das Magnetfeld erzeugt in dem Atomdampf eine optische Anisotropie, denn infolge der Zeemanaufspaltung erhält man für Licht, das senkrecht und parallel zum Magnetfeld B polarisiert ist, verschiedene komplexe Brechungsindizes n_σ und n_π. Doppelbrechung und Dichroismus, die den Polarisationszustand des Lichtes im Bereich der Resonanzlinien ändern, führen zur Aufhellung zwischen den gekreuzten Polarisatoren.

Für die transmittierte Lichtintensität I ergibt sich

$$I(\omega) = I_o(\omega)\frac{1}{4}\left| e^{i\omega(\frac{n_\sigma}{c_o} \cdot L - t)} - e^{i\omega(\frac{n_\pi}{c_o} \cdot L - t)} \right|^2 \quad \begin{array}{l} L = \text{Länge der Meßzelle} \\ \omega = \text{Frequenz des Lichtes} \end{array} \quad (1)$$

Die Differenz der Brechungsindizes $n_\pi - n_\sigma$ ist proportional zum Produkt aus Teilchendichte N und Oszillatorenstärke f. Für kleine Teilchendichte folgt damit aus (1)

$$I \sim N^2 \cdot L^2 \cdot f^2 \quad (2)$$

Dieser quadratische Zusammenhang bewirkt eine hohe Empfindlichkeit der Methode, denn kleine Änderungen der Teilchendichte verursachen eine relativ größere Änderung der Intensität.

Weiterhin zeigt (2), daß es sich bei der Vorwärtsstreuung um eine Nullmethode handelt, d.h. ohne nachzuweisendes Element ergibt sich auch kein Signal. Dies bedeutet gegenüber der AAS einen meßtechnischen Vorteil. Dort hat man ohne Probe ein großes Signal und weist geringe Teilchendichten durch kleine Änderungen einer großen Intensität nach. Der besondere Vorteil der Vorwärtsstreuung gegenüber der AAS ist darüber hinaus, daß eine einzige Lampe /3/ für alle nachzuweisenden Elemente benutzt werden kann.

In unseren Experimenten benutzen wir als Kontinuumstrahler sowohl eine 450 W Xenonhochdrucklampe als auch eine Deuteriumlampe. Die Atomisierung der Proben erfolgt in dem Graphitrohrofen bei Temperaturen bis zu 2700°C unter Argonatmosphäre. Die verwendeten Glanpolarisatoren haben ein Löschvermögen von 10^{-5}. Der Spektrumanalysator besteht aus einem Polychromator und einem Diodenarray mit einer Kanalplatte zur Vorverstärkung. Das Target umfaßt 700 Dioden, von denen jede einen Spektralbereich von 0,16 nm überdeckt. In Verbindung mit der Vorwärtsstreuung ermöglicht dieser Spektrumanalysator eine echte simultane Multielementanalyse für alle diejenigen Elemente, die Resonanzlinien im abgebildeten Spektralbereich besitzen und die mit dem Graphitrohrofen atomisierbar sind.

MESSERGEBNISSE

Abb. 2 zeigt eine Trinkwasseranalyse. In dem dargestellten Wellenlängenbereich von insgesamt 110 nm enthält das Spektrum im wesentlichen Peaks der Elemente Na, Ni, Fe, Mg und Mn. Für die Elemente Fe, Mn und Na wurden durch Kalibrierung mit einer Mischung aus verdünnten Standardlösungen die in Tabelle 1 angegebenen Konzentrationen ermittelt.

Probe: 10 µl Trinkwasser

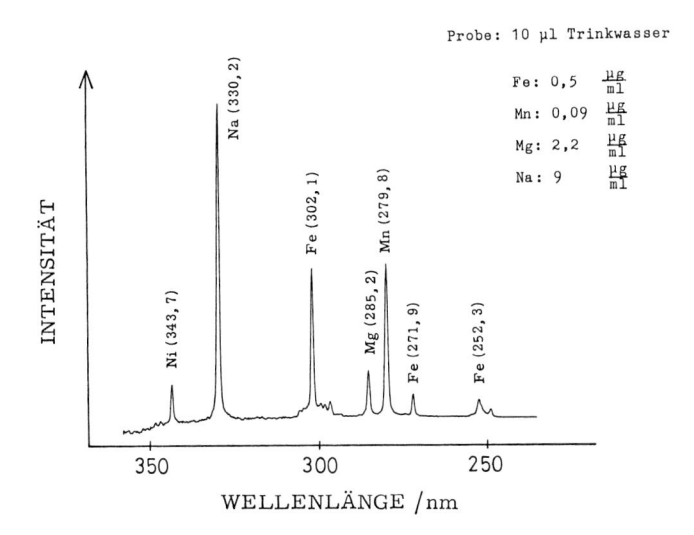

Fe: 0,5 µg/ml
Mn: 0,09 µg/ml
Mg: 2,2 µg/ml
Na: 9 µg/ml

Abb. 2: In Vorwärtsstreuung aufgenommenes Spektrum einer Trinkwasserprobe.
(Oben rechts sind die mit einer ICP-Vergleichsmessung bestimmten Konzentrationen der Elemente angegeben, die die Resonanzlinien im erfaßten Spektralbereich besitzen. Zusammenstellung der Ergebnisse siehe Tab. 1)

Tab. 1: Gegenüberstellung von Meßergebnissen durch Vorwärtsstreuung mit Vergleichsmessungen (ICP)

Element	Vorwärtsstreuung	Emission (ICP)
Fe	0,6 µg/ml	0,5 µg/ml
Mn	0,088 µg/ml	0,09 µg/ml
Na	9,2 µg/ml	9 µg/ml

Aus dem Spektrum wird ein Vorteil der Vorwärtsstreuung gegenüber der Emissionsanalyse sichtbar. Die Spektren bleiben viel übersichtlicher und die spektralen Interferenzen sind von geringer Bedeutung, weil im wesentlichen nur Resonanzlinien auftreten. Insbesondere bei Übergangselementen und hohen Konzentrationen können neben den Resonanzlinien noch weitere Linien auftreten, die zu Übergängen von niedrigen thermisch besetzten Niveaus gehören. Ein solches Beispiel mit Elementen, die auch noch in der Vorwärtsstreuung linienreich erscheinen, ist in Abb. 3 dargestellt. Um einen größeren Spektralbereich zu erfassen, wurden zwei Messungen mit unterschiedlicher Gitterstellung an der gleichen Probe durchgeführt. Der linke Spektralbereich wurde dabei durch einen engeren Spalt um den Faktor 4 abgeschwächt. Obwohl gerade bei den Nebengruppenelementen viele zusätzliche Linien durch thermische Besetzung auftreten, bleibt das Spektrum im Vergleich zu einem Emissionsspektrum doch recht übersichtlich.

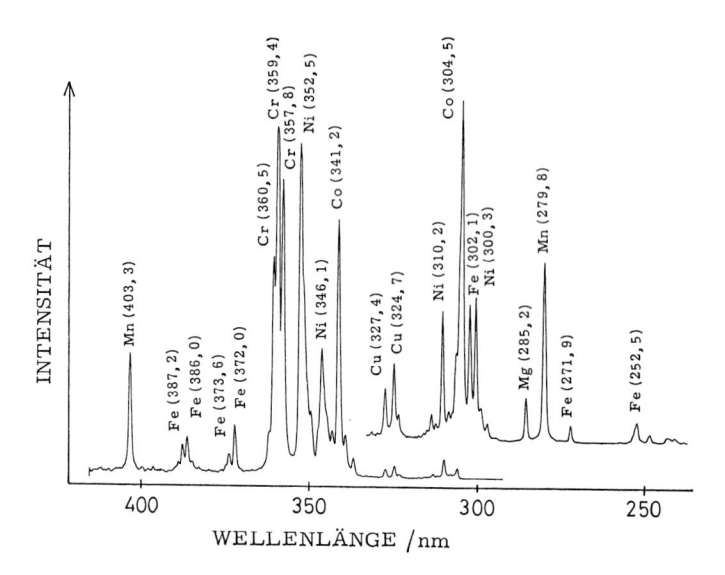

Abb. 3: In Vorwärtsstreuung aufgenommenes Spektrum einer aus verdünnten Standardlösungen zusammengestellten Probe.

Immerhin ist hier ein Wellenlängenbereich von 240 nm bis 420 nm aufgetragen. Die Auswertung der Spektren erfolgt mit einem Rechnerprogramm, das die Linien identifiziert, die entsprechenden Elemente zuordnet und die zugehörigen Intensitäten ermittelt.

Als weiteres Beispiel zeigt Abb. 4 das Spektrum einer aufgeschlossenen Haarprobe, wobei die in der Probe enthaltene Haarmenge 0,3 mg betrug.

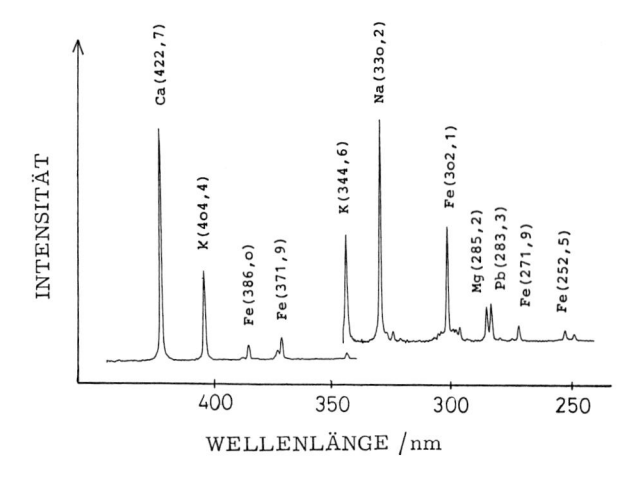

Abb. 4: In Vorwärtsstreuung aufgenommenes Spektrum einer aufgeschlossenen Haarprobe (analysierte Haarmenge 0,3 mg).

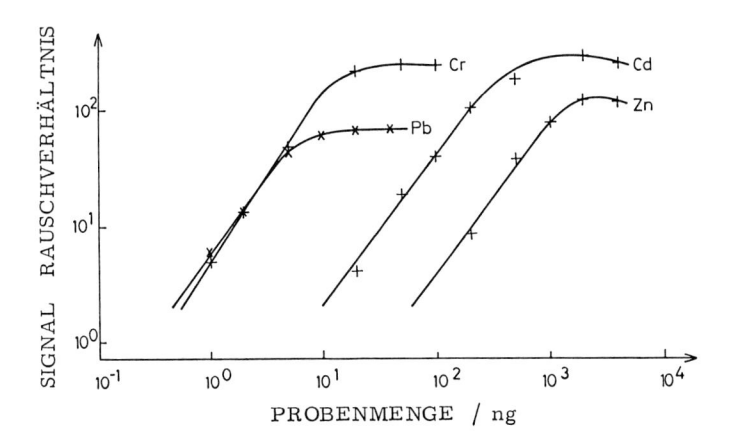

Abb. 5: Auf das Untergrundrauschen normierte Kalibrierfunktionen
der Elemente Pb (283,3 nm), Cr (357,8 nm), Cd (326,2 nm)
und Zn (307,5 nm), gemessen mit dem optischen Spektrumanalysator. Zu beachten ist, daß es sich bei Cd und Zn
um die schwachen Triplettlinien handelt.

Aussagen über den dynamischen Bereich und die Nachweisgrenzen der Analysenmethode liefert die in Abb. 5 gewählte Darstellung. Sie zeigt die auf das Untergrundrauschen normierten Kalibrierfunktionen der Elemente Cr, Pb, Cd und Zn. Diese Kurven wurden mit einer Xenonhochdrucklampe und einem Diodenarray aufgenommen. Im Einzelfall kann es sich als nützlich erweisen, das Diodenarray durch Photomultiplier zu ersetzen.

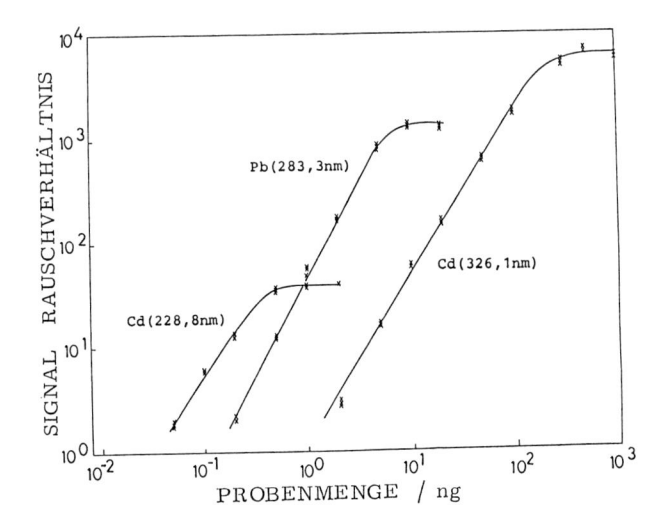

Abb. 6: Auf das Untergrundrauschen normierte Kalibrierfunktionen
der Elemente Cd (228,8 nm und 326,1 nm) und Pb (283,3 nm),
gemessen mit einem Photomultiplier.

Durch die entsprechende Anordnung lassen sich Nachweisgrenze und Dynamik bis zu einer Größenordnung verbessern, wie an den Beispielen der Elemente Cd und Pb in Abb. 6 deutlich wird. Die erste Kurve wurde an der Cd-Singulettlinie 228,8 nm gemessen, die zweite an der Pb-Linie 283,3 nm und die dritte an der Cd-Triplettlinie 326,1 nm. Man sieht, daß die beiden Cd-Linien zusammen einen großen dynamischen Bereich von 50 pg bis 200 ng überdecken. Der abnehmende dynamische Bereich zum UV hin hat seine Ursache zum einen in der spektralen Strahldichteverteilung der Xenonhochdrucklampe und zum anderen in der spektralen Durchlässigkeit der benutzten Polarisatoren. Unterhalb 280 nm ist die Deuteriumlampe der Xenonhochdrucklampe vorzuziehen. Die Nachweisgrenze für Cd läßt sich dadurch auf 15 pg verbessern. Generell läßt sich sagen, daß die Nachweisgrenzen der Vorwärtsstreuung bei Verwendung eines Kontinuumstrahlers als Lichtquelle für einige Elemente etwa in der gleichen Größenordnung liegen wie bei der AAS, für andere Elemente etwa eine Größenordnung schlechter sind /2/.

/1/ Ito, M., Murayama, S., Kayano, K., and Yamamoto, M.: Spectrochim. Acta 32 B, 347 (1977)

/2/ Ito, M.: Anal. Chem. 52, 1592 (2980)

/3/ Debus, H., Scharmann, A., and Wirz, P.: Spectrochim. Acta 36 B, 1015 (1981)

/4/ Wirz, P., Debus, H., Hanle, W., and Scharmann, A.: Spectrochim. Acta 37 B, 1013 (1982)

INTENSITY AND SPECTRAL DISTRIBUTION OF THE RESONANCE RADIATION OF SODIUM IN FORWARD
SCATTERING MEASURED WITH A CW-DYE-LASER

S. Ganz, M. Gross[*], W. Hanle, G. Hermann, A. Scharmann

I. Physikalisches Institut der Justus-Liebig-Universität Giessen, FRG

Am Beispiel der D_1-Resonanzlinie des Natriums soll gezeigt werden, daß Dichroismus und Dispersion spektral unterschiedliche Beiträge zur Vorwärtsstreuungsintensität liefern. Die Natrium D_1-Linie entspricht dem Übergang aus dem $^2S_{1/2}$-Grundzustand in das $^2P_{1/2}$-Niveau (Abb. 1a). Im Magnetfeld spaltet die Feinstrukturlinie in 4 Zeemankomponenten auf, die sich symmetrisch zum Linienschwerpunkt bei B = 0 gruppieren.

Bei transversaler Beobachtung sind die beiden σ-Komponenten senkrecht zum Feld, die beiden π-Komponenten parallel zum Feld linear polarisiert. Durch das Magnetfeld wird innerhalb des atomaren Dampfes eine optische Anisotropie erzeugt; das Medium wird doppelbrechend. Den beiden Polarisationsrichtungen transversal und parallel zum Feld entsprechend, muß man nun dem Medium zwei verschiedene komplexe Brechungsindizes n_π und n_σ zuordnen. Der Realteil des Brechungsindex beschreibt die Dispersion im Bereich der Resonanzlinie, der Imaginärteil den spektralen Verlauf der Absorption (Abb. 1b).

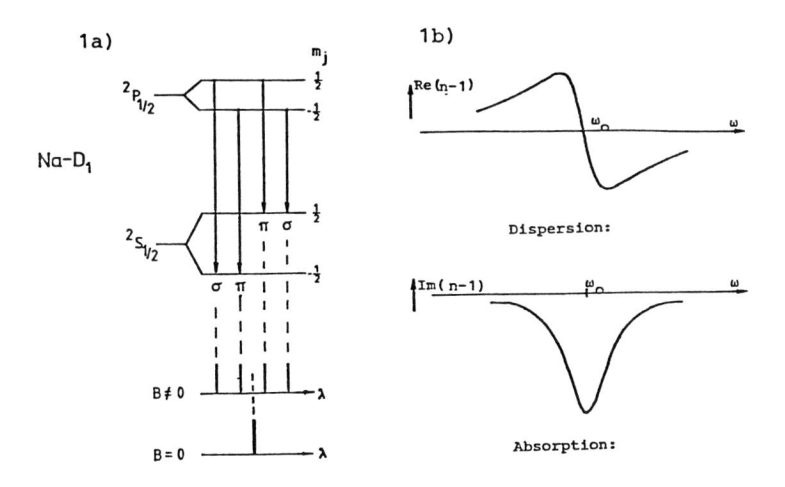

Abb. 1: a) Termschema und b) spektraler Verlauf des Real- und Imaginärteils des Brechungsindex.

---

[*]Part of thesis (D 26)

Bei einer Anordnung mit gekreuzten Polarisatoren, deren Schwingungsrichtung unter jeweils 45° zur Magnetfeldrichtung stehen, wird vom Analysator folgende spektrale Intensität $I(\omega)$ durchgelassen.

$$I(\omega) = I_o(\omega)\frac{1}{4} \left| e^{i\omega(\frac{n_\sigma}{c_o} L - t)} - e^{i\omega(\frac{n_\pi}{c_o} L - t)} \right|^2$$

$I_o(\omega)$ = Intensität vor Eintritt in den Polarisator; c_o = Lichtgeschwindigkeit
n_σ, n_π = Brechungsindizes; L = Länge der Meßzelle

Die transmittierte Intensität setzt sich aus zwei Anteilen zusammen. Der dichroitische Anteil wird bestimmt durch den Imaginärteil κ des Brechungsindex, also durch die Absorption der einzelnen Zeemankomponenten. Der spektrale Verlauf der Dispersion wird dagegen von dem Realteil n' wiedergegeben.

Dichroismus: $$I_A(\omega) = \frac{I_o(\omega)}{4} \cdot \left[e^{-\frac{2\omega L}{c_o}\kappa_\pi} - e^{-\frac{2L}{c_o}\kappa_\sigma} \right]^2$$

Dispersion: $$I_B(\omega) = \frac{I_o(\omega)}{2} \cdot e^{-\frac{\omega L}{c_o}(\kappa_\sigma + \kappa_\pi)} \left[1 - \cos\frac{\omega L}{c_o}(n'_\sigma - n'_\pi) \right]$$

Beide Anteile addieren sich zur Gesamtintensität $I(\omega) = I_A(\omega) + I_B(\omega)$

$$I(\omega) = \frac{I_o(\omega)}{2} \cdot e^{-\frac{\omega L}{c_o}(\kappa_\sigma + \kappa_\pi)} \left[\cosh\frac{\omega L}{c_o}(\kappa_\sigma - \kappa_\pi) - \cos\frac{\omega L}{c_o}(n'_\sigma - n'_\pi) \right]$$

Mit Hilfe eines Computers wurde der spektrale Verlauf der Vorwärtsstreuintensität berechnet (Abb. 2). Aufgetragen ist die relative Intensität gegen die Frequenz des eingestrahlten Lichtes. Zum Vergleich sind die Intensitäten zweier verschiedener Teilchendichten dargestellt. Die Pfeile deuten die Lage der einzelnen Zeemankomponenten an.

Die Anteile aus Dispersion und Dichroismus ergänzen sich spektral, sowohl bei kleinen als auch bei großen Teilchendichten. Insbesondere verschwindet der dispersive Beitrag gerade dort, wo sich die Linienschwerpunkte der Zeemankomponenten befinden (Nulldurchgang der Dispersionskurve; siehe Abb. 1b). Bei großen Teilchendichten kommt es zu Phasenverzögerungen von Vielfachen von 2π. Deshalb zeigt die spektrale Intensität besonders an den Flügeln und in der Linienmitte ein stark oszillatorisches Verhalten. Im Bereich starker Absorption durch die Zeemankomponenten bestimmt der Dichroismus die Vorwärtsstreuungsintensität.

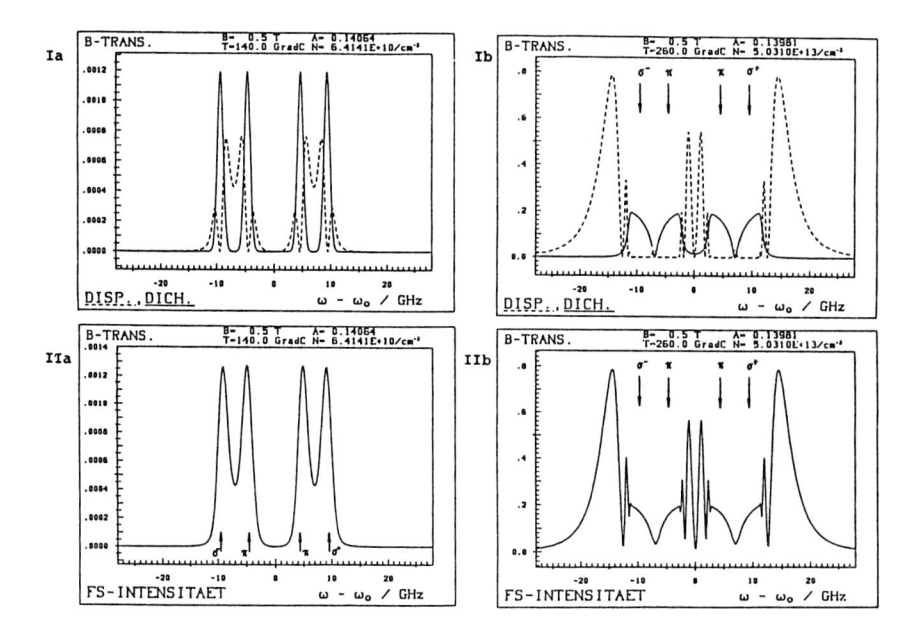

Abb. 2: Ia,b) Intensitätsbeiträge des Dichroismus (——) und Dispersion (---) und
IIa,b) zugehörige Gesamtintensität
a) für kleine Teilchendichten, b) für große Teilchendichten

Die integrale Gesamtintensität (Abb. 3) zeigt, daß bei kleinen Teilchendichten sowohl Dispersion, als auch Dichroismus gleiche Beiträge zur Gesamtintensität liefern. Bei hohen Teilchendichten erreicht der Dichroismus zuerst einen Sättigungswert und die Gesamtintensität wird von der Dispersion bestimmt.

In einem entsprechenden Experiment wurden diese Verhältnisse für Natrium verifiziert. Den dazu notwendigen Versuchsaufbau zeigt Abb. 4. Eine Quarzzelle wurde mit Natrium gefüllt und in einem elektrisch beheizten Ofen auf Temperaturen von ca. 100 - 300°C gebracht. Ofen und Zelle befinden sich in einem statischen Magnetfeld der Stärke 0,5 Tesla zwischen gekreuzten Polarisatoren. Mit einem hochauflösenden Laserspektrometer (Farbstofflaser, Linienbreite 1 MHz) wurde die spektrale Intensität der Vorwärtsstreuung im Bereich der Natrium D_1-Linie in Abhängigkeit der Teilchendichte aufgenommen. Abb. 5 zeigt die relative Intensität in Abhängigkeit der Laserfrequenz für eine große Teilchendichte. Man beobachtet wie erwartet bei hohen Teilchendichten starke Oszillationen am Flügel und in der Linienmitte.

Ein wesentliches Ergebnis ist, daß sich die Intensität in der Vorwärtsstreuung aus einem großen spektralen Bereich ableitet. Im Falle des Natriums liegt dieser Bereich in der Größenordnung 30-60 GHz, das entspricht ca. 20-30 Dopplerlinienbreiten oder etwa 0.5 Å.

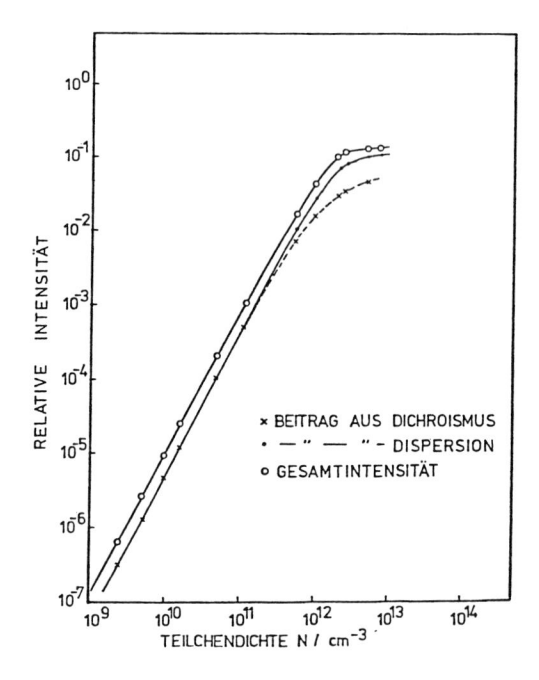

Abb. 3: Berechnete Gesamtintensität und die Beiträge aus Dispersion und Dichroismus.

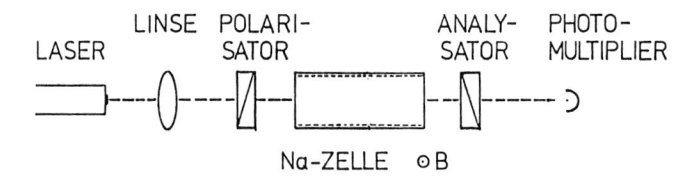

Abb. 4: Versuchsaufbau zur Messung der spektralen Intensität in der Vorwärtsstreuung

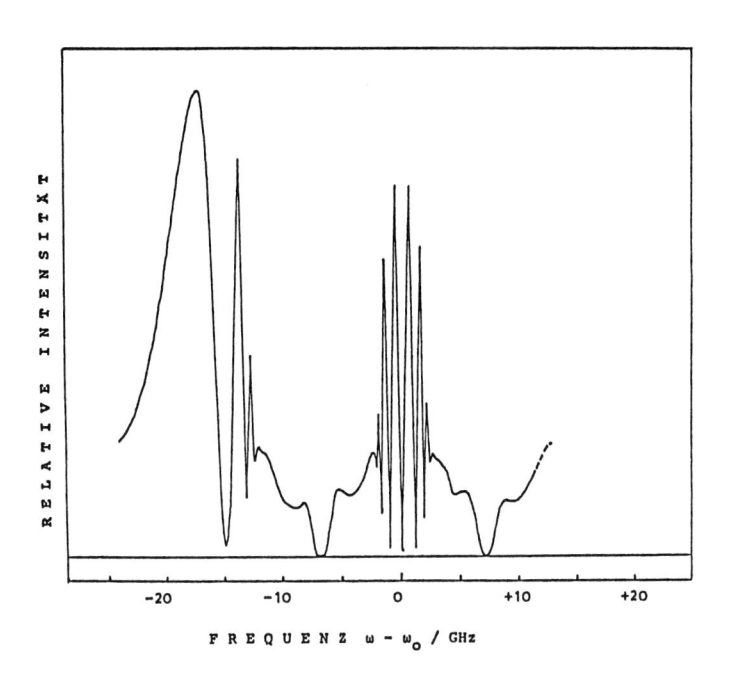

Abb. 5: Gemessene spektrale Intensität bei großen Teilchendichten.

---

*Part of thesis (D 26)

/1/ Corney, A., Kibble, B.P., and Series, G.W.: Proc. Roy. Soc. A 293, 70 (1966)

/2/ Kersey, A.D., Dawson, J.B., and Ellis, D.J.: Spectrochim. Acta 35 B, 865 (1980)

RECENT DEVELOPMENTS AND APPLICATIONS IN HOLLOW CATHODE LAMP-EXCITED
ICP ATOMIC FLUORESCENCE SPECTROMETRY

Donald R. Demers, Elisabeth B.M. Jansen

Baird Corporation, 125 Middlesex Turnpike, Bedford, MA 01730 USA
Baird Europe B.V., Produktieweg 30, Zoeterwoude, The Netherlands

SUMMARY--Recent developments and operating conditions in hollow cathode lamp-excited ICP atomic fluorescence spectrometry (HCL-ICP-AFS) are presented. Direct spectral line overlap interferences actually observed are given and are shown to be few in number. Use of an 80-mm focal length scanning monochromator is shown to provide sufficient flexibility to avoid at least some of these interferences. Detection limits are retained in concentrated matrices for elements whose atomic fluorescence wavelengths are below 350 nm, but are degraded about one order of magnitude for wavelengths above 400 nm. To determine refractory elements, it is necessary to introduce a carbon containing gas, such as propane, via the sample introduction system to prevent reformation of oxides in the plasma tailplume where measurements are carried out. With organic solvents, a small amount of oxygen must be introduced to promote combustion of the organic solvent in the plasma tailplume, and dioxane or a dioxane-organic solvent mixture is the recommended solvent. It is shown that sulfur and phosphorus can be determined at the 1- and 20-ug/mL level, respectively, in the vacuum ultraviolet, provided a purged optical system is employed.

Introduction--In this communication various aspects of hollow cathode lamp-excited ICP atomic fluorescence spectrometry (HCL-ICP-AFS) will be discussed. The data to be reported applies to the HCL-ICP-AFS technique in general and to the Baird Plasma/AFS^{TM} spectrometer in particular. This spectrometer uses an ICP operated at a nominal frequency of 40 MHz for sample atomization and pulsed HCLs for excitation. Up to 12 independent and rapidly interchangeable element

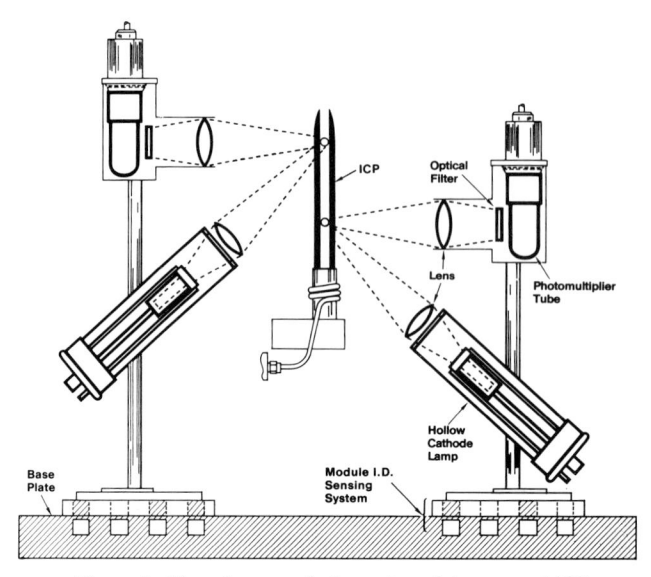

Figure 1. Line diagram of element modules around ICP

modules (channels) encircle the ICP (Figure 1). Each module comprises an HCL, a photomultiplier detector, an optical interference filter and lenses. A detailed description of this spectrometer has been given elsewhere (1).

2. Operating Conditions--The operating conditions found optimal with this spectrometer are listed in Table I. It is seen that, compared to

Table I

Plasma/AFS™ Operating Conditions

Nominal RF powers	400 watts (alkali elements)
	700-850 watts (Al, Si, B, Be)
	500-600 watts (all others)
Observation heights	80-100 mm (refractory elements)
(above torch central tube tip)	100-120 mm (Cr, Al, alkaline earths)
	120-140 mm (all others)
Plasma gases flows	
Coolant (Ar)	8-10 L/min
Carrier (Ar)	2 L/min
Auxiliary	none
Propane	5-40 mL/min (for refractory elements)
Oxygen	50-200 mL/min (for organic samples)
Sample uptake	1 mL/min (aqueous)
	0.3 mL/min (organic)
Nebulizer	Crossflow pneumatic
Aspiration chamber	Conical-shaped
Torch	Fassel-type with outermost (coolant) tube extending 60 mm relative to central tube tip

ICP-atomic emission spectrometry (ICP-AES), the RF powers used are lower, while the observation heights in the plasma and the carrier gas flow rate are higher in HCL-ICP-AFS. All these parameter settings make for a cooler plasma axial channel (between $3200°$ and $3800°$ K, depending on the observation height and the RF power(2)) than that customarily present in ICP-AES systems. However, the use of a torch with an extended coolant tube retards air entrainment and results in a taller plasma discharge. This increases sample residence time in the plasma and largely compensates for the lower axial temperature. For example, at 550 watts nominal and a 100-mm observation height, the classic solute vaporization interference in atomic spectroscopy, the depression of the calcium signal by the presence of phosphorus (3), is absent when using a torch with an extended coolant tube, but is present using a conventional short ICP-AES torch. Only when the RF power is increased to about 900 watts does this interference finally disappear with the short torch.

The coolant gas flow rate is at least 50 percent less than that of commercial ICP-AES systems. The lower flow produces a tall, laminar plasma tailplume needed for low noise atomic fluorescence measurements in this region. Finally, the torch in HCL-ICP-AFS is positioned well inside the load coil--the central tube tip is about halfway between the second and third bottom turns of the three-turn load coil. This position does not harm the central tube tip, provided a carrier gas flow is always present, and it enables introducing propane or oxygen gas (to be discussed later) into the plasma without extinguishing it.

3. Spectral Interferences--The most attractive characteristic of HCL-ICP-AFS vis-a-vis ICP-AES is its virtual freedom from spectral interferences. Only direct spectral line overlap and particulate light scattering can cause spectral interferences in HCL-ICP-AFS. The latter interference has been observed in samples high in aluminum, calcium, lanthanum, and silicon (in the 1 percent range), but is absent in samples high in most other elements, such as sodium, iron, and nickel. The rarity of this type of interference results from the use of an ICP for atomization and from the long residence time of the sample in the tall ICP discharge. Approximate matrix matching usually suffices to reduce this interference to insignificance. Particulate light scattering from silicon is completely eliminated by making up samples to be about 5 percent in hydrofluoric acid, which forms volatile fluorides of silicon.

Table II lists the relevant observed direct spectral line overlap interferences in atomic absorption spectrometry (AAS)(4) and in

Table II

Relevant Observed Spectral Line Overlap Interferences in AAS and in HCL-ICP-AFS

In AAS			In HCL-ICP-AFS		
Analyte	Interferent	Line Separation (nm)	Analyte	Interferent	Line Separation (nm)
Al	V	0.004	V	Al	0.004
Sb	Pb	0.024	Cu	Eu	0.001
Pb	Sb	0.024	Hg	Co	0.003
Ni	Sb	0.05	Co	Hg	0.003
Cd	As	0.01	Pt	Fe	0.001
Ca	Ge	0.016			
Hg	Co	0.003			
Si	V	0.000			
Zn	Cu	0.003			
Zn	Fe	0.003			
Cu	Eu	0.001			

HCL-ICP-AFS. "Relevant" interferences refer to those that can occur at the usually used, most sensitive wavelength by each technique--the small number of other such interferences found at alternate, infrequently used wavelengths are not listed. It is seen from the table that HCL-ICP-AFS is indeed virtually free of spectral line interferences, even freer than AAS.

4. Scanning Monochromator Element Module--To provide the user with the flexibility to analyze for a wider range of elements, an optional universal element module has been developed. This module consists of a scanning mini-monochromator of 80-mm focal length instead of, and at the same location as, the fixed wavelength optical interference filter of a conventional element module, and a universal hollow cathode lamp holder so that the user can employ any HCL, including AA-style lamps.

The universal element module has been used to circumvent the direct line overlap interference of iron on platinum listed in Table II. This interference is common in precious metals analyses, an application for which the Plasma/AFS^{TM} is particularly well suited otherwise. As seen in Figure 2, the platinum wavelength at 271.903 nm from the platinum HCL overlaps the iron absorption wavelength in the plasma at 271.904 nm, only 0.001 nm away. Once the latter energy level is excited, signals at all literature-reported iron atomic fluorescence wavelengths are observed simultaneously with the platinum signals. Given the large spectral bandpasses of the optical interference filters used in the conventional element modules, it is impossible to find an alternate platinum wavelength that does not also pass some iron line. The monochromator element module, however, enables a much narrower spectral

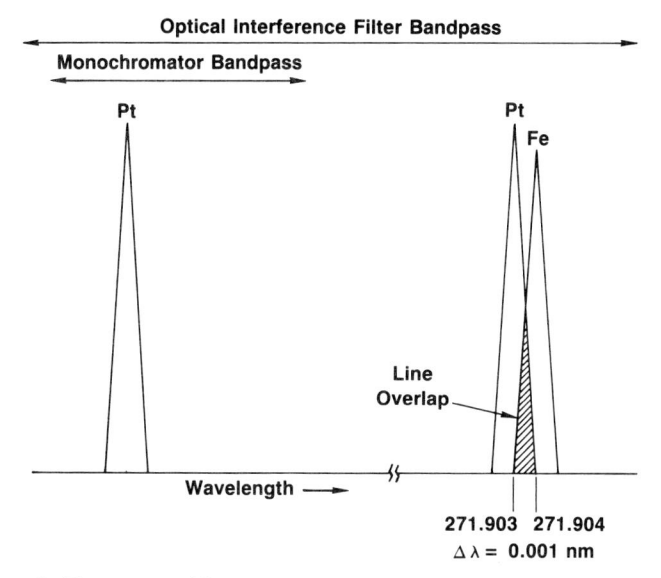

Figure 2. Direct spectral line overlap interference of Fe on Pt in HCL-ICP-AFS

bandpass to be used, which makes it easy to isolate the principal platinum wavelength at 265.9 nm from any iron line and thereby avoid the Fe-Pt interference.

5. Detection Limits In Complex Samples--Underneath the atomic fluorescence signals are also the atomic emission signals of the respective elements. In view of the fact that the ICP is a relatively high background source, even at high observation heights, and that a wide spectral bandpass (several nanometers) is allowed to reach the photodetector through the optical filter, it is reasonable to wonder if the detection limits observed in simple water solutions are retained in concentrated matrices. Table III compares the detection limits for ten elements whose atomic fluorescence wavelengths range from 196 to 670 nm

Table III

Plasma/AFS™ Detection Limits Measured in Absence and Presence of High Concentrations of Matrix Material

Element	λ_{AFS} (nm)	H_2O Solution (μg/l)	3.5% NaCl Solution (μg/l)	1.0% Ni Solution (μg/l)
Se	196.0	250	125	200
Cd	228.8	0.3	0.15	0.2
Si	251.6	300	200	225
Mn	279.5	3	0.8	1
Bi	306.8	200	175	220
Ag	328.1	1	1	1
Al	396.2	40	65	100
Ca	422.7	0.4	0.9	2
Sr	460.7	2	8	20
Li	670.8	0.4	1	3.5

in water, in 3.5 percent sodium chloride, and in 1.0 percent nickel solutions. The sodium chloride solution simulates a seawater sample, while the nickel solution gives many intense atomic emission lines throughout the UV-visible region.

It is seen that those elements whose principal atomic fluorescence lines lie below 350 nm, and this is the large majority of elements, exhibit no degradation in the presence of the concentrated matrix. In the case of manganese, the matrix had the favorable effect of removing the hydroxyl emmision band atop the ICP background at 281 nm and, thereby, reduced the ICP background noise.

Above about 350 nm the atomic emission of the elements becomes increasingly intense and the detection limits are noticeably degraded, especially in the nickel solution, but are generally still acceptable. In short, the importance of Table III is that it establishes the HCL-ICP-AFS technique possesses good sensitivity even in concentrated, real-world samples.

6. Special Considerations for Determination of Refractory Elements-- In HCL-ICP-AFS, the atomic fluorescence must be measured from the ICP tailplume where the ICP background and the atomic emission atop that background are low enough so as to not saturate the photodetector, especially since optical interference filters with relatively wide spectral bandpasses are used. To measure the refractory elements, some means must be employed to maintain these elements as free atoms, instead of as metal oxides, in the plasma tailplume. An effective and simple means to achieve this is to introduce a small amount of a carbon-containing gas, such as propane, via a second port in the sample aspiration chamber. The gas is carried from the aspiration chamber along with the sample aerosol into the plasma. At a very low propane flow, about 10 cc/min, a narrow "sliver" of green emission is observed that extends 1-2 cm beyond the end of the outermost tube of the torch. At a higher propane flow, about 40 cc/min, the green emission expands into a wide, triangular-shaped plume that extends 4-5 cm beyond the end of the outermost tube of the torch. The green emission is due to C_2 emission from the Swan bands at 473 and 516 nm (5). The presence of carbon species in the tailplume prevents the free atoms of the refractory elements from forming oxides. The addition of hydrogen, on the other hand, has no effect.

Elements with only a moderately strong oxide-forming tendency, such as Al, Ba, Be, Ca, Cr, Sn, and Sr exhibit enhanced signals and improved detection limits in the presence of only enough propane to produce a green "sliver" in the plasma tailplume. Atomic fluorescence from these elements is best measured from about 1 cm above the green emission (be it only a "sliver" or a tall, green plume). In contrast, elements with a strong oxide-forming tendency, such as B, Si, Ti, V, W, etc., require enough propane to produce a tall, green plume and measurements must be made inside the green plume. This phenomenon is similar to the need to use a fuel-rich (i.e., carbon-rich) nitrous oxide-acetylene flame in atomic absorption. Here the refractory elements cannot be detected if one views above the red "feather" in the flame, or if it is absent (6).

The value of adding propane on detection limits for both moderate and strong oxide-forming elements is shown in Table IV. Without

Table IV

HCL-ICP-AFS Detection Limits (μg/mL) of Refractory Elements With and Without Propane Gas Addition

Element	With Propane	Without Propane
Al	0.02	0.04*
B	0.5	ND
Be	0.0008	0.003*
Cr	0.006	0.04*
Ge	0.2	ND
Mo	0.1	ND
Sn	0.2	8
Ta	2	ND
Ti	0.3	ND
V	0.2	ND
W	2	ND

ND = Not Detectable
*High RF Power Required

propane, strong oxide-forming elements cannot be detected, while moderately strong oxide-forming elements can still be determined, but with reduced sensitivity, and higher RF powers are usually required.

The addition of propane also serves as a good visual indicator of nebulizer performance because the intensity, height, and stability of the green emission is very sensitive to fluctuations in aerosol transport. The addition of a carbon-containing gas is never required with organic samples. These samples are already carbon-based and furnish directly the necessary green (reducing) region in the plasma tailplume.

7. Special Considerations For Determinations In Organic Solvents--
When an organic solvent is nebulized into an ICP operated at low RF power (i.e., under conditions compatible for atomic fluorescence

measurements), there occurs in the tailplume the desired green emission along the outer edges of the axial channel, as well as an intense yellow-colored "sliver" at the center of the axial channel that spans the whole length of the tailplume. The latter emission, due to incomplete combustion of the organic solvent, causes very noisy baselines (i.e., degraded detection limits), or worse, can saturate the photodetector, precluding any determination of some elements. By introducing enough oxygen (about 300 cc/min), again via the sample aspiration chamber, complete combustion of the organic solvent occurs. Under these conditions, the plasma tailplume takes on an appearance similar to when water is nebulized (i.e., no color is observed, except for a bluish tinge) and non-oxide forming elements can be determined with good sensitivity, but refractory elements reform oxides and cannot be determined at all.

This dilemma is completely avoided if the organic solvent is dioxane, or if the sample is a mixture of a given organic solvent and dioxane. Dioxane (also called diethylene dioxide, $C_4H_8O_2$) mixes well with oils and organic solvents and it contains a high percentage of oxygen in its molecular structure. When dioxane is nebulized into an ICP, the oxygen from it, along with about 100 cc/min of oxygen introduced via the sample introduction system, result in a plasma tail plume devoid of any yellow "sliver", while the rest of the carbon-based molecule produces the tall, green (reducing) tailplume necessary for the determination of the refractory elements.

Detection limits observed in dioxane are generally comparable to those in aqueous media, except for molybdenum, titanium, and vanadium where the detection limits (50, 80 and 100 ug/L, respectively) are about three times better in dioxane.

If other organic solvents must be used, the organic solvent should be mixed with dioxane until a mixture is found at which the yellow "sliver" disappears. Typically, one part of the organic solvent diluted with two or three parts of dioxane will suffice.

The introduction of a small amount of oxygen via the sample introduction system also serves to prevent any carbon buildup at the tip of the central and intermediate tubes of the torch, even when no auxiliary gas flow is used.

8. Applications in Multielement analysis--The advantages of HCL-ICP-AFS will be discussed with the help of application examples.

Water-- The suitability of the Plasma/AFS™ for trace element analysis is illustrated by an analysis of a standard reference material water sample for a number of trace metals (table V). The calibration standard for this analysis was a multielement aqueous solution in 2% nitric acid. No effort was made to match the major constituents of the unknown, Na, Ca, K and Mg. The correlations with the certified values and the precisions are excellent in every case.

TABLE V: Trace elements in water

Element	SRM value (ppb)	Results with Plasma/AFS™ (ppb)	Detection Limits (ppb)
Ag	2.8	3.4 ± 0.5	1
Be	19	18.6 ± 1.0	1
Cd	10	10.4 ± 0.5	1
Co	19	18.5 ± 1.8	10
Cr	17	20 ± 2.5	4
Cu	18	19.5 ± 1.3	2
Fe	88	88.0 ± 2.5	10
Mn	31	31.3 ± 0.8	3
Ni	55	56 ± 1.5	8
Sr	239	240 ± 2.5	2
Zn	72	70.2 ± 0.6	1

Operating Conditions:
Power   500W
Argon flow rates
    Coolant   10L/min
    Auxiliary   none
    Carrier   2L/min
Propane flow rate   20mL/min
Solution uptake rate   1mL/min
Integration time   20 seconds

Steel-- Because of the absence of spectral interferences, the Plasma/AFS™ is suitable for the analysis of steel. Table VI shows the detection limits of some elements with and without Fe-matrix. Only aluminium and chromium show a deterioration in detection limits. Strong atomic emission from Fe at 404.5 and 357.0 nm passes the optical filters of these element modules and causes a strong increase in detector noise. This table also shows the results of analysis of low alloy standard reference material steel sample.

TABLE VI: Steel Analysis; detection limits and analytical results

Element	Wavelength (nm)	Detection Limits		SRM Steel Sample	
		H_2O mg/L	0.5% Fe solution mg/L	SRM value wt %	Plasma/AFS™ wt %
Al	396.2	0.025	0.080	0.095	0.095±0.002
Co	240.7	0.002	0.005	0.30	0.288±0.003
Cr	357.9	0.005	0.075	0.30	0.331±0.005
Cu	324.9	0.001	0.001	0.50	0.483±0.003
Mn	279.5	0.001	0.001	1.04	0.994±0.010
Mo	313.3	0.100	0.100	0.068	0.071±0.004
Ni	232.0	0.003	0.003	0.59	0.598±0.006
Si	251.6	0.300	0.300	0.39	0.38 ±0.02

Operating Conditions:
Power   700W
Argon flow rates
    Coolant   10L/min
    Auxiliary   none
    Nebulizer   2L/min.
Propane flow rate   60mL/min
Solution uptake rate   1.5mL/min
Integration time   20 seconds

Biological Material--As an example for the analysis of biological materials blood is taken. Table VII shows a comparison of Plasma/AFS™ and AAS on the analysis of blood. It follows that the precision of AFS is a factor 4 better than with AAS. The accuracy of the measurements at concentrations above about 100 times the detection limit is approximately 1% relative.

TABLE VII: Analysis of Blood (mg/L)

Element	Plasma/AFS™[1)	AAS
Na	2094±21	2077±78
K	1711±25	1710±110
Fe	496± 7	463±36
Ca	65.7±0.4	57± 4
Mg	40.4±0.6	33.7± 2
Zn	5.3±0.2	5.2±0.8
Cu	1.30±0.05	1.05±0.21

1) 100 times diluted

Oil-- The incomplete combustion of the organic material on the plasma causes a very bright plasma tailplume, which increased the shot noise at the detector and causes the detection limits to deteriorate. Addition of oxygen to the carrier gas solves this problem by enabling complete combustion of the organic matrix. Figure 3 and 4 show the influence of oxygen on the detection limits of some elements.

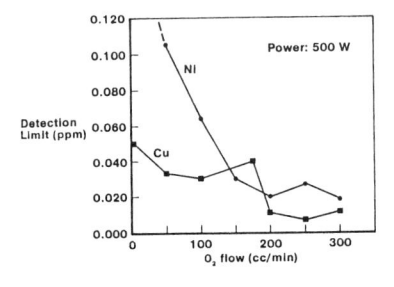

Figure 3. Effect of oxygen on copper and nickel detection limits.

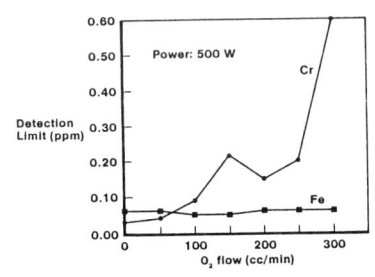

Figure 4. Effect of oxygen on iron and chromium detection limits.

Optimum sensitivity for copper and nickel are found with addition of approximately 300 ml O_2 per minute. The influence of iron is negligeable. Chromium detection limits deteriorate with higher oxygen flow because of the reformation of oxides. Simultaneous analysis of these elements can be performed when dioxane is used as dilutant and approximately 100 ml O_2 per minute. The accuracy and precision obtained with this method for our elements in fuel oil is given in table VIII.

TABLE VIII: Determination of trace elements in fuel oil (NBS SRM 1634a)

Element	Found (ppm)	Given (ppm)
Na	87 ± 4	89 ± 3
Ni	30 ± 1	29 ± 1
V	55 ± 3	56 ± 2
Zn	2.7 ± 0.1	2.7 ± 0.2

Table IX compares the results of analysis of six elements in samples of used engine oil from dieseltruck obtained with Plasma/AFS and DCP atomic emission spectrometer.

Table IX Comparison of AFS and AES* for the Analysis of Used Engine Oil

Sample No.	Cr (ppm)		Cu (ppm)		Fe (ppm)		Mg (ppm)		Mo (ppm)		Pb (ppm)	
	AFS	AES	AFS	AES	AFS	AES	AFS	AES	AFS	AES	AFS	AES
1	30	34	18	24	62	78	17	22	2	6	21	21
2	29	22	10	13	88	111	17	24	12	21	13	14
3	44	44	33	41	160	160	17	17	45	51	50	42
4	50	66	13	17	187	178	18	23	44	45	18	18
5	29	32	45	58	203	198	18	23	15	13	117	110
6	26	29	63	82	204	204	19	24	10	14	247	209

*DC Plasma Echelle (DCP)

Typical operating conditions for oil analysis are:

Power	600W
Argon flow rates	
Coolant	10L/min
Auxiliary	none
Nebulizer	1.2L/min
Oxygen flow rate	100mL/min
Sample uptake rate	0.5mL/min
Diluent	dioxane or dioxane/kerosine
Integration time	20 seconds

Precious Metals-- The Plasma/AFS™ method is well-suited for the detection of precious metals traces and is especially suited for their determination in the presence of high concentrations of matrix elements like Fe, Ni, Cr and Al, again because of its virtual freedom from spectral interferences.

With the usual emission technique it is hardly possible to satisfactorily determine Gold (242,8 nm) in presence of high iron concentrations due to severe spectral line interference from an intense iron line at 248,3 nm. In contrast, the atomic fluorescence spectrum of gold is not influenced by iron. In table X the detection limits of precious metals with Plasma/AFS™ are shown. The linearity ranges from 4 to 6 decades.

TABLE X: Detection limits of precious metals

Element	Detection Limit
Ag	0.001
Au	0.020
Ir	0.500
Os	0.180
Pd	0.005
Pt	0.075
Rh	0.005
Ru	0.050

In table XI a comparison between Plasma/AFS™ and the much slower and tedious classical methods is shown.

TABLE XI: Analytical results with Plasma/AFS™

Element	Plasma/AFS™ (wt %)	Fire Assay (wt %)
Au	95.3	95.5
Pd	2.98	2.94
Ag	1.28	1.28
Pt	0.24	0.23

Sulfur and Phosphorus-- The principal atomic fluorescence lines of sulfur and phosphorus are in the vacuum UV around 180 nm . In this wavelength region, ambient oxygen absorbs the radiation from the excitation source as well as any atomic fluorescence produced. To access this region, a V-shaped chamber has been added to a conventional element module of the spectrometer, and the chamber is flushed with argon. The chamber extends from the front face of the HCL to about 5 mm from the outside wall of the coolant tube and back to the photodetector.

Stable signals for both sulfur and phosphorus are obtained after only 4 minutes of purging with argon at a rate of 1.5 L/min. Detection limits in both water and dioxane are approximately 1 and 20 ug/mL for sulfur and phosphorus, respectively.

For phosphorus, a tall green plume (high propane flow) and viewing into the plume is necessary, while sulfur requires only a green "sliver" (low propane flow) and gives best sensitivity if its atomic fluorescence is measured just above the "sliver". Linearity with both elements extends to about 10,000 ug/mL. The detection limits are much inferior to those of ICP-AES, but would be adequate for many types of samples, such as sulfur in oils and phosphorus in oil additives, foods and fertilizers.

To test for matrix effects, sulfur signal intensities from compounds having different sulfur bonds, all at the same concentration with respect to sulfur, have been compared. All compounds studied give identical signal intensities, indicating complete dissociation of all compounds by ICP.

9. Future Developments-- Despite possessing important inherent advantages, the major drawback to the wider acceptance of HCL-ICP-AFS to date has been the fact that its detection limits for the refractory elements have been inferior to those of ICP-AES systems by one to two orders of magnitude. Recently, a full order of magnitude improvement for all elements has been observed in our laboratory by replacing the pneumatic nebulizer with an ultrasonic nebulizer. With this nebulizer the Plasma/AFS^{TM} detection limits now range from equal to ten times better than those of ICP-AES systems for the non-refractory elements and from equal to about one order of magnitude worse for the refractory elements. Table XII gives a comparison of detection limits for both nebulisation systems.

TABLE XII: HCL-ICP-AFS detection limits with pneumatic and with ultrasonic nebulisation

Element	Detection Limits (ppb)	
	ultrasonic nebulisation	pneumatic nebulisation
Ag	0.2	0
Al	5	40
B	60	200
Cd	<0.1	0.7
Co	1	10
Cr	1	8
Cu	0.3	4
Fe	1	10
Mn	0.4	3
Mo	20	200
Ni	0.8	7
Pb	20	200
Se	25	300
Sn	30	300
W	200	5000
Zn	<0.1	0.5

Moreover, the reproducibility observed with this nebulizer is easily twofold better (less than 0.3 percent for some elements).

Another probable development will be the adaptation of boosted-output HCLs and/or electrodeless discharge lamps. The expected increase in intensity from these sources should result in an important additional decrease in detection limits. The combined result may be a detection capability which would obviate the need for electrothermal atomization methods for many analyses. Such a development would provide a major boost to the method's desirability and acceptance.

REFERENCES

1 - D.R. Demers, D.A. Busch, and C.D. Allemand, Am. Lab., No. 3, (1982) 167

2 - M.A. Kosinski, H. Uchida, and J.D. Winefordner, Talanta, 30 (1983) 339

3 - G.F. Kirkbright and M. Sargent, Atomic Absorption And Fluorescence Spectroscopy, Academic Press, London, 1974, Chap. 12

4 - J.D. Norris and T.S. West, Anal. Chem., 46, (1974) 1423

5 - M.W. Blades and B. Caughlin, ICP Inform. News., 9 (1983) 386

6 - G.F. Kirkbright, M.K. Peters, and T.S. West, Talanta, 14 (1967) 789

7 - K.C. Thompson and R.J. Reynolds, Atomic Absorption, Fluorescence And Flame Emission Spectroscopy, John Wiley & Sons, New York, 1978, Chap. 7, p. 213

8 - R.K. Winge, V.J. Paterson, and V.A. Fassel, Appl. Spectros., 33 (1979) 206

ONE SET OF CONDITIONS: PREREQUISITE FOR MULTI-ELEMENT DETERMINATION USING
ELECTROTHERMAL ATOMIZATION

G. Schlemmer and B. Welz

Department of Applied Research, Bodenseewerk Perkin-Elmer & Co GmbH,
Postfach 1120, 7770 Überlingen, Federal Republic of Germany

Graphite furnace atomic absorption spectrometry is one of the most sensitive element analytical techniques. However, only one element at a time can be determined during a pretreatment/atomization cycle, and it seems to be very attractive to combine the low detection limit of the graphite furnace with a spectrometric technique capable of multi-element determination.

The short residence time of atoms in a graphite furnace requires a spectrometer capable of simultaneous multi-element detection, and among the principally applicable techniques which may provide detection limits similar to AAS are Carbon Furnace Atomic Emission Spectrometry and Coherent Forward Scattering.

One of the prerequisites for multi-element determination is that all elements under consideration have to be pretreated and atomized with one single set of conditions. It becomes apparent from numerous publications, however, that an individual set of conditions including thermal pretreatment, atomization temperature and a possible matrix modification is typically applied for each element.

Nevertheless, groups of elements are handled very similarly and a single set of conditions could be applied for their determination. One set of conditions means that a common thermal pretreatment temperature has to be found that allows to remove as much of the matrix as possible but is low enough to avoid preatomization losses. Important temperatures that have to be applied for removing common matrices are:

600 °C to remove organic matrix;
1000 °C to remove salts like NaCl, KCl and K_2SO_4;
1300 °C to remove phosphates.

Most effective for stabilizing a lot of elements is matrix modification. A common modifier has to be found for as many elements as possible. A common atomization temperature has to be found that renders highest atomization efficiency. This optimum temperature, however, will change with the spectrometric technique used (absorption, emission or fluorescence) and with the method of signal evaluation.

Multi-element determination is attractive only if the elements under consideration can be determined free from matrix interferences. Direct determination against dilute acid solutions should be possible. To fulfill these requirements the graphite furnace and the spectrometer have to meet a number of specifications:

1. directed purge gas stream to avoid condensation of matrix during thermal pretreatment procedure,
2. fast heating rate and L'vov platform to avoid vapor phase interferences,
3. signal integration to avoid interferences caused by kinetic effects during atomization,
4. accurate background correction.

The elements that can be determined with AAS can be classified into six groups with respect to the items discussed above:

> Group A elements (volatile elements)
> platform atomization without modifier
> thermal pretreatment: 600 °C; atomization 2000 °C
> Ag, Au, Bi, Cs, K, Mg, Na, Pb, Sb, Te
>
> Group B elements (volatile elements)
> platform atomization; modifier: 200 µg $NH_4H_2PO_4$
> thermal pretreatment: 700 °C; atomization: 1800 °C
> Cd, Pb, Zn
>
> Group C elements (medium volatile elements)
> platform atomization; modifier: 10 µg Ni (as the nitrate)
> thermal pretreatment: 1000 °C; atomization: 2200 °C
> As, Sb, Se, Te
>
> Group D elements (medium refractory elements)
> platform atomization; modifier: 50 µg $Mg(NO_3)_2$
> thermal pretreatment: 1200 °C; atomization: 2500 °C
> Al, Be, Co, Cr, Mn, Ni, Cu, Fe
>
> Group E elements (refractory elements)
> platform atomization without modifier
> thermal pretreatment: 1000 °C; atomization: 2650 °C
> Ir, Pd, Pt, Si
>
> Group F elements (refractory elements)
> wall atomization without modifier
> thermal pretreatment: 1000 °C; atomization: 2650 °C
> Ba, Ca, Ir, Mo, Pd, Rh, Ru, Si, Sr, Ti, U, V, Lanthanoids

Group E elements can as well be determined under the conditions of group F elements without the benefits of the L'vov platform, however. From the remaining 7 elements B, Ge, Hg, Li, P, Sn, Te two elements can be classified as group F elements (Ge, Li) but with a lower thermal pretreatment temperature. Hg requires a very low thermal pretreatment temperature and B, P, Sn and Te require special matrix modifiers.

As an example, the recovery of As, Sb and Te was tested in Mediterranean Sea water (diluted 1+1) under identical furnace parameters (table 1). While antimony and arsenic were recovered completely the common modifier used $(NH_4)_2Cr_2O_7$ was not optimum for tellurium. This element was recovered completely by using copper as matrix modifier.

A second example is the determination of lead and cadmium on the one hand and of selenium and arsenic on the other hand in marine biological tissue. Under the conditions listed in table 2 the determination could be performed directly against the reference solutions.

Table 1. Recovery of As, Sb and Te in Mediterranean Sea water (diluted 1+1) under identical furnace parameters.

thermal pretreatment temperature: 1000 °C; atomization temperature: 2000 °C; matrix modifier: 100 µg $(NH_4)_2Cr_2O_7$; Zeeman Background Correction

Element	As	Sb	Te	Te with Cu modifier
reference solution char. mass (pg/0.0044 A.s)	16	14	11	12
seawater	15	15	18	11
HGA cookbook	17	15	11	-

Table 2. Determination of lead and cadmium and of selenium and arsenic in marine biological tissue using Zeeman Background Correction

Element		Cd	Pd	As	Se
Conditions		modifier 200 µg $(NH_4)_2H_2PO_4$ therm. pretr. 750 °C atomization 1600 °C		modifier 10 µg Ni, 10 µg $Mg(NO_3)_2$ therm. pretr. 1000 °C atomization 2100 °C	
char. mass (pg)	refer. sol.	0.54	13	14	22
	tomalley	0.58	14	16	23
	scallops	0.62	13	16	31
	plaice	0.70	13	17	24
	HGA cookb.	0.35	13	17	23

POTENTIALITIES OF VOLTAMMETRY IN ENVIRONMENTAL OLIGO-ELEMENT ANALYSIS
OF TRACE METALS

Hans Wolfgang Nürnberg

Institute of Applied Physical Chemistry, Nuclear Research Center
(KFA), P.O. Box 1913, D-5170 Jülich

SUMMARY

Voltammetry offers particular favourable potentialities for the analysis of ecotoxic heavy metals in all types of specimens from aquatic and terrestrial ecosystems, from the atmosphere, from food and human material. For a number of heavy metals oligo-element determinations are provided whereas for some metals only single element determination, yet with very high sensitivity, is possible. The most significant mode, usually applied, is the differential pulse mode either in connection with stripping voltammetry or with adsorption voltammetry. Recently also the square wave mode in connection with both approaches is gaining growing significance. The major methodological and practical analytical aspects are presented and the manifold analytical potentialities are featured by a brief treatment of various application areas.

1 INTRODUCTION

1.1. General Ecochemical Aspects

Among the environmental chemicals emitted and discharged from various anthropogenic sources (fossil energy consumption, industrial and technological activities, traffic, modern agriculture and generally waste and garbage from urban agglomerations) into the atmosphere, aquatic and terrestrial ecosystems, a number of heavy metals has gained particular significance, due to their ecotoxicity and their toxicity for man and animals. These heavy metals are in the first place the per se toxic metals Cd, Pb, Hg and As(III).

A further group, e.g. Cu, Zn, Ni, Co, Cr, is also of particular ecochemical and ecotoxicological interest, although they have for man, animals and plants below respective threshold levels, which are specific and different for the various types of organisms, indispensable essential functions and turn over to toxic actions only well above their essential doses. Nevertheless, it has to be borne in mind, that levels of these metals which might be essential or at least not yet hazardous for a variety of types of organism can exert for other categories of organisms already adverse or toxic effects. The actually concerning levels will differ according to the respective metal and the respective type of organism and will furthermore depend on the strength of simultaneously acting other stress factors. Thus, also the second group of potentially toxic heavy metals is in general of particular ecochemical and ecotoxicological interest (1,2).

A particular common feature of the ecotoxic heavy metals is, that they are in contrast to the numerous organic environmental chemicals not biologically or physically and chemically degradable in the ecosystems, but they undergo a biogeochemical cycle with quite different residence times in the various environmental compartments. Within this cycle the metals undergo a number of transformations with respect to their chemical species forming metal complexes and compounds of higher or lower toxicity. The major pathways of the metals from the sources lead through the atmosphere and by discharge of waste water and run off water through the rivers. There is a definite tendency of the sea to act as a pseudo-sink for heavy metals introduced by deposition from the atmosphere, fluvial discharge and run off water from land. Particularly in the shallow coastal waters and estuaries metal pollution causes the built-up of metal depots at the bottom sediments from which ecotoxic metal amounts can be remobilized under certain conditions. Another concerning depot function for ecotoxic heavy metals has the soil. This depot is continuously loaded by wet and dry deposition of heavy metals from the atmosphere and to a certain extent also by direct input with artificial fertilizers or sewage sludge in the course of agricultural activities or by waste and garbage disposal. From this soil depot a considerable amount of toxic metals is taken up by plants and exerts in this manner not only ecotoxicological stress to a diffe-

ring extent on various types of the vegetation, e.g. the forests, but
introduces also continuously heavy metals into the food chains. Of
particular significance are in this context grass as fodder for the
cattle and with respect to human nutrition grains and vegetables. In
addition to this uptake from the soil the vegetation experiences further a significant burden of ecotoxic heavy metals by deposition from
the atmosphere onto the leaves of plants or the foliage of forests.

For man the common predominant exposure to toxic heavy metals occurs
by consumption of food. The degrees differ according to the type and
the pollution at the cultivation site of the food. Exposure via respi-
ration is on a global scale negligible and gains only significance
under special mostly occupationally caused conditions.

1.2. General Analytical Aspects

It is obvious, that investigations on the ecochemical behaviour and
fate of ecotoxic heavy metals in the environment, as well as, with
respect to environmental protection the continuous control of their
levels in aquatic and terrestrial ecosystems and in food and on their
effects and metabolisation in man, animals and plants, have become
tasks of high scientific and public significance and priority. Analytical chemistry fulfills here a key function. In this context a further common feature of ecotoxic heavy metals has to be emphasized.
These environmental pollutants with sometimes severe and insidious,
because in most cases for man chronical, toxic effects (1,2) occur
commonly at trace levels and sometimes even in ultratrace amounts in
environmental matrices. Therefore, suitable and reliable trace analytical methods are needed which combine high determination sensitivity
with good precision and accuracy. Furthermore, as particularly in environmental surveillance, monitoring and banking large numbers of samples have to be analysed, the applied method should provide a reasonably high determination rate and the investment and running costs
should be moderate. These latter budgetary requirements act somewhat
prohibitive on the large scale application of sophisticated and powerful but with respect to investment and running costs demanding multi-
element methods in routine environmental monitoring. On the other hand
concerning many ecochemical research problems, particularly in still
rather unpolluted ecosystems and generally in certain ecosystem types
and their components, as rivers, lakes, the sea and atmospheric precipitates, but also in certain samples of biotic or human origin, the
determination sensitivity of the in principle always very attractive

multi-element methods, e.g. atomic emission spectroscopy or X-ray fluorescence, is often not sufficient for the existing low trace levels of the considered ecotoxic heavy metals. This limitation exists with respect to certain metals, e.g. Cd and Pb, even for neutron activation analysis being a method restricted anyway to a limited number of special laboratories (3,4).

Thus, still atomic absorption spectrometry (AAS), according to the respective to be analysed ecotoxic heavy metal in the graphite tube, hydride or cold vapor mode, and advanced modes of voltammetry, i.e. the differential pulse and the square wave mode, remain the determination methods predominantly applied in environmental analysis of ecotoxic heavy metals (3-6). None of these two analytical procedures is a multi-element method.

Table 1. Costs of instrumental methods for trace metal analysis

Method	Costs 10^3 US	Remarks
Voltammetry (mannual operation)	8 - 10	
Voltammetry (microprocessor controlled automated devices)	20	contains also SW-mode
Graphite tube AAS	20 - 60	costs depend on automation level (see ref. 83)
Graphite tube AAS with Zeeman-compensation	40 - 60	
Atomic emission spectrometry with ICP excitation	50 - 200	costs depend on automation and data processing level
Neutron activation analysis	60 - 100	costs depend on size and potentialities of counting equipment

AAS is only a single element method. It has the advantage that besides the considered ecotoxic heavy metals it can be applied to a large number of further metals. Yet on the other hand, it has to be borne in mind, that with respect to ultra trace levels of certain toxic heavy metals, e.g. Cd, Pb, Ni and Co, the determination power of the to be applied graphite furnace mode (GFAAS) is not fully sufficient. At any rate it remains under such extreme circumstances a rather difficult and demanding method yielding reliable results only if substantial efforts are made and great experience and expertise exist. This is connected with the to experts well known substantial accuracy risks inherent to GFAAS, particularly in the range close to its determinaion limits for the respective trace metal. An advantage of AAS is, that it is with respect to solid samples less demanding concerning complete mineralisation. Commonly AAS is often regarded as a method with a high determination rate. This remains, however, restricted to the in practice less frequent cases where one is interested only in the determination of one metal. Frequently the practical requirements in environmental analysis, food control or investigations of human tissue and body fluids are quite different and demand the determination of several toxic or potentially toxic heavy metals in the sample. Also investment and running costs for AAS are by no means low (see table 1).

2 VOLTAMMETRIC METHODOLOGY

2.1. General Aspects

Between the true multi-element methods and the single element approach by AAS stand the voltammetric methods. They permit frequently the simultaneous determination of several heavy metals, usually a group of 2 to 5, and voltammetry should be therefore aptly considered as an oligo-element method.

Voltammetry uses the electrochemical reduction or oxidation of the to be analysed heavy metals, dissolved in a suitable analyte, in an electrode process at a suitable test electrode, mostly mercury, sometimes gold, for the determination measuring the electrical current flowing due to the electrode process. The method is based on the law of Faraday according to which the reaction of 1 mole substance in an electrode process would be equivalent to the transfer of the large electrical charge of $n \times 96500$ C through the interface electrode/solution. Usually the number n of the electrons transferred in the elementary step is 2 for heavy metals, for some it can be even 3. The favourable equivalence between heavy metal concentration and electrical charge is the reason of the extraordinary high determination potentiality reaching 10^{-11} moles/liter or 1 ng/l. For some metals, as Cd, it can be even about one order of magnitude less. In this context it is to be emphasized that these are practical determination limits set by the blanks but not by the basic methodological limits or instrumental noise levels of modern voltammetry. Nevertheless, they are at present sufficient to cope with all ultra trace analytical problems of ecotoxic heavy metals, even if they occur to an extreme extent in base-line studies of unpolluted regions, as e.g. in remote areas in the Arctic or Antarctica. The amount of heavy metal consumed for the measurement is negligible. Therefore the bulk concentration of the heavy metals in the analyte remains practically unaltered by the recording of the voltammogram. This can be therefore repeated several times to check reproducibility. In this sense the method is virtually destructionless.

In principle there is today a number of different voltammetric modes available. As in trace metal analysis, however, concentrations below 10^{-5}M or 1000 µg/l have to be determined the originally by Heyrovský (7), the founder of polarography and voltammetry, introduced conventional dc-polarography at the dropping mercury electrode (DME) had to be substituted by more advanced voltammetric modes providing by many orders of magnitude improved determination sensitivities.

2.2. Differential Pulse Voltammetry

2.2.1. Basic features

The most important and versatile advanced mode for voltammetric trace analysis has become the differential pulse mode, the for this largest application area of modern voltammetry most significant achievement from the pioneering methodological work of Barker (8,9) on the development of advanced voltammetry. The differential pulse mode is now incorporated into all commercial voltammetric devices and constitutes the for analytical use most important function. The recording of the signals in these instruments is performed according to the concept of Parry and Osteryoung (10). A detailed description of the principles and properties of the various versions of differential pulse voltammetry has been given elsewhere (11).

In the differential pulse mode the test electrode is polarized by a dc-voltage ramp (5-10 mV/s) on which rectangular voltage pulses (height 25 - 50 mV; duration 20 - 60 ms) are superimposed.

In this manner a very substantial improvement of the signal-to-noise ratio can be achieved, because use can be made of the different time laws for both components over each pulse duration. The during an electrode process flowing current consists always of two components. One is the faradaic current i_F, due to the reduction or oxidation of the respective trace metal at the electrode and is consequently proportional to its bulk concentration in the analyte solution. This faradaic current is usually controlled by diffusion of the trace metal towards the test electrode interface and decays therefore with the square root of the time during each pulse. The electrochemical noise is the charging current component i_c of the total current flowing. This component i_c is connected with the necessary change in the charge of the double layer capacity at the test electrode interface when the pulsed voltage alteration occurs. The charging current i_c decreases, however, much faster according to an exponential time law during the pulse. Therefore the recording of the current i during each pulse is restricted to the later fraction of the pulse duration when the noise component i_c has decayed to negligible amounts and the recorded current i corresponds consequently practically to the for analysis relevant current component i_F. Although this has also decayed to a certain extent according to its square root dependence on time since the begin of the pulse, there remains now a clean current signal. Sufficient electronic amplification of this clean signal creates today no problems at all. The ultimate methodological determination limit of voltammetry is set by the ratio of both current components i_F/i_c and is reached when this ratio approaches 1.

At trace metal concentrations in the analyte solution below 10^{-6}M or in other words typically below 100 or 50 µg/l this determination limit is approached for direct measurements by differential pulse polarography (DPP) at the DME. As a rule, however, the levels of ecotoxic heavy metals in samples from the terrestrial environment, in food and in human material are so low that after digestion analyte concentrations below 10^{-6}M frequently result. The same is the case for all types of natural waters where always levels of ecotoxic heavy metals are significantly lower. Therefore, preconcentration before the determination becomes necessary. A basic advantage of the voltammetric approach is, that this preconcentration can be achieved at the interface of the test electrode without introducing any additional accuracy risk by two different electrochemical approaches. Always a stationary test electrode instead to the dynamic DME characteristic for polarographic methods is applied in these techniques with electrochemical preconcentration.

2.2.2. Differential pulse stripping voltammetry

The first already since about 30 years in voltammetry used approach are the stripping methods also termed inverse voltammetry. There are two versions. In the more frequently to be used anodic stripping a rather cathodic potential (-1,0 or -1,2 V at the stationary mercury electrode) is applied for some minutes and a certain amount of the trace metals present in the analyte is plated at the electrode and thus accumulated in its interfacial zone. If the stationary test electrode is, as in most cases, a mercury electrode this preconcentration method is restricted to stable amalgams forming heavy metals, i.e. Cu, Pb, Cd, Zn among the here interesting ecotoxic heavy metals, whereas important other metals of ecotoxic relevance, as Ni, Co, Cr and As are excluded for this reason.

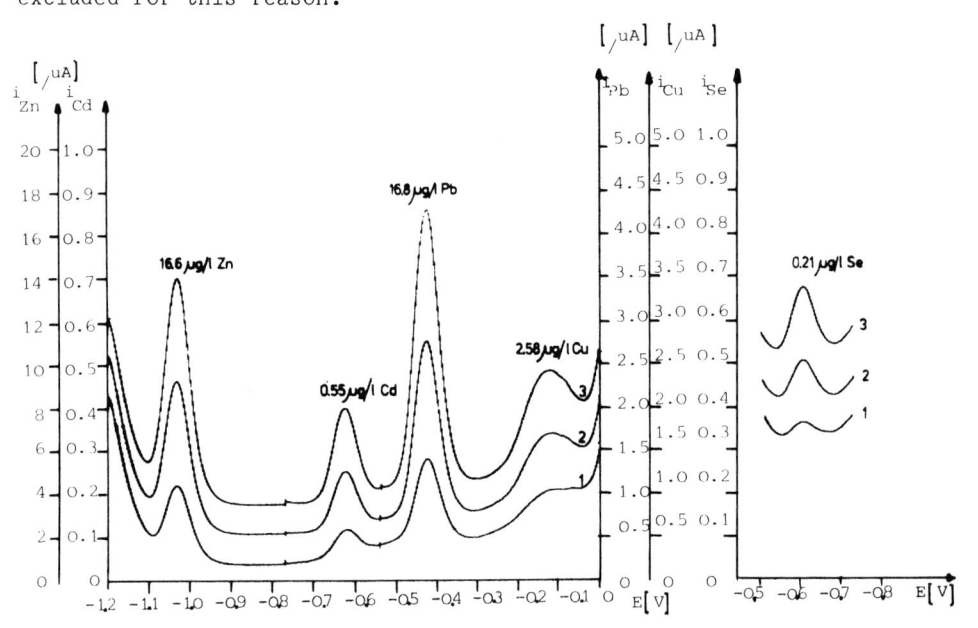

Fig. 1. Simultaneous oligo-element determination of heavy metals in rain water by DPASV at HMDE at pH 2; preconcentration time 3 min at -1,2 V (Ag/AgCl) and subsequently 5 min for DPCSV of Se(IV); 1 original analyte, 2 and 3 after first and second standard addition; total analysis time 30 - 40 min.

To speed up mass transfer towards the test electrode from the bulk of the analyte solution this is stirred or a rotating stationary test electrode is used. After the termination of the preconcentration step the stirring or rotation is stopped and in resting solution the in the interfacial zone of the test electrode accumulated trace metals are reoxidized during a voltage scan into anodic direction. The respective faradaic current (i_F) signals, due to these electrode processes of re-oxidation of the respective trace metals, are recorded and yield the peaks in the stripping voltammogram (see fig. 1). The peak heights are proportional to the concentrations of the respective trace metals in the analyte solution. As each metal has at a test electrode of given material, e.g. mercury, and for a given composition of the analyte solution a characterisitic redox potential simultaneous determinations for certain groups of ecotoxic heavy metals are possible (see table 3) and this property makes voltammetry an oligo-element method.

In the past the mentioned reoxidation had been performed by a linear voltage scan into anodic direction. During the last 15 years more and more the application of the differential pulse mode has become common use instead, turning this stripping version into differential pulse anodic stripping voltammetry (DPASV). The principle is depicted in fig. 2.

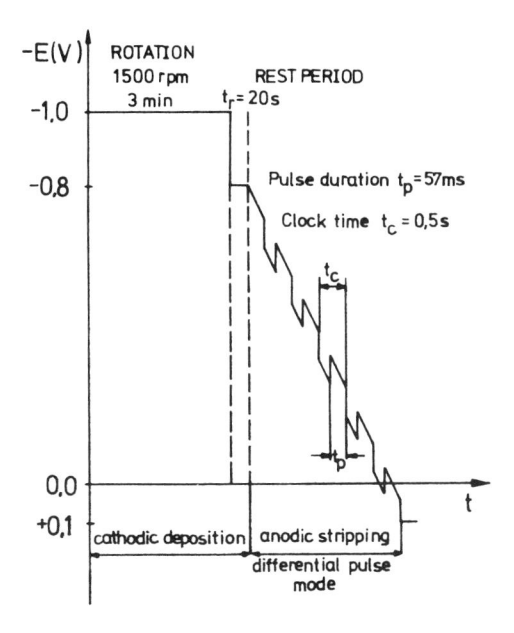

Fig. 2. Principle of differential pulse anodic stripping voltammetry (DPASV)

This combination of the differential pulse mode with its excellent signal-to-noise ratio with the powerful preconcentration concept of inverse voltammetry has opened to voltammetry the potentialities to satisfy for important heavy metals of ecotoxic relevance all analytical requirements down to the lowest ultra trace levels occurring in the environment. This is achieved with a determination limit, accuracy degree and convenience hitherto not exceeded by any other method in instrumental trace element analysis.

The application of the differential pulse mode provides also several further significant advantages. Preconcentration times can be restricted severely compared to the conventional ASV with linear dc-voltage scan. Often 1 or 2 min suffice and even at the lowest ultra trace levels 15 min are not exceeded. In this manner determination time is saved, an important practical aspect in routine analysis. Fundamentally even more important is, however, that the more moderate trace metal accumulation required, due to the high determination sensitivity of the differential pulse mode, eliminates potential sources of interferences causing accuracy problems. Thus, the moderate accumulation of trace metals retards to negligence the formation of intermetallic compounds between the plated trace metals in the mercury.

These aspects of potential interferences require particular attention if the more frequently as test electrode used hanging mercury drop electrode (HMDE) is substituted by the mercury film electrode (MFE) consisting of a thin mercury film, with a thickness of only several hundred $\overset{\circ}{A}$-units, formed on a glassy carbon substrate. The application of this MFE should be restricted to the true ultra trace range typically between 0,1 and 1000 ng/l. The fabrication of high performance mercury film electrodes providing glassy carbon disc electrodes, now also commercially available under the name Rotel-2 from EG & G, Princeton Applied Research, Munich and the handling of these electrodes in ultra trace DPASV is described in detail elsewhere (12,13).

Certain ecotoxic metals require a gold disc electrode instead of mercury. Thus, a high performance DPASV-method for the simultaneous determination of Cu and Hg has been developed (14) and successfully applied to ecochemical investigations in natural waters (14,15) and in fish (16). It is, however, added that for Hg the cold vapor mode of AAS is an easier and for routine analysis more convenient approach. The same holds for the determination of As(III) in natural waters (17) in comparison to the applicaton of the hydride AAS mode. Nevertheless, the existence of high performance voltammetric alternatives remains significant, because they provide possibilities for analytical quality control by independent analytical procedures, a basic necessity in trace element analysis to establish accuracy (18).

For some metals or metalloids of ecotoxic relevance differential pulse cathodic stripping voltammetry (DPCSV) has to be applied. An important example is Se(IV). Preconcentration is achieved at the HMDE by reduction of Se(IV) to Se^{2-} which reacts with at the same time, due to the adjusted rather positive potential, with the formed Hg^{2+} to a film of HgSe on the electrode surface. During the subsequent cathodic stripping step the electrode potential is altered in the differential pulse mode into more negative direction. The Hg(II) in the HgSe-film is reduced to elemental Hg and the Se^{2-} goes into solution. In practical analysis this DPCSV-determination of Se(IV) can be performed in the same run subsequent to the prior DPASV-determination of Cu, Pb, Cd and Zn. In this manner a simultaneous oligo-determination of 5 ecotoxic metals is obtained (19). This method has been extensively used for the investigation of rain water in studies of the wet deposition of ecotoxic metals from the atmosphere (20). It should be added, that if samples contain sufficient levels of Bi, as it can be the case in certain coastal water regions, this metal can be determined simultaneously with Cu, Pb, Cd and Zn by DPASV (21).

2.2.3. Square wave stripping voltammetry

Recently the square wave mode, a further significant achievement from the fundamental methodological key contributions of Barker (9) to modern advanced voltammetry, has been successfully introduced into practical trace analysis (22,23). In this mode the test electrode is polarized by a sequence of small square wave voltages (pulse height 20 mV; frequency 100 Hz) alternating around a mean potential growing in negative or positive direction with 2 mV steps in the stair-case manner. The method is essentially a version of ac-voltammetry but with square wave instead of sinusoidal polarization pattern. Due to the use of square wave pulses, as in the differential pulse mode, again the different time laws for the faradaic current, i_F, constituting the signal relevant for analysis, and the capacity current i_c, being the electrochemical noise component, over each square wave interval can be used and an excellent signal-to-noise ratio is achieved. As already Barker has pointed out almost 30 years ago (9), the square wave mode can be very advantageously combined with the concept of inverse voltammetry yielding square wave stripping voltammetry (SWSV). As recent investigations (23,24) have revealed, the use of SWSV is connected with several distinct advantages in practical trace metal analysis. These are related to specific properties and potentialities of this mode. The scan rate in the stair-case manner can be elevated to 200 mV/s which leads to substantial savings in determination time. For example the simultaneous oligo-element determination of Cu, Pb, Cd and Zn requires with two standard additions only 15 min in the SWSV-mode compared with 40 min by DPASV (23). As a version of ac-voltammetry the square wave mode is selectively sensitive to reversible electrode processes, a property which acts for instance favourably to achieve a good discrimination of the reversible Zn-peak from the following irreversible hydrogen evolution response (23). The SW-mode is also less sensitive to interferences by dissolved organic matter in the analyte. This makes square wave voltammetry somewhat less demanding with respect to the mineralisation completeness of digestion procedures for solid samples of biotic origin or from food (23,24). It has, however, to be emphasized that with respect to the ultimate determination sensitivity (see table 2), usually needed in the investigation of natural waters, the differential pulse mode is still superior and has then to be applied.

2.3. Adsorption Voltammetry

An in principle much wider application potentiality to many more metals has the second more recent voltammetric approach with preconcentration. Although it was designed originally for trace metals forming no stable amalgams, it can be also applied to those heavy metals, which do form stable amalgams. This approach has been termed aptly adsorption voltammetry (AV).

Hitherto it has been very successfully applied to the simultaneous determination of Ni and Co in natural waters, environmental biotic materials, food and human body fluids (25-27), see also (20,24,71) for applications in natural waters. The principle consists in the formation of trace metal chelates with a suitable organic ligand added to the analyte. In the case of Ni and Co this is dimethylglyoxime (DMG). The formed chelates are adsorbed at the surface of the HMDE adjusting the potential in the range of the zero charge potential, which is most favourable for adsorption but is situated outside the potential range of reduction of the trace metals. In this manner preconcentration is achieved. The adsorption time is adjusted to such small intervals, according to the concentration of the trace metals in the analyte, that a full coverage of the electrode surface is avoided. As mass transfer can be speeded up by stirring (900 rpm) adsorption times of a few minutes suffice. Subsequently the electrode potential is scanned in resting solution towards the more negative reduction potentials of Ni and Co and both metals are determined simultaneously. Here at higher concentration levels just a linear dc-voltage ramp can be applied permitting to use a higher scan rate and saving thus determination time. At ultra trace levels the differential pulse mode has to be used instead providing then by adsorption differential pulse voltammetry (ADPV) an extreme determination sensitivity down to 1 ng/l. In this manner new potentialities for reliable ultra trace determinations of Ni-base line levels in body fluids and human tissue of occupationally not exposed persons have been opened up (26) overcoming the hitherto existing limitations, due to the sensitivity restrictions of graphite furnace AAS.

The ADPV-method has been also used in connection with the MFE (28), although this test electrode provides no particular advantages with this mode compared to the easier manipulable HMDE. For higher Ni and Co levels, above 1 µg/l, also direct measurements by differential pulse polarography (DPP) at the DME can be used advantageously (29-31). As the preconcentration step is not necessary, analysis time is saved and the high reproducibiliy of the DME provides a good precision. A particularly high precision and reliability is provided by combining the square wave mode with the concept of adsorption voltammetry to square wave adsorption voltammetry (SWAV), as extended studies on Ni and Co in soil, biotic and human materials have demonstrated (23, 24,32).

New procedures with other appropriate chelators by linear scan adsorption voltammetry for Zn (33), Cu (34), Fe (35), V (36) and U(VI) (37) have been also reported recently. The not ecotoxic heavy metals are mentioned as well to emphasize the wide scope of adsorption voltammetry in trace metal analysis still to be explored to a much wider extent.

Very recently also a new high performance method based on ADPV has been introduced in aquatic trace metal chemistry for the determination of dissolved Cr down to 20 ng/l in all types of natural waters and in rain and snow (38). As chelator diethylenetriamine pentaacetic acid (DTPA) is applied and in addition use of the catalytic action of added nitrate is made. This new method completes the availability of high performance voltammetric methods for all ecotoxic heavy metals existing in natural waters. For a comprehensive treatment of the other ecotoxic metals in aquatic systems see also the following reviews (6,23,39,40).

2.4. Determination Limits and Precision

The concentration evaluation has to be attained in voltammetry as a rule by standard additions as it is usually the case with other instrumental methods in high performance trace element analysis. Usually two standard additions are sufficient.

Table 2. Ultimate practical determination limits in µg/l analyte solution for differential pulse voltammetric modes with preconcentration for a precision of 20% RSD (in brackets ≦10% RSD)

Metal	DPSV/HMDE	DPSV/MFE	ADPV/HMDE
Cu	0,05 (0,10)	0,007 (0,05)	–
Pb	0,02 (0,30)	0,001 (0,0015)	–
Cd	0,02 (0,10)	0,0003 (0,0015)	–
Zn	0,02 (0,50)	n. d.	–
Ni	–	–	0,001 (0,02)
Co	–	–	0,001 (0,02)
Cr	–	–	0,02 (0,1)
Se(IV)	0,1 (1,0)	n. d.	–
As(III)	0,1 (2,0)		–
Hg	0,04 (0,20)	gold disc	–
Cu	0,02 (0,10)	electrode	–

The at present attainable determination limits and the corresponding precision are listed in table 2. These are practical determination limits set by the attainable minimisation of the blanks according to the degree of efforts which will considerably increase the closer one has to come to these limits. For a detailed discussion of the necessary precautions and efforts concerning cleaning of labware and sampling equipment, purification of reagents, water quality and requirements on cleanness and permissible dust levels in the laboratory and particularly at the bench where samples are manipulated in preparatory steps and during the determination stage the reader is referred to the following references (39-41). It is further emphasized, that the given determination limits in the low ultra trace range will be only attainable by laboratories with staff having special expertise and longer lasting experience as well in voltammetry as generally in ultra trace metal analysis. The ultimately possible methodological potentialities in determination sensitivities of the stripping or the adsorption mode of voltammetry have been, however, hitherto not yet exhausted, as there was no analytical necessity for that in the environmental chemistry of ecotoxic heavy metals. From the methodological viewpoint there is still space for further lowering the determination limits by even more stringent and more cumbersome efforts to lower the blanks if this should become necessary for future problems.

2.5. Determination Rate and Other Practical Aspects

The groups of ecotoxic heavy metals for which voltammetry provides simultaneous oligo-element determinations are listed in table 3. Compared with the hitherto in trace metal analysis extensively applied and in many laboratories unduely overfavoured single element modes of AAS, voltammetry offers, if more than one metal has to be determined, distinct advantages with respect to determination rate (see table 4).

Table 3. Combinations of heavy metals determinable simultaneously by oligo-element procedures in the differential pulse or square wave mode

---
Cu, Bi, Pb, Cd, Zn and subsequently Se(IV) at the HMDE according to the stripping approach

Cu, Pb, Cd at the MFE by the stripping approach

Ni, Co at the HMDE or MFE according to concept of adsorption voltammetry

Cu, Hg at the gold disc electrode by the stripping approach

Meanwhile the convenience has been considerably increased by the availability of automated devices with microprocessor control. At present devices are available with different degrees of automation. At any rate the recording of the voltammogram occurs in an automated manner but more advanced versions perform also automatically the standard additions. Furthermore, even more advanced concepts, to be introduced, will be able to include certain preparatory steps before the voltammetric determination stage, e.g. in natural water analysis filtration, UV-irradiation of the filtrate and acidification (42,43). Obviously these automated devices will further increase the determination capacity of laboratories charged with routine analysis. Moreover, the data processing and data storage potentialities of these automated voltam-

Table 4. Comparison of analysis rate between GFAAS and voltammetric modes

Graphite furnace AAS
One measurement: 2 min

Determination with 2 standard additions: 6 min

Per working day (8 h) for 1 metal result ca. 45 determinations, minus 15 - 30 % for within-run tests to establish accuracy yields 30 - 38 determinations of 1 metal in reality

If for example 4 metals have to be determined, 7 - 9 analyses can be carried out per working day.

Differential pulse stripping voltammetry
Determination time for 4 metals simultaneously with 2 standard additions 40 min.
Per working day 10 - 12 analyses can be carried out

If, as frequently possible, the faster square wave mode is applied, the number of analyses will be doubled to 20 - 25

---

metric analysers contribute to the improvement of the precision by providing smooth and better defined base lines against which the peak heights can be evaluated with greater precision.

A not unimportant practical aspect is also the compactness of modern voltammetric instrumentation. This property is favourable in field studies with mobile laboratories and on research vessels in missions at sea or on waterways where always laboratory space is limited.

With respect to the costs voltammetry holds compared with other methods for trace element analysis a most favourable position as table 1 reflects.

Although the number of metals is restricted for which voltammetry provides a determination method with high potentialities, this number comprises all heavy metals of ecotoxic significance. Therefore the applications of voltammetry in environmental chemistry of ecotoxic heavy metals have become manifold and extensive and are further expanding rapidly.

3 APPLICATIONS

3.1. Natural Waters

For investigations on ecotoxic heavy metals in all types of natural waters and in rain and snow voltammetry is doubtless in many cases the method of first choice (6,39,40). A scheme of the complete analytical procedure with the sample pretreatment steps adapted to subsequent voltammetric determination is presented in fig. 3. The major problems

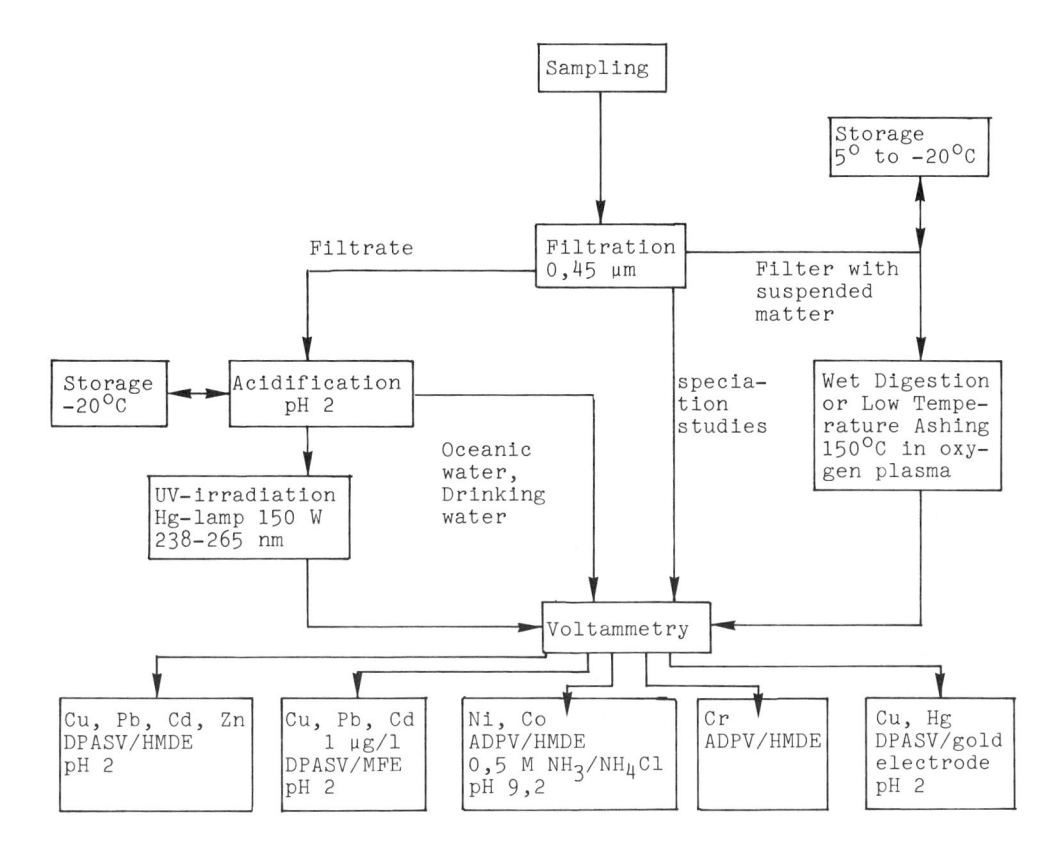

```
                          ┌──────────┐
                          │ Sampling │
                          └────┬─────┘
                               │                    ┌──────────────┐
                               │                    │   Storage    │
                               │                    │ 5° to -20°C  │
                               │                    └──────▲───────┘
                               ▼                           │
                       ┌───────────────┐    Filter with    │
         Filtrate      │  Filtration   │    suspended      │
       ┌───────────────┤   0,45 µm     ├───  matter  ──────┘
       │               └───────┬───────┘
       ▼                       │
┌───────────────┐              │          specia-       ┌─────────────────┐
│ Acidification │              │          tion          │ Wet Digestion   │
│     pH 2      │              │          studies       │ or Low Tempe-   │
└───┬───────────┘              │                        │ rature Ashing   │
    │    ▲                Oceanic                       │ 150°C in oxy-   │
    │    │                water,                        │ gen plasma      │
┌───┴────┴──┐              Drinking                     └─────────────────┘
│  Storage  │              water
│   -20°C   │
└───────────┘
    │
    ▼
┌────────────────┐
│ UV-irradiation │
│ Hg-lamp 150 W  │
│  238-265 nm    │
└────────────────┘
```

Cu, Pb, Cd, Zn | Cu, Pb, Cd | Ni, Co | Cr | Cu, Hg
DPASV/HMDE | 1 µg/l | ADPV/HMDE | ADPV/HMDE | DPASV/gold
pH 2 | DPASV/MFE | 0,5 M NH_3/NH_4Cl | | electrode
 | pH 2 | pH 9,2 | | pH 2

Fig. 3. Flow chart of analytical procedure for natural waters

and difficulties to be overcome in minimising to a satisfactory extent contamination are predominantly comprised in the stages of sampling and sample pretreatment. Only if there the with respect to the often occurring very low ultra trace levels of ecotoxic heavy metals existing very stringent requirements are met to a satisfactory degree, voltammetry can fully display its high potentialities to yield reliable results. For instance errors made in inappropriate sampling will lead to the paradox situation, that with precision highly inaccurate and consequently irrelevant data are determined. The various aspects and precautions to be considered in aquatic trace metal analysis have been the subject of various recent reviews (6,23,39,40) to which the reader is referred for further details.

Extensive applications of DPASV have provided based on reliable analytical data extensive contributions on the distribution of Cd, Pb, Cu, Zn, Ni and Co, their depth profiles and their vertical cycling in the oceans (23,44-48), on their levels and concentration regulation in coastal waters (23,49,50) and in different types of estuaries (51-55) as well as in major rivers (54-58) and lakes (59-61).

Voltammetry is a particularly suitable and efficient tool to investigate and characterize the speciation of dissolved ecotoxic heavy metals in all types of natural waters. This stems to a significant extent from the fact, that voltammetry is basically a substance specific and not just an element specific method. The potentialities for elucidating diagnostic and fundamental speciation studies have been treated in various recent reviews (62-64). In fundamental investigations on the interactions of defined ligands with various heavy metals often simultaneous oligo-measurements are possible. Extensive applications on the speciation of Cd, Pb and Zn have yielded a number of findings of general validity and significance (65-67). The kinetics and mechanism of the formation of inert strong complexes with organic ligands have been clarified (68) and the competitive role of Ca and Mg in sea water and fresh waters was elucidated (65-68). Knowledge on the significance of various organic ligand types for the speciation of the aforementioned three heavy metals in sea water has been provided and the speciation pattern of Pb and Cd with inorganic ligands prevailing in the oceans has been established (63,69), to mention some of the major recent achievements.

3.2. Atmospheric Precipitation

Particular significance has gained the simultaneous oligo-element determination of the ecotoxic heavy metals Cu, Pb, Cd, Zn and Se(IV) (19,20) and to a certain extent also of Ni and Co (20) in rain and molten snow water. These precipitates provide for these heavy metals the predominant deposition mode from the atmosphere onto the vegetation and soils as well as into lakes and into the sea. Since 1980 several thousand samples have been analyed in a program, by which the situation in the Federal Republic of Germany is observed with a network of automated wet deposition samples. Based on reliable analytical data the basic contours of the heavy metal burden caused in the various regions of the country by this deposition with precipitates as well as its seasonal fluctuations and its general trend over the last semidecade could be established (20). Special studies have elucidated the situation in the vicinity and the ambient region of a strong point source represented by a large lead smelter (70).

A fundamental study in the remote Arctic has yielded for the first time informations on the seasonal dependence of the atmospheric long distance transport of ecotoxic heavy metals from Europe into the Arctic (71). It is to be emphasized, that these findings for the heavy metals have pilot character for the atmospheric long distance transport also of other pollutants to the Arctic. The by voltammetry established base line values for ecotoxic heavy metals in the Arctic are regarded as the presently most reliable data for polar regions (72).

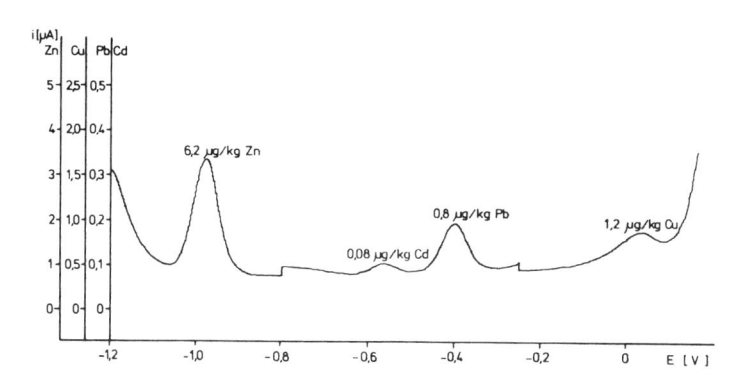

Fig. 4. Simultaneous oligo-element determination of heavy metals in drinking water at HMDE by DPASV

3.3. Drinking Water

Particularly suitable is voltammetry for the in many countries regulated control of drinking water and mineral waters on the levels of toxic heavy metals (73). Usually the only sample pretreatment required is the adjustment of the pH optimal for subsequent simultaneous oligo-element determination by voltammetry (see fig. 4). Drinking water control for toxic heavy metals will be of course a very suitable application area for fully automated voltammetric analysers (42).

3.4. Matrices Requiring Digestion

3.4.1. General Aspects

The potentialities for oligo-element determinations of certain groups of ecotoxic heavy metals (see table 3) make voltammetry also a very attractive and high performance trace analytical alternative for the investigation of all types of biotic materials from the environment, of soil, all kinds of food, human material and body fluids and of municipal waste water.

Of course, a basic prerequisite is an efficient to mineralisation leading complete digestion of the matrix. In this respect voltammetry is more demanding than AAS. This may be one of the reasons why there is in a number of laboratories still a certain reluctance to apply voltammetry in the trace metal analysis of biotic and human materials and food.

3.4.2. Waste Water

Meanwhile the favourable potentialities of DPASV subsequent to an efficient wet digestion have been convincingly demonstrated for the trace metal analysis of municipal waste water (74).

3.4.3. Food

To cope with the special demands in routine heavy metal control of food a streamlined analytical procedure (see fig. 5) with voltammetric oligo-element determinations of toxic heavy metals has been developed and its applicability to all components of the food basket has been demonstrated by an extensive study (24). For the efficient mineralisation a universally applicable wet digestion with a $HNO_3/HClO_4$ mixture is applied. With slight modifications this convenient and cheap wet digestion procedure works also satisfactory for fats, meat (75) and fish. As only small amounts of food sample are taken and therefore small amounts of HNO_3 and $HClO_4$ have to be applied, explo-

Primary Sample Treatment:
Dissection, Homogenisation, Portioning

↓

Universal Wet Digestion:
0.5 g sample in quartz dish, 0,5 ml 70 % $HClO_4$; 0.5 ml 65 % HNO_3; 100 °C until NO_2-fumes cease; 200 °C, for fats 300 °C, until light yellow colour; evaporation to white residue. Duration 0.5 - 2 h according to matrix

↓

Dissolution to analyte with $HClO_4$, pH 2

↓

Voltammetric Determination

↓ ↓ ↓

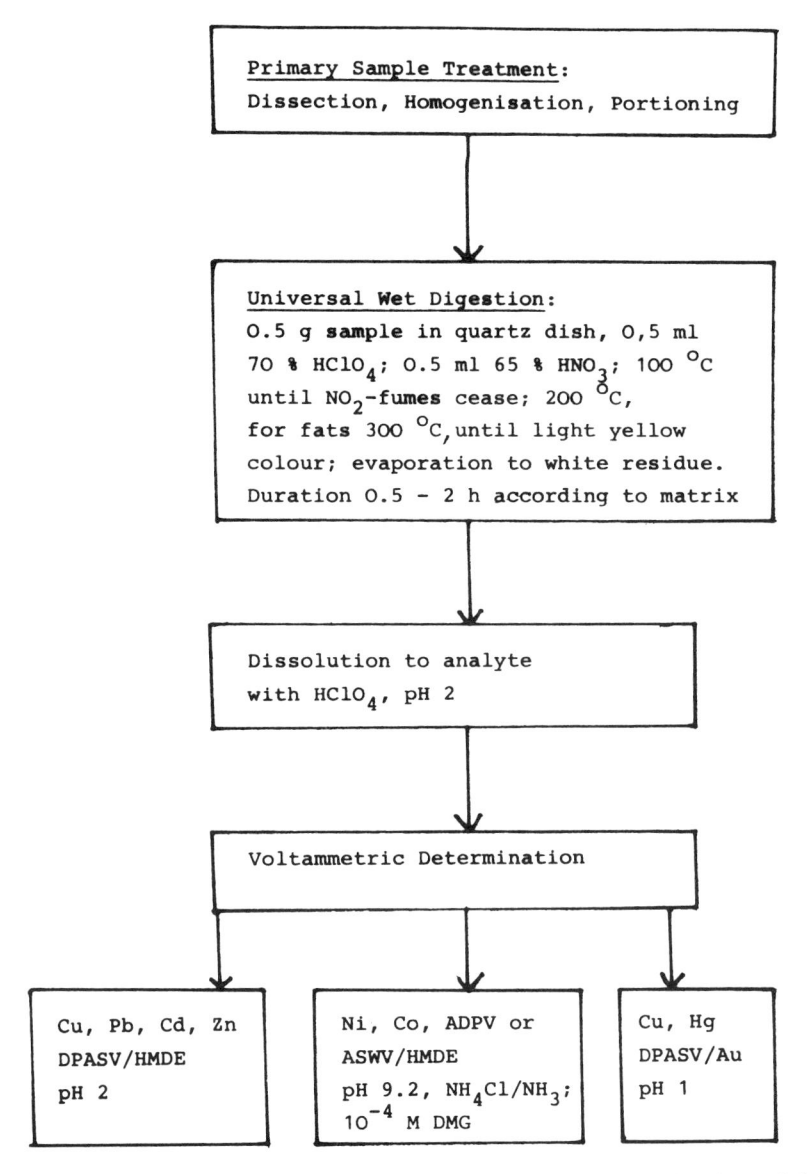

Cu, Pb, Cd, Zn	Ni, Co, ADPV or	Cu, Hg
DPASV/HMDE	ASWV/HMDE	DPASV/Au
pH 2	pH 9.2, NH_4Cl/NH_3; 10^{-4} M DMG	pH 1

Fig. 5. Flow chart of analytical procedure for food, biotic and human materials

sion risks are excluded. The argument that the small sample amount of
0.5 g might cause homogeneity problems is not relevant, as this problem cannot be settled by a factor 10 or 100 larger samples but requires before digestion appropriate homogenisation procedures.

For wines and fruits juices containing not elevated sugar levels the
wet digestion of the here dissolved organic matter can be substituted
by the more convenient UV-irradiation of the sample under addition of
some H_2O_2 (76,77). Beer, however, requires wet digestion (24).

It can be thus easily foreseen, that voltammetry will become soon an
important routine tool in the control of food for toxic heavy metals.

3.4.4. Biotic materials from the environment

The in fig. 5 presented analytical procedure has been also successfully applied to a number of biotic environmental materials in the German
Environmental Specimen Banking Program (27,78).

Particularly for the in biotic materials and food compared to natural
waters higher levels of toxic heavy metals the substitution of the
differential pulse mode by the square wave mode in form of the function incorporated into new advanced microprocessor controlled automated voltammetric analysers will provide advantages with respect to
determination rate, discrimination from interfering irreversible
responses and precision. A very important advantage is furthermore
that the square wave mode is less sensitive to interferences by
dissolved organic traces in the analyte than the differential pulse
mode. Thus, the square-wave mode makes permissible also the use of
somewhat less efficient wet digestion procedures, e.g. the commonly in
combination with AAS common pressurized wet digestion with HNO_3 (23).

3.4.5. Human materials

Growing significance is also gaining the voltammetric oligo-analysis of the considered toxic and potentially toxic metals in samples
from human organs and tissues in environmental epidemology and occupational medicine (79). For exposure base-line studies in environmental
hygiene and in case studies of risk groups blood is the usually investigated matrix. The above mentioned universal digestion procedure for
biotic materials and food can be also applied in human material in

connection with differential pulse voltammetric measurements, substituting with respect to blank levels somewhat less favourable though useful digestion procedures applied previously (26). If, however, the square wave mode is used, the popular pressurized wet digestion with HNO_3 yields also satisfactory results (23). An efficient and very reliable but not inexpensive high performance digestion alternative is low temperature ashing of blood in a micro wave induced oxygen plasma (80).

For low levels of Cd and Pb in human material oligo-determinations by voltammetry in the DPASV or SWASV mode have meanwhile become a strong competitor of graphite furnace AAS in routine analysis, due to their greater sensitivity, high reliability and favourable determination rate. Certainly the simultaneously possible determination of Ni and Co by adsorption voltammetry in the differential pulse (26) or even better in the square wave mode (23,32,81), has become the superior and preferential analytical approach. A further very important application area of voltammetry in occupational medicine and toxicology is the particularly in these fields very important analytical quality control of AAS by an independent method to prohibit the risk of wrong diagnostic medical conclusions caused by irrelevant inaccurate analytical data (18,82).

4. CONCLUDING REMARKS

In general conclusion it can be stated, that the application of advanced voltammetry, mainly in the differential pulse and square wave mode, has provided and will provide significant potentialities and major contributions to deepen the knowledge in the ecochemistry and ecotoxicology of heavy metals in manifold aspects and constitutes an efficient and reliable tool for environmental monitoring and surveillance of toxic and potentially toxic heavy metals. Doubtless voltammetry is to be regarded, therefore, as an indispensable analytical method of the basic instrumental equipment of all laboratories concerned with research and/or routine tasks in trace chemistry and trace analysis of ecotoxic heavy metals.

5 REFERENCES

(1) E. Merian (Hrsg.), Metalle in der Umwelt, Verlag Chemie, Weinheim 1984

(2) L. Friberg, G.F. Nordberg, B. Vouk (eds.), Handbook on the Toxicology of Metals, Elsevier/North Holland Biomedical Press, Amsterdam, New York, Oxford 1979

(3) M. Stoeppler, H.W. Nürnberg in ref. (1), p. 45-104

(4) M. Stoeppler, H.W. Nürnberg in A. Vercruysse (Ed.), Hazardous Metals in Human Toxicology, Elsevier, Amsterdam, Oxford, New York, Tokyo 1984, pp. 95-149

(5) H.W. Nürnberg in W.F. Smyth (Ed.), Electroanalysis in Hygiene, Environmental, Clinical and Pharmaceutical Chemistry, Elsevier, Amsterdam 1980, pp. 351-372

(6) H.W. Nürnberg, Pure Appl. Chem. 54 (1982) 853-878

(7) J. Heyrovsky, Philos. Mag. 45 (1923) 303

(8) G.C. Barker, I.L. Jenkins, Analyst 77 (1952) 685

(9) G.C. Barker, A.W. Gardner, Fresenius Z. Anal. Chem. 173 (1960) 79-82

(10) E. Parry, R.A. Osteryoung, Anal. Chem. 37 (1964) 1634-1637

(11) H.W. Nürnberg in R. Bock, W. Fresenius, H. Günzler, W. Huber, G. Tölg (Hrsg.) Analytiker-Taschenbuch, Springer-Verlag, Berlin, Heidelberg 1981, Bd. 2, pp. 211-230

(12) L. Mart, H.W. Nürnberg, P. Valenta, Fresenius Z. Anal. Chem. 300 (1980) 350-362

(13) L. Mart, H.W. Nürnberg, H. Rützel, Fresenius Z. Anal. Chem. 317 (1984) 201-209

(14) L. Sipos, J. Golimowski, P. Valenta, H.W. Nürnberg, Fresenius Z. Anal. Chem. 298 (1979) 1-8

(15) L. Sipos, H.W. Nürnberg, P. Valenta, M. Branica, Anal. Chim. Acta 115 (1980) 25-42

(16) R. Ahmed, P. Valenta, H.W. Nürnberg, Mikrochim. Acta (1981) 171-184

(17) F.G. Bodewig, P. Valenta, H.W. Nürnberg, Fresenius Z. Anal. Chem. 311 (1982) 187-191

(18) M. Stoeppler, P. Valenta, H.W. Nürnberg, Fresenius Z. Anal. Chem. 297 (1979) 22-34

(19) V.D. Nguyen, P. Valenta, H.W. Nürnberg, Sci. Tot. Environm. 12 (1979) 151-167

(20) H.W. Nürnberg, P. Valenta, V.D. Nguyen, M. Gödde, E. Urano de Carvalho, Fresenius Z. Anal. Chem. 317 (1984) 314-323

(21) P. Valenta, H.W. Nürnberg, Gewässerschutz-Wasser-Abwasser 44 (1980) 105-204

(22) J.G. Osteryoung, R.A. Osteryoung, Anal. Chem. 57 (1985) 101A-110A

(23) P. Ostapczuk, M. Froning, M. Stoeppler, H.W. Nürnberg, Fresenius Z. Anal. Chem. 320 (1985) 645

(24) P. Ostapczuk, P. Valenta, H.W. Nürnberg, Z. Lebensm. Unters. Forsch. (1985), in press

(25) B. Pihlar, P. Valenta, H.W. Nürnberg, Fresenius Z. Anal. Chem. 307 (1981) 337-346

(26) P. Ostapczuk, P. Valenta, M. Stoeppler, H.W. Nürnberg in S.S. Brown, J. Savory (Eds.) Chemical Toxicology and Clinical Chemistry of Metals, Academic Press, London, New York, Paris, San Diego, San Francisco 1983, pp. 61-64

(27) P. Ostapczuk, M. Gödde, M. Stoeppler, H.W. Nürnberg, Fresenius Z. Anal. Chem. 317 (1984) 252-256

(28) H. Braun, M. Metzner, Fresenius Z. Anal. Chem. 318 (1984) 321-326

(29) Hongyuan Chen, R. Neeb, Fresenius Z. Anal. Chem. 314 (1983) 657-659

(30) A. Meyer, R. Neeb, Fresenius Z. Anal. Chem. 315 (1983) 118-120

(31) K. Torrance, Analyst 109 (1984) 1035-1038

(32) H.F. Hildebrand, B. Roumazeille, J. Decouix, H.C. Herlant-Peers, P. Ostapczuk, M. Stoeppler in Proc. 3rd Int. Conf. Nickel Metabolism and Toxicology, Paris, Sept. 1984, Blackwells, Oxford, in press

(33) C.M.G. van den Berg, Talanta 31 (1984) 1069-1073

(34) C.M.G. van den Berg, Anal. Chim. Acta 164 (1984) 195-207

(35) C.M.G. van den Berg, J. Electroanal. Chem. 56 (1984) 2383-2386

(37) C.M.J. van den Berg, Zi Qiang Huang, Anal. Chim. Acta 164 (1984) 209-222

(38) J. Golimowski, P. Valenta, H.W. Nürnberg, Fresenius Z. Anal. Chem. (1985), in press

(39) H.W. Nürnberg, Sci. Tot. Environm. 37 (1984) 9-34

(40) H.W. Nürnberg, Anal. Chim. Acta 164 (1984) 1-21

(41) L. Mart, Talanta 29 (1982) 1035-1040

(42) P. Valenta, L. Sipos, I. Kramer, P. Krumpen, H. Rützel, Fresenius Z. Anal. Chem. 312 (1982) 101-108

(43) W. Dorten, P. Valenta, H.W. Nürnberg, Fresenius Z. Anal. Chem. 317 (1984) 264-272

(44) L. Mart, H. Rützel, P. Klahre, L. Sipos, U. Platzek, P. Valenta, H.W. Nürnberg, Sci. Tot. Environm. 26 (1982) 1-17

(45) H.W. Nürnberg, L. Mart, H. Rützel, L. Sipos, Chem. Geology 40 (1983) 97-116

(46) L. Mart, H.W. Nürnberg, D. Dyrssen in C.S. Wong, E. Boyle, K.W. Bruland, D. Burton, E.D. Goldberg (Eds.), Trace Metals in Sea Water, Plenum Press, New York, London 1983, pp. 113-130

(47) L. Mart, H.W. Nürnberg, D. Dyrssen, Sci. Tot. Environm. 39 (1984) 1-14

(48) L. Mart, H.W. Nürnberg, Experienta (1985) in press

(49) L. Mart, H.W. Nürnberg, P. Valenta in M. Branica, Z. Konrad (Eds.), Lead in the Marine Environment, Pergamon Press, Oxford 1980, pp. 155-179

(50) L. Mart, H.W. Nürnberg, Mar. Chem. (1985), in press

(51) R. Breder, R. Flucht, H.W. Nürnberg, Thalassia Jugosl. 18 (1982) 135-171

(52) P. Valenta, H.W. Nürnberg, P. Klahre, H. Rützel, A.G.A. Merks, S.J. Reddy, Mahasagar 16 (1983) 109-126

(53) P. Valenta, H.W. Nürnberg, H. Rützel, A.G.A. Merks, Vom Wasser 62 (1984) 295-306

(54) L. Mart, H.W. Nürnberg, H. Rützel, Sci. Tot. Environm. (1985) in press

(55) P. Valenta, E.K. Duursma, A.G.A. Merks, H. Rützel, H.W. Nürnberg, Sci. Tot. Environm. (1985) in press

(56) R. Breder, H.W. Nürnberg, J. Golimowski, M. Stoeppler in H.W. Nürnberg (Ed.), Pollutants and their Ecotoxicological Significance, J. Wiley, Chichester, New York 1985, pp. 205-225

(57) M. Weidenauer, K.H. Lieser, Fresenius Z. Anal. Chem. 320 (1985) 550-555

(58) J. Golimowski, A. Sikorska, Chem. Analityczna 28 (1983) 411-420

(59) L. Sigg, M. Sturm, W. Stumm, L. Mart, H.W. Nürnberg, Naturwissensch. 69 (1982) 546-548

(60) R. Breder, P. Klahre, Proc. 3rd Int. Symp. Interactions between Sediments and Water, Geneva, Aug. 1984, CEP Consult., Edinburgh 1984, pp. 113-116

(61) L. Sigg in W. Stumm (Ed.), Chemical Processes in Lakes, J. Wiley, New York, Chichester, Brisbane, Toronto, Singapore 1985, pp. 283-310

(62) H.W. Nürnberg, P. Valenta in C.S. Wong, E. Boyle, K.W. Bruland, D. Burton, E.D. Goldberg (Eds.), Trace Metals in Sea Water, Plenum Press, New York, London 1983, pp. 671-697

(63) H.W. Nürnberg, Fresenius Z. Anal. Chem. 316 (1983) 557-565

(64) H.W. Nürnberg in C.J.M. Kramer, J.C. Duinker (Eds.), Complexation of Trace Metals in Natural Waters, Martinus Nijhoff/W. Junk Publ., The Hague, Boston, Lancaster 1984, pp. 95-115

(65) H.W. Nürnberg, B. Raspor, Environm. Technol. Lett. 2 (1981) 457-483

(66) B. Raspor, H.W. Nürnberg, P. Valenta, M. Branica, Mar. Chem. 15 (1984) 217-230; 231-249

(67) P. Valenta, M.L.S. Simoes-Goncalves, M. Sugawara in C.J.M. Kramer, J.C. Duinker, Complexation of Trace Metals in Natural Waters, Martinus Nijhoff/W. Junk Publ., The Hague, Boston, Lancaster 1984, pp. 357-366

(68) B. Raspor, H.W. Nürnberg, P. Valenta, M. Branica, J. Electroanal. Chem. 115 (1980) 293-308

(69) L. Sipos, P. Valenta, H.W. Nürnberg, M. Branica in M. Branica, Z. Konrad (Eds.), Lead in the Marine Environment, Pergamon Press, Oxford 1980, pp. 61-76

(70) P. Valenta, V.D. Nguyen, R. Flucht, H.W. Nürnberg, Wissenschaft und Umwelt (1984) 211-220

(71) L. Mart, Tellus 35B (1983) 131-141

(72) E.W. Wolff, D.A. Peel, Nature 313 (1985) 535-540

(73) P. Klahre, P. Valenta, H.W. Nürnberg, Vom Wasser 51 (1978) 199-219

(74) B. Pilhar, P. Valenta, J. Golimowski, H.W. Nürnberg, Z. Wasser Abwasser Forsch. 13 (1980) 130-138

(75) H.D. Narres, P. Valenta, H.W. Nürnberg, Z. Lebensm. Unters. Forsch. 179 (1984) 440-446

(76) J. Golimowski, P. Valenta, H.W. Nürnberg, Z. Lebensm. Unters. Forsch. 168 (1979) 333-359

(77) J. Golimowski, P. Valenta, M. Stoeppler, H.W. Nürnberg, Z. Lebensm. Unters. Forsch. 168 (1979) 439-443

(78) M. Stoeppler, H.W. Dürbeck, H.W. Nürnberg, Talanta 29 (1982) 963-972

(79) H.W. Nürnberg in S. Facchetti (Ed.) Analytical Techniques for Heavy Metals in Body Fluids, Elsevier, Amsterdam 1983, pp. 209-232

(80) P. Valenta, H. Rützel, H.W. Nürnberg, M. Stoeppler, Fresenius Z. Anal. Chem. 285 (1977) 25-34

(81) P. Ostapczuk, M. Froning, M. Stoeppler, H.W. Nürnberg, Proc. 3rd Int. Conf. Nickel Metabolism and Toxicology, Paris, Sept. 1984, Blackwells, Oxford, in press

(82) M. Stoeppler, C. Mohl, P. Ostapczuk, M. Gödde, M. Roth, E. Waidmann, Fresenius Z. Anal. Chem. 317 (1984) 486-490

(83) M. Stoeppler, Nachr. Chem. Techn. Labor 33 (1985) M1-M19

Multi-Element-Analyse mittels Ionen-Chromatographie

Georg Schwedt
Institut für Lebensmittelchemie der Universität,
Pfaffenwaldring 55, D-7000 Stuttgart 80

Bernd Rössner und Da-ren Yan
Institut für Anorganische Chemie der Universität,
Tammannstr. 4, D-3400 Göttingen

ZUSAMMENFASSUNG

Für die Simultanbestimmung anorganischer Anionen werden verschiedene ionen-chromatographische Systeme (Ionenaustausch- und Ionenpaar-RP-Chromatographie) vorgestellt. Als Trennmaterialien werden polymere Ionenaustauscher, Ionenaustauscher auf Kieselgelbasis mit chemisch-gebundenen austauschfähigen Gruppen sowie ein RP-18-Material, als Eluenten wäßrige Lösungen der Salicyl- und Phthalsäure als anorganische bzw. als Tetrabutylammoniumsalze (für das RP-System) eingesetzt. Je nach Eluens erfolgt die Detektion mittels Leitfähigkeits-, RI-oder UV- Detektion (indirekte UV-Detektion).
Für die Simultanbestimmung von Kationen wurden je ein Kationenaustauscher niedriger und höherer Austauschkapazität eingesetzt. Für die Elution erwiesen sich Wein-, Oxal- und Citronensäure sowie komplexierende Gemische mit Ethylendiamin als geeignet. Die Detektion der Kationen kann nach post-chromatographischer Umsetzung mit dem Reagenzgemisch PAR-ZnEDTA sehr empfindlich photometrisch durchgeführt werden. Am Kationenaustauscher NUCLEOSIL SA-10 (chemisch-gebundene Sulfonsäure-Gruppen am Kieselgel) lassen sich mit einer Tatratlösung pH 2,75 13 Metallionen innerhalb von 34 Minuten trennen und quantitativ mit Nachweisgrenzen bis in den ppb-Bereich analysieren. Durch eine gezielte Auswahl des Eluenten kann für Analysen z.B. in Lebensmitteln die Elutionsreihenfolge so geändert werden, daß die Elution der Erdalkali- nach den Spurenelement-Ionen erfolgt.

1. EINLEITUNG

Im Unterschied zu den typischen spektrometrischen Multi-Element-Analysenmethoden haben sich bisher chromatographische Verfahren trotz vorhandener Trennmöglichkeiten für Simultanbestimmungen mehrerer Elemente kaum durchgesetzt. Die für eine empfindliche Detektion meist erforderlichen prä-chromatographischen Derivatisierungen bringen zahlreiche Probleme durch Blindwerte, unvollständige Umsetzungen und auch instabile Derivate mit sich (s. in 1).
Neue chromatographische Möglichkeiten hat dagegen die Ionen-Chromatographie gezeigt, welche in der ersten Arbeit von SMALL/STEVENS/BAUMAN 1975 (2) als Koppelung einer Ionenaustausch-Trennsäule mit einer Suppressor-Säule zur Neutralisation des Elektrolyten im Elutionsmittel und der direkt anschließenden Detektion mittels Leitfähigkeitsmessungen beschrieben wird. Die Unterdrückung der Grundleitfähigkeit des Eluenten (Säure bzw. Base oder deren Salze für Kationen bzw. Anionen-Trennungen) durch einen Ionenaustausch- bzw. Neutralisationsvorgang in einer zweiten Ionenaustausch-Säule (OH^- bzw. H^+-Form) ohne eine Verringerung der in der ersten Säule erzielten Auflösung bildet den entscheidenden Schritt für den direkten Einsatz eines Leitfähigkeits-Detektors (s. auch 3).
Heute werden unter dem Begriff "Ionen-Chromatographie" alle flüssigkeits-chromatographischen Methoden zusammengefaßt, welche die Trennung von Kationen oder Anionen ohne prä-chromatographische Derivatisierungen in unterschiedlichen chromatographischen Systemen (mit Trennmaterialien geringer Teilchendurchmesser = HPLC-Materialien) und eine direkte Detektion (UV/VIS-, RI-, Leitfähigkeits-, elektrochemische Detektion) miteinander verbinden (3).

2. ANIONEN-CHROMATOGRAPHIE

Die Ionen-Chromatographie anorganischer Anionen hat sich bereits bis in die tägliche Praxis der Routinelaboratorien durchgesetzt. Bisher standen im Gegensatz zur Kationenanalytik für Anionen-Bestimmungen nur wenige leistungsfähige Einzelbestimmungs-Methoden zur Verfügung. Die Ionen-Chromatographie ermöglicht heute je nach chromatographischem System und Detektionsmethode Simultanbestimmungen von 6 bis 9 Anionen in 10 bis 20 Minuten mit Nachweisgrenzen auch unter 1 ppm (z.B. in Wasserproben).

2.1 ANIONEN-CHROMATOGRAPHIE MIT HOHLFASER-SUPPRESSOR

Für Anionen-Trennungen werden als Trennmaterialien Harze mit quarternären Ammoniumgruppen, als Eluenten Natriumcarbonat/-hydrogencarbonat-Gemische eingesetzt. Das Verhältnis von Carbonat- zu Hydrogencarbonat-Ionen bestimmt die Elutionsstärke. Hinter der Trennsäule erfolgt nach der klassischen Art der Ionen-Chromatographie (2) in einer zweiten Ionenaustausch-Säule (H^+-Form) ein Austausch der Natrium- gegen Wasserstoffionen, wodurch die in wäßriger Lösung kaum dissoziierte Kohlensäure gebildet wird. Anstelle einer Ionenaustausch-Säule, die periodisch regeneriert werden müsste, werden heute in der kommerziellen Ionen-Chromatographie Hohlfaser-Suppressoren eingesetzt: Nach dem Gegenstromprinzip kommt der Eluent hier mit verdünnter Schwefelsäure in Kontakt, so daß ebenfalls ein Austausch von Na^+- gegen H^+- Ionen an einer Membran erfolgen kann. In beiden Fällen wird eine niedrige Grundleitfähigkeit erreicht, so daß die Leitfähigkeit getrennter Anionen gemessen werden kann.

Mit diesem Analysensystem lassen sich z.B. folgende Anionen in etwa 25 Minuten vollständig in der angegebenen Reihenfolge trennen (4) und im ppm-Bereich quantitativ analysieren:

$$\boxed{F^- - Cl^- - NO_2^- - HPO_4^- - Br^- - NO_3^- - SO_4^{2-}.}$$

Durch die Anreicherung aus Wässern in einer Vorsäule kann die Nachweisgrenze bis etwa 50 ppb (je nach Ion) verringert werden.

2.2 ANIONEN-AUSTAUSCH-CHROMATOGRAPHIE UND RI-DETEKTION

Ohne Suppressor-System lassen sich ebenfalls ionen-chromatographische Bestimmungen von Anionen durchführen, wenn anstelle der Leitfähigkeits-Detektion z.B. Änderungen des Brechungsindexes (RI) gemessen werden. Ein für Anionenanalysen geeignetes System setzt sich aus einem stark basischen Ionenaustauscher und einer Natriumsalicylat-Lösung (0,025 mol/l, pH 6,1) als Eluenten und einem RI-Detektor zusammen. Das Ionenaustauscher-Material (NUCLEOSIL SB-5) besteht aus 5 µm-Teilchen auf Kieselgelbasis mit chemisch-gebundenen quarternären Ammoniumgruppen (Austauschkapazität etwa 1 meq/g).

Folgende Anionen lassen sich voneinander trennen und mit Nachweisgrenzen zwischen 1 und 10 ppm bestimmen:

$$\boxed{HCO_3^- - H_2PO_4^- - Cl^- - NO_2^- - Br^- - NO_3^- - SO_4^{2-} - I^-.}$$

Die Gesamtanalysenzeit beträgt etwa 25 Minuten.

2.3 ANIONEN-AUSTAUSCH-CHROMATOGRAPHIE MIT INDIREKTER UV-DETEKTION

Verwendet man Ionenaustauscher geringerer Austauschkapazität (0,1 meq/g und weniger), so kann die Salzkonzentration im Eluenten verringert werden. Bei Salzen organischer Säuren, die im UV-Bereich absorbieren, besteht dann die Möglichkeit einer indirekten UV-Detektion; es wird die Abnahme der UV-Absorption infolge der Anwesenheit eines nicht UV-absorbierenden Anions gemessen. Auf diese Weise werden ionen-chromatographische Analysen mit einer 3 mmol/l Kaliumhydrogenphthalat-Lösung pH 4,5 an einer WESCAN-Anionenaustauscher-Säule (Nr. 269.001) und einer UV-Detektion bei 254 nm (z.B. mittels Interferenzfilter) folgender Anionen mit niedrigeren Nachweisgrenzen als bei 2.2 (etwa 0,5 bis 2 ppm) möglich:

$$\boxed{H_2PO_4^- \ - \ Cl^- \ - \ NO_2^- - Br^- \ - \ ClO_3^- \ - \ NO_3^- \ - \ SO_4^{2-} \ - \ I^- \ - \ HCO_3^{2-}.}$$

Bei Durchflußraten von 3 ml/min für die mobile Phase beträgt die Gesamtanalysenzeit etwa 20 Minuten. Mit einem empfindlicheren, rauschärmeren UV-Detektor lassen sich Nachweisgrenzen von 0,1 ppm erreichen (siehe z.B. 6).

2.4 ANIONEN-AUSTAUSCH-CHROMATOGRAPHIE MIT LEITFÄHIGKEITS-DETEKTION

Das unter 2.3 beschriebene Analysensystem eignet sich aufgrund der niedrigen Grundleitfähigkeit des Eluenten auch für die Leitfähigkeits-Detektion (siehe z.B. 7). Der optimale pH-Wert für eine Multi-Anionen-Analyse (bei geringstmöglicher Dissoziation des Kaliumhydrogenphthalats und bestmöglicher Auftrennung möglichst vieler Anionen) liegt bei pH 5,15.

Mit einer Gesamtanalysenzeit von 15 Minuten sind folgende Anionen trenn- und mit Nachweisgrenzen von etwa 1 ppm bestimmbar:

$$\boxed{H_2PO_4^{-*} \ - \ Cl^- \ - \ NO_2^- - Br^- \ - \ ClO_3^- \ - \ NO_3^- \ - \ SO_4^{2-} \ - \ I^- \ - \ HCO_3^- \ .}$$

* (negatives Signal)

2.5 IONENPAAR-CHROMATOGRAPHIE MIT LEITFÄHIGKEITS-DETEKTION

An Reversed-Phase-Materialien sind Trennungen anorganischer Anionen auch über eine Ionenpaarbildung, z.B. im System Tetrabutylammoniumsalicylat (pH 3,4) möglich. Auf dem RP-Material bildet sich eine Schicht aus dem Salz der mobilen Phase, zwischen dieser Schicht und den Ionen in der mobilen Phase findet ein Anionenaustausch statt (siehe z.B. 8). Geringe Konzentrationen an Tetrabutylammoniumsalicylat mit geringer Leitfähigkeit der wäßrigen Lösung ermöglichen ionenchromatographische Analysen in sehr kurzen Zeiten von etwa 12 Minuten mit Nachweisgrenzen von 1 bis 5 ppm von folgenden Anionen:

$$\boxed{HAsO_4^{2-} - H_2PO_4^- - Cl^- - Br^- - NO_3^- - ClO_3^- - I^- - SO_4^{2-}\ .}$$

Die Trennungen der ersten drei Anionen reichen jedoch für quantitative Analysen auch unter optimierten Bedingungen nicht aus.

2.6 ANIONEN-CHROMATOGRAPHISCHE SYSTEME FÜR MULTI-ELEMENT-ANALYSEN

Die besten Trennergebnisse (Basislinien-Trennungen) werden nach unseren eigenen Untersuchungen und Optimierungen am WESCAN-Anionenaustauscher (s. 2.3) mit einer 3 mmol/l Kaliumhydrogenphthalat-Lösung pH 4,5 als mobiler Phase (und indirekter UV-Detektion oder auch Leitfähigkeitsdetektion - 2.3 bzw. 2.4) mit 9 Anionen in weniger als 20 Minuten erreicht. Ausführliche Vergleiche der verschiedenen ionenchromatographischen Analysensysteme für Wasseranalysen werden von uns demnächst veröffentlicht werden (9).

3. KATIONEN-CHROMATOGRAPHIE

In der Ionen-Chromatographie von Kationen wurden zunächst ähnliche Wege wie bei der Anionen-Chromatographie beschritten: Analysen mit Suppressor-System (2), Kombinationen aus Kationenaustauscher mit niedriger Austauschkapazität und Leitfähigkeitsdetektor mit z.B. Ethylendiamintartrat als Eluens (10). Eine andere Möglichkeit besteht im Unterschied zu den Anionen in der post-chromatographischen Derivatisierung nach Trennungen an Kationenaustauschern auch mit Eluenten höherer Leitfähigkeit.
Der Metallindikator 4-(2-Pyridylazo)resorcinol (PAR) reagiert mit zahlreichen Metallionen zu gefärbten, wasserlöslichen Komplexverbindungen. An Kationenaustauschern wie Aminex A-5 (11) oder Vydac-401 (12) mit niedriger Austauschkapazität lassen sich zahlreiche Metallionen mit Hilfe wäßriger Lösungen organischer, komplexierender Säuren als Eluenten trennen und photometrisch nach post-chromatographischer Umsetzung mit PAR oder dem Ligandenaustausch-Reagenz PAR-ZnEDTA (siehe 13) bis in den ppb-Bereich bestimmen.
Unsere Arbeiten hatten zum Ziel, die Möglichkeiten einer Multi-Element-Analytik auf ionen-chromatographischem Wege an verschiedenen

Ionenaustauschern mit unterschiedlichen Eluenten und post-chromatographischer Umsetzung festzustellen.

3.1 KATIONENAUSTAUSCHER MIT NIEDRIGER KAPAZITÄT (DIONEX CS-2)

Am Kationenaustauscher DIONEX CS-2 (inerter Polymerkern mit einer Latexhülle, an welche Sulfonsäuregruppen gebunden sind) können nach einer "Technical Note" (10/July 1982) der Firma mittels eines Oxalat-/Citrat-Eluenten pH 4,35 in 48 Minuten folgende Metallionen getrennt und nach post-chromatischer Umsetzung mit PAR kontinuierlich photometrisch bestimmt werden:

$$\boxed{Fe^{3+} - (Cu^{2+}) - Ni^{2+} - Zn^{2+} - Co^{2+} - Pb^{2+} - Fe^{2+}.}$$

Wir ermittelten als optimalen Eluenten für eine Multi-Element-Analyse einen Eluenten mit 10 mmol/l Oxalat und 7,5 mmol/l Citrat pH 4,0 für folgende Ionen:

$$\boxed{Fe^{3+}/Cu^{2+} - Ni^{2+} - Zn^{2+} - Co^{2+} - Pb^{2+} - Fe^{2+}.}$$

Die Trennzeit konnte auf etwa 12 Minuten verringert werden.
Die Anwendung eines Tartrat-Eluenten (0,1 mol/l, pH 2,90) ermöglichte auch die Basislinien-Trennung von Fe^{3+}- und Cu^{2+}-Ionen, die Analysenzeit erhöhte sich hier auf etwa 20 Minuten. Anstelle des PAR-Reagenzes wurde von uns das Gemisch PAR-ZnEDTA (s. 13) verwendet.
Am gleichen Ionenaustauscher lassen sich mit Ethylendiamin-Tartrat (jeweils 2 mmol/l an Citrat bzw. Ethylendiamin) pH 4,0 in Kombination mit einem Leitfähigkeitsdetektor folgende Metallionen trennen und quantitativ analysieren:

$$\boxed{Ni^{2+} - Zn^{2+} - Mg^{2+} - Mn^{2+} - Ca^{2+} - Sr^{2+}.}$$

Die Nachweisgrenzen liegen z.B. für Ca- und Mg-Ionen bei 20 bzw. 10 ppb (Probevolumen 20 μl). Über die Optimierung dieses Analysensystems, quantitative Ergebnisse sowie Anwendungen für Ca- und Mg-Bestimmungen wird demnächst ausführlich berichtet (14).

3.2 KATIONENAUSTAUSCHER HÖHERER KAPAZITÄT (1 meq/g) UND POST-CHROMATOGRAPHISCHE UMSETZUNG

Die am Kationenaustauscher DIONEX CS-2 optimierte Tartratlösung wurde auch für Trennungen am Ionenaustauscher NUCLEOSIL SA-10 (chemisch-

gebundene Sulfonsäure-Gruppen am Kieselgel, Teilchendurchmesser 10 μm)
eingesetzt. Mit einer 0,14 mol/l Tartrat-Lösung pH 2,75 lassen sich
nach post-chromatographischer Umsetzung mit PAR-ZnEDTA insgesamt 13
Metallionen erfassen:

$Fe^{3+} - Cu^{2+} - Pb^{2+} - Zn^{2+} - Ni^{2+} - Co^{2+} - Cd^{2+} - Fe^{2+} - Ca^{2+} - Mn^{2+} - Mg^{2+} - Sr^{2+} - Ba^{2+}.$

Die Gesamtanalysenzeit beträgt 34 Minuten. Die Trennungen von Mg/Sr
und Ca/Mn sind jedoch nicht voll befriedigend. Im Hinblick auf Anwendungen für Umwelt- und Biomatrices mit hohen Konzentrationen vor
allem an Erdalkalien ist eine andere Reihenfolge wünschenswert.
Mit Hilfe des Eluenten Citrat/Ethylendiamin konnte dieses Ziel erreicht werden, wobei in der Umwelt in Spuren vorkommende Metalle vor
den Erdalkalien eluiert werden:

$Fe^{3+} - Ni^{2+} - Zn^{2+} - Fe^{2+} - Cd^{2+} - Mn^{2+} - Ca^{2+} - Mg^{2+} - Sr^{2+}.$

Der optimale Eluent enthält nach unseren Untersuchungen 20 mmol/l
Citrat und 2 mmol/l Ethylendiamin bei pH 3,8, die Analysenzeit beträgt etwa 22 Minuten.
Eine selektive Spurenanalyse essentieller Metalle neben den Erdalkalien ermöglicht ein Elutionsgemisch aus jeweils 10 mmol/l Citrat
und Oxalat sowie mit 2 mmol/l Ethylendiamin pH 3,4. Die Elutionsreihenfolge mit einer Gesamtanalysenzeit von 22 Minuten lautet:

$Zn^{2+} - Fe^{2+} - Mn^{2+} - Mg^{2+} - Ca^{2+} - Sr^{2+} - Ba^{2+}.$

Die Eichmessungen ergaben für alle Metallionen Linearitäten über mehr
als eine Zehnerpotenz der Konzentration für den Bereich von 0,5 bis
20 ppm.
Die relativen Standardabweichungen betrugen für Konzentrationen
zwischen 1 und 10 ppm je nach Konzentration und Element der zuletzt
aufgeführten Metallionen-Reihe 1 bis 2 %.
Die Bestimmung anderer essentieller Elemente neben den Erdalkalien
ist mit Hilfe eines Elutionsgemisches aus 1,5 mmol/l Oxalsäure und
2 mmol/l Ethylendiamin pH 3,95. Die Elutionsreihenfolge mit einer
Gesamtanalysenzeit von 24 Minuten lautet:

$Fe^{3+} - Cu^{2+} - Zn^{2+} - Ni^{2+} - Fe^{2+} - Mn^{2+} - Mg^{2+} - Ca^{2+}.$

Erste Anwendungsbeispiele für Getränke (Apfelsaft, Rotwein), Haferflocken, Tee und Wasserproben zeigen die prinzipielle Anwendbarkeit dieser ionen-chromatographischen Verfahren für eine gezielte, der Element-Zusammensetzung mit Hilfe des Eluenten angepaßten Analysensystems aus HPLC-Gerät, HPLC-Kationen-Austauscher-Trennsäule, peristaltischer Pumpe für die post-chromatographische Derivatisierung (mit Glasmischspiralen zum Transport des luftsegmentierten Flüssigkeitsstromes) und einem Durchfluß-Photometer. Die Probenvolumina können von 20 µl auf mindestens 200 µl ohne wesentliche Veränderung in der Auflösung erhöht werden - damit werden auch Spurenanalysen im ppb-Bereich möglich. Eine ausführliche Arbeit über diese Untersuchungen wird von uns demnächst erscheinen (15).

4. AUSBLICK

Ionenaustauscher auf der Basis von HPLC-Trennmaterialien, selektive und variable Eluentensysteme sowie verschiedene Detektionssysteme vom Leitfähigkeits-, über den RI-Detektor bis zum UV-/VIS-Spektralphotometer ermöglichen sowohl Anionen- als auch Kationen-Simultanbestimmungen. Die Anionen-Chromatographie hat bereits einen festen Platz in der Routineanalytik der Wasseruntersuchungen gefunden, für die Kationen-Chromatographie zeichnen sich ähnliche Anwendungsmöglichkeiten aufgrund der bisher vorliegenden Untersuchungen auch für Geo- und Biomatrices ab. Die Kationen-Chromatographie mit post-chromatographischer Derivatisierung besitzt vor allem auch eine orientierende Funktion in Form einer Übersichtsanalyse. Über die Zuverlässigkeit in den Spurenanalysen lassen sich aufgrund der bisher vorliegenden Ergebnisse noch keine sicheren Aussagen machen. Vergleiche mit Analysen, die mit der AAS und Polarographie/Voltammetrie durchgeführt wurden, sind geplant.

LITERATUR

(1) Schwedt, G., Chromatographic Methods in Inorganic Analysis. Hüthig: Heidelberg, Basel, New York 1981
(2) Small, H., Stevens, T.S., Bauman, W.C., Anal.Chem. $\underline{47}$ (1975), 1801 - 1809
(3) Schwedt, G., LaborPraxis $\underline{8}$ (1984), 30-42
(4) Williams, R.J., Anal. Chem. $\underline{55}$ (1983), 851 - 854
(5) Buytenhuys, F. A., J. Chromatog. $\underline{218}$ (1981), 57 - 64
(6) Cochrane, R.A., Hillman, D.E., J. Chromatog. $\underline{241}$ (1982), 392 - 394
(7) Jupille, T.H., Togami, D.W., Burge, D.E., Chromatographia $\underline{16}$ (1982), 312 - 316
(8) Iskandarani, Z., Pietrzyk, D.J., Anal. Chem. $\underline{54}$ (1982), 2427 - 2431
(9) Rössner, B., Schwedt, G.,   Anal. Chim. Acta in Vorb.
(10) Sevenich, G.J., Fritz, J.S., Anal. Chem. $\underline{55}$ (1983) 12 - 16
(11) Cassidy, R.M., Elchuk, S., McHugh, J.O., Anal. Chem. $\underline{54}$ (1982), 727 - 731
(12) Hwang, J.-M., Chang, F.-Ch., J. Chinese Chem. Soc. $\underline{30}$ (1983), 167 - 172
(13) Jezorek, J.R., Freiser, H., Anal. Chem. $\underline{51}$ (1979), 373 - 376
(14) Yan, D.-R., Schwedt, G., Fresenius Z.Anal.Chem. in Vorb.
(15) Yan, D.-R., Schwedt, G., Fresenius Z.Anal.Chem. in Vorb.

Sonstige Analysenschritte

CHEMISCHE MULTI-ELEMENTANREICHERUNG -
PROBLEME DER ANPASSUNG AN PROBENMATERIAL UND BESTIMMUNGSMETHODE

E. Jackwerth

Ruhr-Universität Bochum, Abteilung für Chemie,
Postfach 10 21 48, D-4630 Bochum 1

ZUSAMMENFASSUNG

Verfahren zur chemischen Multi-Elementanreicherung beruhen im allgemeinen auf Trennsystemen, in denen die zu bestimmende Elementpalette in Umfang und Ausbeuten sowohl von der Gleichgewichtslage zahlreicher Komplexreaktionen als auch von der Ausprägung der zur Abtrennung geeigneten physikalischen Eigenschaften der Spuren- oder der Matrixverbindungen abhängt. Solche Eigenschaften sind die Flüchtigkeit, Extrahierbarkeit, Schwerlöslichkeit, Sorbierbarkeit oder Flotierbarkeit der abzutrennenden Spurenkomponenten.

Von Bedeutung für den Verbund einer Anreicherungstechnik mit verschiedenen instrumentellen Multi-Elementmethoden sind neben dem Aggregatzustand des anfallenden Spurenkonzentrats die bei jeder Methode unterschiedlichen Störungen durch nicht abtrennbare Matrixanteile oder durch Reste der zur Anreicherung erforderlichen Reagenzien. Ebenso kann der Aufwand, mit dem Konzentrate von gleichmäßiger und dünner Schicht bzw. von kleinem Volumen erhalten werden, die zudem weitgehend frei von Blindwerten sind, die Anwendbarkeit einer Methode zur Multi-Elementanreicherung beeinflussen. Störungen bei der Spurenanalyse entstehen z. B. durch Elemente oder Verbindungen, deren Signale mit denen der Spuren interferieren oder die einen unkompensierbar hohen und streuenden Untergrund verursachen. Die Mehrzahl der nachweisstarken instrumentellen Bestimmungsmethoden wird außerdem schon durch geringe Gehalte an organischen Stoffen, die z. B. von unvollständigen Aufschlüssen,

Extraktionsphasen, Ionenaustauschern oder Komplexbildnern herrühren, empfindlich gestört.

Wesentlich für die Richtigkeit der Analyse ist eine zuverlässige Kalibrierung von Anreicherungs- und Bestimmungsprozeß. Da gut analysierte Kalibrierproben für die spurenanalytische Multi-Elementbestimmung nur für wenige Materialien verfügbar sind, werden Analysenverfahren oft unter Einsatz von Modell-Lösungen entwickelt, die man dem zu analysierenden Probenmaterial nachgebildet hat. Allerdings zeigt es sich häufig, daß Probe und Modell mit abnehmenden Spurengehalten einander immer unähnlicher werden, so daß systematische Fehler entstehen und eine Übertragung der Ergebnisse schließlich nicht mehr möglich ist. Durch mangelnde Artgleichheit bedingte Probleme entstehen auch bei der Ermittlung von Anreicherungsausbeuten und Nachweisgrenzen sowie bei der Anwendung von Standard-Additionstechniken.

ALLGEMEINE GESICHTSPUNKTE

Verfolgt man die analytische Literatur der vergangenen zwei Jahrzehnte, so erkennt man einen starken Anstieg in der Zahl von Veröffentlichungen zur Entwicklung und Anwendung spurenanalytischer Multi-Elementverfahren. Sichere Auslöser für die Häufung solcher Arbeiten sind jeweils Publikationen über neuartige instrumentelle Bestimmungsmethoden und die nachfolgende Verfügbarkeit entsprechender Geräte im Handel. Deutliche Beispiele hierfür sind die Entwicklung der Atomabsorptions-Spektrometrie mit ihren zahlreichen Varianten und in neuerer Zeit die der Plasma-Emissionsmethoden. In gleicher Weise können gesetzliche Auflagen die analytische Produktivität erheblich verstärken, wie dies z. B. nach der Verabschiedung der Trinkwasserverordnung und des Abwasserabgabengesetzes in der Bundesrepublik erkennbar wurde.

Die wohl wichtigsten Einsatzgebiete für Multi-Elementverfahren, bei denen bereits routinemäßig bis zu 10 oder sogar mehr Spurenbestandteile quantitativ erfaßt werden müssen, sind neben der Reinststoffanalyse die Durchschnittsanalyse metallischer und nicht-metallischer Werkstoffe. Hier kann man feststellen, daß der Umfang der zu bestimmenden Elementpalette in dem Maße anwächst, wie neue Erkenntnisse einen erwünschten oder unerwünschten Einfluß von Spurenverunreinigungen auf die Materialeigenschaften aufzeigen. Ebenso bewirkt der zunehmende Einsatz von Recycling-Produkten eine Ausweitung des Analysenumfangs auf die Bestimmung eingeschleppter Fremdelemente, die häufig genug für die jeweiligen Werkstoffe als exotisch anzusehen sind. Auch die systematische

Prospektierung von Lagerstätten, die landwirtschaftliche Bodenuntersuchung und das überaus vielfältige Gebiet der Umweltkontrolle verlangen heute Multi-Elementanalysen an zumeist äußerst umfangreichen Probenserien. Ein vollständiger Katalog aller Disziplinen, in denen spurenanalytische Methoden notwendig oder wenigstens hilfreich sind, ist kaum noch zu erstellen; die Bandbreite erstreckt sich von Fragestellungen der Archäometrie über praktisch alle naturwissenschaftlichen und technischen Fachgebiete bis hin zu speziellen Problemen im Bereich der medizinischen Diagnostik.

Ziel der Entwicklung der modernen Multi-Elementanalytik ist zweifellos die von jeder chemischen Probenvorbereitung freie und von Instrumenten weitgehend selbsttätig durchgeführte Spurenbestimmung. Sofern die zu bestimmenden Elementkonzentrationen den unteren $\mu g \cdot g^{-1}$-Bereich nicht wesentlich unterschreiten, kommen einige atomspektroskopische und radiochemische Methoden diesem Wunsch bereits sehr nahe. Für Durchschnittsanalysen in noch geringeren Konzentrationsbereichen sowie bei vielen komplex zusammengesetzten Probenmaterialien ist es heute oft jedoch noch unverzichtbar, im Verbund mit der eigentlichen Spurenbestimmung zusätzliche Analysenschritte wie Homogenisieren, Aufschließen und Lösen, Trennen und Anreichern vorzuschalten. Solche Schritte haben den Zweck, störende Begleitstoffe zu entfernen, die Proben in eine für den Analysenprozeß geeignete Form zu bringen und dadurch ein für das Analysenergebnis notwendiges Maß an Richtigkeit, Präzision und Nachweisvermögen zu gewährleisten. Am ehesten werden alle Anforderungen an die Leistungsfähigkeit eines Verbundverfahrens erfüllt, wenn es gelingt, von einer genügend großen Probeneinwaage ausgehend, möglichst viele der analytisch interessanten Spuren mit hohen Ausbeuten abzutrennen und sie in ein weitgehend matrixfreies Spurenkonzentrat zu überführen. Erfahrungsgemäß können dabei nur in wenigen Fällen alle für die Spurenanalyse eines Materials wichtigen Elemente aus einer einzigen Einwaage heraus gleichzeitig angereichert werden. Dazu unterscheidet sich die chemische Reaktivität der Spuren im Anreicherungsprozeß meist zu sehr von der gemeinsamen Stoffeigenschaft, die das analytische Interesse hervorruft, wie z. B. ihrer Toxizität im Abwasser, ihrer Farbgebung bzw. Trübung in einem Lichtleiter oder ihrer die Korrosion eines Metalls fördernden Wirkung.

Als besonders nützlich erweisen sich erprobte Verbundverfahren zur gelegentlichen Richtigkeitskontrolle der mit Analysenautomaten gewonnenen Daten. Wegen ihrer oft unproblematischen Kalibrierung sind sie vor allem auch für die Ermittlung der Elementgehalte von Referenzmaterialien geeignet. Schließlich eröffnet eine leistungsfähige Multi-

Elementanreicherung im Verbund z. B. mit der Photometrie, der Polarographie oder der Flammen-AAS auch vielen kleineren, einfach ausgestatteten Laboratorien den Zugang zur sequentiellen Bestimmung von Elementen in geringer Konzentration.

Zur chemischen Multi-Elementanreicherung haben sich eine Reihe meist schon aus der klassischen Analyse her bekannter Trenntechniken bewährt: Methoden wie die Extraktion, Fällung, Verflüchtigung oder der Ionenaustausch herrschen in der Häufigkeit ihrer Anwendung vor. Die Bewertung der für eine definierte Analysenaufgabe vorgesehenen Methode kann dabei von mehreren Randbedingungen mitbestimmt werden, die mit ihrer analytischen Leistungsfähigkeit nur wenig zu tun haben. Dies gilt etwa für betriebsinterne Kosten-Nutzen-Vereinbarungen oder für die personelle und apparative Ausstattung des Laboratoriums. Im folgenden soll auf einige bei der Auswahl von Anreicherungsverfahren zu bedenkende Wechselbeziehungen eingegangen werden, die durch die chemischen Eigenschaften verschiedenartiger Probenmaterialien sowie durch die individuellen Anforderungen entstehen, die jedes instrumentelle Bestimmungsverfahren an die Beschaffenheit der durch Anreicherung gewonnenen Spurenkonzentrate stellt. Überlegungen dieser Art müssen natürlich nicht nur bei der Anwendung, sondern bereits bei der Entwicklung chemischer Analysenschritte angestellt werden, damit sie später problemlos an ein leistungsfähiges Bestimmungsverfahren angepaßt werden können.

Eine schon zu Beginn jeder Durchschnittsanalyse zu stellende Frage betrifft die Größe der Mindesteinwaage, die erforderlich ist, um Inhomogenitäten des Materials nach Auflösen der Probe sicher auszumitteln und um das Nachweisvermögen des Verbundverfahrens für alle interessierenden Spuren genügend weit zu überschreiten. Ganz abgesehen davon, daß eine entsprechende Probenmenge überhaupt verfügbar sein muß, ist zu überprüfen, ob und in welcher Weise sie im Analysenprozeß verarbeitet werden kann. Es gibt Beispiele dafür, daß selbst Teilmengen von 100 g nach ihrer durch Auflösen erzwungenen Homogenisierung noch nicht ausreichen, um für das Gesamtmaterial repräsentativ zu sein (1).

Auf die Notwendigkeit zur Beachtung einer besonderen Regel bei der Bestimmung von Restverunreinigungen in präparativ aufgereinigten Materialien hat ZOLOTOV hingewiesen. Danach sollte das zur Spurenanreicherung angewandte Verfahren in seinem Trennmechanismus sich deutlich von dem des Reinigungsprozesses unterscheiden: Bei einem durch Zonenschmelzen gereinigten Einkristall ist es z. B. nur wenig effektiv, die durch diese Methode nicht entfernten Spuren durch Anwendung eines weiteren Zonenschmelzverfahrens analytisch anzureichern (2).

Als selbstverständlich sollte man schließlich voraussetzen, daß der durch Blindwerte und deren Streuungen verursachte Einfluß auf Richtigkeit und Nachweisgrenze eines Analysenverfahrens sorgfältig beachtet wird. Für die Auswahl von Anreicherungsverfahren bedeutet dies, daß bei abnehmender Konzentration der zu bestimmenden Spuren die Ansprüche an die Begrenzung der Blindwerte soweit steigen können, daß damit zugleich ein Ausschluß aller Anreicherungstechniken verbunden ist, bei denen größere Reagenzmengen, etwa zur Maskierung der Matrix, unvermeidbar sind.

EINFLUSS DER MATRIXKOMPONENTEN AUF DIE ANWENDBARKEIT
VON ANREICHERUNGSMETHODEN

Jede Art von Probenmaterial ist in ihrem spurenanalytischen Verhalten ein Individuum. Das ist der Grund dafür, daß die problemlose Übertragung spurenanalytischer Arbeitsvorschriften von einem Material auf ein anderes fast nie gelingt. Diese Erfahrung hat zu einer Regel geführt, nach der Aussagen über die analytischen Eigenschaften von Spurenkomponenten ausschließlich auf dasjenige System aus definiertem Probenmaterial und vollständiger Arbeitsvorschrift zu beschränken sind, für das diese Eigenschaften experimentell ermittelt wurden und reproduziert werden können. Jegliche Verallgemeinerung oder Extrapolation auf andere Materialien oder Analysenbedingungen sind also unzulässig, und zwar umso mehr, in je geringeren Konzentrationsbereichen zu analysieren ist. Dabei gilt diese Regel nicht nur für alle Bereiche der "naßchemischen" Probenvorbereitung, sondern mit gleicher Strenge auch für jedes instrumentelle spurenanalytische Bestimmungsverfahren.

In der Praxis bedeutet das, daß die Übertragung einer erprobten Methode zur Multi-Elementanreicherung auf ein in seiner Zusammensetzung verändertes Probenmaterial jeweils mit der Entwicklung einer neuen Arbeitsvorschrift und damit eines neuen Verfahrens verbunden ist. Ist in der neuen Probe lediglich das Konzentrationsverhältnis der Hauptkomponenten über den in der alten Arbeitsvorschrift tolerierten Umfang hinaus geringfügig verändert, so kann dies ohne Leistungsverlust des bisherigen Verfahrens oft allein durch erneutes Optimieren der Parameter korrigiert werden. Auch bei der Spurenanreicherung aus organischen bzw. biologischen Matrices kann man häufig gleichartige Anreicherungsverfahren bei verschiedenartigen Probenmaterialien anwenden: Erfolgt die Anreicherung aus organischem Material z. B. lediglich durch Druckaufschluß und Einengen der resultierenden Spurenlösung, so ist prinzipiell die gleiche Methodik anwendbar, ob es sich um Mehl oder Fleisch handelt, bzw. ob im Probenmaterial bestimmte aufschließbare organische

Verbindungen enthalten sind oder nicht. Allenfalls müssen die Aufschlußparameter, wie Art und Menge des Aufschlußreagenzes, Temperatur und Reaktionszeit, dem veränderten Material angepaßt werden. Ebenso ist es meist ohne Belang für die Funktionstüchtigkeit eines Anreicherungsverfahrens, ob eine Probe Natrium- oder andere Alkaliionen enthält usw.

Werden allerdings Haupt- oder Nebenbestandteile eines Probenmaterials durch andere Komponenten ersetzt, bzw. werden einem Material neue Bestandteile hinzugefügt, die aufgrund ihres individuell ausgeprägten Reaktionsverhaltens im Anreicherungsprozeß das gesamte Trennsystem verändern, so ist eine gleichwertige Multi-Elementanreicherung meist nur noch nach Auswechseln der Methode oder sogar des Trennprinzips möglich. Besonders deutlich wird dies jeweils, wenn als zusätzliche Matrixkomponente ein Element aufgenommen wird, das - entsprechend der AHRLAND-CHATT-DAVIES-Klassifizierung (3 - 5) - einer Gruppe mit völlig anderer Komplexbildungstendenz zuzuordnen ist. Fälle dieser Art sind für die Spurenanalyse von Legierungssystemen besonders typisch: So können in Legierungen des Zinks mit den Nebenbestandteilen Kupfer oder Cadmium die Matrixkomponenten als Amminkomplexe maskiert und die Spuren z. B. durch Kollektorfällung angereichert werden. Innerhalb der obigen Klassifizierung gehören Zn, Cu und Cd gemeinsam zur Gruppe der "b-Kationen". Dotiert man eine dieser Zinklegierungen zusätzlich mit Aluminium oder Titan ("a-Kationen"), so kann das Verfahren nicht mehr angewandt werden, weil die Hydroxide dieser Elemente in ammoniakalischer Lösung ausfallen und einen Teil der Spuren mitreißen. Trotzdem ist es noch möglich, die ursprüngliche Methodik der Anreicherung - die Kollektorfällung der Spuren - auch bei dem veränderten Legierungstyp anzuwenden, wenn ein Verfahren gewählt wird, bei dem die Spuren aus saurer Lösung abgetrennt werden können.

Anders ist dies bei der Multi-Elementanreicherung aus Antimon bzw. aus Blei-Antimon-Legierungen. Während es bei reinem Antimon noch gelingt, eine Reihe von Elementspuren als Dithizonchelate extraktiv abzutrennen (6), verhindert die Gegenwart von Blei in höherer Konzentration praktisch jede Möglichkeit der Anwendung von Chelatbildnern zur Multi-Elementabtrennung der Spuren. Obwohl Sb^{3+} und Pb^{2+} zu derselben Gruppe der "b-Kationen" gehören, ist die allgemeine Tendenz des Bleis zur Bildung stabiler Chelatkomplexe wesentlich stärker ausgeprägt, und zwar so sehr, daß keines der gebräuchlichen analytischen Gruppenreagenzien mehr geeignet ist, um Spuren in größerer Palette für eine Anreicherung zu komplexieren, ohne daß das Blei störend einwirkt. Eine Multi-Elementanreicherung im System Blei-Antimon gelingt gegenüber reinem Antimon also nur noch durch Wechsel des Trennprinzips: Anstelle der Spuren müssen jetzt die Matrixkomponenten selektiv chemisch umgesetzt und abgetrennt werden. Dies ist z. B. durch die Ausfällung in Form der Bromide bzw. Oxibromide möglich, wobei zahlreiche, für die Qualitätskontrolle solcher Legierungen wichtigen Elemente mit hohen Ausbeuten angereichert im Filtrat verbleiben (7).

Wegen der erwähnten allgemeinen Tendenz mancher Elemente zur Bildung stabiler Komplexe ist das Prinzip der Matrixabtrennung durch Fällung in Form einfacher anorganischer Verbindungen oder durch Extraktion als Halogeno-Solvens-Mischkomplexe u. dgl. bei Probenmaterialien mit Hauptbestandteilen wie Ag, Au, Bi, Fe, Hg, In, Tl usw. oft die einzig sinnvolle Möglichkeit zur Multi-Elementanreicherung. Solche Methoden können in bezug auf ihre einfache Handhabung und hohe Selektivität von großem Wert sein; von der Blindwertbelastung her sind sie den Chelatreaktionen der Spuren allerdings oft unterlegen.

Unter den Schwermetallen gibt es eine Reihe von Elementen, die als Hauptkomponenten ebenfalls die Chelatbildung vieler Spuren empfindlich stören, selbst aber nicht alternativ durch Matrixfällung oder -extraktion entfernt werden können, weil selektiv abtrennbare schwerlösliche bzw. extrahierbare Verbindungen bei diesen Metallen fehlen. Dies gilt z. B. für Cd, Co, Cu, Ni und Zn. Chelatbildner kommen für die Matrixabtrennung dieser Elemente allein schon aus Kostengründen nicht infrage, es sei denn, man fällt einen geringen Teil der Matrix als Kollektor für alle Spuren aus, die unter den Fällungsbedingungen ebenfalls schwerlösliche Verbindungen bilden oder die aufgrund anderer, meist nicht näher bekannter Sorptionsmechanismen mitgerissen und dadurch angereichert werden. Für die Analyse von Feinzink können z. B. mehr als 10 Elementspuren abgetrennt werden, indem man nach Lösen der Metallprobe einen geringen Teil des Zinks als Dithiocarbamidatkomplex ausfällt (8). In ähnlicher Weise kann als Kollektor auch ZnS benutzt werden. Da die Löslichkeitsprodukte vieler Metalldithiocarbamidate bzw. -sulfide wesentlich kleiner sind als die der entsprechenden Zinkverbindungen, ist die bei der Anreicherung durch "Anfällen" der Matrix erfaßbare Zahl von Elementspuren verhältnismäßig groß. Benutzt man zum Anfällen von Matrixelementen das starke Reduktionsmittel Natriumtetrahydridoborat $NaBH_4$, mit dem aus einer Reihe von Metallsalzlösungen Kollektoren in feinverteilter metallischer Form bzw. als Borid ausgeschieden werden, so erfaßt man jeweils alle Spuren, die elektrochemisch edler sind als der Kollektor und zusätzlich solche Elemente, die schwerlösliche Boride bilden. Aus Lösungen mit 10 g Ni^{2+} können auf diese Weise z. B. wenige μg Ag, Au, Bi, Cd, Cu, In, Pb, Pd, Pt, Sn und Tl mit Ausbeuten > 95 % angereichert werden. Anwendbar ist diese Technik der partiellen reduktiven Matrixfällung auf Proben mit Hauptbestandteilen von Ag, Bi, Co, Cu, Ni, Pb und Sn (9).

Probleme mit der Auswahl geeigneter Gruppenreagenzien und dementsprechend mit der Trenn- und Anreicherungstechnik entstehen auch dann, wenn das Probenmaterial Matrixkomponenten enthält, die unter den für viele Chelatbildner optimalen neutralen oder schwach ammoniakalischen Reaktionsbedingungen bereits hydrolysieren. Vor allem die Ausfällung der Hydroxide höherwertiger Metallionen wie Sb, Sn, Ti usw. beginnt schon bei pH-Werten, bei denen eine Chelatbildung der anzureichernden Spuren oft noch nicht nachzuweisen oder zumindest nicht vollständig ist. Mit zunehmender Konzentration solcher hydrolyseempfindlichen Komponenten verschiebt sich der Fällungsbeginn zudem zu kleineren pH-Werten. Von besonderer Bedeutung für zahlreiche Analysenaufgaben sind deshalb Gruppenreagenzien, die schon in sehr sauren Lösungen zur Spurenanreicherung eingesetzt werden können. Hierzu gehört etwa das Hexamethylenammoniumsalz der Hexamethylendithiocarbaminsäure (HMA-HMDTC), mit dem selbst in Lösungen, die mehr als 1 M an Säure sind, eine große Gruppe von Metallionen zu schwerlöslichen und extrahierbaren Chelaten reagiert (10). Auch in der Säurebeständigkeit des Chelatbildnermoleküls ist dieses Reagenz dem bekannten Natrium-diethyldithiocarbamidat und analogen Verbindungen stark überlegen.

Insgesamt bietet die Anwendung chelatbildender Gruppenreagenzien die vermutlich vielfältigsten Möglichkeiten für eine gewollt unselektive Multi-Elementanreicherung. Zu den wegen ihrer großen Bandbreite an komplexierbaren Elementen wichtigsten Verbindungen gehören die schon erwähnten, in den Substituenten sich unterscheidenden Dithiocarbamidate, weiterhin die Chelatbildner 8-Hydroxychinolin, Dithizon, Cupferron und 1-(2-Pyridylazo)-2-naphthol (PAN) sowie die 1,3-Diketone Acetylaceton, Trifluoracetylaceton und Thenoyltrifluoraceton. In der Literatur werden mit solchen Reagenzien Verfahren zur gemeinsamen Abtrennung von 20 und mehr Spuren aus zahlreichen Probenmaterialien beschrieben. Dies gilt insbesondere für Stoffe, die komplexchemisch unproblematische Hauptkomponenten enthalten, also z. B. für Alkali- und Erdalkaliverbindungen, für die Metalle und Metallsalze des Magnesiums, Aluminiums, Mangans und Chroms, ferner für zahlreiche natürliche und künstliche Silicate sowie für Wässer. Häufigste Trennmethoden für die jeweilige Gruppe der komplexierten Spuren sind neben der Extraktion die Kollektorfällung, die Spurensorption an Aktivkohle und anderen Sorptionsmitteln sowie die Mikrofiltration der Spurenchelate mit Hilfe feinporiger Membranfilter (11 - 14). Für die Spurenanreicherung aus großvolumigen Wasserproben empfiehlt MIZUIKE die Anwendung einer Flotationstechnik

(15) nach Zerstören der in natürlichen Wässern enthaltenen organischen Stoffe durch UV-Licht und Überführen der Spuren in flotierbare Verbindungen.

Tatsächlich läßt sich der Anreicherungsprozeß an Proben mit komplexchemisch unproblematischen Matrixelementen durch eine geeignete Auswahl von Chelatbildnern sowie durch Einstellen optimaler Versuchsparameter meist ausgezeichnet anpassen. Die mit solchen Verfahren erfaßbare Elementpalette hängt allerdings nicht allein davon ab, wie viele der im Probenmaterial enthaltenen Spuren mit dem Chelatbildner reagieren. Wichtig ist vielmehr auch, daß die entstehenden Chelatkomplexe der Elemente die für die Abtrennung jeweils erforderlichen physikalischen Eigenschaften wie Extrahierbarkeit, Schwerlöslichkeit, Flüchtigkeit, Flotierbarkeit usw. besitzen. Leider sind sowohl die Selektivität chemischer Reaktionen als auch das Trennverhalten der entstehenden Verbindungen bis heute einer theoretischen Behandlung oder Vorplanung kaum zugänglich. Die Kenntnisse über Trennsysteme beruhen vielmehr auf umfangreichen empirischen Untersuchungen und Erfahrungen, die ihren Ursprung zumeist in der klassischen "Naßchemie", vor allem in der Gravimetrie und Photometrie, haben; oft sind die dort bereits als Fällungsmittel oder Farbreagenzien eingesetzten Chelatbildner auch die heute in der Spurenanalyse gebräuchlichen Verbindungen.

Art und Zusammensetzung der Probenmatrix bestimmen schließlich auch darüber, wie groß der tolerierbare Spielraum beim Einstellen der Arbeitsparameter ist, über den ein Multi-Elementverfahren verfügt, ohne daß die analytische Leistungsfähigkeit, etwa der Umfang der anzureichernden Elementgruppe, der Trennfaktor oder die Anreicherungsausbeuten, beeinträchtigt wird. Bei Vorhandensein mehrerer, in ihrem chemischen Verhalten sehr unterschiedlicher Matrixelemente gelingt eine Multi-Elementanreicherung - wenn überhaupt - nur bei strengem Einhalten der Versuchsparameter wie Arbeitsvolumen, Reaktionstemperatur und -zeit, pH-Wert usw.: Das individuelle Verhalten der Matrixelemente und auch deren gegenseitige Beeinflussung im chemischen Trennprozeß einer Multi-Elementanreicherung verursachen fast immer eine so starke Einengung der Arbeitsbedingungen, daß die Anwendung solcher Verfahren zu einer Gratwanderung über die Kommastellen der Parameterwerte werden kann. Für den routinemäßigen Einsatz im Betriebslabor sind Verfahren dieser Art dann nur noch wenig geeignet.

Noch schwieriger und zeitaufwendiger wie die Anwendung ist natürlich die Entwicklung von Verfahren, bei denen eine größere Zahl von Parametern überprüft und optimiert werden muß. Die bekannten Methoden zur

Versuchsplanung und systematischen Optimierung von Systemzuständen -
etwa die Simplexmethode (16) - helfen hier leider nur wenig. Schon die
Festlegung eines zu optimierenden Gesamtkriteriums ist bei einem Multi-
Elementverfahren problematisch: Ist es z. B. "optimaler", von insgesamt
10 interessierenden Spuren 8 zu jeweils 60 % anzureichern oder davon
nur 5 zu erfassen, diese aber zu 95 %? Außerdem weiß man zu wenig darüber, ob und in welcher Weise Arbeitsparameter sich gegenseitig beeinflussen, bzw. in welchem Ausmaß noch unerkannte, das Verhalten der Spuren beeinträchtigende Faktoren im Trennsystem enthalten sind und die
Optimierung erschweren: Vorgänge wie die Spurensorption an Gefäßwandungen, Emulsionsbildungen bei Extraktionsprozessen, Übersättigung und
verzögerte Niederschlagsbildung sowie andere Ungleichgewichtszustände
sind meist unreproduzierbar, und ihr Einfluß verfälscht die einer Optimierung zugänglichen gesetzmäßigen Abhängigkeiten der Zielgröße von den
Systemfaktoren. Es ist also nicht überraschend, daß Intuition und Fleiß
bei der Entwicklung von Verfahren zur Multi-Elementanreicherung heute
eher anzutreffen sind als eine geplante systematische Optimierung der
Verfahrensparamter.

KALIBRIERPROBLEME BEI ANREICHERUNGSVERFAHREN

Die Richtigkeit eines spurenanalytischen Verbundverfahrens hängt u. a.
davon ab, wie genau die Anreicherungsausbeuten R (Wiederfindungsraten)
für die abgetrennten Spuren bekannt sind und am Analysenergebnis korrigiert werden können. Am einfachsten ist die Beurteilung eines Verfahrens, wenn für alle Elemente eine quantitative Anreicherung
(R > 95 %) erzielt wird. Bei Minderausbeuten sind Analysen nur dann
sinnvoll, wenn sichergestellt ist, daß die ermittelten Ausbeutewerte
durch die unvermeidbaren Unreproduzierbarkeiten bei der Einstellung der
Verfahrensparamter nicht wesentlich beeinträchtigt werden.

Eine wichtige Methode zur Bestimmung und Kontrolle von Anreicherungsausbeuten ist der Zusatz geeigneter Radiotracer zur Probe und die
Ermittlung der Aktivitätsverluste durch Analyse des Spurenkonzentrats
(17). In ähnlicher Weise ist die Überprüfung der Anreicherung möglich,
indem man einer Reihe aliquoter Probenanteile ansteigende Gehalte der
anzureichernden Elemente zusetzt und die Proben dem Anreicherungsverfahren unterwirft. Man vergleicht - für jedes Element gesondert - die
Steigung der Kalibriergeraden für die wiedergefundenen Elementanteile
mit der Steigung einer entsprechenden Geraden, die durch Analyse einer
analog dotierten Reihe von Spurenkonzentraten gewonnen wird. Die Ausbeuten ergeben sich also jeweils als Quotient der Steigungen zusammen-

gehöriger Geraden, die an Proben mit Spurenzusätzen unter Einschluß
bzw. unter Umgehen des Anreicherungsverfahrens gemessen wurden.

Diese Technik der Ausbeutebestimmung entspricht in der Funktionsweise der Standard-Additionsmethode; sie kann also durch dieselben Arten von Fehlern verfälscht sein: Erkannt bzw. korrigiert werden ausschließlich die auf die Steigung der Kalibriergeraden einwirkenden "multiplikativen" systematischen Fehler, zu denen auch die Minderausbeuten bei der Spurenanreicherung zählen. Voraussetzung für eine richtige Anwendung ist, daß die den Proben bzw. Konzentraten zugesetzten Elemente bzw. Radiotracer sich im Analysenprozeß verhalten wie die entsprechenden als Eigengehalte vorhandenen Spuren. Falls diese "Artgleichheit" nicht gegeben ist, muß mit einer Verfälschung der Steigungen und der Ausbeutewerte gerechnet werden. Ebenso bleiben alle "additiven" systematischen Fehler wie Blindwerte oder der apparativ bedingte Signaluntergrund bei der Standard-Additionsmethode unerkannt; auch sie verfälschen also das Ergebnis. Verstöße gegen die Forderung nach Artgleichheit bei Kalibriervorgängen sind vermutlich die wichtigste Quelle für falsche Spurenanalysen. Im folgenden sollen einige Beispiele hierzu angeführt werden.

Gut analysierte Kalibrierproben für die spurenanalytische Multi-Elementbestimmung sind nur für wenige Materialien im Handel. Die Arbeitsparameter von Verbundverfahren werden deshalb oft unter Einsatz von Modell-Lösungen optimiert, die man dem später zu analysierenden Probenmaterial in der Zusammensetzung möglichst gut nachgebildet hat. Tatsächlich zeigt sich, daß eine Übertragung der an solchen Modellen gewonnenen Ergebnisse auf das "echte" Probenmaterial nur in Ausnahmefällen problemlos gelingt; je geringer die zu bestimmenden Spurengehalte sind, als desto unähnlicher erweisen sich meist Modell und Probe. So führt die Anreicherung von Spuren aus Legierungen vielfach zu völlig anderen Werten als mit demselben Verfahren an entsprechenden Mischungen der metallischen Legierungskomponenten oder gar an Metallsalzlösungen der Matrixelemente gefunden wird (18). Selbst durch Zusammenschmelzen der Legierungsbestandteile sind die Unterschiede zum technischen Probenmaterial nicht immer zu eliminieren, da auch das Kristallgefüge einen deutlichen Einfluß auf das Analysenergebnis ausüben kann (19, 50). Zusätzlich entstehen hier sogar die Gefahren der Spurenverluste durch Verflüchtigung sowie der Kontamination durch Tiegelmaterial und Ofen. Auch das in der Literatur beschriebene Mischen von Standardproben unterschiedlicher Konzentrationen bzw. das Verschneiden solcher Proben mit reinen Hauptkomponenten mit dem Ziel, Zwischenwerte für die Spurengehalte einzustellen, kann unter diesen Gesichtspunkten bedenklich sein (50).

Es gibt sicher eine Vielzahl von Ursachen für die Probleme bei der Herstellung und Analyse von Materialmodellen. Zum Beispiel ist denkbar, daß entweder die Proben oder die Modelle zu geringen Anteilen unerkannte und daher unberücksichtigt gebliebene Bestandteile enthalten, welche die Spurensignale meßbar beeinflussen. Für Unterschiede im Verhalten der Spuren in Legierungen bzw. Modellen aus Metallmischungen können auch die verschiedenen elektrochemischen Potentiale der Proben bei ihrem Auflösen in Säure verantwortlich sein: Legierungspotentiale werden jeweils stark von einer Metallkomponente allein bestimmt, während bei Metallmischungen die Partikel jeder einzelnen Komponente individuelle Potentiale besitzen. Dementsprechend unterschiedlich kann der Redox-Einfluß der jeweils noch ungelöst vorhandenen Metallreste auf die Elementspuren und z. B. deren Wertigkeit sein. In drastischer Weise erkennt man die Potentialunterschiede meist schon daran, daß manche Legierungen und die entsprechenden Metallmischungen unter sonst gleichen Bedingungen von Säure in sehr unterschiedlicher Weise aufgelöst werden. Natürlich sind Erklärungen dieser Art für die Verhaltensunterschiede von Spuren in Probe und Modell häufig nur spekulativ; meist bleiben die vielfältig zu beobachtenden Effekte unerklärbar.

Bekannt ist auch, daß die zur Kalibrierung zugesetzten Spuren im Anreicherungsprozeß sich sehr verschieden verhalten können, je nachdem, ob man sie der Probe vor oder nach dem Lösen in Säure zudotiert hat. Besonders Edelmetalle in Form verdünnter Spurenlösungen können beim Zusatz zur Metalleinwaage im Löseprozeß irreversibel zementiert werden; der bereits gelösten Probe zudotiert, werden sie dagegen bei der nachfolgenden Anreicherung mit erfaßt.

Probleme mit der fehlenden Artgleichheit von Probe und Modell sind für die Spurenanalyse natürlicher Wässer besonders typisch: Hier weiß man seit langem, daß allein durch Auflösen von NaCl und anderer Elektrolyte in destilliertem Wasser und durch Zudotieren der Spuren in Form verdünnter Metallsalzlösungen ein Meerwasser nicht simuliert werden kann. Unberücksichtigt bleibt dabei nämlich, daß wesentliche Anteile der Spuren in Komplexen (z. B. Carbonatokomplexen) und hydroxidischen Kolloiden oder in organischen Verbindungen sowie in Mikroorganismen gebunden sein können. Auf Zusatz von Chelatbildnern bzw. mit Ionenaustauschern reagieren aber jeweils nur die ionogen vorliegenden Spurenanteile; bereits anderweitig gebundene Elemente bzw. Spurenanteile bleiben unberücksichtigt und können der Analyse verloren gehen. Die als verdünnte Salzlösungen dotierten Spuren des Wassermodells werden dagegen durch diese Art Verbindungsbildung in ihrer Reaktionsfähigkeit nicht beeinträchtigt. Durch Ansäuern von natürlichem Wasser kann man derartig verursachte systematische Fehler kaum vermeiden; organisch bzw. biologisch fixierte Elemente werden allenfalls durch einen Probenaufschluß, z. B. mit Peroxid unter UV-Bestrahlung, freigesetzt (20, 21). Probleme dieser Art treten aber nicht nur bei Legierungen und Wässern auf; man findet sie mehr oder weniger deutlich ausgeprägt bei praktisch allen Probenmaterialien.

Zum Kalibrieren lösungsspektrometrischer Verbundmethoden sind nur in Ausnahmefällen Lösungen geeignet, die ausschließlich die zu bestimmenden Spuren in geeigneter Verdünnung enthalten, da die in den Spurenkonzentraten meist noch vorhandenen Matrixreste starke additive und multiplikative Fehler verursachen können. Die in solchen Fällen häufig angewandten Korrekturverfahren benutzen einen inneren Standard, oder sie schreiben vor, die im Spurenkonzentrat zu erwartenden Matrixbestandteile nach Art und Konzentration den Kalibrierlösungen zuzusetzen ("Matrixausgleich"). Eine Kombination aus beiden Methoden ist die bereits erwähnte Standard-Additionsmethode, bei der die im Spurenkonzentrat noch enthaltenen Matrixreste selbst zum Matrixausgleich und die zudotierten Spuren als eine Art innerer Standard verwendet werden. In allen Fällen führt die Kalibrierung auch unter den so "verbesserten" Bedingungen nur dann zu richtigen Werten, wenn die Forderung nach Artgleichheit bei Spuren- und Matrixzusätzen streng erfüllt und additive systematische Fehler abwesend bzw. korrigierbar sind.

In diesem Zusammenhang soll schließlich auf einige Schwierigkeiten hingewiesen werden, die bei der Bestimmung der Nachweisgrenzen spurenanalytischer Verbundverfahren entstehen können. Nach der Definition von KAISER und SPECKER (22) erhält man den Meßwert an der Nachweisgrenze $\underline{x}$ aus dem mittleren Blindwert $\bar{x}_{Bl}$ und dem dreifachen Wert der Standardabweichung der Blindwerte s_{Bl}: $\underline{x} = \bar{x}_{Bl} + 3\,s_{Bl}$. In strenger Auslegung benötigt man zur Erstellung der erforderlichen Daten ein Material, welches in Art und Zusammensetzung den "echten" Proben so weit entspricht, daß lediglich das für die Bestimmung der Nachweisgrenze interessierende Element fehlt. Dieses Material muß in mindestens 20 Parallelversuchen der vollständigen Arbeitsvorschrift des Analysenverfahrens unterworfen werden - angefangen von der Probenzerkleinerung und -einwaage bis hin zur Auswertung der Meßdaten. Als "Blindwerte" gehen alle Signalanteile in die Rechnung ein, die an der Meßstelle des in Frage stehenden Elementes ermittelt werden. Ursache für solche Signale sind u. a. die aus den Reagenzien, Gefäßmaterialien, der Laborluft usw. eingeschleppten Verunreinigungen ("echte Blindwerte"), die als Interferenzen mit den Signalen anderer Probenbestandteile "vorgetäuschten Blindwerte" sowie der verfahrens- bzw. apparatebedingte Meßuntergrund ("Störpegel").

Die auch von KAISER (23) diskutierte wesentliche Schwierigkeit in praktisch allen Anwendungsbereichen der Spurenanalyse besteht darin, daß solche für die Bestimmung der Nachweisgrenze erforderlichen Proben fast nie verfügbar sind ("Doch sollte man nicht zu sehr beklagen, daß es noch Bereiche gibt, in denen Nachdenken, Phantasie und kritisches Urteilsvermögen gebraucht werden" (23)). Man hilft sich, indem man entweder für sein Verfahren eine andere, experimentell besser zugängliche Nachweisgrenze definiert (24) oder indem man lediglich Teilschritte des Verfahrens berücksichtigt. Bei Verbundverfahren weicht man häufig auf Probenmodelle aus, um die vollständige Arbeitsvorschrift "leer" durchführen zu können. Hierbei kommt jedoch zusätzlich zu den diskutierten Problemen mit der Artgleichheit noch die Schwierigkeit mit den meist unbekannten Eigenverunreinigungen der Modellsubstanzen hinzu, welche die durch das Modell simulierten Blindwerte in Höhe und Streuung verfälschen. Übrig bleibt in solchen Fällen die Beschränkung auf einen groben Schätzwert für die Nachweisgrenze des Verfahrens, ermittelt aus den Reagenzienblindwerten und aus dem Störpegel des Bestimmungsverfahrens, also unter Verzicht auf den Einsatz von Probenmaterial und auf die Anwendung der vollständigen Analysenvorschrift. Je nachweisstärker spurenanalytische Verbundverfahren sind und je komplexer das zur Ermittlung der Nachweisgrenze erforderliche Probenmaterial zusammengesetzt ist, desto schwieriger ist es, brauchbare Bedingungen zu finden, unter denen die tatsächlichen Blindwerte eines Verfahrens und damit zuverlässige Daten für dessen Nachweisgrenze ermittelt werden können.

ANFORDERUNGEN DER BESTIMMUNGSMETHODEN
AN DIE BESCHAFFENHEIT VON SPURENKONZENTRATEN

Spurenkonzentrate sind die Schnittstellen zwischen Anreicherungs- und Bestimmungsmethoden. Ähnlich wie das Probenmaterial einen wesentlichen Einfluß darauf haben kann, ob eine Anreicherungstechnik mit Erfolg zur Multi-Elementabtrennung anzuwenden ist, so bestimmt auch die Beschaffenheit des Spurenkonzentrats die Anwendbarkeit einer Bestimmungsmethode. Ganz allgemein kann man feststellen, daß für den Verbund mit einem für die Spuren gewollt unselektiven Anreicherungsverfahren eine Bestimmungsmethode von hoher Selektivität erforderlich ist. Einzel-Elementverfahren (z.B. die Photometrie) verlangen dagegen meist eine selektive Abtrennung einzelner Spuren. Von einiger Bedeutung für jeden Methodenverbund ist der Aggregatzustand der in den Bestimmungsprozeß einzubringenden Probe: Viele Varianten der optischen Emissionsspektralanalyse sowie die Röntgenspektrometrie und Aktivierungsanalyse bevorzugen z.B. feste Proben bzw. feste Spurentargets; die Atomabsorptions-Spektrometrie und Plasma-Emissionsspektralanalyse sowie die Methoden der Elektroanalyse gehen überwiegend von flüssigen Proben bzw. von Spurenkonzentratlösungen aus. Zwar macht es im allgemeinen keine grundsätzlichen Schwierigkeiten, feste Konzentrate, etwa Kollektorniederschläge mit den anhaftenden Spuren, in Lösung zu bringen, und ebenso kann man aus Eluaten oder aus Extraktionsphasen durch Eindampfen einen festen Rückstand erhalten; Prozesse dieser Art sind aber immer mit einem zusätzlichen Aufwand und einer erhöhten Blindwertbelastung verbunden. Eleganter ist es daher, die Konzentrate zur Analyse so zu verwenden, wie sie im Anreicherungsverfahren anfallen.

Eine besondere Bedeutung für die Kombinierbarkeit von Anreicherungs- und Bestimmungsverfahren haben die ins Spurenkonzentrat gelangenden Störelemente. Verursacher von Störungen beim Einsatz instrumenteller Bestimmungsmethoden können z. B. noch vorhandene Matrixanteile sowie Reste von Reagenzien oder deren Zersetzungsprodukte sein. Weitgehend frei in der Wahl der Bestimmungsmethode ist man deshalb nur, wenn solche Beimengungen im Spurenkonzentrat völlig fehlen. ZOLOTOV hat darauf hingewiesen, daß Anreicherung und Bestimmung in Genauigkeit und Nachweisvermögen einander entsprechen sollten (2). Er unterscheidet dabei Verbundmethoden danach, wie eng Anreicherung und Bestimmung im methodischen Ablauf des Verfahrens miteinander verknüpft sind. Besonders häufig anzutreffen sind die mehr zufälligen Kombinationen aus zwei in ihrer Leistungsfähigkeit voneinander unabhängigen Analysenschritte, etwa die photometrische Spurenbestimmung im eingeengten wäßrigen Eluat einer chromatographischen Matrixabtrennung. In anderen Verbundmethoden sind Anreicherung und Spurenbestimmung so aufeinander abgestimmt, daß allein durch den Verbund eine deutliche Leistungsverbesserung resultiert. Dies gilt z. B. für die Kombination von Extraktion und Flammen-AAS der organischen Phase, wo infolge einer erhöhten Aerosolausbeute die Empfindlichkeit erhöht wird. Als "Hybrid-Methode" werden schließlich Analysenprozesse bezeichnet, bei denen Anreicherung und Bestimmung gewissermaßen untrennbar in einem Analysenschritt, oft sogar in derselben Apparatur ablaufen, etwa bei der Gaschromatographie oder der Inversen Voltammetrie.

Im folgenden soll für eine Auswahl instrumenteller Bestimmungsmethoden untersucht werden, welche allgemeinen Anforderungen an ein Spurenkonzentrat zu stellen sind, damit eine unmittelbare Elementbestimmung möglich ist und Störungen durch Fremdbestandteile vermieden werden. Weitere Informationen hierzu findet man unter (2, 11 - 14). Nicht diskutiert werden soll über die Vorzüge und Nachteile der sequentiellen oder simultanen Analyse von Spurenkonzentraten. Hier soll nur daran erinnert werden, daß die vor allem in der Routineanalytik gefragte zeit-

und kostensparende Simultanbestimmung aller interessierenden Elemente meist mit einem Kompromiß über wichtige Leistungsdaten des Verfahrens und mit einem Verlust an Flexibilität in der Anwendbarkeit erkauft werden muß.

RÖNTGENSPEKTROMETRIE. Eine umfangreiche Übersicht über Möglichkeiten der Kombination von Anreicherungsmethoden mit der Röntgenfluoreszenzanalyse (RFA) und der teilcheninduzierten Röntgenemissionsspektrometrie (PIXE) ist in einer Arbeit von VAN GRIEKEN enthalten (25), der auch zusätzliche Literaturhinweise über die hier diskutierten Techniken entnommen werden können.

Das höchste Nachweisvermögen erzielt man bei röntgenspektrometrischen Methoden mit festen Proben in dünner Schicht. Spurenkonzentrate sollten von schweren Matrixelementen weitgehend frei und homogen verteilt auf einem Träger fixiert sein, der aus Elementen mit niedrigen Ordnungszahlen besteht. Nahezu ideale Bedingungen erhält man, wenn die Spuren in schwerlösliche Verbindungen überführt, ohne Zusatz eines Kollektors durch Filtration über ein feinporiges Membranfilter isoliert und dabei mit hohen Ausbeuten angereichert werden können. In ähnlicher Weise gilt dies für Papierfilter, die mit einigen Mikrolitern Spurenlösung getränkt und nach dem Trocknen als Target verwendet werden. Auch die elektrolytische Abscheidung von Spuren kann einer nachweisstarken röntgenspektrometrischen Spurenbestimmung vorgeschaltet werden. Sobald allerdings größere Volumina eines flüssigen Probenmaterials, etwa von Wässern oder Extraktionsphasen, durch Verdampfen in eine die Spuren enthaltene dünne und feste Schicht überführt werden sollen, entstehen Probleme durch unvermeidbare Unregelmäßigkeiten in der Schichtdicke und in der Elementverteilung des Rückstandes. Hinzu kommen die mit dem zeitaufwendigen Eindampfprozeß verbundenen Blindwertprobleme.

Elegant zur direkten Erzeugung fester Spurenkonzentrate sind Methoden, bei denen die Spuren aus einer geeignet vorbereiteten Probenlösung durch Ionen- oder Fällungsaustausch, durch Adsorption oder durch andere Sorptionsmechanismen an einen Träger gebunden werden. Hierfür werden im Probendurchfluß z. B. Kationen- bzw. Anionen-Austauschermembranen sowie imprägnierte Filter verwendet. Im "batch-Verfahren" werden feinkörnige Austauscher in geringer Menge der Probenlösung zugesetzt, nach der Reaktion abfiltriert und auf dem Filter fixiert. Dabei spielen chelatbildende Austauscher mit unterschiedlichen Gerüstsubstanzen und Ankergruppen eine zunehmende Rolle in der röntgenspektrometrischen Wasseranalyse (25). Unter den Adsorptionsmitteln ist vor allem die Anwendung von Aktivkohle zur Anreicherung von Spurenchelaten und anderer

Verbindungen bei nachfolgender Bestimmung durch RFA und PIXE beschrieben worden (26, 27): Zur Spurenanreicherung muß die mit einem Fällungsreagenz versetzte Probenlösung lediglich durch ein mit einer dünnen Aktivkohleschicht bedecktes Filter filtriert werden. In vielen Fällen lassen sich auf diese Weise auch in Wasser leicht lösliche Spurenverbindungen sowie Kolloide anreichern. Die Kohle kann mit einem Kleber, Fixativ u. dgl. abriebfest auf dem Filter gebunden werden. Hinderlich für diese vielfältig einsetzbare Technik sind vor allem die in handelsüblichen Aktivkohlen bisher unvermeidbaren Verunreinigungen (28). Ähnliche Ergebnisse wie mit Sorptiosmitteln erhält man mit reaktionsfähigen Schichten, die einen Fällungsaustausch zulassen. Filtriert man z. B. Lösungen mit Schwermetallionen durch eine Schicht mit Zinksulfid, so bleiben alle Elemente haften, deren Sulfide ein kleineres Löslichkeitsprodukt besitzen als ZnS (29).

Von wesentlicher Bedeutung unter den Anreicherungsmethoden, die unmittelbar ein festes Spurenkonzentrat lierfern, sind die verschiedenartigen Kollektor- und Spurenfängerreaktionen. Hierbei fällt man die anzureichernden Elementspuren aus der Probenlösung aus, wobei ein bei der Bestimmung nicht interessierendes Element, zu wenigen Milligramm zugesetzt, einen filtrierbaren Niederschlag bildet, der die Spurenverbindungen einschließt. In günstigen Fällen kann auch ein geringer Anteil eines Matrixbestandteils als Kollektorelement ausgefällt werden. Metallsulfide, -hydroxide oder -chelate sind die bevorzugten Kollektoren; als fällende Gruppenreagenzien haben sich das 8-Hydroxychinolin und vor allem die Dithiocarbamidate gut bewährt. Das Kollektorelement kann gleichzeitig als innerer Standard für die Korrektur von Schichtdickenunterschieden verwendet werden. Solche Ungleichmäßigkeiten in der Filterbedeckung wirken sich besonders in der PIXE-Analyse und bei ähnlichen Techniken mit geringem Strahlendurchmesser auf die Höhe der Spurensignale aus, wenn nicht auf die gleichermaßen verfälschten Signale des inneren Standards bezogen wird (28). Ist ein Kollektorelement zugleich auch innerer Standard, so muß auf spektrale Interferenzen mit den Röntgenlinien der Spuren, auf die Zählratenbegrenzung bei energiedispersiven Methoden und auf Interelementeffekte wie Sekundäranregung oder Strahlungsabsorption geachtet werden. Durch die hohe Energie des auftreffenden Mikrostrahls und die dadurch verursachte Erwärmung des Targets können im Bereich des Brennflecks Spurenverluste durch Verflüchtigung entstehen (28). Die Güte der Filterbedeckung durch das Spurenkonzentrat ist bei der Röntgenfluoreszenzanalyse mit meist wesentlich größerem Strahldurchmesser weniger kritisch; allerdings sollte man hier versuchen, die mit dem Spurenkonzentrat belegte Targetfläche möglichst gut durch die anregende Strahlung auszuleuchten. Von den geringen Ordnungszahlen der enthaltenen Elemente her sind als Kollektor für die Röntgenanalyse die in Wasser wenig löslichen chelatbildenden Gruppenreagenzien wie z. B. das PAN besonders geeignet (30): Das Reagenz wird, in Alkohol gelöst, der Probenlösung zugesetzt, wobei der ausfallende überschüssige Anteil als Spurenfänger für die gebildeten Spurenchelate dient.

Besondere Anforderungen an die Beschaffenheit der Proben stellt die Röntgenfluoreszenzanalyse mit totalreflektierendem Probenträger. Diese Methode liefert im Vergleich zur "konventionellen" RFA zwar ein wesentlich verbessertes Signal/Untergrund-Verhältnis, sie besitzt zugleich aber auch eine höhere Störempfindlichkeit gegenüber Matrixanteilen (31). Das Probentarget muß hier aus einem möglichst dünnen Film bestehen, z. B. aus der auf einem Quarzträger eingedampften wäßrigen oder organischen Spurenlösung. In merklicher Konzentration vorhandene Matrixbestandteile, also auch Kollektoren, verringern die Empfindlichkeit der Methode. Für die Multi-Elementanalyse erhält man geeignete Targets z. B. durch eine Dithiocarbamidat-Fällung der Spuren mit nachfolgendem Lösen in Chloroform und Eindampfen des über Membranfilter isolierten Konzentrats (32).

Für die nahezu universell anwendbare Technik der Kombination von Kollektorfällung und Röntgenspektrometrie hat LUKE 1968 den Namen "coprex"-Methode (coprecipitation and X-ray analysis) geprägt (33).

AKTIVIERUNGSANALYSE. Wichtigste Meßtechnik für die durch Bestrahlen einer Probe mit Neutronen, Photonen und geladenen Teilchen erzeugte Radioaktivität ist die Gammaspektroskopie. Wegen der Ähnlichkeit zu den röntgenspektrometrischen Methoden sind die an die Beschaffenheit der Spurenkonzentrate zu stellenden Forderungen weitgehend übertragbar. Auch hier bieten feste Proben in dünner Schicht eine Reihe von Vorteilen; die reproduzierbare Einstellung der Meßgeometrie verlangt dabei ebenfalls eine homogene Bedeckung des Targets. Im Verbund mit der Multi-Elementanreicherung sind also alle Methoden von Interesse, bei denen die Spuren durch Ionenaustausch, Adsorption, Kollektorfällung, Mikrofiltration u. dgl. von der Matrix getrennt und an einem festen Trägermaterial, welches die Bestimmung selbst möglichst wenig beeinträchtigt, fixiert werden (34).

Zu unterscheiden bei der Kombination von Spurenanreicherung und Aktivierungsanalyse ist, ob die Anreicherung vor der Aktivierung oder im Anschluß an die Bestrahlung der Probe durchgeführt werden muß. Eine Vorabtrennung ist z. B. zweckmäßig, um hohe Matrixaktivitäten bei besonders störenden Elementen (Na, K, Br, P) zu vermeiden und um das Nachweisvermögen für geringe Spurengehalte zu erhöhen. Auch die Bestimmung sehr kurzlebiger Nuklide erfordert es, zeitaufwendige Trennoperationen bereits vor der Aktivierung durchzuführen.

In solchen Fällen ist die Auswahl der geeigneten Anreicherungsmethode - von der Kombinierbarkeit mit der Gammaspektroskopie her - unkritisch. Prinzipiell können also alle Methoden verwendet werden, mit denen das in der Analysenaufgabe vorgegebene Trennproblem gelöst werden kann. Da eine längere Bestrahlung von Lösungen im allgemeinen nicht möglich ist, benutzt man bei Methoden, die flüssige Spurenphasen liefern, wie Extraktion und Ionenaustausch, oft den Eindampfrückstand als zu aktivierende Probe. Bei allen Verbundverfahren mit einer Vorabtrennung der Spuren verschenkt man allerdings den für die Aktivierungsanalyse einzigartigen Vorteil der Ausschaltung von Blindwerten: Die im Trennvorgang eingeschleppten Verunreinigungen werden - wie die zu bestimmenden Spuren - mitaktiviert.

Für die Spurenanreicherung aus bereits aktiviertem Probenmaterial wird die Auswahl der Methoden neben der Trennleistung vor allem durch die Halbwertzeiten der zu bestimmenden Spuren und - aus Gründen des Strahlenschutzes - durch die Handhabbarkeit bzw. Mechanisierbarkeit bestimmt. Außerdem ist die Strahlungsresistenz der zur Trennung erforderlichen Reagenzien, organischen Phasen, Ionenaustauscher usw. zu beachten.

Zur Abtrennung von Radiokliden sind seit der Entdeckung der Radioaktivität die verschiedenartigsten Spurenfänger-Reaktionen in Gebrauch, weil hiermit durch Ausnutzen der vielfältigen Sorptionseffekte Elemente auch in extrem geringen Konzentrationsbereichen abgetrennt werden können. Für die Einzel-Elementanreicherung beliebt ist der Zusatz eines mit dem Radionuklid identischen inaktiven Elements als Kollektor, dessen Verbleib im Trennprozeß leicht überprüft werden kann und der eine unmittelbare Aussage über die Vollständigkeit der Spurenanreicherung ermöglicht. Auch Aktivkohle ist mit Erfolg zur Abtrennung chelatisierter aktiver Nuklide verwendet worden (35).

Ist eine sehr rasche Trennung aus bestrahltem Material erforderlich, so sind Extraktionsverfahren anderen Techniken häufig überlegen; der manuelle Aufwand kann hierbei allerdings relativ hoch sein. Die oft eingesetzte Ionenaustauschtrennung auf der Säule ist dagegen einfach zu handhaben und auch zu mechanisieren; die Methode ist oft aber sehr zeitaufwendig und für die Abtrennung kurzlebiger Nuklide dann ungeeignet. Insgesamt ergibt sich auch hier, daß ein optimaler Methodenverbund erst unter Berücksichtigung zahlreicher von den Eigenschaften des jeweiligen Probenmaterials ausgehender Gesichtspunkte geplant werden kann.

OPTISCHE ATOMSPEKTROMETRIE. Im Anwendungsbereich der optischen Elementspektroskopie spielen Anreicherungsverfahren seit Jahrzehnten eine bedeutende Rolle. Es gibt kaum Trenn- und Anreicherungstechniken, die nicht bereits erfolgreich mit Verfahren der Absorptions- und Emissionsspektralanalyse kombiniert worden sind. Zur Einteilung der vielfältigen Verbundmöglichkeiten kann man grob unterteilen zwischen der Spektralanalyse fester Proben z. B. mit Lichtbogen, Funken- oder Glimmentladung als spektrochemischer Lichtquelle und den Methoden der Lösungsspektroskopie unter Verwenden chemischer Flammen bzw. von Plasmen. Voraussetzung für einen Gewinn bei der Anwendung von Trennverfahren in Kombination mit einer dieser Bestimmungsmethoden ist, daß durch den Anreicherungsprozeß ein Konzentrat geschaffen wird, in dem störende Matrixbestandteile fehlen, und die Spuren problemlos thermisch verdampft und angeregt werden können. Feste Spurenkonzentrate erhält man auch hier am einfachsten durch die verschiedenen Spurenfängerreaktionen sowie durch Einengen von Eluaten und Extraktionsphasen.

Für die Emissionsspektralanalyse ist es wichtig, daß besonders solche Matrixelemente sauber abgetrennt sind, die ein linienreiches Emissionsspektrum besitzen und spektrale Koinzidenzen verursachen können. Dementsprechend müssen auch Kollektorelemente neben ihrer Funktion als Anreicherungshilfe unter dem Gesichtspunkt eines störenden Linienreichtums ausgewählt werden. Die Spurenkonzentrate werden meist mit Kohle- bzw. Graphitpulver, einem inneren Standard und - wenn erforderlich - mit einem spektrochemischen Puffer vermischt und z. B. im Lichtbogen abgebrannt. Zu beachten sind die Beiträge aller dieser Zusätze zum Gesamtblindwert der Proben. Bei der Anwendung von Kollektorfällungen dient das Kollektorelement oft gleichzeitig als innerer Standard zur Korrektur von Verdampfungs- und Anregungsstörungen. Zu beachten ist auch hier die Forderung nach Artgleichheit beim Kalibrieren der Analyse. Wegen des geringen Probenbedarfs der Bogenanalyse sollten die eingesetzten Kollektormengen niedrig gehalten werden; auch Elektrolytballast im Spurenkonzentrat sollte man möglichst vermeiden. Falls im Spurenkonzentrat größere Gehalte an Chelatbildnern oder an Ionenaustauschern u. dgl. enthalten sind, ist im allgemeinen eine Veraschung erforderlich, um die Brenneigenschaften des Bogens nicht zu beeinträchtigen. Auch der direkte Einsatz mit Spuren beladener Aktivkohle zur Bogenanalyse gelingt wegen der in den Poren vorhandenen hohen Gasanteile nicht problemlos: Beim Zünden des Bogens quillt die Aktivkohle -

gelegentlich unter Verpuffen - aus der Becherelektrode heraus. Flüssige Spurenlösungen wie Ionenaustausch-Eluate, Extraktionsphasen oder Rückstände aus der Naßveraschung werden häufig in Gegenwart von Spektralkohlepulver sowie einem spektrochemischen Puffer eingedampft, um eine gleichmäßige Verteilung der Spuren in Kohle bzw. Salz zu erreichen.

Für die Spurenanalyse mit der Glimmentladungslampe nach GRIMM müssen die Konzentrate mit einem den elektrischen Strom leitenden Material verpreßt werden; meist nimmt man dazu pulverförmige Kohle, Silber oder Kupfer (36, 37). Hinderlich ist vor allem in Kupfer der für einige Elemente hohe Eigengehalt, der wegen des erforderlichen etwa 4fachen Überschusses im Preßling entsprechend hohe Blindwerte verursacht. Ähnliches gilt auch für die Herstellung von Elektroden für die optische Funken-Spektralanalyse sowie für die Funken-Massenspektrometrie, wo Spurenkonzentrate ebenfalls als Preßlinge in einer Matrix aus Ag, Cu oder Kohle eingesetzt werden. Für den Verbund der nachweisstarken Massenspektrometrie mit Anreicherungsverfahren gibt es zahlreiche Anwendungsbeispiele (38).

Probleme gibt es auch bei der Anwendung lösungsspektroskopischer Methoden zur Analyse flüssiger Konzentrate. Von den zahlreichen Varianten solcher Methoden sollen hier lediglich die Atomabsorptions-Spektrometrie sowie die Plasma-Emissionsspektralanalyse diskutiert werden. Bei der Elementbestimmung durch Flammen-AAs bewirken alle Fremdbestandteile eines Spurenkonzentrats Störungen, die gegenüber wäßrigen Kalibrierlösungen die Flammeneigenschaften wie Oxidations- oder Reduktionsvermögen, Temperatur und Transparenz verändern. Auch Änderungen in der Aerosolausbeute als Folge veränderter Dichte, Viskosität und Oberflächenspannung der eingesprühten Lösung (39) sowie in der Atomisierbarkeit der Spurenverbindungen wirken auf die Signale ein. Schließlich gehören Stoffe dazu, die einen unkompensierbar hohen oder einen strukturierten Signaluntergrund erzeugen. Auf die zahlreichen Möglichkeiten für additive und multiplikative systematische Fehler, die aus mangelnder - oft aber nicht einfach zu erreichender - Artgleichheit herrühren, wurde bereits hingewiesen.

Die Flammen-AAS ist eine probenverbrauchende Sequenzmethode, d. h. der Gesamtbedarf an Meßlösung steigt mit der Zahl der zu bestimmenden Elemente an; gleichzeitig nimmt die Empfindlichkeit für alle Elemente ab. Deshalb sollte das Volumen der Spurenkonzentratlösung für eine nachweisstarke Multi-Elementbestimmung möglichst klein gehalten werden und zur Bestimmung aller interessierenden Elemente gerade ausreichen. Benutzt man die Injektionstechnik und bestimmt jedes Element in Ali-

quoten von 20 - 50 µl, so reicht 1 ml Gesamtvolumen für die Bestimmung
von mehr als 10 Elementen aus (40). Zur Anreicherung für diese Art der
Analyse sind Sorptionsreaktionen geeignet, bei denen ein von Matrix-
bestandteilen praktisch freies Spurenkonzentrat erhalten wird. Hierzu
gehören Austauscher, die im "batch-Verfahren" eingesetzt werden sowie
Aktivkohle und andere Sorbentien, von denen die Spuren leicht wieder ab-
gelöst werden können. Dasselbe gilt für Spurenfängerreaktionen. Kollektorelemente dürfen hier jedoch nur in geringer Menge eingesetzt werden,
um keine störende neue Matrix im Konzentrat zu erzeugen. Ebenfalls geeignet sind Extraktions- und Chromatographie- bzw. Ionenaustausch-Ver-
fahren, sofern die Spurenphasen und Eluate weitgehend rückstandsfrei
zu einem geringen Meßvolumen eingeengt werden können.

Besonders problematisch bei der Kalibrierung von AAS-Verfahren sind
organische Anteile aus nicht vollständig entfernten Extraktions- und
Elutionsphasen sowie Zersetzungsprodukte unvollständig aufgeschlosse-
ner organischer Probenmaterialien, da sie die Empfindlichkeit der Spurenbestimmung durch Verändern der Ansaugrate und Aerosolausbeute stark
verfälschen können. Manche organischen Stoffe beeinflussen außerdem
die Bindungsverhältnisse der Spurenelemente in Meßlösung und Flamme und
tragen so ebenfalls zu einer meist unreproduzierbaren Veränderung der
Empfindlichkeit bei.

Interessant ist die direkte Analyse organischer Extraktionsphasen
durch Flammen-AAS, da hierbei die Empfindlichkeit gegenüber Bestimmungen in wäßriger Lösung oft um das 3- bis 5fache erhöht wird. Dementsprechend nehmen Extraktionsverfahren - meist unter Verwenden von
Methylisobutylketon (MIBK) und Ammonium-pyrrolidindithiocarbamidat
(APDTC) als Gruppenreagenz - eine herausragende Stellung unter den publizierten Anreicherungstechniken im Verbund mit der Flammen-AAS ein
(41). MIBK erweist sich dabei als nahezu optimal in bezug auf seine
Extraktionsfähigkeit für Chelate und andere Verbindungen, gleichzeitig
aber auch aufgrund seiner guten Brenneigenschaften in der Flamme. Nachteilig ist die merkliche Mischbarkeit von MIBK und Wasser, was sich besonders bei sehr unterschiedlichen Phasenverhältnissen und auch bei der
Kalibrierung störend bemerkbar macht. In Kombination mit APDTC können
sehr umfangreiche Elementpaletten angereichert und bestimmt werden. Da
beim Zerstäuben des organischen Lösungsmittels die Flammenstöchiometrie
wesentlich verändert wird, empfiehlt sich bei der Analyse von Extrakten
eine Verminderung der Brenngaszufuhr. Die Anwendung der Injektionstechnik wird hierbei im allgemeinen jedoch durch Veränderungen der Basis-
linie in den injektionsfreien Zwischenzeiten erschwert.

Schließlich soll auch in diesem Zusammenhang noch einmal auf die Kalibrierprobleme hingewiesen werden: In Carbamidatkomplexen oder in anderen Chelaten gebundene Spuren verhalten sich deutlich anders in der Flamme als entsprechende Metallsalze; das gilt selbst für die "öllöslichen Standards", die für die Kalibrierung in organischen Lösungsmitteln im Handel sind. Systematische Fehler schaltet man am besten aus, wenn man zum Kalibrieren entsprechend der Arbeitsvorschrift des Analysenverfahrens gewonnene organische Phasen verwendet, deren Spurengehalte durch ein unabhängiges Vergleichsverfahren ermittelt wurden.

Die für die Flammen-AAS von Spurenkonzentraten geltenden Probleme sind im wesentlichen auch für Flammen-Emissionsverfahren gültig. Durch organische Stoffe verursachte Temperaturänderungen wirken sich hier zusätzlich auch auf die Anregungsverhältnisse und damit in verstärktem Maße auf die Spurensignale aus. Auch bei allen Ofenmethoden der AAS muß man mit erhöhten Störungen rechnen, wenn im Spurenkonzentrat Bestandteile wie Matrixreste, Kollektorelemente oder organische Stoffe enthalten sind, welche die Verdampfbarkeit der Spuren oder deren Signaluntergrund beeinflussen. Zusätzliche Effekte können bei der Temperaturprogrammierung der Küvette entstehen, z. B. wenn durch Fremdbestandteile im Konzentrat Spurenverbindungen gebildet werden, deren Signale im Atomisierungsschritt des Heizprogramms gegenüber wäßrigen Kalibrierlösungen verfrüht oder verzögert mit jeweils veränderter Empfindlichkeit erscheinen. Schließlich sollte man darauf achten, daß die gelösten Spurenkonzentrate möglichst keine aggressiven Säuren (H_2F_2, HNO_3, HBr) in höheren Konzentrationen enthalten, da hierdurch die Lebensdauer des Zerstäubers und auch der Graphitrohr-Küvetten ganz wesentlich beeinträchtigt wird.

In den Anwendungsmöglichkeiten ergänzt, zum Teil auch bereits ersetzt, werden die AAS-Verfahren durch die Emissionsspektralanalyse mit Plasmen, vorzugsweise mit dem ICP (Inductively Coupled Plasma). Vorteile ergeben sich wegen der Möglichkeit zur simultanen Multi-Elementbestimmung unter Einschluß einiger sonst nur schwierig zu erfassender refraktärer Elemente und wegen des über mehrere Größenordnungen sich erstreckenden dynamischen Bereichs. Bei dieser Methode ist auch die Analyse injizierter geringer Probenvolumina möglich (42, 43). Grundsätzlich wirken bei der ICP-Spektrometrie alle Bestandteile von Spurenkonzentraten auf die Analysensignale ein, welche die Aerosolausbeute des Zerstäubersystems beeinflussen. Insbesondere organische Verbindungen haben, wie bei der Flammen-AAS bereits diskutiert wurde, einen Einfluß auf die Probenzerstäubung (39), zugleich aber auch auf die Generatorleistung und die Entladungsform und damit auf das optische Signal

(44,45). Zwar können organische Extrakte, z. B. in MIBK, unmittelbar im Plasma zerstäubt werden, man erhält jedoch als Folge von Bandenemission und erhöhter Streulichtintensität vielfach einen gegenüber wäßrigen Lösungen erhöhten Untergrundpegel, ohne daß die Empfindlichkeit wesentlich verbessert wird (45). Eine Reextraktion der Spuren und die Analyse der resultierenden wäßrigen Spurenlösung ist deshalb im allgemeinen vorzuziehen. Organische Restgehalte jeglicher Herkunft sollten dabei vor der Bestimmung möglichst gut entfernt werden, um Probleme beim Kalibrieren zu vermeiden. Wegen der bei der ICP-Spektrometrie erforderlichen Begrenzung der Salzkonzentration in der Meßlösung auf etwa 20 mg·ml^{-1} müssen Kollektorreaktionen in Kombination mit der Probeninjektion aus geringem Gesamtvolumen so ausgewählt werden, daß dieser Wert durch das Kollektorelement nicht überschritten wird. Zu vermeiden sind Elemente in höherer Konzentration, die ein sehr linienreiches Emissionsspektrum besitzen. Unter den Spurenfängerreaktionen ist dem Einsatz von Sorptionsmitteln, von denen die Spuren in einfacher Weise wieder abzutrennen sind, der Vorzug zu geben.

VOLTAMMETRIE. Unter den elektroanalytischen Bestimmungesmethoden haben Verfahren der Voltammetrie (Polarographie, Chronopotentiometrie, Inverse Voltammetrie) wegen ihres hohen Nachweisvermögens in der Spurenanalyse die größte Bedeutung. Im Vergleich zu den spektrochemischen Methoden können sie allerdings nur eingeschränkt zur Multi-Elementanalyse verwendet werden, da die aus einer Lösung jeweils erfaßbare Spurenpalette nur selten über vier Elemente hinausgeht. Die ebenfalls sehr nachweisstarke Coulometrie erlaubt sogar nur die Bestimmung einer einzigen reaktionsfähigen Komponente in der Probenlösung.

Die durch voltammetrische Verfahren an die Beschaffenheit von Spurenkonzentraten gestellten Anforderungen entsprechen in mancher Hinsicht denen der Ofen-AAS. Störungen verursachen Matrixreste und Kollektorelemente, deren Peak- bzw. Halbstufenpotentiale benachbart zu denen der Spuren liegen und Signalkoinzidenzen herbeiführen. Störungen anderer Art bewirken die aus Probenaufschlüssen, Ionenaustausch-Eluaten oder Extraktionsphasen herrührenden, in ihrer Zusammensetzung meist undefinierten organischen Zersetzungsprodukte: In Gegenwart solcher Stoffe entstehen im Voltammogramm "Geisterpeaks" sowie unreproduzierbare Spurensignale als Folge einer Hemmung der zugehörigen Elektrodenreaktion (46, 47). Bekannt sind solche Effekte besonders bei empfindlichen inversvoltammetrischen Analysen im Anschluß an den Druckaufschluß organischer Probenmaterialien. Insbesondere bei biologischen Proben konnte nachgewiesen werden, daß eine hinreichende Oxydation durch Druckaufschluß unter den üblichen Bedingungen praktisch nie gelingt (48). Auch die zur Maskierung oder Anreicherung verwendeten Chelatbildner sowie andere organische Reagenzien und deren Zersetzungsprodukte gehören bei allen voltammetrischen Verfahren für Elementspuren zu den problematischen Stoffen. Oft werden bereits durch äußerst geringe Gehalte an organischen Verbindungen erhebliche Störungen verursacht, z. B. nach Kontakt der Meßlösung mit organischen Gefäßmaterialien oder nach einer Filtration durch Papierfilter. Selbst das nach der Reinigung durch Ionenaustausch und durch nachfolgendes mehrfaches Destillieren aus Quarz erhaltene Wasser enthält noch störende Stoffe, obwohl durch coulometrische Bestimmung ein Gesamt-Kohlenstoffgehalt von nur wenigen $\mu g \cdot l^{-1}$ ermittelt wurde (49). Am leichtesten vermeidet man die durch organische "Artefakte" verursachten Probleme, wenn man die Spurenkonzentrate vor der Bestimmung mit einem starken Oxidationsmittel, z. B. mit Perchlorsäure, mehrfach abraucht.

LITERATUR

(1) Berndt, H., Jackwerth, E., Kimura, M: Anal. Chim. Acta 93 (1977) 45
(2) Zolotov, Yu.A.: Pure Appl. Chem. 50 (1978) 129
(3) Ahrland, S., Chatt, J., Davies, N.R.: Quart. Rev. 12 (1958) 265
(4) Ahrland, S.: Structure and Bonding 1 (1966) 207; 5 (1968) 118
(5) Umland, F., Janssen, A., Thierig, D., Wünsch, G.: Theorie und praktische Anwendung von Komplexbildnern, Akad. Verlagsges. Frankfurt/M. (1971)
(6) Häberli, E.: Fresenius Z. Anal. Chem. 160 (1958) 15
(7) Danz, J., Jackwerth, E.: unveröffentlicht
(8) Jackwerth, E., Mittelstädt, H.: Mikrochim. Acta (Wien), Suppl. 10, (1983) 325
(9) Jackwerth, E., Musaick, K.: Mikrochim. Acta (Wien), Suppl. 9, (1981) 71
(10) Busev, A.I., Byrko, V.M., Tereschtschenko, A.P., Novikova, N.N., Naidina, V.P., Terentev, P.B.: Zh. analit Khim. 25 (1970) 665; Ref.: Analyt. Abstr. 21 (1971) 1705
(11) Mizuike, A.: Enrichment Techniques for Inorganic Trace Analysis, Springer-Verlag, Berlin, Heidelberg, New York (1983)
(12) Zolotov, Yu.A., Kuzmin, N.M.: Konzentrirovanije Mikroelementov, Khimija, Moskau (1982)
(13) Bächmann, K.: CRC Crit. Rev. Analyt. Chem. 12 (1981) 1
(14) Minczewski, J., Chwastowska, J., Dybczinski, R.: Separation and Preconcentration Methods in Inorganic Trace Analysis, Ellis Horwood Series in Analytical Chemistry, John Wiley & Sons (1982)

(15) Mizuike, A., Hirade, M.: Pure Appl. Chem. 54 (1982) 1555
(16) Retzlaff, G., Rust, G., Waibel, J.: Statistische Versuchsplanung, Verlag Chemie, Weinheim, New York (1978)
(17) Krivan, V.: Talanta 29 (1982) 1041
(18) Doolan, K.J., Belcher, C.G.: Prog. Analyt. Atom. Spectrosc. 3 (1980) 125
(19) Thierig, D., Unger, H., Dehrendorf, H., Theiß, H.J.: Fresenius Z. Anal. Chem. im Druck
(20) Mart, L.: Talanta 29 (1982) 1035
(21) Burba, P., Willmer, P.G.: Fresenius Z. Anal. Chem. 311 (1982) 222
(22) Kaiser, H., Specker H.: Fresenius Z. Anal. Chem. 149 (1956) 46
(23) Kaiser, H.: Fresenius Z. Anal. Chem.: 216 (1966) 80
(24) Ebel, S., Kamm, K.: Fresenius Z. Anal. Chem. 316 (1983) 382
(25) Van Grieken, R.: Anal. Chim. Acta 143 (1982) 3
(26) Vanderborght, B.M., Verbeek, J., Van Grieken, R.E.: Bull. Soc. Chim. Belg. 86 (1977) 23
(27) Johansson, E.M., Akselsson, K.R.: Nucl. Instr. Meth. 181 (1981) 221
(28) Brüggerhoff, S., Jackwerth, E., Raith, B., Divoux, S., Gonsior, B.: Fresenius Z. Anal. Chem. 311 (1982) 252; 316 (1983) 221
(29) Disam, A., Tschopel, P., Tölg, G.: Fresenius Z. Anal. Chem. 295 (1979) 97
(30) Püschel, R.: Talanta 16 (1969) 351
(31) Knoth, J., Schwenke, H.: Fresenius Z. Anal. Chem. 291 (1978) 200
(32) Knöchel, A., Prange, A.: Fresenius Z. Anal. Chem. 306 (1981) 252
(33) Luke, C.L.: Anal. Chim. Acta 41 (1968) 237
(34) Krivan, V.: Angew. Chemie 91 (1979) 132
(35) Vanderborght, B.M., Van Grieken, R.E.: Anal. Chem. 49 (1977) 311
(36) El Alfy, S.: Dissertation Dortmund 1978
(37) El Alfy, S., Laqua, K., Massmann, H.: Fresenius Z. Anal. Chem. 263 (1973) 1
(38) Beske, H.E., Gijbels, R., Hurrle, A., Jochum, K.P.: Fresenius Z. Anal. Chem. 309 (1981) 329
(39) Lemands, A.J., McClellan, B.E.: Anal. Chem. 45 (1973) 1455
(40) Berndt, H., Slavin, W.: At. Absorpt. Newsl. 17 (1978) 109
(41) Welz, B.: Atom-Absorptions-Spektroskopie, Verlag Chemie, Weinheim, 1983
(42) Broekaert, J.A.C., Leis, F.: Anal. Chim. Acta 109 (1979) 73
(43) Boumans, P.W.J.M.: Fresenius Z. Anal. Chem. 299 (1979) 337
(44) Sommer, D., Ohls, K.: Laborpraxis (1982) 598
(45) Broekaert, J.A.C., Leis, F., Laqua, K.: Talanta 28 (1981) 745
(46) Kotz, L., Henze, G., Kaiser, G., Pahlke, S., Veber, M., Tölg, G.: Talanta 26 (1979) 681
(47) Stoeppler, M., Valenta, P., Nürnberg, H.W.: Fresenius Z. Anal.

Chem. 297 (1979) 22
(48) Stoeppler, M., Müller, K.P., Backhaus, F.: Fresenius Z. Anal. Chem. 297 (1979) 107
(49) Jackwerth, E.: unveröffentlichtes Ergebnis
(50) Ohls, K., Sommer, D.: Fresenius Z. Anal. Chem. 312 (1982) 195

ANREICHERUNG VON SELTENEN ERDEN, URAN UND THORIUM UND BESTIMMUNG DURCH
RÖNTGENFLUORESZENZANALYSE

G. Hartmann, B. Sarx, H. Klenk, K. Bächmann

Fachbereich Anorganische Chemie und Kernchemie
Technische Hochschule Darmstadt

1. Einleitung

Die Anzahl der Elemente, die man mit der Röntgenfluoreszenzanalyse direkt bestimmen kann, ist relativ gering. Um das Spektrum der analysierbaren Elemente zu erweitern, müssen diese abgetrennt, angereichert und in eine geeignete Form (homogene Schicht) für die Messung mit der RFA gebracht werden. Ein Beispiel für die Notwendigkeit einer Abtrennung sind die S.E., Th und U, deren L-Röntgenlinien in den meisten Proben von den K-Röntgenlinien häufiger vorkommender Elemente (Übergangselemente) überlagert werden. Die Bestimmung von S.E., Th und U ist für folgende Anwendungen wichtig (1):
1. Radioökologische Untersuchungen (S.E. sind Ersatzelemente für die Untersuchung des Actinidenverhaltens)
2. Untersuchung des Reinheitsgrades von einzelnen S.E.
3. Untersuchung der Ausgangsprodukte für S.E. (z.B. Monazitsand)
4. Geologische Proben

Im Hinblick auf die Radioökologie kann die chemische und physikalische Ähnlichkeit einiger S.E. - so besitzt Nd^{3+} mit 995 pm fast den gleichen Ionenradius wie Am^{3+} (990 pm) - ausgenutzt werden, um Analogieschlüsse zu Elementen zu ziehen, die nur aufwendig und kostenintensiv nachweisbar sind. Für solche Einsatzgebiete werden Analysenmethoden benötigt, die schnell und möglichst simultan verschiedene S.E. bestimmen können. Solchen Anforderungen kann nach einer entsprechenden Probenaufbereitung die RFA genügen.

2. Experimenteller Teil

Zur Abtrennung und Herstellung geeigneter Festproben aus wäßrigem Medium sind Fällungen, Adsorptionsverfahren und Ionenaustausch üblich. Für S.E., Th und Uran läßt sich eine selektive Abtrennung nur durch Ionenaustausch erzielen. Dabei wird so vorgegangen, daß Proben mit unterschiedlichen Matrices (Boden-, Erz, getrocknete Pflanzenproben, Luftfilter) einem sauren Aufschluß unterworfen werden und so in eine einheitliche Form überführt werden. Aus der Aufschlußlösung lassen sich dann verschiedene Elementgruppen abtrennen. In der vorliegenden Arbeit wurden S.E., Thorium und Uran mit einem Ionenaustauscher nach einem von Korkisch (2) beschriebenen Verfahren, das dem Analysenproblem entsprechend modifiziert und optimiert wurde, abgetrennt.

Der mit den Elementen beladene Ionenaustauscher wurde getrocknet und gemahlen. Danach wurde das Bindemittel "Somarblend" zugemischt und das Gemisch zu einer Tablette gepreßt, die durch eine energiedispersive RFA mit Sekundärtarget analysiert wurde. Die Energien der charakteristischen Analysenlinien (L-Linien)der Seltenen Erden Lanthan, Cer und Neodym liegen zwischen 4 und 6 keV. In diesem Energiebereich ist die Anregung durch ein Kupfer-Sekundärtarget am günstigsten. Für die Anregung der höherenergetischen Linien von Uran und Thorium wurde die Sekundäranregung durch ein Molybdäntarget vorgenommen.

Die Auswertung erfolgte mit einem Computerprogramm (3), das nach Abzug des Untergrunds die einzelnen Peakgruppen auf Überlagerungen hin untersuchte und nach Zuordnung der Elemente eine Compton-Korrektur durchführte.

Abb.1 zeigt eine Monazitsandprobe, die direkt gemessen wurde. Die Linien der S.E. sind von den Linien des in der Probe enthaltenen Bariums und Eisens überdeckt. Abb. 2 zeigt dieselbe Probe nach Abtrennung durch den Ionenaustauscher.

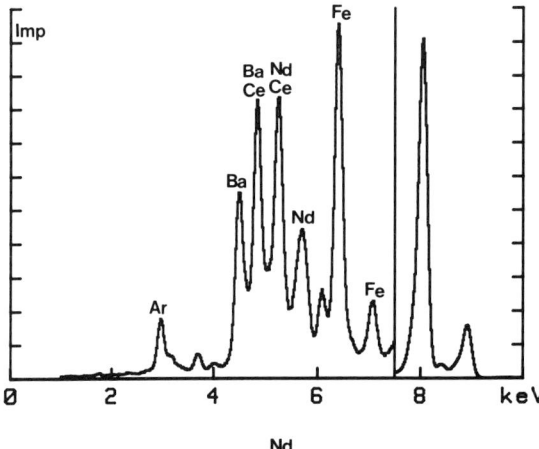

Abb.1
Röntgenfluoreszenzspektrum von Monazitsand ohne Abtrennung und Messung mit Cu-Target

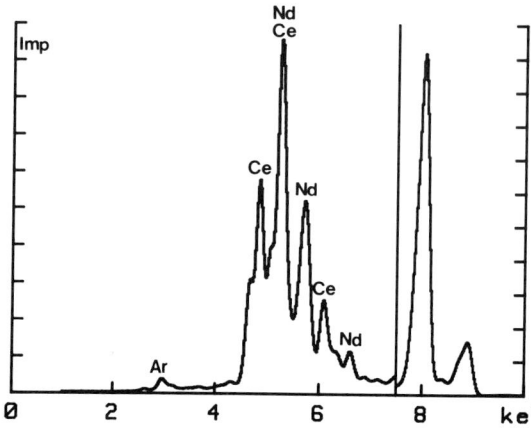

Abb. 2
Röntgenfluoreszenzspektrum von Monazitsand nach Abtrennung und Messung mit Cu-Target

In Abb. 2 sind nach der Abtrennung, insbesondere von Fe und Ba, die Linien des Cers
und Neodyms auswertbar. Wenn man dieselbe Probe über ein Molybdän-Target anregt, ist
auch die Auswertung von Thorium möglich (Abb. 3).

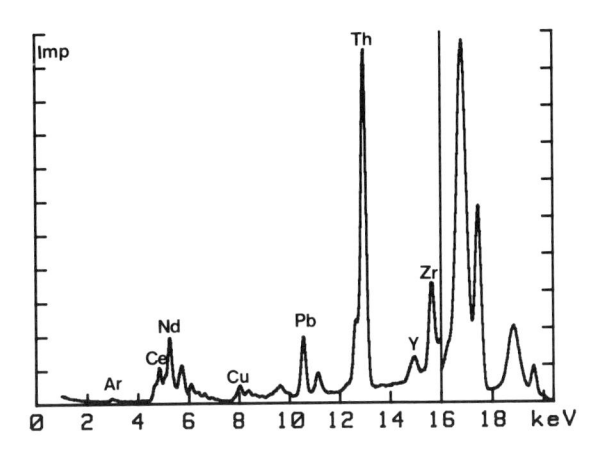

Abb. 3
Röntgenfluoreszenzspektrum
von Monazitsand nach Abtrennung und Messung mit
Mo-Target

Die Nachweisgrenzen der Seltenen Erden sind in Abb. 3 um den Faktor 150 höher als
in Abb. 2 bedingt durch die Sekundäranregung über das Molybdän-Target. In Mineralien
wie Monazit sind sowohl die S.E. als auch Thorium im Prozentbereich enthalten. Wenn
man biologische Proben untersucht, kann man von durchschnittlichen Gehalten im unteren ppm-Bereich für S.E. und im ppb-Bereich für Thorium und Uran ausgehen. Allerdings sind die sogenannten "Seltenen Erden" immer noch häufiger in der Erdrinde und
in biologischen Proben anzutreffen als z.B. Cd, Sb oder Ag. In Tab. 1 ist eine
Zusammenfassung über die Häufigkeiten, Nachweisgrenzen und Störungen der Analyse
der Seltenen Erden, Uran und Thorium gegeben. Außerdem sind die Gehalte der von
uns analysierten Boden- und Pflanzenproben aufgeführt. Die Reproduzierbarkeit der
Messung liegt bei allen analysierten Elementen bei 2 %.

Tabelle 1:

	Konzentration in der Erdrinde ug/g	Nachweisgrenze RFA ug	Stoerungen	Konzentrationen in Umweltproben	
				Boden	Pflanze
				ug/g Trockensubstanz	
57 La	30	2	Ti,V,Mn,Te,J,Cs	56	0.4
58 Ce	60	2	Ti,V,Cr,J,Cs,Ba	116	0.5
59 Pr	9	6	Ti,V,Cr,Fe,Ba		
60 Nd	28	8	Mn,Cs	46	0.2
61 Pm	–	–	–		
62 Sm	7	400	Fe,Ba		
63 Eu	1	400	Cr,Mn,Fe,Ni		
64 Gd	5	400	Cr,Co		
90 Th	12	52	Bi,Pb	20.5	–
92 U	4	28	Rb	3.7	–

In Abb. 4 ist ein Röntgenfluoreszenzspektrum einer Bodenprobe dargestellt.

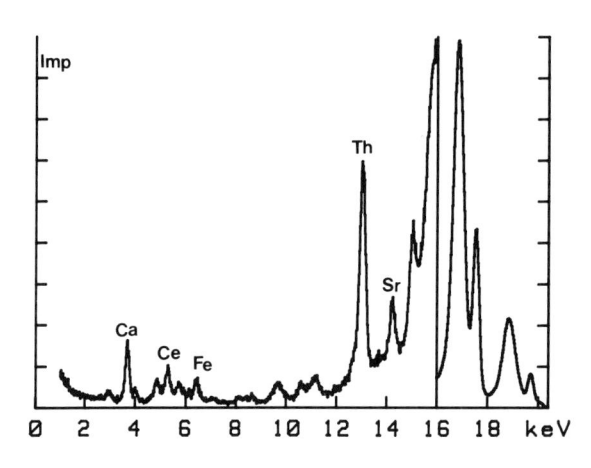

Abb. 4
Röntgenfluoreszenzspektrum
einer Bodenprobe.
Anregung über ein Mo-Sekundär=
target

Auch nach Abtrennung durch den Ionenaustauscher ist das Calcium noch deutlich zu sehen. Im Vergleich zu dem Spurenelement Thorium besitzt das Matrixelement Calcium zwar einen um den Faktor 1000 geringeren Verteilungskoeffizienten; es ist jedoch in Bodenproben um den Faktor 10^6 angereichert. Eine vollständige Abtrennung des Calciums nach einem einstufigen Ionenaustausch ist deshalb nicht möglich. Da die Analysenlinien des Calciums jedoch in einem Energiebereich liegen, die weder die Seltenen Erd- noch die Thorium- oder Uran-Bestimmung stören, kann auf eine nochmalige Aufarbeitung und Abtrennung verzichtet werden. Größere Schwierigkeiten treten bei dem Element Titan auf. Titan kann in seiner dreiwertigen Form als TiO_2 oder als Ti-Acetatkomplex vorliegen, das auf dem Austauscher zurückgehalten wird und somit in die Probe gelangen kann. In Abb. 5 ist das Röntgenfluoreszenzspektrum einer Fichtennadelprobe dargestellt.

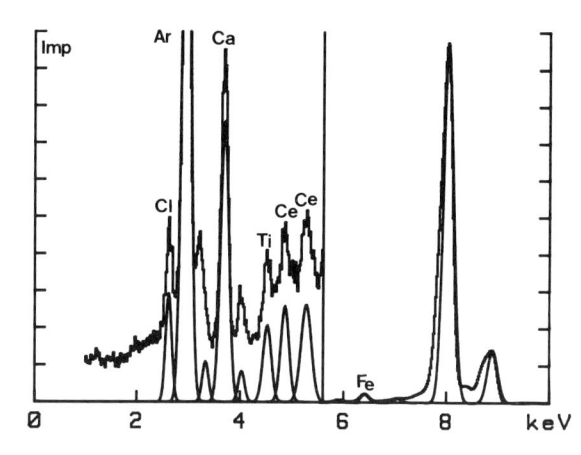

Abb. 5
Röntgenfluoreszenzspektrum von veraschten Fichtennadeln und Messung mit Cu-Target

In Abb.5 ist aus Gründen der Übersichtlichkeit eine etwas geänderte Form der Darstellung gewählt. Von 0 bis 5,6 keV ist das Spektrum mit einer größeren Empfindlichkeit gemessen, so daß die L-Linien der Seltenen Erden deutlich zu erkennen sind. Das durch das Computerprogramm errechnete und korrigierte Spektrum ist nach Abzug des Untergrunds in das Originalspektrum eingezeichnet.

Das Titan konnte nicht vollständig abgetrennt werden, obwohl die Aufschlußlösungen, bevor sie in den Ionenaustauscher gegeben werden, mit H_2O_2 versetzt werden, um das Titan in die vierwertige Form überzuführen. Ti^{4+} wird vom Ionenaustauscher nicht zurückgehalten. Die Konzentration des Cers in der Probe wurde zu 0,5 µg/g Trockengewicht bestimmt.

Neben den Störungen durch Elemente, die durch Ionenaustausch abgetrennt werden können, treten noch Überlagerungen der Seltenen Erden untereinander auf. In Abb. 6 ist reines Dysprosiumoxid vermessen, das Verunreinigungen von Europium, Holmium sowie Yttrium enthält. Die L-α und L-β-Linie des Holmiums wird von den Dysprosium-Linien vollständig überdeckt.

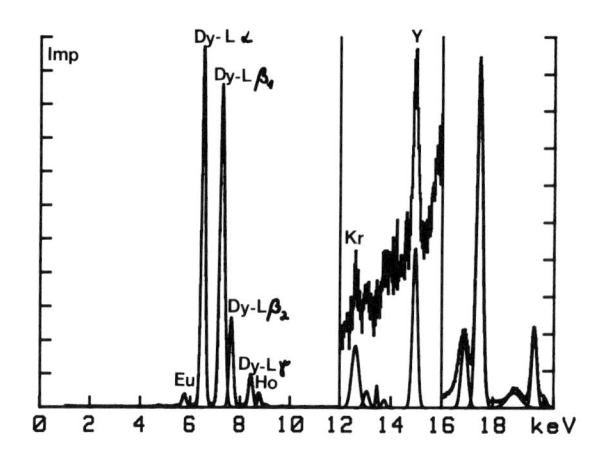

Abb. 6
Röntgenfluoreszenzspektrum von kommerziell erhältlichem Dysprosiumoxid (Fa.Fluka) und Messung mit Mo-Target

Auch in Abb. 6 ist eine Darstellung gewählt, die es erlaubt, das in der Probe enthaltene Yttrium zu sehen. In dem Energiebereich von 12 bis 16 keV wurde mit einer hundertfachen Verstärkung gearbeitet, um die Kα-Linie des Yttriums zu analysieren.

Es konnte weder Terbium, das aufgrund seiner chemischen und physikalischen Ähnlichkeit zum Dysprosium als Verunreinigung zu erwarten gewesen wäre, noch Gadolinum, das aufgrund seiner Häufigkeit erwartet werden könnte, nachgewiesen werden.

3. Zusammenfassung

Für die genannten Anwendungen der Bestimmung von Seltenen Erden, Uran und Thorium stellt die RFA eine simultane und schnelle Analysenmethode dar. Da Störungen durch Übergangselemente auftreten, muß eine Abtrennung und entsprechende Aufarbeitung der Proben vor der Analyse stattfinden. Die Nachweisgrenzen der RFA und das Vorkommen von Seltenen Erden, Uran und Thorium im ppb- bis in den unteren ppm-Bereich macht eine Anreicherung erforderlich, die durch Ionenaustausch erzielt wird. Die Überlagerung von Analysenlinien der Seltenen Erden untereinander wird bei der Auswertung der Spektren durch ein speziell für diese Problematik entwickeltes Computerprogramm wesentlich vereinfacht, so daß auch übereinander liegende Analysenlinien noch zugeordnet und ausgewertet werden können.

Literatur

1. K. Reinhardt, Chemie in unserer Zeit $\underline{1}$ (1984), 24-34
2. J. Korkisch, G. Arrhenius, Anal. Chem., Vol. $\underline{36}$ (1964), 250
3. G. Hartmann, Diplomarbeit, TH Darmstadt (1983)

MULTI-ELEMENT STANDARDS

O. Suschny

International Atomic Energy Agency, Laboratory Seibersdorf

SUMMARY

New instrumental methods of analysis have led to an unprecedented increase in sensitivity, while the progress in electronics has allowed the upgrading of these methods for a simultaneous analysis of many elements in one sample, with little processing. The results obtained, however, are frequently difficult to interpret. To determine and control the accuracy of these measurements which are often found wrong by orders of magnitude in intercomparisons, reference materials and, in particular, suitable multi-element standards are required. Reference materials for a few elements in industrial raw products such as metals, glasses or fine chemicals have been available on an increasing scale since many years. Multi-element standards of geological, environmental and biological materials have been developed more recently and not all needs are yet satisfied. Certification of them is frequently based on the evaluation of intercomparisons in which many laboratories participate. Experience in the preparation and use of some of the IAEA's multi-element standards is described by way of illustration.

1. INTRODUCTION

Development of inorganic analytical chemistry during the last decades has taken two main directions: instrumental methods, most of which are being discussed in these Proceedings, have led to an unprecedented increase in sensitivity, pushing elemental detection limits to concentrations which are between 10^4 and 10^6 times lower than those obtainable with classical methods. At the same time the progress in the use of electronics and microprocessing devices has allowed the upgrading of instrumental techniques to permit the simultaneous or the rapid sequential analysis of a large number of different elements in a sample, with need for only very simple processing.

The substitution of physical and electronic apparatus for the burette, the balance and the cuvette of classical chemistry has brought about many advantages; it has, however also led to increased complexity: the new data are easier to obtain but frequently more difficult to interpret in terms of individual element concentrations. The specificity of the measurements and the accuracy of their results have not been able to keep up with the increased sensitivity of the new methods. Instrumental instabilities, matrix and element interferences, cross-contamination of samples, impurities of reagents, adsorption and desorption effects lead to a loss of precision and to bias which latter is difficult to detect and still more difficult to eliminate.

Intercomparisons between laboratories have shown that an individual laboratory using only one preferred method is normally unable to determine its own accuracy. It can only determine its precision, but this has little to do with accuracy[1]. As an example, we may take the results of an intercomparison of multi-element determinations in a lake sediment sample which the Agency has organized some years ago[2]. Fig. 1 shows the results obtained in the case of cobalt which was present in the sample at ppm concentration. If the large mean X is taken to represent the reference value, individual laboratory means, shown with their standard deviations and standard errors, can be seen to be widely scattered around this value; there is little correlation between precision, as indicated by the length of the vertical lines and accuracy, as indicated by the distance of individual means from the large mean.

There are only two ways by which an individual laboratory can really control the quality of its results: by participation in intercomparisons with other laboratories and by the use of standards or, as they are now more correctly called, reference materials (RM's), which have been certified by a competent authority, usually on the basis of analyses carried out by many different analysts using different analytical methods.

2.SUPPLY OF REFERENCE MATERIALS

Reference materials did not become available on any significant scale until the early years of this century when the need for a control of routine analysis, in particular in iron and steel production was felt. One of the largest producers of reference materials today, the US National Bureau of Standards which now supplies about 1000 different materials, only started its work (with four cast iron standards) in 1906[3]. The International Standardization Organization (ISO) has recently issued a "Directory of Certified Reference Materials"[4] which was put together by its Committee on Reference Materials (REMCO). Although still incomplete, it lists 131 suppliers of reference materials, some of which are national or international organizations, some commercial firms. The number of composition RM's listed is 7000, of which about 4500 are for inorganic analysis. A breakdown of them into different cate-

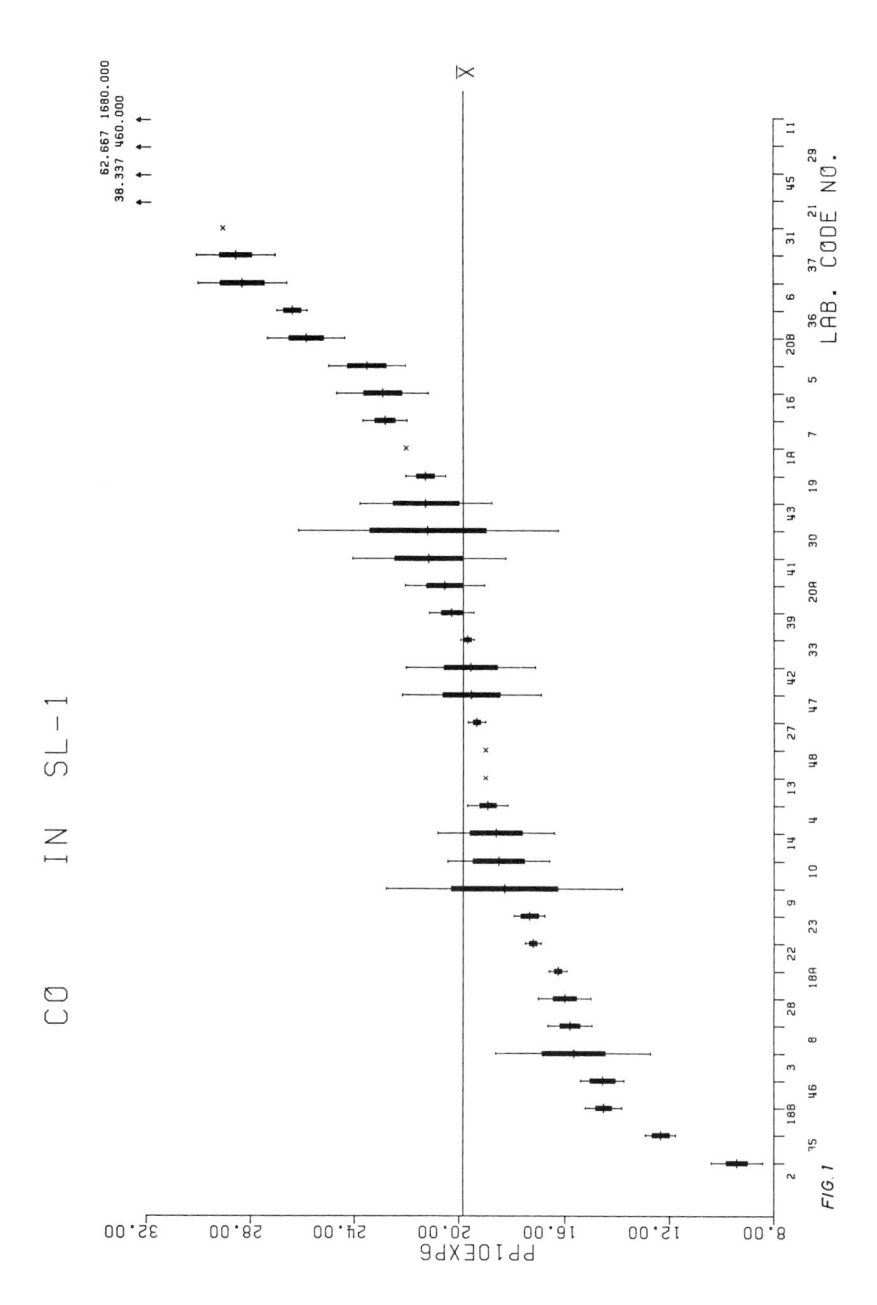

FIG. 1   Intercomparison of cobalt determinations in a lake sediment sample.
         See text for explanation of symbols

gories is given in Table 1. The figures are rounded and approximate only, since frequently an unknown number of individual RM's is listed in a single entry in the directory.

Table 1

Inorganic composition reference materials
in the ISO Directory

Type	Number	
	for all purposes	for multi-elem. anal.
Inorganic chemicals	1000	none
Nuclear and radioactive	600	none
Ferrous metals	1300	some
Non-ferrous metals	1200	some
Glasses, ceramics, etc.	60	20 - 30
Geological	100	20 - 30
Environmental	200	20
Biological	30	10

Most of the RM's listed are certified only for one or a few elements and can certainly not be regarded as multielement standards. This is true in particular for the inorganic chemicals and the nuclear and radioactive materials. The metals and the glasses form an intermediate category, since the present range of certification of them frequently includes a considerable number of elements (up to about thirty) some of which are present at low ppm concentrations. The typical multielement standards, however, are most often found in the categories of geological, biological and environmental materials.

3.SUPPLY OF MULTI-ELEMENT STANDARDS

A reasonable number of multi-element standards is available for the analysis of metals, their alloys and their oxides. Suppliers are the large national standardization and materials testing institutions (e.g BAM in Germany, AFNOR in France, etc). The NBS, in particular, also caters for the international market. Geological materials also are in reasonable supply, e.g. from NBS, the US Geological Survey, the Bureau of Analysed Samples (UK), the Canada Centre for Mineral and Energy Technology, the French Centre National de la Recherche Scientifique, the South African Council for Mineral Technology, the Instituto de Pesquisas Tecnologicas, Sao Paulo, Brazil, and others. Additional information may be found in a publication by Abbey[5] who lists 175 standard samples of silicate rocks and minerals, Flanagan[6] and Govindaraju[7].

Information on reference materials for the determination of minor and trace elements in biological materials has been assembled by Parr[8]. His listing, updated to include recent additions, amounts to 31 materials which are provided by only six laboratories as shown in Table 2: the Community Bureau of Reference (BCR) of the Commission of the European Communities, Brussels (3 materials), the Behring Institute, Marburg, FRG (5) the University of Reading (Dr. Bowen, 1), the International Atomic Energy Agency, IAEA (12), the NBS (9) and the National Institute for Environmental Studies (NIES) of the Japanese Environment Agency (1). Internationally known environmental multielement standards have still fewer suppliers as shown in Table 3: BCR (1), IAEA (6), NBS (8) and NIES (1). The highest number of elements listed is in IAEA's SOIL-5 (34 element concentrations certified and information values provided for additional 23).

Table 2

Biological oligo- and multi-element standards

Supplier	Material	Code	No. of elements listed*
BCR	Aquatic plant	CRM 60, CRM 61	6
BCR	Olive leaves	CRM 62	6
Behring	Blood	OSSD 21, OSSE 21	3
Behring	Urine	OSSA 52, OSSB 52 and OSSC 52	7
Bowen	Kale	–	27
IAEA	Milk powder	A-11	12
IAEA	Animal blood	A-13	10 + 4
IAEA	Animal muscle	H-4	14
IAEA	Animal bone	H-5	12
IAEA	Copepod	MA-A-1/TM	14
IAEA	Fish flesh	MA-A-2/TM	14
IAEA	Rye flour	V-8	11 + 10
IAEA	Cotton cellulose	V-9	13 + 15
IAEA	Hay	V-10	av.from 1985
IAEA	Mixed human diet	H-9	av.from 1985
IAEA	Horse kidney	H-8	in prep.
IAEA	Human hair	HH-2	av.from 1986
NBS	Oyster tissue	SRM 1566	19 + 9
NBS	Wheat flour	SRM 1567	10 + 6
NBS	Rice flour	SRM 1568	12 + 5
NBS	Orchard leaves	SRM 1571	25
NBS	Citrus leaves	SRM 1572	22
NBS	Tomato leaves	SRM 1573	14 + 12
NBS	Pine needles	SRM 1575	15 + 11
NBS	Bovine liver	SRM 1577a	15
NBS	Freeze-dried urine	SRM 2670	13 (in prep.)
NIES	Pepperbush	SRM No.1	16

* Where two numbers are shown, the second refers to elements for which only non-certified information values are provided.

Table 3

Non-biological environmental multi-element standards

Supplier	Material	Code	No. of elements listed*
BCR	Fly ash	SRM 038	10
IAEA	Air filters	Air-3/1	13 + 4
IAEA	Soil	SOIL-5	34 + 23
IAEA	Soil	SOIL-7	in prep.
IAEA	Lake sediment	SL-1	28 + 36
IAEA	Marine sediment	SD-N-1/2	in prep. for 85
IAEA	Fresh water	W-4	19 (in 1985)
NBS	Coal	SRM 1632a	18 + 12
NBS	Coal fly ash	SRM 1633a	19 + 15
NBS	Fuel oil	SRM 1634	in prep.
NBS	Coal (subbituminous)	SRM 1635	14 + 10
NBS	Water	SRM 1634a	17 + 1
NBS	River sediment	SRM 1645	18 + 1
NBS	Estuarine sediment	SRM 1646	16
NBS	Urban particulate	SRM 1648	14 + 19
NIES	Pond sediment	SRM ...	information incomplete

* Where two numbers are shown, the second refers to elements for which only non-certified information values are provided.

4. PROBLEMS OF PREPARATION AND USE OF MULTI-ELEMENT STANDARDS

The present trend goes towards certification of a large number of elements, including trace elements, in natural matrices which may be quite heterogeneous to start with. Homogeneity is produced by careful grinding to very small particle size and thorough mixing. Also, while in the past most standards were certified on the basis of in-house analysis carried out by a few analysts using well-established reference methods, many present-day multi-element standards are products of interlaboratory testing, frequently involving the co-operation of thirty to fifty laboratories, which are using many different methods[9]. Certification based on analysis distributed over many laboratories lacks the simplicity of individual laboratory certification and raises administrative as well as statistical problems. It is, however, the more robust method, since small deviations will tend to be distributed statistically and any gross errors may be detected and eliminated as outliers.

As an example we may take the determination of elements at ppb concentrations in an artificial water sample[10]. In spite of considerable scatter of the individual values, there was no statistically significant difference between the large means and the known input values of 14 out of the 17 trace elements put into and analysed in

this sample. Fig. 2 gives the results for lead. It shows intralaboratory means with
their standard deviations and standard errors, single values (x), outliers (arrows),
the large mean X and the input value X. This proves that with the aid of quite simple
statistical methods even relatively heterogeneous data can be used to arrive at estimates of concentrations which are close to the "true" value in the certification of
reference materials. Fig. 3 shows an example taken from the preparation of a multi-
element standard based on a biological material, freeze-dried animal blood[11]. The
data shown were obtained in the determination of magnesium in this material. They are
presented in the form of a "density plot" in which the frequency of results falling
within a certain concentration interval is plotted against concentration. The original data are shown as circles at the bottom of the figure, on top are shown the mean
and the median with their confidence intervals and the mode. Details can be found in
the original publication.

6.CONCLUSION

A thorough calibration of multi-element determinations is obviously needed. Such a
calibration can be achieved by participation in an intercomparison between many laboratories using different methods, or by the use of competently certified reference
materials ("multi-element standards") which most frequently are prepared in such
intercomparisons. Supply of multi-element standards is still not abundant, in particular there is a lack of such materials for determinations in organic and environmental matrices.

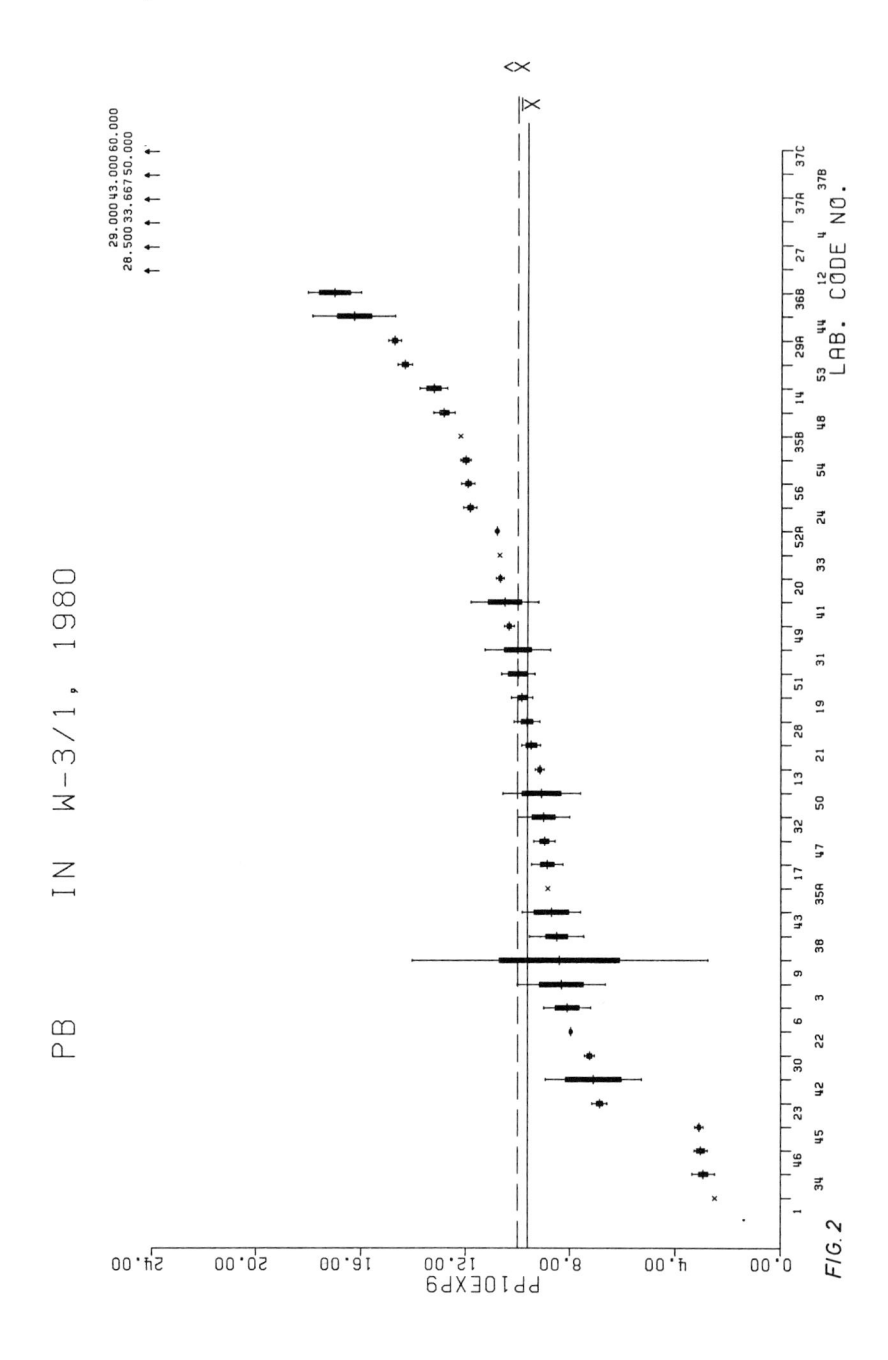

FIG. 2 Intercomparison of lead determinations in an artificial water sample. See text for explanation of symbols.

MG   IN A-13, 1981

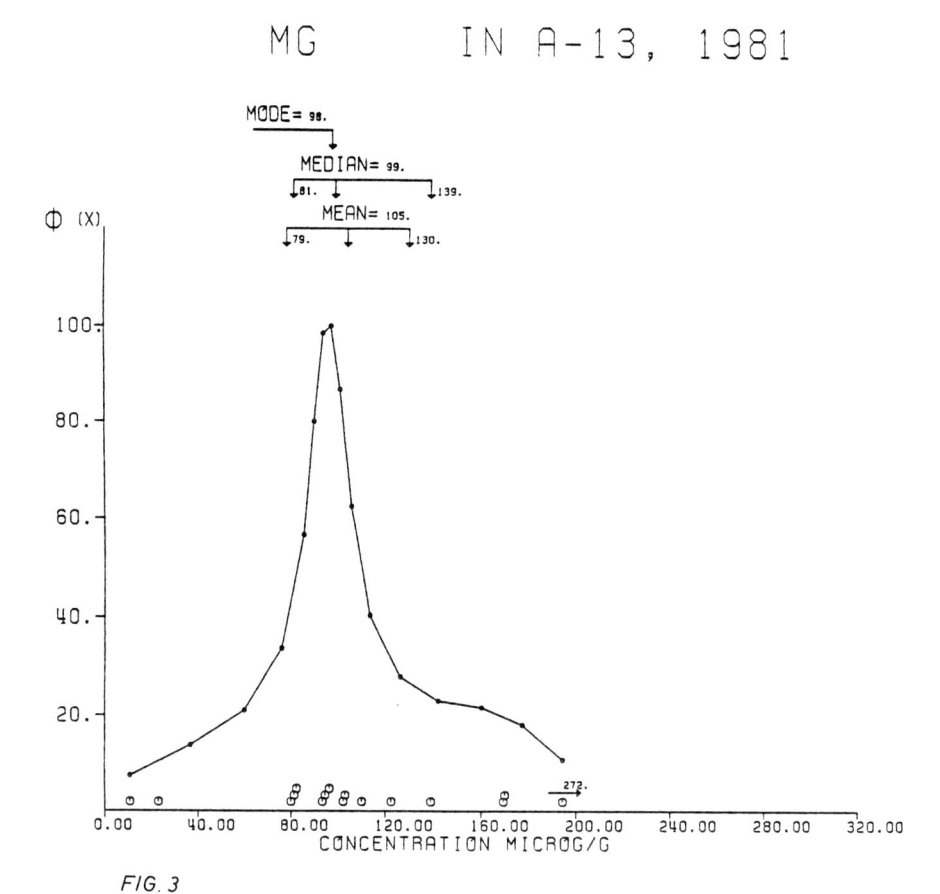

FIG. 3 Intercomparison of magnesium determinations in animal blood. The curve shows the results in the form of a density plot (see text for explanation).

REFERENCES

[1] R. Dybczynski, A. Tugsavul and O. Suschny, Analyst 103, (1978), 733-744.

[2] R, Dybczynski and O. Suschny, "Final Report on the Intercomparison SL-1 for the Determination of Trace Elements in a Lake Sediment Sample, IAEA/RL/64, International Atomic Energy Agency, Vienna 1979.

[3] US Department of Commerce, NBS Special Publication 260, NBS Standard Reference Materials Catalog, 1981-83 Edition, Washington, 1981.

[4] International Organization for Standardization, Directory of Certified Reference Materials, Geneva 1982.

[5] S. Abbey, Studies in Standard Samples of Silicate Rocks and Minerals 1969-1982, Geological Survey Canada Paper 83-15, 1983.

[6] F.J. Flanagan, Reference Samples in Geology and Geochemistry, in preparation (private communication)

[7] K. Govindaraju, Geostandards Newsletter, in preparation (private communication).

[8] R.M. Parr, Survey of Currently Available Reference Materials for Use in Connection with the Determination of Trace Elements in Biological Materials, IAEA/RL/103, International Atomic Energy Agency, Vienna 1983.

[9] O. Suschny, The Analytical Quality Control Programme of the International Atomic Energy Agency, IAEA/RL-35, International Atomic Energy Agency, Vienna 1976.

[10] L. Pszonicki, A. Veglia and O. Suschny, Report on Intercomparison W-3/1 of the Determination of Trace Elements in Water, IAEA/RL/94, International Atomic Energy Agency, Vienna 1982.

[11] L. Pszonicki, A.N. Hanna and O. Suschny, Report on Intercomparison A-13 of the Determination of Trace Elements in Freeze-dried Animal Blood, IAEA/RL/98, International Atomic Energy Agency, Vienna 1983.

MULTI-ELEMENT STANDARDS FROM OXIDE POWDERS FOR SEQUENTIAL X-RAY SPECTROMETERS

C. Freiburg, W. Reichert, and A. Solomah*

Central Department for Chemical Analysis, Nuclear Research Establishment Jülich, D-5170 Jülich
*Whiteshell Nuclear Research Establishment, Pinawa Manitoba ROE 1LO, Canada

Multi-element standards containing 1 permill of each element are needed for an extensive calibration of X-ray spectrometers. Such working standards are prepared from mixtures of oxide powders. Two different matrices were used: Al_2O_3, 0.3 µm[1], and TiO_2, < 1 µm[2]. Oxides are usually added for the standard elements from Ti to U; one of several exceptions is $NaAsO_2$ for As. The standard compounds are selected according to the following criteria: a) no line interference, b) no secondary enhancement, c) no grain size effects. The last point is the most important one. In order to achieve good accuracy, grain sizes must be less than 1 µm for XRF analysis, especially for rare earth oxides [1]. Such grain sizes and the necessary homogeneity can be obtained by ball-milling[3]. Optimum milling conditions are: 1 l porcelaine jar, 3 different sizes of corundum balls, 1/3 of the volume of the jar full with the balls, powder weight 5 % of the ball weight and 6 hours of milling [2]. This means for 1 l jars: 162 balls with 1 cm diameter, 48 balls with 2 cm diameter and 6 balls with 3 cm diameter, and a charge of 50 g powder. Control of grain sizes with a light microscope in the final pellet is only possible for coloured or dark compounds like PbO_2 or Pr_6O_{11}. After milling methyl/n-butyl-methacrylate copolymer[4] is added to the powder as a binder, typical amounts being 10 g powder, 2 ml of 12 % copolymer solution in acetone, i.e., 2.4 wt% for the binder. After the evaporation of the acetone the mixture is pressed into pellets under 10 to of pressure. This pressure is necessary to produce stable pellets which can be measured in vacuo without contamination of the spectrometer. There is no influence of pressure on the X-ray intensity if the powders are sufficiently fine. The pellets must be dried and evacuated before measurement.

The measurements were performed on two different sequential X-ray spectrometers: 1) PW1410[1], Cr side window tube and 2) S/MAX[2], Rh front window tube with 40, 50, and 60 kV excitation. Only such elements were used which can be measured with the NaJ(Tl)-photodetector and the LiF(200)-analyzing crystal. Elements which are enhanced by the tube lines are avoided. These elements can be divided into 3 groups:

I) Ti to Sr (no 22 to 38), $K\alpha$ lines on the low energy side of the 16.6 keV sensitivity maximum,

II) Sr to Ba (no 38 to 56), $K\alpha$ lines on the high energy side of the sensitivity maximum and on the maximum itself.

III) La to U (no 57 to 92), $L\alpha_1$ lines which are all on the low energy side of the sensitivity maximum.

The results of the elements from groups I) and III) can be represented by linear functions.

(1) $\log I = A - B \cdot \sin \vartheta$ (or $\log I = f(1/E)$),
where I is the Intensity (counts/sec/°/$_{oo}$, ϑ is the Bragg angle, E is the energy (keV), and A and B are calibration constants. The results of the elements from group II) can be represented by the function

(2) $\log I = C - D \cdot (E - E_0)^2$ (or $\log I = F(E^2)$),
where E_0 is the sensitivity maximum near the Tl-L_{III} absorption edge of the photodetector, and C and D are calibration constants. The formulars (1) and (2) allow the interpolation of not measured elements in the specified ranges. Curves for a Cr side-window tube (60 kV, 40 mA) and Al_2O_3 matrix for groups I and II are shown in the computer plot of Fig. 1. Changes in sensitivity by factors of 10 and more are obvious. The stochiometry of the Pd-, Rh-, and Ru-oxides used is not very well known; neither is the value of SeO_2 very reliable. Other compounds will be used instead in the future. The border between the groups and the overlapping in this region has been tested but the point is not very clear yet. From standards with 0.5, 1, 1.5, and 2 per mil content linear calibration curves for single elements may be obtained [3]. Lower limits of detection can be calculated according to (3) L = $3 \cdot \sqrt{b/m}$, where b is the background intensity, m is the slope of the linear calibration curve, and L is the lower limit of detection. The combination of formulas (1), (2), and (3) is not yet possible, mainly because difficulties with the background.

We should like to thank Mrs. M. Melchers for programming and data processing, Miss M. Plum, Mr. T. Pieper, and Mr. M. Römer for experimental help. Thanks are due to Prof. B. Sansoni who supported and discussed this work.

REFERENCES

[1] F. Claisse and C. Samson, Adv. in X-Ray Anal. 5, 335 (1962)
[2] T.M. Hare and H. Palmour III in: Ceramic Processing before Firing, eds. G.Y. Onda jr. and L.L. Hench, Wiley-Interscience New York (1978)
[3] Dataflex 350B, Software Description, Rigaku Ind. Corp., Osaka (1981)

MATERIALS AND INSTRUMENTS

1) Linde A, Union Carbide; 2) RLK Kronos, 99 %; 3) Schwinherr and Haldenwanger; 4) Elvacite 2013; Dupont; 5) Philips; 6) Rigaku.

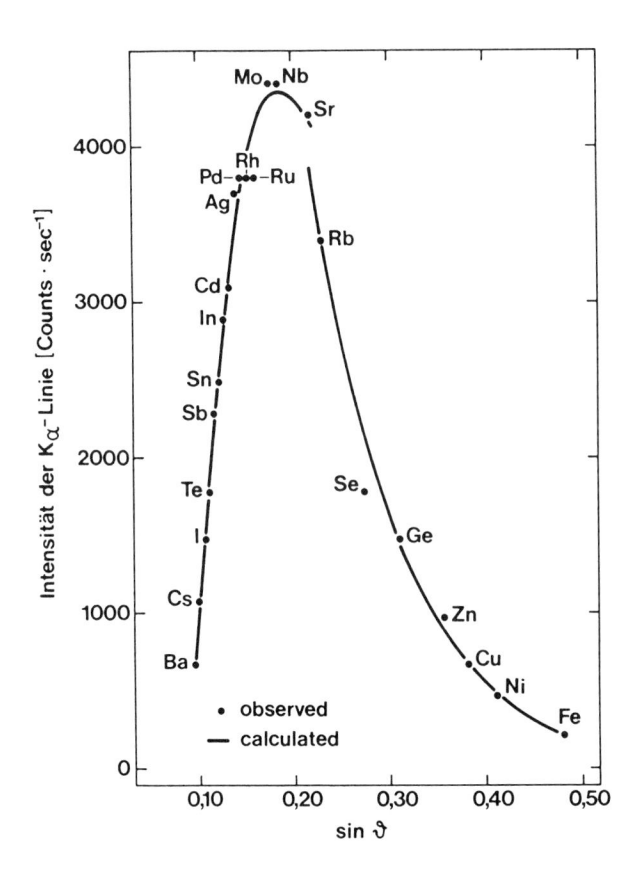

Fig. 1: Curves for the sensitivity performance of a PW 1410 spectrometer, explanations in the text.

DATENBEURTEILUNG IN DER INDUSTRIELLEN MULTIELEMENTANALYTIK

Dr. Karl-Heinz Koch
Hoesch Hüttenwerke AG, Chemische Laboratorien, D-4600 Dortmund 1

ZUSAMMENFASSUNG

Die Beurteilung multielementanalytischer Daten werden hinsichtlich ihrer Güte und ihrer prozeß- oder produkttechnischen Relevanz betrachtet. Die Analysendaten dienen der Prozeßüberwachung und der Charakterisierung von Betriebsstoffen und Produkten oder haben Bedeutung für umwelttechnische Fragestellungen.

Für das genannte Gebiet haben simultan messende atomspektroskopische Multielement-Methoden, wie die Emissionsspektrometrie mittels Funken-, Glimmlampen- und Plasmaanregung, die Röntgenfluoreszenz- und die Massenspektrometrie, überragende Bedeutung erlangt. Sie erlauben die (schnelle) instrumentelle Multielementanalyse einer Vielzahl von Elementen in weiten Bereichen und in zahlreichen Materialien.

Die dem Analytiker in allen Fällen gestellte allgemeine Aufgabe besteht in der Optimierung der Untersuchungstechnik und mit ihrer Hilfe in derjenigen des Herstellungsablaufes von Erzeugnissen. Die Methodenoptimierung ist damit eine strategische Aufgabe, die nur bei kritischer Betrachtung des analytischen Umfeldes gelöst werden kann.

Die Datenbeurteilung ist natürlich nur ein Schritt in der Funktionskette, die mit der Probenahme beginnt und bei der qualitätskontrollierten und gesicherten Datenausgabe endet. Datenbeurteilung bedeutet Interpretation von Untersuchungsergebnissen, die folglich nur von einem Analytiker vorgenommen werden kann. Die Beurteilung von Analysendaten geschieht in verschiedenen technischen Bereichen mit sehr unterschiedlichen Zielsetzungen.

1. EINLEITUNG

Der Titel dieses Beitrages enthält bereits eine programmatische Aussage über den zu behandelnden Problemkreis: Die Beurteilung multielementanalytischer Daten hinsichtlich ihrer Güte und ihrer prozeß- oder produkttechnischen Relevanz; das Adjektiv "industriell" weist aus, daß diese chemisch-analytische Tätigkeit prozeßbezogenen (d. h. verfahrenstechnischen, z. B. zeitlichen) Forderungen genügen und wirtschaftlichen Erfordernissen gerecht werden muß [1].

Die Analysendaten dienen der Prozeßüberwachung und der Charakterisierung von Betriebsstoffen und Produkten oder haben Bedeutung für umwelttechnische Fragestellungen. Diese Aufzählung deutet bereits die vielfältigen Verknüpfungen von Analytik und technischem Geschehen an. Entsprechend vielfältig sind die Beurteilungsprozesse, an denen der Analytiker mit seiner Arbeit beteiligt ist, und es gibt einen umfänglichen Katalog von Beurteilungskriterien für die verschiedenen Stoffe, mit denen der Analytiker umgehen muß. Das können außer dem Hauptgebiet der Analyse von Zwischen- und Fertigprodukten z. B. die Beurteilung verschiedener Analysendaten eines Mineralöles oder eines Lösemittels oder die Interpretation der Prüfergebnisse von Brennstoffen sein. Obwohl hierbei auch im allgemeinen mehrere Eigenschaftswerte ermittelt werden, sind diese Vorgänge natürlich nicht dem Bereich der Multielementanalytik zuzurechnen. Auf dem Gebiet der Wasseranalytik haben sich inzwischen automatisierte fotometrische Verfahren zur Multikomponentenanalyse eingeführt, die ebenfalls nicht als Multielementverfahren im Sinne dieser Arbeit betrachtet werden können.

An typischen Beispielen aus der Stahlindustrie, in der die instrumentelle Multielementanalytik [2] die herausragende Rolle spielt, soll die Thematik der Datenbeurteilung aus chemisch-analytischer und verfahrens- oder produkttechnischer Sicht erläutert werden.

2. VERFAHREN UND AUFGABEN DER MULTIELEMENTANALYTIK

Für das genannte Gebiet haben simultan messende atomspektroskopische Multielement-Methoden, wie die Emissionsspektrometrie mittels Funken-, Glimmlampen- und Plasmaanregung, die Röntgenfluoreszenz- und die Massenspektrometrie, überragende Bedeutung erlangt. Sie erlauben die (schnelle) instrumentelle Multielementanalyse einer Vielzahl von Elementen in weiten Bereichen und in zahlreichen

Materialien (<u>Bild 1</u>).

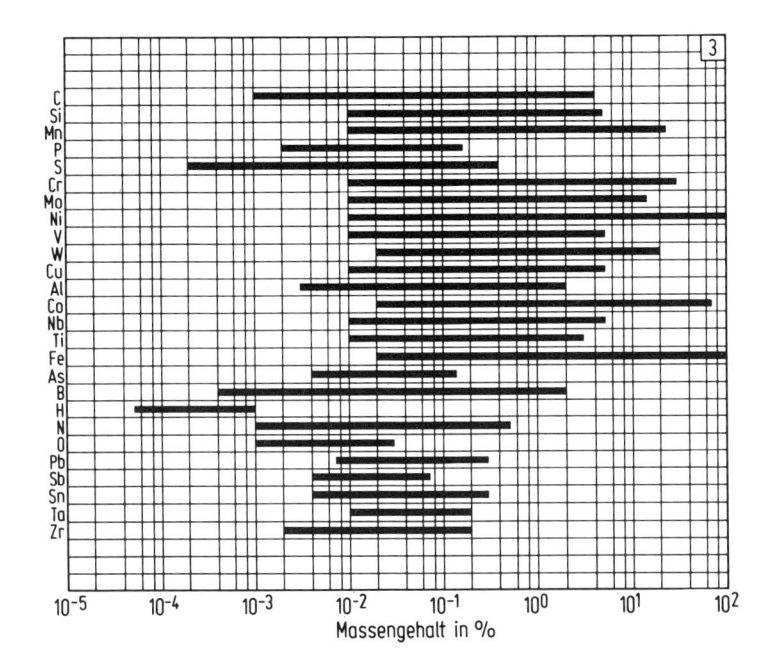

Bild 1: Arbeitsbereiche der instrumentellen Multielementanalytik;
Beispiel: Eisen und Stahl

   Hierbei handelt es sich sowohl um metallische wie oxidische
Stoffgruppen. Als oxidische Roh- und Hilfsstoffe für die verschiedenen Prozeßstufen sind z. B. zu nennen Erze, Sinter, Kalk, Dolomit, feuerfeste Baustoffe und zu den Nebenprodukten zählen die metallurgischen Schlacken und die Schlämme und Stäube, die bei den metallurgischen Prozessen in verschiedenen Aggregaten abgeschieden werden. Die metallischen Stoffe umfassen Ferrolegierungen, verschiedene Metalle und Schrott als Einsatzmaterialien, Roheisen und Rohstahl als Zwischenprodukte und Stahl in vielfältiger Zusammensetzung und Form als Endprodukt. Die zu bestimmenden Anteile reichen über mehrere Zehnerpotenzen hinweg (siehe Bild 1), dabei sind häufig 30 und mehr Elemente in einer Probe zu erfassen [2].

   Die Weiterentwicklung des Werkstoffes Stahl im Hinblick auf die unterschiedlichsten Verwendungszwecke forderte in neuerer Zeit die Entwicklung und Anwendung mikro- und spurenanalytischer Methoden. Letztere wurden notwendig, da bereits Elementanteile im Stahl in der Größenordnung von mg/kg die Werkstoffeigenschaften entscheidend beeinflussen können. Die Lösung dieser Aufgaben gelang außer mit der AAS auch problemabhängig mit multielementanalytischen Methoden der OES, ICP-OES und MS.

Die Entwicklung, die die Analytik von Eisen und Stahl genommen hat, ist symptomatisch für viele Bereiche der angewandten Analytischen Chemie. Die steigenden, aus der Verfahrens- und Werkstofftechnik resultierenden Anforderungen führten zu komplizierterer Analysentechnik und zu erhöhtem Untersuchungsumfang, damit verbunden dann zu höherem Aufwand (Kosten) und erhöhter (spezieller) Qualifikation des analytischen Personals.

Die dem Analytiker in allen Fällen gestellte allgemeine Aufgabe besteht in der Optimierung der Untersuchungstechnik und mit ihrer Hilfe in derjenigen des Herstellungsablaufes von Erzeugnissen. Die Methodenoptimierung ist damit eine strategische Aufgabe, die nur bei kritischer Betrachtung des analytischen Umfeldes (z. B. Zeit- und Genauigkeitsforderungen, Erkennen systematischer Fehler) gelöst werden kann [3]. Im Falle der Untersuchung von Eisen und Stahl und ihrer Ausgangsstoffe waren es die simultanen Methoden der Spektrometrie, die in Verbindung mit dem Einsatz von elektronischen Rechenanlagen zahlreiche Rationalisierungsmöglichkeiten boten und große Rationalisierungserfolge bedeuteten [4,5].

Eine Teilaufgabe bestand dabei in der Anpassung der Analysenverfahren an den prozeßbedingten Datenfluß [6].

Dahinter verbergen sich die Beurteilung der Probenbeschaffenheit im Zusammenhang mit den gewählten Analysenverfahren, sodann die systematische Untersuchung der verschiedenen Verfahrensparameter und ihr Einfluß beispielsweise auf die benötigte Analysenzeit. Das Ergebnis ist ein Verfahren, das den prozeßtechnischen Erfordernissen Rechnung trägt und dabei für den jeweiligen Zweck ausreichend gesicherte Ergebnisse liefert.

Damit sind die Methoden skizziert, mit deren Hilfe die prozeß- und produktrelevanten Analysendaten erstellt werden; Rechenanlagen sorgen für die Datenverarbeitung und -verwaltung und garantieren den prozeßgerechten Datenfluß vom produzierenden Betrieb über das Laboratorium bis zur Produktauslieferung.

3. DATENBEURTEILUNG

Die Datenbeurteilung ist natürlich nur ein Schritt in der Funktionskette, die mit der Probenahme beginnt und bei der qualitätskontrollierten und gesicherten Datenausgabe endet. Datenbeurteilung bedeutet Interpretation von Untersuchungsergebnissen, die folglich nur von einem Analytiker vorgenommen werden kann. Für diesen Teilaspekt, der das meßtechnische Verfahren und die Interpretation der Untersuchungsergebnisse umfaßt, wird neuerlich auch der Begriff "Chemometrie" verwendet, der damit keinesfalls ein Synonym für Analytische Chemie sein kann [7].

Die Beurteilung von Analysendaten geschieht in sehr verschiedenen technischen Bereichen einer Produktionskette mit sehr unterschiedlichen Zielsetzungen.

1. Datenbeurteilung im chemisch-analytischen Bereich:

1.1 Beurteilung von Meßergebnissen hinsichtlich der Probengüte und der "Analysenqualität".

Der hier angedeutete Vorgang ist von besonderer Bedeutung in der Emissionsspektrometrie kompakter Proben, die bei der Analyse von Roheisen und Stahl nach wie vor die am weitesten verbreitete Methode ist. Wegen der an die Analyse gestellten Zeitforderungen ist es wichtig, nur die Zusammensetzung von Proben ausreichender Güte (Homogenität; keine Lunker) zu ermitteln. Unbrauchbare Proben müssen frühzeitig erkannt und verworfen werden. Dieses Problem kann durch die "analytische Beurteilung" der spektrometrischen Vorfunkphase gelöst werden [8]. Dabei wird die Vorfunkphase in 3 Abschnitte unterteilt (<u>Bild 2</u>): Zunächst wird 2 s vorgefunkt, dann während der folgenden 4 s des Vorfunkens die elektrische Ladung des Eisen-Kanals (Referenzkanal) integriert und - während das Vorfunken noch 1 s weiterläuft - der erhaltene Spannungswert mit einem vorgegebenen Wert verglichen. Stimmen Meßwert und Vorgabe überein, wird die Analyse automatisch fortgesetzt, im anderen Falle wird sie sofort abgebrochen und unter Umständen eine neue Probe angefordert.

Diese Verfahrensweise setzt natürlich Voruntersuchungen über die Höhe des Fe-Meßwertes von einwandfreien Proben voraus, um den Vorgabewert festlegen zu können [8]. Hier findet also zunächst die Beurteilung eines analytischen Meßsignals statt, die aber dann für die Präzision und Richtigkeit einer Vielzahl von Elementen von grundsätzlicher Bedeutung ist.

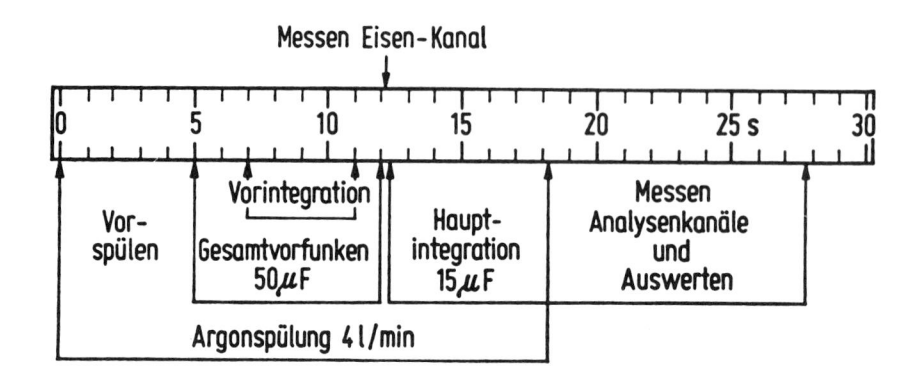

```
              Messen Eisen-Kanal
                    ↓
|0    5    10   15   20   25 s  30|
  Vor-    Vorintegration    Haupt-    Messen
 spülen  Gesamtvorfunken  integration  Analysenkanäle
              50 μF        15 μF      und
                                      Auswerten
         Argonspülung 4 l/min
```

Bild 2: Zeitablauf der spektrometrischen Stahlanalyse mit kontrolliertem Vorfunken

Bei der Datenverarbeitung spektrometrischer Meßsignale erfolgt noch ein weiterer Schritt: Die Beurteilung der Analysendaten hinsichtlich ihrer Qualität (Güte). Darunter ist die Beurteilung von Doppel- oder Mehrfachbestimmungen anhand vorgegebener Werte für die Präzision der Analysendaten zu verstehen. Das setzt die Ermittlung der möglichen Streubereiche für die einzelnen Elemente unter Festlegung der gewünschten statistischen Sicherheit voraus. Im allgemeinen dienen die an homogenen Proben durch Mehrfachbestimmungen erhaltenen doppelten Standardabweichungen (2 s) als Beurteilungskriterium. Abweichungen von diesen Vorgaben werden als Ausreißer ausgewiesen und fordern eine weitere Analyse zur Entscheidungsfindung an. Die am besten im Sinne der genannten Festlegung übereinstimmenden Werte der Multielementanalyse werden dann gemittelt und als Endergebnis weitergeleitet. Führen nun beispielsweise auch 3 oder 4 Untersuchungen zu jeweils voneinander stark abweichenden Analysendaten, so liegt in der Regel eine inhomogene Probe vor und die Analysenwerte werden wegen ihrer mangelnden Aussagefähigkeit verworfen; die Analyse einer neuen Probe ist erforderlich.

Es kann aber auch der Fall vorliegen, daß eine von diesen Mehrfachuntersuchungen durchaus zu richtigen Werten geführt hat. Hier kommt dann unter Umständen eine Form der Datenbeurteilung zum Zuge, die nachfolgend kurz erläutert werden soll.

1.2 Beurteilung von Analysendaten hinsichtlich ihrer Plausibilität.

Trotz aller Automatisierung bei der Analyse und der Datenerstellung sollte der analytisch geschulte Mensch in den Prozeß der Datenbeurteilung eingeschaltet bleiben. Im Falle der instrumentellen Multielementanalyse von Roheisen und Stählen kann das z. B. auf folgende Weise geschehen.

Nachdem der im vorhergehenden Abschnitt geschilderte Prozeß der analytischen Datenverarbeitung abgelaufen ist, hat der Analytiker vor der endgültigen (automatischen) Abgabe der Analysenergebnisse bei entsprechendem Rechner- und Bedienungskomfort des Analysensystems die Möglichkeit, sich die Analysenergebnisse noch einmal auf einem Bildschirm oder Drucker aufzurufen und mit den dem Produktionsbetrieb vorgegebenen Zielgrößen zu vergleichen. Darüber hinaus kann er die Analysendaten mit allen an den im Produktionsablauf voraufgegangenen Proben ermittelten (gespeicherten) Werten im Zusammenhang betrachten. Diese Überprüfung der Plausibilität der Analysendaten durch den Analytiker stellt dann die Endstufe der Datenbeurteilung im Bereich der chemischen Analytik dar. Dabei können unter Umständen auch solche Werte zweckdienlich sein und (mit bestimmten Vorbehalten) weitergeleitet werden, die sich zunächst als nicht ausreichend zuverlässig erwiesen haben (siehe oben unter 1.1).

Selbstverständlich kann man eine derartige Aufgabe einer abschließenden Bewertung analytischer Daten auch einem Rechner übertragen. Der sachkundige Analytiker hat sich jedoch in dem hier dargestellten analytischen Bereich gegenüber dem Rechner als überlegen gezeigt, da sich alle analytischen und technischen Wechselbeziehungen, die dem abschließenden Denk- und Beurteilungsprozeß zugrunde liegen, nur schwer in einem Rechnerprogramm vollbefriedigend niederlegen lassen.

1.3 Beurteilung der Ergebnisse bei zeitaufgelösten Untersuchungsabläufen, wie z. B. bei der Oberflächenanalytik.

Bisher war von Multianalysen die Rede, deren Ergebnisse sich in Zahlen vollständig ausdrücken lassen und keiner weiteren Interpretation bedürfen. Außer dieser Elementanalytik haben instrumentelle Methoden Bedeutung erlangt, die ebenfalls

zum Bereich der Multielementanalytik gehören, die aber nicht
nur die globale Zusammensetzung eines Stoffes zu beschreiben
sondern besondere Eigenschaften und Zustände eines Werkstoffes
zu charakterisieren gestatten. Hier ist z. B. die Emissionsspektrometrie mit einer Glimmlichtanregungsquelle(GDOS) zur
Analyse von Werkstoffoberflächen zu nennen [9].

Die GDOS liefert in wenigen Sekunden Multielementinforma-
tionen, die sich als Tiefenprofile der verschiedenen Elemente
interpretieren lassen (Bild 3). Es gibt ferner inzwischen Mög-

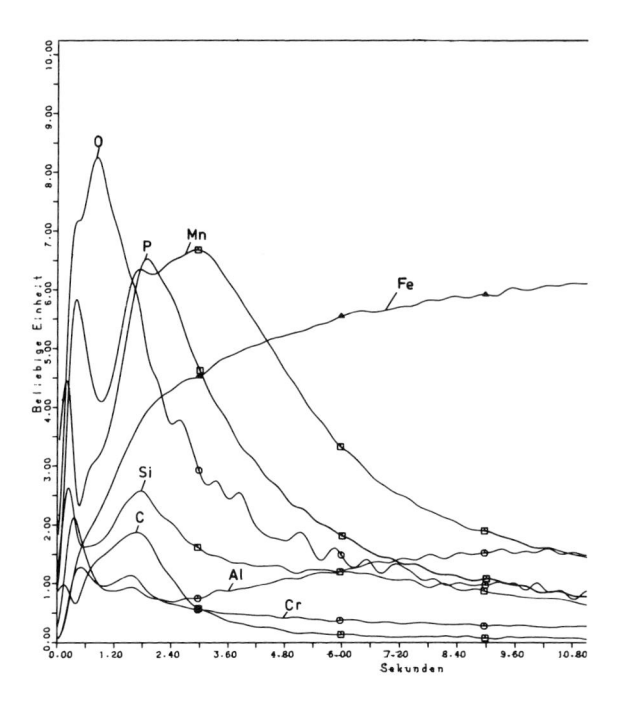

Bild 3: Emissionsspektrometrische Multielementanalyse einer Stahloberfläche

lichkeiten - bei Kenntnis verschiedener Untersuchungsparame-
ter - die gemessenen Linienintensitäten in quantitative Analysenergebnisse umzusetzen [10]. Letzteres entspricht dem üblichen Verfahren der Datenverarbeitung und -beurteilung spektrometrischer Analysen. Hier kommt aber folgender Aspekt
hinzu. Die Untersuchungsergebnisse werden in Form von zeitaufgelösten Kurvenzügen erhalten, die sich auf einem Bild-

schirm oder Plotter unter Vorgabe verschiedener Auswahlkriterien darstellen lassen. Durch die Interpretation dieser Kurvenzüge können qualitative und quantitative Aussagen über die verschiedenen "Schichten" eines Werkstoffes, ihre Dicke und ihre chemische Zusammensetzung, gemacht werden. Der Vorgang der Datenbeurteilung ist damit wesentlich umfassender als bei anderen analytischen Abläufen.

1.4 Beurteilung von Analysendaten hinsichtlich der Einhaltung oder Überschreitung von festgelegten Grenzwerten.

Im Bereich der Wasseranalytik oder der Umweltanalytik gilt es häufig, Analysendaten hinsichtlich der Einhaltung oder Überschreitung von festgelegten Grenzwerten zu beurteilen. Das kann nur unter Berücksichtigung der Charakteristik des Analysenverfahrens (also nur durch den sachkundigen Analytiker) erfolgen.

Auf dem Gebiet der Wasser- und Abwasseranalytik gibt es Grenzwerte für bestimmte Inhaltsstoffe [11] und Bewertungsgrößen [12], deren Einhaltung durch die Analytik festzustellen ist bzw. deren Festlegung analytische Untersuchungen zur Voraussetzung hat. Sofern hier die Bestimmung der Kationen angesprochen ist, handelt es sich meistens um eine Aufgabe für die instrumentelle Multielementanalytik, z. B. die ICP-Spektrometrie.

Im Bereich der Umweltanalytik, die fast immer Spurenanalytik bedeutet, kommt der Beurteilung der Analysenergebnisse besondere Bedeutung zu, da an diese Daten oft weitreichende Entscheidungen für die Technik und die Gesellschaft geknüpft werden [13]. Durch einfache Umrechnung von möglicherweise mit systematischen Fehlern behafteten Analysenergebnissen auf z. B. Staubemissionsraten für eine Region oder Belastungen von Gewässern in einem bestimmten Zeitraum würden unter Umständen falsche Schlüsse gezogen und nicht hinreichend begründete Entwicklungen eingeleitet werden [14]. Datenbeurteilung heißt hier vor allem Kenntnis über die zufälligen Fehler aller Untersuchungsschritte, insbesondere derjenigen der Probenahme und Probenvorbereitung. Ferner muß sie viele weitere Faktoren miteinbeziehen, wie z. B. die Probenhäufigkeit innerhalb eines bestimmten Zeitraums und die Frage der Repräsentativität der Proben. Nur in den seltensten Fällen werden on-line-Messungen möglich sein, so daß einem aufzustellenden Prüfplan grundsätzliche Bedeutung zukommt. Schließlich ist die Qualität des Analysenverfahrens (Präzision und Richtigkeit) in die Datenbeurteilung miteinzubeziehen.

Wegen der Vielfalt und Vielschichtigkeit der offenen Fragen ist die Diskussion auf diesem Gebiet noch lange nicht als abgeschlossen zu betrachten [14].

2. Datenbeurteilung im Produktionsbereich.

Auftraggeber für die analytischen Industrielaboratorien sind vor allem die Produktionsbetriebe. Der Datenfluß von den Laboratorien zu diesen Betriebspunkten und die verschiedenen Beurteilungsprozesse innerhalb des analytischen Bereiches wurden im Vorstehenden aufgezeigt. Die beim Hersteller eines Produktes eintreffenden Analysenergebnisse werden nun als Beurteilungskriterium für die Feststellung der Übereinstimmung dieser Daten mit den Zielvorgaben, z. B. der chemischen Zusammensetzung einer Schmelze, genutzt. Auftretende Abweichungen führen sofort zu Maßnahmen, die eine zielgerichtete Beeinflussung des Prozeßablaufes bewirken.

3. Datenbeurteilung bei der Endprüfung von Erzeugnissen.

Der letzte Beurteilungsvorgang auf dem Wege vom Rohstoff zum Endprodukt findet vor der Auslieferung des Erzeugnisses statt. Bei metallischen Werkstoffen wird anhand der Ergebnisse einer Endprüfung beurteilt, ob die im Rahmen von Lieferbedingungen oder -verträgen festgelegten Anteile, meist einer Vielzahl von Elementen, eingehalten worden sind.

Die analytische Beurteilung geschieht anhand der bekannten Kriterien [15], wie Standardabweichung, Wiederholbarkeit, Vergleichbarkeit, Vertrauensintervall u. ä., so daß an dieser Stelle auf die entsprechende Literatur verwiesen werden kann [16].

4. Sicherstellung von Analysenergebnissen.

Nach Ausarbeitung vollständiger Analysenvorschriften ist ein reproduzierbarer Formalismus eine notwendige Voraussetzung für eine gesicherte Analytik ("Gute-Analytische-Praxis" - GAP -) [17]. Hinzu kommen als weitere Maßnahmen zur Sicherstellung von Analysendaten [18] die Durchführung von Ringuntersuchungen [19], die problemgerechte Verwendung von Referenzmaterialien und die Anwendung unabhängiger Verfahren. Im Bereich der Umweltanalytik stellen sich in diesem Bereich besondere, in vielen Fällen bisher ungelöste, Probleme.

LITERATUR

1) K. H. Koch, Arch. Eisenhüttenwes. 53 (1982) 97 - 100.
2) K. Ohls, Vortrag, Symposium Instrumentelle Multielementanalyse, Jülich, 2. - 5. April 1984.
3) G. Tölg, Vortrag, 8. Internationales Symposium für Mikrochemie, Graz 1980.
4) K. H. Koch u. W. Loose, Radex-Rundschau 1979, 931 - 939.
5) K. H. Schmitz, E. Thiemann, Vortrag, Symposium Instrumentelle Multielementanalyse, Jülich, 2. - 5. April 1984.
6) K. H. Koch, K. Ohls, G. Becker, Arch. Eisenhüttenwes. 41 (1970) 25 - 28.
7) E. Bayer, H. Kelker, Naturwissenschaften 70 (1983) 473 - 479.
8) K. H. Koch, K. Ohls, G. Becker, Arch. Eisenhüttenwes. 45 (1974) 17 - 22 .
9) M. Kretschmer, K. H. Koch, D. Grunenberg, Z. anal. Chem. 314 (1983) 226 - 234.
10) K. H. Koch, M. Kretschmer, D. Grunenberg, Mikrochim. Acta 1983 II 225 - 237.
11) F. K. Ohnesorge, Gas- und Wasserfach (GWF) 121 (1980) 515 - 522.
12) WLB (Wasser, Luft u. Betrieb) 1979, 16 - 18.
13) B. Sansoni, Sicherheit Chem. Umwelt 1 (1981) 195 - 198.
14) A. Scholz, Staub - Reinhalt. Luft 41 (1981) 304 - 309.
15) Anal. Chem. 55 (1983) 2210 - 2218.
16) K. Doerffel u. K. Eckschlager, "Optimale Strategien in der Analytik", VEB Deutscher Verlag f. Grundstoffind., Leipzig 1981/ Verlag H. Deutsch, Thun 1981.
17) J. Brauner, K. H. Koch, G. Staats, E. Thiemann, H. Wünsch, Stahl u. Eisen 102 (1982) 1047 - 1051.
18) K. H. Koch, Stahl u. Eisen 103 (1983) 449 - 452.
19) K. Ohls u. D. Sommer, Z. anal. Chem. 312 (1982) 195 - 220.

ANALYTISCHE UND GEOCHEMISCHE KONTROLLE DER MULTIELEMENTANALYTIK GEOLOGISCHEN MATERIALS AUF STATISTISCHER GRUNDLAGE

D.SAUER

Bundesversuchs- und Forschungsanstalt Arsenal Wien,
Geotechnisches Institut   POB 8, A-1031 Wien

1. EINLEITUNG

Die Anwendung der Multielementanalytik in der geochemischen Analyse verlangt eine andere Form der Qualitätskontrolle als sie bisher für Einelementmethoden üblich gewesen ist, umso mehr als auch eine höhere Anforderung an die Genauigkeit und an die Vergleichbarkeit des Datenmaterials gestellt wird (1).

2. BEGLEITENDE QUALITÄTSKONTROLLE

Die Multielementanalyse von Großserien bedarf einer begleitenden Qualitätskontrolle, die bereits bei der Kalibrierung der Analysenverfahren einsetzen muß. Sie wird in Probenpaketen zu 100 bis 200, maximal zu etwa 1000 Proben abgewickelt. Voraussetzung ist eine Kalibrierung mit SRM's (Standard Reference Materials). Systematische Fehler sind mit Sicherheit nur in Serienanalysen erkennbar. So können bei der Kalibrierung mit Hilfe von Prozeßrechnern zum Ausgleich von Matrixeinflüssen Überkorrekturen vorkommen. Derartige systematische Fehler und mögliche Koinzidenzen lassen sich mit Hilfe einer Korrelationsmatrix daran erkennen, daß ungewöhnlich gute Korrelationskoeffizienten auftreten.

2.1 QUALITÄTSKONTROLLE FÜR ELEMENTE MIT DOPPEL- UND MEHRFACHANALYTIK

Die Doppel- und Mehrfachanalytik erlaubt die Auswahl der besten Methode für die Probenmatrix, die Nachweisgrenze, den Konzentrationsbereich, die Wiederholbarkeit, Genauigkeit und Zuverlässigkeit.

Die Beurteilung kann durch Laborstandards, Testproben und Testprobenserien oder durch den Vergleich des Datenmaterials erfolgen. Dazu kann eine einfache statistische Auswertung mit arithmetischen und geometrischen Mittelwerten einschließlich deren Standardabweichungen, Mittelwertvergleichen, Maximal- und Minimalwerten, Parametern der Häufigkeitsverteilung ebenso dienen wie die Darstellung der Häufigkeitskurven, Korrelationskoeffizienten bis zur Anwendung der Hauptkomponentenanalyse. Die Doppelanalytik hat auch den Vorteil, Probenverwechslungen, Ausreißer,

Koinziden etc. rasch erkennen zu können. Externe Kontrollanalytik ist kaum erforderlich.

2.2 QUALITÄTSKONTROLLE FÜR ELEMENTE OHNE INTERNE KONTROLLANALYTIK

Abgesehen von externen Kontrollanalysen gibt es auch die Möglichkeit, statistische Methoden unter Verwendung geochemischer Beziehungen, zum Beispiel geochemische kohärenter Elementpaare, wie La/Ce, La/Y, Th/Zr, Th/Ce, K/Rb u.a. zur Überprüfung des Datenmaterials einzusetzen.

3. QUALITÄTSKONTROLLE NACH ABSCHLUSS DER GROSSSERIENANALYSE

Wenn die Großserienanalyse über einen größeren Zeitraum erfolgt ist, in dem Änderungen in der personellen oder methodischen Abwicklung eingetreten sein können, ist auch eine retrospektive Kontrollanalytik, aufschlußreich, die wenigstens 2 % der Proben erfaßt. Man kann so Rückschlüsse auf die langfristige Genauigkeit des Analysenprozesses ziehen.

Bei Anwendung der Doppelanalytik ist eine gute langfristige Wiederholbarkeit dann gegeben, wenn die beiden Häufigkeitsverteilungen einander entsprechen.

Die Erfolgskontrolle, ob das geochemische Datenmaterial der geologischen Wirklichkeit entspricht und vor allem der von der Natur vorgegebenen geochemischen Elementverteilung nahe kommt, ist am besten an Hand der Betrachtung des Einzelelementes und der Interelementbeziehungen in Variationsdiagrammen oder mittels der Multivariateanalyse festzustellen Vereinzelte, nicht erklärbare Anomalien, können nur durch Nachprobung verifiziert werden.

Generation of accuracy in multielement systems by reconstitution of the sample (II).*)

Gotthard Staats, AG der Dillinger Hüttenwerke, Laboratorium, Postfach 15 80, D-6638 Dillingen, Federal Republic of Germany

Only after the analytical process can an opinion be formed about the quality of the accuracy of the results obtained. They may contain systematic errors exceeding the repeatability of the method. These errors are not predictable and for their part, are subject to statistic partitions, limiting the quality of accuracy unnecessarily.

This state of affairs does not comply with the scientific and legal criteria of Quality Assurance. This problem is solved by incorporating the active generation of accuracy into the analytical method, in order to permit the specification, provide evidence for, and to control, the accuracy.

Evidence for analytical results is provided only by their interference - free deduction from accepted reference materials: i.e. from a defined number of atoms of the analyte in the reference sample preferably in the same amount as present in the analytical sample.

The condition for an interference - free transfer of the confidence in the reference to that in the analytical sample is realized by removing all differences of interference between the two kinds of samples.

This means identity of the reference sample and the analytical sample, and is produced deliberately by noting all possible technical variables as follows,

- performing multielement analysis for all the relevant elements.
- reconstituting the sample weight by its compounds in their pure form (sample matching), following procedures to be described.
- accomplishing identical physical and chemical conditions for the reference sample and the analytical sample (isoformation).
- applying no mathematical matrix correction.

*) Part I : Staats, G : Fresenius Z. anal. Chem. (1983) 315: 1 - 5

Provided these conditions are observed, the interval of uncertainty of the analytical results will be a direct measure of the accuracy.

Using the procedure described, the difference between absolute and relative methods becames meaningless.
Obviously, the historical princible of the analytical mode of operation will be abolished in favour of a synthesizing one. This contradiction is well justified technically, semantically clear but may be mentally difficult to accept.

Results

This method permits the production of a secondary reference material even of a very complex matrix in amounts of some tens of kilograms, comprising homogenization and analysis of all analytically relevant elements with proved accuracy. This operation takes about 60 hours to complete. As compared to commercially available certified reference materials, these samples have the advantage of defined accuracy and even of superiority by their abundance of information, availability and adaptability to production programs.

The means of intervals of uncertainty of the accuracy were determined on 56 of about 300 secondary reference samples after an effort of work, of which a medium-sized industrial laboratory is capable.

The results in Table 1 were obtained mostly by X-RFA (X-ray fluorescence analysis). The dependence observed of these results on the ranges of concentration is determined by the repeatability of this analytical method. When applying methods with better efficiencies at concentrations < 1 %, better results should be expected as indicated by Table 1.

As expected and according to the 10 years experience with this method of reconstitution, the repeatability of the analytical method proved to be the limiting factor for the accuracy of the analytical results. Thus the excellent repeatability of X-RFA at high concentrations permits control of the homogeneity of the sample material, within the limits of the mass of analytical sample applied.

Table 1. Intervals of uncertainty of the accuracy, calculated from 4 to 14 independent reconstitutions; level of probability P = 95 %.
Means of all elements analyzed in 56 of 300 secondary reference samples within one range of concentration.
26 types of material: steel, ferro-alloys, copper-alloys, lime, dolomite, gravel, feldspar, fluorspar, fire brick, iron ores, slags,

Range of concentration [%]	Intervals of uncertainty P = 95 %		n
	m $[\% \cdot 10^4]$	M $[\%]$	
100 - 10	1253	0,25	11
10 - 1	342	0,68	15
1 - 0,1	138	2,80	16
0,1 - 0,01	61	12,00	22
0,01 - 0,001	14	28,00	12
0,001 - 0,0001	1	20,00	3

m = average of the intervals of uncertainty of the accuracy of all analytes in all samples in one range of concentration.

M = relative average of the intervals of the accuracy, referred to the mean of the concentration range.

n = partial number of elements analysed

Total number of all analytes: 24; C,Mg,Al,Si,P,S,Ca, Ti,V,Cr,Mn,Fe,Co,Ni,Cu,Zn,Zr,Nb,Mo,Sn,Ba,W,Pb.

MULTI-ELEMENT TRACE ANALYSIS OF GEOTHERMAL WATERS : PROBLEMS, CHARACTERISTICS AND APPLICABILITY

R. Vandelannoote, W. Blommaert, L. Van't dack, R. Van Grieken and R. Gijbels

Department of Chemistry, University of Antwerp (U.I.A.), B-2610 Wilrijk, Belgium

SUMMARY

Geothermal water samples were collected from numerous sources in France and, after pretreatment, analyzed by NAA, SSMS, XRF, AAS, potentiometry. A data set with over 5000 trace element results was obtained for the dissolved and particulate phases, offering a unique opportunity to compare the precision, accuracy and elemental range of the various techniques. Procedures to avoid contamination, storage and filtration problems were developed. The applicability of panoramic multi-element analysis in geothermal water research was thoroughly evaluated and illustrated.

INTRODUCTION

Trace and ultra trace data are only seldomly applied in geothermal energy research due to difficulties in interpreting data, to sampling and storage problems and the necessity to invoke sophisticated analysis techniques of high sensitivity. In an effort to evaluate the usefulness of multi-element trace data in this research field, water samples from some 60 sources in the French E. Pyrenees and Vosges were analysed by various techniques. This was done in the framework of EEC-research projects, in collaboration with French teams of the Bureau de Recherches Géologiques et Minières (Orléans) and the Département de Géochimie des Eaux, Université de Paris-7 (Paris) who also performed major component, dissolved gas and isotopic composition analyses. A total of well over 5000 elemental determinations in the dissolved and suspended phase were carried out. All the measured data are available in final EEC-reports (1-5).

The thermal sources all had emergence temperatures between 22 and 76°C. Most showed sulfide species in solution and exhibited the typical characteristics for so-called nitrogen-alkaline thermal springs, e.g. high pH (8.0-9.7), low total dissolved solid content (< 1g/l), trace elements typical for a granito-gneissic environment.

EXPERIMENTAL

Figure 1 illustrates schematically the different overall sampling, preparation and analysis schemes followed by the Antwerp and Paris teams. As indicated, the invoked physical preconcentration step was either freeze-drying or evaporation to small volumes or to dryness on graphite or on Whatman-541 filter paper. Chemical preconcentration involved co-crystallization with 1-pyridylazo-2-naphtol (PAN) and, for Br^-, precipitation with $AgNO_3$. The analysis techniques were : neutron activation analysis (NAA) via an ultra-short, short and long cycle, atomic absorption spectroscopy (AAS), energy-dispersive X-ray fluorescence (XRF), spark source mass spectrometry (SSMS) with photoplate detection, and potentiometry. Details of the analytical procedures have been reported earlier (1, 3, 6-9).

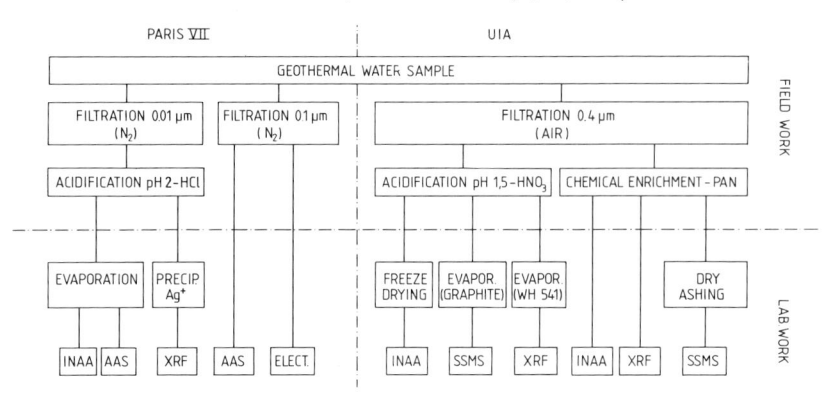

Figure 1 : Representation of the overall sampling and analysis scheme.

SAMPLING AND SAMPLE TREATMENT

Most of the studied thermal waters are used for medical treatments, they are brought to the surface through different types of tubings or conducts, often since ancient times. When sampling from taps, the water was allowed to flow for at least 30 min. No evidence of contamination from tubings was ever found, probably since most waters are oversaturated with respect to metal sulfides and have little affinity for additional metal uptake, or since the inside of the tubings is usually covered with precipitated secondary minerals which prevent direct contact.

New 1 l polyethylene bottles are to be preferred for storage; their excellent properties are well documented in the literature. A thorough cleaning of the containers appeared mandatory, in order to minimize leaching of mainly zinc. The procedure implying 1 week soaking in analytical grade 1/1 HCl, 1 week in analytical grade

1/1 HNO$_3$ and several weeks in bidistilled water proved to be satisfactory.

The pore size of the filter used for the separation of suspended matter from dissolved elements proved to have a major influence on the final results. Although using a 0.4 µm pore-size is an accepted procedure for differentiating between "dissolved" and "suspended" matter, in many waters clay colloids having sizes from 0.5 to 0.1 µm introduce serious discrepancies, especially for elements such as Al, Mn, Fe, Ti. The French group applying 0.01 µm pore size filters found appreciably lower amounts of "dissolved" trace elements than the UIA group with 0.4 µm filters (e.g. 3-5 times for Fe, Cu, Eu,...). Colloids such as amorphous sulfides and claylike particles (such as kaolinite, chlorites and illites) were indeed detected to be present.

The storage time prior to analysis should be controlled. Although the waters were acidified to pH 1-1.5 and stored in the dark, white precipitates were formed after a few months and a decrease in trace element concentration by up to a factor of 10 was observed.

EVALUATION OF THE ANALYSIS TECHNIQUES

In spite of the favourable enrichment factors, the chemical procedure involving co-crystallization was not found to be satisfactory for this type of analyses : many geochemically interesting elements were not collected, and the collection yields were found to vary significantly (up to a factor of 10) for water types with different major ion composition.

Table I lists typical _detection limits_ for INAA and SSMS after freeze-drying or evaporation as a physical preconcentration step. About 40 elements were usually assessed in geothermal waters by combining both techniques.

Table I : Detection limits in low salinity geothermal waters.

Element	Detection limit (µg l^{-1}) INAA	SSMS	Element	Detection limit (µg l^{-1}) INAA	SSMS	Element	Detection limit (µg l^{-1}) INAA	SSMS
F	80	-	Cu	30	0.1	Cs	0.07	0.1
Na	20	-	Zn	0.80	0.15	Ba	25	0.4
Mg	15	-	Ga	-	0.2	La	0.05	0.4
Al	5	-	Ge	-	0.4	Ce	0.50	0.5
Cl	20	-	As	0.80	0.15	Sm	0.06	1.5
K	1500	-	Br	0.50	-	Eu	0.02	1
Ca	20	-	Rb	4	0.06	Lu	0.01	1
Ti	20	0.08	Sr	40	0.07	Hf	0.02	1.5
V	0.10	0.08	Zr	20	0.7	W	2.5	4
Cr	0.20	0.15	Mo	2	1.3	Au	0.003	1
Mn	0.09	0.07	Ag	0.9	0.1	Hg	0.3	-
Fe	50	0.1	Cd	20	0.6	Pb	-	1.3
Co	0.25	0.1	Sb	0.02	0.4	Th	0.08	1.5
Ni	-	0.1	I	1	-	U	0.12	1.5

In spite of the low concentration levels involved and the sampling problems, the <u>accuracy</u> of NAA was acceptable since the results agreed quite well with those of the other techniques. The average ratios of the results for 8 identical geothermal sources, obtained via SSMS and NAA were e.g. 1.13 for As, 1.04 for Mo, 1.07 for Cs, 0.78 for W. For elements not likely to be initially affected by the pore-size of the filter used, like F, Cl, Ca, Rb and Cs, the results of AAS or potentiometry as found by the French group for 15 identical sources differed from those of NAA by a factor of 0.92, 1.17, 0.94, 0.85, 0.76, respectively.

Typical <u>standard deviations</u> per measurement for the concentration levels encountered in sulfide geothermal waters were e.g. in the case of NAA : 7 % for F, 5 % for Na, 9 % for Ca, 9 % for As, 22 % for Mo, 10 % for Cs, 7 % for W; in the case of SSMS : 68 % for As, 33 % for Mo, 23 % for Cs, 32 % for W; in the case of AAS : 4 % for Ca, 15 % for Rb, 6 % for Cs.

APPLICATIONS OF TRACE ELEMENT DATA IN GEOTHERMAL RESEARCH

Most characteristics of geothermal waters and of their deep reservoir can be assessed directly, at very large expense, by drilling a borehole at a location of possibly economic geothermal energy interest. This study has shown, however, that trace element data of the emerging thermal waters can be a very valuable and economic alternative in a number of cases :
- The deep reservoir temperature is one of the main parameters for a successful exploitation of geothermal energy. Classical empirical chemical geothermometers based on one or two elements, often lead to incoherent results especially for temperatures below 150°C, which are most common. Trace elements such as Li, Rb, Cs, Ti, Sr, Mo and W were found to reflect still the deep reservoir temperature conditions, even at the emergence of a source. A combination of many element concentration data in a chemical multi-element trace geothermometer yields more acceptable results, since it is less sensitive towards some secondary effects.
- Pressure at depth can be calculated when the pressure dependence of equilibria is known. Especially Mg-minerals equilibria are sensitive indicators.
- The temperature and depth of the oxidation of e.g. sulfides can be assessed using dissolved arsenic levels, and assuming mineral equilibria towards As_2S_3.
- Mixing of thermal water with shallow ground water can be detected not only through conventional tritium measurements but also by using data for dissolved Li, Rb, Cs, As, W. A better collection of the hottest water component is thus possible, leading to a more efficient exploitation of the heat source.
- The nature of the geological setting is clearly reflected in thermal water by the suite of dissolved elements present : a multi-element trace analysis of the water will not only yield general information on the geological structure, but also detailed information on the extent of a given formation at depth, and on hidden mineralisation. E.g. in granite, valuable rare metal mineralizations of Sn, W, Mo, Ta are

difficult to localize; a panoramic trace element analysis of water, that cross cuts such veins, may be a good indicator.
- Problems concerning scaling,corrosion and disposal of geothermal fluids can only be correctly solved after trace element analysis of the thermal water. Precipitation of secondary minerals can be foreseen and adjustment of a parameter (e.g. pH) can decrease the amount of scaling drastically. The optimal way of disposal of cooled thermal fluids (directly in the river net or e.g. through reinjection) merely depends of the composition of the primary water.

In addition to these practical applications of multi-element analysis of thermal water, other more fundamental fields (thermodynamics, mineralogy, ore formation, etc.) may also benefit from this type of work.

Since trace element data are at present still too scarce, it is often felt that the resulting conclusions are still rather speculative. We feel,however, that more attention should be paid to this promising field.

REFERENCES

1. R. Gijbels, R. Van Grieken, W. Blommaert, R. Vandelannoote and L. Van't dack
    Final Report, EUR 6440 (1979),120 pp.
2. R. Gijbels, R. Van Grieken, W. Blommaert, R. Vandelannoote and L. Van't dack
    Final Report, EEC-contract nr. 576-78-4 EGB (1981), 43 pp.
3. R. Gijbels, R. Van Grieken, W. Blommaert, R. Vandelannoote and L. Van't dack
    Final Report, EUR 8871 (1983), 149 pp.
4. J.C. Baubron, B. Bosch, P. Desgranges, J. Halfon, M. Leleu, A. Marcé and C. Sarcia
    Final Report, EEC-contract nr. 178-77 EGF (1977), 31 pp.
5. G. Michard, C. Fouillac, G. Ouzounian, M. Evrard, C. Gourmelon, M. Demuynck, D. Lavergne and M. Javoy
    Final Report : EEC-contract nr. 147-76 EGF (1976), 40 pp.
6. G. Ouzounian, Ph.D. thesis, Paris-7-University (1978), 87 pp.
7. W. Blommaert, R. Vandelannoote, L. Van't dack, R. Gijbels and R. Van Grieken
    Journ. Radioanal. Chem., vol. 57 (1980), pp. 382-400.
8. R. Vandelannoote, W. Blommaert, R. Gijbels and R. Van Grieken
    Frezenius Zeitschrift Anal. Chem., vol. 309 (1981), pp. 291-294.
9. M. Vanderstappen and R. Van Grieken
    Talanta, vol. 25 (1978), pp. 653-658.

INTERDEPENDENCE OF SELECTIVITY AND PRECISION IN MULTIELEMENT ANALYSIS

Matthias OTTO[1] and Wolfhard WEGSCHEIDER[2]

[1] Department of Chemistry, Bergakademie Freiberg,
9200 Freiberg, G.D.R

[2] Institute for Analytical Chemistry, Micro- and Radiochemistry,
Technical University Graz, Technikerstraße 4,
A-8010-Graz, AUSTRIA

Quantitative multielement analysis requires analytical sensors of sufficient selectivity to operate for samples of varying complexity. Not infrequently in the emergence of multielement techniques it have been limitations in selectivity that eventually led to the replacement of one method by another one. Yet, to date neither on the exact definition of selectivity behold on its influence on precision and accuracy analytical chemists seem to have agreed upon. This contribution is to demonstrate that at any level of reproducibility the obtainable precision with non-specific sensors is determined by the condition number of the calibration matrix.

Multielement methods that give linear signal/concentration (or mass) relationships can be quantified by solving the system

$$\underline{S}\,\underline{c} = \underline{Y} \qquad /1/$$

for the concentration vector $\underline{c}$. In equ. /1/ $\underline{S}$ stands for the sensitivity matrix and $\underline{Y}$ for the signal matrix. The calibration step carried out beforehand is to determine $\underline{S}$ from a suitable set of standards. Numerical methods for handling equ./1/ are given in (1) and are fairly well established in Analytical Chemistry (see e.g. 2,3). The total error of such an analysis cannot be stated explicitely but can only be bounded by some maximum level depending on the condition number of $\underline{S}$ and the relative errors in $\underline{S}$ and $\underline{y}$, the signal vector of the unknown (1). Problems arise in cases of high levels of noise, time drifts, deviations from linearity or miscalibration along the wavelength (or mass) axis (4). Under such circumstances the calibration model /1/ would need to be extended by further terms to correct for those effects. Lacking the knowledge of the appropriate model extension it was found better feasible to rotate the signal and concentration matrices by principal component analysis such that discrimination against all deviations from the original model /1/ becomes possible.

Of the factors affecting the results, the condition number of the calibration matrix $\underline{S}$ appears to be a numerically consistent measure of the degree of collinearity of a multielement method and has been shown (5) to be at least qualitatively related to Kaiser's original definition (6) of selectivity. The condition number is calculated from the norms of the sensivity matrix $\underline{S}$ and of its inverse $\underline{S}^{-1}$ as

$$\text{cond } (\underline{S}) = //\underline{S}// \; //\underline{S}^{-1}// \qquad /2/$$

$//\underline{S}//$ denoting the norm of $\underline{S}$ and calculated from square roots of the highest and the reciprocal smallest Eigenvalue (7). The condition number of a fully selective system is unity.

A large number of computer simulations were run to evaluate the influence of selectivity in combination with variables such as noise level, number of sensors, component number, signal strength ratios and band separation on accuracy of the analytical results. Adjusting the levels of those variables within limits met in practical spectroscopic analysis the relative analytical error was found to be less than 20 % as long as the condition number is lower than 150.

An experimental test was conducted for a simultaneous multielement determination for Ni^{2+}, Co^{2+}, Cu^{2+}, Fe^{3+}, and Pd^{2+} by UV/VIS spectroscopy of their diethyldithiocarbamate complexes. Measuring 14 wavelengths between 320 and 500 nm the system gave a cond $(\underline{S}) = 50.6$. The prediction error at the 10^{-5} molar level ranged for 4 analyses between 2.4 and 5.8 % which agrees favorably with the predictions derived from computer simulations.

References

(1) Stoer, J. "Einführung in die Numerische Mathematik I", Springer: Berlin-Heidelberg-New York 1976, 2nd.ed.

(2) Jochum, C.; Jochum, P.; Kowalski, B.R. Anal. Chem. 1981, 53,85.

(3) Haaland, D.M.; Easterling, R.G. Appl.Spectr. 1982, 36,665.

(4) Otto, M.; Wegscheider, W.; Lankmayr, E.P. Techn. Univ., Graz, 1983; unpublished work.

(5) Kalivas, J.H. Anal.Chem. 1983, 55, 565.

(6) Kaiser, H. Fresenius Z. Anal.Chem. 1972, 260, 252.

(7) Lawson, L.; Hanson, R. "Solving Least Squares Problems"; Prentice Hall: Englewood Cliffs, NY, 1974.

EIN NEUES DATENBANK-, INFORMATIONS- UND AUSWERTUNGSSYSTEM CHEMOMETRIE

J. Bürstenbinder

3S Statistik Software Systemtechnik GmbH
D - 1000 Berlin 45, Dürerstr. 28 a

Die Interpretation von Ergebnissen der chemischen Analytik verlangt oft außer den Analysenergebnissen mehrerer Methoden Angaben über die Herkunft der Probe, die Art der Probennahme und die Beschreibung der Probe nach medizinischen, mineralogischen oder anderen Kriterien. Diese Informationen müssen von den Beteiligten zusammengeführt werden, um eine systematische Untersuchung zu gewährleisten.

Mit dem Programmsystem DIAS-Chemie wird eine Datenbank verwaltet und mit den Informationen aus dieser Datenbank werden mathematische und statistische Untersuchungen durchgeführt.

1 Das Informationssystem

Die Angaben zu einer Probe bilden bei DIAS-Chemie die kleinste logische Einheit. Eine Probe in diesem Sinn kann z. B. ein Handstück, ein Bohrkern, eine Serumprobe oder ein Versuchstier sein. Zu den Proben werden deskriptive Informationen, z. B. Lokalität der Probennahme, Probenmaterial, Geologie, Petrologie, und Analysenergebnisse gespeichert. Der Benutzer von DIAS-Chemie entwirft die Struktur der Datensätze und legt den Detaillierungsgrad der deskriptiven Informationen fest. Einzelproben können zu Datengruppen zusammengefaßt werden, z. B. alle Einzelproben für sulfidische Erze aus Spanien, o. ä. Variable können durch Funktionen zusammengefasst werden, z. B. As * Sb / Na * Na.

2 Das Auswertungssystem

Die folgenden Darstellungs- und Auswertungsverfahren werden eingesetzt:
- Ausgabe von Tabellen
- Statistische Analyse einer Verteilung:
  Die Verteilung einer Variablen wird durch verteilungsunabhängige Parameter und bezüglich einer (Log-)normalverteilung beschrieben, Tests über die Art der Verteilung werden durchgeführt (Kolmogoroff - Smirnov).
- Statistische Untersuchung des Zusammenhanges zwischen Variablen:
  Die Verteilungen mehrerer Variabler werden auf Abhängigkeiten untersucht
  z. B. durch Korrelogramme, Regressionsrechnungen, Kontingenztests, multivariate Methoden.
- Lokationsvergleiche abhängiger und unabhängiger Stichproben:
  Testmethoden, die auf Ranginformationen beruhen, werden eingesetzt, z. B.
  U -Test von Mann-Whitney, Vorzeichenrangtest von Wilcoxon.

- Vergleich von Datensätzen mit Normierungssystemen:
  Die Daten von Einzelproben werden auf Soll- oder Vergleichswerte bezogen, die grafische Darstellung ermöglicht eine mehrdimensionale Untersuchung zur Gruppierung von Proben.
- Zweidimensionale grafische Darstellungen:
  Zeichnungen zur grafischen Clusteranalyse werden ausgegeben.
- Spezialanwendungen:
  Für die Geowissenschaften werden Karten der räumlichen Verteilung von Proben und Darstellungen der Konzentrationen längs eines Profils ausgegeben.

3 Anwendungen

DIAS-Chemie wird für Dokumentation, Methodenvergleich, Gruppierung von Proben und systematische statistische Reihenuntersuchungen eingesetzt. Die Benutzer von DIAS-Chemie sind in der Lage auch ohne Programmierkenntnisse nach einer kurzen Einweisung mit dem System zu arbeiten. Seit 18 Monaten arbeiten Chemiker und Geologen selbständig mit DIAS-Chemie. Zur Zeit wird das Programmsystem für Arbeiten in der Biomedizin eingerichtet.

Anwendungen

VOR- UND NACHTEILE DER ICP-ATOMEMISSIONSSPEKTROSKOPIE UND DER ATOMABSORPTIONSSPEKTROSKOPIE BEI DER ANALYSE GEOCHEMISCHER UND BIOLOGISCHER MATERIALIEN

H. Heinrichs, H.J. Brumsack und K.H. Wedepohl

Geochemisches Institut der Universität Göttingen

Beim Vergleich der Bestimmungsgrenzen (3s der Untergrundvariation) der ICP-AES mit der Flammen-AAS zeigt die ICP-AES für die meisten Elemente günstigere Bedingungen. Eine Ausnahme besteht bei Kalium und Rubidium. Bei der Verwendung hochauflösender ICP-Vakuumspektrometer kommen die Elemente Lithium, Natrium und Cäsium noch hinzu. Im Vergleich zur flammenlosen AAS zeigt die ICP-AES bei den Elementen B, Ba, Ce, Gd, La, Lu, Nb, Os, P, Pr, S, Sc, Sr, Ta, Ti, Th, U, W, Y und Zr niedrigere Bestimmungsgrenzen. Die ICP-AES bietet gegenüber den AAS-Methoden einen wesentlich größeren Meßbereich mit Linearität zwischen Konzentration und Intensität.

Dampfphasenstörungen wie z.B. die Bildung von Aluminaten, Titanaten, Phosphaten, Silikaten, Oxiden und Hydoxiden in der Flamme bei Temperaturen von 2000-2500°C treten aufgrund der hohen Temperaturen von 5000-8000°C im Plasma nicht auf. Ein weiterer Vorteil der hohen Plasmatemperaturen sind hohe Elektronendichten, die sich als gute Ionisationspuffer auswirken und Anregungsunterschiede zwischen den Elementen vermindern. In der Lachgas/Azetylen-Flamme kann die Ionisations-Interferenz bei Temperaturen von 2900°C dagegen bis zu 75 % betragen. Im Plasmavolumen besteht ein Gradient der Elektronendichten. Nur bei größeren Beobachtungshöhen treten deutliche Ionisationsstörungen auf. Dieser Bereich (>18 mm über der HF-Spule) ist wegen einer gleichzeitigen Abnahme der Bestimmungsgrenzen für die Messung der meisten Elemente uninteressant. Bei der AAS-Anregung mit der Graphitrohrküvette treten vor allem Dampf- und Festphasenstörungen z.B. durch die Bildung von Monohalogeniden, Monocyaniden, Dicarbiden, Hydroxiden, Oxiden und Einlagerungsverbindungen mit dem Graphit auf. Einen Überblick über die Störungen und deren Vermeidung bei der ICP-AES und der AAS gibt die Tabelle 1.

Bei der Analyse geochemischer und biologischer Proben spielen die Präparationstechniken, insbesondere die Aufschlüsse, eine entscheidende Rolle. Ein normaler $HF/HClO_4$- oder HF/H_2SO_4-Aufschluß für silikatische und karbonatische Matrizes stellt eine Verdünnung um den Faktor ≥250 dar. Organische Materialien lassen sich mit einem Verdünnungsfaktor ≥50 in einem $HClO_4/HNO_3$-Aufschluß aufschließen. Anhand der Bestimmungsgrenzen

Tabelle 1: Störungen und deren Vermeidung bei der ICP-Atomemissionsspektroskopie und der Atomabsorptionsspektroskopie

ICP - AES		GRAPHITOFEN - AAS		FLAMMEN - AAS	
STÖRUNGEN	VERMEIDUNG DER STÖRUNGEN	STÖRUNGEN	VERMEIDUNG DER STÖRUNGEN	STÖRUNGEN	VERMEIDUNG DER STÖRUNGEN
Spektrale Interferenzen Überlagerung von Linien, Überlappung nahe beieinanderliegender Linien, Koinzidenz mit Absorptions- und Emissionsbanden	Ausweichen auf störungsfreie, unempfindliche Linien, Störungen durch Banden können durch den Aufsatz eines Quarzrohres auf die Fackel und durch einen erhöhten Kühlgasdurchsatz reduziert werden.	Dampfphasenstörungen Bildung von Monohalogeniden, Monocyaniden, Dicarbiden, Hydroxiden, Oxiden, Dimeren, Clustern etc.	Verwendung von Matrix-Modifikationszusätzen (z.B. von $(NH_4)_2HPO_4$, $(NH_4)_2SO_4$, H_2SO_4, Ni, La etc.), Plattformrohren, bestimmten Inert- und Reaktionsgasen	Dampfphasenstörungen Bildung von Aluminaten, Titanaten, Silikaten, Phosphaten, Oxiden, Hydroxiden u.a.	Einsatz einer heißeren Flamme (z.B. der Lachgas/Azetylen-Flamme, Zusatz eines Befreiungsagens (z.B. von La bei der Bestimmung der Erdalkalimetalle
		Festphasenstörungen Bildung von Einlagerungsverbindungen im Graphit mit Fluor, Brom, Jod, Alkalien, Halogeniden von Be, Al, Tl und Sb, Halogeniden und Sulfiden der meisten Übergangselemente, Oxychloriden, HNO_3, $HClO_4$, H_2SO_4, NH_3 etc. Bildung von CaF- biden der Elemente $Hf, La, Mo, Nb, Ta, Ti, Zr, W, Y, Cr, V, B, Si, Ba$ etc. Abhängigkeit des Atomisierungsprozesses von der Größe und Form der beim Trocknen entstandenen Kristalle	Zusatz von Befreiungsagenzien (z.B. von K bei der Bestimmung von Rb oder von La bei der Bestimmung von Mo. Einsatz pyrolytisch beschichteter Rohre, Aerosolbeschichtung heißer Flächen	Ionisationsstörungen Anregungsunterschiede der Elemente in Proben- und Eichlösungen aufgrund unterschiedlicher Elektronendichte in der Flamme	Zusatz von Ionisationspuffern (z.B. von Cs)
Physikalische Störungen Unterschiede in der Viskosität und Oberflächenspannung der Lösungen	Arbeiten mit innerem Standard, Additionsverfahren und Angleichung von Matrizes	Unspezifische Lichtverluste Lichtstreuung an Partikeln in der Gasphase	Kompensation (z.B. mit Kontinuumstrahlern)	Unspezifische Lichtverluste Lichtstreuung an Partikeln in der Flamme	Kompensation (z.B. mit Kontinuumstrahlern)
Ionisationsstörungen Minimale Anregungsunterschiede der Elemente in Abhängigkeit vom Gradienten der Elektronendichte im Plasma	Veränderung der Beobachtungshöhe in Richtung höherer Elektronendichte			Physikalische Störungen Unterschiede in der Viskosität und Oberflächenspannung der Lösungen	Arbeiten mit innerem Standard, Additionsverfahren und Angleichung von Matrizes

für die ICP-AES läßt sich leider nachrechnen, daß z.B. folgende Elemente nicht direkt in Aufschlußlösungen von Gesteinen und Pflanzen gemessen werden können: Ag, As, Au, Bi, Br, Cd, Cl, Cs, F, Ga, Ge, HF, Hg, Ho, In, Ir, J, Li, Lu, Os, Pd, Pr, Pt, Rb, Re, Rh, Ru, Sb, Se, Sn, Ta, Tb, Te, Th, Tl, Tm, U und W. Hinzu kommen auch noch Gd und Pb, da bei diesen beiden Elementen alle nachweisstarken Linien erheblich gestört sind. Die Analyse mit der ICP-AES beschränkt sich in Flußwässern auf die Elemente Al, B, Ba, Ca, Fe, Mg, Mn, S, Si, Sr, (Ti) und Zn, in Pflanzen auf Al, B, Ba, Be, Ca, Co, Cr, Cu, Fe, Mg, Mn, Ni, P, S, Sc, Si, Sr, Ti, Y, Yb, Zn und Zr und in Gesteinen auf Al, B, Ba, Be, Ca, Ce, Co, Cr, Cu, Dy, Er, Eu, Fe, La, Mg, Mn, Mo, Nb, Ni, Nd, S, Sc, Si, Sm, Ti, V, Y, Yb, Zn und Zr. In den meisten Gesteinsaufschlüssen erfolgt die Messung von Dy, Er, Eu, Mo, Nd, S und Zn dicht an den Bestimmungsgrenzen, so daß wegen der sehr genauen Einstellung des Gerätes die Elemente nur einzeln gemessen werden können. In vielen Fällen reicht die Nachweisempfindlichkeit bei diesen Elementen nicht aus. Berücksichtigt man weiter, daß bei sehr vielen Elementen in Abhängigkeit von ihrer Leichtflüchtigkeit oder Schwerlöslichkeit ganz unterschiedliche Aufschlüsse durchgeführt werden müssen, so wird der begrenzte Einsatz der ICP-AES als Multi-Element-Analysenverfahren bei der Untersuchung von geochemischen und biologischen Proben deutlich.

Die ICP-AES zeigt gegenüber der Flammen-AAS eine deutlich schlechtere Zerstäubercharakteristik. Die Ansaugrate bei einem ICP-Zerstäuber ist 5-10mal kleiner als bei einem AAS-Zerstäuber. Daraus ergibt sich eine größere Abhängigkeit der Emissionsintensitäten von der Viskosität und Oberflächenspannung der Probe. Beim Meinhard-Zerstäuber fallen die Nettoemissionsintensitäten von 100 auf 80 % bei einer kontinuierlichen NaCl-Zugabe bis zu 1 % (w/v).

Die ICP-AES zeigt wie alle Emissionsmethoden spektrale Interferenzen (Tab. 2), d.h. Überlagerung von Linien, Überlappungen nahe beieinanderliegender Linien und Koinzidenz mit Absorptions- und Emissionsbanden. Insbesondere kann Überlappung mit O_2-Absorptionsbanden mit Wellenlängen <2000 Å, mit NO-Emissionsbanden zwischen 2000-2500 Å, mit OH-Emissionsbanden zwischen 2950-3200 Å und CN-Emissionsbanden zwischen 3800-4000 Å auftreten. Darüber hinaus gibt es eine Vielzahl von Rekombinationskontinua. Das Ausmaß spektraler Interferenzen hängt von der Auflösung des verwendeten Monochromators ab. Jedoch können Störungen durch Banden auch durch einen erhöhten Kühlgasdurchsatz oder durch den Aufsatz eines Quarzrohres auf die Fackel reduziert werden.

In der AAS spielen spektrale Interferenzen praktisch keine Rolle, da modulierte, elementspezifische Lichtquellen und ein auf diese Modulationsfrequenz abgestimmter Verstärker die Interferenzen verhindern.

Tabelle 2: Nachweisstarke Linien bei der ICP-AES und mögliche spektrale Interferenzen in geochemischen und biologischen Materialien

Element	Wellen-länge Å	Bestimmungs-grenzen ng/ml	mögliche spektrale Interferenzen	Element	Wellen-länge Å	Bestimmungs-grenzen ng/ml	mögliche spektrale Interferenzen
Ag	3280.68	5.	Mn, Fe, Al, Ti	Ni	2216.47	12.	Co, W
Al	3092.71	16.	V, Fe, Mg, OH-Bande	Os	2255.85	0.6	Fe
Al	3082.16	55.	V	P	1782.87	500.	I
As	1890.42	65.	O_2-Bande	P	2136.18	90.	Cu, Al
As	1936.96	70.	Al, O_2-Bande	Pb	2203.53	50.	Al, Sn, Pd, Fe, Ti
B	2497.73	5.	Fe, NO-Bande	Pd	3404.58	40.	V, Fe, Mo, Zr
Ba	4554.03	0.5	Zr, Sr	Pr	3908.44	35.	
Be	3130.42	0.3		Pr	4222.98	40.	Ce, U
Be	2348.61	0.5	V, OH-Bande	Pt	2144.23	35.	
Bi	2230.61	70.	NO-Bande, Fe, Al, Ti	Re	2214.26	10.	Os, Pt, Pd
Ca	3933.66	0.3	Sc	Rh	2334.77	40.	Sn
Cd	2288.02	5.	As	Ru	2402.72	35.	Fe
Ce	4186.60	80.	Zr, V, Y	S	1807.31	200.	Al
Co	2286.16	10.	Ti, Fe, V	Sb	2068.33	50.	Cr, Ge, Mo, Fe, Al, Ti
Cr	2677.16	7.	V, Mn, Mo, U	Sc	3613.84	1.0	
Cu	3247.54	8.	OH-Bande, Fe, Mn, Ti, Nb	Se	1960.26	70.	Al, Fe, Mn
Dy	3531.70	13.	Mn, Nd, Sm, Ti	Si	2516.11	15.	V, Mo
Er	3372.71	2.	Ti	Sm	3592.60	20.	Nd, Gd, V, Ni
Eu	3819.67	3.	Gd, Nd, Fe, Ca	Sn	1899.76	40.	Ti, Mg, Al, Fe, V, Zr, As
Fe	2599.40	5.	Ta, Mo	Sr	4077.71	0.2	
Ga	2943.64	45.	Ni, V	Ta	2262.30	50.	
Gd	3422.46	15.	Cr, Ni, Fe, Ce, Dy, Sm	Tb	3509.17	30.	Ru, V
Hf	2773.36	15.	Cr, Fe	Te	2142.81	60.	
Ho	3456.00	7.	Er, Mo, Zr	Th	2837.30	70.	Fe
Ir	2242.68	35.	Cu	Ti	3349.41	5.	Cr, Nb, Cu, Co
La	3794.78	10.	Fe	Tl	1908.64	70.	Mo, V
La	3988.52	20.	Ca	Tm	3131.26	6.	Be, OH-Bande
Lu	2615.42	1.	Fe, V, Ni, Er	U	3859.58	200.	CN-Bande, Nd, Fe, Ca, Ti
Mg	2795.53	0.15	Fe, Cu	V	3093.11	5.	Al, Mg, OH-Bande
Mn	2576.10	1.	W	V	2908.82	6.	Nb
Mo	2020.30	15.	Ca	W	2079.11	30.	Ni, Cu
Nb	3094.18	10.	Fe, Al, Ca	Y	3710.30	3.	
Nb	2950.88	10.	V, Mg, Al, Ti, OH-Bande	Yb	3289.37	1.	Fe, Al, V
Nd	4061.09	30.	Hf	Zn	2025.48	5.	Mg, Cu, Nb, Al, Ca
Nd	4303.58	30.	Ni, Sr, Gd, Sm, Ca, Fe, Ti	Zn	2138.56	9.	Ni, V, Fe, Cu
Ni	2316.04	15.	Fe, Zn, Co, Sr, Sc	Zr	3496.21	2.	Y, Mn

MÖGLICHKEITEN DER INSTRUMENTELLEN MULTI-ELEMENT-ANALYSE IN DER WASSER-
CHEMIE

Karl-Ernst Quentin
Institut für Wasserchemie und Chemische Balneologie der Technischen
Universität München, Marchioninistraße 17, 8000 München 70

ZUSAMMENFASSUNG
Identifizierung und Konzentrationsbestimmung der Wasserinhaltsstoffe
sind für die Bewertung der Wassergüte und die Beurteilung der Wasser-
nutzungs-Möglichkeiten unabdingbare Voraussetzung. Unter diesen Gegebenheiten hat sich auch die Wasserchemie vermehrt mit Multi-Element-
Methoden befaßt und Erfahrungen gesammelt. Die insbesondere für die
Wasserchemie in Frage kommenden Methoden-Entwicklungen lassen sich in
6 Hauptgruppen gliedern, die im einzelnen erläutert werden. Die bislang
bekannten Anwendungsmöglichkeiten werden an Hand von Beispielen beschrieben und die Perspektiven dargelegt. Einige Methoden haben sich bereits bewährt und sind im laufenden Einsatz einzelner Wasserlaboratorien.
Die generelle Verwendung von Multi-Element-Methoden ist aber in der
Wasserchemie noch nicht vollzogen.

―――――――――

Identifizierung und Konzentrationsbestimmung der Inhaltsstoffe sind unabdingbare Voraussetzungen für die Bewertung der Güte des Wassers und die
Beurteilung seiner Verwendungsmöglichkeiten; daher ist die Analytik des
Wassers von grundlegender Bedeutung. Ereignisse in der "fließenden Ge-
wässerwelle", Beeinflussungen des Grundwassers, Auswirkungen von Abwas-
sereinleitungen oder Differenzierungen zwischen geogener Herkunft und
anthropogener Sekundärbelastung erfordern aber nicht nur eine genaue
sondern auch eine rasche Analytik zahlreicher Einzelstoffe und Einzelproben, um binnen kurzer Zeit eine informative Gesamtübersicht zu erhalten, Folgerungen für die Aufbereitung und Nutzung des Wassers zu treffen und Gewässerschutz-Maßnahmen einzuleiten. Gerade unter diesen Gesichtspunkten der Proben- und Stoffvielzahl ist die wasserchemische
Einzelbestimmung bis zur Datenauswertung oft zu umständlich und zu
langwierig.

In der Wasserchemie ist zu unterscheiden zwischen den Hauptbestandteilen im Konzentrationsbereich mg/l (z.B. Ca, Mg, Cl) und den Spurenbestandteilen im Konzentrationsbereich µg/l und darunter, zu denen beispielsweise die Metalle rechnen; sie werden allgemein als Schwermetalle bezeichnet und liegen als Kationen im Wasser vor. Als Unterscheidungsmerkmal zwischen gelösten und ungelösten Stoffen gilt die Vereinbarung, daß gelöste Stoffe ein Membranfilter der Porenweite 0,45 µm passieren.

Ein solches Konzentrationsraster der Hauptbestandteile und Spurenstoffe im Wasser ist aber nur als Anhalt zu verstehen, da die Wässer schon nach Herkunft recht unterschiedliche Inhaltsstoffe und Konzentrationen aufweisen können, die auf den Fließwegen weitere Veränderungen erfahren.

Beispielsweise liegt Bromid in Grund- oder Oberflächenwässern üblicherweise als Spurenstoff vor, im Meerwasser oder in bestimmten Tiefenwässern ist aber mit Milligramm-Mengen zu rechnen; in verschiedenen Abwässern zeigen Metalle erhebliche Konzentrationserhöhungen. Ein weiteres Beispiel ist das Zink als Spurenstoff der Oberflächen- und Grundwässer, das im Trinkwasser material- und leitungsbedingt höhere Konzentrationen aufweisen kann.

Schließlich spielen auch die Wechselbeziehungen zwischen Wasser und Sedimenten oder Schlämmen, zu denen auch der Klärschlamm gehört, im Hinblick auf den Stoffaustausch eine bedeutende Rolle. Zur Kenntnis der Zusammenhänge bedarf es oftmals der Untersuchung dieser Materialien nach entsprechendem Aufschluß.

Die Wasserchemie hat es aber nicht nur mit einer Vielzahl von Stoffen und erheblichen Konzentrationsunterschieden zu tun, sie muß als Folge der Umweltgesetzgebung in der Analytik auch eine Fülle nationaler und supranationaler Vorschriften mit Parametern und Grenzwerten berücksichtigen. Ihr obliegt es daher, Analysenmethoden zu erarbeiten und für justitiable Meßergebnisse zu sorgen.

Wenn auch in der heutigen Zeit die organischen Belastungsstoffe einen erheblichen Teil der Wasseranalytik in Anspruch nehmen, bilden doch nach wie vor die anorganischen Inhaltsstoffe das Grundgerüst der Wasserbeschaffenheit und charakterisieren den Wassertyp. Unter diesen Gegebenheiten sind Multi-Element-Methoden für die Wasserchemie von großem Interesse; allerdings wird sich stets die Frage erheben, bei welcher Institution und für welchen Zweck diese Methodik eingesetzt werden soll und ob der apparative Aufwand im richtigen Verhältnis zum Nutzen steht.

Die bekanntesten Methoden-Entwicklungen lassen sich in 6 Hauptgruppen gliedern: 1. Atomspektroskopie; 2. Photometrie mit Multikanal-Detektion; 3. Flüssigkeitschromatographie mit verschiedener Detektion; 4. Gaschromatographie mit spezifischer Detektion; 5. Polarographie-Voltammetrie; 6. Sonstige Methoden.

Methodisch ist die Atomabsorptionsspektrometrie von Haus aus eine Einzelbestimmung, die mit ihren verschiedenen Probeneinbringungstechniken heute ihren festen Platz zur Metallbestimmung in der Wasseranalytik hat. Möglichkeiten einer sequentiellen oder auch gleichzeitigen Bestimmung von zwei Metallen bestehen zwar, haben aber keine besondere Bedeutung erlangt.

Demgegenüber ist seit einigen Jahren die Atomemissionspektroskopie mit Plasmaanregung (ICP - AES) als Multi-Element-Methode verschiedentlich in der Wasserchemie im Einsatz (1,2). Die Entscheidung für ein Simultangerät oder ein sequentielles Gerät wird von Fall zu Fall zu treffen sein. Bei hohem Probendurchsatz und einer größeren Anzahl von Elementen ist die Simultanbestimmung vorteilhaft. Vorzüge sind die einfache Probenvorbereitung, die Zeitkürze und die Möglichkeit, auch Borat und Phosphat als Elemente einzubeziehen. Die Methodik läßt sich vor allem zur Schwermetallbestimmung in Oberflächenwässern und zur Überwachung industrieller oder kommunaler Abwässer einsetzen, wenn ein rascher Überblick über das Vorkommen einzelner Metalle gewonnen und die Einhaltung der zulässigen Konzentrationen nach nationalen und supranationalen Festlegungen laufend geprüft werden soll. Auflistungen in den verschiedenen Vorschriften zeigen, daß gegebenenfalls mehr als 20 Metalle zu bestimmen sind.

Tab. 1. Auflistung von Metallen nach verschiedenen Richtlinien
A Übereinstimmung

EG-Richtlinie Gewässerschutz Liste II	Arbeitsblatt A 115 der ATV	EG-Richtlinie Oberflächenwasser für Trinkwassergewinnung	EG-Trinkwasserrichtlinie
Arsen; Blei; Cadmium; Chrom; Eisen; Kobalt; Kupfer; Nickel; Selen; Quecksilber; Zink			
Summe 11			

Bei der Ausbaufähigkeit der ICP-Methode könnte man durch eine Auswahl von etwa 30 Elementen die Bestückung des Gerätes so vornehmen, daß mit Ausnahme der Stickstoffverbindungen, des Kohlenstoffes und der meisten Anionen in einem Analysenschritt eine weitgehende Wasseranalyse durchführbar ist. Bei Grund- und Trinkwässern wird allerdings für verschiedene Metalle die Nachweisgrenze der ICP-Methode nicht genügen, um die erforderlichen Konzentrationsangaben zu machen (3).(Tab. 3, S. 5).

Tab. 2. Auflistung von Metallen nach verschiedenen Richtlinien
B Unterschiede

EG Liste II	A 115 der ATV	EG-Oberfl. W. Trinkwasser	EG-Trinkwasser-Richtlinie
Antimon	-	-	Antimon
-	Aluminium	-	Aluminium
Barium	-	Barium	Barium
Beryllium	-	Beryllium	Beryllium
-	-	Mangan	Mangan
Molybdän	-	-	-
Silber	Silber	-	Silber
Vanadium	-	Vanadium	Vanadium
Zinn	Zinn	-	-
Tellur	-	-	-
Thallium	-	-	-
Titan	-	-	-
Uran	-	-	-
21	14	15	18 (Gesamtsumme)

Schon länger wird in der Wasserchemie die Photometrie mit Multikanal-Detektion in Geräten angewandt, die automatisch Probenahme, Reagenzzugabe und photometrische Detektion vornehmen (4). In der Regel erfolgt kein Trennschritt. Mit einer Mehrkanal-Dosier-Schlauchpumpe könne mehrere Wasserproben und Reagenzlösungen gleichzeitig gefördert werden. Das Mischen der Untersuchungswässer mit den Reagenzlösungen ist nahezu in jedem Verhältnis möglich. Sequentiell läßt sich das gepumpte Untersuchungswasser z.B. mit Luft in definierte Volumen unterteilen, in denen dann verschiedene Inhaltsstoffe photometriert werden.

Tab. 3. Werte der EG-Trinkwasser-Richtlinie (EG) und Nachweisgrenzen der AES/ICP (AES) sowie der AAS mit elektrothermaler Anregung (AAS)

	EG µg/l	AES µg/l	AAS µg/l
Antimon	10	30	2,0
Arsen	50	30	0,1*
Barium	100	0,2	10,0
Beryllium	+	1	1,0
Blei	50	20	1,0
Bor	1000	5	1000
Cadmium	5	2	0,1
Chrom	50	8	2,0
Eisen	50	2	2,0
Kobalt	+	3	2,0
Kupfer	100	2	5
Mangan	20	0,5	1
Nickel	50	6	5
Phosphor	175	13	–
Quecksilber	1	30	0,05**
Selen	10	25	0,05*
Silber	1	6	2
Vanadium	+	2	2
Zink	100	2	0,1

*Hydridmethode
** Kaltdampfmethode

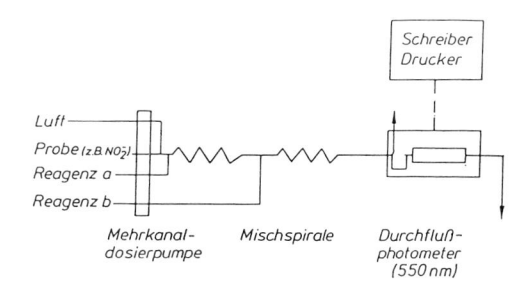

Geeignet für: NO_2^-, NO_3^-, NH_4^+, Cl^-, SO_4^{2-}, PO_4^{3-}, SiO_3^{2-}, Ca^{2+}, Mg^{2+}, Fe^{2+}, Mn^{2+}

Abb. 1. Photometrie mit Multikanaldetektion (Auto-Analyzer mit einer Analysenstraße)

Simultan ist es möglich, durch Verteilung des Untersuchungswassers
auf verschiedene Analysenstraßen mehrere Bestandteile des Wassers
gleichzeitig zu photometrieren. Diese automatische Methode hat sich
hauptsächlich dort eingeführt, wo kontinuierlich in einer Vielzahl
von Wasserproben recht einheitlicher Grundzusammensetzung die Konzentration bestimmter Inhaltsstoffe überprüft werden soll. Die weite
Verbreitung der Methodik in klinischen Laboratorien zur Analysierung
einer Probenvielzahl annähernd gleicher Zusammensetzung war seinerzeit
der Anstoß, solche Autoanalysatoren auch für die Wasseranalytik nutzbar
zu machen. Für die Analytik einer größeren Anzahl von Einzelstoffen
in Wässern unterschiedlicher Zusammensetzung ist die Verfahrensweise
weniger geeignet.

Während bei den Kationen Multi-Element-Analysen schon früher in der
Wasserchemie erprobt wurden, war diese Analytik bei den Anionen mit
weitaus größeren Schwierigkeiten verbunden. In den letzten Jahren ergaben sich allerdings Möglichkeiten, die auch für die Wasseranalyse
von Bedeutung sind. Ausgehend von konventionellen HPLC-Geräten machte
die Flüssigkeitschromatographie erhebliche Fortschritte. Die wesentlichsten Verbesserungen betrafen das Ionenaustauschmaterial. Verwendet
werden 3 verschiedene Trennverfahren (HPIC, HPICE und MPIC). In Verbindung mit einer hochempfindlichen Leitfähigkeitsdetektion wurden die
ersten Geräte unter der Bezeichnung "Ionenchromatograph" geliefert (5).
Abgesehen von den Kosten wiesen sie erhebliche Schwächen auf und
eigneten sich noch nicht für den Routinebetrieb; sie ermöglichten die
simultane Bestimmung zahlreicher Anionen (z.B. Cl^-, F^-, Br^-, NO_3^-,
NO_2^-, SO_4^{2-} und PO_4^{3-}). Da nur geringe Wassermengen benötigt werden,
ist das Verfahren besonders dann von Interesse, wenn nur kleine Probenmengen zur Verfügung stehen (z.B. Sickerwässer). Die heutigen Geräte der 2. Generation sind bedeutend verbessert; die bestimmbare
Anionenzahl wurde erweitert und die Ermittlung von Kationen ermöglicht.
Man benutzt verschiedene Trennungs-, Suppressor- und Detektionsverfahren. Neben der Leitfähigkeitsdetektion wird die Fluoreszenzdetektion
sowie die direkte und indirekte UV-Detektion eingesetzt. Grundsätzlich
kann hinter die Auftrennungsstufe nahezu jede Detektion geschaltet
werden. Die Bestimmung der Ionen ist bis in den Spurenbereich (µg/l)
möglich. In der Wasserchemie wird bislang die Ionenchromatographie
nur für spezielle Untersuchungen verwendet. Die weitere Entwicklung
der Gerätetechnik, aber auch der Gerätekosten wird zeigen, ob dieses
Verfahren als wasserchemische Multi-Element-Analyse allgemeine Verbreitung findet.

Auch die Gaschromatographie hat Möglichkeiten der Multi-Elementbestimmung, die wasserchemisch zu nutzen sind. Nach Derivatisierung lassen sich anorganische Stoffe als metall- oder metalloid-organische Verbindungen, als flüchtige Metallchelate, als Komplexe mit gemischten Liganden oder in Form anderer organischer Verbindungen simultan ermitteln. Ein Beispiel ist die gaschromatographische Trennung und Bestimmung von Metallen nach ihrer Überführung in Dithiocarbamate und deren Extraktion aus dem Wasser (6). Bei den Anionen im Wasser lässt sich gaschromatographisch die wichtige Spurenermittlung von Bromid und Iodid vereinfachen. Wenn man die Halogenide mit Ethylenoxid in 2-Brom- bzw. 2 Iodethanol überführt und diese Derivate extrahiert, können anschließend gaschromatographisch mit ECD an Hand der unterschiedlichen Retentionszeiten simultan Bromid und Iodid im Mikrogrammbereich bestimmt werden (7).

Die unter den Begriffen Polarographie-Voltammetrie zu summierenden Verfahren gehören zweifellos zu den leistungsstarken Methoden bei der Bestimmung aquatischer Inhaltsstoffe, vor allem der Schwermetalle (8). Wenn sich auch der Multi-Elementcharakter häufig nur auf die gleichzeitige Ermittlung von 3 - 5 Metallionen erstreckt, sind doch alle Vorteile einer Simultanbestimmung gegeben. Meist können die angesäuerten Wasserproben direkt analysiert werden. Meerwasser mit seiner dominierenden Salzmatrix, die bei anderen Methoden Schwierigkeiten bereitet, ist gerade wegen des hohen Leitsalzgehaltes für die klassische Polarographie geeignet. Bei sauer konservierten Trinkwasserproben ist es oft angezeigt, den pH-Wert anzuheben, um Messungen im negativen Potentialbereich zu erleichtern.

Das wasserchemisch am häufigsten angewandte voltammetrische Verfahren ist die differentielle Pulspolarographie mit der anodic Stripping Methode; differentielle Pulsvoltammetrie wäre die exaktere Bezeichnung, da mit einer stationären Elektrode gearbeitet wird, oftmals mit hängendem Hg-Tropfen oder Hg-Film. Die inverse Voltammetrie, wie der Stripping-Prozeß auch genannt wird, verzeichnet als wesentlichen Schritt eine dem eigentlichen Meßvorgang vorgeschaltete Anreicherung der Metallionen. So gelingt es, mit den Bestimmungsgrenzen deutlich die Trinkwassergrenz- oder Richtwerte zu unterschreiten. Als Beispiele der praktischen Anwendung sind auf den folgenden Abbildungen verschiedene Schwermetalle im Münchener Leitungswasser und unterschiedliche Zinkgehalte in Abhängigkeit von der Standzeit des Leitungswassers in der Hausinstallation dargestellt.

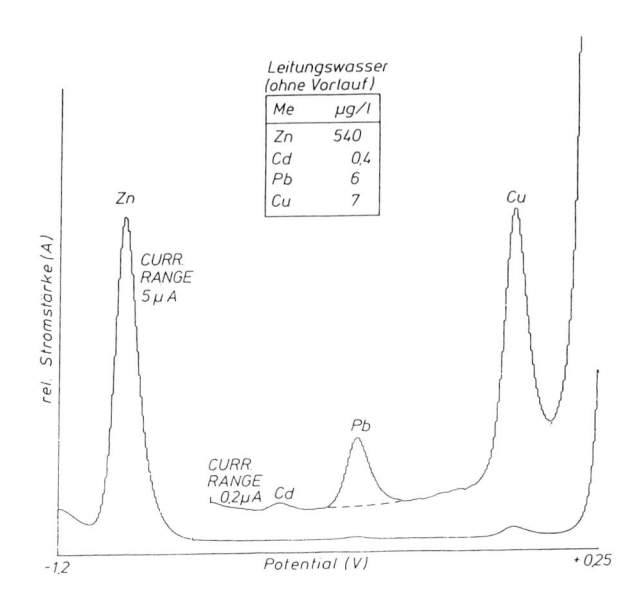

Abb. 2. Stromspannungskurve (DPASV) für Leitungswasser
(Institut, München, 23.3.1984, 8.30 Uhr)

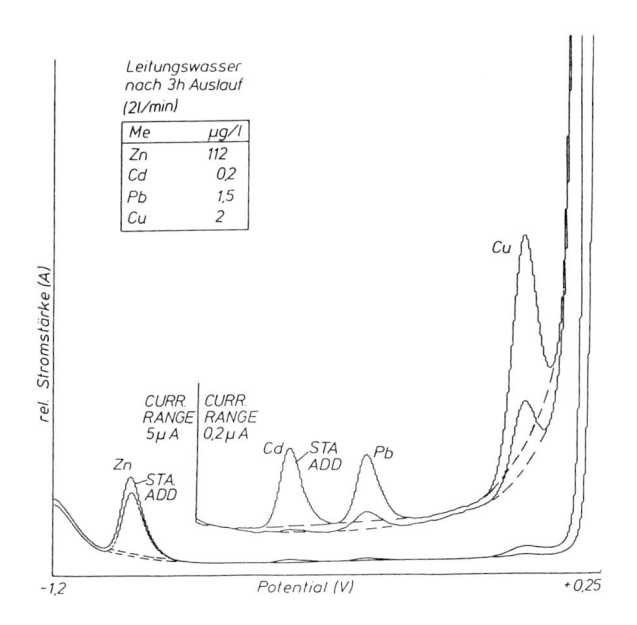

Abb. 3. Stromspannungskurve (DPASV) für Leitungswasser
(Institut München, 23.3.1984, 11.30 Uhr) mit
Standard Addition (Sta.Add.)

Über die zahlreichen Qualitätskontrollen des Wassers hinaus befaßt sich die aktuelle Wasserchemie auch mit der Wechselwirkung kationischer und anionischer Wasserinhaltsstoffe. Auf Metalle bezogen, handelt es sich um eine Komplexierung oder bei mehrzähnigen Liganden um eine Chelatisierung. Die polarographischen bzw. voltammetrischen Methoden sind besonders geeignet, diese Reaktionen aufzuklären, da die Messung nur unwesentlich in die naturgegebenen Gleichgewichte eingreift und die Wasserproben unverändert beläßt. Es können daher Reaktionen zwischen gleichzeitig vorhandenen Metallionen und verschiedenen Liganden verfolgt werden. Mit praxisorientierten Modellen lassen sich wichtige Einblicke in die komplizierten aquatischen Systeme erzielen.

Unter den sonstigen Methoden einer Multi-Element-Analyse des Wassers ist an erster Stelle die Röntgenfluoreszenzanalyse zu nennen, die schon seit längerer Zeit mit Erfolg benutzt wird. In speziellen Fällen ist auch die Neutronenaktivierungsanalyse eingesetzt worden. Zu den neueren Methoden gehört die Felddesorptions-Massenspektroskopie, die ursprünglich für die Analyse organischer makromolekularer Stoffe entwickelt wurde, mit der aber auch Metalle simultan und quantitativ bestimmbar sind. Allerdings ist eine besondere Probenvorbereitung notwendig (9).

Zusammenfassend ist festzustellen, daß für die Wasserchemie gerade in der heutigen Zeit der Umweltbelastung und des Umweltschutzes instrumentelle Multi-Element-Analysen von großem Nutzen sein können. Daß ihr Einsatz noch nicht weit verbreitet ist, hat seine Gründe im instrumentellen Aufwand, in den praktischen Gegebenheiten kleinerer Laboratorien vor Ort und in mancher noch laufenden Weiterentwicklung der Geräte-Generationen. Die erfolgreiche Umsetzung wissenschaftlicher Erkenntnisse auf diesem Sektor in die Praxis der Routinemessung hängt auch von der Gerätezuverlässigkeit und der oftmals notwendigen Methodenmodifizierung auf die speziellen Erfordernisse der Wasseranalytik ab.

REFERENCES

(1) Hoffmann, H.-J.: Labor-Praxis 4, H. 4 (1980) 18 - 25.

(2) Huber, L.: Vom Wasser 58 (1983) 173 - 185.

(3) Quentin, K.-E.: Schriftenreihe Gewässerschutz-Wasser-Abwasser (1984) im Druck

(4) Grasshoff, K. et.al. (Ed.): Methods of Seawater Analysis. Verlag Chemie, Weinheim; Deerfield Beach, Florida, Basel 1983

(5) Weiß, J.: Chemie für Labor und Betrieb 34 (1983) 293 - 297, 342 - 345, 494 - 500; 35 (1984) 59 - 66

(6) Tavlardis, A., Neeb, R.: Fresenius Z.Anal.Chem. 292 (1978) 199 - 202

(7) Grandet, M., Weil, L., Quentin, K.-E.: Z. Wasser- Abwasser- Forsch. 16 (1983) 66 - 71

(8) Nürnberg, A.W.: Fresenius Z. Anal.Chem. 316 (1983) 557 - 565. - Frimmel, F.H., Immerz, A.: Fresenius Z. Anal. Chem. 302 (1980) 364 - 369

(9) Schulten, H.-R., Bahr, U., Palavinskas, R.: Fresenius Z. Anal. Chem. 317 (1984) 497 - 511

ANALYSE VON SELTENERDELEMENT-MINERALEN MIT HILFE DER ICP-OES

J. Luck

Hahn-Meitner-Institut für Kernforschung Berlin GmbH
D-1000 Berlin 39, Glienickerstr. 100

Die quantitative Bestimmung der Seltenerdelemente (SEE) ist von großer Bedeutung zur Aufklärung geochemischer Prozesse (1), die zur Bildung unterschiedlicher SEE-Minerale führten. Für diese Untersuchungen ist es notwendig, die SEE-Gehalte richtig und mit hoher Reproduzierbarkeit im Konzentrationsbereich von % bis zu wenigen µg/g zu bestimmen. Mit Hilfe der ICP-OES können diese Elemente z.T. nach dem chemischen Aufschluß ohne weitere Trennungen direkt gemessen werden. Die Nachweisgrenzen für die eizelnen Elemente bei dieser Methode entsprechen den Anforderungen für petrogenetische Studien.

Es wurden verschiedene Aufschlußmethoden wie Lithiummetaborat-Schmelze, HNO_3/HCl Aufschluß und Aufschlüsse mit Säuregemischen "Spektrosolv PS" und "Spektrosolv AS" (Merck) eingesetzt.

1. Experimentelles

Alle Einzelelementstammlösungen mit jeweils 1 000 ug/g stammen von Ventron GmbH. Die Eichlösungen wurden durch Verdünnen mit den jeweiligen Substanz-Säure-Mischungen erstellt.

Zur Verfügung stand ein 3520 ICP-Spektrometer (sequentiell) von ARL mit einem DEC PDP 11/03 Rechner zur Steuerung und gleichzeitigen Auswerten der Meßergebnisse. Da die Salzgehalte und damit die Dichten der Probelösungen über einen großen Bereich variierten. wurde eine peristaltische Pumpe eingesetzt, um eine konstante Flußrate für den Zerstäuber zu erzielen. Es wurden im wesentlichen die Wellenlängen der ARL - Programmbibliothek benutzt.

Zur Analyse von La, Ce, Sm und Eu in Bastnäsit kann ein HNO_3/HCl Aufschluß angewendet werden. Bei den sehr schwer aufschließbaren Monazitproben wurde eine Boratschmelze durchgeführt, die aber zu sehr hohen Salzkonzentrationen führte. Gute Erfolge zeigte der Aufschluß mit dem Säuregemisch "Spektrosolv PS" von Merck für beide Mineralarten.

2. Ergebnisse und Diskussion

Ein Bastnäsit aus Burundi und drei verschiedene Standard-Proben (2,3) IGS-36: Monazit, NIM 66/69:"high grade" Monazit sowie NIM 70/71:"medium grade" Monazit wurden analysiert. Die Messungen erfolgten gegen flüssige Eichstandardlösungen mit simulierter Matrixzusammensetzung und für einzelne Elemente zusätzlich nach dem Standardadditionsverfahren. Die Ergebnisse zusammengefaßt in Tab.1 sind in guter Übereinstimmung mit den Literaturdaten. Wenn die Analysenergebnisse chondritnormiert werden, kann man auch für alle SEE die Kohärenz überprüfen. Für die Abtragung der normierten Werte gegen die Ionenradien ergibt sich für Eu ein deutlich unterhalb des Kurvenzuges liegen-

	Bastnäsit	IGS-36	Lit	NIM66/69	Lit	NIM50/71	Lit
La	$14,30\pm0,2\%$	$10,10\pm0,2$	$10,2\%$	$7,68\pm0,2$	$7,8\%$	$3,32\pm0,2$	$3,44\%$
Ce	$23,50\pm0,2\%$	$18,46\pm0,2$	$20,2\%$	$16,96\pm0.1$	$16,2\%$	$7,45\pm0,1$	$7,1\%$
Pr		$3,07\pm0,2$	$2,3\%$	$2,82\pm0,1$	$2,1\%$		
Nd		$8,84\pm0,1$	$9,0\%$	$6,80\pm0,1$	$6,5\%$	$3,20\pm0,2$	$3,0\%$
Sm	6520 ± 90	$1,31\pm0,1$	$1,3\%$	$1,08\pm0,1$	$1,0\%$	4700 ± 230	5200
Eu	1488 ± 53	350 ± 25	301	269 ± 30		105 ± 10	
Gd	1120 ± 20	4006 ± 200	6700	2433 ± 235	7500	1085 ± 50	
Dy	696 ± 25	3530 ± 240	2700	4164 ± 222	3800	1800 ± 43	
Ho		378 ± 7	347	580 ± 100		209 ± 40	
Er		145 ± 32	535	734 ± 28		336 ± 35	
Yb	32 ± 5	435 ± 33	259	380 ± 30		177 ± 10	182
Lu		55 ± 4	51	42 ± 6		23 ± 4	

Tab. 1 Bestimmung der Gehalte der SEE im Bastnäsit und Monaziten (2,3)
Konzentrationen in μg/g wenn nicht anders angegeben

der Wert, der durch den Wertigkeitswechsel des Eu von 3+ zu 2+ erklärt werden kann.

Die Analysenwerte in Tab.1 sowie eine Chondritnormierung zeigen, daß die Daten die notwendige Reproduzierbarkeit haben.

Die Verteilungsmuster der SEE in Mineralen lassen auf einfache Art die Fraktionierung der SEE vor und während der Mineralisation erkennen. Damit stellen die Verteilungsmuster ein wesentliches Hilfsmittel für petrographische Untersuchungen dar.

3. Referenzen

(1) Möller,P.,Morteani,G.,and Schley,F.:Lithos 13 (1980) 171-179
(2) Lister,B.:Geostand.Newsl. 5 ,1 (1981) 75-81
(3) Stoch,H.,and Ring,E.J.:MINTEK-Report No. M 104 (1983)1-45

VERFAHREN ZUR BESTIMMUNG DER SELTENEN ERDEN (SE) IN GEOLOGISCHEN MATRIZES MIT HILFE DER ICP-OES

Jörg Erzinger

Mineralogisch-Petrologisches Institut der Justus-Liebig Universität Giessen
Senckenbergstrasse 3 , D-6300 Giessen

Da die Seltenen Erden eine für petrogenetische Aussagen wichtige Elementgruppe ist, wurde ein entsprechendes ICP-OES-Verfahren optimiert und zur Routinereife weiterentwickelt, um die Gehalte an SEE in silikatischen Gesteinen quantitativ bestimmen zu können. Da sie im 0.x bis 10x mg/kg-Bereich liegen, sind in vielen Fällen Abtrennungs- und Anreicherungsschritte notwendig. Das hier beschriebene Verfahren ist in (1) und (2) ausführlicher dargestellt, einzelne inzwischen optimierte Parameter werden hier erstmalig mitgeteilt.

Das analysenfein gemahlene Probenmaterial wird mit einem $HF-HClO_4$-Aufschluß nach (2) oder (3) in Lösung gebracht. Die Lösungen müssen danach 1n HCl enthalten. Eventuelle unlösliche Rückstände werden mit einem NaOH-Schmelzaufschluß (siehe (2) oder (3)) gelöst und mit der Säureaufschlußfraktion vereinigt.

Die Abtrennung und Anreicherung der SEE erfolgt wie in Abbildung 1 dargestellt.

Ionenaustauschersäule 20 mm ⌀, 300 mm lang, 100 mm hoch, gefüllt mit Kationenaustauschharz (z. B. DOWEX AG 50 W-X8)
↓
Auswaschen mit 4n HCl
↓
Spülen mit 1n HCl
↓
Aufschlußlösung aufgeben
↓
Auswaschen mit 400 ml 1.4n HCl
↓
Eluieren der SEE (+ Ba) mit 500 ml 4n HCl
↓
Filtrieren des Eluats
↓
Einengen des Filtrats bis zur Trockene
↓
Aufnehmen des Rückstands mit 5 ml 3n HCl
↓
Aufbewahren in Kunststoffflaschen

Abb. 1: Flußdiagramm zur Abtrennung und Anreicherung der SEE

Außer Ba, Sr und Y sind nun alle Hauptbestandteile der Aufschlüsse nahezu quantitativ abgetrennt, sodaß die notwendige Wellenlängenauswahl dadurch relativ einfach wird und die möglichen Interelement- und Matrixstörungen leichter überschaubar werden. Gegenüber der Aufschlußlösung ist die Meßlösung jetzt um Faktor 20 angereichert.

DieAbtrennung der SEE von der Matrix wurde bezüglich Wiederfindung und Selektivität eingehend untersucht, die Wiederfindungsraten betragen für alle SE 98% bis 102%.

Für alle SE konnten optimale Geräteparameter (Tabelle 1), Meßwellenlängen und Untergrundkorrekturlinien (Tabelle 2), sowie Interferenzkorrekturen gefunden werden.

Eingehende Untersuchungen ergaben für dieses Verfahren eine befriedigende Reproduzierbarkeit für alle SE besser als 10%, nur für Tb und Tm sind die Werte etwas schlechter.

Brenner : Quarz-Teflon	Zerstäuber	: Cross-Flow mit Probenpumpe
Plasmagas : Argon - 14 l/min	Hilfsgas	: Argon - 0.8 l/min
Zerstäuberdruck : 32 psi	Beobachtungshöhe	: 16 mm über HF-Spule
	Angelegte HF-Leistung : 1.25 kW	

Tabelle 1 : Geräteparameter für ICP-OES "ICP 5500" (Perkin Elmer)

Die Richtigkeit der gefundenen Werte wurde durch die Analyse vieler internationaler Referenzgesteine bestätigt, der Vergleich mit Literaturdaten zeigt gute Übereinstimmungen. In Tabelle 3 sind unsere Ergebnisse (d.A.) für zwei Referenzproben mit den Literaturmittelwerten aufgelistet, weitere Daten können beim Autor angefordert werden.

Element	(nm)	Untergrund unten	Untergrund oben	Element	AGV-1 $\bar{x}$ (Lit.)	AGV-1 $\bar{x}$ (d.A.)	BHVO-1 $\bar{x}$ (Lit.)	BHVO-1 $\bar{x}$ (d.A.)
La	408.67	0.08	-	La	37	35	16.7	15
Ce	418.66	0.05	0.05	Ce	66	61	41	38
Pr	390.84	0.06	0.05	Pr	6.5	6.7	5.6	5.3
Nd	406.11	0.08	-	Nd	34	33	24	24
Sm	359.26	-	0.06	Sm	5.9	6.4	6.1	5.9
Eu	381.97	0.05	-	Eu	1.67	1.53	2.0	1.81
Gd	335.05	-	0.08	Gd	5.1	4.4	7.0	6.2
Tb	350.92	-	0.06	Tb	0.7	0.5	1.0	0.9
Dy	353.17	0.05	-	Dy	3.4	3.5	4.8	5.0
Ho	345.60	-	0.08	Ho	0.67	0.66	0.94	0.9
Er	369.27	0.06	-	Er	1.61	1.8	2.0	2.3
Tm	313.12	-	0.07	Tm	0.33	0.3	0.31	0.3
Yb	328.94	0.06	-	Yb	1.72	1.8	2.1	1.9
Lu	261.14	0.06	-	Lu	0.28	0.23	0.32	0.3

Tab. 2: Wellenlängen mit Untergrundintervallen (in nm)

Tab. 3: Literaturdatenvergleich an internationalen Referenzgesteinsproben (Angaben in mg/kg)

Nachdem die Aufschluß- und Abtrennungsverfahren optimiert und die Methodenentwicklung am Spektrometer abgeschlossen waren, kann heute eine angelernte Laborkraft mit entsprechender Ausstattung etwa 7 bis 10 Proben pro Tag zur Messung vorbereiten. Die Bestimmungvon 14 SEE an 10 Probenlösungen kann mit allen Nebenarbeiten in drei Stunden durchgeführt werden.

Es steht damit eine Multielementmethode für die Gesteinsanalytik routinemäßig zur Verfügung, die kostengünstiger und apparativ wesentlich weniger aufwendig ist, als z.B. INAA oder andere bisher verwendete Verfahren.

REFERENZEN

(1) Walsh,J.N., Buckley,F. & Barker,J.: Chemical Geology, 33, 141-153, 1981.
(2) Erzinger,J., Heinschild,H.-J. & Stroh,A.: Vorträgesammelband des 2. Colloquium Atomspektrometrische Spurenanalytik, Verlag Chemie, im Druck.
(3) Herrmann,A.G.: Praktikum der Gesteinsanalyse.- Springer-Verlag Berlin-Heidelberg-New York, 1975.

DETERMINATION OF ANTIMONY, INDIUM, RHENIUM, SELENIUM, TELLURIUM AND TIN IN GEOCHEMICAL SAMPLES BY RNAA.

J. Nonaka

Max-Planck-Institut für Chemie, Saarstr.23, D-6500 Mainz

In order to supplement the data for primitive spinel lherzolites of Jagoutz et al. (1), the RNAA method of our Institute (2) has been extended for the determination of poorly known elements such as In, Re, Sb, Se, Sn and Te. As a standard sample the meteorite Murchison, a C 2-chondrite, was analysed. The samples and standards were irradiated for 7 days at the FR-2 reactor in Kernforschungszentrum Karlsruhe at a thermal neutron flux of about 9×10^{13} ncm^{-2}s^{-1}. The amount of carriers added were as follows (in mg): In 1; Re 0.72; Sb 10; Se 10; Sn 30 and Te 30. The γ-counting was performed with coaxial Ge(Li) detectors in conjunction with a multi-channel-analyser and magnetic tape. Separated sample solutions had volumes of 20 ml and were placed in a counting vessel with defined geometry.

1. Separation scheme

Fig. 1 shows an abbreviated flow diagram of the separation scheme. Transfer the irradiated sample to zircon crucible containing the carriers. Fuse the mixture with 5 g of Na_2O_2 for about 20 min. with occasional swirling. After cooling, transfer the crucible to a beaker containing 50 ml of water. Centrifuge the solution and decant the supernate S1. Dissolve the precipitate P1 with 30 ml of 5.5 N HBr and centrifuge the unsoluble part. Combine the supernate S1 with the supernate S2. Adjust the solution to pH 8 with 10 % NaOH solution and centrifuge. Decant the supernate S3 into a beaker.

Adjust the supernate S3 to 0.5 N HCl. Transfer the solution to a separatory funnel counting 100 ml of diethylether and shake. The organic layer O1 contains Au and Os. Drain the aqueous phase A1 and add at first 60 ml of ethanol, then alkalify the solution with 10 % NaOH solution and centrifuge. Boil the supernate S4 until the volume is reduced to 50 ml and add 1 g of Na_2CO_3. Centrifuge the carbonate of Ba and Sr. Adjust the supernate S5 to pH 2 with HCl. Add about 20 ml of 3 % sodiumtetraphenylboron solution (3) drop by drop and collect the precipitate on the filter paper. The filtrate F6 contains Se and Re, the precipitate P6 Ag, Cs and Rb. Add a few ml of KCl solution to the filtrate F6 to remove the excess of sodiumtetraphenylboron from the solution and filter the precipitate. Prepare an anion exchange

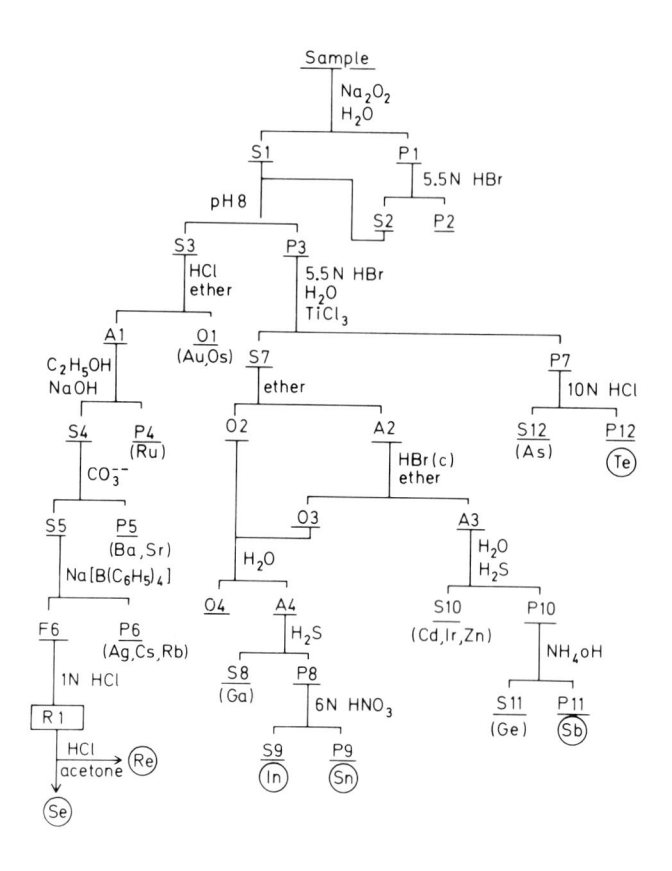

Fig. 1. Flow diagram of the radiochemical separation procedure. Letters and numbers designate various phases to which the text refers: A, aqueous layer; F, filtrate; O, organic layer; P, precipitate; R, resin and S, supernate. Elements within parentheses were not determined in this work.

column with 1 g of Dowex 1-x8 resin, 100-200 mesh, chloride form, which has been pre-soaked in 1 N HCl. Adjust the filtrate to 1 N in HCl and load onto the column. The effluent contains Se. Rhenium is retained on the resin (4).

Dissolve the precipitate P3 in 20 ml of 5.5 N HBr. Add the 20 ml of 15 % $TiCl_3$ solution and adjust to 2 N in HBr. The precipitate P7 contains Te and As. Wash the precipitate with 20 ml of 10 N HCl and centrifuge. The supernate S12 contains As, the precipitate P12 Te.

Transfer the supernate S7 to a separatory funnel containing equal volume of diethylether and shake. Drain the aqueous layer A2 into another separatory funnel and adjust the solution to 4 N in HBr. Add equal volume of diethylether and shake. Drain the aqueous layer A3 and dilute to 1 N in HBr. Saturate the solution with H_2S to precipitate sulfide and centrifuge. The supernate S10 contains Zn, Cd and Ir, the precipitate P10 Sb and Ge. Dissolve the precipitate P10 in 20 ml of NH_4OH and centrifuge. The supernate S11 contains Ge, precipitate P11 Sb.

Combine both ether solution O2 and O3 and transfer into a separatory funnel containing 50 ml of water. Shake the funnel and drain the aqueous layer A4. Saturate A4 with H_2S to precipitate sulfide and centrifuge. The precipitate P8 contains In and Sn, supernate S8 Ga. Wash the precipitate P8 with 20 ml of water and dissolve it in 20 ml of 6 N HNO_3. Heat the solution on a hot plate until metastannic acid precipitates completely. Cool and centrifuge. The supernate S9 contains In, precipitate P9 Sn.

2. Individual purifications

Rhenium: Wash the resin R1 on which Re is retained with 20 ml of a mixed solution from CH_3OH (90 % v/v) and 6 N HCl (10 % v/v). Elute Re with 50 ml of mixed solution from acetone (90 % v/v) and 6 N HCl (10 % v/v) (5). Heat the eluate to dryness on a water bath. Dissolve the residuum with 20 ml of 6 N HCl. Count the 137 keV-γ-peak of ^{186}Re.

Selenium: Transfer the effluent to a separatory funnel. Make the solution 7 N in HCl, add 1 ml of ethylmethylketone and shake. Wait for 15 min.(6). Add 20 ml of $CHCl_3$ and shake. Drain the aqueous layer to another separatory funnel and shake with 20 ml $CHCl_3$. Combine the both $CHCl_3$ layer and add 10 ml of 3 N HNO_3. Reduce the volume of the solution to 10 ml by heating on a water bath. Dilute it to 20 ml with water and count the 136, 265 and 280 keV-γ-peaks of ^{75}Se.

Indium: Prepare an anion-exchange column with 2 g of Dowex 1-x8 resin, 100-200 mesh, chloride form, which has been pre-soaked in 8 N HCl. Add 5 ml of 6 N HCl to the supernate S9 and heat to dryness. Dissolve the residuum with 50 ml of 8 N HCl and load onto the column. The main interfering element Fe retains on the resin. Reduce the volume of effluent to 20 ml on a water bath. Count the 190 keV-γ-peak of ^{114m}In.

Tin: Transfer the precipitate P9 to a porcelain crucible. Add 100 mg of sulphur and 100 mg of Na_2CO_3 and fuse it for 10 min.(7). After cooling dissolve the fusion cake with 20 ml of water. Count the 158 keV-γ-peak of ^{117m}Sn.

<u>Antimony</u>: Prepare an anion-exchange column with 2 g of Dowex 1-x8, 100-200 mesh, chloride form, which has been pre-soaked in 4 N HCl. Wash the precipitate P11 with water several times and dissolve it with 20 ml of HCl (c). Centrifuge the solution. Make the supernate 4 N in HCl and load onto the column, collect Sb with 20 ml of 2 N HNO$_3$, count the 603 and 1691 keV-γ-peaks of ^{124}Sb.

<u>Tellurium:</u> Dissolve the precipitate P12 with 4 ml of aqua regia and add 16 ml of 6 N HCl. Transfer the solution to a separatory funnel containing 40 ml of isobuthyl-methylketone and shake (8). Transfer the organic layer into another separatory funnel. Back-extract Te with 20 ml of water and count the 159 keV-γ-peak of ^{123m}Te.

3. Recovery

Dilute the sample solution exactly to 25 ml with water. Transfer 100 ul of this solution to a small PVC vial and evaporate to dryness. The vials are irradiated in Mainz at the TRIGA rector with a thermal neutron flux of about 7×10^{11} ncm^{-2}s^{-1}. Irradiation time is 10 min. for Sn (count the 158 keV-γ-peak of ^{123}Sn) and Te (count the 150 keV-γ-peak of ^{131}Te). For the other elements it is 6 h. Typical recoveries for the elements are as follows: In 40-60 %; Sb 20-30 %, Se 40-70 %, Sn 10-20 % and Te 20-80 %.

4. Results

The results of the analyses are listed in Tab. 1, together with the results of Morgan et al. (9,10). The precision of the method was checked by an analysis of the meteorite Murchison, a C 2 chondrite (Tab. 2). Our data for Murchison agree well with the data of other authors. Tin has not been previously determined in Murchison, but the mean value of tin in other C 2 chondrites is 0.85 ppm (11).

Acknowledgments - I am indebted to Mr. B. Spettel for advice and technical assistance. I wish to thank the staff of the TRIGA research reactor of the Institut für anorganische Chemie und Kernchemie der Universität Mainz.

Tab. 1 Concentration of 6 trace elements in 6 spinel lherzolites collected by Jagoutz. For comparison the data of 14 spinel lherzolites of Morgan et al. (9,10) are given.

	In (ppb)	Re (ppb)	Sb (ppb)	Se (ppb)	Sn (ppb)	Te (ppb)
SC1	18a+	0.23b	4.5b	13a	99b	20d
FR1	19a	0.44	2.9b	15b	230b	14d
D1	9.5b		3.6b		30c	7.6d
Ka168	14b	0.08c	15b	5.9b	53b	3.8d
PO1	17b	0.07c	4.0b	7.0b	83c	7.0d
KH1	15b	0.05c	1.7d	9.3b	36c	11d
Morgan et al.	5-25	0.01-0.26	0.7-4.9	1.3-63		2-14

+Errors based on counting statistics: a, 5 %; b, 5-10 %; c, 10-25 %, d, 25-50 %.

Tab.2 Comparison of 6 trace elements in the meteorite Murchison: A = Krähenbühl et al. (12), B = Kallemeyn & Wasson (13), C and D = unpublished data of our laboratory, E= this work.

	In (ppb)	Re (ppb)	Sb (ppb)	Se (ppm)	Sn (ppm)	Te (ppm)
A	46	43	107			1.8
B	48		120			
	48		130			
			125			
C		51	100	12.1		
D		50		13.1		
E	49	45	123	11.2	0.84	1.6

REFERENCES

1. Jagoutz E., Palme H., Baddenhausen H., Blum K., Cendales M., Dreibus G., Spettel B., Lorenz V., Wänke H., Proc. Lunar Planet. Sci. Conf. 10 (1979) 2031-2050.
2. Baddenhausen H., private communication.
3. Merck-Bericht "Natriumtetraphenylborat".
4. Korkisch J., Modern methods for the separation of rarer metal ions (1969) p.304, Pergammon Press.
5. Korkisch J., Feik F., Analyt. Chim. Acta 37 (1967) 364-369.
6. Jardanov N., Mareva St., Talanta 13 (1966) 163-168.
7. Willand H.H., Furman N.H., Grundlage d. quantitativen Analyse (1950) p.350, Springer Verlag, Wien.
8. Hayashi K., Ogata T., Japan. Analyt. Chem. 15 (1966) 1120-1124.
9. Morgan J.W., Wandless G.A., Petrie R.K., Irving A.J., Proc. Lunar Planet. Sci. Conf. 11 (1980) 213-233.
10. Morgan J.W., Wandless G.A., Petrie R.K., Irving S.J., Tectonophys. 75 (1981) 47-67.
11. Mason B. (ed.), Handbook of Elemental Abundances in Meteorites (1971) p.378, Gordon and Breach Science Publishers, New York.
12. Krähenbühl U., Morgan J.W., Ganapathy R., Anders E., Geochim. Cosmochim. Acta 37 (1973) 1353-1370.
13. Kallemeyn G.W., Wasson J.T., Geochim. Cosmochim. Acta 45 (1981) 1217-1230.

FEASIBILITY OF BETA-RAY SPECTROMETRY IN I N A A:
APPLICATIONS IN GEO- AND COSMOCHEMISTRY

Gerd Weckwerth
Max-Planck-Institut für Chemie, 6500 Mainz

1. INTRODUCTION

Instrumental neutron activation analysis (INAA) of silicate rocks have almost exclusively been carried out by gamma-ray spectrometry, using the high energy resolution. The continuous energy distribution of the beta decay leads simultaneously to an extensive overlapping of beta spectra from different isotopes. Therefore radiochemical separation steps are usually necessary, before quantitative determinations with beta counting are possible. However, if the composition of samples is varying only in a limited range - as in silicate rocks - direct beta counting proves to be feasible.

1.1 Potassium
The first reported analysis of silicate minerals by direct beta counting was the determination of potassium (Winchester, 1961). The beta radiation of ^{42}K used in this case has the highest $ß_{max}$-energy of all relevant isotopes with half-lifes more than an hour. Therefore interferences from other beta-ray emitters can be reduced to a negligible amount, using an absorber of suitable thickness (800 mg/cm^2, Figure 1). The main difficulty of this technique is the gamma background of ^{24}Na. Even by using a very thin detector with the lowest available gamma effiency, we find 50% to 95% of the counting rate, produced from irradiated basalts, to be derived from gamma radiation. This portion can be calculated from an additional measurement with a thick absorber, allowing only gamma radiation to pass. But the statistical error caused by subtracting this background compensates the advantages of higher counting rates in comparison to measurements with gamma-ray spectrometry.

1.2 Phosphorus
A second application of direct beta counting in geochemistry is the determination of phosphorus, first used by Steinnes(1971). The beta radiation of ^{32}P is the only emitted radiation usable for an analysis of P in silicates. Its $ß_{max}$-energy of 1.71 MeV is somewhat lower than that of ^{42}K, requiring a smaller absorber thickness (170 mg/cm^2). The longer half-life of ^{32}P (14.28 days) gives the possibility to wait, until isotopes with higher beta energies have decayed to a suffient extent.
Figure 2 shows that more than 14 days after irradiation, ^{32}P, produced

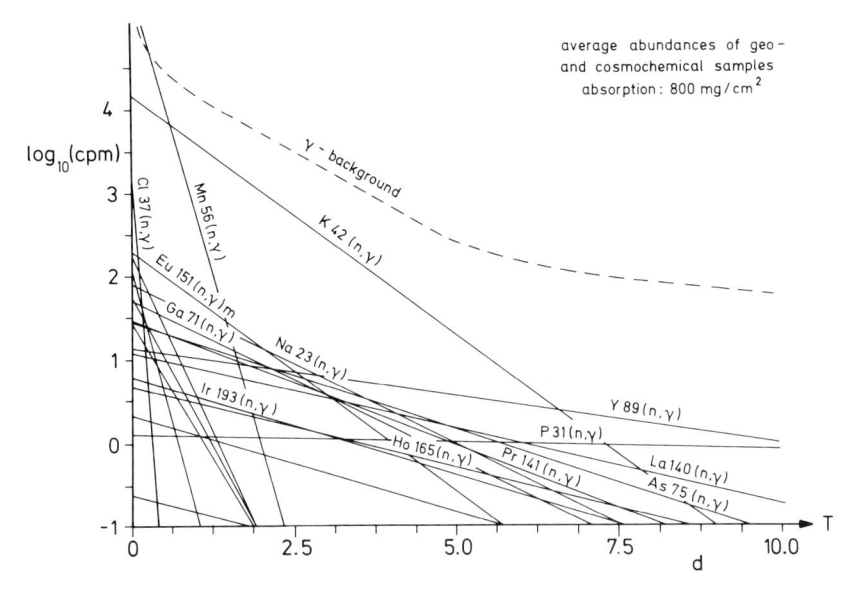

Beta-ray spectrometry in geo- and cosmochemistry 559

Fig. 1 Calculated counts per minute of a surface-barrier detector (500μ) derived from varies nuclear reactions during the first 10 days after irradiation

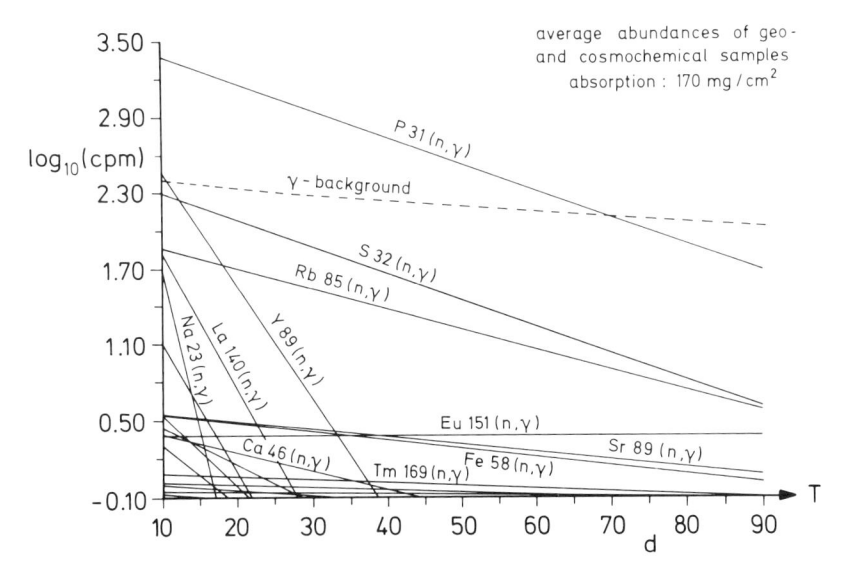

Fig. 2 Calculated counts per minute of a surface-barrier detector (500μ) derived from varies nuclear reactions between the 10th and 90th day after irradiation

from ^{31}P in silicates, on the average has a far dominant counting rate. This does not change even at low P-contents, because the main interferences come from incompatible elements showing correlated abundances. We intended to use the technique of Steinnes as part of a multi-element analysis. That means to do P-measurements of the same samples, irradiated for gamma-ray spectrometry, by an additional beta counting (3). The advantages would be:

1. Simultaneous determinations of the abundances of interference elements needed for corrections
2. Improvement of element ratios by elimination of otherwise possible variations in sampling and irradiation parameters.
3. No additional consumption of rare material, e.g. meteorites.

2. EXPERIMENTAL

The pulverized samples were filled in small capped polyethylene cylinders with a base of 0.5 cm^2. These were exposed for 6 hours to a thermal neutron flux of $7*10^{11}$cm^{-2}s^{-1} in the TRIGA-reactor of the University of Mainz. During the first days after irradiation, samples were counted on a Ge(Li)-detector for analysis by gamma-ray spectrometry. Before starting beta counting, about 14 days after irradiation, the mass of the samples, if not already in this range, was reduced to 100mg. This was necessary to keep the self-absorption of the samples and the geometry of all beta measurements constant. Variations of these parameters proved to be the main source of error.

For beta counting a surface-barrier detector with an area of 100 mm^2 and 0.5 mm sensitive thickness was used. We chose an absorption thickness of 170 mg/cm^2, given by the bottom of the cylinder, the sample holder, and a thin plastic foil. To correct for interference from gamma-rays, all samples were counted a second time with a 4 mm thick Al-absorber. The correct counting rate of beta radiation was obtained with:
$B = B_1 - G_1*1.1$; B_1 and G_1 are the counting rates with and without the Al-absorber. The factor 1.1 on average gives the influence of the gamma absorption. Standards for P and for the main interfering elements were treated in the same way as the samples. High activation-rates of some elements required the use of only a small amount of material (< 1 mg). In those cases specific counting rates were recalculated to a self-absorption of 100 mg. Table 1 shows specific activities of interference elements relative to P for measurements 14 days and 21 days after irradiation. These values are used as interference factors (I) in corrections calculated with:

$$C° = C^+ - \sum_n I_n * C_n$$

$C°$ = corrected-, C^+ = uncorrected P-content
C_n = content of interference-element n

element	radio-isotope	half-life (days)	E_{max} (MeV)	(spec. activ.)$_{el}$/(spec. activ.)$_p$	
				14 days after irradiation	21 days after irradiation
phosphorus	P 32	14.28	1.71	1	1
sulfur	P 32	14.28	1.71	0.042	0.042
rubidium	Rb 86	18.65	1.78	0.91	0.99
yttrium	Y 90	2.67	2.27	0.90	0.2
antimony	Sb 122	2.74	1.97	1.98	1.45
	Sb 124	60.2	2.31		
europium	Eu 152	4529	1.46	1.69	2.34
	Eu 154	3104.5	1.85		
arsenic	As 76	1.09	2.96	0.08	0.0014
calcium	Ca 47	4.54	1.98	0.000065	0.000031
chlorine	P 32	14.28	1.71	0.0069	0.0069
iron	Fe 59	45	1.57	0.000013	0.000017
strontium	Sr 89	50.5	1.46	0.008	0.01
lanthanum	La 141	1.68	2.18	0.3313	0.0261
samarium	Sm 153	1.95	0.8	0.038	0.0045
terbium	Tb 160	72.1	1.74	0.33	0.44
holmium	Ho 166	1.12	1.86	0.22	0.004
thulium	Tm 170	128	0.97	2.45	3.33
sodium	Na 24	0.63	1.39	0.000014	$9*10^{-9}$

Tab. 1 Specific activities of the most important beta-ray emitters normalized to P, 14 and 21 days after an irradiation of 6 h. thermal neutron flux: $7*10^{11}$ fast neutron flux: $7.5*10^{10}$ absorber used: 170 mg/cm^2

Sample (type)	P(ppm)	accuracy	interference elem. (port.)		lit. values. (ppm)	(ref)
Standard rocks						
BCR-1 (basalt)	1580	11 %	Rb, S	4 %	1570	(4)
AGV-1 (andesite)	2245	11 %	Rb, Sb	3.5 %	2140	(4)
W-1 (diabase)	585	12 %	Rb,S,Ca	5 %	610	(4)
PCC1 (peridotite)	10	25 %	Sb, Ca	50 %	12	(4)
AN-G (anorthosite)	120	14 %	Ca,S,Rb	10 %	50 - 140	(5)
BE-N (basalt)	4630	10 %	Rb	1 %	4585	(5)
MA-N (granite)	5600	14 %	Rb	37 %	6066	(5)

Sample (type)	P(ppm)	accuracy	Sample (type)	P (ppm)	accuracy
Perdotites			Lunar samples		
SC1 (sp.lherz.)	64	12 %	66095 (matrix)	1125	14 %
KH1 (sp.lherz.)	24	15 %	66095 (basalt)	1140	20 %
J4 (gr.lherz.)	162	13 %	66095 (anorth.)	172	20 %
Ka167 (harzburgite)	33	13 %	67539.7 (impact)	8	30 %
SC1 (olivine)	46	20 %	67567.3 glass)	390	16 %
KH1 (olivine)	18	23 %	67627.10 "	980	16 %
J4 (olivine)	34	20 %	Meteorites		
Ka167 (olivine)	24	18 %	AH 81005 (Moon)	90	25 %
SC1 (orthopyrox.)	16	26 %	Murchison (C2)	1135	18 %
KH1 (orthopyrox.)	12	25 %	Allende (C3)	1270	17 %
J4 (orthopyrox.)	32	20 %	Stannern (Eu)	440	16 %
SC1 (clinopyrox.)	40	26 %	Shergotty (SNC)	3470	13 %
KH1 (clinopyrox.)	20	35 %	AH 77005 (SNC)	1520	13 %
J4 (clinopyrox.)	30	32 %	Nakhla (SNC)	450	15 %
Q1 (quartz, fig. 3)	2.1	25 %	Chassigny (SNC)	250	16 %

Tab. 2 Results of instrumental P-mesurements with beta counting

Except for S and Y, we got the abundance of interference-elements from gamma-ray spectrometry. The S-content was taken from varies techniques (e.g. X-ray flourescence). The interference of S can be reduced using a reactor with a lower ratio of fast to thermal neutrons. This would also reduce Rb-interferences, because the fraction of epithermal neutrons decreases. Y-interferences are observable because of the short half-life of 2.7 days. When appearing we prefer an additional waiting-time.

3. RESULTS AND APPLICATIONS

P-measurements of a few well known standard rocks proved to be in good agreement with literature values (Table 2). Multiple analysis of samples with a high P-content showed variations up to 6%, mainly caused by variations of the sample position relative to the detector. Taking into account a similar error of the standard value, the accuracy of the technique is about 10%. The error increases for samples with a different density or mass, for higher contents of interfering elements, and for P-abundances less than 10 ppm, because of the increasing statistical error. The counting rate of the background (4 min^{-1}) was eliminated when subtracting the measurements with and without Al-absorbers. The detection limit, defined by this background, was calculated to 0.1 ppm. About 1 ppm is required to ascertain that the measured activity was due to ^{32}P, by determining the half-life. Figure 3 shows counting rates of a cleaned quartz-sample. The log. scale (base 2) gives the possibility to a fast determination of the half-life. The counting rate recalculated to the end of irradiation, is about 9 min^{-1}, meaning approximately 2 ppm of P.

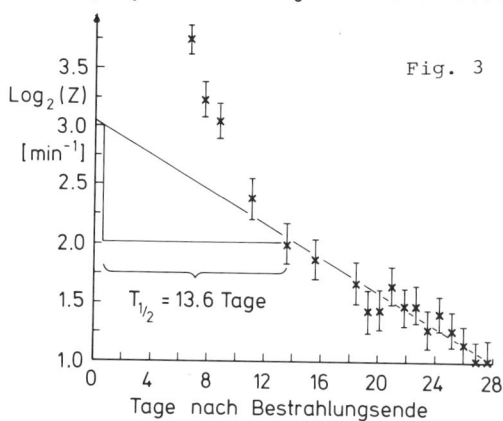

Fig. 3

The method proved to be suitable for ultramafic samples and rock-forming minerals such as olivine (3). This was used for studying the behaviour of P in magmatic processes (6). From P-abundances in comparison to other incompatible elements in lunar and meteoritical samples, we estimated the P-content in the mantle of other planetary bodies (7).

REFERENCES: (1) WINCHESTER J.W., Anal.Chem. 33 (1961) 1007
                  (2) STEINNES E., Anal.Chim.Acta 57 (1971) 451-456
(3) WECKWERTH G., Diplomarbeit, Universität Mainz, 1983
(4) FLANAGAN F.J., Geochim.Cosmochim. Acta 37 (1973) 1189-1200
(5) GOVINDARAJU K., Geostand.Newslet. 4 (1980) 49-138
(6) BLUM K., WECKWERTH G., PALME H., Fortsch.Mineral. 60 (1982) 51-52
(7) WECKWERTH G., SPETTEL B. and WÄNKE H., Terra Cognita 3 (1983) 79-80

OPTIMIERUNG DER MULTIELEMENTANALYTIK GEOLOGISCHEN MATERIALS IN PULVER-
FORM FÜR WELLENLÄNGENDISPERSIVE RÖNTGENFLUORESZENZ

N.MÜLLER

Bundesversuchs- und Forschungsanstalt Arsenal Wien,
Geotechnisches Institut   POB 8,   A-1031 Wien

1. EINLEITUNG

Die routinemäßige Multielementanalytik von Bach- und Flußsedimenten
verlangt die Untersuchung von Proben mit verschiedenartiger Matrixzusammensetzung von sauer bis ultrabasitisch. Es wurde ein Verfahren entwik-
kelt, das ohne Vorselektion auskommt. Möglichkeiten und Grenzen werden
im Folgenden behandelt.

2. METHODIK UND INSTRUMENTATION

Das Probenmaterial wird bei 60° C getrocknet und anschließend auf
80 mesh abgesiebt. Die analysenfeine Aufmahlung erfolgt in Scheiben-
schwingmühlen mit Sinterkorundmahlgefäßen, wobei eine mittlere Korngröße
50µ erreicht wird. Etwa 2 g Probensubstanz werden auf einen Träger aus
Borsäure mit 20 Tonnen Druck aufgepreßt. Es entstehen dabei überaus wi-
derstandsfähige und reproduzierbare Preßtabletten mit einem Durchmesser
von 40 mm.

Die Analyse selbst erfolgt mit einem Röntgen-Sequenzspektrometer PW 1400
(Philips). Zur Probenanregung dient eine 3 kW Röntgenröhre mit Rh-Target.
Die Konversion der Intensitäten in Konzentrationen erfolgt in einem Rech-
ner DEC PDP 11/23.

3. KALIBRIERUNG

Zur Eichung wird eine große Zahl SRMs herangezogen (bis 50). Einen
Versuch, die Matrixschwankungen zu erfassen, stellt die Einbeziehung des
Comptoneffekts dar. Dabei werden die Intensitäten der Elemente mit einer
charakteristischen Energie, die größer als die der Fe-Absorptionskante
ist, durch die Intensität des compton-gestreuten Anteiles der Rhodium Linie der Röntgenröhre dividiert. Die Berechnung der Interelementeinflüsse
durch Verwendung des Korrekturmodells von Rasberry- Heinrich ermöglicht
die Analyse mit einer für geochemische Analysen ausreichender Genauigkeit.

4. ERGEBNISSE

4.1 REPRODUZIERBARKEIT

In der vorliegenden Arbeit wurden drei Untersuchungsreihen durchgeführt. 10 Messungen an einem Preßling unmittelbar hintereinander geben ein Maß für die Gerätestabilität über etwa 12 Stunden. Die relative Standardabweichung vom Mittelwert liegt bei den Hauptkomponenten unter 0,5 % (Mg 1,97 %) bei Spuren über 30 ppm unter 6,5 %. Bei Berücksichtung der Subprobennahme durch Messung von 4 Preßlingen liegen die Werte unter 0,8% (Mg 2,65 %) bzw. unter 10 %, bei Langzeitmessungen (45 Messungen in 6 Monaten) zwischen 1,5 % und 15 % (Mg 22,4 %).

4.2 GENAUIGKEIT

Ein Maß für die Genauigkeit stellte die Differenz zwischen referiertem und errechnetem Wert in der Eichkurve dar. Auch hier kann eine mittlere statistische Schwankung der Differenz bestimmt werden. Dieses Verfahren wurde sowohl für SRMs aus einem breiten Matrixbereich als auch für Granite und Böden angewendet. Es zeigt sich, daß die Werte bei granitischen Gesteinen bei einzelnen Elementen um bis zu einen Faktor 10 besser werden, im allgemeinen nur um den Faktor 2 bis 5. Bei den Bodenproben ist das Verhältnis für das allgemeine Eichprogramm noch günstiger.

5. ARBEITSBEREICHE FÜR AUSGEWÄHLTE ELEMENTE

Mg	0,02	-	30,0 %	Cr	3	-	2.900 ppm
Al	0,03	-	20,8 %	Co	1	-	150 ppm
Si	6,08	-	35,4 %	Ni	1	-	2.300 ppm
P	0,001	-	0,5 %	Zn	10	-	4.000 ppm
K	0,05	-	10,0 %	Rb	40	-	2.300 ppm
Ca	0,07	-	15,0 %	Sr	10	-	5.000 ppm
Ti	0,01	-	2,5 %	Y	4	-	750 ppm
Fe	0,9	-	24,5 %	Zr	3	-	3.000 ppm
Mn	0,01	-	1,5 %	Pb	10	-	1.300 ppm

6. SCHLUSSBEMERKUNG

Es ist möglich, Proben eines breiten Matrixbereiches relativ genau zu analysieren. An der unteren Grenze des Arbeitsbereiches beträgt die relative Standardabweichung (S) zwischen 10 und 50 % und verbessert sich ab dem 10-fachen Wert auf 2 - 10 % und liegt ab dem 100-fachen Wert zwischen 1 und 3 %.

MULTI-SPURENELEMENT-ANALYSE IN GNEISEN UND DEREN SCHWERMINERALFRAKTIONEN MITTELS INAA

W. Kiesl und F. Kluger
Institut für Analytische Chemie der Universität Wien

ZUSAMMENFASSUNG

Die zerstörungsfreie Bestimmung einer Anzahl von Spurenelementen in vorwiegend sauren Gesteinen (Graniten, Gneisen, Quarziten) bzw. deren Schwermineralfraktionen wurde zur routinemäßigen Durchführung adaptiert und optimiert. Dies war insbesondere im Hinblick auf unterstützende Untersuchungen bei Prospektionsarbeiten notwendig, da hier ein größerer Probenanfall bewältigt werden konnte.

Darüber hinaus wurde der Einfluß der Neutronenabsorption durch die in den genannten Gesteinen auftretenden Minerale berücksichtigt, da dieser Effekt nahezu die einzige Quelle möglicher Fehlanalysen darstellt, insbesondere dann, wenn nach der Probenaufarbeitung die Korngrößefraktion 0,25 - 0,06 mm unmittelbar der Analyse zugeführt wird.

EINLEITUNG

Geochemische Untersuchungen eines vorwiegend durch Gesteine granitischer Zusammensetzung dominierten Gebietes, sowie die Unterstützung von Prospektionsarbeiten machten eine routinemäßige Bearbeitung einer großen Anzahl von Proben erforderlich.

Zu diesem Zweck wurde ein zerstörungsfreies Verfahren erarbeitet, welches die Erfassung von 8 Seltenen Erdelementen, nämlich La, Ce, Nd, Sm, Eu, Tb, Yb und Lu gestattet, womit genetische Gesichtspunkte geochemisch zu interpretieren sind. Darüber hinaus konnten Th- und U-Konzentrationen bis in den unteren ppm-Bereich ermittelt werden, was für Prospektionsarbeiten eine durchaus ausreichend niedere Konzentration darstellt. Neben Sc, Ta und Zr diente vorwiegend Hf zum Erkennen von Anreicherungen des Minerals Zirkon. Die Elemente Cr, Co und Au sowie As, W, Rb, Cs und Ba runden das Spektrum der bestimmbaren Elemente ab. Womit die Prospektion auf mit diesen Elementen assoziierte Lagerstätten ausgedehnt erscheint.

Die chemische Variationsbreite der Matrix erfährt nur bei gewissen Elementen eine Einschränkung. FeO-Gehalte > 10 % sowie Sc-Gehalte von > 40 ppm wirken nachhaltig auf den Untergrundbeitrag der Probe und haben einen negativen Einfluß auf die von uns angegebene Erfassungsgrenze bei 25 % rel. Standardabweichung. Schließlich wirkt sich ein Na_2O-Gehalt > 3 % auf die nach der Bestrahlung einzuhaltende Abkühlzeit aus. Alle anderen Haupt- und Nebenbestandteile wirken sich kaum aus, sie streuen bei den Gesteinsproben in weiten Grenzen. SiO_2 etwa von einem Paragneis mit ca. 63 % bis 75 % bei einem Disthenquarzit oder CaO von < 0,10 % in dem Disthenquarzit bis ~51 % im Apatit von Grobgneisen.

Die Schwermineralfraktionen wiesen sehr unterschiedlichen Mineralbestand auf. Dies war davon abhängig, ob die Schweretrennung mittels Clerici-Lösung (Dichte 3,5 - 3,8) oder mit Tetrabromäthan (Dichte) 2,94) durchgeführt wurde. Erze, Granat, Rutil und Zirkon setzten die Schwerfraktion nach der Trennung mit Clerici-Lösung zusammen, nach der Tetrabromäthantrennung stiegen bisweilen die Anteile der Minerale Apatit, Disthen, Titanit, Epidot und Turmalin beträchtlich. Daneben enthielt die Schwerfraktion gelegentlich Anteile der Hauptminerale, etwa Glimmer, Quarz usw.

ANALYSENSCHEMA

20 - 300 mg Probenmaterial wurden in PVC-Röhrchen eingewogen und gemeinsam mit den entsprechenden Primärstandards im Core des Reaktors 6 h bei $2.10^{12} n.cm^{-2}.sec^{-1}$ bestrahlt. Nach einer Abklingzeit von ca. 6 d (abhängig vom Na_2O-Gehalt) wurde die 1. Messung durchgeführt, nach weiteren 14 d die 2. Messung des γ-Spektrums vorgenommen. Die Tabelle 1 enthält die entsprechenden Nuklide und die quantitativ ausgewerteten γ-Peaks. Die Tabelle enthält schließlich auch die jeweilige Erfassungsgrenze bei 25 % rel. Standardabweichung, die im übrigen von der Zusammensetzung der Probe abhängig ist.

Die Meßstrecke bestand aus einem 70 cm^3 Ge-Li-Detektor (Auflösungsvermögen 2,22 keV am 1,33 MeV Co-60-Peak) in Verbindung mit einem Canberra 8180-PDP-11/05 Vielkanalanalysator.

Da für die Analyse üblicherweise die Korngrößefraktion von 0,25 - 0,06 mm verwendet wurde, kann eine Verfälschung der Analysenresultate dadurch eintreten, daß stark neutronenabsorbierende Minerale vorliegen. Deshalb wurde für die in Frage kommenden Minerale (Apatit, Monazit, Xenotim bzw. Turmalin) nach Bartels (1950) der maximale Korngrößeradius für 10 %ige zulässige Neutronenflußdepression berechnet.

Tabelle 1: Meßplan für die bestrahlten Proben

Bestrahlungsbedingungen: 6h Core bei 2.10^{12} n . cm^{-2} . sec^{-1}

Einwaagen: 20 - 300 mg

1. Messung: 6d Abklingzeit, 1h Meßzeit

Nuklid	Ausgewertete γ-Energien (keV)		Erfassungsgrenze (ppm)*
As - 76	559		1
La - 140	487	1596	0,5
Sm - 153	103		0,1
Yb - 175	396	283	0,07
Lu - 177	208		0,04
W - 187	686		3
Au - 198	412		0,035
U(Np-239)	228	278	2

2. Messung: 20d Abklingzeit, 5 - 12h Meßzeit

Nuklid	Ausgewertete γ-Energien (keV)		Erfassungsgrenze (ppm)*
Sc - 46	889		0,7
Cr - 51	320		10
Co - 60	1332		1
Rb - 86	1077		50
Zr - 95	757		500
Ba - 131	496		25
Cs - 134	796		1
Ce - 141	145		3
Nd - 147	531		10
Eu - 152	122	1408	0,03
Tb - 160	299	879	0,07
Yb - 169	177		0,1
Lu - 177	208		0,03
Hf - 181	482		0,5
Ta - 182	1221	1231	1
Th(Pa-233)	312		2

*Erfassungsgrenze bei 25 % rel. Standardabweichung, abhängig v.d. Zusammensetzung der Probe.

Wie aus den Tabellen 2 und 3 ersichtlich, ist der Effekt nur bei
Anwesenheit von Xenotim, Monazit bzw. Turmalin relevant. Bei Anwesenheit dieser Minerale (optische Inspektion der Gesteinsproben) müßte
die Proben analysenfein vermahlen und entsprechend verdünnt (Stärkepulver oder dgl.) werden, bevor die Bestrahlung durchgeführt wird.

Den hierfür nötigen Berechnungen wurden die in Tabelle 2 wiedergegebenen Analysen zugrundegelegt. Für den durchschnittlichen Einfangquerschnitt der Minerale sind insbesondere die Elemente Sm, Eu, Gd,
Dy und B relevant. Tabelle 3 enthält schließlich jene Parameter, die
für die Berechnung des maximalen Korngrößenradius benötigt werden.
Dabei wurden jene Elemente, deren Beitrag < 1 % im Vergleich zum durchschnittlichen Einfangsquerschnitt ist, nicht berücksichtigt. Im wesentlichen wird bei den Phosphaten der durchschnittliche Einfangquerschnitt vom Gehalt an Gd dominiert, bei Turmalin von B. Man erkennt

Tabelle 2: Pauschalchemismus und für die Neutronenflußdepression relevante Spurenelementgehalte einiger Mineralphasen (Angaben in Gew.-%)

Apatit		Zirkon		Monazit		Xenotim		Turmalin (Var. Elbait)	
H	0.015	O	34.43	O	26.44	O	30.15	H	0.34
O	38.05	Si	15.11	F	0.74	P	14.59	Li	1.02
F	3.37	Zr	47.96	Si	0.20	Y	27.00	B	3.44
Na	0.12	Sm	0.0105	P	12.46	La	0.17	O	51.33
P	18.42	Gd	0.017	Y	0.71	Ce	0.80	F	1.59
Cl	0.20	Dy	0.020	La	10.84	Pr	0.14	Na	2.55
Ca	38.87	Hf	1.96	Ce	30.11	Nd	1.01	Mg	0.57
Fe	0.34	Th	0.07	Pr	3.29	Sm	0.77	Al	21.14
Sr	0.10	U	0.158	Nd	10.16	Eu	0.16	Si	17.90
Sm	0.0805			Sm	1.28	Gd	2.62	Fe	0.12
Gd	0.0657			Gd	0.54	Tb	0.67		
Dy	0.0392			Dy	0.21	Dy	6.18		
				Th	0.07	Ho	1.76		
				U	1.89	Er	5.60		
						Tm	0.89		
						Yb	4.81		
						Lu	0.49		
						Th	0.51		
						U	0.77		

Tabelle 3: Molare Elementhäufigkeiten der in Tab. 2 gelisteten Minerale und zur Abschätzung der Neutronenabsorption nötige physikalische Parameter; es sind nur diejenigen Elemente aufgeführt, die mehr als 1 % zur gesamten Neutronenflußdepression beitragen.

Element σ[1] (barn)	Apatit Atom-% σ' [1]		Monazit Atom-% σ'		Zirkon Atom-% σ'		Xenotim Atom-% σ'		Turmalin Atom-% σ'	
Hf 102					0.34	0.35				
Nd 50.5			2.78	1.4						
Sm 5800	0.0129	0.75	0.34	19.7	0.002	0.12	0.181	10.5		
Eu 4600	0.0027	0.12								
Gd 49000	0.0101	4.95	0.13	63.7	0.003	1.47	0.59	289.3		
Dy 930					0.004	0.04	1.325	12.3		
Li 70.7									2.59	1.8
B 759									5.64	42.8
$\bar{\sigma}$[1]		5.82		84.8		1.98		312.1		44.6
Dichte [g.cm^{-3}]		3.22		5.2		4.6		4.8		3.02
Mittleres Atomgewicht		23.8		38.6		30.9		35.05		17.7
Makroskopischer Absorptionskoeffizient [cm^{-1}]		0.48		6.9		0.18		26		4.6
Kugelradius[2] bei max. 10% Absorption [mm], Σ_a		2.7		0.19		7.6		0.052		0.30

[1] Die angegebenen Einfangquerschnitte gelten für monoenergetische Neutronen der Geschwindigkeit 2200 cm.sec^{-1}; die anteiligen Einfangquerschnitte σ' ergeben sich als Produkt von σ und der jeweiligen Atomhäufigkeit, die mittleren Einfangquerschnitte $\bar{\sigma}$ als Summe der elementaren Komponenten.

[2] nach BARTELS (1950) r = 1,36/Σ_a

aus Tabelle 3, daß die Anwesenheit von Xenotim in den Gesteinsproben den größten Einfluß bezüglich einer Neutronenflußdepression ausübt, gefolgt von Monazit, Turmalin und Apatit. Für alle anderen genannten Minerale ist der Beitrag zur Selbstabschirmung unbedeutend, selbst für durchschnittliche SEE-Gehalte des Zirkon, der bekanntlich die schweren Seltenen Erdelemente anreichert.

LITERATUR

Bartels, W.J.C., Self-Adorption of Monoenergetic Neutrons, KAPL - 336, 1 - 25, 1950

OPTIMIERUNG DES ICP-OES-SPEKTROMETERS FÜR DIE ANWENDUNG AUF GEOCHEMISCHE MULTIELEMENTANALYSENVERFAHREN

P. DOLEZEL

Bundesversuchs- und Forschungsanstalt Arsenal Wien,
Geotechnisches Institut   POB 8, A-1031 Wien

1. EINLEITUNG

Seit vier Jahren ist ein ICP-OES-Spektrometer im Betrieb. Die erste Aufgabenstellung war, die Multielementanalysen von Fluß- und Bachsedimenten für die geochemische Bestandsaufnahme im österreichischen Bundesgebiet auf Al, Ba, Be, Ca, Ce, Co, Cr, Cu, Fe, Ga, K, La, Li, Mg, Mn, Na, Nb, Ni, P, Pb, Sc, Sr, Ti, V, Y und Zn durchzuführen.

2. METHODIK UND INSTRUMENTATION

Aufschlußmethode.
500 mg analysenfein gemahlener Probe werden in Teflonschalen (30 ml) mit 6 ml $HClO_4$ 70 % : HNO_3 65 % 1:1 und ca. 20 ml H_2F_2 48 % versetzt und über Nacht kalt stehen gelassen und danach auf Sandbad bei 170 - 180° C zur sirupösen Konsistenz eingedampft. Die abgekühlte Probe wird nochmals mit der Säuremischung (6 ml, aber ohne H_2F_2) versetzt und mit etwas H_2O bis zur vollständigen Auflösung auf dem Sandbad (ca. 160 - 180° C) aufgewärmt. Die gelöste Probe wird in 50 ml Plastikfläschchen umgefüllt und mit H_2O auf Volumen ergänzt. Das verschlossene Fläschchen wird geschüttelt, 2 Stunden stehengelassen, und ein Anteil davon gelangt im Probenwechsler zur Messung.

Analysengerät.
Luftsimultanspektrometer
Philips PV 8210 mit ICP-Source PV 8490 mit
autom. Untergrundkorr. Einheit PV 8260/51
49 fixe Detektoren und 1 beweglicher Detektor

Plasmabrenner:	Universaltype
Plasma-Argon:	21 l/min
Carrier-Argon:	1,2 l/min
Zerstäuber:	Pneumatischer Querstromzerstäuber mit perist. Pumpe
Probenaufgabenmenge:	1,9 ml/min
Plasmaleistung:	1,6 KW
Beobachtungshöhe:	12 - 16 mm über der Spule
Integrationszeit:	15 sek
Spülzeit:	50 sek
Probenwechsler:	SP 4 - 1 PYE-UNICAM für 100 Proben

3. KALIBRIERUNG

Die Eichung erfolgt mittels SRMs, die den Bereich der basischen bis sauren Gesteinstypen (inklusive Böden und Sedimente) überdecken. Der Aufschluß erfolgt genauso wie bei den Routineproben. Die Eliminierung der verschiedensten Interferenzeffekte wird entweder durch die im Softwarepaket enthaltene Routine oder über Messung von Blindprobe mit Störelementzusatz zur Errechnung der Korrekturfaktoren durchgeführt.
Zur Nacheichung dient die Blindlösung als tiefer und ein homogener, sorgfältig aufbereiteter und homogenisierter Diabas als hoher Standard.

4. ERGEBNISSE

4.1 Reproduzierbarkeit:

Bei genauer Einhaltung der vor und bei Eichung ermittelte Parameter liegen die statistischen Kennzahlen bei Mehrfachaufschluß einer Probe für alle analysierten Elemente im Durchschnitt bei $\pm$ 5,5 %, bei 60-facher Messung einer Probe (2^h) bei $\pm$ 7,2 % und bei 160 Aufschlüssen einer Probe über ein Jahr bei $\pm$ 9,2 %.

4.2 Analysenleistung:

Probenwechslerfassungsvermögen 100 Proben in 2 Chargen:
je 2 Nacheichstandards (Lo/Hi) und 48 Proben
Analysenzeit: 3 1/2 Stunden für 100 Proben
Durchschnittliche Nettoprobenzahl pro Tag: 192 Proben

5. SCHLUSSBEMERKUNG

Die Analysenleistung von ca. 200 Proben täglich auf 26 Elemente kann mit ausreichender Genauigkeit im Rahmen des durch die SRMs vorgegebenen nutzbaren Arbeitsbereiches zufriedenstellend erreicht werden. Hauptaugenmerk ist auf die Einhaltung der Geräteparameter, der Nacheichstandards und die Eliminierung von Ausreißerproben wie Karbonate und Ultrabasite zu legen, die gesondert analysiert werden müssen.
Der Aufmahlungsgrad der Probe ist daher von außerordentlicher Bedeutung. Die SRMs, mittels derer die Kalibrierung erstellt wird, sind sehr feinkörnige Pulver, wie sie nur nach etwa halbstündiger Aufmahlung einer Probe erreicht werden. Bei einer Kapazität von 200 Proben pro Tag muß die Mahlzeit so kurz wie möglich gehalten werden, sodaß u. U. die Proben variierende Kornverteilungskurven aufweisen können.
Die Schlußfolgerungen aus der vierjährigen Erfahrung mit dem Silikatprogramm sind folgende:
1) Probenfeinheit ist wesentlich für den raschen und möglichst vollständigen Aufschluß.
2) Wegen der Schwierigkeiten beim Aufschluß tendieren vor allem Chrom in höheren Konzentrationen (Spinelle), Kalium (teilweise wegen unlöslichem Perchlorat, aufschlußtemperaturabhängig) und Aluminium

zu Minderbefunden. Nicht hinreichend aufgemahlene Proben können ebenfalls zu niederen Werten führen.

3) Besondere Sorgfalt ist bei der Herstellung des Laborstandards erforderlich. Die Probe muß sehr fein aufgemahlen werden, sich reproduzierbar im Aufschluß verhalten und vor allem zuverlässig homogenisiert sein.

4) Die Geräteparameter inklusive der Raumklimabedingungen müssen striktest eingehalten werden, ebenso die gesamte Probenmanipulation beim Aufschluß.

5) Eine gruppenweise (je 96 Proben) statistische Überprüfung der Mittelwerte, der Standardabweichung, der Maximal- und Minimalkonzentrationen, eventuell der Schiefe und Wölbung können auch Auskunft über Unstetigkeiten geben.

Bei Einhaltung dieser Parameter sind die statistischen Kennzahlen bei Mehrfachaufschluß einer Probe: durchschnittliche Varianz $\pm$ 5,5 %, bei 60-facher Messung einer Probe: $\pm$ 7,2 % und bei 160 Aufschlüssen einer Probe $\pm$ 9,2 % über einen Zeitraum von einem Jahr Zahlen, die realistisch die tägliche Routinearbeit kennzeichnen.

INSTRUMENTAL NEUTRON ACTIVATION ANALYSIS
OF SMALL SPHERES OF VARIOUS ORIGINS

H.T. Millard Jr., P. Englert[*] and U. Herpers[*]

United States Geological Survey, Denver, Colorado 80225, USA
[*] Institut für Kernchemie der Universität zu Köln,
D 5000 Köln 1, FRG

1. INTRODUCTION

Small spherical objects are often found in recent and older geological sediments. It is generally assumed that magnetic spherules from such sites are of extraterrestrial origin [1]. However, systematic and accurate elemental and trace elemental analyses of such small samples are difficult for several reasons and have not frequently been performed. No attempts have been made to systematically study collections of small silicate spheres by instrumental methods other than semi-quantitative microprobe studies [1]. This, however, yields only major elemental concentrations or concentration ratios. The goal of this study is to reveal the extent to which instrumental neutron activation analyses is capable of establishing major, minor or trace elemental patterns of such small spherical objects that would be sufficient to give clues on their origins. Analysing small particles, one faces several problems, a major one of which is the sample preparation, which should not contaminate the small objects on the trace elemental level. Another problem, which requires much experience, is the handling of small samples during all procedures performed with them without destroying or losing them. The preparation of suitable standards on a microscale is another important point to be addressed.

2. EXPERIMENTAL

Spherical particles of different collections were analyzed during this study. These include: 1.) Deep sea spherules from the "Millard collection", which were prepared from 750 kg of Pacific red clay by means of magnetic separation, density separation with heavy liquids and/or handpicking under the microscope. A detailed description of the collection procedures and of the physical properties of the spherules is given by Millard and Finkelman (1970) [2] and Murell et al. (1980) [3]. The majority of these stony spherules, which have diameters of approximately 100 µm are considered to be products of meteorite ablation [2]. 2.) Transparent, glassy spherical objects from 400 years old Central European Jungfraujoch glacier ice (~700 kg) were collected on membrane filters and handpicked under the microscope. Their average diameter is ~100 µm and the main constituent of a large fraction of these objects is carbon, as determined by Laser Ion Microprobe Mass Analyses. A description of the collection is given by Englert and Herpers [4]. 3.) In addition to objects from these collections, spherules from Mount Kilauea (Hawaii, Pele's tears) and those from cave sediments were considered for analysis. The cave spheres were taken from residues of HCl-dissolved $CaCO_3$-containing layers of West German and Greek cave floors, investigated by Hennig [5]. Most of the spherules involved in this study were weighed on microbalances prior to analysis, except for the glacier spherules, whose low density of ~2g/cm^3 did not allow this. Their weights were estimated using their radii and density.

Standard preparation is a major problem for instrumental neutron activation analysis of small spheres. Several approaches are possible: The ideal case is to find or produce a multielement standard material which is homogeneous within the weight and size ranges of the spherules. Another approach is the use of single element standards, (10^5-10^7 times the expected concentration in the spherules) which have to be dissolved and diluted after irradiation to match the activities induced in the spherical objects. A third possibility is to use a series of certified multielement rock standards at the 100-600 µg level (10-1000 times the spherule mass) and to rely on precisely determined counting rate dependent pulse pile-up corrections for a given counting system [6]. We used mg-chips of a meteorite (Filomena) for the irradiation of deep sea spheres. The single-element standard approach provides reliable results; it is, however, impracticable when more than 20 elements are to be considered. Single-element standards exclusively were applied for all elements during the glacier ice, volcanic and cave spheres irradiation and for the determination of noble metals in the deep sea spherules. The use of certified rock standards is the most convenient technique and has been applied in all irradiations.

Because of the considerable interest in determining major elements in small spherical objects, the particles from the Millard collection were analysed for Mg, Al, Ti, V and Mn by the respective short-lived neutron capture products. Each individual sphere was irradiated for 4 minutes at a flux of 6 x 10^{12} n/cm^2s and measured 4 minutes after end of irradiation and removal from the irradiation container by standard γ-ray counting techniques. For a long-term irradiation the same spherules were individually packed into a special Suprasil quartz container with standards directly adjacent to them. The assembly was irradiated at the Omega-West reactor of the Los Alamos National Laboratory for 8h, and after a cooling time of 16h; for an additional 15 minutes at a flux of 8×10^{13} n/cm^{2}s. Some shorter-lived isotopes were measured by standard γ-ray counting techniques at Los Alamos. Further counting was done at the United States Geological Survey in Denver, using high resolution γ-ray spectrometry with coaxial and well-type detectors Compton suppression- and triple-coincidence-techniques, the latter for iridium [6,7].

Glacier ice spherules and those from other sources were irradiated for 3 days at a flux of 6×10^{13} n/cm^{2}s in the Merlin reactor of the KFA Jülich, accumulating a total neutron dose of 1.8×10^{19} n/cm^2. Counting of samples and diluted standards was done in low-level shielded well-type detectors and by anticompton spectrometry [8].

3. RESULTS

Table 1 contains results of 8 of the 22 spherical objects, which were the subject of this study. Only elemental concentrations with uncertainties of 20% or less are listed here. In addition the elements Br, Ta, Al, Mg, Ti and Zn were detected in some of the spherules. Otherwise detection limits are given. More than 10 elements could be determined in individual small spherical objects by instrumental NAA only. For spherules subjected to sequential irradiations, more elements (up to 18) were detectable. The few elements found in glacier spheres and their low concentrations are due to their high carbon content [4]. Iridium was not found in glacier ice, volcanic and cave spheres. However, iridium could be determined by triple-coincidence-techniques in deep sea spherules. The detection limit was as small as 10^{-14} g/g.

As this paper is predominantly dealing with analytical problems of instrumental neutron activation analysis of small objects, a detailed discussion of the results is not intended here. However, a few remarks should be made to demonstrate the utility of the data in determining the origins of the objects studied. The volcanic spheres and two of the cave spheres follow the K/La trend line for terrestrial material [9]. The K/La ratios of the glacier ice spheres do not indicate terrestrial origin. All deep sea spherules have similar elemental distribution patterns, but the trace element iridium, which is used as an indicator of extraterrestrial origin was only detected in three of them, and, contrary to expectation, not predominantly in the especially large ones. However, it has been shown that Ir is very inhomogeneously distributed in small spherical objects and is mostly related to metallic inclusions [10]. Whether the elemental patterns analysed thus far justify the assumption of extraterrestrial origin for the deep sea spheres which do not contain ppm-amounts of Ir, is presently under investigation. It has been shown, that up to 20 elements can be detected in small objects by INAA. Further development of instrumental techniques for routine spherule and radiochemical NAA for detailed analyses is needed [11].

4. ACKNOWLEDGEMENTS

The authors wish to thank M.Murell (Caltech) and J.R.Arnold (UCSD) for providing samples, the reactor crews of the U.S.G.S., Denver, of LANL and the KFA Jülich for providing irradiations, and D.McKown of the U.S.G.S., Denver, B.Dropesky and R.C.Reedy (LANL) for the use of their counting equipment.

5. TABLE 1: Elemental concentration in small spherical objects

Spherule	Source	Weight [µg]	K [%]	Sc [ppm]	Cr [ppm]	Fe [%]	Co [ppm]	La [ppm]	Eu [ppm]	Au [ppm]	Ir [ppm]
USGS-D-A	(1)	440	-	4.9	3270	23.1	415	3.8	0.15	-	<0.005
LJ-437	(1)	5.4	-	6.6	1170	23.1	240	<2	<0.4	-	2.8
K-123	(2)	2.6	<1	60	276	11.3	<50	20	5.7	0.4	-
K-62	(2)	1.7	1.5	<0.2	4.9	<0.1	<50	20	<3	4.8	-
Pele's Tear	(3)	25.1	0.43	30	44	8.2	55	3.6	4.5	0.09	-
Cave 1	(4)	7.5	4.3	39	70	11.2	340	97	13.5	21.0	-
Cave 2	(4)	22.5	0.6	49	<28	5.8	90	123	9.6	2.0	-
Cave 3	(5)	23.5	<1	<28	<10	0.1	<50	4042	110	0.4	-

(1) Deep sea spherules from the "Millard collection" [2,3]; (2) Glassy glacier ice spherules [4]; (3) Volcanic spherules from Mount Kilauea; (4) Spherules from a West German cave [5]; (5) Spherule from a Greek cave [5];

Elemental concentration presented here have uncertainties less than 20 %. Otherwise detection limites are given.

6. REFERENCES

[1] R.Ganapathy, D.E.Brownlee, and P.W.Hodge (1978), Science 201, 1119-1121; [2] H.T.Millard, jr., and R.B.Finkelman (1970), J.Geophys. Res. 75, 2125-2134; [3] M.T.Murell, P.A.Davis, K.Nishiizumi and H.T. Millard, jr., (1980), Geochim.Cosmochim.Acta 44, 2067-1074; [4] P.Englert and U.Herpers (1984), submitted to: J.Radioanal.Nucl.Chem.; [5] G.J.Hennig (1979), Dissertation, Universität zu Köln; [6] H.T. Millard, jr., (1984), Nucl.Instr.Meth., in press; [7] U.Herpers and R.Wölfle (1974), Messtechnik 82, 111-118; [8] H.G.Riotte, U.Herpers, and E.Weber (1980), Radiochim.Acta 27, 209-211; [9] H.Wänke (1974), in Analyse extraterrestrischen Materials, W.Kiesl and H.Malissa Jr., Eds., Springer Verlag, Wien, pp. 183; [10] J.Czajkowski and P.A. Farnsworth (1983), Meteoritics 18, 287-288; [11] R.Ganapathy and D.E.Brownlee (1979), Science 206, 1075-1077.

APPLICATION OF MULTI-ELEMENTAL NEUTRON ACTIVATION ANALYSIS IN ENVIRONMENTAL RESEARCH

R.DAMS

Institute Nuclear Sciences, University Gent, Proeftuinstraat 86, B-9000 Gent, Belgium

SUMMARY

Instrumental neutron activation analysis coupled to high resolution Ge(Li) γ-spectrometry and computer assisted data reduction, is for most environmental and biological matrices extremely well-suited. The major elements of these matrices, C, O, H, S, P, Si do in fact not produce intensive γ-emitting radioisotopes after neutron irradiations. Besides the intrinsic sensitivity, reproducibility and accuracy the main advantage of this technique consists in its non-destructive and instrumental character. Analytical blanks and uncertainties resulting from sample dissolution and chemical separations are avoided. The sample remains available for other analyses and the procedure as a whole can be largely automated. With an appropriate irradiation-counting scheme and sometimes with epithermal neutron irradiations, also a number of not so very sensitive elements can be determined. Today with the utmost refinement of the art one succeeds in determining up to 45 elements in a variety of environmental matrices, such as atmospheric particulates, stack emissions, coal, fly ash, liquid fuel, shales, oil sands, sewage sludge, soil, road dust, workroom dust, incineration waste, rain, surface- and waste-water and in a variety of biological materials.

In certification of environmental and biological reference materials, instrumental neutron activation analysis has played a major role and still does, owing to its sensitivity and freedom of contamination.

Also in industrial hygiene studies in workrooms this multipurpose instrumental technique can be used to analyse stationary samples, source samples and personal samples. Computer programs for classification and clustering of elements or samples are applied for the interpretation of the wealth of information obtained.

1. Introduction

During the past two decades, increasing attention has been focussed on pollution of the natural environment. The abatement of contamination of the air, of the surface-, sea- and drinkingwaters and of the soils is an important and complex task, involving the cooperation of specialists of various disciplines including analytical chemists. It became soon obvious that identification and estimation of the potential hazard of air, water and soil pollution with particulate matter are only possible after a thorough chemical analysis. The threat pollution by heavy metals can be to the human environment is well-known by now. All possible sources of metal pollution such as combustion of fossil fuels, municipal waste incineration and waste water treatment, industrial processes, transport, and all kinds of waste disposal should be investigated, starting with an elemental analysis of the material involved.

An important question still is, which elements or compounds should be determined and which not. It is at the present state of our knowledge, a very difficult task to rank pollutants according to their importance. Most elements present in the environmental samples may in one sense or the other be more or less important either as an indicator of pollution sources, as an element of general environmental interest or as one of toxicological interest. Table 1 illustrates this point (1). The safest approach is therefore not to limit the scope of the analysis, but to analyse for as many elements as possible, preferably with a multi-elemental technique. Owing to the complex nature and the variety of the environmental samples involved and the low concentrations of most elements the method of analysis should be sensitive, specific, accurate and applicable to major, minor and trace constituents. Before meaningful conclusions can be drawn concerning sources, pathways and environmental impact large numbers of analytical data are required. Therefore a technique which is entirely instrumental and easily automated is highly desirable.

2. Instrumental Neutron Activation Analysis

Most analytical techniques routinely applied for this purpose only partly meet these requirements. When solid samples are to be handled most techniques involve a dissolution step of the sample and subsequent chemical separation steps with the inherent danger for losses and contamination. Instrumental neutron activation analysis (INAA) coupled to high resolution Ge(Li) γ-spectrometry and computer assisted data reduction, is for most solid environmental and biological matrices extremely well-suited. In a number of textbooks (2-8), conference proceedings (9-12) and overview articles (13-16) there is a wealth of information on this technique available. The major elements of these matrices, C, O, H, S, P, Si, do in fact not produce intensive γ-emitting radioisotopes after neutron irradiation. Today with the utmost refinement of the art, one succeeds in determining more than 45 elements by INAA (1, 17-22). Table 1 shows that in atmospheric aerosols with an appropriate irradiation-counting

TABLE 1 : Elements of Interest in Environmental Work (1)
* Priority grading : first (1) ; second (2) ; third (3)

Element	Interest**			Prior.*	Element	Interest**			Prior.*
	Env.	Ind.	Tox.			Env.	Ind.	Tox.	
Al	0			2	I	0			2
Ag	0			2	K		0		2
As	0	0		2	Mg				2
B	0			3	Mn				2
Ba	0	0		2	Mo				2
Be	0		0	3	Ni	0	0	0	2
Bi				2	P				3
Br	0	0		3	Pb	0	0	0	1
Ca		0		1	S		0		1
Cd	0		0	1	Sb	0	0		2
Cl	0	0		2	Sc		0		3
Co	0			2	Se	0		0	2
Cr	0		0	2	Si		0		2
Cu		0		1	Sn		0		2
F	0	0		2	Ti				2
Fe		0		1	V	0	0	0	1
Hg	0	0	0	1	Zn		0	0	1

** : Env. : of environmental interest
Ind. : of interest as an indicator of a pollution source type
Tox. : of toxicological interest.

— Element determined by INAA.

scheme in routine the large majority of elements of interest with Z > 8 can be detected purely instrumentally. In spite of its multi-elemental character some shortcomings are shown in the same table. No information about the chemical state of the elements is provided. Speciation studies must thus involve chemical separations. Some possibly important elements cannot (Be, Bi, B, P, Tl) or hardly (Cd, Mo, Ni, S, Sn) or by applying special techniques such as fast (Si), epithermal (F, U) or cyclic short activation (Pb), be determined.

2.1. Applicability

The availability of high neutron flux reactors, the increased detection efficiency and resolution of nowadays Ge(Li) and intrinsic Ge-detectors and the complete computerization of calculations and reductions of complex data from the multichannel analyzers have largely extended the possibilities of the method during the last 5 years. Besides the intrinsic sensitivity, reproducibility and accuracy of the method its main advantage consists in its non-destructive character. Analytical blanks and uncertainties resulting from sample dissolution and chemical separations are avoided.

The sample remains available for other analyses and the procedure as a whole can largely be automated. Aqueous samples must often be pretreated by freeze-drying, evaporation or any other form of concentration which implies an additional difficulty and chances of errors, making the technique less popular for liquid samples. The technique has found large application for the analysis of a variety of environmental samples such as atmospheric particulates, stack emissions, coal, lignite, fly ash, liquid fuel, shales, oil sands, sewage sludge, compost, soil, road dust, workroom dust, incineration waste, sediments, rain, snow, surface-, sea-, ground-, drinking- and industrial water.

2.2. Accuracy - Certification of Reference Materials

Not only the sensitivity but especially the accuracy with which one may observe minor and trace elements in these samples has greatly increased during the past years. Because, also in environmental research, it is always mandatory to question the accuracy of the analytical result, a great need has risen for a wide variety of environmental reference materials, studied for the very small amounts of elements or components present. In the last decade several national and international institutes and organisations have started certification campaigns often leading to the issuing of certified reference materials (NBS, IAEA, BCR) (23). In the certification of many elements INAA has been a pillar and still plays a major role owing to its inherent sensitivity and freedom of contamination. This point can be illustrated with several ex-aamples(24). Figure 1 shows the results of 13 laboratories taking part with 5 different analytical techniques in the certification campaign of As in fly ash by BCR. Primarily based on the INAA and HAAS results the As content was certified with 95% confidence limits of less than 5%. A similar figure is shown for Mn in fly ash (Fig.2).

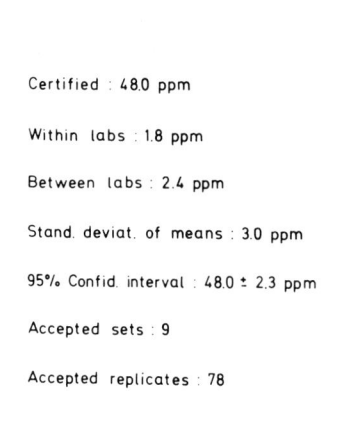

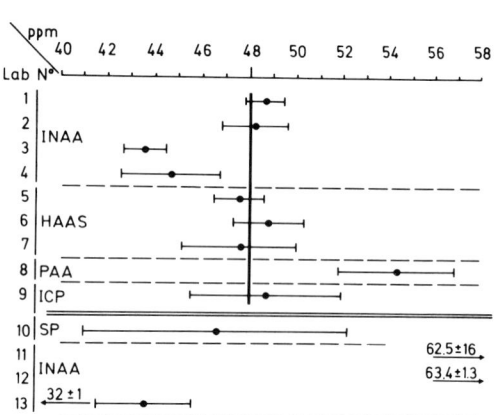

Certified : 48.0 ppm

Within labs : 1.8 ppm

Between labs : 2.4 ppm

Stand. deviat. of means : 3.0 ppm

95% Confid. interval : 48.0 ± 2.3 ppm

Accepted sets : 9

Accepted replicates : 78

As in Fly Ash

Fig. 1 : Determination of As in fly ash (23, 24) by different laboratories

Again INAA and AAS are the major techniques sustained by ICP and SP results. It is interesting to see that AAS, PAA and ES results had to be discarded for straggling mean or for outlying variances. Again certification with narrow 95% confidence limits (< 4%) was possible.

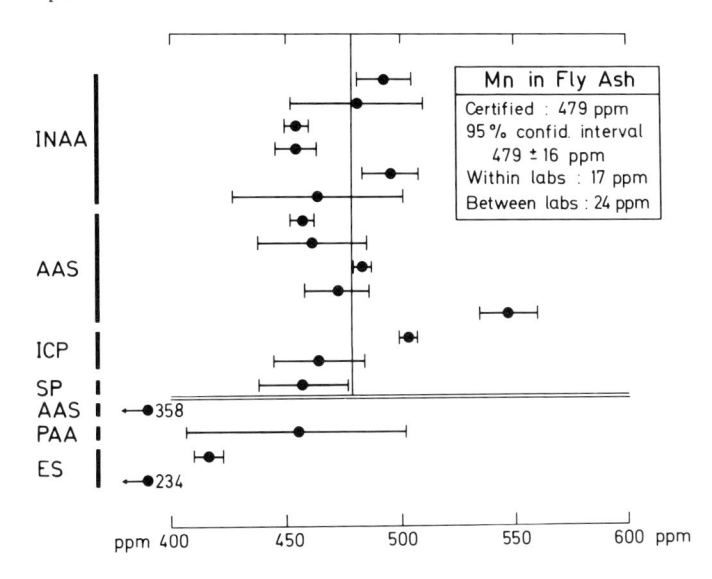

Fig. 2 : Determination of Mn in fly ash (24) by different laboratories

3. INAA of Atmospheric Aerosols

Instrumental multi-elemental neutron activation analysis without post-irradiation chemical separation is only feasible provided the activity induced in the matrix is not prohibitively high and no single major activity is produced which overshadows the other radioisotopes. Atmospheric particulate matter, or aerosols, collected on a clean substrate are ideally suited to fulfill these requirements. Therefore the literature now contains an overwhelming number of papers on INAA of airborne particulates. Fields of investigation include urban, industrial, rural, marine and remote background aerosols.

3.1. Sampling

Although there are several techniques to collect aerosols the most obvious way is by filtration. The selection of the optimum filter requires simultaneous consideration of physical and chemical properties. As filter materials with sufficiently low blank values the following types have been used : cellulose (Whatman, Schleicher & Schull), Microsorban (Dexter), membrane (Millipore, Gelman), Nuclepore (Shandon), Quartz-fiber (Gelman) etc. (1). By means of cascade impactors the particles can also be collected as a function of their particle size on clean impaction surfaces. This surfaces are generally thin sheets of organic polymers (polyethylene, teflon, polycarbonate, etc.). To reduce the chances for contamination the pre-irradiation manipulations can be restricted to a minimum. A representative fraction is cut of the filter or the impaction surface and sealed into a small polyethylene bag or in order to obtain a more reproducible geometry for irradiation and counting the fractions can be wrapped in a piece of the same clean material and pressed into a pellet.

3.2. Procedure

It is not feasible to choose for each element the optimal irradiation, cooling and counting conditions : a compromise must be found. The balance between the working-up of a reasonable number of samples and the desired quality of the analytical results will finally dictate which irradiation-counting scheme should be adapted. The scheme shown in figure 3 consists of two irradiations and four to five γ-spectrometric measurements (14). The short irradiation together with a flux monitor allows the detection of 10 to 15 isotopes. Up to 15 samples are irradiated simultaneously with one or two standards for at least 7 hours. Correction for flux gradients in the irradiation capsule can be made by means of a monitor (Fe-sheets). The measurements after one to three days and especially the measurement after at least 10 days, for the decay of ^{82}Br and ^{24}Na, give a wealth of information. In favourable conditions 42 elements could be detected by applying this irradiation-counting scheme. In general practice this figure often reduces to some 30 elements, while for the other elements only upper concentration limits are obtained. The scheme can of course be simplified, generally at the

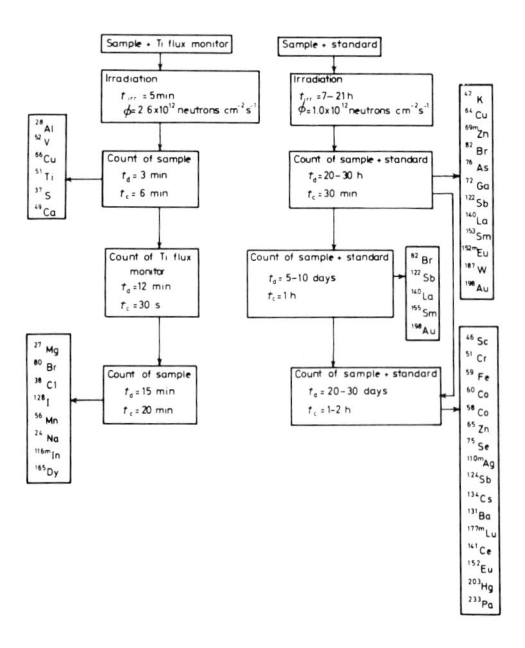

Fig. 3 : Irradiation counting scheme for multi-elemental instrumental neutron activation analysis of environmental samples

expense of the precision and the sensitivity. Standards are prepared by spotting on filter paper know amounts of solutions in which the elements are combined in proportions roughly simulating the aerosol composition. Important is that the shape of the standard spectrum, the deadtime of the spectrometer and its decrease can be made to resemble that of the sample.

Figure 4 shows part of a spectrum recorded after a decay time of 4 days of a fly ash sample collected at a coal fired power plant. It can be seen that a careful net peakarea calculation is required for the triplet 554.2 keV (^{82}Br), 559.1 keV (^{76}As) and 564.0 keV (^{122}Sb). The highest resolution obtained is desirable. A choice of the most suited decay time is often helpful.

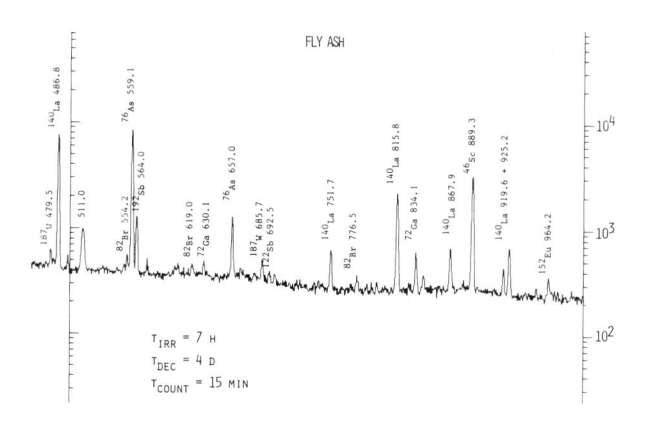

Fig. 4 : Part of gamma spectrum of fly ash sample, obtained after a decay time of four days

Figure 5 illustrates part of a spectrum recorded after 14 days decay time. In this sample the intense photopeak of ^{46}Sc at 1120.5 keV, overshadows its neighbour at 1115.5 keV of ^{65}Zn. The determination of Zn thus becomes difficult.

The multielemental character of INAA, the volume of data resulting from γ-spectrometry and the routine application to a large number of samples have made the intervention of a computer for data reduction a necessity (16).

The programmes LAMAMA and OLIVE are routinely used for spectrum analysis of short- and long-lived isotopes. They automatically locate the peaks, calculate net peak areas, convert areas to masses of elements, subtract the contribution of filter materials, calculate concentrations (ng.m^{-3}) and uncertainties on these concentrations, signal to blank ratios and other quantities which aid the analyst in interpreting the results. Although the routine use of these programmes increases accuracy and speed and eliminates human errors, it is always mandatory to examine the data carefully before accepting them. Figure 6 compares the sensitivity obtained by this routine procedure using

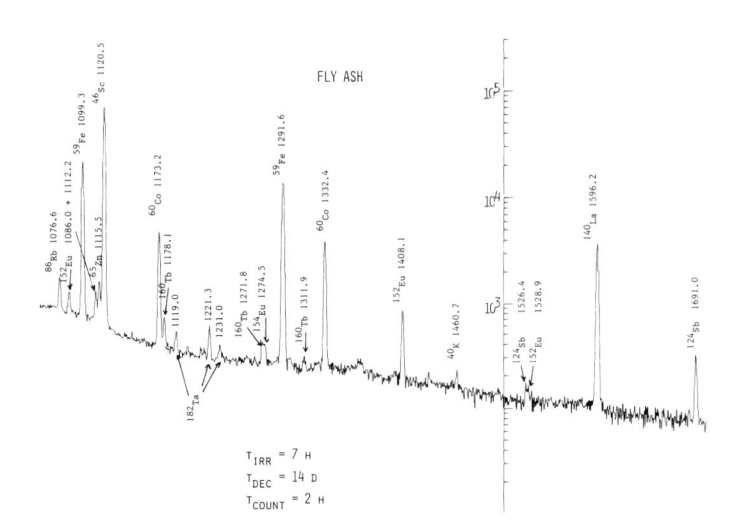

Fig. 5 : Part of gamma spectrum of fly ash sample, obtained after a decay time of fourteen days

short- or long-lived isotopes with the typical concentrations of 39 elements in a typical urban aerosol. The detection limits are calculated after a 24 hour sampling time. It can be seen that for the large majority of the elements the sensitivity is largely sufficient, exceptions being, S, Ni, Cu, Rb, Mo, Ag, Cd, I, Lu and W.

3.3. Special Irradiation-counting Systems

In some cases irradiation with epithermal neutrons under Cd-cover may present advantages. Also use can be made of the extremely high resolution obtainable with a LEPD in the low energy region. Janssens et al. (25) showed that after a short irradiation under Cd-cover a 20 min measurement with a LEPD enables the detection of the isotopes ^{80m}Br, ^{82m}Br, ^{60m}Co, ^{122m}Sb, ^{116m}In and ^{239}U. Also the X-rays of Br, Se, Te, Sn and Sb were seen. Similar work was performed more recently by Nakanishi and Sansoni (26). An even shorter irradiation (30 s) and a fast count (t_d = 25 s) gives some additional information. Short irradiations with and without Cd-cover were applied to detect isotopes with half-lives of less than 2 minutes. The Cd-cover enhances the detection of some isotopes such as ^{19}O (from F(n,p)), ^{116m}Hf, ^{23}Ne (from Na(n,p)), ^{82}Br, ^{80}Br, ^{110}Ag, ^{116m}In. Sometimes F can also be detected from the (n,γ) reaction producing ^{20}F ($t_{1/2}$ = 11 s). Owing to the interference of the (n,p) reaction on Na a correction is required.

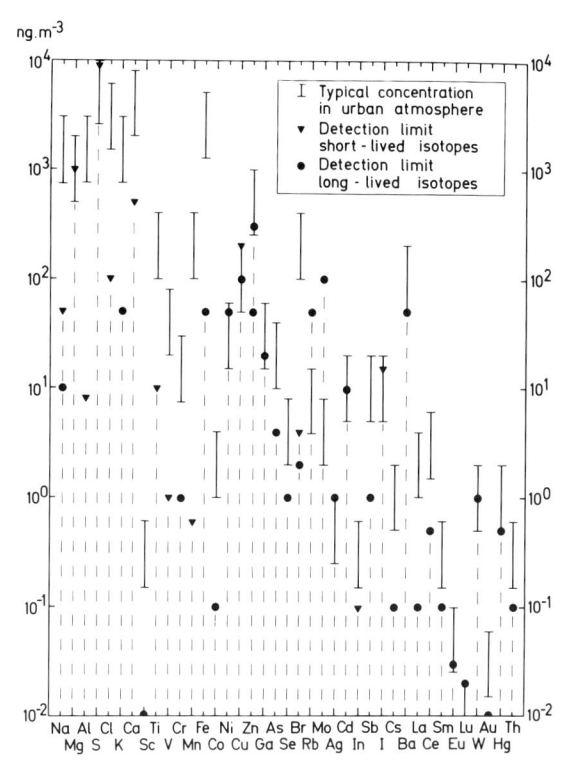

Fig. 6 : Sensitivity of routine instrumental neutron activation analysis of atmospheric aerosols, collected on cellulose filter paper, as compared to typical concentrations in an urban atmosphere

4. Elemental Composition of Dust in an Iron Foundry

A workshop of an iron foundry is a multi-source area of particulate contaminants. Owing to the variety of sources in such an area no valid assessment of the impact of the particulate contaminants on the workers health can be made without a knowledge of the chemical composition and the aerodynamic size of the particles suspended in the air. During a two week period particulate samples were collected daily with a network of 16 stationary filter samples, with three cascade impactors and with three personal samplers carried by workers active in the metal pouring-(PT1), the core-making-(PT2) and the shake-out department (Fig. 7) (28-30). Also some samples were taken of the major emission sources in the foundry. All samples were analysed by instrumental neutron activation analysis. Data were obtained for 21 elements in all samples, generally with reproducibilities better than 10%, while 15 additional elements were detected in the large majority of the samples, but not in all of them.

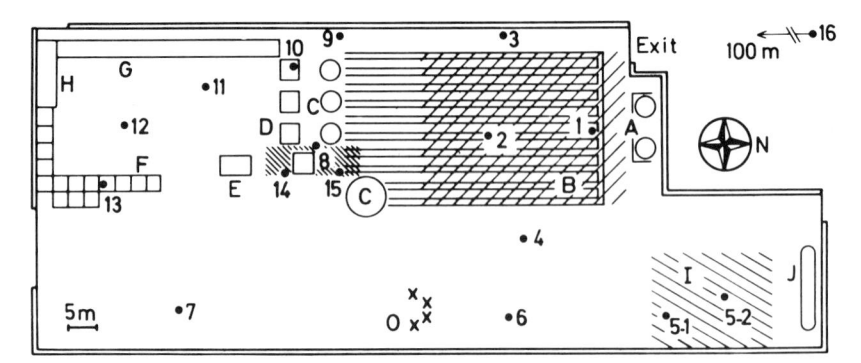

- Stationary sampling sites
× Sampling site for comparison of stationary samplers
▨ Personal sampling (station PT1, PR1)
▧ Personal sampling (station PT2, PR2)
▨ Personal sampling (station PT3, PR3)

Fig. 7 : Scheme of the workshop and sampling sites. With A : melting and metal treatment department ; B : pouring lines ; C : molding machines ; D : shaking-out grid ; E : shot-blasting ; F : fettlingshop ; G : sand preparation ; H : sand preparation and return sand ; I : core-shop ; J : drying oven

The day by day variations of the concentrations in the air were typically a factor of 2.5 but will not be discussed. Instead a time-averaged mean concentration was calculated for all stations and for 21 elements. For visualizing and for interpretation of the data a computer program, called CONTOR-TRIDIM, was applied (16, 31, 32). Figure 8 shows the distribution of La, where isoconcentration lines have been drawn from calculated intermediate concentrations. This distribution is typical for a first group of lithophilic elements namely Na, K, Cs, Mg, Ca, Al, Sc, La, Ce and Eu. These elements have their maximum in the fettling shop and a minor maximum at the pouring department. The element Fe is indicative for a second group of elements with V, Cr, Mn, Fe, Co, Zn, Ba, As and Sb. Typical is a very sharp peak at fettlingshop. The third group of volatile elements consists only of Br and Cl. The concentration variations of these elements are very smooth. This grouping of elements already sketches out the origin and some physical properties of the particles involved. Since the total weight of the samples was known the analytical data were expressed in terms of concentrations in the dust and again time-averaged mean values were calculated for all 16 stations and 21 elements. Again contour diagrams were drawn from which the same three groups of elements resulted but a better association with the sources became obvious. The lithophilic elements (La-group) are primarily generated by the pouring, shake-out and molding process. The Fe- group originates primarily during fettling and shake-out. The elements Br and Cl are produced from burning pitch, used in the preparation of the molding sand. The particle size distributions measured in the pouring- core-making and shake-out

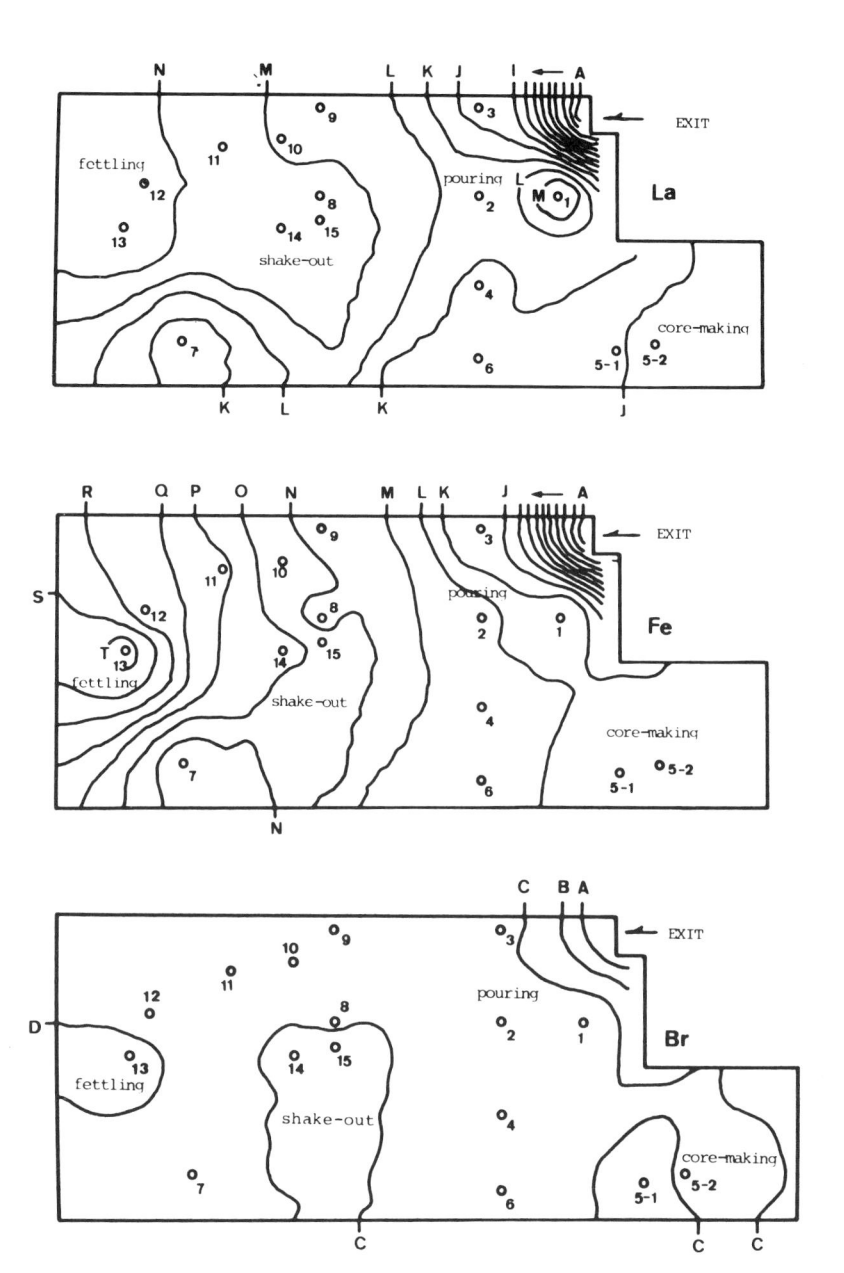

Fig. 8 : Contour plots of the elements in the workshop (ng.m^{-3}).

La : A : 0.735 ; I : 31.5 ; J : 50.4 ; K : 80.5 ; L : 129 ; M : 206 ; N : 3
Fe : A : 1.45 ; J : 61.9 ; K : 159 ; L : 254 ; M : 415 ; N : 649 ; O : 1040 ;
P : 1160 ; Q : 2660 ; R : 4250 ; S : 6790 ; T : 10900
Br : A : 41.6 ; B : 66.5 ; C : 106 ; D : 209

departments (Fig. 9) indicates that the elements of the La-group and most of those of the Fe-group are at all locations associated with coarse particles. A mass median diameter larger than 11 μm is found. Elements such as As, V, Zn and Sb have distribution curves with almost equivalent peaks on either side of the distribution curves. In the pouring- and core-making departments the volatile elements Cl and Br are preferentially associated with submicrometer particles. Manganese relates both with coarse and fine particles as was shown with a scanning electron microscope. In order to obtain a more mathematical basis for the grouping of elements an element clustering program was applied. The classification is carried out with a computer program, named HADCLU, which features a Hadamard transformation and a pattern matching routine, resulting in similarities between patterns (32). Fifteen stations (excluding the fettling shop) and 3 personal samplers are included. Figure 10 illustrates the result in a dendrogram.

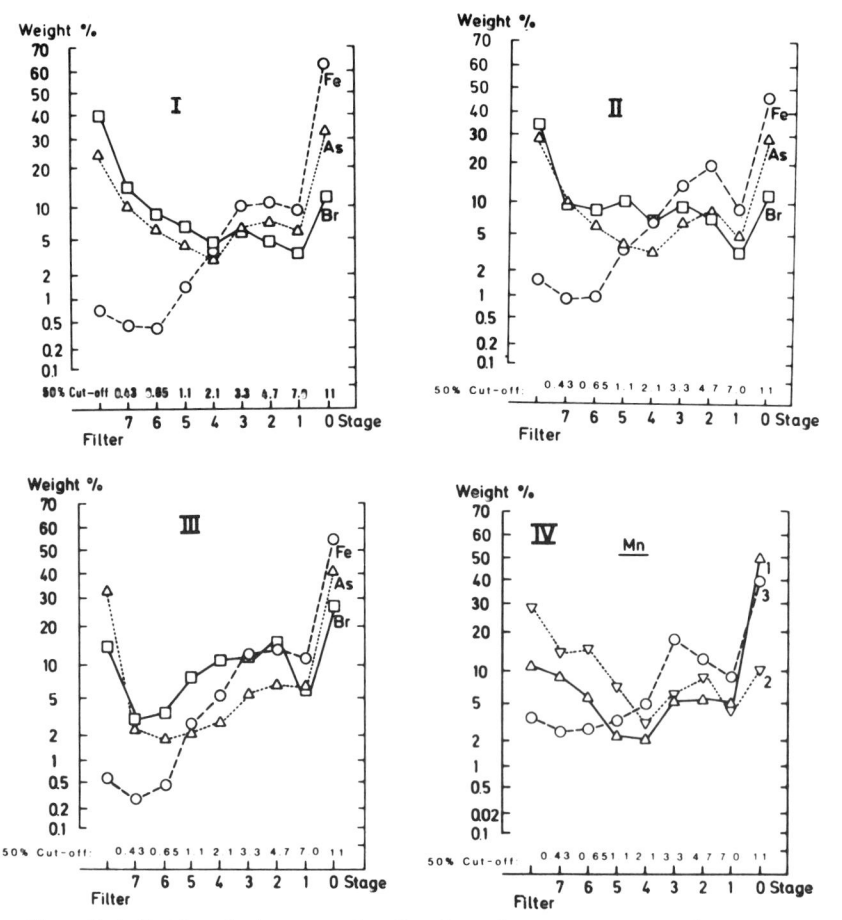

Fig. 9 : Size distribution (μm) curves for Fe, As and Br in : I : pouring department ; II : core-making department ; III : shake-out department ; IV : For Mn in 1. pouring department ; 2. core-making department ; 3. shake-out department

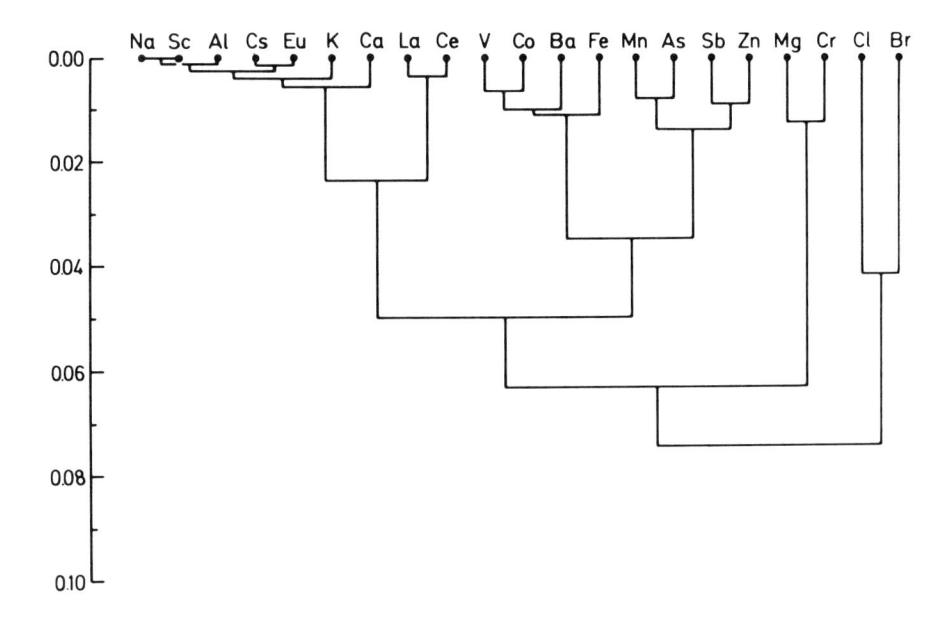

Fig. 10 : Element clustering among time averaged samples and fall-out dust except the sample from the fettling shop

All elements are placed equidistant at the top. Vertical distances indicate dissimilarities between elements as calculated by the program. Two major clusters, with small within-group dissimilarities are produced, namely the La-group and the Fe-group. The elements Br and Cl form also a separate cluster. Supprising is the behaviour of Mg and Cr. By analyzing the materials used in the shop it appeared that the core-wash material brushed onto the surfaces of the cores by the coremaker has a very high content of both elements. The coremaker cleans the cores with compressed air in such a way that he is a direct receiver of the dust generated by this process. The set of data were thus re-run, deleting personal sample PT2, worn by the coremaker. The result is shown in Figure 11. Three clusters are now produced with exactly the same composition as those obtained by comparing the contourplots. At lower dissimilarity levels subgroups of As, Sb, Zn and Mn (submicrometer particle association) and of La and Ce (nodulization elements) can be seen. Easily to be determined elements can be selected as indicators for each group.

In order to compare the composition of the air samples, with the composition of some materials handled in the shop a classification program, called DISSIM, was used. Again a dendrogram is produced (Fig. 12) from which clusters of samples are obtained (10, 11). Besides the air samples the following compositions were included : pitch (PI), molding sand (MS), shake-out sand (SS), ground-dust collected in the fettling shop (FD) and between the pouring lines (PD), core-wash (CW) and finally fall-out

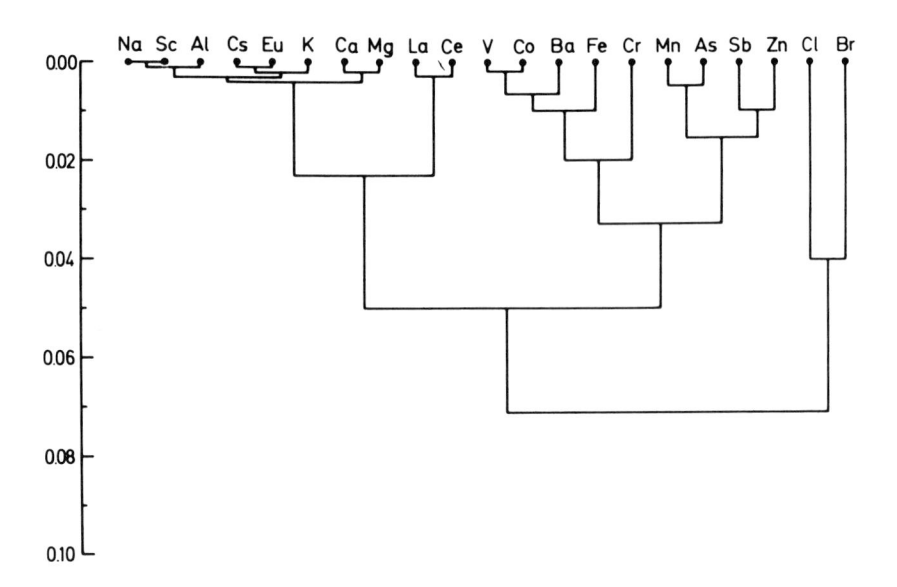

Fig. 11 : Element clustering among time averaged samples and fall-out dust except the samples from the fettling shop and personal sample PT2

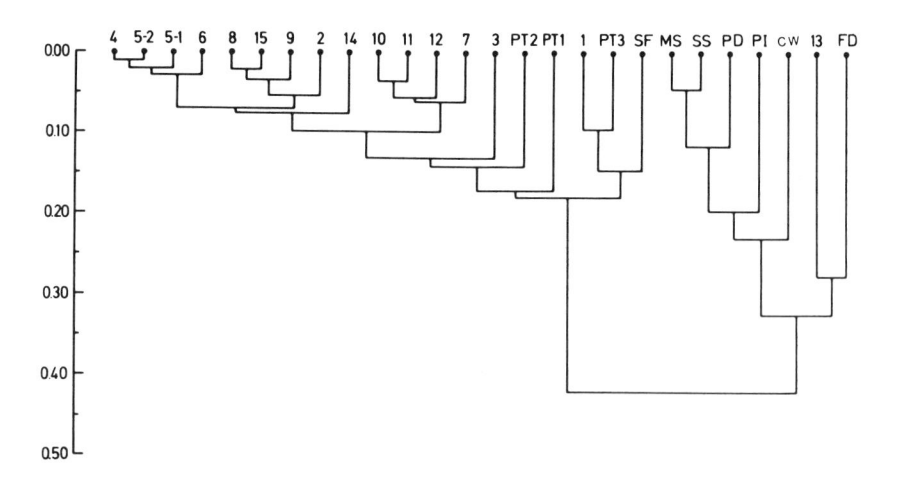

Fig. 12 : Clustering of time averaged stationary and personal samples, fall-out dust, ground dust and materials handled in the shop

dust collected in the vicinity of the shake-out machine (SF). At the dissimilarity level 0.2, 2 major clusters are obtained. The largest group includes all air samples and the fall-out sample (SF). It is concluded that shake-out, metal melting and pouring are the most significant area-wide dust generators encompassing the entire workshop. The fettling shop is a very intense but localized source owing to the extremely large particle size. The ground dust does not reflect the composition of the airborne particles. Also the clusters obtained at lower dissimilarity levels can easily be explained by logical deductions.

5. Conclusion

It is obvious that multi-elemental instrumental neutron activation analysis can be a powerful tool in the study of particulate pollution of atmospheric and indoor air. Its non-destructive and multi-elemental character together with the high reproducibility, sensitivity and especially the accuracy form a guarantee for a bright future for this technique not only in purely scientific research but also in monitoring and abatement of environmental pollution. For reducing and interpreting the wealth of multielemental information obtained, mathematical techniques featuring concentration profile delineation with contour plots and classification of elements or samples with cluster analysis are very useful and give additional merits to the multi-elemental technique.

REFERENCES

1. Heindryckx R. and Dams, R., Progress in Nuclear Energy, 3 (1979) 219-252

2. De Soete D., Gijbels R. and Hoste J., "Neutron Activation Analysis - Chemical Analysis", (Eds. Elving P.J. and Kolthoff I.M.) Vol. 34, Wiley Interscience, London (1972)

3. Amiel S., "Nondestructive Activation Analysis" Elsevier, New York (1981)

4. Erdtman G. and Petri H., "Nuclear Activation Analysis : Fundamentals and Techniques" in "Treatise on Analytical Chemistry" Sec. Edit., Part I, Vol. 9 (Eds. Elving P.J. and Krivan V.) Wiley Interscience, New York (1984) in press

5. Adams F. and Dams R., "Applied Gamma-ray Spectrometry", Pergamon Press, Oxford (1970)

6. Prepper G., Görner W. and Niese S., "Spurenelementbestimmung durch Neutronenaktvierung" in "Moderne Spurenanalytik", Band 6 ; Akademische Verlaggesellschaft Geest und Portig, K.G.Leipzig (1981)

7. Kruger P., "Principles of Activation Analysis" Wiley Interscience, New York, (1970)

8. Erdtman G., "Neutron Activation Tables", K.H.Lieser (Ed.) Verlag Chemie, Weinheim (1976)

9. Braun T. and Bujdoso E. (Ed.). Proceed. "Intern.Conf. 1981, Modern Trends in Activation Analysis", J.Radioanal.Chem., 69, 70, 71 and 72 (1982)

10. Henkelman R., Kim J.I., Lux F., Stärk H. and Zeisler R. (Ed.) Proceed. "Intern. Conf. 1976, Modern Trends in Activation Analysis" Vol.I and II, München (1976)

11. IAEA (Ed.) Proceed. "Measurement, Detection and Control of Environmental Pollutants 1976", Vienna (1976)

12. IAEA (Ed.) Proceed. "Nuclear Techniques in the Life Sciences 1978", Vienna (1979)

13. Cornelis R., Hoste, J., Speecke A., Vandecasteele C. and Versiek J., "Activation Analysis - Part 2" Physical Chemistry Series. Vol. 2 Analytical Chemistry Part 1 (Ed. West T.), Butterworths, London (1976) 71-150

14. Dams R., De Corte F., Hertogen J., Hoste J. and Maenhaut W., "Activation Analysis Part 1", Physical Chemistry Series. Vol. 2 Analytical Chemistry Part 1 (Ed. West T.), Butterworths, London (1976) 1-70

15. Cornelis R., Dams R., De Corte F., Hertogen J., Hoste J., Maenhaut W., Op de Beeck J. and Vandecasteele C. "Nuclear Activation Analysis Applications" in "Treatise in Analytical Chemistry" Sec. Edit., Part I, Vol. 9 (Eds. Elving P.J. and Krivan V.) Wiley Interscience, New York (1984) in press

16. Op de Beeck J. and Hoste J. "The Application of Computer Techniques to Instrumental Neutron Activation Analysis", Physical Chemistry Series 2, Vol. 2, Analytical Chemistry, Part 1 (Ed. West T.) Butterworths, London (1976) 151-180

17. Dams R., Robbins J.A. and Winchester J.W., Anal.Chem., 42 (1970) 861-868

18. Zoller W.H. and Gordon G.E., Anal.Chem., 42 (1970) 257-264

19. Lyon W.S., "Trace Element Measurement at the Coal-fired Steamplant", CRC Press Uniscience West Palm Beach, Florida (1977)

20. Malissa H., "Analysis of Airborne Particles by Physical Methods" CRC Press, Uniscience Analyt.Chem. for Environm. Control, West Palm Beach, Florida (1978)

21. Oikawa K., "Trace Analysis of Atmospheric Samples" Halsted Press Book, John Wiley, New York (1977)

22. Valkovic V., "Trace Elements in Coal", Volume II, CRC Press Inc., Boca Raton, Florida (1983) 228-246

23. Dams R., Pure and Applied Chemistry, 55 (1983) 1957-1968

24. Colinet E., Griepink B., Guzzi G. and Haemers L., "The certification of the mass-fractions of As, Cd, Co, Cu, Fe, Mn, Hg, Na, Pb and Zn in fly ash obtained from combustion of pulverized coal. BCR n° 38", report EUR 8080 En, Commission of the European Communities, Luxemburg (1982)

25. Janssens M., Desmet B., Dams R. and Hoste J., J.Radioanalyt.Chem., 26 (1975) 305-315

26. Nakansishi T. and Sansoni B., J.Radioanalyt.Chem., 37 (1977) 945-955

27. Dams R., Billiet J. and Hoste J., Intern.J.Environ.Anal.Chem., 4 (1975) 141-153

28. Zhang J., Billiet J. and Dams R., Staub Reinhalt. Luft, 41 (1981) 381-386

29. Zhang J., Billiet J., Nagels M. and Dams R., Science Total Environm., 30 (1983) 167-180

30. Zhang J., Billiet J. and Dams R., Science Total Environm., submitted (1984)

31. Op de Beeck J., J.Radioanalyt.Chem., 37 (1977) 213-221

32. Op de Beeck J., "Program Packages DISSIM and HADCLU" Unpublished work, University of Gent, Belgium (1980)

INSTRUMENTELLE ANALYSE VON LUFTSTAUB DURCH AKTIVIERUNG MIT PHOTONEN UND PHOTONEUTRONEN

B.F. Schmitt, C. Segebade
Bundesanstalt für Materialprüfung (BAM), Berlin

Bei der Analyse von umweltrelevanten Proben wird Auskunft über eine große Anzahl von Elementen, z.B. über den Gehalt an Schwermetallen, an Halogenen oder auch an Hauptbestandteilen gewünscht. Bei Luftstaubimmissionen, Schwebstaub, Bodenproben oder Sedimenten erschweren silikatische oder organische Matrices die Analyse, sofern Trennungsoperationen zur Erfassung einzelner Komponenten erforderlich sind. Bei Luftstaubproben liegen die Massen des abgeschiedenen Staubes etwa zwischen 0.1 und 1.0 mg/cm^2, d.h. die Staubmasse ist sicher kleiner als die Filtermasse, so daß die Reinheit des Filters das Ergebnis beeinflußt. Bei täglicher oder wöchentlicher Probenahme an mehreren Meßstellen fallen viele Proben in relativ kurzer Zeit an. Unter diesen Bedingungen sind durch Photonen- und Photoneutronenaktivierung mit Hilfe eines Elektronenlinearbeschleunigers insgesamt 34 Elemente instrumentell bestimmbar. So wurden mehrere hundert Luftstaubanalysen im Luftstaubverbundprogramm der AFR (Arbeitsgemeinschaft zur Förderung der Radionuklidtechnik), Karlsruhe, und im Stichprobenmeßprogramm des Senators für Stadtentwicklung und Umweltschutz, Berlin, durchgeführt.

Bei einer Langzeitaktivierung (4 bzw. 5 Stunden mit einer maximalen Photonenenergie von 30 MeV und einem durchschnittlichen Strahlstrom von 150 µA) können durch Photonenaktivierung die Elemente Na, Mg, Ca, Ti, Cr, Ni, Cu, Zn, As, Se, Br, Rb, Zr, Nb, Mo, Cd, Sn, Sb, J, Cs, Ba, Ce, Nd, Sm, Tl, Pb und U sowie gleichzeitig durch Photoneutronenaktivierung Mn analysiert werden. Die Photonenaktivierung findet dabei in dem eng gebündelten Photonenstrom statt, der an einem Schwermetalltarget austritt. Der Photonenstrahl erscheint als nahezu gradlinige Fortsetzung des beschleunigten Elektronenbündels. Außerhalb dieses Photonenbündels ist die Dosisleistung der Photonen nur gering. Im Gegensatz dazu breiten sich die Neutronen, die vor allem durch die γ,n-Reaktion schon im Target entstehen, isotrop aus. Seitlich vom Target, im Winkel von 90° zur Strahlrichtung, befindet sich daher eine Bestrahlposition für die Neutronenaktivierung /1/.

Bei einer zweiten Aktivierung von 45 min. Dauer werden durch Photonenaktivierung die Elemente Si, Cl, K und Fe analysiert sowie durch Photoneutronenaktivierung die Elemente Al, V und I. Alle genannten Ele-

mente liegen bei üblichen Bedingungen der Staubsammeltechnik, d.h. bei
Benutzung einer Filterfläche von ca. 18 cm^2, einer Sammelzeit von 8 h
bis 7 d, einem Staubgehalt der Luft zwischen 10 und 150 µg/m^3 und einer dabei erreichbaren Filterbelegung von 3 bis 20 mg Staub in Mengen
deutlich über den Bestimmungsgrenzen (0,01 bis 5 µg) vor, lediglich
beim Co gehen die Gehalte bis an die Bestimmungsgrenze von 0,01 µg
herunter. Aus Gründen des Arbeitsaufwandes wurden im AFR-Luftstaubverbundprogramm nur die Elemente Ca, Cr, Mn, Co, Ni, Zn, As, Br, Cd, Sn,
Sb und Pb analysiert (1981 - 1982) /2/, in Berlin fiel Co heraus und
Cl, Fe, Se, Mo, Tl, U, Sr und Ba werden zusätzlich bestimmt (1984).
Als Referenzmaterial wurde ein hausinterner Standard "Müllverbrennungsasche" URM 1 /3/ sowie das NBS-Referenzmaterial SRM 1648 "Urban Particulate" verwendet. Einen Auszug aus den Ergebnissen zeigt Abb. 1.

Für die instrumentelle Analyse sind Filter aus Glasfasern nicht geeignet, da sowohl Hauptbestandteile (Na, Ca, Al, Si) und Nebenbestandteile (Zn) den Untergrund der Gammaspektren derart erhöhen, daß keine
Spurenelementanalysen mehr möglich sind. Geeignet erwiesen sich Celluloseazetat, welches aber feuchteempfindlich ist, und Polycarbonat, bei
dem jedoch dessen Spurenverunreinigungen berücksichtigt werden müssen.

Referenzen

/1/ Reimers, P., BAM-Berichte Nr. 17, Berlin, August 1972
/2/ Neider, R., Schmitt, B.F., Segebade, C., Kühl, M., KfK-AFR 006
(1983) und KfK-AFR 007 (1983)
/3/ Schmitt, B.F., Segebade, Chr., Fusban, H.-U., J.Radioanal.Chem.
60 (1980), 99-109

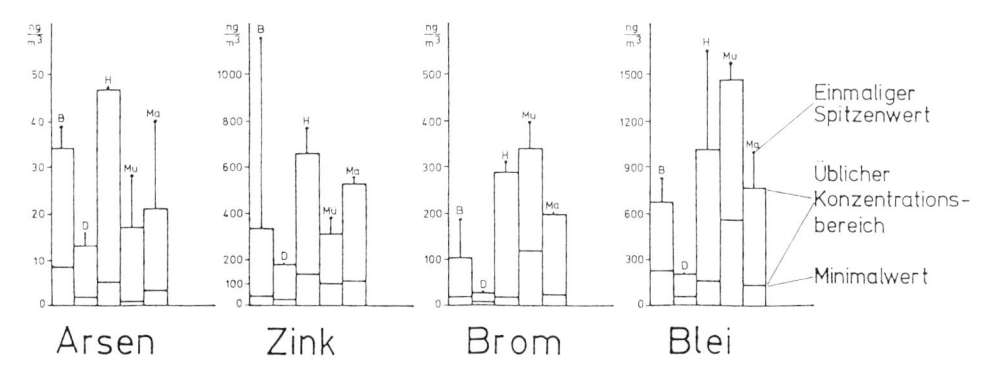

Abb. 1: Schadstoffe in Luft (auszugsweise). Sammelorte: B Berlin,
D Deuselbach (Hunsrück), H Hamburg, Mü München, Ma Mannheim

FLY ASH OF A WASTE INCINERATION FACILITY AS A REFERENCE MATERIAL FOR INSTRUMENTAL MULTIELEMENTANALYSIS

C. Segebade, B. F. Schmitt
Bundesanstalt für Materialprüfung Berlin

Fly ash produced during incineration of city waste (industrial and private household) has proved to be a suitable basic material for preparation of a multi-purpose analytical reference material[1,2,3]. Two batches of the fly ash were taken from the waste incineration facility in Berlin Ruhleben. The first batch served for preliminary investigations; twenty kilograms of the ash were ground to an average particle size of about 60 micrometers, homogenised and then analysed multiply using several analytical methods, primarily activation analysis. pure elements or their simple, stoichometrically well-determined and high-purity compounds were used as primary standards in the most cases. However, this was, as mentioned above, a preliminary investigation and therefore performed on a comparably small scale. Nonetheless, this material has been re-analysed frequently and the concentration data have been updated very often. It has been in use in the authors' laboratory successfully hitherto.

The other batch of fly ash was prepared in the same manner to serve as a candidate material to pass the certification procedure within the multielement reference material program of the European Community Reference Bureau (BCR). In the table below, concentration data for both material batches are given. Each value was obtained by between six and more than a hundred replicate determinations. Since the BCR certification procedure is not completed as yet, only preliminary data can be given as they were obtained in the authors' laboratory.

Both ashes can be used for a large variety of materials to be analysed, but the overall composition suggests primarily environmental application. The main matrix consists of alumino-silicate with high contents of calcium and iron. The large concentrations of several components known as severe environmental pollutants are remarkable, e.g. Cr, Cu, Zn, As, Cd, Sn, Pb. The overall composition is somewhat similar to that of city air-dust[4]. This kind of material is particularly useful for analytical methods which do not require any chemical treatment, e.g. activation analysis, because the decomposition of the ash entails difficulties and sometimes might not be complete.

References

[1] Schmitt, B. F., Segebade, C., Fusban, H.-U., Journal of Radioanal. Chem. 60 (1980), 99-109

[2] Fusban, H.-U., Segebade, C., Schmitt, B. F., ibidem 67 (1981), 101-117

[3] Schmitt, B. F., H.-U. Fusban, C. Segebade, Proceed. Internat. Symps. on the Production and Use of Reference Materials, BAM Berlin, Nov. 13 - 16, 1979

[4] Neider, R., Schmitt, B. F., Segebade, C., Kühl, M., KfK-AFR 006 (1983) and KfK-AFR 007 (1983)

Table 1: Element contents of the waste incineration fly ash batches URM 1 and URM 2; given in µg/g or as indicated.

Element	URM 1	URM 2	Element	URM 1	URM 2
C %	2,5	1,77	Y	34	27
F %	0,15	---	Zr	152	140
Na %	1,65	4,30	Nb	15,9	28
Mg %	1,80	2,18	Mo	24	---
Al %	8,28	10,16	Ag	42	65
Si %	14,00	14,04	Cd	250	474
S %	5,80	4,46	In	---	2,4
P %	0,3	0,55	Sn %	0,324	0,62
Cl %	1,28	4,21	Sb	220	417
K %	2,3	4,5	I	24	---
Ca %	10,16	8,80	Cs	16	14,7
Sc	20	4,1	Ba %	0,441	0,43
Ti %	0,71	0,85	La	34	25
V	117	47	Ce	56	43,4
Cr	410	832	Nd	44	---
Mn %	0,102	0,14	Sm	6	4,3
Fe %	7,89	2,20	Eu	0,23	0,44
Co	38	32	Tb	0,59	---
Ni	126	120	Yb	1,6	---
Cu	769	1305	Lu	0,6	0,4
Zn %	1,30	2,574	Hf	3,6	3,4
Ga	261	28	Ta	1,85	4,52
Ge	16,4	0,4	W	36	30
As	93	88	Au	0,6	1
Se	22	39	Hg	0,27	30,7
Br	136	293	Tl	3,9	3,1
Rb	139	110	Pb %	0,635	1,098
Sr	778	43	Th	11,3	11,2
			U	4,2	---

MULTI-ELEMENT ANALYSIS OF SINGLE DUST GRAINS
IN THE µg-MASS RANGE

K. Thiel, J. Peters, and W. Schröder[*]

Institut für Kernchemie der Universität zu Köln
[*] Institut für Neurobiologie der KFA / Jülich

1. INTRODUCTION

For the investigation of potentially extraterrestrial dust particles from polar ice and deep sea sediments LAMMA, EDAX, and INAA were applied to individual grains with the aim of a rapid chemical classification. By hand-picking under an optical microscope approximately 550 spherical objects in the range of 10 - 300 µm in diameter were extracted from the water-insoluble residue of ~ 4 tons of shelf ice recovered at Atka Bay / Antarctica, near the German Georg-von-Neumayer Station. The deep sea spherules (DSS) originated from ~ 400 kg of Pacific clay (1). Such microspheres are known to be candidates of having cosmic origin (2).

2. LAMMA

Laser microprobe mass analysis (LAMMA) provided a quick method for the semi-quantitative determination of all elements and main isotopes up to uranium. Using the Leybold-LAMMA 500 instrument of the Institute of Neurobiology / KFA Jülich more than 1000 spectra of ~ 50 antarctic spherules and selected DSS could be recorded and evaluated.

By a high energy laser pulse ~ 10^{-13} g of material of an individual spherule can be evaporated for analysis in the LAMMA time-of-flight mass spectrometer. Multiple laser shots (up to 100) focussed on the same surface spot allowed to establish a rough depth profile of the elemental composition in the upper surface layer (0...$\leq$ 5µm) of a single dust grain.

In order to avoid complex mutual interference of negative molecular ion species, preferably positive ion spectra were recorded. In the case of positive ions mainly SiO_2^+ and isotopic species of FeO^+ interfered with ^{60}Ni and the Ge-isotopes of masses 70...74. Except for Ni which is commonly used as an indicator for material of extraterrestrial origin this interference is of minor importance. Ni was subsequently determined by EDAX and NAA.

Using the NBS trace element glass SRM 610 as a reference, it was shown that for the sample geometry of dust grains only semi-quantitative element data can be derived, the reproducibility being about 30 %. The detection limit for many elements lies in the sub-ppm range, corresponding to $< 10^{-19}$ g.
The LAMMA technique despite its restrictions with respect to quantitative analysis provided a quick means of chemically classifying the dust grains and of achieving relative concentration depth profiles not accessible to the other methods. Thus for certain deep sea spherules a clear correlation of Fe, Si, and Mg and of Ca and K within the upper ~ 5 μm of the surface layer of the particle could be established on a sub-micron scale. Such element layering in DSS is believed to be caused by sea water leaching.

3. EDAX

Electron microprobe analysis (EDAX) turned out to be especially suitable for quantitative <u>bulk element</u> determination of the dust grains. Similarly to the LAMMA technique the spherules were mounted at the edge of a special double glueing tape on a single-hole copper grid. Either the original surface or a polished section of a spherule was then measured at a voltage of 120 kV and a TEM magnification of typically 10 000 x. The chemical homogeneity of a specific sample was checked by multiple spot scanning.

In view of accumulating statistically relevant data of a large number of individual grains within reasonable measuring times the ZAF-correction method without matrix standards was applied (3). A careful test of this method using the NBS standard SRM 361 and a commercial steel of known composition showed that the certified values were reproduced within ~ 5 - 10 % depending on the concentration of the respective element.

EDAX allowed a rapid determination especially of the elements Mg, Al, Si, S, Cl, K, Ca, Ti, Cr, Mn, Fe, Ni, and Zn in the 0.1 to 100 % range. The approximate detection limit for most of these elements was ~ 500 ppm if the counting time was 100 s / spectrum.
Spectral interferences mainly occurred due to the extremely high Fe-content (up to ~ 70 %) of a great number of spherules. Particularly small amounts of Co could not reliably be determined by the $K_{\alpha 1}$-line because of the broad tail of the Fe-$K_{\beta 1,3}$-line. These and other spectral interferences (e.g. Ti - La) were overcome by applying INAA.

4. INAA

Instrumental neutron activation analysis provided an independent control of the EDAX-results e.g. for K, Cr, Mn, Fe, and Zn. Since the INAA-signal is representative of the whole volume of a dust grain it may give more reliable average concentration values in the case of chemically inhomogeneous samples. INAA additionally allowed the measurement of <u>trace elements</u> present in the ppm range (Sc, Cr, Mn, Co, Ni, Zn, K) and in the ppb range (Sb, La, As, Ir, Au, W).

For n-activation the spherules were individually placed in ~ 300 µm deep holes in a fused quartz plate of 1 mm thickness. A second quartz plate served as protective cover, Al-clamps holding both plates together. Defined volumes of aqueous solutions of NBS-standard SRM 361 and of combinations of elements not interfering with each other during γ-spectrometry were prepared as standards in quartz tubes, IR-dried, and sealed.

The n-irradiations were carried out in the FRJ 1 reactor of the KFA, Jülich. The samples were exposed to fluences of typically 3.5×10^{19}, 3.5×10^{18}, and 1.6×10^{19} n cm^{-2} for thermal, epithermal, and fast neutrons, respectively.

After a cooling time of 5 - 10 hrs a first 30 min-γ-count yielded the approximate element composition of each sphere which allowed optimum coordination of the subsequent long-term (1 - 3 d) measurements. Except for Ni which was determined via its (n,p)-product ^{58}Co, all elements were measured via (n,γ)-activation. The detection limits for Sc, Cr, Mn, Ni, Zn, and K lie in the lower ppm range, for Co in the upper ppb range, for Sb, La, As, Ir, and W in the lower ppb range, and for Au in the sub-ppb range ($< 10^{-15}$ g). Depending mainly on the element concentration the accuracy of the measurements varied from $\leq$ 5 % (Fe) to $\leq$ 50 % (K, As, Au).

5. IMPLICATIONS

Multi-element analysis of approximately 100 antarctic spherules revealed the existence of at least 3 chemically different groups:
- spherules of chondritic-like element pattern
- 2 groups of spheres of distinctly non-chondritic composition.

The vast majority of all spherical dust particles investigated show an element pattern which does not match with the chemistry of average crustal rocks, volcanic fly ash, satellite ablation debris or weld-spatter from commercial steel.

It is concluded that at least the spherules of chondritic-like composition which make up only ~ 1 wt-% of all particles studied are of extraterrestrial origin and have been formed of small dust grains entering the earth's atmosphere or by melt ablation of larger meteorites (4). Whether the 2 non-chondritic dust components originate from chondritic-like material that suffered element fractionation is presently being investigated.

REFERENCES

(1) DSSs placed at our disposal by Prof. J.R. Arnold, Dept. of Chemistry, UCSD, San Diego, USA

(2) D.E. Brownlee, Chap. 19 in: The Sea, Ed.: C. Emiliani, John Wiley & Sons, New York 1981

(3) J.A. Chandler, X-Ray Microanalysis in the Electron Microscope, North-Holland Publ. Co., Amsterdam 1977

(4) K. Thiel and J. Peters (in preparation)

This work was financially supported by Deutsche Forschungsgemeinschaft, Bonn-Bad Godesberg under contract TH 320/1-3.

Neutron activation analysis for the modern electronics industry

M.L. Verheijke, J. Hanssen, H. Jaspers,
L. Steuten and P. Wijnen

Philips Research Laboratories
Eindhoven, The Netherlands

In the organization of our research laboratories the Radiochemical Department is a part of a rather large analytical group, in which almost all analytical techniques are available. The consequence of this arrangement is that we only get those analytical problems which cannot be solved better with other methods. The question is, however, what is better? Faster? In general, NAA is not a fast method, and certainly not if there is no nuclear reactor near the laboratories. Furthermore, often the matrix activity needs time to decay. No, the powers of NAA are: little sample preparation and a combination of the multielement character and the lowest limits of detection for a large number of elements. The former is important, because we have several times found unexpected impurities in materials which were thought to be very pure.

For the irradiation of our samples we can make use of four nuclear reactors: BR1 and BR2 at Mol, Belgium, FRJ1 at Jülich and HFR at Petten, at distances of 45, 140 and 180 km respectively. The last three of them are high flux reactors with about 2×10^{14} n/cm^2.s, and the shortest times between end of irradiation and arrival in our laboratories are 4 hours for the BR2 reactor and one hour for the low flux BR1 reactor. In order to obtain the lowest possible detection limits we need a rather large (high efficiency) Ge-detector and measuring times of hours. The consequence of this is that the supply of samples determines the number of detectors one needs: we use eight Ge(Li) detectors, connected to a complete Nuclear Data 6620-system with two terminals, two hard disks, a magnetic tape unit and a line printer.

We use the monostandard method (Co as flux monitor), determined the thermal/epithermal flux ratios for the irradiation facilities we use and have a nuclear library with 127 nuclides, in which the k_o- and Q_o-values of De Corte and Simonits[1] are used. Measuring conditions are standardized[2] and pulse pileup correction[2] as well as

subtraction of the annihilation background peak is carried out. The
software is organized as follows:

a) Peak search etc. is carried out with the ND system.

b) If only one spectrum of a sample is measured, the evaluation
can be performed with the ND system as well, because all detector
efficiency data and the nuclear library are stored on a disk. All
these programs are written in FORTRAN.

c) If we want to make the most of an analysis, however, we
measure a sample 3, 4 or even 5 times during a month after irradiation. The reduced spectra (peak data etc.)
are sent to an IBM computer. On this computer we have a large
program in PASCAL, which can evaluate up to 5 reduced spectra
of a sample simultaneously, which means that the half lives are also
used for the identification. With this program a number of corrections are also performed: cascade summation, burn-up, U-235 fission
interference, Cu-Na and Pt-Au interference[2]. In the final table
the contents, the limits of detection and the possible interfering
nuclides are printed out.

In the next part of this paper we describe how NAA is used for
the determination of very low amounts of material in the fields of
silicon technology, glasses for optical fibres, III-V compounds and
materials for optical recording.

For the moment the analyses in the area of <u>Integrated Circuit</u>
technology take up by far the greatest part of our time. It may be
evident that nature is on our side in this case, because the activity
of the silicon matrix decays rather rapidly and moreover Si-31 is almost a pure beta-emitter. The last few years the starting material,
i.e. the silicon wafers of 3 and 4 inches diameter, have been no
problem from an analytical point of view: they are sufficiently
pure. But sometimes additional contamination occurs during the many
process steps or cleaning procedures, and it may be well known that
nanograms of some metals on a wafer may be disastrous. For instance
we have found contaminations from stainless steel (Fe, Cr, Ni, Co,
Mo, W) caused by the tubing of the equipment.
Another example: for very high temperature treatments the ovens are
made of silicon carbide instead of quartz but, yet we found 20-100
ppb Cu in the silicon which came from this oven material.

We also investigated materials transport from graphite suscep-
tors into silicon wafers by irradiating a piece of susceptor mate-

rial, and after some processing on a high temperature we measured the wafer. About 10 ppb of some metals in the graphite resulted in 1 ng of these metals in a 2 inch wafer.

In many cases we locate the impurities in the slices by removing the oxide layer, the nitride layer, the epitaxial layer or the back layer from the wafers by means of a well chosen etching procedure and by measuring gamma spectra between these steps. Of course the etching solutions, to which carriers are added, can be measured as well.

The last example of analysis in the silicon technology field we will describe here is testing the effectiveness of cleaning methods. There are many solutions and mixtures of strong acids for cleaning the surfaces in some stage of the process. By contaminating the surfaces with radioactive tracers we can see how these solutions work. We improved the detection limit of iron and chromium by irradiating enriched isotopes, Fe-58 and Cr-50. In this way we were able to detect about 0.1 pg of these elements.

The second group of samples came from the investigators who are working on optical fibres. We analyzed fibres (quartz as well as soft glass) and, going back into the fabricating process, also samples of core glass, cladding glass and starting materials such as SiO_2, $NaNO_3$, KNO_3, Na_2CO_3, Li_2CO_3, $CaCO_3$, Al_2O_3, GeO_2 and Y_2O_3. It must be emphasized that one ppb of a transition metal leads to unacceptable light losses. It may be clear that in a number of these materials the activity of the matrix is such that either a chemical separation or a decay period is necesasry. But even in the latter case a sufficient number of relevant impurities can be determined in the ppb range.

The next type of samples are the III-V compounds which are materials in the Light Emitting Diode technology: GaAs, Al_xGa_yAs, Ga, As etc. but also the starting materials TMA (trimethyl aluminium), TMG (trimethyl gallium) and AsH_3. These substances are dangerous gases or liquids which must be converted into more manageable solids like Al_2O_3, Ga_2O_3 and As_2O_3, respectively, before irradiation. All these materials are not yet as pure as silicon, so the impurity contents range from the ppb to the ppm level. The determination of Cr dopes in GaAs could be done easily after sufficient decay of As-74 from the (n,2n) reaction[3].

The last field of research we want to mention briefly is that which we call Optical Recording. It concerns the Compact Disk and the Video Long Play. These disks are made of PMMA (Polymethylmetacrylate) and PC (polycarbonate) and much effort has been put into the research of adhering the metal mirror on these materials. The mirror is brought onto the surface by means of electroless plating but for a good adhering a nucleation with a monolayer of some metal is necessary. Without going into details of this process we will only mention the analytical contribution to this subject: we determined Ag, Cu, Cr, Sn, Mn, Zn, Pt, Pd, Br and Cl on surfaces in the amount of about 10^{14} atoms/cm^2. We used only low flux irradiations (BR1) because PMMA and PC do not withstand high fluxes.

References

1. A. Simonits, L. Moens, F. de Corte, A. de Wispelaere, A. Elek, J. Hoste, J. Radioanal. Chem. 60 (1980) 461 and more articles from these authors in this journal.
2. M.L. Verheijke, J. Radioanal. Chem. 35 (1977) 79.
3. G.M. Martin, M.L. Verheijke, J.A.J. Jansen and G. Poiblaud, J. Appl. Phys. 50 (1979) 467.

CURRENT ASPECTS OF MULTIELEMENT ANALYSIS IN THE LIFE SCIENCE

R.Michel[*], G.V.Iyengar[**], R.Zeisler[***]

[*]Institute for Nuclear Chemistry, University of Cologne
[**]Institute of Medicine, Juelich Nuclear Research Center, Jülich
[***]Center for Analytical Chemistry, National Bureau of Standards, Gaithersburg, MD

SUMMARY

The general problems and trends of multielement analysis in the life sciences are exemplarily discussed on the basis of three applications to nutritional, biomedical and environment investigations. The key problem of trace element analysis, accuracy, is described by the results for the analysis of IAEA A-11 milk standard. Here interlaboratory comparisons provide a basis for unbiased worldwide investigations of such an important nutrient as human milk. Studies of the corrosion of medical implants in the human body have shown the importance of blanks, the need to determine interelement relationships and the capability to detect unexpected effects by a large elemental coverage. The need for multi element analysis in environmental sciences is described using the National Environmental Specimen Bank Pilot Program at the National Bureau of Standards as an example. The methods chosen to determine more than 30 elements in human tissues are surveyed and the results obtained in the analysis of human liver are described. On the basis of these experimental data a method to compare different techniques is proposed and some outlines for further developments are indicated.

1. INTRODUCTION

Nearly 30 elements are known to be essential for good health of living organisms (1). Several others, mostly heavy elements, must be regarded as toxic. In order to quantitatively determine all these elements nearly all analytical techniques available have been used. The analytical schemes include real multielement techniques such as X-ray fluorescence analysis, activation analysis, ICP atomic emission spectroscopy and spark source mass spectrometry as well as combinations of several single element techniques such as atomic absorption spectrometry, isotope dilution mass spectrometry and the electrochemical methods, selective ion electrode, differential pulse polarography and anodic stripping voltametry. Moreover, techniques for the investigation of the spatial distribution of trace elements such as electron microprobe, proton induced X-ray emission with microbeams and laser induced mass spetrometry have been applied. For a systematic survey on all these methods see ref. (2).

In spite of the fact that an enormous amount of analytical data has been accumulated in the life sciences during the last decades, the knowledge about the action of trace elements in living organisms is far from being complete. Even worse, a considerable part of the existing data must be regarded as heavily biased and suffering from a considerable lack of accuracy. Consequently, the most important goal in trace element analytical chemistry in the life sciences is to overcome these difficulties and to provide a reliable and comprehensive data base describing the behaviour of trace elements in biological media.

In order to compare the various analytical techniques several analytical and practical criteria (2) have to be considered. Generally, the most important ones are sensitivity, accuracy and presision. In the life sciences, however, all these criteria are governed by a particular problem arising from the extremely low trace element levels to be determined. In many cases contamination and high blank values seriously affect sensitivity and accurcy resulting in biased and faulty data. An outstanding example of this problem is the analysis of human serum (3), which still is not resolved for most trace elements.

The need for obtaining "analytically valid samples" emphasizes the role of sampling for trace element analysis. In principle, most of the basic aspects of specimen collection are independent of the analytical techniques while the subsequent preparation steps are to some extend method oriented. Also, there are distinct differences in the importance of blank problems which are related to the analytical techniques employed. It may be mentioned that generally the instrumental techniques have considerable advantage over all others. The advantages of a method such as neutron activation analysis is obvious in this context. Such improvements in the analytical methodology form the back bone of success of most trace element studies in the life sciences. A survey of the problems particulary with respect to specimen collection and handling are presented elsewhere (4).

Since it is not possible to give a complete survey on the immense number of analytical applications in the life sciences, only three typical investigations will be decribed here, with discussions of some of the current aspects of multielement analysis.

2. INVESTIGATIONS OF NUTRIENTS

The trace element status of an individual is essentially determined by the elemental composition of the diet, which is crucial as the primary source of input. For various purposes it becomes necessary to monitor a large number of elements in the basic food as well as in tissues and body fluids. Under these circumstances, the quality of results is of particular significance since analytical data often may be used as basis for legal recommendations or in some cases to impose restrictions.

Regarding the importance of human milk for the growth of infants, an international collaborative program for the investigation of human milk was coordinated and supported by United Nation's Agencies (5). In the course of this collaboration a review of the literature on human (and animal) milk (6) and a laboratory intercomparison on the IAEA A-11 milk standard (7) were performed.

As a general result of this comparison, it turned out that much more analytical work is needed before a reliable degree of consistency can be attained for the results. There are certain elements such as Co and Zn (fig.1) for which the reported results fall into a narrow range, though there are still some extreme outliers pointing to severe contamination problems. Other elements such as Cu (fig.2) and Mn showed results that were dependent on the analytical technique used. Thus, AAS (54 % of all results) gave mean values of 1140 ± 100 ng/g (range of means 540 - 1872) for Cu and 480 ± 60 ng/g (range of means 198 - 733) for Mn. These values are higher than those obtained by NAA (35 % of the results), which are 550 ± 60 ng/g (range of means 302 - 1050) for Cu and 300 ± 40 ng/g (range of means 120 - 584) for Mn. This observation reveals serious methodological flaws. The extremely low blanks and the practical absence of matrix interferences of NAA are of particular advantage here. However, the problems encountered in this intercomparison do not necessarily reveal the problems of particular analytical methods, but those of analysts and their skill. Actually, it has been demonstrated (9) that NAA and AAS are suitable for the analysis of Cu and Mn in such a matrix and that their results are in good agreement.

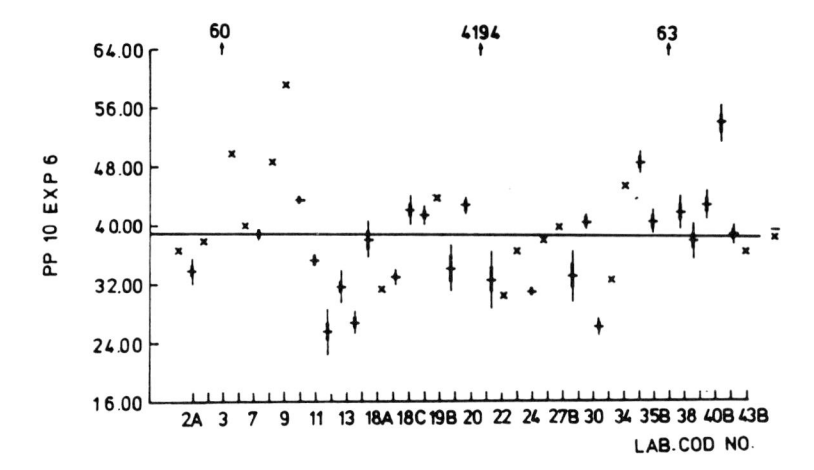

Fig. 1: Zn in milk powder, A-11, as an example for good agreement of results of the IAEA laboratory intercomparison.

As a result of such intercomparisons (5,6) the final set of analytical techniques used for the further investigation of human milk is: ion selective electrode for F, AAS for Ca, K, Mg, Na, Ni, ICP-AES for Pb, NAA for Cl, Cu, Fe, Zn, Cd, Co, Cr, Hg, I, Mn, Mo, Sb, Se, Sn and V, and calorimetry for P. With the experience gained from intercomparison it

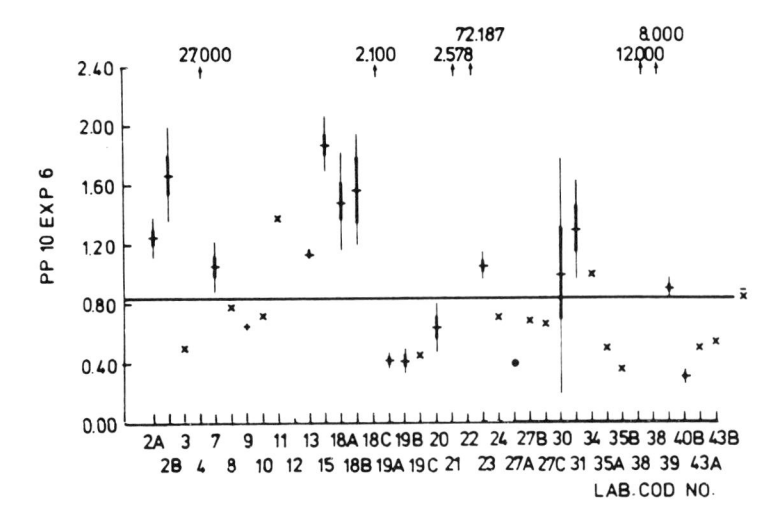

Fig. 2: Cu in milk powder, A11, showing non acceptable discrepancies between many laboratories of IAEA laboratory intercomparison.

was then possible to analyse over 400 samples of human milk from different countries under well defined conditions (e.g.9) and to compare populations living under quite different geographical and socioeconomic conditions. It turned out, that there are e.g. distinct differences in the Mn contents of human milk ranging from 4 µg/l (Guatemala, Hungary, Sweden) up to 40 µg/l (Philippines), while other elements as Cu, Fe, Se amd Zn are fairly uniform. This study was also used to asses the question of daily intakes of trace elements through mother's milk (10). The observed daily intakes are significantly lower for Cu, Cr, F, Fe, Mn and Zn than the recommended WHO/NAS values (11). An important question is whether or not infant formulas should be based on these recommandations or should reflect more natural conditions. This question can only be answered by an improvement in the accuracy of the trace element analyses, since the differences under discussion are smaller than the ranges of nonevaluated experimental results.

3. MEDICAL APPLICATIONS

Anomalies in the trace element balance in man can either cause or result from diseases. Also in the course of medical treatment intentional and unintentional changes in the trace element levels are observed for more than 30 elements (12). In many fields of medicine, however, the knowledge about action of trace elements is still insufficient and often it is impossible to completely define an analytical task a priori. Consequently, multielement techniques have to be applied which can detect as many elements as possible to cover the entire trace element status. An example of this type of analysis is the investigation of the corrosion of orthopaedic implant materials in the human body. For 10 constituent elements of nails, plates, prostheses and bone cements, the effects of corrosion and dissolution have been observed in the analysis of soft human tissues (13). But only stainless steel and Co-Cr alloys have been investigated in some detail (13-18). For all these investigations, INAA was used to determine up to 18 elements in the respective tissues (fig.3). NAA is particularly well suited for this type of investigation, since the search for trace element burdening by corrosion products requires an analytical technique with extremely low blank values (17). For the investigation of human tissues originating from surgical interventions a particular problem arises from the limited sample amount available. Moreover, the sampling is governed by the needs of medical care. Therefore one has to deal with extremely inhomogenous sample materials and often the entire sample is just enough for one analysis.

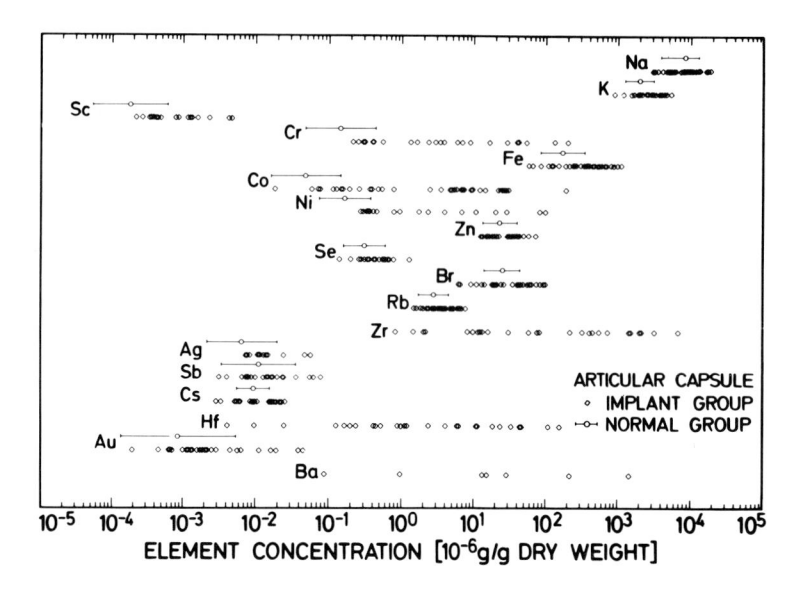

Fig. 3: Results of INAA of human articular capsules after removal of hip joint prostheses made of Co-Cr-alloys. For comparison the mean trace element concentrations of normal articular capsule and their variances are given.

On the other hand one has not only to determine the absolute amounts of the corrosion products, but one has also to look for interelement relationships. Considering the small amounts and the inhomogeneity of the available materials, only real multielement techniques can be applied.

The investigation of interelement relationships generally is of importance in medical applications because interactions between trace elements and even synergistic effects are known. A comparison of different elements also facilitates the extraction of valuable information from heterogenous tissues such as those coming from orthopaedic surgery. Though the concentrations of elements specific for the alloys such as Co, Cr, Fe, Mo and Ni vary over several orders of magnitude (fig.4) due to an immense variation in the degree of corrosion, the ratios of the concentrations of corrosion products can be used to investigate the details of the corrosion process (17,18) and to learn about the transport phenomena in the body (13). The contents of Cr and Ni in tissues after application of stainless (13,14; fig.4a) or of Co-Cr alloys (17,19; fig.4b) do not match the composition of the implants. Combining NAA with radiotracer studies and ESCA (18,19) it has been shown that the abundance ratios of Cr and Ni of fig.6 are mainly determined by selective solution processes in the course of passivation of the implant materials. For non-passivating materials and for implants undergoing more severe types of corrosion, the Cr-Ni-ratios in the tissues reveal the abundance ratios of the alloys used (18). Moreover, NAA as a real multielement techniques allows one to discover unexpected effects for elements not originally searched for, which would be easily overlooked with single element techniques. For example the high abundances of Ba, Hf and Zr in the tissues of patients with hip prostheses (fig.3) were identified in a recent study (17). These elements originated from additives to the bone cements, which were so far regarded as inert in the body. This investigation showed that in contrast the cements undergo distruction and transport phenomena. Moreover, the Au data in fig.3 are extremely high (up to 100 ng/g) for some patients as a consequence of past Au therapy for chronic polyarthritis, demonstrating an additional unforeseen body burdening. The corrosion of the implants does not only affect the ele-

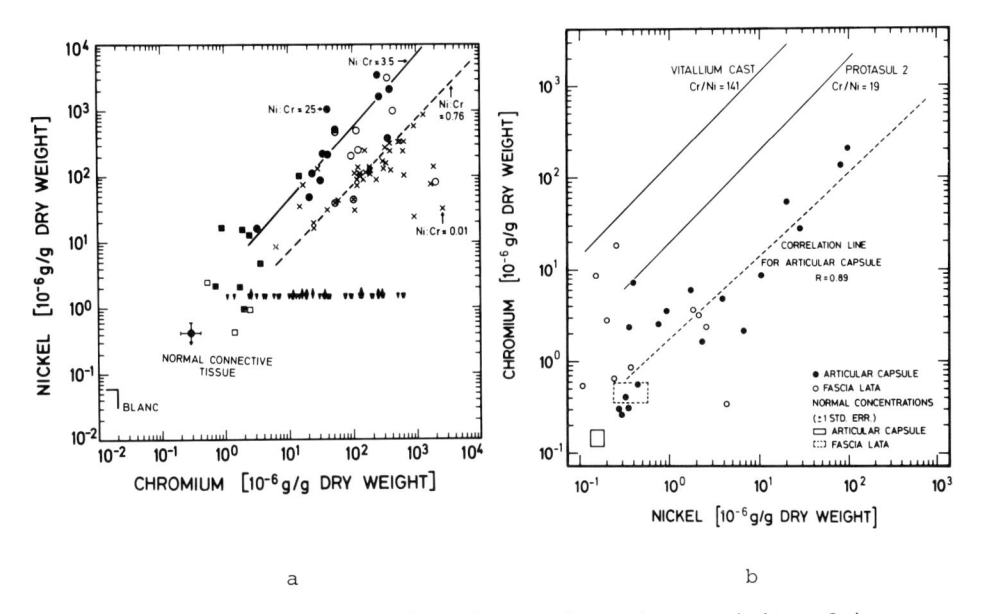

Fig. 4: Cr and Ni in connective tissues from the proximity of implants made of stainless steel (a) and Co-Cr alloys (b), the element ratios revealing the complexity of the corrosion process in the human body. See (18,19) for a detailed discussion.

ments specific to the implant materials, but also gives rise to changes in essential trace elements not contained in the metals. So a decrease in Zn concentration with increasing abundance of corrosion products (13) and an anticorrelation of Ni and K (20) were observed.

Still the knowledge of the fate of corrosion products in the human body, their transport, excretion and storage and their possible impact on the patient's health is insufficient. But multielement analysis promises an approach to a complete description of the complex interaction of implant materials with the human body.

4. ENVIRONMENTAL STUDIES

The number of elements which need to be analyzed is continually increasing to meet the needs of todays environmental studies. The programs for environmental specimen banks (21,22) demand for a nearly complete description of the sampled materials, which can only be achieved by a combination of various analytical techniques. The current trends in this field will be described using the analysis of human liver specimen for the U.S. National Environmental Specimen Bank Pilot Program as an example (23). The U.S. Pilot Program for a National Environmental Specimen Bank (21) was initiated to develop protocols for the contamination-free sampling and long term storage of environmental specimens to elaborate analytical methods for a complete compositional description, in particular for the inorganic (and organic) pollutants. An analytical protocol combining AAS, NAA, IDMS and VOL was established (24) allowing for the determination of 31 elements in human liver specimens. Table 1 gives a survey on the elements determined and the detection limits achieved in the analysis of liver. ICP and XRF could not be readily applied to the liver specimen because of the limited amount of material available and a lack of sensitivity for many elements of interest. However, another technique was applied successfully to the analysis of liver specimen, namely neutron capture prompt gamma activation analysis (PGAA;25). This techniques allows for the determination of B, C, Cd, Cl, H, K, N, P and S in liver specimens, adding 6 more elements to those 31 mentioned above (26).

The findings from an investigation of a first set of 36 human liver specimens (fig.5) pointed out some new facts about the use of different analytical techniques in the life sciences. One significant result is that many essential trace elements (Co, Cu, Mg, Se, Zn) show very narrow ranges of concentrations while pollutant elements such as Al, As, Cd, Hg, Pb and Tl vary widely. For example the concentrations of Se vary by only a factor of 1.8 and other essential traces mentioned above by factors of only 2-3.5, whereas the pollutants exhibit differences of a factor of 100.

However, many pollutant trace elements were found at concentration levels which are on the low end or below previously reported data (29).

	Priority	Method	Detection Limit (µg/g)
Be	1	AAS	0.0006
Na	3	NAA	1.0
Mg	3	NAA	2.0
Al	2	AAS/NAA	0.1/0.5
Cl	3	NAA	20
K	3	NAA	100
Sc	4	NAA	0.0001
V	1	NAA	0.02
Cr	1	NAA	0.02
Mn	1	AAS/NAA	0.2/0.05
Fe	3	NAA	1.0
Co	1	NAA	0.0005
Ni	1	AAS	0.01
Cu	2	AAS/IDMS	0.2/0.1
		NAA/VOL	1.0/0.2
Zn	1	NAA/VOL	0.1/1.0
As	1	AAS/NAA	0.001/0.02
Se	1	AAS/NAA	0.05/0.01
Br	3	NAA	0.1
Rb	3	NAA	0.1
Mo	1	NAA	0.2
Ag	2	NAA	0.02
Cd	1	AAS/NAA/VOL	0.01/0.1/0.05
Sn	1	NAA	0.005
Cs	4	NAA	0.01
La	4	NAA	0.05
Ce	4	NAA	0.05
Pt	1	NAA	0.000002
Au	4	NAA	0.0001
Hg	1	NAA	0.005
Tl	1	IDMS	0.0001
Pb	1	IDMS/VOL	0.01/0.01

Table 1: Analytical protocol for human liver specimen used in the U.S. Pilot National Environmental Specimen Bank. Priority scaling: 1) first priority elements according to (27), 2) additional pollutant elements according to (28), 3) biological and trace elements, 4) elements possibly suitable for monitoring contaminations. For the detection limits samples of 1 g wet weight were assumed.

Specifically, levels of Al, As, Hg, Tl and Pb are significantely lower than previously reported, pointing to possibly insuffucient sampling techniques in many earlier investigations (30). These observations emphasize the importance of state-of-the-art sampling and storage techniques for trace element analysis in the life sciences. It also strengthens the statement that a considerable part of the existing data is heavily biased.

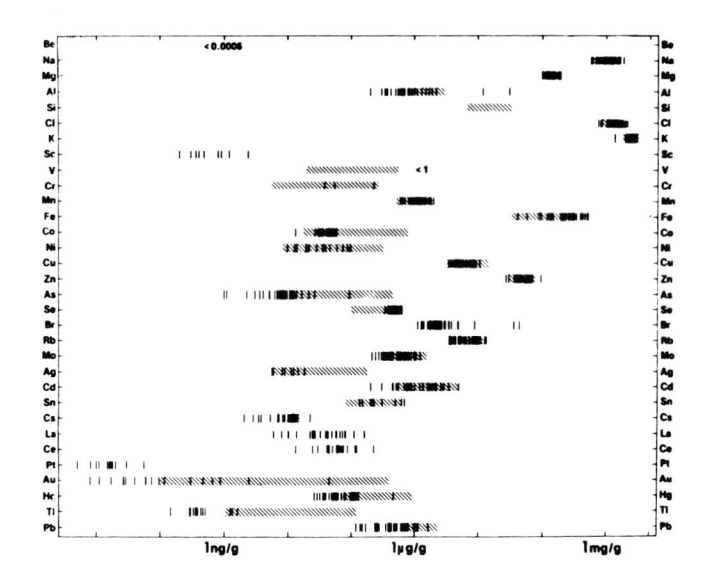

Fig. 5: Elemental composition of 36 individual liver specimen
(One line per each data point) and concentration
ranges in literature (29) represented by the shaded
areas.

Element	Concentration Range (μg/g wet weight)	Percent Imprecision			
		AAS	IDMS	NAA	VOL
Al	0.3 - 2(31)	22- 5(1.6)	---	>30 -9(1.5)	---
Mn	0.6 - 2	25- 5	---	4.5-1.5	---
Cu	3.5 -11	7- 2	0.5	20 -6	6-1
Zn	28 -96	---	---	0.4-0.2	25-2
Se	0.4 -0.65	18-14	---	1.2-0.9	---
Cd	0.3 - 5	8- 6	---	23 -2	>30-1.5
Pb	0.12- 1.7	---	0.2	---	15-3

Table 2: Comparison of dynamic imprecision of four analytical
techniques observed during the analysis of 36 human
liver specimen.

One further result was derived from the analysis of liver specimens with regard to the comparison of analytical techniques. Due to the multiple observation of some elements by the analytical scheme used, conclusions were derived regarding accuracy and precision of the different techniques. The results of a statistical analysis of the data showed that there is a strong dependency between the imprecision of an analytical technique and the concentrations of the element analyzed. Table 2 shows the result obtained for seven elements by four techniques involved in the analytical protocol (24). It turns out, that there is no general statement possible in favour of one particular technique. The precision of a single determination is a function of the concentrations for all techniques except for the single element technique, IDMS, with its outstanding precision. The data demonstrate that for each particular element the appropriate technique has to be evaluated carefully. They illustrate a problem encountered in environmental trace analysis where large concentration ranges exist. Only a combination of different techniques appears to provide a promising approach to reliable and accurate data.

5. CONCLUSION AND PROSPECT

The present development of multielement analysis in the life sciences is characterized by intensive efforts to increase the accuracy of the analytical data and to minimize the bias due to contamination. Here laboratory intercomparisons and the experiences of the specimen bank programmes have to be applied in "normal" analytical laboratories.

The gain of knowledge about the action of trace elements in biological systems has still increased the requirements for highly sensitive multielement techniques with large elemental coverage. The required analytical schemes have to result in accurate data, thus allowing for the investigation of interelement relationships. They must be able also to detect unexpected effects and ideally should describe the complete trace element composition of a sample. This can be achieved by elaborating analytical schemes in which nuclear methods play an important role. Full effectiveness, however, can only be obtained by a combination of several analytical techniques. The particular choice will strongly depend on the individual element and the matrix to be analyzed.

However, in addition to determining of the elemental concentrations attempts to determine the chemical form of the traces become more and more important. Such speciation studies are still relatively rare (31-35). The future of trace element investigations will not be limited to bookkeeping of concentrations, but will require the knowledge of the chemical form and of the biochemical action, even at the extrem trace element levels. Already the amount of literature (36) dedicated to studies of the functional behaviour of trace elements and to dose/response relationships is increasing, showing that the area of trace element analytical chemistry in the life sciences has become a real interdisciplinary field, the tools of which are modern multielement techniques.

REFERENCES

(1)   W.Merz, Science 213 (1981) 1332-1338.

(2)   J.S.Hislop, in: P.Brätter, P.Schramel (eds.), Trace element analytical chemistry in medicine and biology (1980) 747-767, W. de Gruyter, Berlin.

(3)   J.Versieck and R.Cornelis, Anal.Chim.Acta 116 (1980) 217-254.

(4)   B.Sansoni and G.V.Iyengar, Sampling and sample preparation methods for the analysis of trace elements in biological materials, Jül-Spez-13 (1978).

(5)   IAEA Research Coordination meeting RC/75-2A (1977) Vienna.

(6)   G.V. Iyengar, Elemental composition of human and animal milk: A Review, IAEA TEC-DOC Series (1982) Vienna.

(7)   R.Dybczynski, A.Veglia and O.Suchny, Report on the intercomparison run A-11 for the determination of inorganic constituents of milk powder, IAEA/RL/68, July 1980.

(8)   G.V.Iyengar, K.Kasparek, L.E.Feinendegen, Y.X.Wang and H.Weese, The science of the tot. Environm. 24 (1982) 267-274.

(9)   R.M.Parr, IAEA Bulletin 25, No 2 (1983) 7-15, and WHO Report (in preparation) 1984.

(10) G.V.Iyengar, Z.Ges.Hyg. 30 (1984) 88.

(11) WHO Report 522 (1973), NAS Washington (1980).

(12) R.Michel, Fresenius Zeitschr.Anal.Chem. 317 (1984) 451-460.

(13) F.Lux and R.Zeisler,Z.Anal.Chem. 261 (1972) 314-328.

(14) R.Michel and J.Zilkens, Z.Orthop. 116 (1978) 666-674.

(15) J.Zilkens, F.Löer, R.Michel and J.Hofmann, Z.Orthop. 119 (1981) 760-763.

(16) F.Löer, J.Zilkens, R.Michel, G.Freisem-Broda and K.H.Bigalke, Z.Orthop. 121 (1983) 255-259.

(17) J.Hofmann, N.Wiehl, R.Michel, F.Löer and J.Zilkens, J.Radioanal. Chem. 70 (1982) 85-107.

(18) J.Hofmann, R.Michel, R.Holm and J.Zilkens, Surf.Interf.Anal. $\underline{3}$ (1981) 110-117.

(19) R.Michel, J.Hofmann, F.Löer and J.Zilkens, Traumatology (1984) accepted.

(20) R.Michel, J.Hofmann and J.Zilkens, in P.Brätter and P.Schramel (eds.), Trace element analytical chemistry in medicine and biology (1980) 137-157, W.de Gruyter, Berlin.

(21) H.Harrison, R.Zeisler and S.A.Wise, Pilot program for the National Environmental Specimen Bank - Phase I, EPA-600/1-81-025, U.S. Environmental Protection Agency (1982).

(22) M.Stoeppler, H.W.Durbeck and H.W.Nürnberg, in Jahresbericht 1979/80 der Kernforschungsanlage Jülich GmbH (1980).

(23) R.Zeisler, S.H.Harrison and S.A.Wise, The Pilot National Environmental Specimen Bank - Analysis of human liver specimen, NBS Spec.Pub. (U.S.) 656 (1983).

(24) R.Zeisler, NBS Spec.Pub. (U.S.) 656 (1983) 35-38, and R.Zeisler, S.H.Harrison and S.A.Wise, Biol.Trace Elem.Res. $\underline{6}$ (1983) 31-49.

(25) D.F.Anderson, NBS Spec.Pub. 656 (1983) 123-126.

(26) J.K.Langland, S.H.Harrison, R.Zeisler and S.A.Wise, NBS Spec. Pub. 656 (1983) 21-34.

(27) N.P.Luepke, Monotoring environment materials and specimen banking (1979) Martinus Nijhoff Publishers, The Hague.

(28) U.S. Environmental Protection Agency, Quality criteria for water (1976) U.S. Environmental Protection Agency, Washington.

(29) G.V.Iyengar, W.E.Kollmer and H.J.M.Bowen, The elemental composition of human tissue and body fluids (1978) Verlag Chemie, Weinheim.

(30) R.Zeisler, NBS Spec.Pub. (U.S.) 656 (1983) 81-90.

(31) F.Girardi, E.Marafante, R.Pietra, E.Sabbioni and A.Marchesini, J.Radioanal.Chem. $\underline{37}$ (1977) 427-440.

(32) M.Bonardi, C.Birattari, M.C.Girardi, R.Pietra and E.Sabbioni, J.Radioanal.Chem. $\underline{70}$ (1982) 337-348.

(33) E.Sabbioni, R.Pietra, E.Marafante, J.Radioanal.Chem. $\underline{69}$ (1982) 381-400.

(34) E.Sabbioni, in: P.Brätter and P.Schramel (eds.), Trace element analytical chemistry in medicine and biology (1980) 407-426, W.de Gruyter, Berlin.

Instrumentelle Multi-Element-Analyse in der Lebensmittelanalytik

Georg Schwedt
Institut für Lebensmittelchemie der Universität,
Pfaffenwaldring 55, D-7000 Stuttgart 80

1. EINLEITUNG

Elementgehalte in Lebensmitteln sind zunächst unter den Gesichtspunkten Mineralstoffe, essentielle und toxische Elemente von Interesse. Den Mineralstoffen werden in der menschlichen Ernährung folgende Elemente zugeordnet:
Na - K - Ca - Mg - P - Cl .
Der tägliche Bedarf liegt zwischen 0,5 und 5 g je nach Element. Andere Elemente werden vom Menschen nur in Milligramm-Mengen pro Tag benötigt, sie kommen außerdem auch nur in geringen Konzentrationen in Lebensmitteln vor. Sie werden als Spurenelemente bzw. essentielle Elemente bezeichnet. Einige von ihnen weisen jedoch in höheren Konzentrationen bzw. bei höherer täglicher Aufnahme auch toxische Wirkungen auf, so daß eine Zuordnung von der Menge bzw. Konzentration abhängt.
Folgende Elemente werden heute als essentiell (in Klammer "möglicherweise essentiell") bezeichnet:
(As) - (Cd) - Co - Cr - Cu - F - Fe - I - Mn - Mo - Ni - (Pb) - Se - Si - Sn - (Ti) - V - Zn.
Als toxisch oder carcinogen gelten folgende Elemente (unterstrichene Elemente sind auch unter "essentiell" aufgeführt):
Ag - As - Au - Ba - Be - Bi - Cd - Co - Cr - Cu - F - Hg - Li - Mn - Ni - Pb - Pd - Pt - Sb - Se - Sn - Te - Tl - V - Zn.   (nach 1 u. 2)
Einer Multi-Element-Analyse in Lebensmitteln kommt auch im Hinblick auf mögliche synergistische Wirkungen (z.B. zwischen Hg und Se) für Beurteilungen physiologischer und toxikologischer Wirkungen, weiterhin für Untersuchungen in der Lebensmitteltechnologie, der Herkunft pflanzlicher Lebensmittel (zur Frage nach den Anbaugebieten),

der Kontamination von tierischen und pflanzlichen Lebensmitteln u.ä.
Gesichtspunkten eine große Bedeutung zu.
Orientierende Übersichtsanalysen und quantitative Spurenanalysen
möglichst vieler Elemente sind Forderungen der Lebensmittelanalytik,
die nur mit leistungsfähigen Methoden der instrumentellen Multi-
Element-Analyse erfüllt werden können.

2. METHODEN DER MULTI-ELEMENT-ANALYSE IN LEBENSMITTELN

Die Auswertung der "Chemical Abstracts" seit 1972 im "General Index"
unter "Food Analysis" vermittelt ein Bild der Entwicklungen und
Anwendungen von Multi-Element-Analysenmethoden. Aus der Zahl der
aufgeführten Publikationen ergibt sich folgende Reihenfolge:
Atomabsorptions-Spektrometrie (AAS) -
Polarographie und Voltammetrie -
Neutronen-Aktivierungsanalyse (NAA) -
Optische Emissions-Spektrometrie (OES-ICP) -
Röntgenfluoreszenz-Spektrometrie (RFA).
In den Routinelaboratorien der Lebensmitteluntersuchung steht zur
Zeit die AAS noch mit weitem Abstand an der Spitze, die optische
Emissions-Spektrometrie (ICP) gewinnt parallel zu den gerätetechni-
schen Entwicklungen an Bedeutung. Die anderen Methoden werden wegen
der meist erforderlichen speziellen, matrixabhängigen Probenvorberei-
tung oder apparativen Voraussetzungen (siehe NAA) nur in Forschungs-
instituten eingesetzt.
Im Hinblick auf die Auswahl und gezielte Anwendung einer Methode wird
zunächst einmal ein orientierender Überblick über die durchschnittlichen Gehalte an Elementen in verschiedenen Lebensmittelgruppen gegeben (Tab. 1 und 2).

Tab. 1: Durchschnittliche Gehalte an Mineralstoffen in ausgewählten
        Lebensmitteln (in mg/kg Trockensubstanz)

Mineralstoff	Lebensmittel				
	Fleisch	Mehl	Kartoffeln	Gemüse	Obst
Na	800	20	30	200	20
K	3000	1550	4400	3000	2000
Ca	100	250	100	400	150
Mg	200	250	250	200	100
P (Phosphat)	200	2000	500	400	200
Cl (Chlorid)	500	3000	450	450	40

Tab. 2: Durchschnittliche Gehalte an essentiellen Elementen in einigen ausgewählten Lebensmitteln (mg/kg Trockensubstanz)

Essentielles Element	Fleisch	Mehl	Kartoffeln	Gemüse	Obst
Co	0,05	0,01	0,03	0,2	0,1
Cr	0,3	0,2	0,2	0,15	0,15
Cu	2	3	2	1	1
Fe	25	20	8	9	5
Mn	0,5	20	1,5	3	4
Mo	1	0,4	0,2	0,5	0,1
Ni	0,1	0,2	0,2	0,6	0,2
Se	0,8	0,4	0,2	0,8	0,1
V	0,9	0,2	0,3	0,3	0,7
Zn	25	20	3	2	1

Anschließend an diese Gehaltsangaben wäre eine Gegenüberstellung von Nachweisgrenzen für die einzelnen Elemente und Bestimmungsmethoden sinnvoll. Dieser Vergleich müßte jedoch unter anderem auf der Basis identischer Probenvorbereitung mit jeweils für die Matrix optimierten Meßbedingungen durchgeführt werden. Solche Angaben sind jedoch in der Literatur nicht zu finden. Eine Tabelle ("Nr. 4") in der Übersichtsarbeit von Morrison 1979 (1), stellt Empfindlichkeiten ("Sensitivity") und Nachweisgrenzen ("Detection limit") aus verschiedenen Veröffentlichungen für acht Multi-Element-Methoden in Bezug zu Serumgehalten der Elemente. Die Aussagefähigkeit eines solchen Vergleichs ist jedoch sehr beschränkt; er kann auch für die Lebensmittelanalytik nur sehr allgemeine Anhaltspunkte für Methodenvergleiche liefern.

2.1 ATOMABSORPTIONS-SPEKTROMETRIE (AAS)

Die Übersichtsarbeit von Crosby 1977 (3) zeigt deutlich die dominierende Rolle der AAS in der Analytik anorganischer Lebensmittelinhaltsstoffe. Bis heute haben sich noch keine wesentlichen Verschiebungen zu einer anderen Methode in der Routineanalytik ergeben. Mit der Flammen-AAS werden auch diejenigen Elemente erfaßt, deren Bestimmung bisher der Flammenphotometrie vorbehalten war (Na, K, Ca) (4). Die Graphitofen-Technik (Flammenlose AAS) wird vor allem zur Bestimmung von Schwermetallen wie Cd, Pb, Cu eingesetzt; um Matrixeffekte gering zu halten, werden diese und andere toxische Elemente (wie As, Hg, Sb, Se) häufig zunächst als Dithiocarbamate extrahiert (5). Diese Dithio-

carbamat-Extraktion stellt auch in der Lebensmittelanalytik einen
wichtigen Vorbereitungsschritt für Multi-Element-Analysen mittels
AAS dar. Weiterhin findet die Hydrid-Technik für die Elemente As,
Se, Te, Sb, Sn (6) verbreitet Anwendung. Je nach Matrix (fett-, pro-
tein-, kohlenhydratreich - pflanzlich/tierisch) und nach den zu bestimmenden Elementen (-gehalten) ist eine überlegte Auswahl des Auf-
schlußverfahrens erforderlich. Diese Verfahren reichen von einfachen
"Extraktionsschritten" (Erhitzen einer Probe mit einem HCl/HNO_3-Ge-
misch auf dem Wasserbad) oder der trockenen Veraschung (4) bis zur
Plasmaveraschung bei niedriger Temperatur in einem Sauerstoff-Fluor-
Plasma (7) für die Bestimmung der Elemente Sn, Fe, Pb und Cr.

2.2 EMISSIONS-SPEKTROMETRIE MITTELS ICP (ES-ICP)

Mehr als die AAS stellt die ES-ICP in simultaner oder sequentieller
Ausführung eine Multi-Element-Bestimmungsmethode dar. Auf der einen
Seite liegen die Anschaffungs- und Betriebskosten höher als bei der
AAS, auf der anderen Seite ergeben sich einige Vorteile im Hinblick
auf Matrixeffekte (infolge hoher Salz-, vor allem auch Phosphatgehalte
in Lebensmitteln) und wegen der hohen Zahl an bestimmbaren Elementen
und somit im Durchschnitt vergleichbaren oder sogar niedrigeren
Kosten je Elementbestimmung. Die Zahl der in Lebensmitteln bestimmbaren Elemente wird von Boyer et al. 1979 (8) mit 25 - 30 angegeben.
Ausführliche Angaben über Probenaufschluß, Extraktion von Metallspuren als Dithiocarbamate, Kalibrierung und Interferenzen sind in
der Arbeit von Evans u. Dellar 1982 (9) über multiple Elementspuren-
analysen in Lebensmitteln enthalten. Über Einflüsse des Probenaufschlusses, des Untergrundes und der Matrix von Lebensmitteln auf die
Ergebnisse von ES-ICP-Analysen berichten Jones u. Boyer 1978 (10).
Stehen bei der AAS Matrixeffekte im Vordergrund, so müssen hier besonders spektrale Interferenzen berücksichtigt werden. Analysen von
Pflanzenaschen zeigen nach Cottenie/Verloo/Velghe 1980 (11), daß
richtige Ergebnisse ohne Interferenzen nur für eine eingeschränkte
Zahl von Elementen zu erhalten waren. Ein Weg zur Beseitigung von
Störungen führt über die genannte Dithiocarbamat-Extraktion, ein an-
derer über das in der AAS bekannte Hydrid-System für die Elemente As,
Bi, Ge, Sb, Se, Sn, Te (12/13). Insgesamt zeigen die bisher vorliegen-
den Untersuchungen für Lebensmittel eine breite Anwendbarkeit trotz
der genannten Einschränkungen, auch wenn für einige Elemente in
Lebensmitteln wie Cr, Cd, Ni und Pb die Nachweisgrenzen für normale
Gehalte nicht ausreichen.

2.3 RÖNTGENFLUORESZENZ-SPEKTROMETRIE (RFA)

Spezielle, für die Meßtechnik erforderliche Probenvorbereitungen charakterisieren diese emissions-spektrometrische Analysenmethode. Für Metall-Spurenanalysen in Lebensmitteln wurde z.B.die Extraktion mit geschmolzenem 8-Hydroxychinolin (aus Zucker) (14), die Flüssig-flüssig-Extraktion mit 1-(2-Pyridylazo)-2-naphtol (PAN) aus Wein mit anschließendem Eindampfen der Extraktionslösung auf einem Probenträger (15) sowie auch die Ausfällung als Dithiocarbamate und Filtration über ein Membranfilter als Probeträger (16) beschrieben. Eine elegante und schnelle Probenvorbereitung für Lebensmittel tierischer Herkunft besteht im Gefriertrocknen mit Mahlen und Pressen einer Tablette (17). Die RFA zeigt sich vor allem hier als schnelle und relativ empfindliche Multi-Element-Methode mit Nachweisgrenzen auch unter 1 Mikrogramm auf dem Probenträger bzw. unter 1 ppm in der Probe (16).

2.4 POLAROGRAPHIE/VOLTAMMETRIE

Wesentlich empfindlicher, jedoch weniger universell sind diese elektrochemischen Bestimmungsmethoden. In der Literatur werden Anwendungen der Kathodenstrahl-, Square-wave-Puls- sowie Differential-Puls-Polarographie vor allem für Spurenanalysen von Cd, Cu, Pb, Sn und Zn in Lebensmitteln beschrieben (18 - 20). Die Voltammetrie (als DPASV) hat sich als besonders leistungsfähig für die simultane Bestimmung von Cd, Cu, Ni, Pb und Zn in sehr unterschiedlichen Lebensmitteln erwiesen (21). Die Nachweisgrenzen reichen bis unter 1 µg/kg Probenmaterial. Für die genannten toxischen Metalle ist die Voltammetrie den vorher behandelten Multi-Element-Methoden vor allem hinsichtlich der Empfindlichkeit und der günstigen Kosten überlegen.

2.5 NEUTRONEN-AKTIVIERUNGSANALYSE (NAA)

Noch mehr Elemente als mit der ES-ICP (2.2) (bis zu 31 in Lebensmitteln) lassen sich mittels der NAA bestimmen. Es werden Nachweisgrenzen von 10^{-6} bis 10^{-11} g (bei Probenmengen um 150 mg) je nach Element erreicht (22). Interferenzen im Gamma-Spektrum durch die Nuklide der in hohen Konzentrationen vorkommenden Elemente Na, K und P verlangen jedoch zum Teil aufwendige Probenvorbereitungs- oder radiochemische Trennschritte (22,23). Auf der anderen Seite bietet sich diese Methode für breite Untersuchungen über die tägliche Aufnahme von Elementgehalten aus der Nahrung an. Die Anwendbarkeit ist jedoch auf Einrichtungen beschränkt, denen Neutronenquellen zur Verfügung stehen.

3. AUSBLICK

Neben der AAS wird für die Routineanalytik sehr wahrscheinlich die ES-ICP zunehmend für die Analytik anorganischer Lebensmittelinhaltsstoffe an Bedeutung gewinnen. Die chemische Probenvorbereitung besitzt trotz der sehr leistungsfähigen instrumentellen Meß- und Analysentechnik einen hohen Stellenwert innerhalb der Verfahren der Multi-Element-Analytik; hier sind immer wieder Verbesserungen und Anpassungen an die Lebensmittel-Matrix erforderlich. Bei extrem niedrigen Gehalten z.B. toxischer Elemente ist eine gezielte Auswahl der Bestimmungsmethode erforderlich. Multi-Element-Analysen insgesamt erfordern eine der Aufgaben- bzw. Problemstellung angepaßte Analysenstrategie sowie die Entwicklung von Analysenschemata (s. z.B. 24). Sinnvoll geplante und durchgeführte Multi-Element-Analysen sind nur in enger interdisziplinärer Zusammenarbeit des Analytikers mit Lebensmittelchemikern, Lebensmitteltechnologen, Ernährungsphysiologen, Toxikologen und Wissenschaftlern anderer verwandter Bereiche wie der Agrarwissenschaften möglich. Die Datenfülle verlangt den Einsatz von Computern aber auch die ständige Kontrolle der Zuverlässigkeit der Verfahren z.B. durch Methodenvergleiche und Ringversuche.

LITERATUR

(1) Morrison, G.H., CRC Crit. Rev. Anal. Chem. $\underline{8}$ (1979) 287 - 320
(2) Wood, J. M., Science $\underline{183}$ (1974) 1049 - 1052
(3) Crosby, N. T., Analyst $\underline{102}$ (1977) 225 - 268
(4) Maurer, J., Z. Lebensm. Unters.-Forsch. $\underline{165}$ (1977) 1 - 4
(5) Müller, H., Siepe, V., Deutsche Lebensm.-Rdsch. $\underline{77}$ (1981) 392 - 400
(6) Evans, W.H., Jackson, F.J., Dellar, D., Analyst $\underline{104}$ (1979) 16 - 34
(7) Williams, E.V., Analyst $\underline{107}$ (1982) 1006 - 1013
(8) Boyer, K.W. et al., Food Process. $\underline{40}$ (1979) 72 - 73
(9) Evans, W.H., Dellar, D., Analyst $\underline{107}$ (1982) 977 - 993
(10) Jones, J.W., Boyer, K.W., in: Appl Inductively Coupled Plasma Emission Spektroscop. p. 83 - 106, Franklin Inst. Press: Philadelphia 1978
(11) Cottenie,A., Verloo, M., Velghe, G., Spectra 2000 $\underline{8}$ (1980) 69 - 74
(12) Hahn, M.H., Wolnik, K.A., Fricke, F.L., Caruso, J.A., Anal. Chem. $\underline{54}$ (1982) 1048 - 1054
(13) Wolnik, K.A., Fricke, F.L., Hahn, M.H., Caruso, J.A., Anal.

Chem 53 (1981) 1030 - 1035
(14) Magyar, B., Lobanov, F.I., Talanta 20 (1973) 55 - 63
(15) Bergner, K.G., Lang, B., Deutsche Lebensm.-Rdsch. 66 (1970) 157 - 164
(16) Grote, B., Montag, A., Lebensmittelchemie u. gerichtl. Chemie 33 (1979) 89 - 92
(17) Rethfeld, H., Fresenius Z. Anal. Chem. 310 (1982) 127 - 130
(18) Kapel, M., Komaitis, M.E., Analyst 104 (1979) 124 - 135
(19) Borus-Böszörmenyi, N., Nahrung 24 (1980) 295 - 302
(20) Paolo, B., Ind. Aliment. (Italy) 22 (1983) 15 - 20
(21) Valenta, P., Ostapczuk, P., Pihlar, B., Nürnberg, H.W., Proc. Int. Conf. Heavy Metals in the Environment, Amsterdam 15. - 18.8.1981, S. 619 - 621, CEP Consult., Edinburgh 1981
(22) Schelenz, R., J. Radioanal. Chem. 37 (1977) 539 - 548
(23) Yeh, S.J. et al., Anal. Chim. Acta 87 (1976) 119 - 124
(24) Jones, J.W., Capar, S.G., O'Haver, T.C., Analyst 107 (1982) 353 - 377

H. Eschnauer, H. Meierer, R. Neeb
Institut für Anorganische Chemie und Analytische Chemie
der Johannes Gutenberg-Universität, D-65 Mainz/Rhein

Simultane Multi-Element-Bestimmung in Wein durch
ICP-Plasma-Emissionsspektralanalyse

Im Wein sind bis jetzt etwa 50 anorganische und weniger als 1000 organische Inhaltsstoffe nachgewiesen.

Die anorganischen Inhaltsstoffe im Wein bestehen aus Mineralstoffen, Spurenelementen und Ultra-Spurenelementen.

Mineralstoffe sind Elemente, die im Wein in grösseren Gehalten (bis zu 1 mg/l, selten mehr) enthalten sind. Es handelt sich um die vier kationischen Elemente Kalium, Magnesium, Calcium und Natrium sowie um die vier anionischen Elemente Kohlenstoff, Phosphor, Schwefel und Chlor.

Die Konzentration der Spurenelemente im Wein schwankt in weiten Grenzen zwischen 0,001 - 10 mg/l (ppm-Bereich). Im Wein sind etwa 25 Spurenelemente vorhanden.

Die Konzentration von Ultra-Spurenelementen im Wein liegt um oder unter 1 Mikrogramm pro Liter Wein (1 μg/l oder ppb-Bereich). Bisher konnten in diesem Bereich ca. 20 Elemente nachgewiesen werden. Die Bestimmung der Ultra-Spurenelementgehalte in Wein erfordert spezielle Methoden mit anspruchsvollen Messgeräten.

Um die Gehalte an Spurenelementen und Ultra-Spurenelementen in Wein besser beurteilen zu können, wird zwischen primären und sekundären Gehalten unterschieden.

Unter einem primären Gehalt an Spurenelementen und Ultra-Spurenelementen in Wein wird nur der Gehalt verstanden, der von Natur aus in Wein vorhanden ist (natürliches Vorkommen in Wein). Dieser primäre Gehalt kommt einmal aus dem Weinbergsboden über Wurzel, Rebe, Traube und zum anderen durch natürliche oder anthropogene Kontamination (Meer, Vulkan u. a.) über Blatt und Traube in dem Traubenmost und in dem fertigen Wein zustande. Der primäre und somit natürliche Gehalt an Spurenelementen und Ultra-Spurenelementen ist in der Regel nur ein Teil des gesamten Gehaltes normaler Weine des Handels.

Unter einem sekundären Gehalt an Spurenelementen und Ultra-Spurenelementen in Wein wird nur der Gehalt verstanden, der durch kellertechnisch-beabsichtigte und/oder durch unbeabsichtigte (künstliche Kontamination) Einflüsse und Massnahmen verursacht wird.

Durch kellertechnisch-beabsichtigte Massnahmen, die dazu dienen, eine Konservierung, eine Klärung, eine Korrektur der Farbe, des Geruchs oder des Geschmacks zu erreichen, können Spurenelemente und Ultra-Spurenelemente in den Wein gelangen. Hierunter fallen eine grosse Zahl in der Kellerwirtschaft geübter Praktiken, die alle gesetzlich verboten sind (Zugabe von Aluminium-, Barium-, Wismut-, Brom-, Blei-, Fluor-, Mangan-, Strontium- oder Zinkverbindungen).

Sekundäre Gehalte an Spurenelementen und Ultra-Spurenelementen durch unbeabsichtigte Einflüsse und Massnahmen, also künstliche Kontamination im weitesten Sinne, umfassen eine ganze Reihe von Einfluss-Faktoren.

Zur künstlichen Kontamination im engeren Sinne gehören standortbedingte Umwelt-Faktoren, wie Fabrikemissionen (Pb, Zn, Cd aus Hüttenbetrieben; F aus Aluminiumwerken und Ziegeleien; Pb, Cd, F, Se aus Pigment- und Keramikfabriken; Tl aus Zementwerken u. a.), Bergbauhalden (über Wurzelaufnahme und/oder anhaftende Bodenteilchen) oder Autoabgase (Pb, Cd).

Zur künstlichen Kontamination im weiteren Sinne gehören vor allem die Weinbau- und Kellerei-Faktoren, wie Schädlingsbekämpfungsmittel (Cu), Korrosion von Lagerbehältern und Kellereimaschinen (Cu, Zn, Sn, Fe, Cr, Ni u. a.) sowie Weinbehandlungsmittel.

Der Gesamt-Gehalt an Spurenelementen und Ultra-Spurenelementen in normalen Weinen des Handels besteht also immer aus dem primären, natürlichen Gehalt und dem sekundären Gehalt (aus künstlicher Kontamination).

Unabhängig von der Herkunft und der Zufuhr an primären und sekundären Gehalten an Spurenelementen und Ultra-Spurenelementen wird im Verlauf der Weinherstellung und insbesondere während der Gärung der grösste Teil wieder ausgeschieden (abgereichert).

Unter Abreicherung versteht man die Konzentrationsabnahme eines Spurenelementes oder Ultra-Spurenelementes aus dem Traubenmost und Wein in den Hefetrub während der Gärung und des nachfolgenden Weinausbaus. Diese Abreicherung wird quantitativ verfolgt und kann durch den Abreicherungs-Faktor (f) oder die Abreicherungs-Rate (%) angegeben werden.

Mit der gesamten Mengenbilanz eines Spurenelements oder Ultra-Spurenelements erfasst man quantitativ den Gesamt-Gehalt im Wein und seine Abreicherung während der Gärung in den Hefetrub. Die Abreicherungsrate ist spurenelementspezifisch und erfolgt in unterschiedlichen Mengen. Extrem hohe Abreicherungsraten werden für einige toxische Metalle festgestellt (durch Metallsulfid-Bildung und -Ausscheidung bei der Gärung). Die Fermentation hat also in hohem Maße einen Selbstreinigungs-Effekt für viele Spurenelemente und Ultra-Spurenelemente. Eine weitere Verminderung tritt durch eine Reihe von kellertechnischen Massnahmen ein, so vor allem durch die Möslinger-Schönung mit Kaliumferrocyanid.

In soweit ist es berechtigt, von Wein als dem reinsten Getränk überhaupt - mindestens aber von dem reinsten alkoholischen Getränk - zu sprechen. Schon im alten Carthago wurde Wein als reinstes Produkt gepriesen!

Tabelle 1   (2)

Mineralstoff- und Spurenelementgehalte in Wein

Definition	Mineral Elements		Trace Elements					
Range (mg/l)	1000–100	100–10	10–1	1–0,1	0,1–0,01	0,01–0,001	≳0,001	≪0,001
Elements	K	Na	Fe	Al	Ba	Cr	Sb	(Bi)
	Mg		Mn	F	Pb	Co	Cs	(W)
	Ca		B	J	As	Mo	Cd	(Ga)
	. . . . .	. . . . .	Si	Rb	Br	Ag	Hg	(In)
	C	Cl	Zn	Sr	Li	Ni	Se	. . . . .
	P		Cu	Ti	V	Nb	Hf	(U)
	S				Sn		Ta	(Ra)
							Au	(Rn)
							Tl	(K40)
							La	(C14)
							Sc	(T)
	H	N O					Eu	
							Be	

ppm → ppb

Range of Mineral Elements and Trace Elements in Wine

Tabelle 2

Konzentrationsbereiche (mg/l) der untersuchten Elemente

Ia																	VIIIb
H																	
	IIa											IIIb	IVb	Vb	VIb	VIIb	
Li	Be											B	C				
Na	Mg											Al	Si				
0	0											0,5–					
	60	150	IIIa	IVa	Va	VIa	VIIa	VIIIa	VIIIa	VIIIa	Ib	IIb	5,3				
K	Ca	Sc	Ti	V	Cr	Mn	Fe	Co	Ni	Cu	Zn	Ga	Ge				
0–	0–			0,05	0,06–	1,2–	5,1–	0,01–	0,01–	0,3–	0,5–						
700	270			0,5	0,5	1,8	16,1	0,5	0,8	5,1	6,1						
Rb	Sr	Y	Zr	Nb	Mo	Tc	Ru	Rh	Pd	Ag	Cd	In	Sn	Sb			
										0,1–	0,07–						
										1,0	0,7						
								Pt	Au	Hg	Tl	Pb	Bi				
										0,01–		0,1–					
										1,3		0,8					

Probenvorbereitung

Aufschluß des Weines und des Mostes

Je nach Schwermetallgehalt werden zwischen 10 und 50 cm^3 Probe mit 3,5 bis 20 cm^3 Perhydrol im Kjeldalkolben unter gleichzeitigem Eindampfen über dem Bunsenbrenner aufgeschlossen. Zu der Probe werden 0,2 cm^3 konz. H_2SO_4 (suprapur) hinzugegeben, um zu vermeiden, daß der Aufschluß bis zur Trockene eindampft und sich dabei selbst entzündet. Der nach dem Aufschluß verbleibende Rückstand wird mit 5 bzw. 10 cm^3 2,2 mHNO_3 aufgenommen. Die so erhaltenen Lösungen werden mit der OES-ICP direkt analysiert.

Aufschluß der Hefe

Zum Hefeaufschluß stehen zwei Methoden zur Verfügung:

Methode 1: Von der getrockneten Hefe wird zwischen 500 und 1500 mg eingewogen und mit 5 bis 10 cm^3 Perhydrol plus 1 bis 2 cm^3 konz. HNO_3 (suprapur) aufgeschlossen. Anschließend wird der Rückstand mit 5 bzw. 10 cm^3 2,2 m HNO_3 aufgenommen und analysiert.

Methode 2: Es werden zwischen 150 und 400 mg getrocknete Hefe eingewogen. Dann wird 1,0 cm^3 konz. HNO_3 (suprapur) hinzugegeben und in einer Teflonbombe auf 180° erwärmt. Bei dieser Temperatur wird die Aufschlußbombe 90 Minuten lang gehalten und dann abgekühlt. Nach Zugabe von 5 cm^3 H_2O (bidest.) wird die Lösung analysiert.

Ergebnisse

Tabelle 3 (Vergleiche (3))
Nachweisgrenzen in reinen Lösungen (ppb) und
zur Bestimmung verwendete Linien (nm)

Element	Nachweisgrenze (ppb)	Linie (nm)	Element	Nachweisgrenze (ppb)	Linie (nm)
Ag	21,8	328,07	K	1647,0	766,49
Al	105,0	396,15	Mn	0,9	257,61
Cd	6,1	228,80	Mg	4,9	279,55
Ca	336,4	422,67	Na	46,0	588,99
Co	5,7	238,89	Ni	126,4	341,44
Cr	7,6	267,71	Pb	48,3	220,35
Cu	10,3	324,75	V	7,2	282,46
Fe	8,8	259,94	Zn	16,5	213,85
Hg	54,4	253,65			

Die Mineralstoffe in den aufgeschlossenen Proben verursachen im untersuchten Konzentrationsbereich eine Abnahme der Intensität der Meßsignale für die Schwermetalle bis max. 10 % (Vergleiche (4)). Ausnahme: Ca/Al-Einfluß. Die Intensität des Meßsignales für Al steigt bei hohen Ca-Gehalten auf über 800 % an.

Tabelle 4

Abreicherungsraten (%) nach der Gärung und nach der Blauschönung

Element	Gärung (%)	Blauschönung (%)
Al	20-30	0
Cd	29-41	ca. 80
Co	17-24	77-100
Cr	24-30	0
Cu	88-90	ca. 100
Fe	0-19	63- 80
Mn	0- 6	46- 73
Pb	32-40	0- 12
V	0-20	50- 80
Zn	0-31	78- 89

Tabelle 5

Methoden-Vergleich mit OES-ICP, G.C. und DPASV von Gehalten an Cu, Pb, Zn, Co und Ni in Wein und Most

Methode	Cu (ppb) Wein	Cu (ppb) Most	Pb (ppb) Wein	Pb (ppb) Most	Zn (ppb) Wein	Zn (ppb) Most	Co (ppb) Wein	Co (ppb) Most	Ni (ppb) Wein	Ni (ppb) Most
OES-ICP	30	310	20	20	540	540	10	10	-	-
G.C. (5)	41	317	-	-	611	-	12	-	62	73
DPASV	38	357	31	40	624	-	-	-	-	-

(- = nicht bestimmt)

Literatur

(1) H. Meierer, Dissertation, Universität Mainz, 1984 (im Druck)
(2) H. Eschnauer, UCD Symposium Proceedings, University of California, Davis, 1980
(3) P. Schramel, B.-J. Klose, S. Hasse, Fresenius Z. Anal. Chem., 310, 209 (1982)
(4) J. H. Kalivas, B. R. Kowalski, Anal. Chem. 53, 2207 (1981)
(5) G. Hartmetz, G. Scollary, H. Meierer, A. Meyer, R. Neeb, Fresenius Z. Anal. Chem., 313, 309 (1982)

STUDIES ON THE CHEMICAL NATURE OF HIGHLY RADIOACTIVE MICROPARTICLES BY
SEM-EDX, AES, ESCA

F. Baumgärtner, A. Huber, R. Henkelmann
Institut für Radiochemie, Technische Universität München

SUMMARY

SEM-EDX, AES und ESCA are applied to radioactive materials with dose rates up to
100 rad/h (at 1 cm distance). In heterogeneous mixtures (fission-product-elements
and transuranics) many micro-particles are successively analyzed with semiquantitative results leading to conclusions about composition and origin of various microcomponents.

INTRODUCTION

The multiemental analysis of heterogeneous samples composed of a large number of
highly radioactive microparticles, possibly heterogeneous, such as residues,
aerosols, precipitates, etc., present a different analytical situation than is
usually encountered in the measurement of homogeneous samples. Problems arise not
only from the analysis of particles in microscopic and submicroscopic dimension,
but also from the heterogeneity in the chemical composition of microparticles and
from the radiation from samples. Meaningful results for such a heterogeneous system
can be obtained only by the multielemental analysis of many individual particles
with the help of an advanced evaluation method and proper interpretation of large
data sets. Suitable instrumental methods for this analytical application appear to
be Scanning-Electron-Microscopy (SEM) combined with Energy-Dispersive-X-ray analysis
(EDX), Auger-Electron-Spectroscopy (AES) and Photoelectron-Spectroscopy (ESCA).

Under normal operation, the resolution of our instrumentations * is demonstrated to
be 10 nm for electron-optical observation, 100 nm for Auger-Electron-Spectroscopy
and 1 μm for X-ray analysis.

* JEOL JSM-35CF Perkin Elmer SAM-595, ESCA-500

INFLUENCE OF DOSE RATE

The samples investigated originated from fields of nuclear fuel cycle and consist of highly radioactive material. First we have to answer the applicability of the chosen methods to samples with high ionizing dose rates. The Auger-Electron-Spectrometry is not disturbed visibly by radioactive radiation, at dose rates up to 100 R/h at a distance of 1 cm. For X-ray analysis using a Silicon-semiconductor detector, a limitation is imposed only by the resolution time of the instrumentation because of the high count rates due to the sample's radiation. At a dose rate of 100 R/h at 1 cm distance resulting in an equivalent detector count rate of 50.000 cps, the measuring time war extended by a factor of 100.

The Auger-Spectra of some elements, like technetium, plutonium and other transuranics, have been measured for the first time in these experiments /1/.

PROCEDURE AND RESULTS

For the characterization of heterogeneous systems of microparticles we applied the following procedures and methods:

1. The zone of microscopic observation is divided into equally sized squares and a fixed number of particles are measured in each square. For every particle, size, shape, elemental composition are analyzed. At the same time, the coordinates of each particle are recorded, which facilitates locating the particles again for further measurements. At this stage, light and heavy elements can already be distinguished with the aid of backscatter-electron images.

2. The table of elemental frequencies gives a first hint if the sample is of homogeneous or heterogeneous nature.

   For example a filtered residue from a nuclear process solution, shows microparticles with 26 elements detectable. This particles can be classified into three size ranges and into five elemental groups depending on their origin:

10 - 100 µm	1.	light elements	Mg, Al, Si, P, S, Cl, K, Ca
	2.	transition metals	Ti, V, Cr, Mo, Fe, Ni, Cu, Zn
	3.	heavy metals	W, Pt, Pb
0.5 - 5 µm	4.	fission products and nuclear fuel particles	Zr, Mo, Tc, Ru, Rh, Pd, Pu
< 0.5 µm	5.	fission products	Zr, Pd

This classification is derived from the assessment made for elemental frequencies, which is based in this case on analyses of only 64 particles. The use of back-scatter electron image facilitates the recognition of the five elemental groups mentioned above.

3. A histogram of elemental combinations is used to recognize chemical compounds. Additonally for particles with diameters greater than 1 µm, the homogeneity of elemental composition can be confirmed by means of linescans.

Radioactive aerosols from a high-temperature experiment, for instance, exhibit crystalline or sperical shape, depending on the applied temperature and contains only cesium and technetium. Both samples are found to be homogeneous in spite of completely different microscopic appearance. Linescans over both crystalline and spherical particles show a homogeneous elemental distribution and a constant concentration ratio between Cs and Tc, suggesting stoichiometric relations of the two elements.

Additionally, ESCA-analysis has verified the chemical compound to be $CsTcO_4$, showing this technique to be another useful microchemical method.

In another aerosol experiment, analyses of 100 particles show the presence of 10 elements in nearly every particle, but the elemental distribution in each particle is not so uniform as can be expected for chemical compounds. The analytic results lead to the conclusion that these aerosols originated form liquid drops, with an elemental composition consistent with that in liquid but heterogeneously assembled within each particle.

As a further example of analysis, the residue from a LWR nuclear fuel is presented. The measurements of 700 particles show the presence of nine elements: Zr, Mo, Tc Ru, Rh, Pd, Ag, U and Pu. The histogram of elemental combinations directs the attention to a group of five elements, Mo, Tc, Ru, Rh and Pd, which have been known up to this time only in fast breeder reactor fuels forming an alloy with ε-structure. Oxygen-analysis with AES confirms the metallic character of the particles containing these elements.

4. A quantitative analysis of microparticles cannot be expected as customary in microprobe analysis because of the irregular surface of the samples. A semiquantitative analysis, however, proved to be feasible. The relative differences of the K_α X-ray intensities from all elemental combinations in question are plotted according to their frequency, Fig.1. (In the case of the above mentioned ε-phase containing Mo, Tc, Ru, Rh and Pd, no correction for excitation and absorption effects is needed, because these elements have adjacent atomic numbers.) Distinct peaks show the range of the quantitative elemental relation respectively, in

rough agreement with microprobe measurements from the ε-phase.

	sample F 4	Kleykamp /2/
Mo	23.0	24
Tc	10.9	8
Ru	46.4	44
Rh	10.3	8
Pd	9.6	10

5. In particular cases elemental distribution maps may be additionally a useful means of investigation, although this technique cannot be applied routinely because of the long measuring times involved.

For a sample constisting of highly radioactive colloidal particles adsorbed at silica grains, the combination of backscatter images and elemental distribution maps revealed unequivocally the presence of ε-phase material in colloidal form.

As another example, the unknown precipitation on the surface of an ε-phase is given. The trigonal crystalline structures of about one micron dimension on the surface of the cake is shown to consist of Ag, Br and I. Even at relatively high magnifications, the elemental maps exhibit a clear picture of the structures, Fig 2.

This sample also demonstrates, how bulk analysis methods may lead to erroneous results, if applied to microscopically heterogeneous samples. The originally homogeneous surface of the metallic ε-phase becomes partially oxidized in time, presumably through ozone generated by radiation. A bulk analysis like X-ray diffractometry cannot realize that the oxidized phase is limited to separated microscopic aereas caused by a secondary oxidation process and it is not precipitated from solution /3,4/.

6. In addition to the elemental composition, the size distribution is an important parameter for characterizing particles.

By means of automatic particle sizing /5/ it is possible to determine particle size spectra with sufficient statistical accuracy and resolution. As an example, the distribution of spherical aerosols is studied. The number of particles measured in two days is 4500.

We are about to combine automatic particle recognition with an elemental analysis.

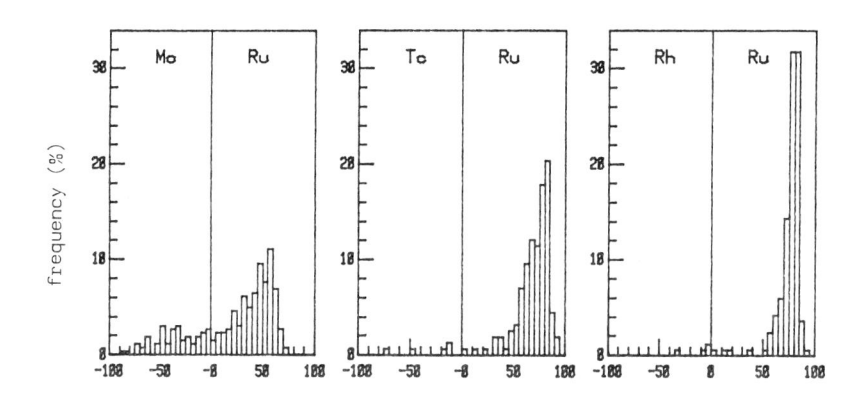

Fig. 1: Relative differences (x-y)max(x,y) (%) of Ru, Mo, Tc and Rh in an alloy of fission products resulting in a relative composition of Ru:Mo:Tc:Rh = 1 : 0.4 : 0.15 : 0.15.

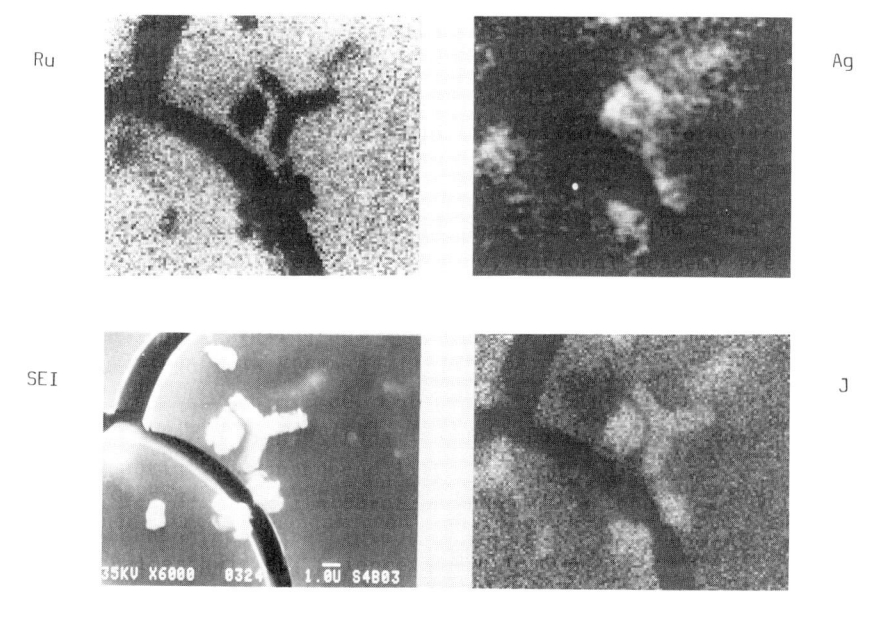

Fig. 2: Elemental distribution maps for Ru, Ag and J of a precipitate of fission products.

Acknowledgement - This work is financially supported by the Bundesministerium für Forschung und Technologie, Bonn.

REFERENCES

/1/ Baumgärtner, F., Dachsel, C., Henkelmann, R., Fresenius, Z.,
    Anal. Chem. (1983) 314, 348
/2/ Kleykamp, H., Gottschalg, H.D., Prejsa, R., Wertenbach, H.,
    KFK-PWA Bericht 40/82 (1982)
/3/ Kleykamp, H., KfK-Bericht 2665 (1978) und Trans. ANS 31, 1979, 508.
/4/ Kleykamp, H., Atomwirtschaft, 1982, 385.
/5/ Erasmus, S.I., Smith, K.C.A.,
    EMAG 1982, 115

PROBLEME BEI DER ANALYSE VON SIMULIERTEM HOCH-AKTIVEM WASTE (HAW)

W. Coerdt, F. Geyer, E. Mainka, H. G. Müller und S. Weis
Kernforschungszentrum Karlsruhe GmbH
Institut für Radiochemie
Postfach 3640
7500 Karlsruhe, Federal Republic of Germany

1. Aufgabenstellung:

Die Charakterisierung von hochaktivem Waste (HAW) ist im Zusammenhang mit der Abfallbehandlung aus dem nuklearen Brennstoffkreislauf von ausschlaggebender Bedeutung. Die Aufgabe war, für dieses Probenmaterial entsprechende Analysentechniken zu testen. In den Proben waren bis zu 25 Elemente zu bestimmen. Die Arbeiten wurden an simuliertem Material durchgeführt.

2. Arbeitsbericht:

Die anfangs instabilen Proben stabilisierten sich nach 14-tägigem Stehenlassen bei ca. 22°C. Der Niederschlag wurde abfiltriert. Die Analyse des Niederschlages ergab, daß Zirkon und Palladium die Hauptkomponenten waren und außerdem noch Ag und Se nachgewiesen wurden. Detailliert analysiert wurde die klare Lösung. Als bevorzugte Analysentechnik haben wir die optische Emissions-Spektroskopie mit Plasmaanregung (ICP) gewählt.
Die zu bestimmenden Analysenelemente haben sehr linienreiche Spektren, so daß ein hochauflösender Analysator benötigt wird; es sei denn, man verdünnt die Proben stark, so daß auf diese Weise das Spektrum linienärmer wird. Wir haben unser Meßsystem so aufgestellt, daß wir gleichzeitig unser Sequenzspektrometer und unseren 3.5 m Gitterspektrographen mit hochauflösendem Gitter (z.B. 200 000) einsetzen können. Ein Schema der Meßapparatur zeigt Abb. 1
Mit dem 3.5 m Gitterspektrographen wurde eine Verdünnungsreihe aufgenommen und halbquantitativ eine Multielementanalyse durchgeführt. Für die quantitative Analyse mit dem Sequenzspektrometer wurden 3 Verdünnungen mit dem Faktor 1000 der Probe hergestellt und gemessen. (Meßkonzentration im ppb-Bereich). Als Analysenlinien sind soweit möglich die empfindlichsten Linien nach [1] ausgewählt worden.

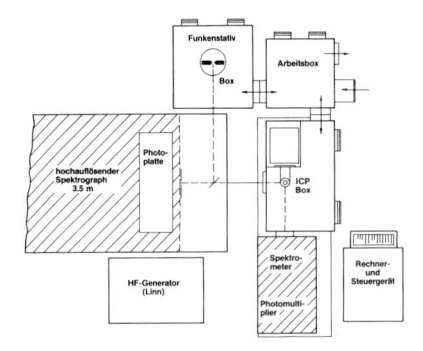

Abb. 1 Schema der Meßapparatur

3. Ergebnisbericht

Die Meßergebnisse faßt Abb. 2 zusammen

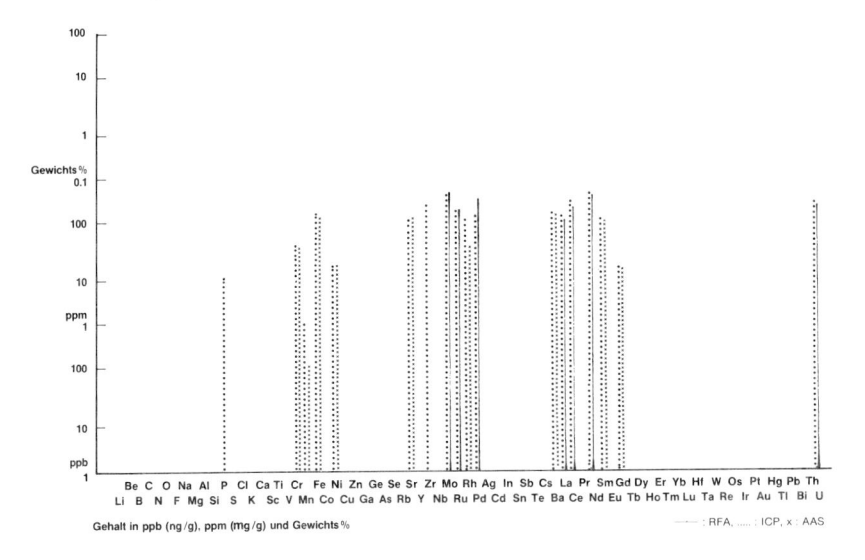

Die Reproduzierbarkeit der Meßwerte war für die Elemente
Ba, Cr, La, Mo, Nd, Pd, Sr und Zr besser 10 %
Ce, Fe, Ru, Sm und U besser 20 %
Gd, Mn und Ni nahe der Nachweisgrenze ca. 25 %.
Die Richtigkeit lag für die einzelnen Elemente in der gleichen Größen-
ordnung. Die Nachweisempfindlichkeit entsprach den in [1] angegebenen
Daten.

1] R. K. Winge, V.S. Peterson and V.A. Fassel: Appl.Spec 33 (1979)
   206 - 19

MULTI-ELEMENT ANALYSIS OF WASTE WATER USING AES-ICP AND XRF METHODS.

P.Hoffmann, K.H.Lieser, R.Speer, R.Pätzold

Fachbereich für Anorganische Chemie und Kernchemie
Technische Hochschule Darmstadt

The central purification plant for the waste water of the city of Darmstadt consists of a mechanical purification step, of an anaerobic sludge decomposition and of an aerobic biological purification step.

The waste water is conducted to the plant in two canals. The "north"-canal leads mainly household waste water, whereas the "south"-canal is contaminated by water from metal manufacturers, printers and electroplating industry.

Many trace elements (toxic heavy metals) disturb the biological purification steps and cannot be separated from the waste water or from the sludge in a practical way. A high content of these elements prevents the agricultural use of the water and of the sludge and therefore they have to be controlled and avoided as far as possible.

2 l-samples were taken each day during the week from 25th to 31st May 1983. For their characterization, the electrical conductivity, the pH-value, the sediment concentration, the water temperature, the rain amount and the waste water amount were measured.

All samples were preserved for 4 weeks by adding 10 ml HNO_3 (65 %) and by storage at $5°C$.

The samples were then worked up in different ways:
- decomposition : adding of 10 ml HNO_3 (65 %) and of 10 ml H_2O_2 (30 %) heating nearly to dryness, taking up the residue with 100 ml HNO_3 (65 %), heating again nearly to dryness, taking up the residue with 50 ml HNO_3 (10 %), and filtration,
- centrifugation: for 2 h at 10,000 rpm, in some cases followed by a filtration step through a 0.4 µm membrane filter,
- filtration : through 0.4 µm membrane filter,
- enrichment : shaking of 1 l solutions at pH 6 with 100 mg of a cellulose ion exchanger with 1,3,5-triazine as connecting group and N,N-diethyldithiocarbaminate as complexing agent for heavy metals.

The analytical determination was performed by atomic emission spectrometry with inductively coupled plasma (ICP) in the liquid samples and by energy-dispersive X-ray analysis (XRF) in the solid samples.

In the "sludge regulation" of the FRG the concentration limits for agricultural use are fixed for the elements Cr, Ni, Cu, Zn, Cd, Hg, and Pb. Hg was found in no sample, whereas the results for the other elements are represented in the figure.

Fig.: Concentration of the elements measured during a week for the "north"- and "south"-canals.

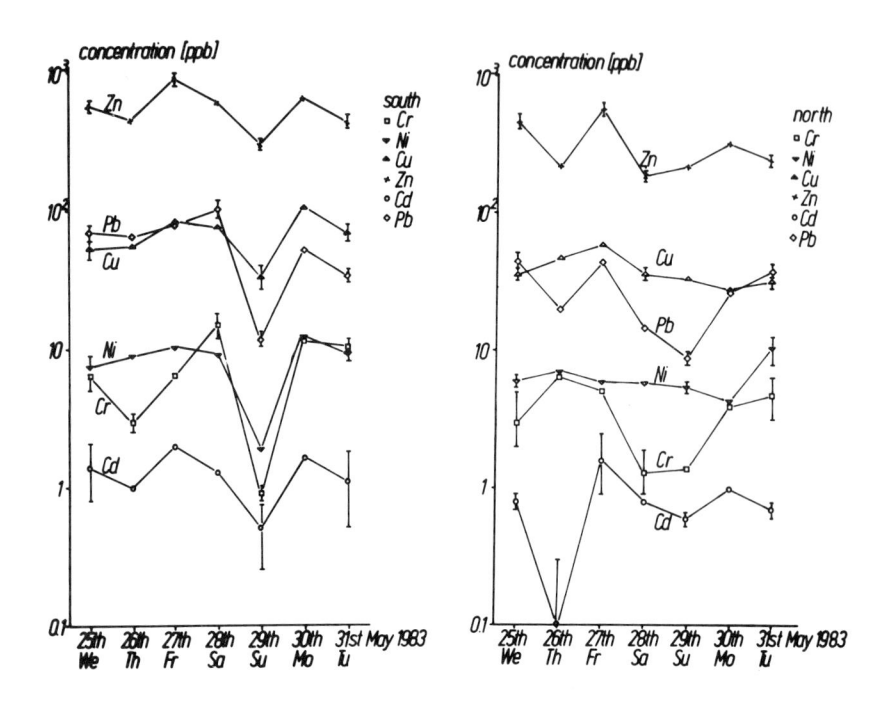

It has been found that the decomposition and the centrifugation are the best methods and nearly equal because, in these cases, there is a maximum trace element concentration in solution.

The contamination of the "south"-canal is for all elements higher, than that of the "north"-canal. The higher values of the electrical conductivity in the "south"-canal have to be explained by neutralization of industrial waters, the higher amount of sediments can be explained by precipitation of hydroxides, sulfides, halogenides, carbonates, phosphates and sulfates at pH-values between 7 and 9. The course of data for the "south"-canal shows a minimum on Sunday, which can be explained by industrial plants which are working on Sunday but not on Saturday. It should be noted that the behaviour of Zn and Cd, of Ni and Cu, and of Pb and Cr resp. is similar. Furthermore, it can be noticed that the concentration of the elements in the solid phase is relatively constant as opposed to the large difference in the liquid.

IN-LINE DETERMINATION OF U, NP, AND PU IN PROCESS STREAMS BY ENERGY-DISPERSIVE XRF.

P.Hoffmann, T.Hofmann, K.H.Lieser

Fachbereich für Anorganische Chemie und Kernchemie
Technische Hochschule Darmstadt

Today the analytical determination of actinides in solutions of nuclear fuel elements is performed by taking single samples, making chemical separations and, in most cases, spectrophotometric measurements. Therefore, an in-line instrumentation in nuclear fuel reprocessing plants is necessary for continuous measurement, for multielement determinations of U, Np, Pu, and other actinide elements, for determinations in the closed system, keeping people away from the toxic solutions, and for regulation of the process streams using the analytical data.

For actinide element analysis, different methods can be used: α-spectrometry, γ-spectrometry, activation analysis, XRF-spectrometry, mass spectrometry, raman spectrometry, voltammetry/polarography, UV/VIS/IR-spectrometry, and atomic absorption and emission spectrometry.

Among these the XRF-spectrometry seems to be the best solution for an in-line determination system, if it is designed as an energy-dispersive arrangement with radionuclide excitation. Such a system has to be computer controlled with an automatic evaluation of complex spectra.

The optimization of the geometrical arrangement of the sample, which is a fluid-leading tube and of the excitation and detection devices gives the following results: as tube material technical glass and stainless steel can be used, the wall thickness of the tube should not exceed 6 mm, the diameter of the sample tube should be about 130 mm, the excitation sources on both sides have to be at an angle of 45° with the tube-detector axis, the detector should be a high-purity germanium diode, the detector should be shielded by lead from the primary radiation of the radionuclide excitation sources.

From different points of view it is preferable to excite and to measure the K-X-ray-lines of the actinides: the penetration through the tube walls is more effective, the resolution of the K-lines is much better than that of the L-lines, and a greater volume of the sample can be analyzed and hence, give more accuracy to the result.

The most usual radionuclide for the excitation of K-lines in heavy elements is ^{57}Co. Taking into account the Compton- and Rayleigh-scattering effects in the sample and in the detector, it is deemed helpful to use a 300 keV-γ-source. A number of suitable radionuclides

Tab.: Radionuclide sources emitting γ-rays of about 300 keV for the excitation of the K-lines of actinide elements.

RADIONUCLIDE	DECAY MODE; γ-energy(keV)/A(%)	HALF-LIFE
^{131}I	β^-, γ; 284/ 6.1 364/81.2	8 d
^{133}Ba	ε, γ; 276/ 7.3 303/18.6 356/62.3 384/ 8.8	10,7 a
^{192}Ir	β^-, γ; 296/30.2 308/31.8 316/87.0 468/51.8 485/ 3.4 589/ 4.6 604/ 8.9 612/ 5.8	74 d
^{231}Pa	α, γ; 284/ 1.6 300/ 2.3 303/ 2.3	$3.27 \cdot 10^4$ a

is summarized in the table. Taking the decay branches and the half-lives into account, ^{133}Ba gives the optimal intensity and stability over a long period. In the case of our laboratory device, an activity of 10-20 mCi is optimal, at higher activities the scattering background is increased and the detection limit becomes poor.

At optimal conditions the lowest detectable concentration of U in acid aqueous solutions is 5 ppm.

EINSATZ EINES SIMULTANEN ICP-SPEKTROMETERS IM ANALYSENDIENST EINES FORSCHUNGSZENTRUMS

G. Wolff, H. Nickel, H. Lippert

Kernforschungsanlage Jülich GmbH, Zentralabteilung für Chemische Analysen

1. EINLEITUNG UND ZUSAMMENFASSUNG

Die Zentralabteilung für Chemische Analysen der Kernforschungsanlage Jülich betreibt seit mehr als 3 Jahren ein Vielkanalspektrometer mit induktiv gekoppeltem Plasma im Analysendienst. Das Gerät ist eine Kombination eines Vakuum-Polychromators, Modell 34000 der Firma ARL, Ecublenz, Schweiz, ausgestattet mit 50 Elementkanälen und einem Luft-Monochromator Modell 35000 als frei wählbarem weiterem Elementkanal. Die Vielzahl der mit dem Gerät bearbeiteten Probleme läßt sich grob in die folgenden Gruppen einordnen:
1. Materialcharakterisierung von metallischen Werkstoffen;
2. Stöchiometriebestimmungen von Sonderlegierungen und Kristallen;
3. Gesamtanalysen von Mineralen, Böden, Gesteinen, Klärschlamm usw.;
4. Multielementanalysen von biologischem Material.

Über die hierbei gemachten Erfahrungen wird hier berichtet.

2. EXPERIMENTELLES

Das für die Routineanalytik eingesetzte Spektrometer wurde mit den folgenden spektralen Daten und Arbeitsbedingungen betrieben:

Spektrometer	:	ARL 3400-ICP Quantovac
Brennweite	:	1 m
Gitter	:	1080 Striche/mm
Blaze	:	600 nm 1. Ordnung
		300 nm 2. Ordnung
		200 nm 3. Ordnung

Spektraler Bereich : 400-820 nm 1. Ordnung
                     230-400 nm 2. Ordnung
                     170-230 nm 3. Ordnung

Plasma   : Ar/Ar-Plasma; 1240 W

Gasströme   : 12 l/min Kühlgas
                   0,8 l/min Plasmagas
                   1 l/min Trägergas

Zerstäuber   : freiansaugender Meinhardzerstäuber, 6bar Druck

Da bei der Geräteanschaffung der größte Teil der damit zu bearbeitenden Fragestellungen bereits festlag, wurde die Elementkanalauswahl unter Berücksichtigung der analytischen Probleme getroffen. Die Ausstattung des Gerätes mit Festkanälen, sowie die damit gemessenen Nachweisgrenzen für die einzelnen Elemente in wässriger Lösung sind in Tabelle I angegeben.

<u>Tabelle I</u>

KANAL N	ELEMENT	WELLENLÄNGE (nm)	ORDNUNG	D.L. (ppm)	KANAL N	ELEMENT	WELLENLÄNGE (nm)	ORDNUNG	D.L. (ppm)
1	LA	379.48	1	0.00208	27	AL2	396.15	1	0.01121
2	U	385.96	1	0.04882	28	CE	413.76	1	0.02062
3	TH	401.91	1	0.00879	29	AR	425.96	1	0.00000
4	SR	407.77	1	0.00007	30	BI	223.06	2	0.01279
5	TE	214.28	2	0.01590	31	NI	231.60	2	0.00871
6	SB	217.58	2	0.01142	32	RU	245.65	2	0.00601
7	CD	226.50	2	0.00081	33	FE	259.94	2	0.01216
8	BA	455.40	1	0.00015	34	P	178.29	3	0.02825
9	W	239.71	2	0.02288	35	S	180.73	3	0.04006
10	B1	249.68	2	0.00178	36	MG	279.52	2	0.00008
11	SN	189.99	3	0.02245	37	AS	189.04	3	0.03188
12	C	193.09	3	0.03568	38	V	292.40	2	0.00176
13	AL1	309.27	2	0.01339	39	NA	589.59	1	0.01627
14	BE	313.04	2	0.00002	40	MO	202.03	3	0.00172
15	ZN	213.85	3	0.00098	41	NB	316.34	2	0.00352
16	CU	324.75	2	0.00118	42	SI1	212.41	3	0.00757
17	LI	670.78	1	0.00120	43	AG	328.07	2	0.00183
18	CS	697.33	1	9.37631	44	PB	220.35	3	0.02418
19	CO	238.89	3	0.00171	45	ZR	343.82	2	0.00086
20	TA	240.06	3	0.00883	46	TL	351.92	2	0.04319
21	TI	368.52	2	0.00186	47	SI2	251.61	3	0.00647
22	HG	253.65	3	0.02713	48	MN	257.61	3	0.00028
23	K	766.49	1	0.03696	49	CR	267.72	3	0.00812
24	RB	780.02	1	0.14675	50	QM+	0.00	0	0.03088
25	CA	393.37	2	0.00003	51	QR+	0.00	0	0.00000
26	SE	196.09	2	0.05081	52	B2	249.68	3	0.00356

Der Kanal 29 für Argon wurde eingebaut, um Änderungen der Gaszufuhr (Druck, Strömungsgeschwindigkeit usw.) während der Analyse feststellen zu können. Der Kanal 12 für Kohlenstoff wird ebenfalls in der Regel nicht analytisch genutzt, sondern dient bei der Analyse von biologischem Material der Überprüfung der Vollständigkeit der Veraschung. Für Cs konnte nur die unempfindliche Linie bei 697, 329 nm gewählt werden,

---

[+] QM und QR stehen für den Meß- und Referenzkanal des im Gerät ebenfalls eingebauten Monochromators. Über einen dieser Kanäle ist eine weitere Wellenlänge frei wählbar.

weil die Hauptnachweislinie bei 852, 111 nm außerhalb des spektralen Bereichs des Gitters liegt und sonst eine erhebliche Einschränkung des kurzwelligen Bereichs hätte in Kauf genommen werden müssen.

3. ERGEBNISSE

3.1 Metallegierungen

Der Schwerpunkt der Analyse von Metallegierungen lag auf dem Gebiet der austenitischen Stähle und der Nickelbasislegierungen. Untersucht wurden die folgenden Materialtypen:
Hastelloy S; Hastelloy X; Inconel 601, 617, 625, X-750; Nimonic 80 A, 86, 105, PE 16; Nimocast 713 LC; Manaurite 36 X; Incoloy 800 H, 802, In-519.

Als Beispiel für die Analyse solcher Legierungen sind in Tabelle II die Werte für Inconel 617 angegeben und mit den mit klassischen Verfahren erhaltenen Werten verglichen.

Tabelle II

Elemente	Inconel 617 Gehalte in Gew. %	
	OES-ICP	klass. Verfahren
Ni	54,6 ± 0,02	54,2 ± 0,02
Cr	22,3 ± 0,02	21,9 ± 0,01
Fe	0,109 ± 0,005	n.b.
Co	11,9 ± 0,01	12,0 ± 0,02
Mo	9,32 ± 0,02	9,75 ± 0,05
Ti	0,39 ± 0,01	0,34 ± 0,02
Al	1,07 ± 0,02	n.b.
Mn	0,034 ± 0,001	n.b.
Cu	0,009 ± 0,001	n.b.
V	0,008 ± 0,001	n.b.
C	n.b.	0,091 ± 0,002
Si	n.b.	0,08 ± 0,02

Die Übereinstimmung der Werte ist gut, mit Ausnahme der Mo-Werte. Diese sind aber beim klassischen Verfahren (photometrische Bestimmung als SCN-Komplex) wahrscheinlich zu hoch, weil bei der Bestimmung kein Fe in größerem Überschuß zugesetzt wurde, um den Thiocyanatkomplex zu stabilisieren.

---

n.b. = nicht bestimmt

3.2 Stöchiometriebestimmungen von Sonderlegierungen und Kristallen

Die Analyse dieser Art Proben stieß mit klassischen Verfahren auf Schwierigkeiten, einmal weil sie vielfach Edelmetalle und Lanthanidenelemente enthielten oder von den Übergangselementen diejenigen, die schwer quantitativ zu trennen waren, zum anderen aber auch deshalb, weil die zur Analyse zur Verfügung stehenden Probenmengen klein waren. In der Regel standen nur Probemengen von 1-100 mg zur Verfügung. Sodann sollte die Stöchiometrie möglichst genau erfaßt werden, mit Absolutfehlern 0,2 %. Beispiele für solche Legierungen sind: AlGa, CuAu, AgPd, IrRh, IrTh, GeGa, Eu-Pd-Si und als Vertreter der Kristalle $GdMo_6S_8$.

Zur Lösung der Fragestellung machten wir uns den Umstand zunutze, daß mit einem Simultanspektrometer gleichzeitig alle in den Proben vorkommenden Elemente gemessen werden können. Somit müssen bei der Analyse 100 % der Einwaage wiedergefunden werden. Bildet man nun die Summe der gefundenen Elementmengen und vergleicht sie mit der Einwaage, so läßt sich aus dem Verhältnis ein Korrekturfaktor errechnen, der größer oder kleiner aber immer nahe bei eins ist, je nachdem, ob die Summe der gefundenen Elementmengen kleiner oder größer als die Einwaage ist. Multipliziert man die Elementsignale mit diesem Faktor, so erreicht man eine höhere Präzision der Analyse, als bei einfacher Auswertung der Elementsignale. Voraussetzung für die Gültigkeit dieser Überlegung ist nur, daß sich während der Analyse alle auftretenden Schwankungen der Signale bei allen gemessenen Spektrallinien gleichsinnig verhalten.

Als Beispiel für Analysen derartiger Substanzen, ist hier die Untersuchung einer Edelmetallegierung angeführt, die die Zusammensetzung Cu_3Au haben sollte. Die Ergebnisse wurden zu Testzwecken von einer größeren Substanzmenge von etwa 200 mg Einwaage erhalten. Die Probelösung wurde geteilt und einmal klassisch chemisch - Cu durch Elektrolyse, Au durch Reduktion und Fällung als Metall -, untersucht, und der andere Teil nach Verdünnung mit dem ICP-Spektrometer gemessen.

Tabelle III Cu_3Au-Legierung

Element	gef. Gew%		ber. Atom%	
	OES-ICP	klass.	OES-ICP	klass.
Cu	48,57	48,8	74,72	74,7
Au	50,94	51,2	25,28	25,3
Summe	99,81	100,0	100,00	100,0

Die hohe, hier mit der OES-ICP erzielte Genauigkeit ist kein Zufallsergebnis, sondern ließ sich auch bei der Analyse von Standardproben erreichen. Die Ergebnisse für zwei Standardproben sind in Tabelle IV angegeben:

Tabelle IV Standardproben

Elemente	AKP NR. 223		BCS Nr. 180/1	
	Zertifikat	OES-ICP	Zertifikat	OES-ICP
Cu	58,74	58,56	67,36	67,29
Zn	38,82	38,96	-	-
Pb	2,13	2,16	-	-
Ni	-	-	30,85	30,92
Fe	-	-	0,82	0,82
Mn	-	-	0,81	0,81

Die Analysen wurden ohne Korrektur der Elementeinflüsse aufeinander durchgeführt. So ergeben z.B. 100 ppm Cu in Lösung auf unserem Zn-Kanal ein Signal, das 0,29 ppm Zn entspricht. Für Zn ergibt sich eine entsprechende Depression des Signals im Cu-Kanal. Werden die Werte damit korrigiert, so erhält man für den Standard AKP Nr.223 Werte von 58,67% Cu und 38,79% Zn.

3.3 Minerale, Böden, Gesteine, Klärschlamm

Proben dieser Art erhält die Zentralabteilung für Chemische Analysen aus dem Institut für Erdöl und Geochemie, aus dem Institut für Radioagronomie, aus den chemischen Instituten und von externen Auftraggebern. Dabei soll entweder die Zusammensetzung der Probe, also die Hauptbestandteile und die Nebenbestandteile oder das Spurenelementspektrum analysiert werden. Die bei weitem schwierigste Aufgabe auf diesem Gebiet war die Analyse von Klärschlamm, weil hierfür keine Standards zur Verfügung standen und außerdem eine Gesamtanalyse einschließlich der toxischen Spurenelemente gewünscht war. Als weitere Schwierigkeit kommt hinzu, daß Klärschlämme je nach Art der Deponie, die unterschiedlichste Zusammensetzung haben. Aus diesen Gründen wird die Analyse eines Klärschlammes von einer kommunalen Deponie aus dem Raum Bonn hier näher beschrieben.

3.3.1 Probenahme und Probenvorbereitung

In dem hier geschilderten Beispiel sollten alle mit der OES-ICP nachweisbaren Elemente in der Probe bestimmt werden. Die wesentliche Voraussetzung für zuverlässige Analysenwerte war ein einwandfreier Aufschluß der koplexen Matrix. Es wurden von uns 3 verschiedene Arten des Probenaufschlusses untersucht.

Boraxaufschluß

Dieser Aufschluß kann angewendet werden, wenn bei der Analyse keine flüchtigen Bestandteile der Probe analysiert werden müssen. Man geht von 200 mg Substanz aus, die durch Glühen der Probe im Muffelofen bei 800-900°C von organischen Bestandteilen befreit wurde. Diese Einwaage wird im PtAu-Tiegel (10% Au), Graphittiegel oder Tiegel aus Glaskohlenstoff mit einer Mischung von 1,8 g $Li_2B_4O_7$ und 0,2 g $LiBO_2 \cdot 2H_2O$ vorsichtig durchmischt und 20-30 min, bei einer Temperatur von 750-850°C über dem Brenner oder im Tiegelofen aufgeschlossen. Gelegentliches vorsichtiges Umschwenken des Tiegels, wobei sich eine Boratperle bildet, beschleunigt den Aufschluß. Die noch flüssige Boratperle wird in etwa 50 ml einer mit einem Teflonstäbchen gerührten, etwa 4%igen Salz- oder Salpetersäure eingegossen. Unter leichtem Erwärmen löst sich die Perle vollständig nach etwa 10 min. Nach dem Abkühlen wird auf 100 ml aufgefüllt. Diese Lösung kann direkt (für Spurenelemente) oder nach Verdünnung (für Hauptbestandteile) mit dem ICP-Spektrometer analysiert werden. Der Nachteil diese Aufschlusses ist, daß zwar die Metalloxide und SiO_2 in Lösung gehen, aber erhebliche Verluste an Hg, As, Pb, Se u.a.m. auftreten.

Aufschluß mit wasserfreier H_3PO_4 unter Zusatz von HNO_3

Dieser Aufschluß hat den Vorteil der geringeren Temperaturbelastung der Probe. Außerdem werden dabei alle kohlenstoffhaltigen Substanzen mit oxidiert. Allerdings können auch bei diesem Aufschluß Verluste an Hg, As, Pb, Sb, Se auftreten. Diese lassen sich aber weitgehend vermeiden, wenn unter Rückfluß aufgeschlossen wird.
200 mg der getrockneten Probe werden mit 10 ml wasserfreier H_3PO_4, die aus reinstem P_2O_5 bereitet wird, im Teflonbecher unter Rückfluß auf ca. 300°C erhitzt. Mit einem Teflonstäbchen wird leicht gerührt. Nach dem Abkühlen der Probe auf Raumtemperatur werden 100 Mikroliter konz. HNO_3 s.p. zugegeben und erneut auf 250-300°C erhitzt. Nach ca. 20 min. ist der Aufschluß beendet. Falls noch dunkle Partikel in der Lösung sind, muß die Prozedur unter erneuter Zugabe von HNO_3 wiederholt werden. Die Aufschlußlösung wird auf 50-500 ml aufgefüllt, je nachdem wie hoch die Konzentration der zu bestimmenden Elemente ist. Anschließend kann mit dem ICP-Spektrometer analysiert werden. Allerdings muß bei einer Endkonzentration von 5% H_3PO_4 oder mehr mit einem inneren Standard gearbeitet werden, weil die Viskosität der Lösung sonst zu hoch ist.
Bei diesem Aufschluß bleibt etwa 2/3 des SiO_2 ungelöst, jedoch konnte kein Einschluß oder eine Adsorption von anderen Elementen am Rückstand festgestellt werden. Al_2O_3, TiO_2, ZrO_2 usw. werden vollständig aufgeschlossen.

Tabelle V

Elemente	1. Probe Gehalt in %	1. Probe Gehalt an Oxiden %	2. Probe Gehalt in %	2. Probe Gehalt an Oxiden %
P	nicht best.	nicht best.	0,131	0,392 (SO_4)
S	0,119	0,357 (SO_4)	0,131	0,392 (SO_4)
Sr	0,0126	0,015	0,013	0,015
Sb	0,0881	0,117	0,087	0,104
Ba	0,0476	0,053	0,047	0,053
B	0,726	2,34	0,521	1,68
Sn	0,102	0,130	0,097	0,124
Al	3,68	6,95	3,66	6,92
Zn	0,0763	0,095	0,077	0,096
Cu	0,0226	0,028	0,022	0,028
Ti	0,276	0,460	0,276	0,460
K	1,31	1,58	1,23	1,48
Ca	2,27	3,18	2,33	3,26
Bi	-	-	0,0030	0,0033
Ni	0,0061	0,0078	0,0046	0,006
Fe	2,05	2,93	2,04	2,92
Mg	0,543	0,900	0,542	0,899
Na	1,53	2,06	1,50	2,02
Mo	0,0044	0,0066	0,0042	0,006
Si	9,96	21,3	10,00	21,5
Pb	0,0038	0,004	0,0040	0,005
Mn	0,0423	0,055	0,042	0,055
Cr	0,103	0,150	0,103	0,151
Co	0,0029	0,004	0,0028	0,004
Cd	0,0019	0,002	0,0015	0,002
Te	0,0160	0,020	0,0100	0,013
V	0,0002	-	0,00015	-
Li	0,0006	-	0,0005	-
C [*]	14,5	-	14,5	-
SiO_2 (Rückstand)		31,3	-	31,3
Glühverlust		30,3	-	30,3

Als Beispiel für eine Vollanalyse nach diesem Aufschluß sind in Tabelle V die Ergebnisse der Analyse von zwei Proben des oben erwähnten Klärschlammes von der gleichen Deponie angegeben.

Die Übereinstimmung der Ergebnisse der Analyse aus den zwei unterschiedlichen Proben ist gut, wenn man die inhomogene Verteilung der Elemente in solchen Proben berücksichtigt. Eine Ausnahme hiervon ist der Borgehalt der Probe. Hier ist es möglich, daß beim Aufschluß Verluste auftreten, weil nach Aussage des Autors, eine der reaktiven Spezies im Aufschlussmittel PH_3 sein soll.

---

[*] C wurde nach Verbrennung im O_2-Strom durch IR-Absorption bestimmt, das SiO_2 im Rückstand und der Glühverlust nach klassisch-chemischen Verfahren.

Aufschluß mit Königswasser

Dieses Aufschlußverfahren entspricht dem Entwurf zur DIN 38 414, Teil 7 vom Februar 1982. Es wurde von uns ebenfalls an den gleichen Proben überprüft. Der Aufschluß eignet sich aber nicht für eine Vollanalyse von Klärschlamm und Böden, sondern nur zur Bestimmung der säurelöslichen Anteile. Es hat aber den Vorteil, daß hierbei keine Elementverluste auftreten. Die speziellen, dafür benötigten Einrichtungen wurden von uns nach den Angaben des DIN-Entwurfs angefertigt. Da man dabei von 3 g Klärschlammprobe ausgeht, erhält man gleichzeitig eine bessere Mittelung der Elementgehalte.

3.4 Multielementanalysen von biologischem Material

Zu dieser Gruppe von Materialien, die von uns untersucht wurden, zählen Pflanzentrockensubstanzen, Algen, Bakterienkolonien aus Biogasfermentern, Konzentrate von Pflanzenviren und Blutserum.

Wiederum als Beispiel soll die Analyse von Blutserum beschrieben werden.

3.4.1 Probenvorbereitung

Ausgangspunkt der Untersuchungen war ein Mischserum humanen Ursprungs aus der Klinik der KFA. Es sollte geprüft werden, in wieweit sich die Metalle sowie P in diesen Seren mit der OES-ICP direkt oder nach Veraschung bestimmen lassen.

Eine Analyse in den Serumproben direkt war nicht möglich, wegen der hohen Viskosität der Proben. Jedoch ergaben sich erste brauchbare Ergebnisse bereits bei einer Verdünnung von 1:10 mit 0,5 n HCL (siehe Tabelle VI).

Die Veraschung der Proben erfolgte nach folgendem Verfahren:
1 ml gefriergetrocknete Serumprobe wurde bei niederer Temperatur mit angeregtem Sauerstoff verascht (Kaltveraschung unter vermindertem Druck mit elektrisch angeregtem O_2). Die Asche wurde mit 0,5 n HCL aufgenommen und auf 10 bzw. 5 ml mit 0,5 n HCL aufgefüllt.

3.4.2 Ergebnisse

Die Ergebnisse, die mit unserem Spektrometer erhalten wurden, sind in der Tabelle VI zusammengestellt.

Tabelle VI   ICP-Analysen von Blutserum

Element	Gehalte in µg/ml				
	Pr. A	Pr. B	Pr. C	Pr. D	Pr. E
Na	3249	3500	3450	3293	3313
K	165	186	177	172	175
Ca	101	110	108	101	102
Mg	20	22	21	20	20
P	149	152	151	151	141
Cu	1,20	1,69	1,48	1,73	1,49
Fe	1,00	1,54	1,20	1,21	1,38
Zn	0,80	1,66	1,40	1,01	0,87

Element	Gehalte in µg/ml		Mittelwerte A-F mit Standardabweichung	
	Pr. F	Lit.		
Na	3431	3313	3373	± 100
K	180	190	176	± 7
Ca	105	97	105	± 4
Mg	21	22	20,7	± 0,7
P	150	141	149	± 4
Cu	1,52	1,19	1,58	± 0,12
Fe	1,34	1,09	1,33	± 0,14
Zn	1,13	1,15	1,21	± 0,32

Bermerkung zu den Proben:
A: 1 ml Serum, 1:10 mit 0,5 n HCL verdünnt
B: 1 ml Serum, verascht (Tieftemp.), Asche gelöst in 10 ml 0,5 n HCL
C: wie B, andere Probe
D: 1 ml Serum, verascht wie B, Asche gelöst in 5 ml 0,5 n HCL
E: 1 ml Serum, verascht wie B, Asche gelöst in 5 ml 0,5 n HCL
F: wie E, andere Probe
Lit: Mittelwerte aus der Literatur nach Iyengar, Kollmann, Bowen
"The Elemental Composition of Human Tissues and Body Fluids"
Verlag Chemie, Weinheim, 1978

Die in der letzten Spalte der Tabelle VI angegebenen Mittelwerte sollen nicht den richtigen Gehalt des Serums repräsentieren, sondern sind nur aus den Ergebnissen der Proben A-F formal berechnet und sollen der Fehlerbeurteilung unseres Probenvorbereitungs- und Analysenverfahrens dienen.

Auch einige Spurenelemente wie Al, Ba, Si, Mn, Sr, und Cd wurden bei diesen Analysen gleichzeitig mit bestimmt. Die Ergebnisse waren aber noch so stark schwankend, daß erst noch weitere Untersuchungen über den Einfluß der Probenvorbereitung auf die Blindwerte und über den Einfluß der Probenviskosität auf die Zerstäubungsrate durchgeführt werden müssen.

Immerhin zeigen die erhaltenen Ergebnisse, daß die OES-ICP auch für die direkte Analyse von Serumproben geeignet ist, wenn man diese nur mit

0,5 n HCL um einen Faktor 10 verdünnt.

4. DISKUSSION

Die hier mitgeteilten Erfahrungen zeigen, daß die Methode der Emissionsspektralanalyse mit Plasmaanregung mit Vorteil für die Lösung unterschiedlichster analytischer Fragestellungen, wie sie in unserem Forschungszentrum im Analysendienst anfallen, eingesetzt werden kann.
Die Hauptprobleme liegen in der Wahl einer geeigneten, für das Verfahren spezifischen Probenvorbereitung. Dies ist bedingt durch die besondere Art der Aerosolerzeugung und ihre Einflüsse auf das Plasma. (3)

Die Analysenzeiten bei Benutzung eines Vielkanalspektrometers sind klein. Die Präzision ist für ein emissionsspektralanalytisches Verfahren hoch und läßt sich in Sonderfällen durchaus mit der Genauigkeit klassisch-chemischer Verfahren vergleichen.

Für die Spurenanalyse im unteren Konzentrationsbereich (1-10 ppb) ist die Analysenmethode nach unseren Erfahrungen nur bei einfachen Matrices (Wasser, Leachlösungen) geeignet, übertrifft aber für einige Elemente die Atomabsorption mit Flammenanregung.
Besondere Anforderungen werden für die Beurteilung der Analysendaten an das Bedienungspersonal gestellt, weil sich dessen Fachkenntnisse sowohl auf das Gebiet der Emissionsspektralanalyse als auch auf das der analytischen Chemie allgemein erstrecken sollten.

5. LITERATUR

(1) L. Morgenthaler, Interner Bericht Fa. Bausch u. Lomb,
    Sunland CA, USA (1982)

(2) P. Scherer, H. Lippert, G. Wolff, Biological Trace Element
    Research 5 (1983) 149-163

(3) Z. Zadgorska, H. Nickel, M. Mazurkiewicz, G. Wolff,
    Z. Anal. Chem. 314 (1983) 356-361

Anmerkung des Herausgebers:
Für die Zahlenangaben in den Tabellen ist der erstgenannte
Autor verantwortlich.

INSTRUMENTAL ANALYSIS AND PROVENANCE OF ARCHAEOLOGICAL ARTIFACTS

E. Pernicka

Max-Planck-Institut für Kernphysik, P.O. Box 103980, 6900 Heidelberg, FRG

SUMMARY

An important aspect of archaeometric research - the application of scientific methods to problems of cultural history - is the determination of the provenance of archaeological artifacts, which demands the use of instrumental methods for multi-element analysis. Thereby the chemical composition and especially trace element contents are used as fingerprint, which can be compared with various production sites.

Relating to the most important groups of materials found in archaeological excavations - metals, pottery, and stone artifacts - examples of provenance determination in the eastern Mediterranean are described: a) Early metallurgy, b) late Bronze Age pottery, and c) prehistoric trade in obsidian.

1. INTRODUCTION

The analysis of archaeological artifacts has a history almost as long and illustrious as analytical chemistry itself. Indeed, the "father of analytical chemistry", Martin Heinrich Klaproth, renowned for his discovery of the element uranium, published analyses of Roman coins already at the end of the 18th century /1/. This first truly archaeometric work of a chemist is also the first account of a quantitative analysis of any copper alloy. Today, archaeometry is an interdisciplinary field between the natural sciences and the humanities. Broadly speaking, it can be described as the application of scientific methods to questions of cultural history. Of course, there exists a large variety of problems from various disciplines of the humanities, but most of them can be classified into areas such as prospection, dating, material identification, technology, provenance, and conservation. In the first two areas physical methods prevail, while analytical chemistry has yielded important contributions to the remaining areas. In the following I shall concentrate on the role of analytical chemistry for determination of the provenance of archaeological objects with some illustrative examples from the Aegean cultural region. In prehistory the Aegean is regarded as mediator between the early civilizations in the Near East and Central Europe. Therefore, contacts of the Aegean with Egypt or Anatolia on the one side and the Balkans and the Western Mediterranean on the other side are of special importance. Especially speed and direction of technological transfer and the extent of prehistoric trade are major issues of controversy. Archaeologists generally demonstrate cultural contacts and trade routes by typological similarities of artifacts. But unfortunately, many typological comparisons are open to discussion.

As early as in 1842 it was suggested that the chemical composition of an artifact could be used as additional criterion for classification /2/ and spread of cultures could be observed by the geographical distribution of certain materials. This method enabled Montelius /3/ to establish a sequence of principal cultural periods based on the material used for tools and weapons. Today we still use the terms Stone Age, Bronze Age, and Iron Age.

Also already in the last century, it was recognized that trace elements would be better indicators for the provenance of materials /4/. This is the "fingerprint" concept of provenance studies when artifacts are compared with raw materials from various regions. This approach requires that the raw materials are either not altered at all by the production and corrosion processes or in a quantitatively predictable way. It also requires that there are not as many sources of raw material as to make the task an impossible one. Very often, however, early artifacts are small, fragile or fragmentary and thus permission to sample them is often only granted, if the samples are very small. In addition, it is necessary for such a study to analyze a large number of samples for several elements at low concentrations, so the work seemed not feasible until the development of sensitive instrumental methods for multi-element analysis. With the advent of optical emission spectroscopy (OES) for chemical analysis /5/ the number of analyses, especially of metal artifacts rose sharply. Other important groups of artifacts studied in this way, comprise pottery and stone implements. These are also the materials found most frequently in archaeological excavations due to their durability.

2. PROVENANCE OF METALS

2.1. Lead and Silver

These two metals have to be considered together because in antiquity silver derived predominantly from argentiferous lead ores. Therefore, lead and silver were first smelted together from the ores and separated from each other by a second process called cupellation, in which molten lead is oxidized together with other impurities and separated as liquid PbO. It is evident that this greatly changes the chemical fingerprint of the ore source. Only Au and the platinum-group elements remain totally in the silver. Cupellation experiments have shown that Cu, Bi, and perhaps Sb are depleted in a predictable way /6/ and may also be used for provenance studies.

Several methods have been used for the chemical characterization of ancient silver including OES /7/, XRF /8/, activation with thermal /9/ and fast neutrons and with protons /10/. Although some of these studies discuss the provenance of silver, no attempt was made to compare the artifacts with analyses of ores so that the results were not conclusive enough. The problem is that the principal impurities in ancient silver are Cu, Au, and Pb. Since lead ores were mostly used for silver production and Cu was frequently added, only Au is a reliable source indicator. Given the large number of lead ore occurrences, many of which argentiferous, there is a danger that there are several occurrences with similar Au/Ag-ratios.

The pioneering works of Wampler and Brill /11/ and Grögler et al. /12/ have shown that lead isotope ratios could provide an additional "fingerprint" to pinpoint ore sources of lead and silver. The main advantage of this method is that the isotopic composition of lead is not changed by extraction or refining processes applied to an ore to produce a metal, nor by fabrication of that metal into an artifact, nor by any subsequent corrosion of that artifact. The lead isotope method is based on the natural radioactive decay of isotopes of uranium and thorium to isotopes of lead. In lead-ore forming solutions U and Th are in association with Pb. While these elements remain associated, the isotopic composition of the lead changes continuously. During lead ore mineralization Pb is separated from U and Th and the momentary lead isotope composition is "frozen in". Hence, the isotopic composition of lead may be unique and different for a number of mines so that the isotopic composition of lead in a silver artifact may be matched with argentiferous lead ores from ancient mining areas.

But even with two methods for comparing artifacts and ores overlap of the geochemical characteristics can occur. The number of possible ore sources is then further reduced considerably when it is investigated by dating methods if a certain lead ored deposit was actually worked in the period of interest. This requires extensive field studies and cooperation of geologists, archaeologists, geochemists, and physicists since physical methods are often involved. Almost ten years ago, such a collaboration was initiated at the Max-Planck-Institut für Kernphysik in Heidelberg to study the archaeometallurgy of the Aegean. Initially, the investigations concentrated on the provenance of silver, but lead and gold were included at an early stage and now copper has been added, too. The methods employed are NAA, AAS, MS, thermoluminescence dating of pottery and slags, and radiocarbon dating of charcoal.

An important aspect, especially with objects made of noble metals, is the danger of re-use of scrap metal. Thereby it would be unavoidable that metal from various sources is mixed and the compositional information on the provenance would be lost. At least for silver there are two periods, in which this danger seems minimal: The first is the beginning of the Early Bronze Age, around 3000 BC in the eastern Mediterranean when the first silver objects appear and the second is the Archaic period in Greece when in the 6th century BC the first silver coins came into use. The fast acceptance of silver coinage in Aegean trade resulted in a sharp rise in the demand for silver and consequently in an increase of the production within a short period.

In contrast, lead does not seem to have been re-used in pre-classical periods. Moreover, there is evidence that most prehistoric lead has not been desilvered /13/ so that more trace elements can be used for provenancing. The following two examples demonstrate that with such an interdisciplinary approach as outlined above it is possible to relate ancient silver and lead objects to certain ore deposits.

2.1.1. <u>Prehistoric lead and silver in the Aegean</u>. The beginning of the Early Bronze Age marks great changes in the nature of early societies. A very important aspect of change is the use of metals to a large extent for the fabrication of tools and weapons. This presupposes the ability to smelt metals from their ores, which, in turn, undoubtedly was an important force leading to craft specialization and social differentiation. In the Aegean this transition did not develop gradually but is characterized by a sudden increase of prosperity in the first half of the third millenium BC. This is indicated by the appearance of a relatively large number of metal artifacts, more than a third of which is made of silver or lead. The simultaneous appearance of lead and silver is certainly not a mere coincidence but rather an indication of the common smelting of the two metals from argentiferous lead ores. As mentioned above, this necessitates knowledge of the cupellation process. Most archaeologists think that this metallurgical technique originated somewhere in Anatolia, and consequently, early lead and silver finds in the Aegean were regarded as imports.

However, on the Cycladic island of Siphnos Wagner et al. /14/ discovered that a lead-silver mine, which was exploited in the archaic period, was already worked in the Early Bronze Age, i.e. in the same period when objects of lead and silver appear. Therefore, it seemed conceivable that the lead-silver metallurgy was independently developed in the Aegean. Analyses of prehistoric Cycladic lead and silver objects indeed show that a large part of the metals was derived from Siphnos or Laurion, in the south of Attika, but also from the north or northeast of the Aegean. Present evidence suggests that the idea of metallurgy and maybe its technology came from the early flourishing urban centers of Troy, Thermi on Lesbos, and Poliochni on Lemnos but that is was soon adopted and developed in the Early Cycladic culture /15/.

2.1.2. <u>Archaic Greek coinage</u>. The so-called Asyut hoard, which consisted of approximately 900 Archaic Greek silver coins, was unearthed in Egypt in 1969 /16/. The importance of this hoard lies both in its size - it is by far the largest hoard find up to now - and the fact that it contains a great variety of coins from mints in mainland Greece and Asia Minor. Although the exact chronology of the hoard is a matter of dispute /17/, it can safely be concluded that most of the coins date from the beginning of the 5th century BC. Since the introduction of silver currency in Greece is thought to have started only in the first half of the 6th century BC /16/, the danger of mixed silver from different ore sources to any large extent is probably precluded.

Another peculiarity of the Asyut hoard is that practically all of the coins are damaged by chisel strokes. This allows the removal of a small sample of about 10 mg from the inner part of a coin without lowering its numismatic value. Tests have shown /18/ that analysis of the surface can be misleading. The samples were grouped according to their concentrations of Cu, Sb, Au, and Bi using average link cluster analysis /19/. This information combined with the lead isotope data resulted in the identification of three major ore sources: Laurion, Siphnos, and an ore deposit in northern Greece, probably the Kassandra district on the Chalkidiki peninsula (Fig.1). All of these were mentioned by ancient authors like Herodotus, but only Laurion had been unequivocally relocated before this study.

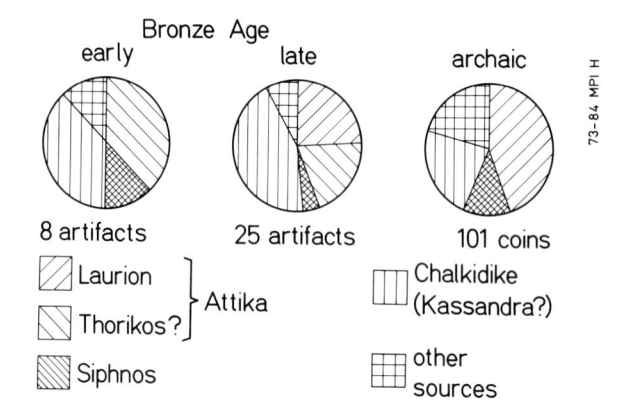

Figure 1: Provenance of ancient silver in the Aegean as deduced from chemical composition and lead isotope ratios.

2.2. Copper and Copper Alloys

Studies on the prehistory of metallurgy have focussed their attention on copper and its alloys, the arsenical and tin bronzes because they are most significant for social change at the transition from the Stone Age to the Metal Age. Therefore, in contrast to lead and silver, a large body of trace element analyses has accumulated, which can already be used for comparison with new analyses of artifacts and of ores. To avoid duplication of work, new analyses need to be compared with published analyses, and particularly with the largest corpus of analyses from Stuttgart, published by Junghans et al. between 1960 and 1974 /20/, who analyzed more than 40 000 European eneolithic and bronze age copper-based artifacts for Fe, Co, Ni, Zn, As, Ag, Sn, Sb, Au, Pb, and Bi. Therefore any new analytical programme should at least assay these 11 elements and Cu, which limits the choice of analytical methods.

In their interpretation of their analyses Junghans et al. /20/ concentrated on the possibilities of identifying compositional groups of artifacts. Their results have raised a controversial discussion on the archaeological significance of their groups and many archaeologists have remained sceptic of the successful application of this method to the provenancing of ancient copper. It now becomes increasingly clear that this scepticism is not justified and that with the use of cluster analysis better sorting of multivariate data can be achieved. However, few attempts have been made to relate copper artifacts to ore sources. The ones reported in the literature /21, 22/ were based on semiquantitative methods. This poses the question of how precise the analysis of copper artifacts should be, given the possible heterogeneity of the artifacts and the ore deposits.

Studies along these lines are now carried out at the MPI Kernphysik in Heidelberg to complement our earlier work on lead and silver in the Aegean. We apply the same approach and the same methods, which have proven successful for the provenancing of lead and silver, to Early Bronze Age copper objects from the Aegean. In a first step we compared the analyses of 47 objects from the Troad, which have been accomplished by the Stuttgart group with OES /23/, with our methods. The result is that - with few exceptions - the data from the two laboratories are comparable within a factor of two /24/. This is in the expected range since the accuracy of the OES method as used in Stuttgart is estimated to be ± 30% for most elements. Since the concentrations range over several orders of magnitude in copper artifacts, this accuracy seems adequate for classification purposes. However, compositional groups become tighter when precise analytical methods are used.

3. POTTERY

Compared with metals determination of pottery provenance is relatively new; first attempts have been reported two decades ago /25/, for two reasons. First, the main constituents of clay vary only slightly and even at the trace level variations are much smaller than in metals. Second, it is necessary to analyze as many elements as possible because, as yet, no group of elements has been singled out, which provides satisfactory group separation. This makes the use of sensitive and accurate multi-element analysis and objective classification methods such as cluster analysis or factor and discriminant analysis mandatory. Of the methods generally used (OES, RFA, NAA), NAA seems to be the best choice because it allows precise determination of up to more than 30 elements in clays /26/.

With the pottery it is usually not important to locate the actual clay beds, from which the raw materials derived. Instead, it is attempted to identify a certain workshop by its characteristic clay composition. This can be achieved by analyzing a range of pottery types from each proven or assumed production center within a geographical region and cultural period. Since the number of workshops can be large, it is evident that the number of analyses needed to establish even the principal groups of pottery in a region like the Aegean exceeds the capacity of a single laboratory. Therefore, it is absolutely necessary that analyses from different laboratories can be compared so that data can be accumulated and eventually a single sherd could be assigned to a certain workshop.

Fortunately, most groups engaged in pottery analysis by NAA use either a clay standard prepared at Berkeley /27/ or rock standards from the U.S. Geological Survey /28/. Both standards have been used to calibrate an in-house clay standard in Heidelberg. The difference between the two calibrations varied between 0.7 and 10% for 19 elements except for Cu, which differs by 34%. However, since already characterized pottery groups show a typical variation of 15 to 20% /29/, the accuracy at present obtainable in pottery analysis seems adequate to deal with many archaeological questions concerning the provenance of pottery. Ultimately, it may be desirable to improve the accuracy to the $\pm$ 5% level, but it certainly seems not necessary that absolute standardizations have to agree down to $\pm$ 1%.

From the Eastern Mediterranean there exist already thousands of pottery analyses, especially of Late Bronze Age pottery /30/. In this period, people of the Mycenaean culture seemingly traded profusely with foreign lands with the result that vast quantities of Mycenaean pottery are found practically all over the Mediterranean. Visual examination reveals a close affinity in style and fabric, implying that all share a common origin. Archaeologists have always been uneasy about accepting such a conclusion and it seems no surprise that the first attempts to use the composition of pottery for discriminating similarly looking wares were made with Late Bronze Age pottery /25/.

In an exploratory study we investigated if Late Bronze Age pottery from the seashore near Troy could be shown to derive from Mycenae. This could provide new evidence for the location of the landing site of the Greek fleet as described in Homer's Trojan War. The results of NAA of several sherds together with the known composition of Mycenaean pottery are given in Table 1. They show that the sherds S15, which were excavated in a Late Bronze Age citadel near the shore, could come from the northeastern Peloponnese, while the remaining samples differ greatly. These sherds were found along the ancient coastal line and would therefore be most indicative for the problem outlined above.

But this example was mainly chosen to demonstrate that many archaeological questions concerning pottery provenance can already be answered with little effort because interlaboratory comparison for pottery analyses is generally good.

Table 1: NAA results of Late Bronze Age pottery sherds from Beşik-Tepe on the coast near Troy. The composition ranges of pottery from Mycenae were taken from /29/. All concentrations are in µg/g, except Na and Fe in percent.

	Mycenae /29/	S15-124	S15-180a	S15-137	Parz.625	OEZ	L17.22
Ba	382-921	426	400	404	1197	608	778
Ce	59.4-67.3	59.2	58.0	59.8	66.2	54.7	58.4
Co	27.5-31.0	24.1	23.7	22.9	39.2	43.1	22.2
Cr	221-286	220	218	218	490	502	174
Cs	5.9-9.2	11.5	11.1	10.7	8.6	11.6	10.3
Eu	1.26-1.40	1.16	1.35	1.25	1.72	1.35	1.28
Fe(%)	4.69-5.52	5.38	5.42	5.25	2.67	4.52	4.09
Hf	3.34-4.27	2.08	1.99	2.25	3.13	2.63	2.53
La	30.8-34.1	32.8	32.0	32.5	37.1	33.0	36.1
Lu	0.37-0.45	0.42	0.42	0.43	0.52	0.53	0.44
Na(%)	-	0.734	0.657	0.738	1.36	1.19	0.99
Rb	115-150	153	148	147	79.0	68.8	114
Sc	19.8-22.7	22.0	21.8	21.5	21.4	16.9	16.2
Sm	-	5.76	5.76	5.69	7.36	5.97	5.69
Ta	-	<0.6	1.13	<0.54	<0.6	1.26	1.17
Th	10.9-12.5	12.4	12.3	12.0	11.9	12.9	15.3
U	-	1.93	1.89	1.89	2.94	3.2	3.57
Yb	-	2.32	2.35	2.36	2.76	2.65	2.11

4. OBSIDIAN

Obsidian is a volcanic glass whose mechanical properties make it suitable for the manufacture of tools. In the Aegean it was a vital raw material through five millenia and large quantities of obsidian can be found at most prehistoric sites. Since there are relatively few sources of the material, it is almost ideal for provenance studies by analytical methods. However, in order to form obsidian, the lava must have a relatively narrow range of compositions. Thus, like with pottery, the main components do not seem to offer a good chance for discrimination of various sources. Nevertheless, Georgiades /31/ was able to distinguish the calcalkaline source of Melos from the alkaline source of Antiparos by wet chemical analysis, but the time and quantity of material required discouraged a more thorough investigation.

Trace element analysis has proven the most certain and general way of characterizing obsidian. OES has been employed to demonstrate the wide extent and early date of prehistoric trade of obsidian in the Aegean /32/. However, an entirely satisfactory separation between all sources of obsidian, which might have been used in the prehistoric Aegean, was not achieved. Among other techniques like fission track analysis /33/ and mass spectrometry /34/ NAA again provided complete discrimination among all relevant sources /35/. Since the authors used element ratios rather than absolute concentrations for differentiation (Fig. 2), their results can be readily compared with precise analyses from other laboratories without any problems of standardization. Also plotted in Fig. 2 are results of NAA on obsidian artifacts from the earliest level of Troy and on natural obsidian from two different sources on Melos. The latter indicate the excellent agreement of our analyses with the published ones and the former document that already at the end of the fourth millenium BC there was contact between Troy and the Cyclades. This is further evidence for early Troy's orientation to the Aegean rather than to central Anatolia.

ACKNOWLEDGEMENT

I thank Professor M. Korfmann (Tübingen) for providing the pottery and obsidian samples from Beşik-Tepe, Professor G.A. Wagner for discussion of the manuscript and W. Bach for technical assistance.

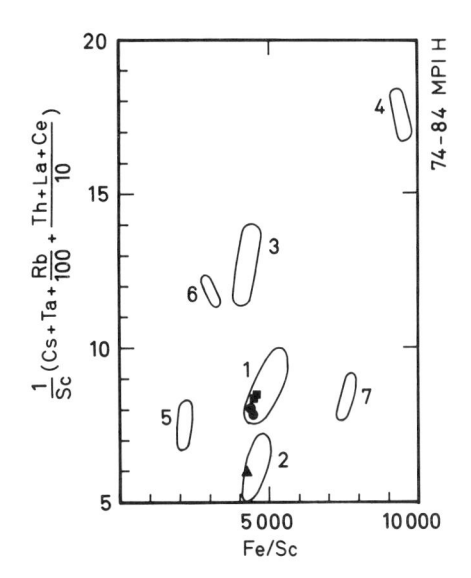

Figure 2: Element concentration patterns for obsidian sources in and near the Aegean; 1, Melos (Adhamas); 2, Melos (Dhemenegaki); 3, Giali; 4, Acigöl; 5, Carpathian; 6, Çiftlik; 7, Antiparos (adapted from /35/). Squares are natural obsidians from Melos (Adhamas), the triangle from Melos (Dhemenegaki) and the dots are two obsidian artifacts from Besiktepe near Troy (Besik-Sivritepe, 4th millenium BC; Besik-Yassitepe, Troy I level).

REFERENCES

/1/ Klaproth, M.H.: Beyträge zur chemischen Kenntniß der Mineralkörper, Bd. 1-6, Stettin 1795-1815.

/2/ Göbel, F.: Über den Einfluß der Chemie auf die Ermittlung der Völker der Vorzeit oder Resultate der chemischen Untersuchung metallischer Altertümer insbesondere der in den Ostseegouvernements vorkommenden, behufs der Ermittlung der Völker, von welchen sie abstammen. Erlangen 1842.

/3/ Montelius, O.: Die älteren Kulturperioden im Orient und in Europa. I. Die Methode. Stockholm 1903.

/4/ Wibel, F.: Die Cultur der Bronzezeit Nord- und Mitteleuropas. 26. Ber. d. Schlesw.-Holst.-Lauenb.Ges. f.d. Sammlung und Erhaltung vaterl. Alterthümer, Kiel 1865.

/5/ Gerlach, W. and Schweitzer, E.: Die chemische Emissionsspektralanalyse, Bd. 1 (1930).

/6/ Pernicka, E. and Bachmann, H.-G.: Archäometallurgische Untersuchungen zur antiken Silbergewinnung in Laurion. Teil III: Das Verhalten einiger Spurenelemente beim Abtreiben des Bleis. Erzmetall 36 (1983) 592-597.

/7/ Allis, E.J. and Wallace, W.P.: Impurities in Euboean monetary silver. Amer.Num. Soc.Mus. Notes 6 (1959) 35-67.

/8/ Klockenkämper, R. and Hasler, K.: Zur Leistungsfähigkeit der energiedispersiven Röntgenspektralanalyse. Multielementanalyse von römischen Silbermünzen. Fresenius Z.Anal.Chem. 289 (1978) 346-352.

/9/ Kraay, C.M. and Emeleus, V.M.: The composition of Greek silver coins. Oxford 1962.

/10/ Meyers, P.: Non-destructive activation analysis of ancient coins using charged particles and fast neutrons. Archaeometry 11 (1969) 67-79.

/11/ Wampler, J.M. and Brill, R.H.: A preliminary investigation of the usefulness of lead isotope analyses in archaeological studies. Trans.Am.Geophys. Union XLV (1964) 109 f.

/12/ Grögler, N., Grünenfelder, M. and Houtermans, F.G.: Isotopenuntersuchungen zur Bestimmung der Herkunft römischer Bleirohre und Bleibarren. Z. für Naturforschung XXIa (1966) 1167-72.

/13/ Wagner, G.A. and Pernicka, E.: Blei und Silber im Altertum: ein Beitrag der Archäometrie. Chemie in unserer Zeit 16 (1982) 46-56.

/14/ Wagner, G.A., Gentner, W. and Gropengiesser, H.: Evidence for third millenium lead-silver mining on Siphnos island (Cyclades). Naturwissenschaften 66 (1979) 157.

/15/ Pernicka, E. and Wagner, G.A.: Die metallurgische Bedeutung von Siphnos im Altertum. In: Silber, Blei und Gold auf Siphnos: Prähistorische und antike Metallproduktion. Wagner, G.A. and Weisgerber, G. (Ed.). Der Anschnitt, Beiheft 2 (1984) in press.

/16/ Price, M. and Waggoner, N.: Archaic silver coinage. The Asyut hoard. London 1975.

/17/ Cahn, H.A.: Asiut - Kritische Bemerkungen zu einer Schatzfundpublikation. Schweiz.Num.Rundsch. LVI (1977) 279-87.

/18/ Schubiger, P.A., Müller, O. and Gentner, W.: Neutron activation analysis on ancient Greek silver coins and related materials. J.Radioanal.Chem. 39 (1977) 99-112.

/19/ Doran, J.E. and Hodson, F.R.: Mathematics and computers in archaeology. Edinburgh 1975.

/20/ Junghans, S., Sangmeister, E. and Schröder, M.: Kupfer und Bronze in der Frühen Metallzeit Europas. Berlin 1968 und 1974.

/21/ Pittioni, R.: Zweck und Ziel spektralanalytischer Untersuchungen für die Urgeschichte des Kupferbergwesens. Arch. Austriaca 26 (1959) 67-95.

/22/ Berthoud, T.: Etude par l'analyse de traces et la modelisation de la filiation entre minerai de cuivre et objets archéologiques du Moyen-Orient. Ph.D. Université Pierre et Marie Curie, Paris 1979.

/23/ Esin, U.: Kuantitatif spektral analiz yardimiyla Anadolu'da başlangicindan asur kolonileri çağina kadar bakir ve tunç madenciligi. Istanbul 1969.

/24/ Pernicka, E.: Instrumentelle Multi-Elementanalyse archäologischer Kupfer- und Bronzeartefakte: Ein Methodenvergleich. Ber.Röm.-Germ.Komm. (im Druck)

/25/ Catling, H.W., Blin-Stoyle, A.E. and Richards, E.E.: Spectrographic analysis of Mycenaean and Minoan pottery. Archaeometry 4 (1961) 31-38.

/26/ Hancock, R.G.V.: Low flux multielement instrumental neutron activation analysis in archaeometry. Anal.Chem. 48 (1976) 1443-1445.

/27/ Perlman, I. and Asaro, F.: Pottery analysis by neutron activation. Archaeometry 11 (1969) 21-52.

/28/ Flanagan, F.J.: Descriptions and analyses of eight new USGS rock standards. Geol.Surv.Prof.Pap. 840, Washington, D.C. (1976).

/29/ Harbottle, G.: Provenience studies using neutron activation analysis: The role of standardization. In: Archaeological Ceramics. Olin, J.S. and Franklin, A.D. (ed.), 67-78, Washington, D.C. (1982).

/30/ Bieber, A.M., Jr., Brooks, D.W., Harbottle, G. and Sayre, E.V.: Application of multivariate techniques to analytical data on Aegean ceramics. Archaeometry 18 (1976) 59-74.

/31/ Georgiades, A.N.: Praktika tis Akademias Athinon 31, 150 f (1962).

/32/ Renfrew, C., Cann, J.R. and Dixon, J.E.: Obsidian in the Aegean. Ann. British School Arch. Athens 60 (1965) 225-247.

/33/ Durrani, S.A., Khan, H.A., Taj, M. and Renfrew, C.: Obsidian source identification by fission track analysis. Nature 233 (1971) 242-245.

/34/ Gale, N.: Mediterranean obsidian source characterization by strontium isotope analysis. Archaeometry 23 (1981) 41-52.

/35/ Aspinall, A., Feather, S.W. and Renfrew, C.: Neutron activation analysis of Aegean obsidians. Nature 237 (1972) 333-334.

INSTRUMENTELLE ELEMENT-ANALYSE - EIN WERKZEUG BEI DER
UNTERSUCHUNG VON KUNSTWERKEN.

Franz Mairinger

Institut für Farbenchemie, Akademie der Bildenden Künste Wien

SUMMARY

Für die Erforschung und Erhaltung von Kunstwerken sind künstlerische, geisteswissenschaftliche, soziologische, handwerkliche und, da jedes Kunstwerk neben diesen Dimensionen auch eine materielle Basis besitzt, naturwissenschaftliche Aspekte von Bedeutung. Die Zusammenhänge all dieser Faktoren soll an Hand eines Flußdiagramms deutlich gemacht werden.

Daraus wird ersichtlich, daß mit naturwissenschaftlichen Methoden primär Aussagen über den materiellen Bestand und über die zeitlichen Veränderungen gewonnen werden können.

Die Erforschung des materiellen Bestandes liefert Informationen über den Materialstil und die kunsttechnologischen Eigenheiten einer Epoche, einer Kunstlandschaft oder einer Werkstätte. Solche Ergebnisse helfen dem Kunstwissenschaftler den historischen Ansatz, der aus der Stilkritik gewonnen wurde, zu sichern oder in manchen Fällen sogar zu verifizieren.

Die Erfassung der zeitlichen Veränderungen, die ein Kunstwerk erlitten hat, wie natürliche Alterung, Verfälschungen oder mechanische Beschädigungen gibt Auskunft über den Erhaltungszustand und damit über die Echtheit. In weiterer Folge sind diese Ergebnisse für die Erstellung eines Restaurierungskonzeptes von Bedeutung.

Für eine systematische Erforschung des materiellen Aufbaus bieten sich zwei zueinander komplementäre Untersuchungstechniken an: Flächenuntersuchungen mit sichtbarer und unsichtbarer Strahlung (IR, UV, X, γ, n, e-) und Punktuntersuchungen. Die Ergebnisse der, im strengen Sinn, zerstörungsfreien Flächenuntersuchungen sind für stilkritische Ansätze vielfach unentbehrlich geworden, liefern aber auch die notwendigen Informationen für eine allfällige problemorientierten Probenentnahme, der kritischste Faktor bei den Punktuntersuchungen. Da die Probenmenge aus verständlichen Gründen in jedem Fall möglichst gering gehalten werden muß oder gar zu unterbleiben hat, sind die Methoden der Instrumentellen Element- bzw. Multielement-Analyse für viele dieser analytischen Probleme optimal geeignet. Dies gilt auch für die Erfassung der Spurenelemente oder der Isotopenverteilung, deren Erfassung Rückschlüsse auf die Lagerstätten mineralischer Pigmente zulassen. Für diese Problemkreise haben sich unter anderem XFR, NAA, PIXE, OES, ICP-AES, LMA und MS sehr bewährt.

Probleme des zweiten Fragenkomplexes, die die zeitlichen Veränderungen von Kunstwerken betreffen, sind analytisch meist wesentlicher schwieriger und aufwendiger zu lösen. Lange Reaktionszeiten von oft vielen hundert Jahren, komplizierte Substanzgemische und verwickelte Reaktionsmechanismen, unübersichtliche Oberflächenverhältnisse bei umweltinduzierten Schäden sorgen für veritable analytische Aufgabestellungen. Hiefür sind, neben den bereits erwähnten analytischen Verfahren, vor allem Methoden wie SEM, EPMA, IPMA und SIMS erfolgreich eingesetzt worden. Die dabei gewonnenen Ergebnisse haben heute bereits für praktische restauratorische Maßnahmen Bedeutung.

Die Aussagefähigkeit solcher Untersuchungen wird nun an Hand von Beispielen aus der Tafelmalerei, Wandmalerei, mittelalterlichen Glasmalerei, sowie der Buchmalerei illustriert.

LITERATUR

F. Mairinger, Untersuchungen von Kunstwerken mit sichtbaren und unsichtbaren Strahlen, Bildhefte der Akademie der Bildenden Künste in Wien, Doppelheft 8/9, Wien, 1977; 96 S.

MULTI-ELEMENT-ANALYSIS IN ARCHAEOMETRY

Josef Riederer

Rathgen-Forschungslabor, Berlin

SUMMARY
Chemical analysis has proved to be of essential importance for the solution of historical problems in the field of the history of art, archaeology and ethnology. The precise defination of the composition of materials of works of art and from that the determination of time and place of their origin, their technique of manufacture their authenticity or methods for their preservation are the main problems where science can provide basic information to history.
The main problems of the analysis of works of art and archaeology are the necessity to work only with very small samples or to use non destructible techniques, further the examination of large series of objects, in order to provide statistically reliable data. Peculiar problems are the analysis of depth profiles, if alterations on the surface have to be detected or scanning analysis, if heterogeneous object are examined.
Modern analytical techniques of multielement analysis have already provide us with essential data on the materials used in earlier periods. Energy-dispersive X-ray fluorescence technique is used as a routine technique on specialized archaeometry laboratories for rapid qualitative informations as well as for quantitative analysis of metals, ceramics, stone and glass. Microprobe techniques are available for small samples. Activation analysis is used first of all for the characterisation of materials by their trace elements. Important results have been obtained from the analysis of ceramics; stone, obsidian, but also for organic materials like paper and bone. During the last years ICP-analysis, mounic X-ray techniques, PIXE and related ion beam analysis have been applied sucessfully for historical problems.

Since some 20 years the term archaeometry is used for the application of science in the field of cultural history. The analysis of materials plays an important role in the description of artefacts since the beginning of analytical chemistry at the end of the 18th century, when Martin Heinrich Klaproth analyzed series of historical objects, made from various materials. He could show that analytical data are not only useful for a precise definition of the material, but also for conclusion on age or proveniance of these objects. Though in the 19th and at the beginning of the 20th century analytical research on art objects provided essential results, not earlier than about 20 years ago archaeometry became a really important instrument for research in art and archaeology. This development is due to the improvement of analytical techniques, which permit data of a high accuracy on a great number of chemical elements on large series of objects without the necessity to take large samples.

By that the main problems of archaeometry are already mentioned. First it is necessary to analyzed large series of objects to provide a reliable fundament on the variation of compositions of materials used for historical objects. For instance, to characterize bronzes used in the bronze age in central Europe, about 22.000 analysis have been done by emission spectroscopy. Second it is necessary to do the analysis on a very small sample or even by nondestructive techniques. This point explains the increased activity of the last 20 years, far earlier too large samples were needed for analysis. Today multielement analysis by various spectrographic techniques, activation analysis or X-ray fluorescence analysis can be done with only minute interventions in to the object's substance. Third a high accuracy is demanded from analysis in archaeometry since it is one of the aims, to distinguish objects of different age and origin by differences of the concentrations of trace elements and these difference usually are very small. Finally the determination of a great number of elements is necessary, for statistical techniques based on a large number of elements permit a precise grouping of objects of different origin. The analysis of pottery has received important support from activation analysis, which provides results on about 30 elements, from which the rare earth elements are of a peculiar importance.

For archaeometry, which is established as well on museum laboratories as on scientific research institutes the whole range of analytical techniques is applied for the analysis of art objects. The smaller laboratories are equipped with instruments for emission spectroscopy and atomic absorption spectrography. ICP techniques gradually come into use. X-ray fluorescence analysis, as well wavelength and energydispersive systems are in use on most of the larger laboratories. Activation analysis, ion beam analysis, as well as gammaspectroscopy have gained increased importance during the last years.

Most of the research on the composition of historical objects actuelly is done on metals, especially gold, silver and copper alloys, on ceramics and glass, while stone, pigments from paintings, and especially organic materials with inorganic compounds like paper or bone have not been studied extensively, though already existing results have provided convincing results.

The aims of archaeometry are first to obtain precise informations on the composition art objects, for until now in many cases erraneous terms for materials are used in museums. Just to cite one example. Metal objects of the roman period or the middleage bronzes, though the amount of copper varies between 60 and 100 %, of tin between 0 and 25 %, of lead between 0 and 30 % and of zinc also between 0 and 30 %. About twenty groups of materials have to be formed to describe the type of alloy precisely. The main reason why precise data on the composition of art objects are needed is the possibility to deduce from analytical data place and date of origin.

Multielement analysis plays an important role, if the proveniance of an art object shall be found out from material properties. The basic idea of this attempt is the experience that on a certain place of manufacture either materials are use which have properties differing from those of other places or techniques of manufacture are applied which again are characteristic for the place of origin.

The first possibility, to find a relation between the composition of an art object and local properties of materials, frequently can be relized when ceramics are analyzed, for the composition of clays varies considerably from place to place because of the heterogenity of the weathering rock material, from which the clay derives. In the case of pottery analysis multielement analysis has proved to be of great importance, for activation analysis permits the determination of about 25 trace elements (As, Ba, Ce, Co, Cr, Cu, Dy, Eu, Hf, Ho, La, Lu, Ni, Nd, Sb, Sc, Sm, Ta, Tb, Th, U, Yb, Zn, Zr), which, when grouped by cluster analysis, usually permit precise differentiations of objects from different origins.

This could be proved for instance on a big number of archaeological objects, like jar handles from Greece, which could be localized on different islands, or with roman terra sigillata, where each center of production can be determined by a certain combination of trace elements. Multielement analysis of the main components (Si, Al, Fe, Mg, Ca, Na, K, Ti, P, Mn) by means of X-ray fluorescence analysis provides further criteria for the characterization of ceramic materials.

The same principle for localizing works of art is possible with metals. It was the aim of the analysis of the 22.000 bronze objects from the bronze age in central Europe to find a link between trace element concentrations and provenance, relying on the idea that local copper ores, which differ considerably in composition due to their origin, were used for making bronzes. And in fact material groups, based on the concentration of As, Sb, Bi, Ni and Ag could be established, when data, obtained by emission spectroscopy were compared. The same was possible with the analysis of gold, where the concentrations of Pt and Sn proved to be charateristic for the provenance of the gold.

From a younger historical period, the end of the middle age and the Renaissance, objects of copper alloys were analyzed in large series and it was found that the lead/silver ratio depends from the place of origin. This observation could be explained first by different concentrations of these elements in the copper ores, but the relatively high concentrations of lead in objects from Nuremberg and the relatively high concentrations of silver in products from Austria could be explained by a technical process. In Nuremberg lead was added to the molten copper to remove the silver by certain technical process, by which a certain portion of the lead remained in the copper. In Austria in earlier periods this process was not known, and by that no lead was added and no silver was removed from the copper. Those relations between the composition of an art object and the technique of manufacture are frequent in historical periods, when the composition of alloys was prescribed by guilds. The place of manufacture of bronze mostars, which cannot be localized because of their stilistic features, can be determined from their composition which was different from one town to the other.

Some important work was done to determine the place of manufacture of pigments used in painting from chemical analysis. Emission spectroscopy and later activation analysis provided interesting data first of all on some pigments. For lead white, the most important historical white pigment, it would be shown that by means of trace elements products from Italy can be distinguished from products from the Netherlands and these too are different again from lead white, manufactured in the 19th century.

Isotope analysis too contributes much to the localization of materials from artefacts. Sulfur isotopes of pigments (cinnabar, ultramarine, cadmium yellow) have been sucessfully studied with the aim to distinguish natural and artificial products, to find differences in the technique of manufacture and to determine the deposits from natural pigments. Another important example of the determination of proveniance from isotope analysis is the localization of marble by means of the determination of $\delta^{13}C$ and $\delta^{18}O$ values. $\delta^{13}C$ varies from -1 to +7, $\delta^{18}O$ from 0 to -10, which is a range, large enough to separate groups of objects of the same origin. Combined with chemical analysis and microscopy a reliable localization of the marble is possible.

From organic materials there are only a few examples of multielement analysis of inorganic compounds. Some convincing results have been obtained on the determination of trace elements in paper. Here too the place of origin can be determined because of differences of the manufacturing process. To early dutch paper blue cobalt pigments were added to increase the whiteness of the paper. The determination of cobalt together with a series of other elements geochemically related with cobalt permits to distinguish this dutch paper from papers of the same period from other countries, which fraudulently were supplied with dutch watermarks to pretend a better quality of the paper.

Like with the deduction of the place of origin from analytical data the time of manufacture can be derived from analysis. Apart from several techniques of absolute dating where methods of chemical multielement analysis are applied as a part of the dating procedure, the age of an art object sometimes can be derived fromits composition. Again the brass objects from the Renaissance period made in Nuremberg provide a striking example of those possibilities. In the period from 1520 to the middle of the 17th century the concentration of six elements shows a characteristic variation in the course of time. While tin and lead increase gradually over these 150 years the concentrations of iron, nickel, arsenic and antimony increase and decrease in shorter periods, depending from the type of ore used at a certain period at Nuremberg. Atomic absorption spectroscopy and a evaluation of the data by cluster analysis provides reliable data on the date of manufacture.

This observation that the composition of copper alloys varies in a certain direction in the course of time because of the use of ores from different mines or because of the improvement of metallurgical techniques can be made with a great number of other groups of historical bronzes, like with chinese mirrors, which can be dated already from variation of their main elements copper, tin and lead.

Also with coins a continuous variation of the composition can be observed, which usually can be explained as a debasement of their value. In gold coins the amount of silver and in silver coins the amount of copper increases. The most convincing example of a coin debasement can be demonstrated by the analysis of sesterzes from the roman empire. During the reign of Augustus (27 BC - 14 AD) these coins were of pure brass, which was invented at that time and at high esteem because of its golden colour. During the following 250 years the amount of zinc decreases continuously from 26 % to 0 %, while on the other hand the less valuable metals tin and lead increased, reducing the value of the coin. Those examples on a certain historical, economical or social situation during a certain period are possible with already all materials.

From the analysis of silver fibulae from the Goths it became obvious that in many cases the alloy was not always very rich in silver. Frequently these fibulae contained up to 60 % of copper and 15 % of zinc. The historical explanation which was found in the technical literature of the 10th century. was not a fraudulent mixing of the silver with a less noble metal, but the way how the silversmith got his metal. The silver had to be provided from the customer in the form of coins and since some customers had not enough silver they gave brass coins for the rest.

An interesting relation between the use of a certain material and the political situation was found, when wall paintings from Central Asia were analyzed. The pigments came either from the west, like natural ultramarine, a green copper chloride and dark red ochres or from the east, like azurite, malachite and cinnabar and it depended only from the more or less peaceful relations between Central Asia and its neighbours from which side the pigments were provided. Those conclusions on the intensity of the trade between different regions frequently are possible by considering the materials of their cultural objects.

Recently multielement analysis of human bone from earlier civilizations has provided important informations on the life of early men. By the analysis of trace elements, especially Sr, Zn, Mg, Na, Cu, Fe, Al, Mn and K it was possible to distinguish population which got their food by hunting or by agriculture. Differences in the trace element concentrations of men and women, elder persons and children could be explained by differences in their diet. Even bones from rich and poor people could be distinguished from differences in the concentrations of trace elements. Of peculiar interest for historians are the concentrations of toxic elements in bone, like lead, mercury or arsenic, which also permit conclusions on the way of life of earlier populations.

Finally one point has to be mentioned, where multielement analysis is of peculiar importance for museums, that is the detection of forgeries of works of art. Due to a notable shortage of antiquities on the art market, forgeries of a high technological and stilistical perfection are produced and analytical procedures can provide essential arguments on their identification.

As shown above, there exist already extensive informations on the properties of materials used in earlier periods for works of art. Even if it is possible for a forger to get informations on the precise types of materials used originally, he will have no chance to provide his imitations with trace element concentrations similar to these of originals. And here again the chance of multielement analysis in archaeometry becomes obvious: it gives us so many detailed informations on the properties of materials which scarcely can be imitated. If by activation analysis from terra sigillata about 25 trace elements are determined or early silver objects can be characterized by about 15 elements, than a forger is confronted with those cifficulties that he concentrates his work on the deception of the eye and in fact, forgeries where it was tried to imitate the material are pretty scarce.

Methodenvergleich

KOMBINIERTE ANWENDUNG SPEKTROCHEMISCHER ANALYSENMETHODEN BEI DER
MULTIELEMENTANALYSE GEOLOGISCHEN MATERIALS AN GROSZSERIEN

E. Schroll und D. Sauer

Bundesversuchs- und Forschungsanstalt Arsenal Wien
Geotechnisches Institut   POB 8, A-1031 Wien

SUMMARY:

   A combination of two multi-element methods, ICP-OES (simultan
quantometry) and WD-XRF (sequential spectrometry) is used as basic
instrumentation for the analysis of geological materials, especially
of stream sediments, but also of soils and of rocks. The sample capacity is 100 up to 150 samples per day. Twenty elements can be analysed
using both methods controling the accuracy. The quantometric analysis
is supplimented by arc-OES, AAS-methods, ED-XRF and other special
methods. The analytical program covers more than 50 elements. Great
importance is attached to automatization and electronic data processing.

1. EINFÜHRUNG

1.1 MULTIELEMENTANALYTIK IN DER GEOCHEMIE

   In den Anfängen der Geochemie dominierte entsprechend der Entwicklung der anorganisch-analytischen Chemie die Monoelementmethodik. Bei
der Anwendung instrumenteller Methoden, wie die der optischen Emissionsspektralanalyse, mußte sich quantitative Analyse auf ein oder
einige Elemente beschränken, um maximale Nachweisgrenzen oder Wiederholbarkeiten zu erreichen. Überdies wäre vor der Einführung der modernen EDV-Techniken die Auswertung eines umfangreichen Multielement-
datenmaterials mit den damaligen rechnerischen Möglichkeiten nicht zu
bewältigen gewesen.

Diese Umstände führten auch zu dem Zustand, daß man über die Interelementbeziehung, abgesehen von geochemisch-kohärenten Elementpaaren oder -gruppen relativ schlecht informiert war oder auch noch ist (8, 11). Erst die Anwendung der Multielementanalytik mit der gemeinsamen Erfassung der Haupt-, Neben- und Spurenbestandteile an jeder untersuchten Mineral- oder Gesteinsprobe schafft die Voraussetzung für die geochemische Charakterisierung. Die chemischen Daten können durch isotopengeochemische Meßwerte ergänzt werden, sodaß auch chemisch ident erscheinende geologische Materialien noch differenziert werden (9).

Die Entwicklung der Multielementanalytik ist daher in allen Anwendungsbereichen der analytischen Geochemie, ob für die Durchschnittsprobe oder für die in-situ-Analyse, von außerordentlicher Bedeutung. Die Multielementanalytik bildet die Grundlage einer besseren genetischen Klassifikation von Gesteinen aller Art einschließlich der für die Rohstoffwirtschaft interessanten Mineral- und Erzlagerstätten. Über die Erforschung wissenschaftlicher Grundlagen hinaus ergeben sich für die Rohstofforschung auch praktische Konsequenzen. Richtige genetische Modellvorstellungen sind für Exploration und Bewertung mineralischer Rohstoffvorkommen unentbehrlich.

Bei der Suche nach mineralischen Rohstoffen hat sich die Anwendung der analytischen Geochemie in Kombination mit geophysikalischen Meßverfahren mit Erfolg bewährt. Die geochemische Prospektionstechnik stützte sich anfangs auf halbquantitative Feldmethoden und emissionsspektrographische Multielementmethoden, dann auf die Absorptionsspektralanalyse, die jetzt durch verschiedene automatisierte instrumentelle Multielementanalysenverfahren ersetzt werden kann. Entscheidend ist der Durchbruch zur Großserienanalytik, mit der anstelle relativer Konzentrationswerte mit weitgehender Annäherung die wahre Elementverteilung erfaßt werden kann.

Dies ist auch die Voraussetzung für die Schaffung international gültigen geochemischen Kartenmaterials der Elementverteilung in Gesteinen, jungen Sedimenten und Böden. Die geochemische Aufnahme ist nicht nur für die Rohstofforschung erforderlich, sondern auch für die Bewältigung der Umweltprobleme, zur Kenntnis des natürlichen geochemischen Untergrundes und zur Beweissicherung.

1.2 SPEZIELLE AUFGABENSTELLUNG

Der Bedarf an Analysen geochemischen Materials für Programme der Rohstoffprospektion und Umweltforschung hat in Österreich zur Schaffung eines zentralen Labors geführt, das die Durchführung von Multielementanalysen in Großserien ermöglicht. Die analytischen Arbeiten wurden zunächst für eine geochemische Basisaufnahme Österreichs aufgenommen, in dessen Rahmen etwa 33.000 Proben von Bach- und Flußsedimenten (stream sediment, d. h. bewegtes Sediment) auf 34 Elemente untersucht worden sind. Als weiteres geologisches Material sind ferner Bodenproben, Gesteinsproben und Schwermineralfraktionen für die geochemische Prospektion und Umweltforschung von vorrangigem Interesse. Im wesentlichen handelt es sich um silikatisches Material. Karbonatgesteine können ebenso gut in das Analysenprogramm einbezogen werden, wenn auch Abänderungen der Methodik zu beachten sind. Unter geochemischer Analyse wird die einfache Bestimmung, etwa nach der Art der technischen Analyse, verstanden, wobei besonderer Wert nur auf gute Wiederholbarkeit gelegt ist. Eine relative Standardabweichung (s) von $\pm$ 10 bis $\pm$ 15 % gilt für Einzelelementanalysen in Konzentrationsbereichen von 10 bis 10000 ppm in Großserien als ausreichend (2). Die Schaffung eines international vergleichbaren Datenmaterials verlangt zusätzlich Genauigkeit, so daß versucht werden kann, die geochemischen Gesetzmäßigkeiten der Elementverteilung auch interkorrelativ zu erfassen. Die Multielementanalytik verlangt außerdem die Entwicklung einer anderen Methodik der Qualitätskontrolle als die Einzelelementanalytik (2).

2. RAHMENBEDINGUNG DER AUFGABENSTELLUNG

Für Planung und Durchführung der analytischen Aufgabenstellung sollten folgende Bedingungen gelten:

1) Probe:
   a) Beschränkung der verfügbaren und dokumentierten Probenmengen auf 50 g.
   b) Möglichst geringe, aber noch repräsentative Subprobenmengen.
2) Analysenmethoden:
   a) Wenig aufwendige Probenvorbereitung.
   b) Möglichst große Bandbreite der Probenmatrix.
   c) Möglichst viele Analysenelemente in simultaner oder sequentieller Multielementanalyse.
   d) Ausreichende Nachweisgrenzen und Konzentrationsbereiche in Bezug zur geochemischen Häufigkeit.
   e) Ausreichende Wiederholbarkeit und Richtigkeit.
   f) Kontrollierbarkeit der Serienanalytik.
   g) Große Analysenkapazität (zumindest 100 Proben pro Tag).
   h) Weitgehende Automatisierung des Analysenvorganges und Auswertung der Analysenresultate.
3) Kosten:
   a) Geringer Personalaufwand.
   b) Möglichst geringer Sachaufwand für Probenvorbereitung und Analyse (Energie, Material, Geräteamortisation).

Die Entwicklung einer geeigneten Analysenmethode muß daher in erster Linie als Optimierungsaufgabe angesehen werden.

3. PROBENMATERIAL

Bach- und Flußsedimente setzen sich aus drei Komponenten zusammen: Gesteinsdetritus, akzessorischen Mineralien (Schwermineralfraktion) und Verwitterungsmaterial (Tonfraktion des Bodens). Konventionell wird die Siebfraktion unter 80 mesh (d.s. 0,02 cm) analysiert, in der die Schwermineralfraktion überproportional (bis fünffach) angereichert werden kann. Die Probennahme selbst ist bei höheren Gehalten an Schwermineralen für vorwiegend monomineralisch gebundene Elemente, wie z.B. Zirkonium im Zirkon, und bei positiver geochemischer Anomalie mit einem größeren Fehler behaftet. Die Konzentrationswerte unterliegen für den Probenahmepunkt einer zeitlichen Varianz, z.B. durch Einwirkung von Hochwässern, die vor allem die Zusammensetzung der Tonfraktion für manche Elemente oder auch anormale Elementkonzentrationen signifikant verändern können, wie dies in Abb. 1 gezeigt wird.

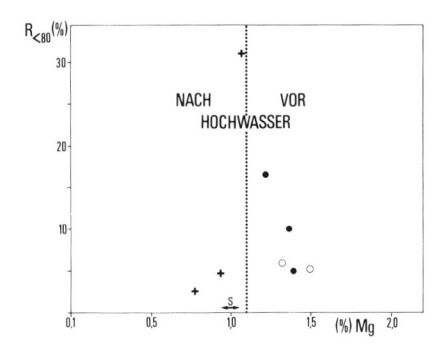

Abb. 1: Beispiel für die Varianz einer Bach- und Flußsedimentprobe, Mehrfachprobung zu verschiedenen Zeiten vor und nach einem Hochwasser.
Lokalität: Etwa 50 m² des Flußbettes der Feistritz, bei einer Brücke 3 km flußabwärts vom Dorf Feistritz am Wechsel/Niederösterreich.
Zeit: 20. und 22. Juni, 02. Juli 1983
Verhältnis Feinanteil/Gesamtprobe unter 80 mesh in R% zur Elementkonzentration Mg. Ähnlich verhalten sich die Hauptelemente Si, Al, Fe, Ca und Spurenelemente, wie Ba, P, Y, Zn, Zr. Jede Probe Mittelwert aus 6 Analysen.

Die Varianz der Probenahme, vor allem in zeitlicher Hinsicht, kann erheblich größer sein als die Varianz der Analysenwerte. Reproduzierbar sind jedoch die lithogeochemischen Beziehungen. Durch statistische Zusammenfassung von mehreren Probenahmepunkten ergibt sich für einen Flächenbereich der geochemischen Karte die hinreichende Aussagewahrscheinlichkeit.

Die chemische Zusammensetzung der Bach- und Flußsedimente ergab für
die Hauptbestandteile folgende Matrixbandbreiten:

Si	1	- 30 %	Al	0,7	- 15 %	Fe	0,01	- 20 %
Mg	0,02	- 21 %	Ca	0,1	- 20 (40) %	K	0,2	- 8 %
Na	0,07	- 5 %	Ti	0,02	- 7,5 %	Mn	0,01	- 5 %

Die Matrix variiert im kristallinen Grundgebirge meist zwischen der chemischen Zusammensetzung saurer und basischer Gesteine. Beiträge von ultrabasitischen Gesteinen sind als lithogeochemische Anomalien zu betrachten. Auch hohe Anteile von Detritus karbonatischer Gesteine sind möglich.

4. PROBENVORBEREITUNG

Für die Probenvorbereitung ist die Herstellung einer genügend homogenisierten, möglichst kontaminationsfreien Analysenprobe geeigneter Korngrößenzusammensetzung erforderlich. Sie ist beim Durchsatz großer Probenmengen, vor allem bei Gesteinsmaterial, das vorher gebrochen werden muß, der arbeitstechnische Engpaß. Das Gewicht liegt bei sorgfältiger manueller Tätigkeit. Die Güte der Aufmahlung entscheidet über die Qualität der Analyse. Für den benötigten Probendurchsatz, der Siebung, Verjüngung und Feinstmahlung verlangt, war die Verdopplung der Einrichtungen nötig. Für die Feinstmahlung haben sich Schwingmühlen mit Mahlgefäßen aus dem relativ teuren Sinterkorund bewährt. Bei einer Mahlzeit von 4 min wird für eine Probenmenge von 50 g eine mittlere Korngröße von 10 bis 25 mμ erreicht, wobei Kornanteile 100 mμ nicht überschritten werden.

5. ANALYSENMETHODIK

Die Multielementanalytik benötigt eine rasche, interne Eigenkontrolle, die durch Parallelanalytik erfüllt werden kann, um Fehler aller Art, von der Probenverwechslung bis zu instrumentellen Störungen, rasch erkennen können. Externe Austauschanalysen sind meist zeitaufwendig und nicht immer zielführend. Als Grundlage der Multielementanalytik wurden zwei komplimentäre instrumentelle Methoden: ICP-OES und WD-XRF (Kurzbezeichnung XRF) gewählt, mit deren Hilfe ein Grundstock an Elementen doppelanalytisch erfaßt und die Zahl der Analysenelemente erhöht werden kann (10).

Für die Analyse mit Hilfe des optischen Spektrums wurde ein Simultanquantometer (49 Kanäle, gleichzeitige Messung von etwa 30 Kanälen) mit einem zusätzlichen beweglichen Spalt ausgewählt. Für die Röntgenspektralanalyse dient ein sequentielles Gerät. Aus Gründen des hohen

Probendurchsatzes liegt die Grenze der Nutzung bei etwa 30 Elementen. Das ICP-Quantometer mit automatisiertem Probenwechsler schafft 100 Proben in einer Meßzeit von 3 1/2 Stunden. Eine automatisierte Abschaltung ermöglicht die Analyse über den achtstündigen Arbeitstag hinaus. Das XRF-Quantometer mit einem 300-Probenwechsler ist rund um die Uhr - auch am Wochenende - im Einsatz, sodaß auch bei Messung von 30 Elementen der Tagesdurchschnitt über 100 Proben liegt.

Die beiden Multielementmethoden sind gesondert vorgestellt worden (1, 4, 5). Die Kombination ICP-OES und XRF hat den Vorteil, zwei grundsätzlich verschiedene Methoden zu verbinden: Lösungs- und Feststoffanalyse, sowie unterschiedliche Spektren. Bis zu einem gewissen Grad können die jeweiligen Nachteile der beiden Methoden ausgeglichen werden.

Der vorteilhafte saure Lösungsaufschluß ($HF + HClO_4 + HNO_3$ 1:1:1) eliminiert aber fluorophile Elemente, wie Si, B, As, und erweist sich für weitgehend monomineralisch gebundene Elemente, wie Zr, Sn oder Cr (bei höheren Gehalten), wegen der Schwerlöslichkeit von Mineralen (Zirkon, Kassiderit, Spinelle) als nachteilig. Beim Säureaufschluß ist auf die Elemente K und Al zu achten, die zu niedrigen Werten tendieren. Die Qualität der Röntgenspektralanalyse ist bei monomineralischer Elementbindung vor allem an den Aufmahlungsgrad der Feststoffprobe gebunden.

Die Subprobenmenge für die instrumentellen Methoden, ICP-OES 0,5 g und XRF 1 - 2 g, ist gering, um Probe- und Standardmaterial zu sparen. Für einige Elemente, bei denen die $L\alpha$-Linie gemessen wird, kann bei der XRF-Analyse eine zu dünne Probenschicht zu fehlerhaften Werten führen.

5.1 ELEMENTAUSWAHL

Die mit beiden instrumentellen Multielementmethoden erfaßbaren Elemente sind in Abb. 2 dargestellt.

5.2 NACHWEISGRENZEN

Für die geochemische Analyse soll die Nachweisgrenze eine halbe bis ganze Konzentrationsdekade unter dem niedrigsten Durchschnittswert (Clark-Wert) der untersuchten Gesteinsart liegen. Bei großer Matrixbreite kann von einer Nachweisgrenze im üblichen Sinn nicht mehr gesprochen werden, sondern eher von einer durchschnittlichen Nachweisgrenze oder "Bestimmungsgrenze". Eine hinreichende Wiederholbarkeit wird erst in einem Konzentrationsbereich erreicht, der um den Faktor Zwei bis Zehn über der Nachweisgrenze anzusetzen ist.

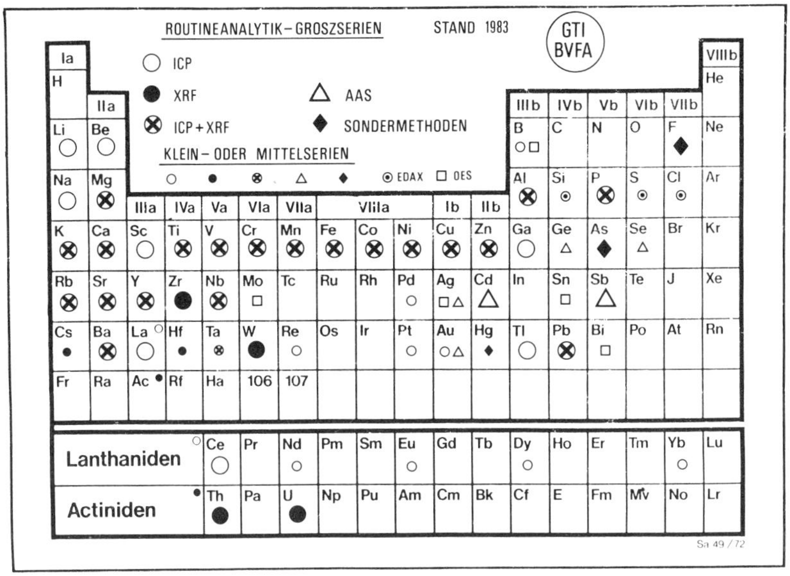

Abb. 2: Elementkombination der instrumentellen Multielementdoppel-
analytik und ergänzender Methoden, die routinemäßig oder
fallweise erfaßt werden.

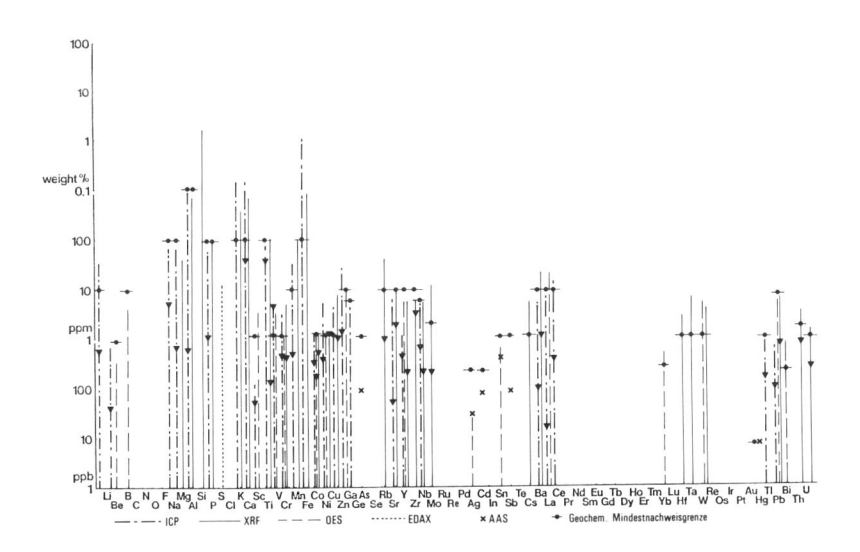

Abb. 3: Nachweisgrenzen der Analysenmethoden und die geochemisch
erforderlichen Mindestkonzentrationswerte. Bei Hauptelementen
ist die niedrigste Konzentration der Kalibrierung gezeichnet.

Die durchschnittlichen Nachweisgrenzen der beiden Multielementmethoden liegen bei 1 ppm, sodaß etwa 30 Elemente in geologischem
Material ohne Voranreicherung mit Erfolg analysiert werden können. Die
idealen analytischen Nachweisgrenzen werden allerdings nicht bei allen
wichtigen Spurenelementen erreicht. Vor allem die XRF-Methode liefert
bei kurzen Meßzeiten für Elemente, wie As, Cs, Hf, Mo, Sb, Sn, W und
U, geochemisch unzureichende Bestimmungsgrenzen. Auch Ag und Cd
scheiden aus. Diese Elemente sind allerdings auch für die ICP-Anregung
zu unempfindlich (Abb. 3).

5.3 KALIBRIERUNG

Es wird eine Art Multi-SRM-Kalibrierungsmethode angewandt. Die
angestrebte analytische Genauigkeit ist von der Existenz einer genügend großen Zahl geeigneter Standardreferenzproben (SRM, standard
reference material) abhängig, die den erforderlichen Meßbereich
abdecken. Die Kalibrationskurven der Multielementmethoden werden mit
Hilfe von bis zu dreißig SRMs verschiedenster Gesteinarten einschließlich Bodenproben aufgebaut. Leider fehlen für eine größere Zahl
seltenerer Elemente noch immer zuverlässig Konzentrationsangaben, so
für Be, Cs, Hf, Nb, Ta, Mo, W, Cd, Tl, Sn, Sb, Bi, Se oder Te. Es
handelt sich dabei meist um jene chemischen Elemente, deren Analytik
bei geologischem Material noch zu wünschen übrig läßt.

Um den Verbrauch der wertvollen SRMs herabzusetzen, weicht man
zweckmäßigerweise auf Laborstandards aus.

5.4 WIEDERHOLBARKEIT UND RICHTIGKEIT
5.4.1 WIEDERHOLBARKEIT

Die kurzzeitige instrumentelle Wiederholbarkeit der Einzelmessung
(s) ist bei beiden Methoden ausgezeichnet, bei vielen Elementen sogar
bei $\pm$ 1 %. Die XRF-Methode zeichnet sich durch eine geringere Geräte-
drift aus (Abb. 4).

Bei langfristigen Messungen weisen beide Methoden je nach Element
und Konzentrationsbereich im Arbeitsbereich der Kalibrationskurven
immer noch Wiederholbarkeiten von s = $\pm$ 3 bis $\pm$ 20 % auf (Abb. 5). Der
Vorteil der großen Gerätestabilität der XRF-Methode wird durch den
Fehler der Preßtablettenherstellung wieder aufgewogen.

5.4.2 VERGLEICHBARKEIT DER DOPPELANALYTIK

Im Konzentrationsbereich über 10 bis 20 ppm ist für die meisten
Elemente eine langfristige Vergleichbarkeit innerhalb der Schranken
der Wiederholbarkeiten (s bis 2s) zu erzielen. Die ICP-OES-Methode ist
wegen besserer Nachweisgrenzen in der Regel vorzuziehen, bei höheren
Konzentrationen die XRF-Methode, so für K oder Cr.

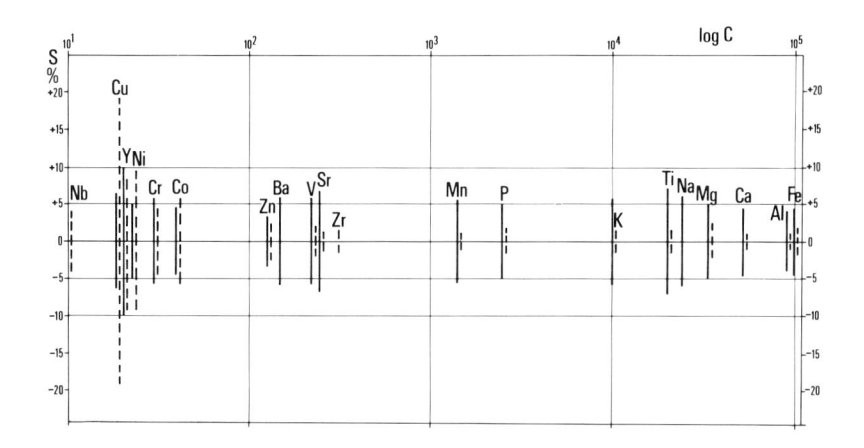

Abb. 4: Standardabweichungen (s) der Doppelanalytik für je 60 Messungen aus ein und derselben Lösung des Laborstandards "Diabas" und ICP-OES (Strich) und 10 Messungen ein und derselben Preßtablette dieser Proben mit XRF (Strichlierung). Bei etwa vergleichbarer Dauer der Meßzeit zeigt das XRF-Quantometer die geringere Drift.

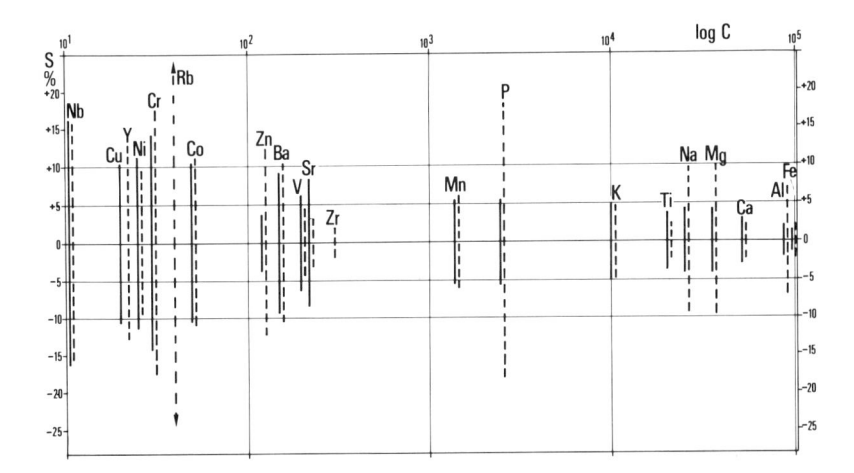

Abb. 5: Standardabweichungen (s) der Doppelanalytik für 160 Messungen, verteilt auf 8 Monate und Subprobenahme für den Laborstandard "Diabas" mit ICP-OES (Strich) und 36 Messungen, verteilt auf 4 Monate mit Subprobenahme mit XRF (Strichlierung).

Im Konzentrationsbereich über 5 Gewichts-% sind beide Analysenverfahren durch eine gesonderte Nachanalyse (mit ED-XRF unter Verwendung der Preßtabletten oder mit WD-XRF unter Einsatz von Schmelztabletten oder nach Verdünnung der Lösungen für ICP-OES), vor allem für die geochemischen Hauptelemente, wie Si, Al, eventuell auch Ca, Mg, Fe oder K, zu korrigieren, wenn Wert auf eine bessere Wiederholbarkeit zu legen ist. Abweichende Analysenergebnisse sind vor allem in der Überschreitung der Grenzwerte der vorgegebenen physikalischen und chemischen Matrixbreite der beiden Analysenverfahren zu suchen. So wirken sich höhere Konzentrationen von Cr, Mn, und Ti ab 1 %, Mg ab 5 % sowie Ca und Fe ab 10 % elementspezifisch aus. Überdurchschnittliche Gehalte anderer Elemente verlangen gleichfalls eine kritische Prüfung auf Koinzidenzen oder Matrixeinflüsse. Vergleiche von Durchschnittswerten, wie sie bei Großserienanalysen angefallen sind, werden in Abb. 6 in einem Variationsdiagramm dargestellt.

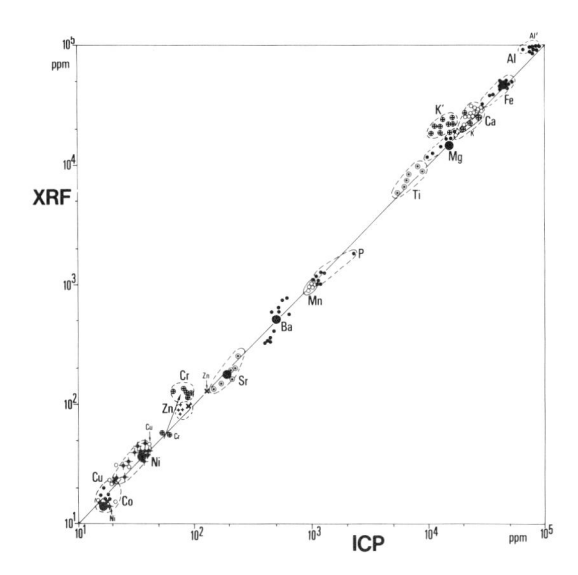

Abb. 6: Vergleichbarkeit der Doppelanalytik für einige Elemente, ausgedrückt durch Mittelwerte für je 1000 Proben. Mittelwerte für je 10.000 Proben größere schwarze Punkte.

5.4.3 RICHTIGKEIT

Die Richtigkeit soll im Idealfall der Wiederholbarkeit entsprechen. Wenn Analysenwerte, die mit wenigstens zwei verschiedenen Analysenmethoden erhalten worden sind, übereinstimmen, dann darf angenommen werden, daß diese Werte richtig sind. Voraussetzung ist aber, daß die Kalibrierung mit Standards erfolgt ist, deren Werte richtig sind. Bei Analysenserien müssen Mittelwertsvergleich, Korrelation und Regression sowie Vergleich der Häufigkeitsverteilungen zur Prüfung herangezogen werden.

Real gesehen ist der wahre Wert der Einzelprobe "verdeckt". Allerdings sind geochemische Richtwerte in Form von Durchschnittsgehalten für die meisten chemischen Elemente bekannt. In Abb.7 wird an einem Beispiel der Versuch dargestellt, zwei Elemente in das geochemische System der Elementverteilung der kontinentalen Kruste im Sinne einer Überprüfung auf generelle Richtigkeit einzuordnen.

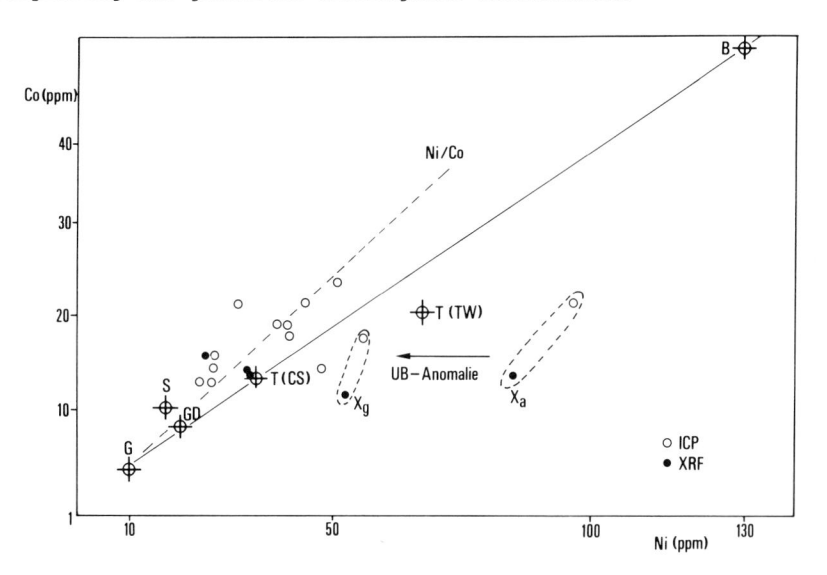

Abb. 7: Ni- und Co-Werte von Bach-und Flußsedimenten Österreichs. Die Clark-Werte von Granit (G), Granodiorit (GD) und Basalt (B) liegen auf einer Entwicklungslinie. Ferner sind die Clark-Werte Böden (S) und von Tongesteinen (T) eingetragen (13). Die ICP-OES- und XRF-Werte von je 1000 Proben folgen einer Regressionsgerade mit einer Steigung von etwa 2. Nur eine Probenserie, die eine lithogeochemische Anomalie ultrabasitischer Gesteine miterfaßt, weicht in ihrem arithmetischen Mittelwert (x_a), weniger im geometrischen Mittelwert (x_g) ab. Einen Hinweis auf die Richtigkeit der Ni/Co-Verhältnisse ergibt sich aus dem Durchschnittswert von 51 Tongesteinsproben (T(CS)) aus dem ostalpinen Karn (Raibler Schichten), die mit derselben analytischen Methodik analysiert worden sind und als gute Durchschnittswerte für die regionale Krustenzusammensetzung zur Zeit der Trias aufzufassen sind.

Als Richtwerte können nicht nur die geochemische Häufigkeit in geochemischen Einheiten, wie Gesteinsarten und -familien, regionale oder geotektonische Provinzen, ozeanische und kontinentale Kruste, sondern auch Korrelationen, Elementverhältnisse u.ä. in Betracht gezogen werden.

In Abb. 8 soll gezeigt werden, daß auch Elementverhältnis und Korrelationskoeffizient zur Entscheidung bei der Auswahl der Analysenmethode herangezogen werden können. Dieses Beispiel ist nur als

Versuch zu werten, da die statistische Grundlage der eingesetzten Richtwerte noch nicht ausreichend erscheint.

Beispiele von Variationsdiagrammen, mit deren Hilfe Analysenwerte überprüft werden können, zeigen Abb. 9 und 10.

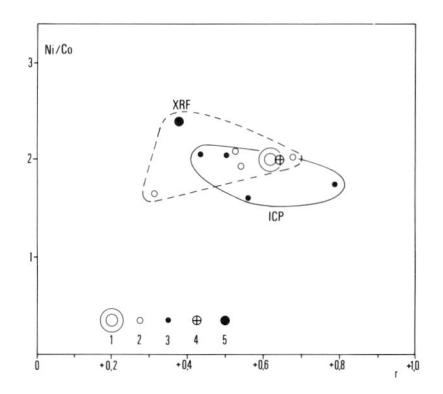

Abb. 8: Kontrolle von Datenmaterial aus der Doppelanalytik Ni und Co durch ein synthetisches System aus SRM-Daten (6 Basite, 14 Granite, 9 Böden und Tongesteine).
Ni/Co-Verhältnisse gegen Korrelationskoeffizienten (r) von Probengruppen N = 1000, bzw. 6000.
Richtbereich - 1, ICP-OES - 3, bzw. 4, XRF - 2, bzw. 5.
Die ICP-OES-Werte liegen näher zur Richtmarke.

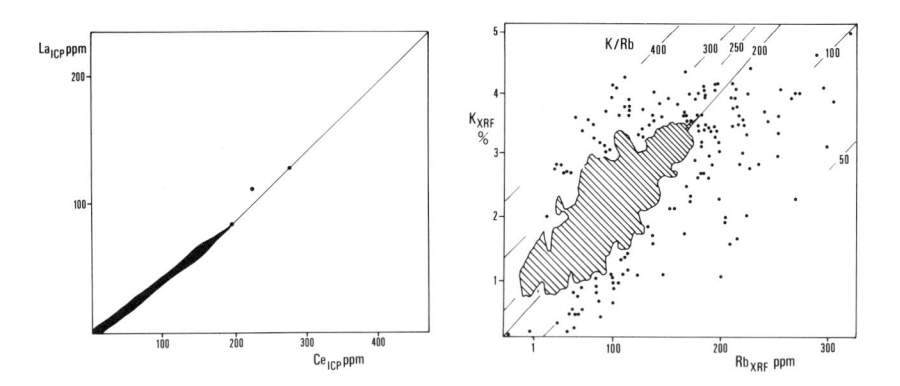

Abb. 9: Variationsdiagramm Ce/La (für 1000 Proben). Die Elementverhältnisse erscheinen kaum variabel (links).

Abb. 10: Variationsdiagramm K/Rb (für 1000 Proben). Die Daten liegen im Erfahrungsbereich des Elementverhältnisses, im Durchschnitt bei etwa 250 (rechts).

Bezüglich weiterer Anmerkungen zur Analysenkontrolle und geochemischen Überprüfungsmethodik sei auf einen weiteren Beitrag verwiesen (7).

6. ANDERE ANGEWANDTE ANALYSENMETHODEN

An anderen Analysenmethoden sind Kohlenbogen-OES, ED-XRF, AAS (Flamme, Hydrid und Kohleofen), Radiometrie (U, Th, K) sowie Sonderanalytik, wie für As, Sb, F, vorhanden, um einerseits für möglichst viele Elemente die interne Kontrolle schnell bei der Hand zu haben und andererseits das Multielementanalysenprogramm element- oder konzentrationsmäßig ausweiten zu können (siehe Abb. 1 und 2). Es ergibt sich auch die Notwendigkeit, den analytischen Konzentrationsbereich der beiden quantometrischen Methoden sowohl in die Bereiche niedriger als auch hoher Gehalte zu erweitern. Auf Überdeckung von Kalibrationsbereichen ist zu achten.

Die klassische OES mit Kohlenbogenanregung und Fotoplattenauswertung, die heute personell und sachaufwandmäßig relativ kostspielig erscheint, ermöglicht unter Verwendung eines Gitterspektrographen mit großer Dispersion (2,5 Å/mm) die Analyse von Spurenelementen in silikatischer Matrix mit Nachweisgrenzen von 100 bis 1000 ppb, wie für Ag, B, Be, Mo oder W. Sie ist für die Bestimmung des Gesamt-Sn die zuverlässigste Methode. Der Vorteil der großen Dispersion wird mit der Verkleinerung des spektralen Arbeitsbereiches erkauft, sodaß damit die Zahl der möglichen Analysenelemente verringert wird. Die verbesserten Nachweisgrenzen werden durch Anwendung der Methode der fraktionierten Destillation und Zusatz thermochemischer Reagentien erreicht (9); dies bedingt eine weitere Selektion der Analysenelemente. Die Anzahl der in einem Analysengang erfaßbaren Elemente erreicht kaum 10. Hiezu kommt, daß der personelle Zeitaufwand bei manueller Messung erheblich ins Gewicht fällt. Bei Automatisierung des Meßvorganges mit anschließender rechnerischer on-line-Auswertung wurde die Anzahl der in einem Arbeitsgang bestimmten Analysenelemente aus Gründen des Bedarfes und der leichteren Manipulierbarkeit auf 4 bis 5 beschränkt.

Diese Methode wird in Routine für vier Analysenelemente verwendet: Ag (Nachweisgrenze 20 ppb), Mo (100 ppb), Pb (3 ppm) und Sn (1 ppm). Das Analysenelement B (Nachweisgrenze 1 ppm) liegt außerhalb des spektralen Arbeitsbereiches. Eine Probeneinwaage von 0,14 g wird nach Zumischung von CdF_2 (1:1) mit einem Gleichstromabreißbogen 30 A/220 V mit einer Belichtungszeit von 25 s angeregt. In ähnlicher Weise kann W (Nachweisgrenze 2 ppm) und mit einer aufwendigeren Sonderbestimmung mit AgCl als thermochemisches Reagens im Doppelbogen mit einer Nachweisgrenze von 0,1 ppm bestimmt werden (3). Es liegen außerdem Erfahrungen über die spektrochemische Spurenanalyse in Großserie von 17 Elementen vor, für die allerdings zwei getrennte Aufnahmeverfahren (A,B) erforderlich sind, um optimale Nachweisgrenzen zu erhalten: A (Ag, B, Be, Ge, Mo, Pb, Sn, V), B (Co, Cr, Cu, La, Ni, Sc, Y, Yb, W). Die Analysenkapazität erreicht nicht die geforderten 100 Proben/Tag.

Die AAS mit Flammenanregung ermöglicht die Kontrollanalytik für zahlreiche Analysenelemente wie Na, Li, Zn u.a., in Form der Hydridmethode die Spurenanalyse der Elemente As, Sb und Se und mit dem Kohlerohrofen werden die tiefen Konzentrationsbereiche, vor allem von Cd, Tl, Pb und Bi zugänglich.

Die ED-XRF dient der schnellen Kontrolle. Die für die WD-XRF hergestellten Preßlinge (H_3BO_3-Basis) können verwendet werden. Die leichten Elemente, auch Nichtmetalle wie S, sind gut bestimmbar. Al und Si können in Silikatgesteinen mit einer Genauigkeit von etwa $\pm$ 0,5 Absolut-% für Al und $\pm$ 1 Absolut-% für Si analysiert werden. Die ED-XRF dient auch einer schnellen orientierenden Analyse bei Kleinserien.

Einen Überblick über die Arbeitskapazität der kombinierten Analysenmethoden gibt Tab. 1.

Tabelle 1: Die angewandte Kombination von Analysenmethoden

Instrumentelle Methode	Anzahl der Analysenelemente	Analysen-kapazität (Probenzahl/Tag)	Mann-jahre
ICP - OES	28 - 30	100 - 200	2 1/2
WD-XRF	30	100	1 1/2
Doppelanalytik	bis zu 20		
AAS (Flamme, Hydridmethode)	1 - 4	100	
AAS (Kohlerohrofen)	4	100	2
OES (z.T. automatische Fotoplattenauswertung)	4 - 10	70	2 1/2
ED-XRF			
manuell	10	10	1
halbautomatisch	12	50	
Sondermethoden	je 1	100 - 300	nach Bedarf je 1

7. SCHLUSSBEMERKUNG

Die Arbeitskapazität des Labors der Abteilung für geochemische Analytik des Geotechnischen Institutes mit bis zu 30.000 Proben pro Jahr, d.s. bis zu 1,000.000 Daten, wird einschließlich Probenvorbereitung von 15 Mitarbeitern, davon 6 akademisch graduierte Fachkräfte, bewältigt; 2 1/2 Arbeitskräfte sind für die EDV-Arbeit, davon 1 1/2 am

...staltsterminal eingesetzt, das mit einer Großrechenanlage in Verbindung steht. Für die Probenaufbereitung und -vorbereitung (1 1/2 Laboranten) sowie Sonderanalytik kann zeitweise zusätzlich der Einsatz von Hilfskräften notwendig sein. (Die Einzelleistung des Analytikers muß sich der Teamarbeit unterordnen, die an alle Mitarbeiter auch erhöhte Anforderungen zur Zusammenarbeit stellt).

Als ideales Ziel ist eine kombinierte und selbstkontrollierte Doppel- oder Mehrfachmethodenanalytik vorstellbar, die simultan abläuft und bei der EDV-gesteuert "Ausreißerdaten" automatisch einer Kontrolle, bzw. einer Wiederholung zugeführt werden. Die zukünftige Entwicklung sollte in letzter Konsequenz zu computergesteuerten Analysengroßeinrichtungen führen.

LITERATUR

(1) DOLEZEL, P.: ibid.
(2) FLETCHER, W.K.: In: FLETCHER, W.K., ed, Analytical Methods in Geochemical Prospecting. Handbook of Exploration Geochemistry, Vol. 2, Elsevier Amsterdam 1981.
(3) JANDA, I.: Mikrochim. Acta (1976 II), 473 - 479.
(4) MÜLLER, N.: ibid.
(5) MÜLLER, N.: X-ray Spectrometry (im Druck)
(6) SAUER, D., SCHROLL, E.: VII. CANAS, Sopron 1982, abstract.
(7) SAUER, D.: ibid.
(8) SCHROLL, E.: In: AHRENS, L.H., ed, Origin and Distribution of the Elements. Pergamon Press, Oxford etc. 1968, 599 - 617.
(9) SCHROLL, E.: In: AHRENS, L.H., ed, Origin and Distribution of the Elements. Pergamon Press, Oxford etc. 1979, 213 - 216.
(10) SCHROLL, E.: Analytische Geochemie, Vol. I, Methodik. Enke Verlag Stuttgart 1975.
(11) SCHROLL, E.: Analytische Geochemie, Vol. II, Grundlagen und Anwendungen. Enke Verlag Stuttgart 1976.
(12) SCHROLL, E.: Fortschr. Miner. 54 (1977), 167 - 191.
(13) THOMPSON, M.: In: HOWARDS, R.J. (ed), Statistics and Data Analysis in Geochemical Prospecting. Handbook of Exploration Geochemistry (FLETCHER, W.K. ed), Vol. 3, Elsevier Amsterdam 1983.
(14) TUREKIAN, K.K., WEDEPOHL, K.H.: Bull. Geol. Soc. Amer. 72 (1961), 172 - 202.

INTERCOMPARISON OF THE MULTIELEMENT ANALYTICAL METHODS TXRF, NAA AND ICP
WITH REGARD TO TRACE ELEMENT DETERMINATIONS IN ENVIRONMENTAL SAMPLES

W. Michaelis, H.-U. Fanger, R. Niedergesäß, H. Schwenke

Institut für Physik, GKSS Forschungszentrum Geesthacht,
D-2054 Geesthacht, Germany

SUMMARY

The paper gives an intercomparison of total reflection X-ray fluorescence analysis, instrumental neutron activation analysis both with thermal and 14 MeV neutrons, and inductively coupled plasma optical emission spectroscopy on the basis of various extensive environmental research and monitoring programmes. The data originate from parallel analyses performed to control precision and accuracy during the studies in addition to interlaboratory tests. Quite different matrices are considered: particulate matter suspended in waters, filtrates of river water, sediment, mussel tissue (*mytilus edulis*), air dust and annual rings of spruce-wood. The data presented give an insight into the reliability of analytical results that is achieved in routine programmes in which large numbers of samples have to be handled. It is not the purpose of this paper to explore the optimum precision, accuracy or detection limit attainable in well-aimed analyses. Finally, data are reported on various interlaboratory tests (Intercomparisons: Water W-3/1, Soil-7, Sediment SD-N-1/2, Mussel tissue Ma-M-2/TM).

1. INTRODUCTION

Environmental research and monitoring programmes, in general, require the quantitative detection of numerous elements in the samples collected. This demand emphasizes the great significance of multielement analytical methods since they offer the possibility to confine the total expenditure. Nevertheless, only in favourable cases a single method will cover

all elements of interest. Another important aspect is that often very
high precision and accuracy are required, for instance, if systematic
local or temporal trends in pollution studies have to be disclosed.
Besides interlaboratory tests the application of different methods to
the analysis of the same samples is the best way to ensure the results.
In extensive sampling programmes, however, the total costs very soon
come to the fore. The methods must, therefore, be fast and economical
and should be combined in a well-considered way.

This paper gives an intercomparison of total reflection X-ray fluorescence analysis (TXRF), instrumental neutron activation analysis both
with thermal (INAA) and 14 MeV neutrons (14 MeV NAA), and inductively
coupled plasma optical emission spectroscopy (ICP) on the basis of extensive data sets obtained in various environmental research and monitoring programmes. These programmes, in particular, comprise studies of
the heavy-metal transport in tidal rivers [1, 2, 3], pollution surveys
of estuaries and shallow waters of the North Sea including organisms
[3 - 5] and contributions to an air-dust monitoring study in collaboration with other laboratories [6, 7]. Thus, quite different matrices
are considered.

The majority of data has been obtained by TXRF since this method has
proved to be particularly efficient in programmes with large numbers of
samples [3, 7]. NAA is used as a reference method and for determining
elements not covered by TXRF. Specific applications of ICP are focussed
on individual elements in particulate matter and on elements in the
liquid phase.

A lot of parallel analyses were performed in the studies mentioned
since it is a general observation (see, e.g., [8]) that from the results
of intercomparison tests alone information about the accuracy in routine
practice can be derived only with reservation. The data which are presented in this paper, therefore, give an insight into the reliability
of analytical results that is achieved in routine measurements with all
the aggravations associated with the handling of large numbers of samples.
It is explicitly not the purpose of this paper to explore the optimum
precision, accuracy or detection limit attainable in well-aimed analyses.

On the one hand, the data support the usefulness of additional intercomparisons in routine applications. On the other hand, they also demonstrate that spectacular analytical errors as repeatedly reported in
the literature [9, 10] can be precluded even in extensive sampling programmes if the possible sources of systematic errors are present to the

analyst's mind and if the operation of the analytical instrument is well understood. Admittedly, this demand can more easily be fulfilled by a laboratory which is also engaged in the development and improvement of analytical methods.

2. ANALYTICAL METHODS AND SAMPLE PREPARATION

2.1 Total Reflection X-ray Fluorescence Analysis

Details of the method have been published elsewhere [7, 11]. It should, however, be supplemented here that the detection limits for the tungsten anode have been improved meanwhile by more than a factor of 5 compared to earlier data given. The specimen to be analysed is preferably processed into a solution or a fine-grained suspension in a liquid with a sufficiently high vapour pressure. This feature makes the technique particularly suited for the analysis of filtrated water, suspended particulate matter (SPM), and air dust.

Sample preparation for river water from the limnetic region is quite simple. After filtration the liquid phase can directly be analysed for the elements S, K, Ca, Mn, Fe, Ni, Cu, Zn, As, Rb and Sr by pipetting 25 µl onto the quartz glass sample support and subsequent drying up. A Co solution is added to the filtrate for standardization (200 ppb). In order to control the results for Ni, Cu, Zn and As part of the samples is, in addition, analysed with increased detection power after matrix separation which is performed on the specimen support following a rather simple, but reliable procedure [12].

For the analysis of SPM the loaded filters are warmed up with 20 ml of conc. HNO_3 to 90 °C for about 2 h. Again, Co is added as an internal standard. After dilution with ultra-pure water to about 40 ml undissolved residua are dispersed in an ultrasonic bath. An aliquot of 20 µl is dried up on the sample support by applying a soft vacuum. Air dust samples are prepared in a similar manner using a temperature of 80 °C. For determining sulphur the method of standard addition is used in order to take into account a possible matrix influence caused by the very low energy of the K radiation.

For sediment analysis 50 to 100 mg of the sample are digested with 3 ml conc. HNO_3 and 0.5 ml HF in a teflon bomb at a temperature of 165 °C for 5 to 6 h. After addition of a Co standard solution the sample is diluted with bidestilled water to about 80 ml. Then, 10 µl of this solution are dried up on the sample support. Organic material and part of the SPM samples were also digested using this procedure.

2.2 Instrumental Neutron Activation Analysis with Thermal Neutrons

Short-term irradiation (3 to 60 min) is performed in the rabbit system at the research reactor FRG-1 (5 MW). The thermal neutron flux ranges from 8.0×10^{12} to 1.4×10^{13} n cm^{-2} s^{-1}. The epithermal and the fast neutron flux are 1.1 to 1.4 % and 12 to 14 % of the thermal flux, respectively. Ni and Fe wires as well as Al, Zr and Au foils are used as flux monitors. Typical sample masses are between 20 and 100 mg. The samples are irradiated in polyethylene capsules which are cleaned with acetone after irradiation.

For long-term irradiation (in most cases between 1 and 3 d) a 'merry-go-round' device at the reactor FRG-2 (15 MW) is utilized. The thermal flux amounts to 4.5×10^{13} to 1×10^{14} n cm^{-2} s^{-1}. Epithermal and fast flux range from 3 to 5 % and 22 to 30 % of the thermal neutron flux, respectively. The fluxes are monitored using Fe wires and Al, Zr, Ag and Au foils. Samples with masses typically between 200 and 500 mg are irradiated in quartz capsules (Suprasilan) etched with HF. After irradiation the capsules are cleaned with HCl and acetone.

Usually, samples are prepared by freeze-drying and, if necessary, sieved, homogenized or milled.

2.3 14 MeV Neutron Activation Analysis

The neutron source is KORONA, a high-intensity sealed neutron tube with an integrated fast pneumatic rabbit system for sample transfer [13]. The maximum source strength is about 5×10^{12} n s^{-1} corresponding to a 14 MeV neutron flux at the sample position of approximately 5×10^{10} n cm^{-2} s^{-1}. The flux is monitored by a long counter. Characteristic features of KORONA are the feasibility of cyclic activation with transfer times of 140 ms to the 16 m distant detector station and the cylindrical target geometry which ensures that the flux variation within the sample volume does not exceed ± 5 % [14]. Depending on their nature the samples are sieved, milled, homogenized and/or freeze-dried. Irradiation is performed in 10 mm diam. × 7 mm polyethylene capsules. The capsules are separated from the carrier rabbit by means of centrifugal force in a curved branch of the pneumatic tube near the detector.

2.4 Inductively Coupled Plasma Optical Emission Spectroscopy

The computer-controlled ICP spectrometer is operated in the sequential mode. Spectra are unfolded by a fitting programme after scanning a high-resolution monochromator in steps of about 0.1 pm over the line profiles and the underlying background. These procedures allow the accurate determination of the net peak areas even in the presence of interferences and, thus, minimize systematic errors [15]. For each element a complete set of optimum operation parameters (emission line, coupled radio-frequency power, gas flows and observation height) has been determined and is adjusted during the measurement. A pneumatic nebulizer is used for injection. Calibration is preferably achieved by means of the standard addition method. Samples are digested by similar procedures as described in subsection 2.1.

3. Applications in Environmetal Research and Monitoring Programmes

3.1 Studies of the Heavy-Metal Transport in Tidal Rivers

The investigation of transport phenomena in tidal rivers is considerably complicated by pronounced spatial heterogeneities and temporal variabilities. To tackle this rather complex problem, a methodology has been developed which combines in an effective way hydrographic field measurements, trace element analyses and mathematical simulation models [1 - 3]. Within this concept, a detailed knowledge of the heavy-metal concentrations in the water, the suspended load and the sediment is required. A number of samples are taken over selected river cross sections at properly distributed locations and at various phases of the tidal period. A theoretical model then simulates the transfer processes for heavy metals through the transport of the liquid phase and the par-

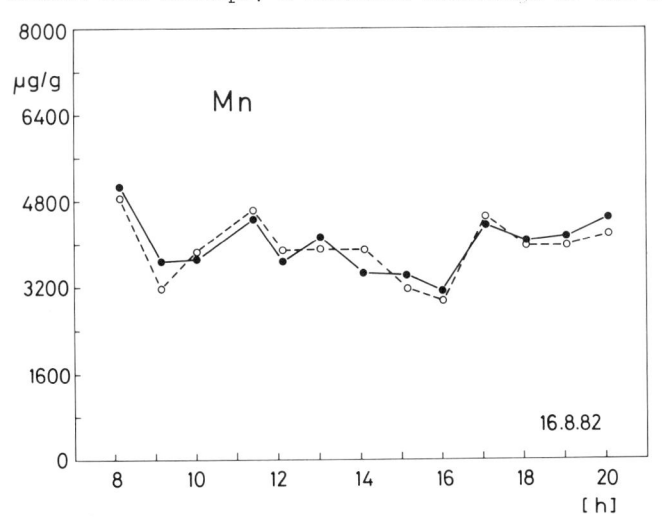

Fig. 1: Specific load of SPM from the Lower Elbe River: intercomparison of NAA (o) and TXRF (●) for Mn and Fe. Dry substance.
a) Mn, sampling on August 16, 1982.

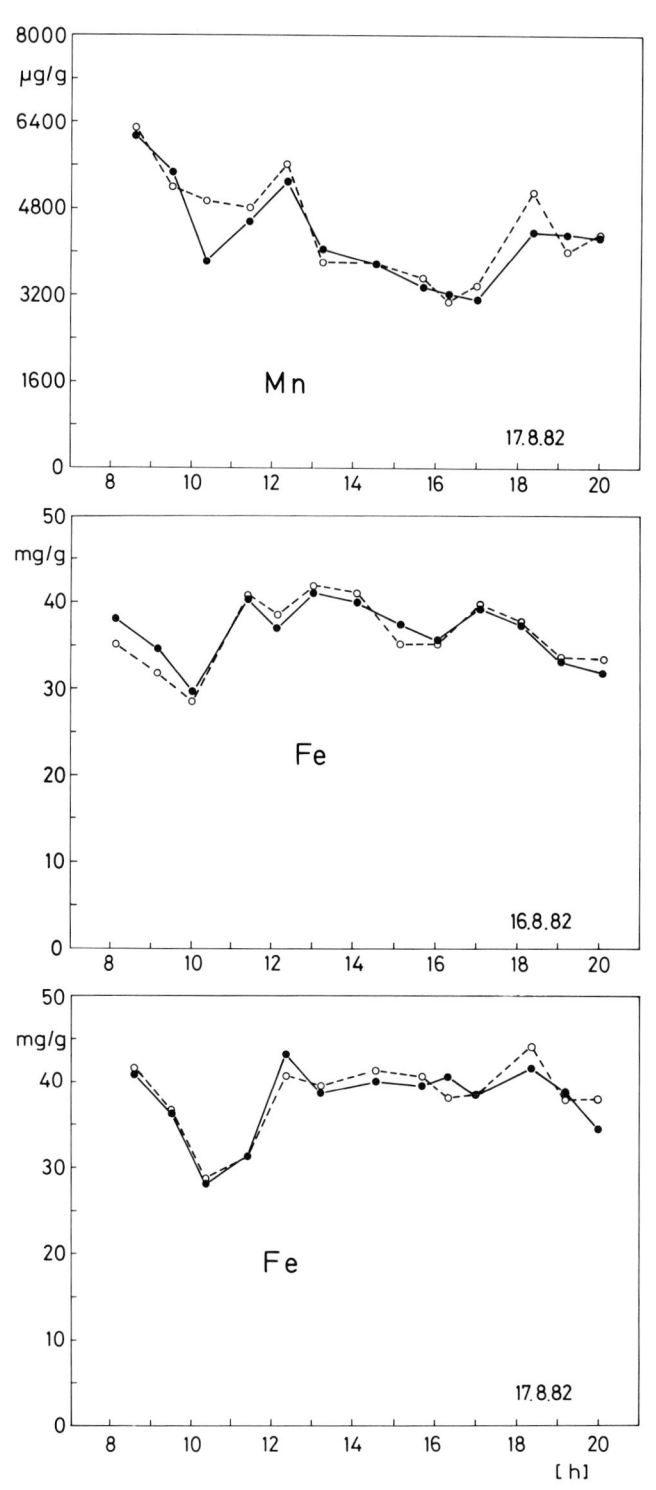

ticulate matter, using the field data and the analytical results as boundary conditions.

During an extensive study performed in 1982 on the Lower Elbe River, a lot of parallel analyses with TXRF, NAA and ICP were carried out in order to assure the observed heterogeneities and variabilities which, in a critical way, influence the transport modelling. Some of the data obtained are displayed in Figs. 1 to 3. Error bars given in these and the subsequent Figs. correspond to the overall estimated error with a 68 % confidence level unless otherwise specified.

Figs. 1 to 3 refer to the specific load of the SPM in the river area investigated. Most of the data are presented as time series, i.e. only results belonging to the same sam-

Fig. 1b - 1d:
b) Mn, sampling on August 17, 1982.
c) Fe, August 16, 1982.
d) Fe, August 17, 1982.

pling time are comparable since the time series include systematic spatial and temporal trends.

The elements Mn and Fe in SPM can reliably be determined in routine measurements both by TXRF and NAA with standard deviations of about 5 % or better (Fig. 1). Only in one case (Fig. 1b) the deviation for Mn was found to be outside the allowed range. Most probably, the source of this discrepancy lies in an erroneous flux correction for the NAA short-term irradiation.

The determination of As by TXRF is complicated by the energetic interference of the L radiation from Pb. Imperfect accuracy in unfolding the spectra may, therefore, induce severe systematic errors.

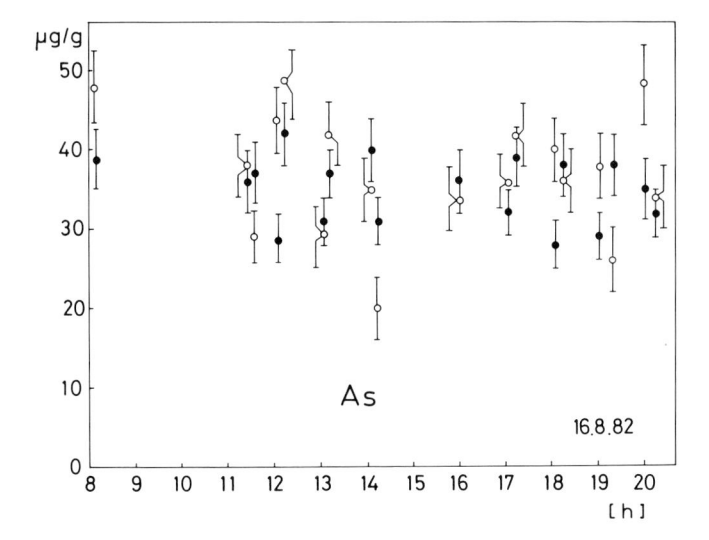

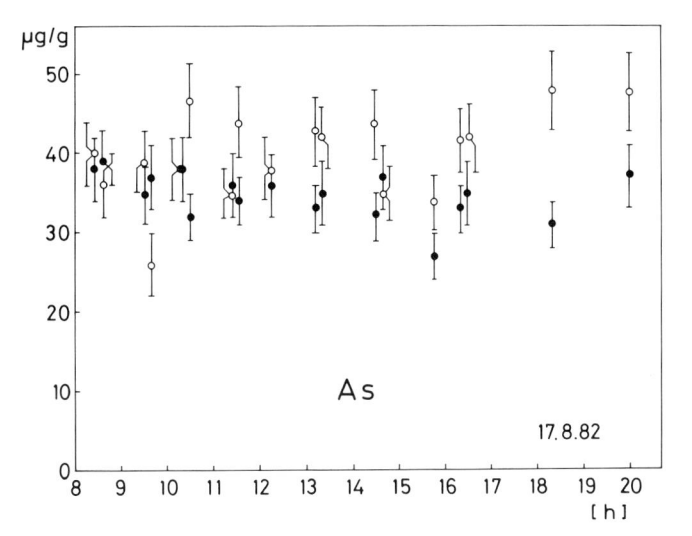

Fig. 2: Specific load of SPM from the Lower Elbe River: intercomparison of NAA (●) and TXRF (o) for As. Sampling on August 16 and 17, 1982. Dry substance.

The comparison with the results of NAA shown in Fig. 2, however, indicates that this problem on the whole can be coped with.

The specific load with Pb was found to be quite high (~ 100 µg/g). Thus, there are no difficulties in the analysis of SPM by TXRF. Nevertheless, 12 arbitrarily selected filter samples were digested and, in addition, analysed by ICP. The comparison is given in Fig. 3. The agreement of the data is pretty good. For most of the samples the mean values are within the single standard deviation of both methods, they are within the double standard deviations for all cases. Altogether, the ICP results seem to be slightly lower than the data obtained by TXRF. A total of 16 elements (K, Ca, Ti, V, Cr, Mn, Fe, Ni, Cu, Zn, As, Rb, Sr, Zr, Ba, Pb) can be determined in SPM by TXRF using a molybdenum anode. Unfortunately, the tungsten anode configuration with improved detection sensitivity was not yet available during this study. Therefore, Cd which by AAS was found to range between 2 and 15 µg/g was below the detection limit at that time. This is also valid for Hg which occurs in SPM of the river section investigated with concentrations between 2 and 3 µg/g as determined by INAA [16]. On the other hand, Cd is near or below the detection limit of INAA. A total of 32 elements are covered by this method: K, Mn, As by short-term irradiation, Na, Ca, Sc, Cr, Fe, Co, Ni, Zn, Se, Rb, Sr, Zr, Ag, Sb, Cs, Ba, Hf, Ta, W, Au, Hg, Th, U and numerous rare earth elements by long-term irradiation.

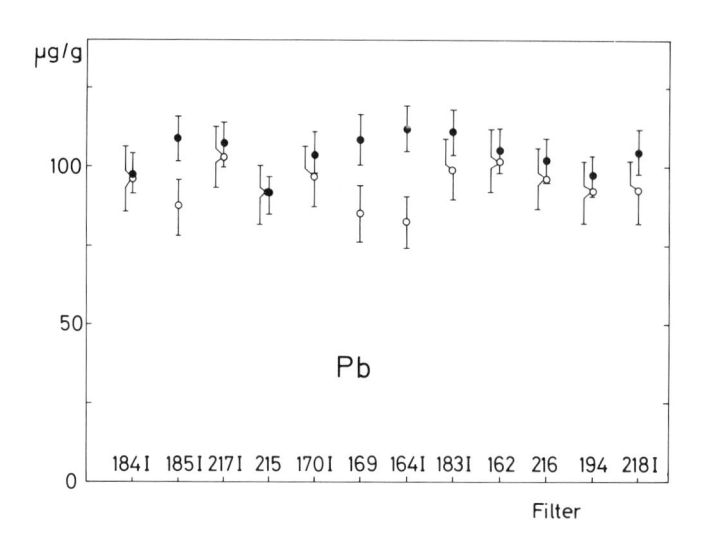

Fig. 3: Specific load of SPM from the Lower Elbe River: intercomparison of TXRF (●) and ICP (o) for Pb. Sampling on August 23, 1982.

3.2 Dissolved Metals in the Mixing Zone of Estuaries

Elements like Mn, Fe, Cu, Zn and Cd in solution show a characteristic behaviour in the mixing zone of estuaries [17]. For instance, in the longitudinal profile the dissolved Mn concentration exhibits a maximum at low salinities. These findings may be explained by removal of dissolved Mn at higher pH and higher dissolved oxygen concentrations in the lower estuary and production of Mn (II) in solution by reduction of higher oxidation states in water with lower oxygen and pH values in the early stages of estuarine mixing [3, 16, 17]. Data on these phenomena are not only of interest with regard to pollution studies, but also provide valuable information on the fundamental processes occurring in estuaries including, in particular, the interaction between suspended matter, the sediment and the liquid phase.

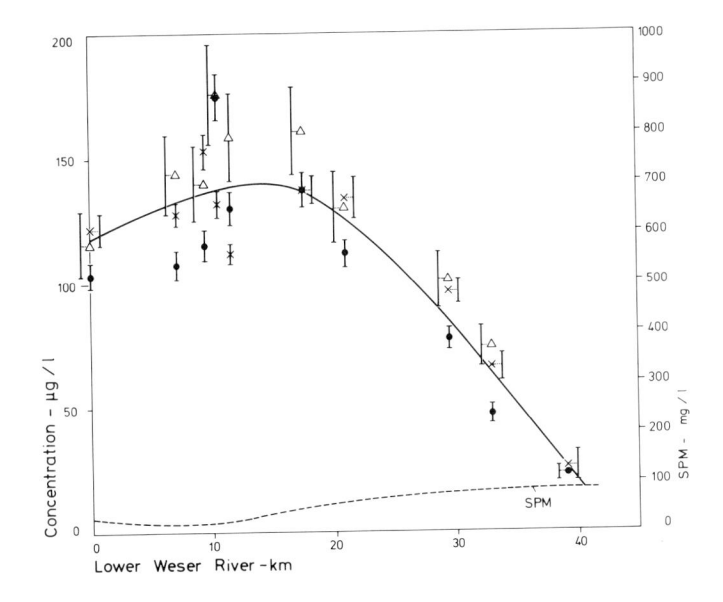

Fig. 4: Dissolved Mn and SPM content in the estuary of the Weser River at flood phase. ● ICP, x INAA, Δ TXRF.

During an extensive field campaign in 1979 on the Lower Weser River [4] besides profiles of the suspension concentration and the specific load of SPM, longitudinal profiles of the dissolved metals were taken at different phases of the tidal period by sampling from a moving boat. The filtrates were analysed by TXRF, INAA and ICP. As an example, Fig. 4 displays the results obtained for the flood phase. For comparison, the SPM concentration has also been included in the diagram. The analytical data presented give a realistic insight into the pattern of the results during routine measurements at the 20 to 200 µg/ℓ level. Meanwhile, the detection limits of TXRF have considerably been improved. Using matrix separation on the sample support (cf. subsection 2.1), they are now as low as 0.3 µg/ℓ (As).

3.3 Pollution Study of the North Frisian Wadden Sea

Under public contract in 1980 a pollution study of the North Frisian Wadden Sea was performed in compliance with the rules of the European Community on the quality standard of mussel culture waters (79/923/EWG). Besides a general pollutant inventory the purpose of the campaign was to explore whether or not there is a downward trend along the coast due to the influx of the Elbe River. Samples of water, sediment and mussels (*mytilus edulis*) were taken at selected sites and analysed by TXRF, INAA, AAS and ASV [5]. Mussels were sized in order to eliminate the influence of age. The soft parts of 20 to 40 individuals were homogenized and freeze-dried in each case. From the sediment samples the grain size fraction < 63 µm was investigated.

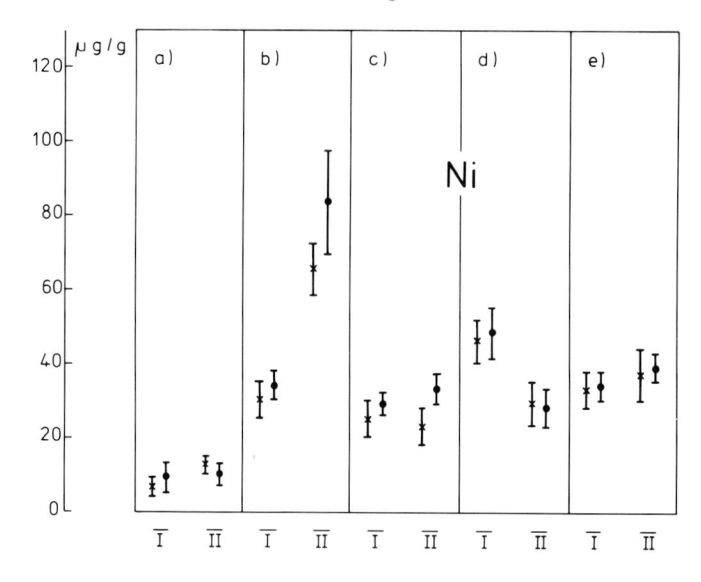

Fig. 5: Ni in sediment samples (< 63 µm) of the North Frisian Wadden Sea. Sampling sites:
a) S of Büsum,   b) SW of Südfall,
c) SE of Föhr,   d) E of Hörnum/Sylt,
e) E of Braderup/Sylt.

x INAA, • TXRF. Error bars: 95 % confidence level.

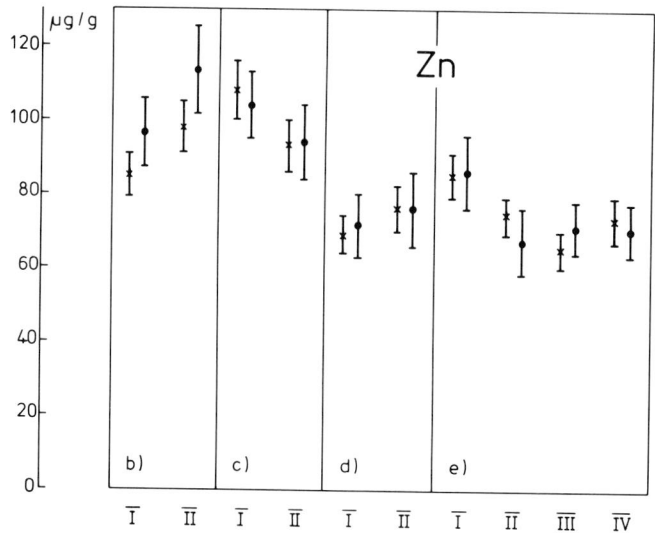

Fig. 6: Zn in mussel tissue (*mytilus edulis*). North Frisian Wadden Sea. Sampling sites: see Fig. 5.   x INAA, • TXRF. 95 % c.l.

Some of the results obtained from parallel analyses by TXRF and INAA are presented in Figs. 5 and 6 which show the Ni content in some sediment samples and the Zn content in some mussel samples, respectively. All data refer to the dry substance. In these diagrams the error bars correspond to the overall estimated error with a 95 % confidence level. The Roman numerals label neighbouring sampling locations in the same area. Associated differences in the heavy-metal concentrations are mainly caused by variations in the mud content of the sediments. Figs. 5 and 6 demonstrate the excellent agreement of the data achieved by TXRF and INAA.

3.4 Air-Dust Monitoring

Precision and accuracy of air-dust analysis have been investigated by performing a variety of comparison measurements and by taking part in numerous interlaboratory tests. A comprehensive presentation of data on Milanese air dust, as obtained by TXRF, INAA and 14 MeV NAA, has been given elsewhere [7]. The last-named method is of particular interest for sulphur since it allows - unlike NAA with thermal neutrons - specific reference measurements for the examination of TXRF results if the source strength is high enough to utilize the $^{34}S(n,p)^{34}P(T_{1/2} = 12.4$ s) reaction.

Within a large-scale air-dust monitoring programme which was performed few years ago in collaboration with ten analytical laboratories [6], TXRF turned out to be particularly suited for routine applications. The reproducibility of the results as to repeated sample preparation and measurement was found to be about 5 % for Fe, Cu, Zn and Pb, 10 to 15 % for S, K, Ca, Ti, Mn, Ni, Br, Sr, and 20 to 35 % for V, Cr, As, Se, Rb, Zr, Nb, Mo, Ag, Cd, Sn, Sb and Ba. The larger scatter for the last-mentioned elements is in part due to overlaps in the fluorescence spectrum (for instance, As in the presence of Pb or Ba and V associated with a high Ti content) and in part caused by the occurrence of very low concentrations (5 to 50 ng/cm^2).

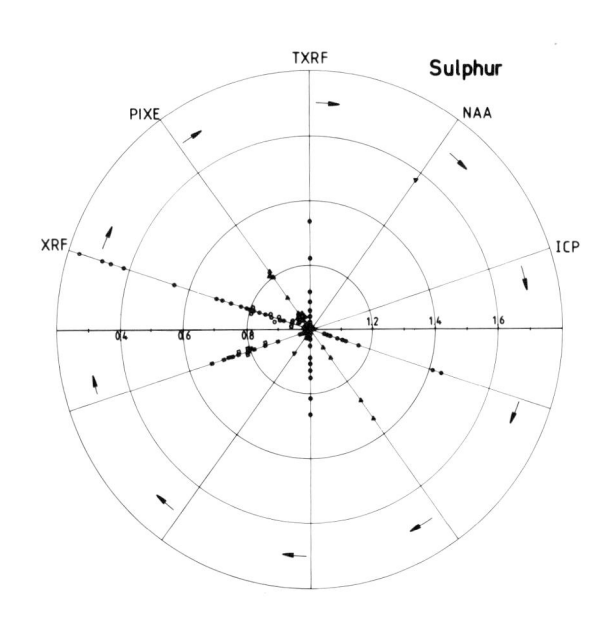

The extensive set of data accumulated in the course of this programme allows a detailed critical evaluation of a variety of analytical methods. Besides TXRF, NAA, and ICP, these include photon activation analysis (PhAA), atomic absorption spectrometry (AAS), proton-induced X-ray emission (PIXE) and conventional types

Fig. 7: Intercomparison of various analytical methods in the analysis of air dust.
a) sulphur

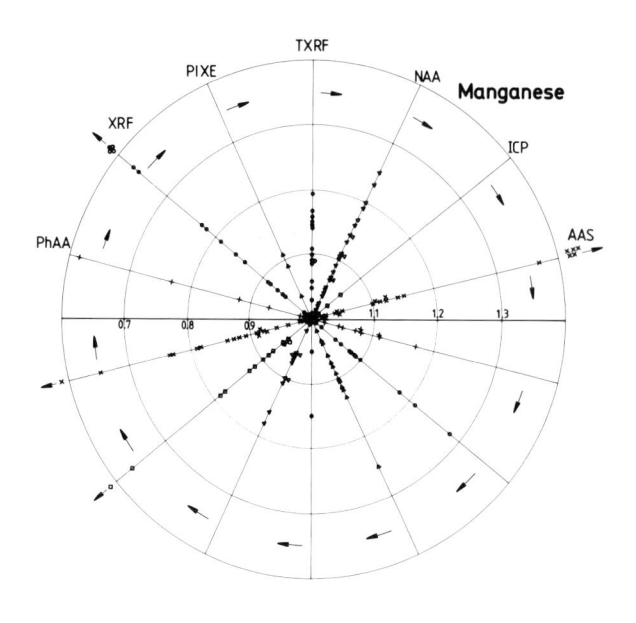

Fig. 7: b) Manganese

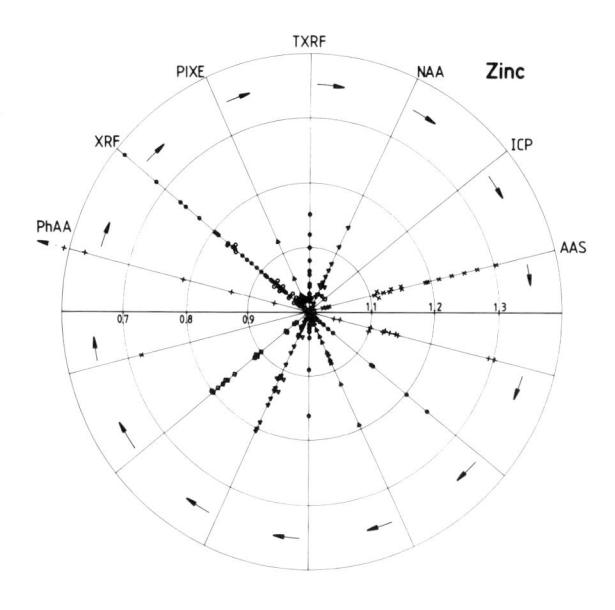

Fig. 7: c) Zinc

of X-ray fluorescence analysis (XRF). Fig. 7 gives a few examples for the elements S, Mn, Zn and Pb. The points plotted in these diagrams stand for the ratios of the analytical results to the mean of the total ensemble as obtained after application of the Nalimov outlier test. This mean may, but must not necessarily coincide with the 'true' value. Ratios > 1 are plotted in the upper half of the 'target' (right-hand scale), those < 1 appear in the lower half (left-hand scale). Values outside the scale are marked by radial arrows. The data were derived from week-averaged results and refer to samples taken at two sites with quite different degrees of pollution over a period of ten arbitrarily selected successive weeks. Only the TXRF and part of the NAA results were obtained in the own laboratory.

The presentation chosen in Fig. 7 allows a direct evaluation of the analytical data with respect to precision and accuracy. As can be seen, TXRF can quite well compete with other methods, including critical elements such as sulphur.

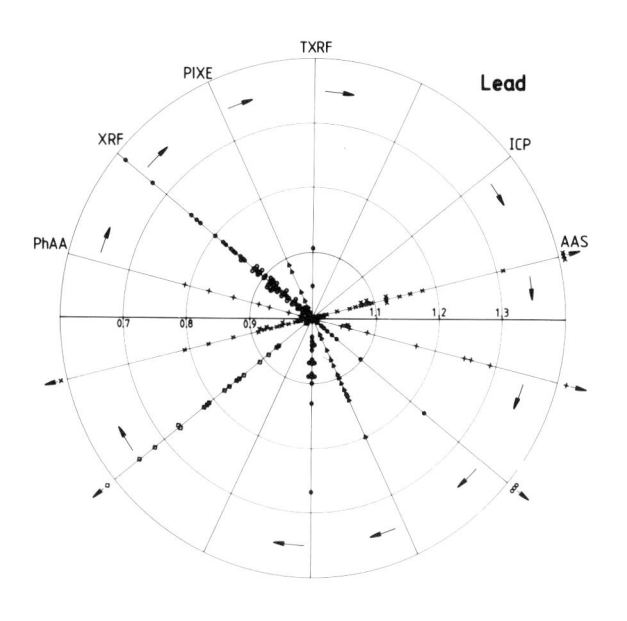

Fig. 7: d) Lead

3.5 Analysis of Wood Samples from Normal and Diseased Trees

In the framework of a coordinated study of potential causes of the rapidly increasing diseases in forests, the analytical capabilities of TXRF, INAA, 14 MeV NAA and ICP have been explored with regard to the most important elements in wood samples. Here, Na, Mg, Al, P, S, K, Ca, Mn, Fe, Zn, Cd, Hg and Pb are of particular interest [18]. It turned out that without using enrichment procedures the concentrations of these elements except P, Cd and Hg are well above the detection limits. Results obtained for a small pulverized sample of a three-year growth ring of a normal spruce stem are summarized in Table I. The error in 14 MeV NAA is mainly due to moderate statistics. Therefore, the counting rate is at present improved by using a bore-hole germanium detector. Possibly, the number of detectable elements may also be increased.

Table I: Some essential elements in a three-year growth ring of a spruce stem. Concentrations in µg/g dry substance.

Element	TXRF	ICP	INAA	14 MeV NAA
Na	n.d.	9.9 ± 1.0	9.0 ± 1.4	n.d.
Mg	n.d.	64.1 ± 3.9	n.d.	61 ± 15
Al	n.d.	2.6 ± 0.4	2.8 ± 0.4	n.d.
S	92 ± 18	n.d.	n.d.	n.d.
K	420 ± 60	446 ± 45	440 ± 66	460 ± 60
Ca	623 ± 60	655 ± 39	n.d.	690 ± 110
Mn	145 ± 7	159 ± 11	156 ± 12	144 ± 21
Fe	49.8 ± 3.0	49.5 ± 5.0	n.d.	57 ± 15
Zn	15.5 ± 1.6	< 17	n.d.	n.d.
Pb	0.40 ± 0.08	n.d.	n.d.	n.d.

Table II: IAEA Intercomparison W-3/1, Trace Elements in Water (1980). Element concentrations in µg/ℓ.

Element	TXRF	Input Value	Overall Mean of Accepted Values	No. of Lab./ Outlier	XRF
As	23.1 ± 2.2	22.4	20.8 ± 7.2	29/2	2.9
Au	2.5 ± 1.1	2.3	2.7 ± 1.2	9/1	n.d.
Ba	122 ± 23	99.6	96.2 ± 31.2	19/0	46-47
Cd	n.d.	4.0	3.9 ± 1.6	44/1	0.8
Co	2.0 ± 0.3	2.24	2.14 ± 0.24	26/8	n.d.
Cr	3.1 ± 1.3	2.0	2.6 ± 0.95	34/5	4.1
Cu	10.0 ± 1.3	9.06	8.74 ± 1.8	47/6	6-11
Fe	118 ± 5.9	125.8	122.3 ± 18.1	49/2	110-288
Hg	1.5 ± 0.3	2.0	2.15 ± 0.6	27/2	17.6
Mn	9.4 ± 1.3	9.6	8.4 ± 2.0	44/7	6.2
Mo	n.d.	2.3	3.6 ± 1.2	9/3	n.d.
Ni	2.0 ± 0.3	2.0	1.95 ± 0.4	22/9	n.d.
Pb	10.7 ± 1.6	10.0	9.6 ± 3.3	45/6	2.9-29
Se	20.9 ± 1.5	21.2	20.0 ± 2.5	19/4	n.d.
U	2.7 ± 1.2	2.04	1.8 ± 0.7	16/4	100
V	1.7 ± 0.4	2.0	1.9 ± 0.4	10/3	n.d.
Zn	22.3 ± 1.0	23.2	22.8 ± 2.4	52/14	22-76

(Total No. of Laboratories = 58; TRFA Code No. 33)

Table III: IAEA Intercomparison Run Soil-7. Element concentrations in µg/g.

	TXRF	INAA	14 MeV NAA
Al	–	44800 ± 3600	47300 ± 4000
As	11 ± 3	12.5 ± 1.0	13 ± 3
Ba	216 ± 76	131 ± 26	127 ± 36
Br	–	–	8.1 ± 0.6
Ca	167000 ± 17000	194300 ± 20000	153000 ± 23000
Ce	–	63.0 ± 6.0	74 ± 15
Co	–	8.89 ± 0.44	–
Cr	58 ± 12	70 ± 6	77 ± 8
Cs	–	5.57 ± 0.05	7 ± 2
Cu	16 ± 3	–	–
Eu	–	1.05 ± 0.05	–
Fe	24500 ± 1300	26300 ± 1300	26300 ± 1600
Ga	10 ± 2	–	–
Hf	–	4.87 ± 0.50	–
K	12500 ± 1900	–	12400 ± 1600
La	–	29 ± 3	–
Mg	–	11700 ± 2300	13900 ± 1800
Mn	632 ± 32	636 ± 51	693 ± 55
Na	–	2360 ± 350	2300 ± 400
Nb	14 ± 4	–	14 ± 4
Ni	27 ± 5	27 ± 5	38 ± 10
Pb	75 ± 5	–	–
Rb	56 ± 6	51.2 ± 4.0	52 ± 7
Sb	–	1.34 ± 0.27	–
Sc	–	8.82 ± 0.44	–
Si	–	–	180000 ± 14000
Sm	–	5.0 ± 0.2	–
Sr	116 ± 12	135 ± 27	104 ± 20
Ta	–	0.66 ± 0.05	–
Tb	–	0.65 ± 0.13	–
Th	11 ± 3	9.1 ± 0.9	–
Ti	3010 ± 210	2670 ± 270	3000 ± 300
U	–	2.21 ± 0.22	–
V	58 ± 12	68 ± 7	–
Y	22 ± 3	–	24 ± 5
Yb	–	1.74 ± 0.35	–
Zn	109 ± 8	116 ± 12	–
Zr	108 ± 32	182 ± 27	173 ± 14

Table IV: IAEA Intercalibration Sample Sediment SD-N-1/2. Element concentrations in µg/g.

	TXRF	INAA	14 MeV NAA
Ag	-	-	-
Al	-	2.0 ± 0.4	38800 ± 1800
As	61.5 ± 6.0	37200 ± 1800	54 ± 5
Au	-	55 ± 5	-
Ba	349 ± 45	0.042 ± 0.010	320 ± 70
Br	-	290 ± 40	-
Ca	56010 ± 1840	59 ± 6	56190 ± 5900
Cd	17.7 ± 1.0	58400 ± 3800	-
Ce	65 ± 4	11 ± 3	-
Co	-	58 ± 6	-
Cr	148 ± 15	12.4 ± 0.8	145 ± 14
Cs	-	158 ± 16	8 ± 1
Cu	75.2 ± 6.0	5.4 ± 0.5	-
Dy	-	-	-
Eu	-	4.5 ± 1.5	-
Fe	37430 ± 870	1.14 ± 0.11	37100 ± 2100
Hf	-	37400 ± 2000	-
Hg	-	8.4 ± 0.8	-
Ga	-	1.52 ± 0.15	-
K	9.9 ± 2.0	15800 ± 2000	14400 ± 1000
La	16460 ± 750	31.6 ± 4.0	-
Mg	-	7600 ± 2000	8200 ± 940
Mn	781 ± 34	794 ± 30	840 ± 40
Na	-	10500 ± 800	10500 ± 1000
Nb	9.5 ± 0.8	-	-
Nd	-	30 ± 3	-
Ni	36.0 ± 2.3	34.3 ± 2.5	-
O	-	-	563700 ± 16900
Pb	132 ± 9	-	78 ± 40
Rb	77 ± 5	90 ± 5	100 ± 60
Sb	-	4.3 ± 0.5	60 ± 22
Sc	-	7.1 ± 0.5	7 ± 5
Se	3.0 ± 1.5	2.9 ± 0.5	-
Si	-	-	285000 ± 10800
Sm	-	6.0 ± 0.6	-
Sn	17 ± 2	-	-
Sr	325 ± 14	325 ± 30	300 ± 40
Ta	-	1.25 ± 0.15	-
Tb	-	0.81 ± 0.09	-
Th	6 ± 1	7.2 ± 1.0	-
Ti	2580 ± 180	2560 ± 400	2650 ± 300
U	-	2.4 ± 0.4	-
V	93.3 ± 11.0	78.8 ± 4.0	-
W	-	2.8 ± 0.8	-
Y	30 ± 4	-	-
Yb	-	4.9 ± 0.5	-
Zn	479 ± 15	464 ± 30	320 ± 15
Zr	145 ± 22	350 ± 50	-

Table V: IAEA Intercomparison Run Mussel Tissue Ma-M-2/TM. Element concentrations in µg/g.

	TXRF	INAA
Ag	-	0.062 ± 0.005
Al	-	359 ± 29
As	13.2 ± 1.3	12.3 ± 1.2
Au	-	0.0162 ± 0.0013
Br	-	417 ± 33
Ca	15 400 ± 1500	16 050 ± 1600
Cd	1.65 ± 0.33	1.9 ± 0.5
Ce	-	0.29 ± 0.04
Cl	-	87 100 ± 9000
Co	-	0.882 ± 0.044
Cr	1.8 ± 0.7	1.23 ± 0.15
Cs	-	0.087 ± 0.009
Cu	8.0 ± 1.0	-
Eu	-	0.0083 ± 0.0008
Fe	282 ± 8	276 ± 14
Hf	-	0.0129 ± 0.0013
Hg	0.98 ± 0.15	0.91 ± 0.07
I	-	17 ± 4
K	12 100 ± 2400	11 700 ± 1800
La	-	0.25 ± 0.04
Mn	69.9 ± 3.5	66.8 ± 5.3
Mo	0.80 ± 0.16	-
Na	-	45 500 ± 7000
Ni	1.20 ± 0.24	1.19 ± 0.18
Pb	2.4 ± 0.5	-
Rb	6.2 ± 0.6	7.20 ± 0.58
S	19 400 ± 4000	-
Sb	-	0.027 ± 0.005
Sc	-	0.044 ± 0.004
Se	2.14 ± 0.32	2.54 ± 0.20
Sr	97.0 ± 8.0	107 ± 16
Ta	-	0.0049 ± 0.0007
Tb	-	0.0065 ± 0.0016
Th	-	0.064 ± 0.006
U	-	< 3
V	-	0.20 ± 0.03
Zn	2.2 ± 0.9	
	169.1 ± 5.1	157 ± 13

4. INTERLABORATORY TESTS

Precision and accuracy of the analytical methods applied are repeatedly examined by taking part in interlaboratory tests for all relevant matrices. Some recent intercomparisons are summarized in Tables II to V. The data refer to the following samples:
- IAEA Intercomparison W-3/1, Trace Elements in Water (1980),
- IAEA Intercomparison Run Soil-7
- IAEA Intercalibration Sample Sediment SD-N-1/2 and
- IAEA Intercomparison Run Mussel Tissue Ma-M-2/TM.

Table II gives an indication of the performance of TXRF. The other Tables also include INAA and 14 MeV NAA and allow an intercomparison of these three methods. While for the water sample input values and overall means of accepted values already exist, data from other laboratories and mean values for the last three samples are not yet available.

REFERENCES

[1] W. Michaelis, Proc. Int. Conf. on Heavy Metals in the Environment, September 6 - 9, 1983, Heidelberg, Germany, Vol. II, p. 972 - 975

[2] W. Michaelis, GKSS 83/E/39

[3] B. Anders, W. Junge, J. Knoth, W. Michaelis, R. Pepelnik, H. Schwenke, Vth Int. Conf. on Nuclear Methods in Environmental and Energy Research, April 2 - 6, 1984, Mayaguez, Puerto Rico

[4] GKSS Research Centre, Gewässeranalytische Untersuchungen auf der Unterweser im Herbst 1979, Teil 2: Hochauflösende hydrographische Messungen und Spurenanalytik, GKSS 80/E/27

[5] GKSS Research Centre, Schadstoffuntersuchungen an ausgewählten Standorten mit Muschelvorkommen im nordfriesischen Wattenmeer

[6] AFR-Berichte, Elementanalyse von Schwebstäuben, KfK-AFR 006

[7] W. Michaelis, H. Böddeker, J. Knoth, H. Schwenke, Vth World Congress on Air Quality, May 16 - 20, 1983, Paris, France, Vol. I, p. 391 - 398

[8] F. Ackermann, H. Bergmann, U. Schleichert, Fresenius Z. Anal. Chem. 296 (1979) 270 - 276

[9] G. Tölg, Naturwissenschaften 63 (1976) 99 - 110

[10] J. Müller, G. Kallischnigg, ZEBS-Berichte 1/1983, Dietrich Reimer Verlag

[11] J. Knoth, H. Schwenke, Fresenius Z. Anal. Chem. 301 (1980) 7 - 9

[12] J. Knoth, H. Schwenke, Fresenius Z. Anal. Chem. 294 (1979) 273 - 274

[13] H.-U. Fanger, R. Pepelnik, W. Michaelis, 4th Int. Conf. on Nuclear Methods in Environmental and Energy Research, April 14 - 17, 1980, Columbia, MO, USA, CONF-800 433, p. 195 - 204

[14] B. Anders, E. Bössow, GKSS 83/E/23

[15] M. Schönburg, GKSS 83/E/57

[16] W. Michaelis, H. Böddeker, K. Kramer, R. Niedergesäß, B. Racky, C. Schnier, K. Weiler, GKSS (to be published)

[17] J.G. Duinker, M.T.J. Hillebrand, R.F. Nolting, S. Wellershaus, Neth. J. Sea Res. 15 (2) (1982) 141 - 169

[18] J. Bauch, private communication

MULTI-ELEMENT-ANALYSE VON VIER GEOLOGISCHEN STANDARD-REFERENZ-PROBEN
MIT HILFE DER ICP-OES, NAS UND SSMS

P. Dulski, J. Luck, W. Szacki

Hahn-Meitner-Institut für Kernforschung Berlin GmbH, D-1000 Berlin 39,
Glienicker Str. 100

Geochemische Untersuchungen erfordern zum Teil umfassende Bestimmungen
von Haupt-, Neben- und Spurenelementen in den unterschiedlichsten
Matrices. Dafür sind geologische Multi-Element-Standards erforderlich,
die in ihrer Hauptzusammensetzung mit den zu analysierenden Proben
vergleichbar sein sollten. Nur dadurch lassen sich Matrix-Effekte
bei einigen Analysemethoden vermeiden.

Der Bedarf an neuen zertifizierten Standard-Referenz-Materialien (SRM)
nimmt zu, da die bekannten Standards zum großen Teil nicht mehr zur
Verfügung stehen und sich das Arbeitsfeld der Geochemie ständig erweitert.

Im Rahmen verschiedener Ringversuche wurden mit Hilfe der drei in
unserem Laboratorium zur Verfügung stehenden Analysemethoden, induktiv
gekoppeltes Plasma mit optischer Emissionsspektroskopie (ICP-OES),
Neutronenaktivierungsspektrometrie (NAS) und Funkenquellen-Massenspektrometrie (SSMS), zwei geologische Multi-Element-Standards
    AL-1 (Albit), GIT (Groupe International de Travaille, Frankreich)
    W-2 (Diabas), USGS (US Geological Survey, USA)
und zwei Standards für Einzelelemente
    TAN-1 (Tantal-Erz)   ) CANMET (Canada Centre for Mineral
    OKA-1 (Niob-Erz, karbonatisch) and Energy Technology/Kanada)
bezüglich ihrer Haupt-, Neben- und Spurenelemente analysiert. Folgende
Elemente wurden in den aufgeführten Standards bestimmt:

AL-1: Na, Mg, Ca, Sc, Ti, V, Cr, Mn, Fe, Co, Cu, Zn, Ga, Ge, As, Rb,
Sr, Y, Zr, Nb, Sb, Cs, Ba, La, Ce, Pr, Sm, Eu, Gd, Tb, Yb, Lu,
Hf, Ta, Th, U.

W-2: F, Na, S, Cl, K, Ca, Sc, Ti, V, Cr, Mn, Fe, Co, Ni, Cu, Zn,
Ga, Ge, As, Rb, Sr, Y, Zr, Nb, Sb, Cs, Ba, La, Ce, Sm, Eu,
Tb, Yb, Lu, Hf, Ta, Th.

TAN-1: Na, K, Sc, V, Cr, Mn, Fe, Co, Ni, Cu, Zn, Ga, Ge, As, Rb,
Sr, Y, Zr, Nb, Cs, Hf, Ta.

OKA-1: Na, Mg, P, K, Ca, Sc, V, Cr, Mn, Fe, Co, Ni, Cu, Zn, Ga, Ge,
As, Rb, Sr, Y, Zr, Nb, Sb, Cs, Ba, La, Ce, Pr, Nd, Sm, Eu,
Tb, Yb, Lu, Hf, Ta, Th, U.

Die Anzahl der mit den eingesetzten Methoden in den vier Standards
bestimmten Elemente ist der Tabelle 1 zu entnehmen.

Standard Methode	AL-1	W-2	TAN-1	OKA-1
ICP-OES	14	11	12	23
NAS	17	20	12	23
SSMS (Platte)	12	24	-	-
SSMS (switch)	12	11	16	12
Anzahl pro Standard	36	37	22	38

Tab. 1: Anzahl der mit den eingesetzten Methoden in den vier Standards
bestimmten Elemente

Für die Standards TAN-1 und OKA-1 liegen bereits Zertifikat-Werte
für Tantal (0.236 %) /1/ bzw. Niob (0.37 %) /2/ vor. Die mit unseren
Methoden ermittelten Werte stimmen im Rahmen der Meßgenauigkeit sehr
gut mit den zertifizierten Werten überein.
Für den USGS Standard W-2 ist eine vorläufige Zusammenfassung der
publizierten Analysenergebnisse erschienen /3/. Die geringe Zahl der
publizierten Daten pro Element (maximal 6) läßt eine genaue Aussage
über die zu empfehlenden Werte noch nicht zu, zumal die Streuung der
Einzelwerte zum Teil beträchtlich ist. Ein Vergleich der mit den uns
zur Verfügung stehenden Methoden ermittelten Elementgehalte mit den
in /3/ angegebenen Mittelwerten, zeigt im allgemeinen eine recht gute
Übereinstimmung.

Für die Elemente, die mit zwei oder mehr Methoden bestimmt wurden,
kann gezeigt werden, daß die eingesetzten Analyseverfahren eine ausreichende Übereinstimmung aufweisen. Außerdem wurde die Zahl der bestimmten Elemente durch den Einsatz mehrerer Elemente beträchtlich
erhöht (Tabelle 1).

REFERENZEN
/1/ Steger, H.F., and Bowman, W.S.: CANMET Report 83-10E (1983)
/2/ Steger, H.F., and Bowman, W.S.: CANMET Report 81-1E (1981)
/3/ Gladney, E.S., Burns, C.E., and Roelandts, I.: Geostandard News-
    letters 7, 1 (1983) 3-226

ATOMMEMISSIONSSPEKTROMETRIE MIT ICP– UND DC–ARC–ANREGUNG:
Ein Vergleich

G. DREWS, Römisch-Germanisches Zentralmuseum, Mainz

Geringe Matrixabhängigkeit, großer dynamischer Meßbereich mit nahezu linearen Eichkurven und niedrige Nachweisgrenzen begründen in der Analytik von Lösungen die eindeutige Überlegenheit der Atomemissionsspektrometrie mit ICP-Anregung über die mit DC-Arc-Anregung. Dies gilt weitgehend auch für die Festkörper-Analytik. Allerdings bedeutet die Notwendigkeit der Überführung eines Festkörpers in eine Meßlösung in vielen Fällen eine erhebliche Verdünnung der zu bestimmenden Festkörperkonzentration in der Meßlösung. Dadurch können Festkörperkonzentrationen, die durchaus im quantitativen Meßbereich der DC-Arc-Technik liegen, in der Meßlösung bis an die Nachweisgrenzen absinken, wodurch auch die ICP-Messungen mit Meßfehlern behaftet werden, wie sie für die DC-Arc-Technik typisch sind (z.B. RSD von ca. 10-20 %). Die wesentlichsten Gründe für hohe Probenverdünnungen sind:

A) Sehr geringe Festkörper-Probemengen (häufig in der Analytik archäologischer Proben)
B) Schwerlösliche Proben, die zur Herstellung einer Meßlösung z.B. einen Schmelzaufschluß notwendig machen (häufig in der Analytik geologischer Proben).
C) Starke Variation interferierender Haupt- und Nebenelement-Konzentrationen bei On-line-Korrekturen der Interferenzen.

Besonders häufig beim Einsatz von Quantometern, die kein Ausweichen auf schwächere Analysenlinien ermöglichen. Hier schlägt oft der Vorteil der hohen Empfindlichkeit der ICP-Technik in einen Nachteil um: während in der DC-Arc-Technik die Variation hoher Elementkonzentrationen infolge der Selbstabsorption in der Lichtquelle nur noch begrenzt für die Interferenzen wirksam wird, macht sie sich in der ICP-Technik gerade wegen des weitgehenden Fehlens der Selbstabsorption voll bemerkbar.

BEISPIEL ZU (A)
Bestimmung von ca. 10 ppm Cr, Ni, Ag in 10 mg Festkörper-Probe.

Lösungsverbrauch:	ca. 2 ml/min
Vorspülzeit:	ca. 1 min
Integrationszeit:	10 sec
Min.Lösungsbedarf:	ca. 2,5 ml
Verdünnungsfaktor:	250; d.h. Elementkonz. in Meßlösung: 0,04 ppm
Nachweisgrenzen (Lit.-Werte):	
DC-Arc-Technik:	ca 1 - 5 ppm
ICP-Technik:	ca 0,003 - 0,01 ppm

Fazit: Trotz der niedrigen Nachweisgrenzen wird auch in der ICP-Technik an der Nachweisgrenze gemessen. Es bietet sich daher an, ein Quantometer mit ICP-Anregung zusätzlich mit einer DC-Arc-Anregung auszurüsten, um — zumindest für halbquantitative Bestimmungen — den Vorteil der einfachen und schnellen Probenpräparation der DC-Arc-Technik nutzen zu können.

Unter diesen Gesichtspunkten wurde der SPEX-COMMON-ELEMENT-STANDARD, ein für die halbquantitative DC-ARC-Analyse auf Li-Karbonat-Basis zusammengestellter Multielement-Festkörperstandard mit den Konzentrationsabstufungen 0,1% (L1); 0,033% (L1A); 0,01 % (L2); 0,003% (L1A) und 0,001% (L3) mit je 0,1% In als Innerem Standard sowohl mit der ICP- als auch der DC-Arc-Technik mit demselben Quantometer gemessen.

EMPFOHLENE UND VERWENDETE LINIEN (Å)

Element	DC-Arc	ICP (Lit)	ICP
	(1)	(2)	(3)
Be	2348,6	2348,6	2348,6 X3
Al	3961,5	3961,5	3961,5 X2
Ti	4981,7	3349,4	3349,4 X2
V	4379,2	3093,1	2924,6 X2
Cr	4254,3	2055,5	2677,2 X2
Mn	4034,5	2576,1	2576,1 X2
Fe	3581,2	2599,4	2599,4 X2
Ni	3414,5	2316,0	3414,4 X2
Cu	3247,5	3247,5	3247,5 X2
Sr	4607,3	4077,7	4077,7 X1
Ag	3280,7	3280,7	3280,7 X2
Cd	2288,0	2288,0	2288,0 X2
Sn	3175,0	1970,8	3034,1 X2
Sb	2068,4	2068,4	2068,4 X3
Bi	3067,7	1953,9	2230,6 X2
In	4511,3	3256,1	4511,3 X1

(1) Empfohlene DC-Arc-Linien nach: Ahrens & Taylor, Spectrochemical Analysis, 2nd ed.
(2) Empfohlene ICP-Linien nach: KONTRON-ICP-Berichte 7 (1979).
(3) In dieser Arbeit für DC-Arc und ICP verwendete Linien in X. Ordnung.

Gerät: Luft-Simulatan-Spektrometer des ARL 3580 mit Int. ICP-Stativ und Spex-Universal-Stativ.
Gitter: Bausch & Lomb 1080 Str./mm
Eingangsspalt: 20μ

Meßbedingungen:
DC-Arc-Anregungen: 10 mg Standard + 20 mg Kohlequlver in Harvey-Elektroden
Probe =Anode; 22,5 Ampere Gleichstrom konst. Elektrodenabstand: 4 mm
Zwischenabbildung; beobachtete Zone: 2-3 mm über Anode Integrationszeit: 99 sec.
Standardabweichung für Intensität Meßelement/Intensität.In:Ag: ca 8 %; Rest: ca. 20 % (n = 10)
ICP-Anregung: Li-Metaborat-Aufschluß: 1 Tl.Standard +2 Tl.Flux
Probenverdünnung: 1: 200 in 2n-HCL
Zerstäubersystem: Babbington GMK
Probenverbrauch: 1,75 ml/min.
Integrationszeit: 10 sec.
Standardabweichung für Intensität Meßelement/Intensität In ca. 2%. (n = 22)

Die erhaltenen Eichkurven (Abb. 1 u.1 A f. DC-Arc ; 2 und 2 A für ICP) zeigen deutlich die Überlegenheit der ICP-Technik hinsichtlich Reproduzierbarkeit und dynamischem Meßbereich, aber auch, daß sich mit der klassischen DC-Arc-Technik bezogen auf den Festkörper für manche Elemente gleich oder bessere Nachweisgrenzen erzielen lassen.

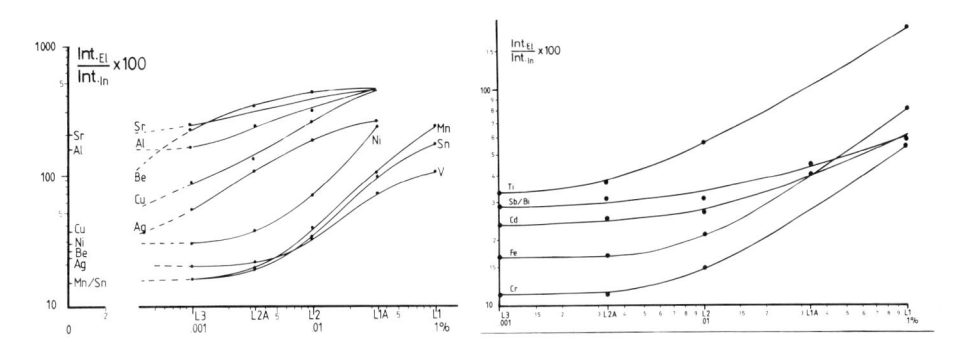

Abb. 1 und 1 A

DC-Arc -Eichkurven; Mittelwerte (n=5) Intensität Meßelemente/Intensität In gegen die Festkörperkonzentrationen.

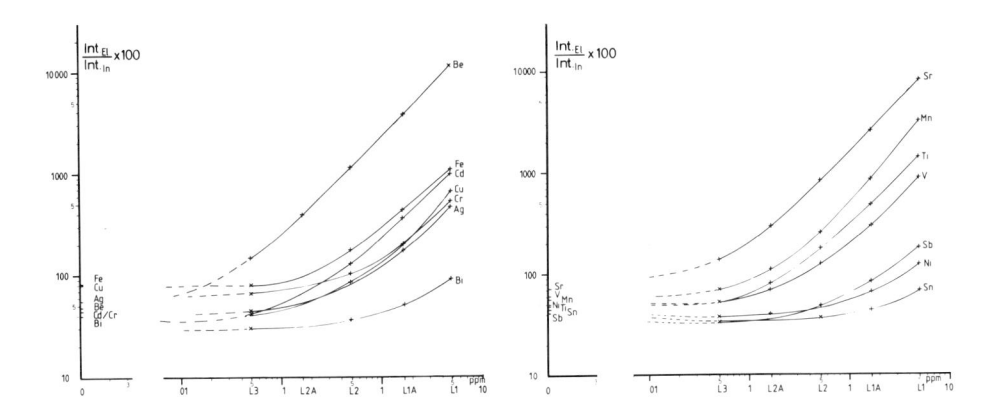

Abb. 2 und 2 A

ICP-Eichkurven; Mittelwerte (n=22) Intensität Meßelement/Intensität In gegen die Elementkonzentrationen in der Meßlösung (L1-L3 = Standard)

COMPARATIVE ANALYSES OF NATURAL ROCK SAMPLES BY MASS SPECTROMETRY, X-RAY FLUORESCENSE, ATOMIC ABSORPTION SPECTROMETRY AND GAMMA-RAY SPECTROMETRY

Hansen, B.T.[1], Henjes-Kunst, F.[2], Baumann, A.[1], Jecht, U.[3]
1) Mineralogisches Institut der Universität Münster, 2) Institut für Petrographie und Geochemie der Universität Karlsruhe, 3) Siemens AG, Karlsruhe

Most methods for analyzing natural rock samples are matrix dependant.
    The aim of this study is to determine some main and trace elements in natural rock samples by a matrix independant method: Mass spectrometric analyses, applying the isotope dilution method, were performed on six igneous rocks (GR, granite; TO, tonalite; GA, gabbro; NP, nosean phonolite; MB, melilite basalt; PE, peridotite) and three rocks of sedimentary origin (SH, shale; GW, greywacke; CG, cordierite gneiss).
The samples have been analyzed for Ca, K, Ba, Sr, Rb, Pb, U, and Th so far.
Different sample aliquots were spiked for the individual elements before dissolution. Element-separations were made by ion exchange techniques. The mass spectrometric analyses were performed with a Teledyne NBS-type 12"90° solid source mass spectrometer using Re-filaments.
The range of concentrations obtained are listed in the following table:

Element	Range of concentrations		Absolute deviations Max.	Average	Methods
CaO (wt%)	0.04	14	0.26	<0.1	MS,AAS,XRF
K_2O (wt%)	0.01	9	0.25	<0.2	MS,AAS,XRF,GRS
Ba (ppm)	3	1300	55	<20	MS,AAS
Sr (ppm)	5	1700	11	<5	MS,XRF
Rb (ppm)	0.3	300	3.2	<1	MS,XRF
Pb (ppm)	0.3	40			MS
U (ppm)	≤0.3	6	1.5	<0.4	MS,GRS
Th (ppm)	≤0.1	30	2	<1.0	MS,GRS

The concentrations of CaO (Fig 1) and K_2O (Fig 2) were determined by three independant methods: mass spectrometry (MS), X-ray fluorescense (XRF) and atomic absorption (AAS); K_2O additionally by gamma-ray spectrometry (GRS).

The average absolute deviations of <0.2 wt% between the values obtained by the different methods correspond to ralative deviations of <5%. This is considered to be a good result for analyses of silicate rocks.

The agreement of the Ba values (Fig 3) beween MS and AAS depends on the concentration of this element in the sample, i.e. concentrations <500 ppm show relative deviations in the order of 5-10%, whereas concentrations >500 ppm are characterized by relative deviations of <5%.

Sr (Fig 4) and Rb (Fig 5) determinations by MS and XRF yielded a very good agreement of the values over a wide range of concentrations. The relative deviations are <2%.

The concentrations for U (Fig 6) and Th (Fig 7) obtained by GRS show average absolute deviations in the order of 0.3 ppm for U and 1 ppm for Th.

These investigations are part of a joint project between Siemens AG, Karlsruhe, and Institut für Mineralogie, Universität Münster, preparing natural rock samples as reference material for XRF.

Acknowledgements

We wish to thank Z. Solyom (Lund), K.H. Becher, L. Richter, E. Gohn (Göttingen) and R. Altherr (Karlsruhe) who kindly provided some of the comparative analyses. The able assistance of B. Borchardt and I. Büning is kindly acknowledged.

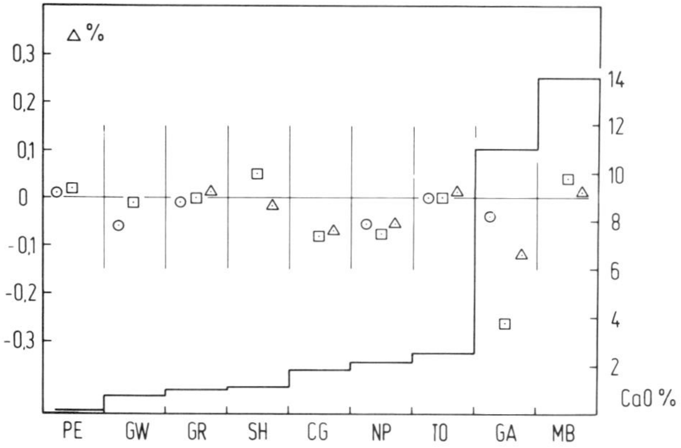

Comparative analyses of rock samples

Fig 1   The figures 1-7 show the element concentrations in the investigated samples and the absolute deviations of values obtained by different methods using matrix corrections as compared to the mass spectrometric data (△).

⊙ XRF (Karlsruhe)   ▫ XRF (Lund)
◐ XRF (Münster)   △ AAS (Münster)
⬡ γ-spectrometry (Göttingen)

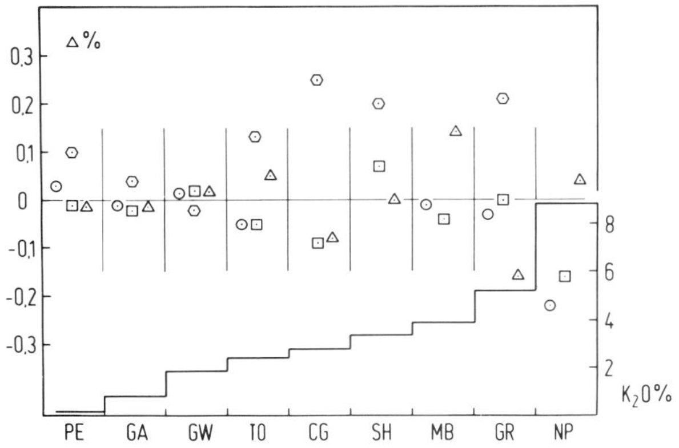

Fig 2

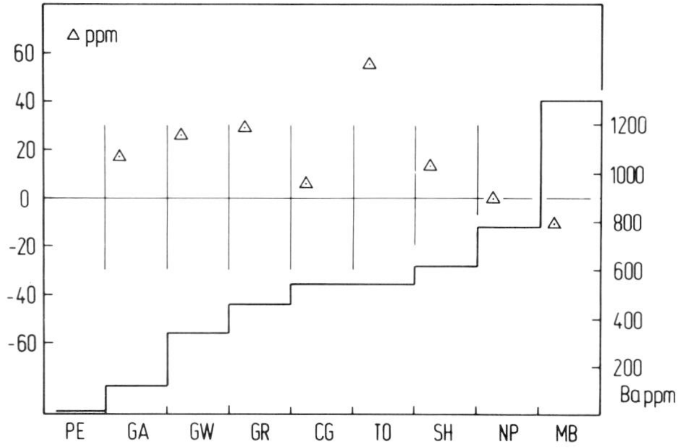

Fig 3

Fig 5

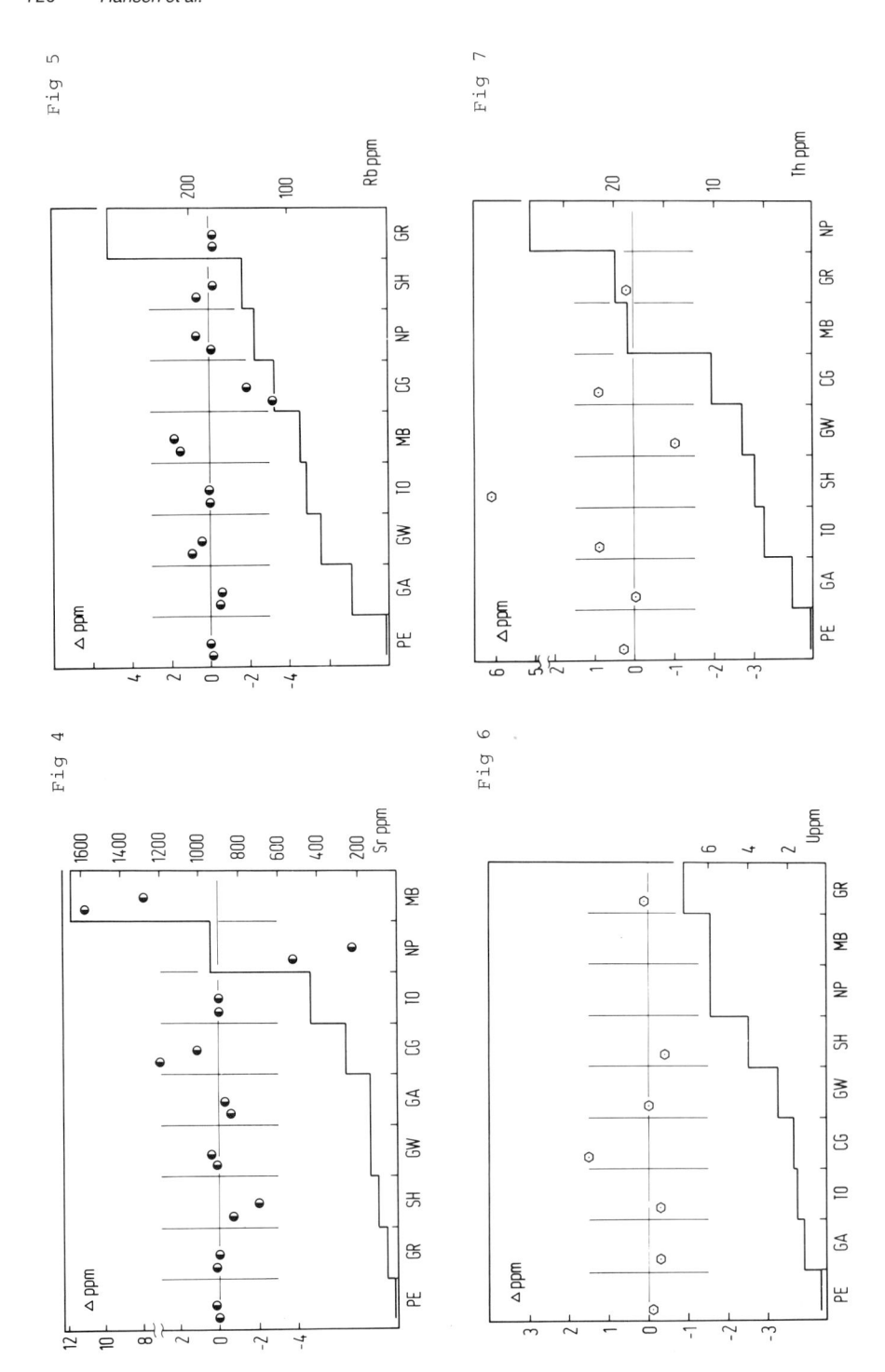

Fig 7

Fig 4

Fig 6

Anhang

1. GROSSGERÄTE ZUR INSTRUMENTELLEN MULTIELEMENTANALYSE

L. Radermacher, B. Sansoni

Zentralabteilung für Chemische Analysen

Kernforschungsanlage Jülich GmbH, D-5170 Jülich-1

Ausgangspunkt jeder Instrumentellen Multielementanalyse ist das dazu erforderliche Gerät. Dieses ist meist relativ aufwendig, was durch weitgehende Computerisierung und Automatisierung noch verstärkt wird. Um eine erste Information zu geben, werden zunächst die wichtigsten derzeit im Handel zugänglichen Geräte aufgelistet. Es folgen Abbildungen entsprechender Geräte aus der dem Symposium angeschlossenen Industrieausstellung. Sie stehen als Beispiel für viele andere ähnliche Geräte. Am Ende des Kapitels werden Anschriften der auf dieser Ausstellung vertretenen sowie weiterer hier genannter Firmen gegeben.

1.1 Zusammenstellung handelsüblicher Spektrometer

Folgende Liste zählt die im Handel befindlichen Spektrometer zur Instrumentellen Multielementanalyse auf. Sie sind nach Spektrometriearten gegliedert. Die Zusammenstellung ist bei weitem nicht vollständig. Auch kann für die Angaben keine Gewähr übernommen werden.

1.1.1 Kernstrahlungsspektrometrie

Ein Meßplatz für Kernstrahlungsspektrometrie besteht im allgemeinen aus einem Detektor mit Hochspannungsversorgung, Impulsverstärker, Vielkanalanalysator, Rechner und Programmen zum Steuern des Analysators sowie zur Auswertung der gemessenen Spektren. Alpha- und Gammaspektrometrie unterscheiden sich wesentlich nur hinsichtlich der Detektoren und der Art der Auswertung der Spektren.

a) Vielkanalanalysatoren

Firmen: Canberra (Modelle: 10, 20, 35, 85, 90),
Nuclear Data (Modelle: 60, 62, 65, 66, 76, 620, 680, 6600, 6700),
Ortec (Modell: ADCAM), Silena (Modell: CATO), Tennelec (Modell: mate 381), Geoscience, Intertechnique, Laben, Nokia.

Die Kosten beginnen für kleine tragbare Geräte in der Größenordnung mit etwa 15.000,-- DM, für große Systeme mit Rechner ab etwa 30.000 und reichen bis 200.000,-- DM.
Wichtigster Beurteilungspunkt des Analysators ist der Analog-Digital-Converter (ADC). Seine Hauptkriterien sind: Geschwindigkeit, Linearität und Stabilität. Die Konvertierung kann nach zwei Prinzipien erfolgen: Wilkinson Typ (die Geschwindigkeit wird durch die Dekodierfrequenz ausgedrückt, übliche Werte liegen zwischen 100 und 200 MHz) und das Wägeprinzip (Impulsverarbeitungszeit etwa 3 - 15 μsec).

b) Detektoren
- Alphaspektrometrie

Hierfür kommen vor allem der Si(Li)-Sperrschichtzähler, das Proportionalzählrohr sowie die Gitterionisationskammer in Frage.

Die Gitterionisationskammer der Fa. Kimmel, Münchener Apparatebau, ist für Probendurchmesser von bis zu 20 cm Ø geeignet. Sie zeichnet sich durch relativ hohe Auflösung von bis zu etwa 25 keV für die 5,15 MeV-Linie des Pu-239 aus.

Si(Li)-Sperrschichtzähler liefern die Firmen Canberra, Nuclear Data, Ortec, Schlumberger.
Proportionalzählrohre sind erhältlich bei den Firmen Canberra, Princeton Gammatech, Getac, Ortec, Schlumberger.

- Betaspektrometrie

Die Betastrahlung besitzt ein kontinuierliches Energiespektrum und erfordert daher für Messung und Auswertung spezielle Geräte und Verfahren. Die am häufigsten angewendete Technik ist die Flüssigszintillationsspektrometrie (liquid scintillation counting). Es gibt nur komplette Geräte, die aus dem Probenwechsler, Zweifotodetektoren, Verstärker, Koinzidenzschaltung, zwei bis sechs Zählkanälen, Rechner und Auswertesoftware bestehen. Lieferfirmen sind Berthold, Kimmel Münchener Apparatebau, LKB, Packard.

- Gammaspektrometrie

Die beiden mit Abstand wichtigsten Detektoren sind der Ge(Li)- sowie der Reinstgermaniumdetektor, wobei der Trend zu letzterem geht. Beide besitzen ein extrem hohes Auflösungsvermögen. Der Reinstgermaniumdetektor zeichnet sich dadurch aus, daß er während der meßfreien Zeit nicht mit flüssigem Stickstoff gekühlt werden muß und bis weit in das Gebiet der weichen Röntgenstrahlung anwendbar ist. Lieferfirmen sind u.a. Canberra, Nuclear Data, Ortec, Schlumberger, Princeton Gammatech, Getac.

c) Verstärker: Canberra (Modelle: GI 2020, 2021, 2022), Ortec (Modelle: 673, 972), Schlumberger (Modell: 7169), Silena (Modelle: 7611, 7612), Tennelec (Modelle: 205A, 243).

1.1.2 Massenspektrometrie (MS)

Hochauflösende Festkörpermassenspektrometer arbeiten zumeist mit doppelt fokussierender Trennung, weniger mit einem Quadrupolsystem. Die wesentliche Untergliederung erfolgt nach der Art der Anregung.

a) Funkenanregung (spark source)
Firma: Jeol (Japan)
      Gerätetyp: JMS-01BM-2
Dieses Gerät ist das einzige zur Zeit verfügbare kommerzielle Gerät. Es eignet sich zur Multilement-Spurenanalyse in Metallen und Isolatoren. Es besitzt sowohl einen Fotoplatten-Ionennachweis als auch einen elektrischen Ionennachweis.

b) Glimmentladung (Glow-discharge, GDMS)
Firma: VG-Instruments
      Gerätetyp: VG-9000
Dieses Gerät eignet sich zur Multielementanalyse für elektrisch-leitende Feststoffe, aber auch für Halbleiter und Isolatoren (Presselektroden).

c) Laseranregung (laser excitation)
Eigenbaugeräte
Die Firmenentwicklungen sind noch nicht abgeschlossen. Diese Geräte werden zur Multielementanalyse an Festkörpern, Metallen, Halbleitern und Isolatoren eingesetzt. Neuerdings bietet die Firma JEOL eine Laseranregung als Option an.

d) ICP-MS (inductively coupled plasma)
Firmen: Sciex, Kanada (Schweiz) Gerätetyp: Elan 250 ICP/MS
        VG-Instruments, (England) Gerätetyp: Plasma Quad
Diese Geräte wurden zur Multielementanalyse in Lösungen entwickelt. Eine ausführliche Gegenüberstellung der technischen Daten beider Geräte wurde von J.A.C. Broekaert (1) gegeben (Literaturangaben in 7.1.1.8).

1.1.3 Röntgenfluoreszenzspektrometrie (RFA)

Nach dem Aufbauprinzip unterscheidet man die wellenlängen-, energiedispersive und totalreflektierende Röntgenfluoreszenzspektrometrie. Hinsichtlich des Multielementcharakters unterscheidet man sequentiellen und simultanen Betrieb. Im wesentlichen kommen folgende Firmen in Frage.

a) Wellenlängendispersive RFA

Firma	Gerätetyp	Meßanordnung	Besonderheiten
Applied Research Laboratories	8400	sequentiell	Stirnfensterröhre, schnelle Meßpositionierung
	8600	simultan	
Asoma	LCA	sequentiell	tragbares Gerät, relativ niedriger Preis
Mitterfellner	Portaspec Rosa 1	sequentiell simultan	tragbares Gerät, relativ niedriger Preis
Philips	1404	sequentiell	100 kV, Multilayer
	1606	simultan	
Rigaku	S/Max	sequentiell	Stirn- oder Seitenfensterröhre, Bestrahlung von oben, B und C-Bestimmung möglich
Shimadzu	VF-320	sequentiell	Stirn- oder Seitenfensterröhre, Bestrahlung von oben
Siemens	SRS 300	sequentiell	Stirnfensterröhre
	MRS 400	simultan	

b) Energiedispersive RFA

Firma	Gerätetyp	Meßanordnung	Besonderheiten
Edax	711		
Getac	Analyst 770		Häufigste
Kevex	770	simultan	Anwendung als Zusatz
Link	860		von Elektronenmikroskopen
Ortec	Tefa 5000		
Tracor	4020		

c) Totalreflektierende RFA

Firma	Gerätetyp	Besonderheiten
Seifert	Extra II	niedrige Nachweisgrenzen bis in den Spurenbereich, flüssige Proben

1.1.4 Optische Atomemissionsspektralanalyse (AES)

Wegen ihrer besonderen aktuellen Bedeutung wird hierbei die Plasmaanregung gesondert behandelt.

a) Plasma-Atomemissions-Spektrometrie

Man unterscheidet die Anregung durch Bogen-, Funken-, Glimmentladung, Laserstrahlung und Plasma. Wegen Einzelheiten sei auf die ausführlichen Besprechungen von Broekaert in Spectrochimica Acta verwiesen.

Firma	Gerätetyp	Besonderheiten (1) bis (8) siehe Literatur in 1.1.8
Applied Research Laboratories	3510 3520 OES	schnelles Sequenz ICP-AES siehe (2)
Baird	Spectromet ICP	siehe (2)
Instrumentation Laboratory	IL Plasma-200	schnelles Sequenz ICP-AES
Jarrell Ash	Atomscan 2000 ICAP-9000	siehe (2) simultanes ICP-AES
Kontron	Plasmakon S 35 ES 750 ICP-ASS 80	siehe (1) siehe (5) siehe (4)
Labtest Equipment	Plasmascan 710	siehe (2)
Leeman Labs	Plasma-Spec	siehe (3)
Perkin Elmer	ICP/5500 ICP/6000	siehe (2) siehe (1)
Spectrametrics	Spectra Span VI	siehe (1), schnelles Sequenz DCP-AES

b) Sonstige Atomemissionsspektrometrie

Firma	Gerätetyp	Besonderheiten
I.S.A. (Jobin Yvon)	JY 32 E	siehe (3)
RSV-Präzisionsmeß- geräte		Glimmentladungs-AES, simultan bis ca. 90 Elemente
Shimadzu	GQM-500	siehe (6)
	GVM-500	siehe (6)
Spectro	Spectrolab	simultan AES (Funken)

1.1.5 Atomabsorptionsspektrometrie

Die Atomisierung erfolgt durch Flammen oder elektrothermal im Graphitrohr, die Kompensation mit Deuteriumlampe, durch Zeeman-Effekt oder nach Hieftje. Im folgenden bedeutet K=Kanal, E/D = Einzel-/Doppelstrahl, M=Monochromator.

Firma	Gerätetyp	Besonderheiten
Baird	A 3400	1 K, E, M=0,25 m Czerny-Turner
Erdman & Grün	ZAAS/SM1	siehe (7)
Hitachi (Colora Meßtechnik)	Z 6000	siehe (8)
	Z 7000	siehe (8)
	Z 8000	siehe (8)
	180	siehe (7), 1 K M=0,45 m Czerny-Turner, D2
Instrumentation Laboratory (IL)	IL Video 12/22	siehe (8), (1)
	IL 440 AVA	siehe (4)
	IL 951 V	2 K, D, M=0,33 m Ebert, D 2
	IL 751	2 K, D, " , D 2
	IL 551 V	1 K, D, " , D 2
	IL 257	1 K, D, " , D 2
	IL 157	1 K, E, " , D 2
Jarrell-Ash/ Fisher	850	1 K, D, M=0,40 m Czerny-Turner, D 2
	810	2 K, D, " , D 2
Maassen	GBC-903/902	siehe (8)
Perkin Elmer	2380	siehe (8)
	3030 b	siehe (8)
	3030	siehe (8),
	5000	siehe (8), (7)
	280	1 K, E, M=Czerny-Turner, D 2
	380	1 K, D, " , D 2
	560	1 K, D, " , D 2
	4000	1 K, D, " , D 2

Philips	SP-9	siehe (8), (4)
(Pye Unicam)	SP 191/192	M=0,33 m Ebert, D 2
	SP 2900	siehe (4)
	PU 9000	siehe (8)
Rank Hilger	H 1551	1 K, E, H 2
Shimadzu	AA-670	siehe (8)
Varian	Spectr.	
	AA 30/40	siehe (8)
	AA 975	siehe (6)
	AA 12/1475	siehe (8)
	AA 775	M=0,33 m Czerny-Turner, D 2

1.1.6 Atomfluoreszenzspektrometrie

Firma	Gerätetyp	Besonderheiten
Baird	Plasma/AFS 2000	Atomisierung mit ICP-Plasma, Anregung durch Hohlkathodenlampen, (ICP-HCL-AFS), siehe (3)

1.1.7 Mikrosonden

a) Elektronenmikrosonde

Firma	Gerätetyp	Besonderheiten
CAMECA		Elementverteilungsanalyse mit
JEOL (Kontron)		wellendispersivem System (Röntgenfluoreszenz)

b) Ionenmikrosonde

Firma	Gerätetyp	Besonderheiten
Atomika	a-Dida	Ionenbeschuß, Sekundärionenmassenspektrometrie (SIMS)
CAMECA	3 f	Elementverteilungsanalyse und Tiefenprofile

c) Lasermikrosonde

Firma	Gerätetyp	Besonderheiten
Leybold-Heraeus	Lamma 500	Ortsanalyse, Durchschuß
	Lamma 1000	Ortsanalyse, Reflexion

1.1.8 Literatur

(1) J.A.C. Broekaert: Instrument Column in:
    Spectrochim. Acta, 39 B (1984) 729-735

(2) Dsgl., 39 B (1984) 589-596

(3) Dsgl., 37 B (1982) 359-364

(4) Dsgl., 37 B (1982) 727-732

(5) Dsgl., 38 B (1983) 533-542

(6) Dsgl., 38 B (1983) 1355-1361

(7) Dsgl., 37 B (1982) 65-69

(8) M. Stoeppler: Marktübersicht Atomabsorptionsspektrometrie,
    Nachr. Chem. Tech. Lab., 33 (1985) M3 - M 19

1.2 Abbildung verschiedener Großgeräte

Als Beispiel für Geräte zur Instrumentellen Multielementanalyse folgen Abbildungen von Großgeräten, die auf der dem Symposium für Instrumentelle Multielementanalyse angeschlossenen Geräteausstellung gezeigt wurden. Aus diesem Grunde ist die folgende Bilderauswahl nur exemplarisch und nicht vollständig.

Der Schwerpunkt der Ausstellung lag naturgemäß bei den ICP-Methoden. Neben mehreren sequentiellen und simultanen ICP-Atomemissionsspektrometern waren, erstmals in Deutschland, ein ICP-Massenspektrometer sowie das erste kommerziell zugängliche ICP-Atomfluoreszenzspektrometer ausgestellt. Die Computerisierung und Automatisierung der Großgeräte ist wiederum deutlich fortgeschritten. Ein sich andeutender Trend liegt in der noch weitergehenden Datenverarbeitung unmittelbar am Gerät, zum Beispiel anspruchsvoller Statistikauswertungen unter Zuhilfenahme von Gesichtspunkten der Chemometrie, Zusammenstellung von Tabellen, Ausgabe abgabefertiger Protokolle. Flexibler wurde gegenüber früher die Bedienung und Datenerfassung der Geräte durch Einsatz integrierter Mikroprozessoren.

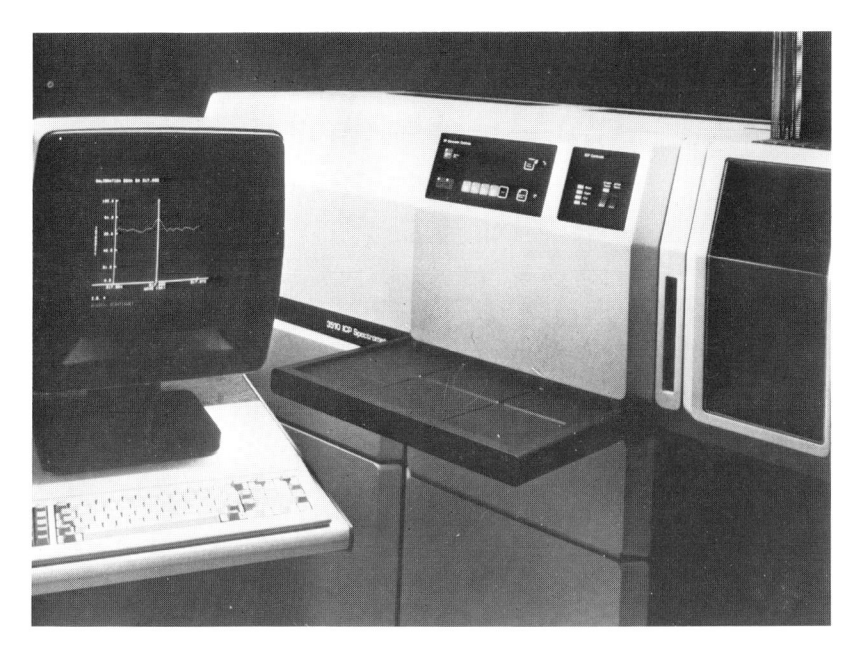

ICP-Atomemissionsspektrometrie, schnelles Sequenzspektrometer Modell 3510
Applied Research Laboratories

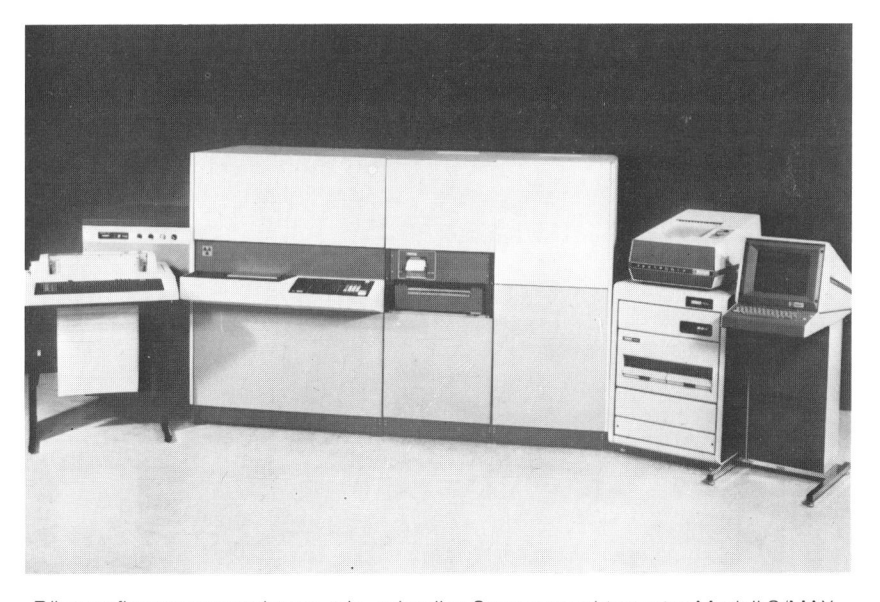

Röntgenfluoreszenzspektrometrie, schnelles Sequenzspektrometer Modell S/MAX
Rigaku Corporation (Atomika)

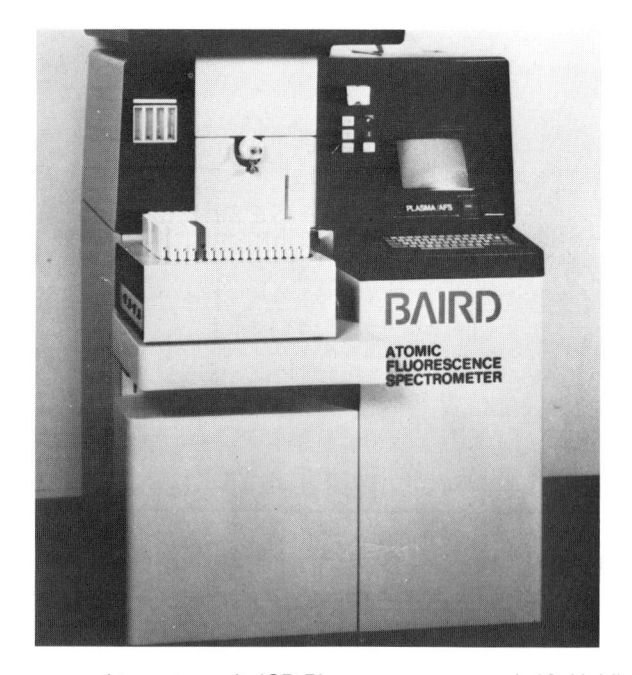

ICP-Atomfluoreszenzspektrometer mit ICP-Plasmaanregung und 12 Hohlkathodenlampen schnelles, Sequenzspektrometer für 12 Elemente, Modell Plasma/AFS 2000
Baird Corporation

DCP-Atomemissionsspektrometer, schnelles Sequenzspektrometer mit Gleichstromplasma (DCP), Echelle Gitter, Modell SpectraSpan VI
Spectrametrics, Inc (Beckman Instruments)

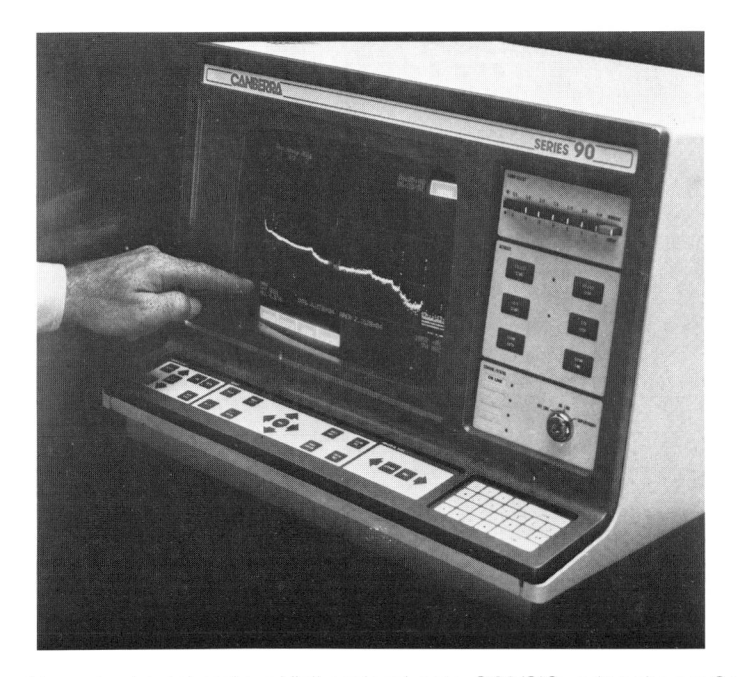

Gammaspektrometer, bestehend aus Vielkanalanalysator S 90/SIO und tragbarem Ge-Detektor Typ MAC
Canberra Industries Inc.

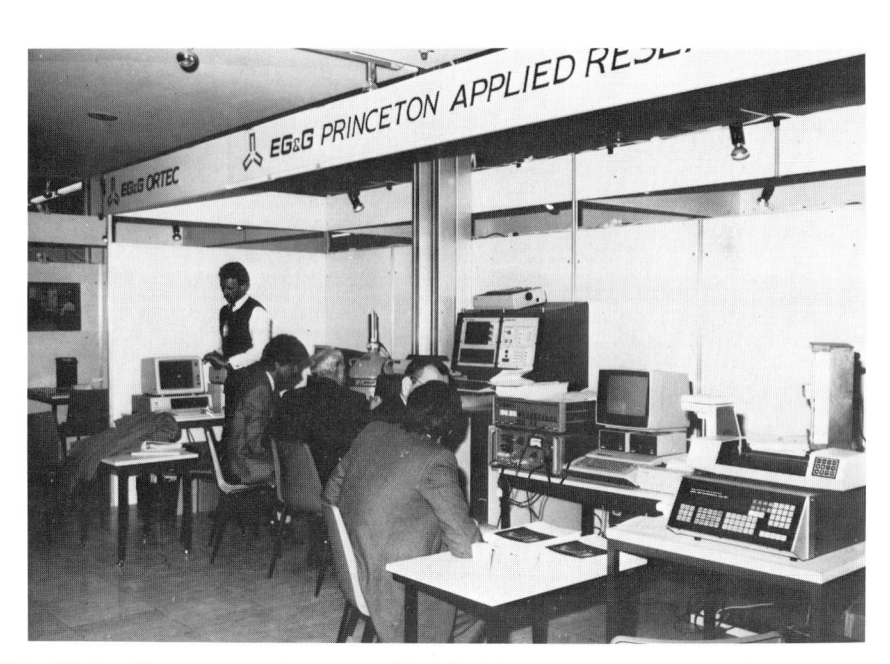

Mitte: Röntgenfluoreszenzspektrometer, Modell 5000, energiedispersiv mit Gammaspektrometer ADCAM,
EG & G Ortec
Rechts: Polarographiesystem 384 B und 264
EG & G, Princeton Applied Research

Wechselstrom-Impedanz Meßsystem für Elektroanalytik mit Potentiostat/Galavanostat, Modell 368-2
EG & G, Princeton Applied Research

Großgeräte 735

Röntgenfluoreszenzspektrometer energiedispersiv, zum Anschluß an elektronenoptische Geräte, Modell KEVEX Analyst 770
Getc Instrumentebau GmbH

Detektoren für Antikompton- bzw. Low-level-α,β-Meßsystem
Harshaw Chemie GmbH

Mechanisches Probenvorbereitungsgerät: halbautomatische Feinmühle (links im Bild), Modell HSM-S
Herzog Maschinenfabrik GmbH + Co

ICP-Atomemissionspektrometer, schnelles Sequenzspektrometer, Modell IL PLasma-200
Instrumentation Laboratory

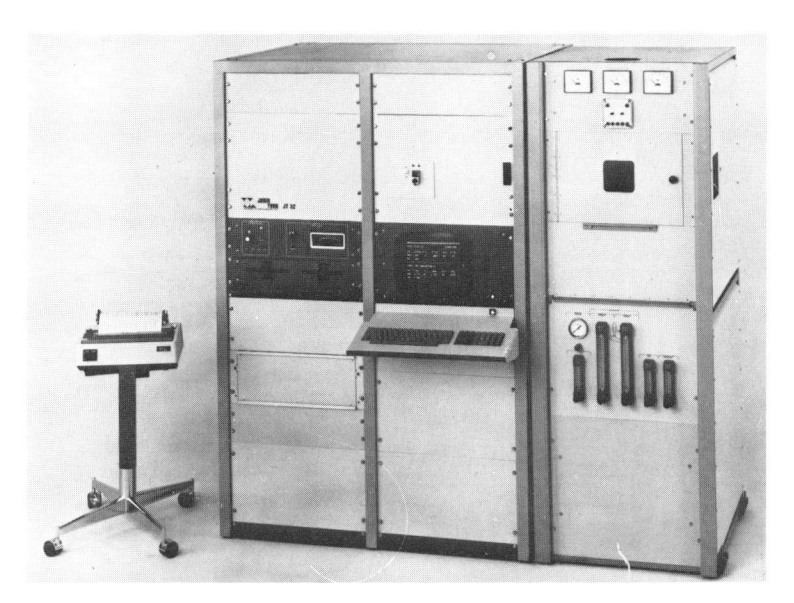

ICP-Atomemissionsspektrometer, Simultangerät, Modell JY 32 P
Instruments S.A. Jobin Yvon

Gammaspektrometrie: Vielkanalanalysator mit gekoppeltem Datenbank-Rechnersystem
Intertechnique Deutschland GmbH

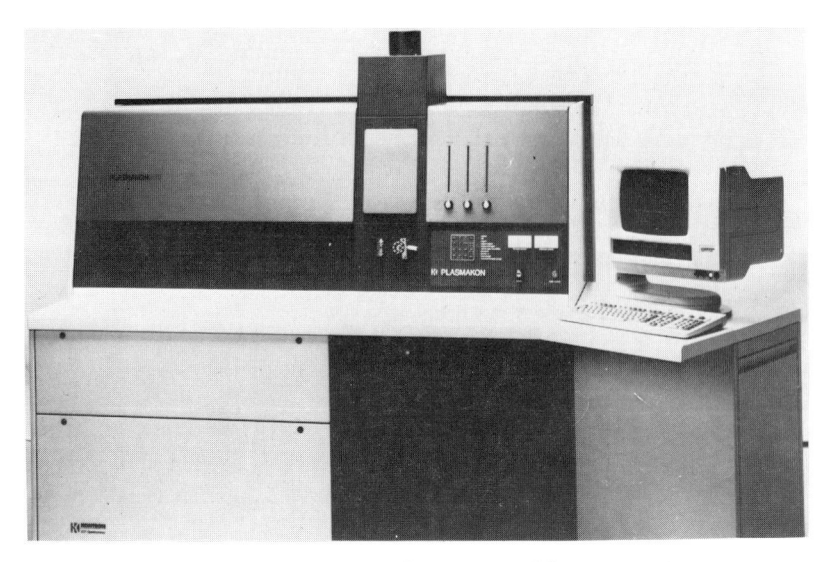

ICP-Atomemissionsspektrometer, Simultan- und Sequenzspektrometer,
Modell Plasmakon S 35
Kontron GmbH, Spektralanalytik

Probenvorbereitungsgeräte
Kürner Anlysentechnik

Großgeräte 739

Tragbare Röntgenfluoreszenzspektrometer:
a) links: Modell: Portaspec
b) rechts: Mini-Vielkanal RFS, Modell Rosa 1
Kurt Mitterfellner GmbH

Gammaspektrometrie, Vielkanalanalysator mit schnellen ADC's und NIM-Technologie
links: Modell ND 66, rechts: Modell ND 76
Nuclear Data GmbH

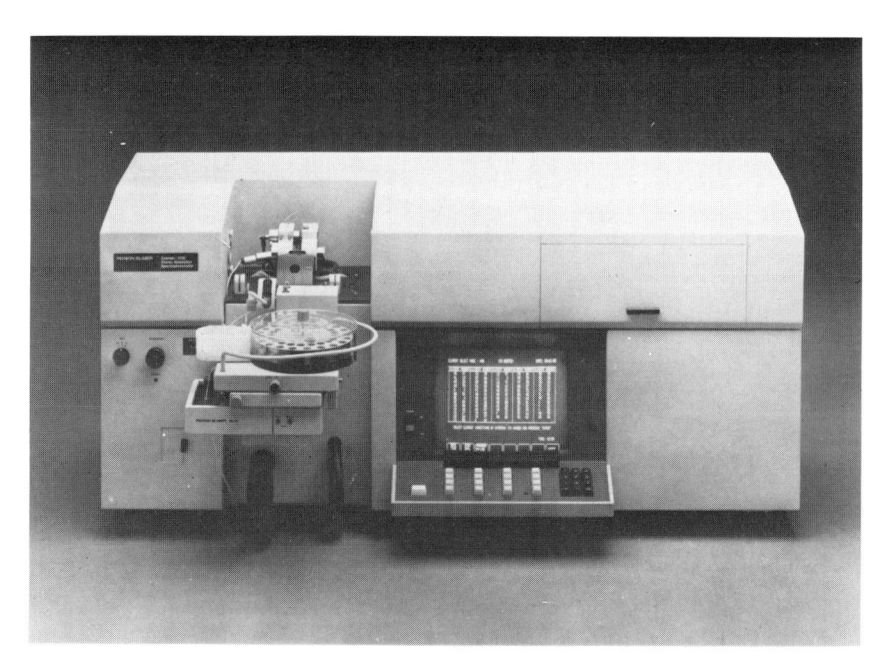

ICP-Atomemissionsspektrometrie, Sequenzgerät mit Zeeman Kompensation, Modell ICP/5500
Bodenseewerk Perkin-Elmer & Co

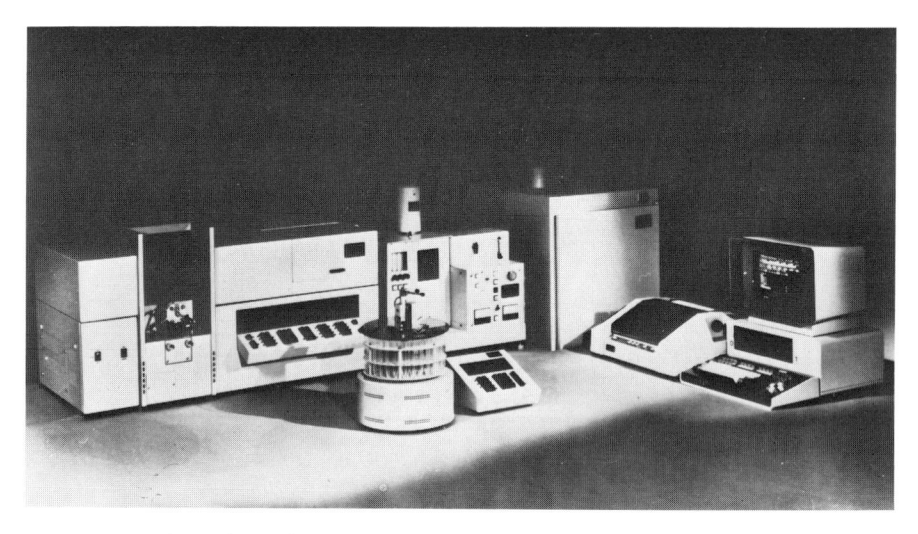

Atomabsorptionsspektrometer 3030, schnelles Routinegerät
Bodenseewerk Perkin-Elmer & Co

Atomabsorptionsspektrometer, langsames sequentielles Multielementgerät für bis zu 16 Elementen
Philips GmbH

Glimmentladungs-Spektrometer für Multielementanalyse
RSV-Präzisionsmeßgeräte GmbH

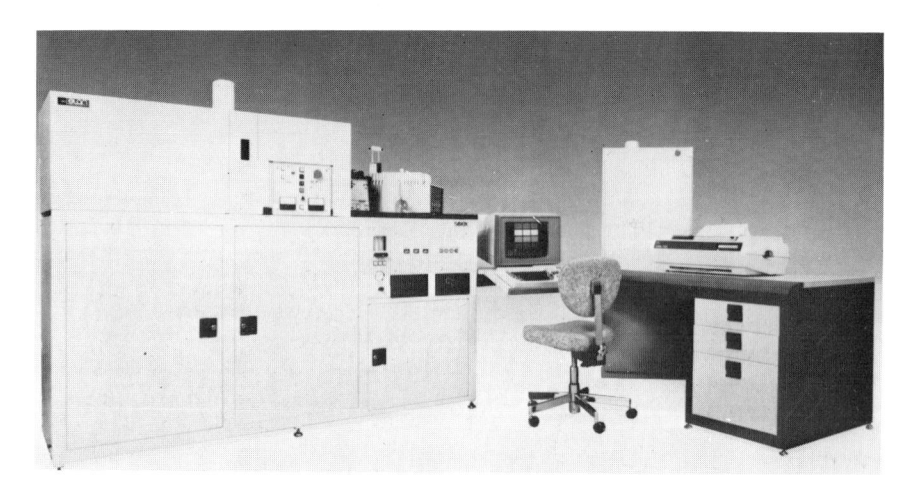

ICP-Massenspektrometer, Modell ELAN ICP/MS
Sciex, Division of MDS Health Group Ltd.

Röntgenfluoreszenzanalyse, energiedispersiv mit Totalreflexion, Modell EXTRA II (Bild: Oberteil)
Rich. Seifert & Co

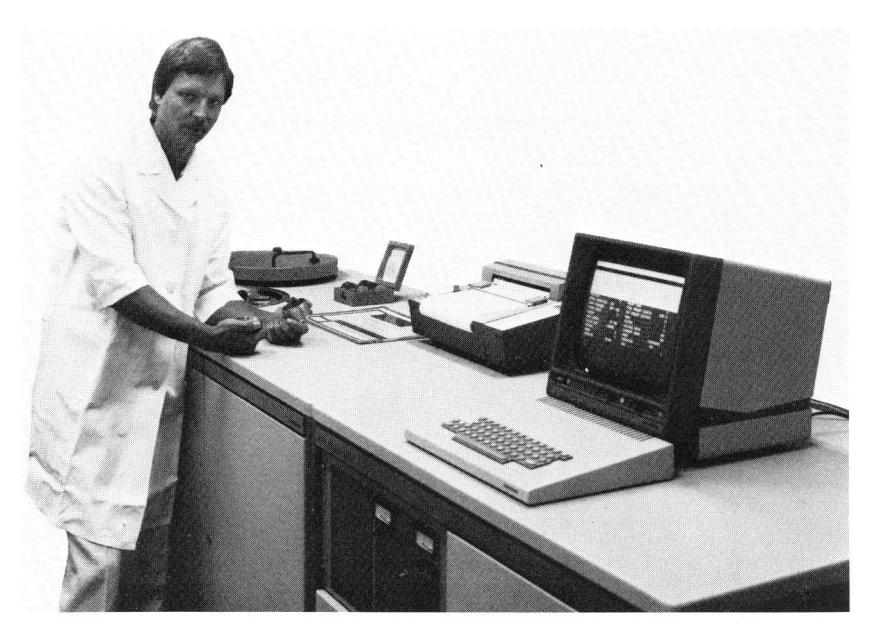

Röntgenfluoreszenzanalyse, Sequenzspektrometer Modell SRS 300
Siemens-AG, Geschäftsbereich Meß- und Prozeßtechnik

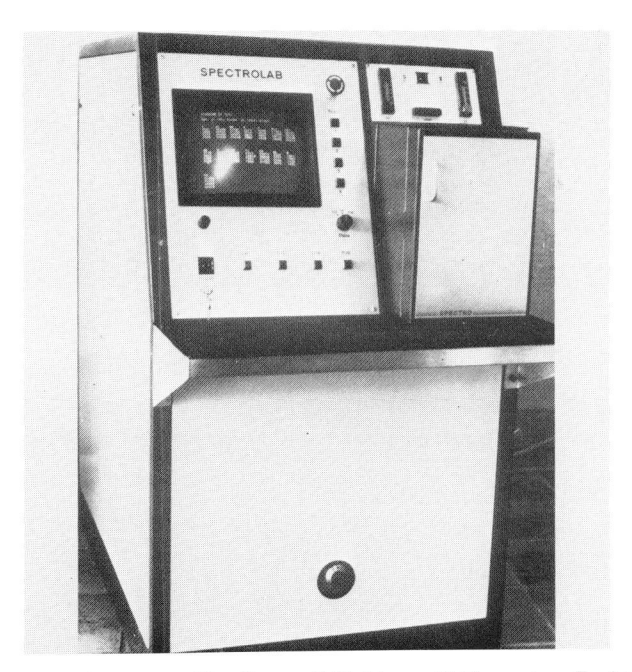

Atomemissionsspektrometer, Simultangerät für bis zu 40 Elementen, Funkenanregung,
Modell: SPECTROLAB
Spectro GmbH

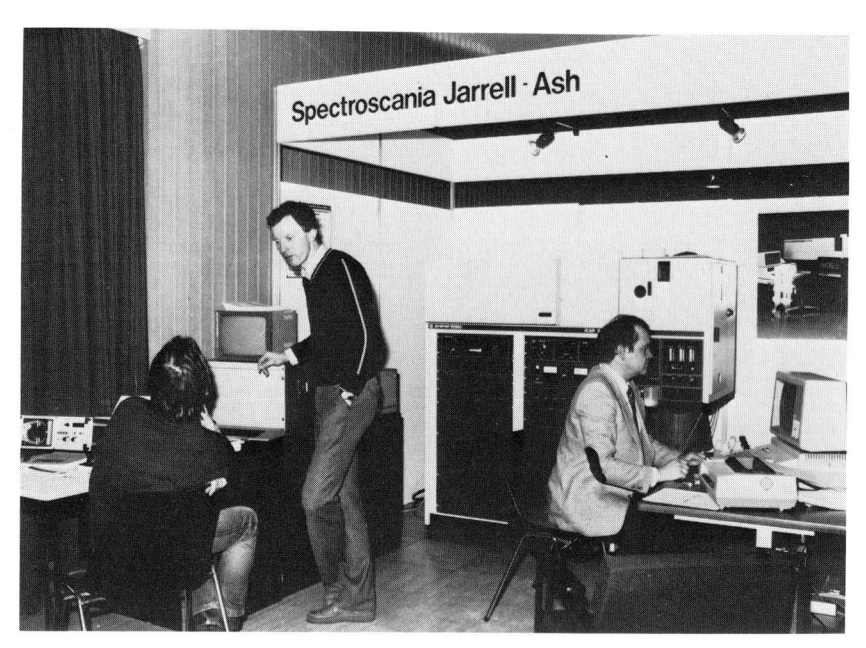

ICP-Atomemissionsspektrometer, Simultangerät mit bis zu 50 Elementen, Modell: ICAP-9000 Jarrell Ash Division (Spectroscania)

Röntgenfluoreszenz-Feststoff-Analysator mit Staubimissionsmeßgerät, BETA-Staubsammler F 703, Modell RD/1000 bzw. F 703
Verewa, Meß- und Regeltechnik Spohr

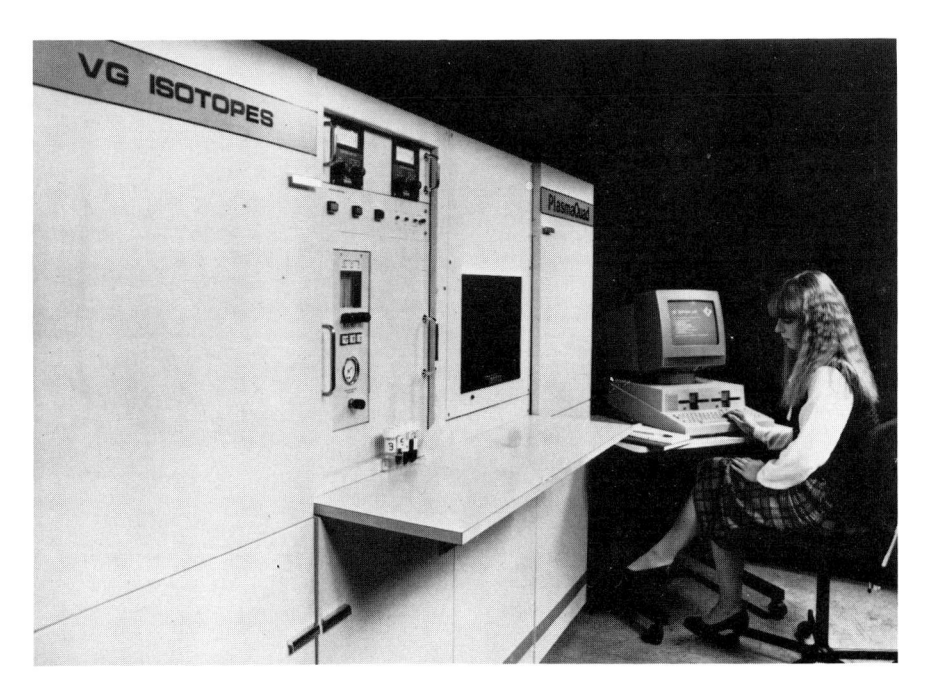

ICP-Massenspektrometer mit ICP-Plasma und Quadrupol-Massenspektrometer,
Modell: Plasma Quad
VG Instruments Ltd.

1.3 Anschriften der Hersteller

Um Kontakte zu erleichtern, folgt eine Zusammenstellung der Anschriften der in der Ausstellung vertretenen sowie im Text zusätzlich genannten Hersteller, jedoch ohne Gewähr.

Applied Research Laboratories SA (ARL SA)
   En Vallaire, CH-1024 Ecublens
   Tel. 021/349701-8

   ARL GmbH, Königstraße 5, D-4000 Düsseldorf
   Tel. 0211/329040-49

Asoma Instruments, 12741 Research Blvd.,
   Suite 501, Austin, Tx 78759, USA
   Tel. (512) 258-6608, Tx 76-7177

Atomika Technische Physik GmbH
   Kuglmüllerstraße 6, D-8000 München 19
   Tel. 089/152031

   Rigaku Corporation
   Segawa Bldg. 2-8 Kandasurugadai
   Chiyoda-ku, Tokyo, Japan
   Tel. (03) 295-3311

   Rigaku/USA Inc., 3 Electronics Ave,
   Danvers, MA 01923, USA
   Tel (617) 777-2446

Beckman Instruments GmbH
   Frankfurter Ring 115, D-8000 München 40
   Tel. 089/3887-1

   Beckman Instruments Inc.,
   204 Andover St., Andover, MA 01810, USA
   Tel. (617) 475 7015, Tx 947 134

   Spectrametrics, Inc. (Subsidary of Beckman Inc.)
   75 Foundation Avenue, P.O. Box 3000
   Haverhill, MA 01831 USA
   Tel. (617) 373-8000, Tx 94-7134

Berthold, Postfach 160, D-7547 Wildbad 1
   Tel. 07081/3981, Tx 724019

Biotronik GmbH
   Postfach 1330, D-6457 Maintal 1
   Tel. 06181/492082-87

   Baird Corporation
   125 Middlesex Turnpike
   Bedford, MA 01730 USA
   Tel. (617) 276-6187, Tx 923 491

   UK. Address: Baird Atomic Ltd., 4 Warner Drive,
   Springwood Industrial Estate, Braintree, Essex CM7 7YL
   Tel. 0376/26560, Tx 987885

Cameca
    103 Boulevard Saint-Denis/B.P. 6, F-92403 Courbevoie Cedex
    Tel. 01 334 30 60, Tx 620 348 F

Canberra Elektronik GmbH
    Hahnstraße 70, D-6000 Frankfurt
    Tel. 069/6666087

    Canberra Industries Inc.
    45 Gracey Ave, Meriden, CT 06450 USA
    Tel. (203) 238-2351, Tx 643251

Colora Meßtechnik GmbH
    Postfach 1240, D-7073 Lorch

    Hitachi Ltd., Nissei Sangyo Co. Ltd,
    15-12 Nushi-Shimbashi, 2-Chome Minato-Ku
    Tokyo (Japan)

Edax International Inc., P.O. Box 135,
    Prairie View, IL 60069, USA

EG & G Instruments, Ortec Division
    Hohenlindenerstraße 12, D-8000 München 80
    Tel. 089/918061

    EG & G Ortec
    100 Midland Rd.
    Oak Ridge, TN 37830 USA
    Tel. (615) 482-4411

EG & G Instruments, PAR Division
    Hohenlindenerstraße 12, D-8000 München 80
    Tel. 089/918061

    EG & G Princton Applied Research (PAR)
    P.O. Box 2565
    Princeton, NJ 08540 USA
    Tel. (609) 452-2111, Tx 843409

Erdmann & Grün KG, Feinmechanik und Optik
    Postfach 1580, Solmserstr. 90, D-6330 Wetzlar

Getac Instrumentebau GmbH
    Am Obstmarkt 32, D-6500 Mainz 21 (Finthen)
    Tel. 06131/40091

Harshaw Chemie GmbH
    Viktoriastraße 5, D-5632 Wermelskirchen 1
    Tel. 02196/4899

Herzog Maschinenfabrik GmbH & Co
    Auf dem Gehren 1, D-4500 Osnabrück-Lüstringen
    Tel. 0541/37370

Hilger Analytical Ltd., Westwood Industrial Estate
    Ramsgate Road, Margate, Kent CT9 4JL
    Tel 0843-25131, Tx 96252

Instrumentation Laboratory GmbH
    Kleinstraße 14, D-5303 Bornheim 2 (Hersel)
    Tel. 02222/8310

Instrumentation Laboratory Inc.
Analytical Instrument Div.
1 Burtt Road, Andover, MA 01870 USA
Tel. (617) 470-1790, Tx 7103471274

Instruments SA GmbH
Hauptstraße 68, D-8025 Unterhaching
Tel. 089/6114077

Instruments SA, Jobin Yvon
16-18 rue du Canal, F-91160 Longjumeau
Tel. 33.6.909.3493

Intertechnique Deutschland GmbH
Postfach 1645, D-6500 Mainz 1
Tel. 06131/234661

Jeol (Europe) B.V.
Building 105, Schiphol-Oost, The Netherlands
Tel. 020/459051

Kevex Corporation
1101 Chess Drive, P.O. Box 4050, Foster City, CA 94404, USA
Tel. (800) 624-7105

Kimmel, Münchener Apparatebau
Hans Stießbergerstr. 2, D-8013 Haar
Tel. 089/462031

Kontron GmbH
Oscar-von-Miller-Straße 1, D-8057 Eching
Tel. 08165/77342, Tx 526710

Kürner Analysentechnik
Herderstraße 2, D-8200 Rosenheim
Tel. 08031/88088

Laben, vertreten durch
Unicom GmbH, Hügelstr. 70, D-6000 Frankfurt/M.
Tel. 069/528022

Labtest Equipment Co (Europe)
Talstr. 35, D-4030 Ratingen

Labtest Equipment Co.,
11828 La Grange Ave, Los Angeles, CA 90025, USA

Leeman Labs Inc.,
540 Main St., Tewksbury, MA 01876, USA

Leybold Heraeus GmbH
Bonner Str. 504, D-5000 Köln 51
Tel. 0221/37011

LKB Produkter AB
Box 305, S-16126 Bromma
Tel. (08) 980040, Tx 909870

Maassen
Hittisauer Str. 13, D-7980 Ravensburg 1

Mitterfellner GmbH
Königsteinerstraße 102, D-6232 Bad Soden
Tel. 06196/25074

Nokia, vertreten durch
    Dr. V. Hornung, Beratungsbüro für Chemische Analytik,
    Jasminstr. 21, D-4600 Dortmund 41
    Tel. 0231/402307

Nuclear Data GmbH
    Bonameserstraße 44, D-6000 Frankfurt 50
    Tel. 069/520152

Packard Instruments GmbH
    Subsidary of AMBAC Industries Inc.,
    Hanauer Landstraße 220, D-6000 Frankfurt/M. 1
    Tel. 069/430171

Perkin-Elmer & Co GmbH, Bodenweewerk
    Postfach 1120, D-7770 Überlingen
    Tel. 07551/811

    Perkin-Elmer Instruments Div.
    Norwalk, CT 06856 USA
    Tel. (203) 762-1000

Philips GmbH, Elektronik für Wissenschaft und Industrie
    Abt. VW, Miramstraße 87, D-3500 Kassel
    Tel. 0561/501413

Princeton Gammatech (PGT)
    Mainzer Str. 103, D-6200 Wiesbaden
    Tel. 06121/79052

Pye Unicam Ltd.
    York Street, Cambridge, CB1 2PX, England

RSV Präzisionsmeßgeräte GmbH
    Hauptstraße 60, D-8031 Seefeld 2 (Hechendorf)
    Tel. 08152/7711

Schlumberger
    Reisstr. 21, D-6200 Wiesbaden
    Tel. 06121/463714

Sciex Europe
    Chemin des Terrasses, CH-1095 Lutry
    Tel. 021/394834

    Sciex, Div. of MDS Health Group Ltd.
    355 Commerce Drive
    Amherst, NY 14150, USA
    Tel. (716) 691-3556

    Sciex, Div. of MDS Health Group Ltd.
    55 Glen Cameron Road, Thornhill, Ontario
    Canada L3T 1P2
    Tel. (416) 881-4646, Tx 06-964722

Seifert & Co
    Bogenstraße 41, D-2070 Ahrensburg
    Tel. 04102/760-1

Shimadzu (Europe) GmbH
    Ackerstr. 111, D-4000 Düsseldorf

    Shimadzu Corp.
    Shinjuku Mitsui Bld 1-1, Nishi-Shinjuku 2-chome,
    Shinjuku-ku, Tokyo 160 (Japan)

Siemens AG
    Bereich: Meß- und Prozeßtechnik, E 689
    Östliche Rheinbrückenstraße 50, D-7500 Karlsruhe 21
    Tel. 0721/595-1

Silena, Wissenschaftliche Instrumente GmbH
    Hänfigstr. 7, D-6467 Hasselroth 2,
    Tel. 06055/4021

    Silena S.p.A.
    Via Negroli, I-20133 Milano
    Tel. (02) 7490565

Spectro GmbH
    Boschstraße 10, D-4190 Kleve
    Tel. 02821/26068

Spectroscania GmbH
    Ambergerstraße 10, D-8036 Herrsching
    Tel. 08152/2021

    Jarrell Ash Division
    590 Lincoln Street
    Waltham, MA 02254 USA
    Tel. (617) 890-4300, Tx 6817015

Tennelec GmbH
    Münchener Str. 50, D-8025 Unterhaching
    Tel. 089/6115060

    Tennelec Inc.
    601 Turnpike, Oak Ridge, TN 37830, USA
    Tel. (615) 483-8405, Tx 810-572-1018

Tracor Instruments
    6500 Tracor Lane, Austin, TX 78721
    Tel. (512) 926-2800

Varian GmbH
    Alsfelder Str. 6, D-6100 Darmstadt

Verewa, Meß- und Regeltechnik
    Eppinghoferstraße 92-94, D-4330 Mülheim/Ruhr 1
    Tel. 0208/472729

VG Instruments GmbH
    Gustav-Stresemann-Ring 12-16, D-6200 Wiesbaden
    Tel. 06121/39131

    VG Instruments Inc. Inorganic Div.
    300 Broad Street, Stamford, CT 06901, USA
    Tel. (203) 329-8050

    VG Isotopes Ltd.
    Ion Path, Road Three, Winsford, Cheshire CW 738 X, England
    Tel. 06065/51121

2. Abkürzungen für die Bezeichnung von Analysenmethoden

B. Sansoni
Zentralabteilung für Chemische Analysen
Kernforschungsanlage Jülich GmbH

D-5170 Jülich-1

2.1 Einleitung

Buchstabenabkürzungen werden in zunehmendem Umfang als Kurzbezeichnung für Analysenmethoden mit langen, komplizierten oder immer wiederkehrenden Namen in Veröffentlichungen verwendet. Dies kann nützlich sein und ist vertretbar, sofern die Abkürzung logisch, zweckmäßig, kurz und möglichst auch geläufig ist. In jedem Fall muß das Kürzel in einer Veröffentlichung aber bei seinem ersten Auftreten in vollem Wortlaut erklärt werden, mit der Abkürzung dahinter in Klammern. In Überschriften sollte es deshalb vermieden werden.

Andererseits kann die Verwendung zu vieler unterschiedlicher Buchstabenkombinationen, die manchmal auch für ganz unterschiedliche Gebiete Verwendung finden — zum Beispiel AA sowohl für Atomabsorption als auch Aktivierungsanalyse — zu Verwirrung führen und einen Veröffentlichungstext schwer lesbar machen. Dies gilt vor allem für Übersichten, in denen mehrere verschiedenartige Methodengruppen behandelt werden, wie in vorliegender Monographie. In solchen Fällen kann die häufige oder gar alleinige Verwendung von Kürzeln zu einer Krankheit werden. Häufig sind die Abkürzungen nicht standardisiert. Sie haben sich aus dem täglichen Gebrauch ergeben und werden in unterschiedlichen Analytikerkreisen und Ländern unterschiedlich verwendet, in den USA häufiger als in Deutschland und Mitteleuropa.

Um das Lesen von Buchstabenkürzeln zu erleichtern, bringt nachstehende Liste eine Zusammenstellung nach Methoden. Danach folgt eine alphabetische Auflistung. In letzterer verweisen die Zahlen in Klammern ( ) jeweils auf Liste 1 mit den voll ausgeschriebenen Namen.

Die Zusammenstellung ist in keiner Weise vollständig oder verbindlich. Es wurden lediglich die dem Herausgeber geläufigeren Abkürzungen zusammengestellt und als Beispiele für die gegenwärtige Umgangssprache die in den Zusammenfassungen der 1269 Vorträge der Pittsburgh Conference 1985 verwendeten Kürzel hinzugefügt (4). Dementsprechend wechseln auch deutsche und englische Bezeichnungen. Es wurde kein Versuch gemacht, diese Bezeichnungen offiziellen Vorschlägen und Vereinbarungen anzupassen. Solche gibt es nur für einen kleinen Teil der aufgeführten Beispiele. Für Oberflächenanalysemethoden hat H. Hantsche eine umfangreiche Zusammenstellung veröffentlicht (3). Sie wird gesondert wiedergegeben, da die Aufnahme in die Hauptliste diese noch komplizierter machen würde.

Nach Abschluß des Manuskriptes wurden zwei weitere Veröffentlichungen bekannt (5) (6), in denen weitere etwa 110 (5) bzw. 500 (6) Abkürzungen enthalten sind. Zahlreiche Abkürzungen in (6) sind jedoch teilweise sehr speziell. Es erschien nicht sinnvoll, diese noch in das vorliegende Manuskript einzuarbeiten, welches etwa 220 Kürzel enthält. Dies soll einer späteren Neufassung vorbehalten bleiben.

Die verwirrende Vielfalt von insgesamt schätzungsweise ca. 750 in der Literatur anzutreffenden Abkürzungen, vorwiegend für Methoden, hat zwei Konsequenzen. Erstens sollte jeder Autor in einer wissenschaftlichen Arbeit oder Veröffentlichung bei der erstmaligen Verwendung der Abkürzung deren vollen Wortlaut gebrauchen und z.B. das Kürzel dahinter in () Klammern setzen. Zweitens wird es sich wohl auf Dauer nicht vermeiden lassen, daß eine internationale Kommission Regeln für zweckmäßige Abkürzungen aufstellt und von ihr eine Liste mit Abkürzungen empfohlen wird.

Für eine spätere Neuauflage dieser Liste bittet der Verfasser freundlich um Hinweise.

Literatur

1) V.A. Fassel (Chairman), Nomenclature, Symbols, Units and their Usage in Spectrochemical Analysis — III. Analytical Flame Spectroscopy and Associated non-flame Procedures, Pure & Appl. Chem., *45*, 105 — 123, 1976

2) M. Grasserbauer, K.F.J. Heinrich, G.H. Morrison, Nomenclature, Symbols and Units Recommended for in situ Microanalysis, Pure & Appl. Chem. *55*, 2023 — 2027 (1983)

3) H. Hantsche, Abbreviations for microscopical and analytical methods, Microscopy Acta *87*, 271 — 276 (1983)

4) 1985 Pittsburgh Conference & Exposition on Analytical Chemistry and Applied Spectroscopy, February 25 — March 1, 1985, 1985 Abstracts (1269 Titles), New Orleans 1985

5) N.N., Akronyme, in: R. Bock, W. Fresenius, H. Günzler, W. Huber, G. Tölg (Hersg.), Analytiker Taschenbuch, Band 5, S. 305 — 310

6) The acronyms used in the world of spectroscopy, microscopy and diffractometry, II. Glossary of abbreviations, Spectrochim. Acta, *36 B* (1981), 361-372

2.2 Nach Methoden geordnete Abkürzungen von Analysenmethoden

1. Aktivierungsanalyse (AA)

1.1 Einzelbezeichnungen

1. **A**   Aktivierung
                aber auch: Atom, Absorption, Analyse
2. **N**   Neutronen
3. **T**   Thermisch(e)
4. **R**   Reaktor (neutronen)
                auch: Radiochemisch(e Trennungen)
5. **E**   Epithermisch(e)
6. **I**   Instrumentell(e)
7. **mono**   Monostandard(methode)
8. **rel**   Relativ- bzw. Multielementstandard(methode)
9. **mono,Ī**   Monostandardmethode mit effektivem Resonanzintegral Ī

1.2 Kombinationen

1. **AA**   Aktivierungsanalyse
                aber auch: Atomabsorption
2. **NAA**   Neutronenaktivierungsanalyse
3. **TNAA**   Neutronenaktivierungsanalyse mit Aktivierung durch thermische Neutronen
4. **RNAA**   Neutronenaktivierungsanalyse mit Aktivierung durch (ungefilterte) Reaktorneutronen
                auch: NAA mit radiochemischen Trennungen
5. **ENAA**   Neutronenaktivierungsanalyse mit Aktivierung durch epithermische Neutronen (z.B. unter Cadmiumabschirmung)
6. **INAA**   Instrumentelle Neutronenaktivierungsanalyse
7. **ITNAA**   dsgl., mit Aktivierung durch thermische Neutronen
8. **IENAA**   dsgl., mit epithermischen Neutronen
10. **NAA, mono** Neutronenaktivierungsanalyse nach Monostandardmethode
11. **NAA, rel**   Neutronenaktivierungsanalyse nach Relativ- bzw. Multielementstandardmethode
12. Analog die Kombinationen TNAA, mono; TNAA, rel; RNAA, mono; RNAA, rel; ENAA, mono; ENAA, rel.

2. Massenspektrometrie

2.1 Einzelbestimmungen

1. **MS**   Massenspektrometrie
2. **FIMS**   Field ion mass spectrometry
3. **FMS**   Funken-Massenspektrometrie (englisch: SSMS)

4. **IMMA** — Ion Microprobe Mass Analyser
5. **LAMMA** — Laser Microprobe Mass Analysis
6. **LIMA** — Laser ionization mass analysis
7. **LMS** — Laser-Massenspektrometrie
8. **QMS** — Quadrupol-Massenspektrometrie
9. **SIMS** — Sekundärionen-Massenspektrometrie
10. **SNMS** — Secondary neutralization mass spectrometry
11. **SSMS** — Spark source mass spectrometry (deutsch: FMS nach 2.1.2)
12. **TMS** — Thermionen-Massenspektrometrie (auch: TIMS)
13. **MCP** — Multi channel plate (detector)

2.2 Kombinationen

1. **DSIMS** — Dynamic secondary ion mass spectrometry
2. **ETV-ICP-MS** — Massenspektrometrie mit ICP-Plasmaanregung und elektrothermaler Verdampfung
3. **FAB-MS** — Fast atom bombardement mass spectrometry
4. **FAB-FTMS** — Fast atom bombardement-Fourier transform mass spectrometry
5. **FDMS** — Field desorption mass spectrometry
6. **FTMS** — Fourier-transform mass spectrometry
7. **INMS** — Ionized neutral mass spectrometry
8. **LD-FTMS** — Laser desorption — Fourier-transform mass spectrometry
9. **MSMS** — Tandem mass spectrometry
10. **NIRMS** — Noble gas ion reflection mass spectrometry
11. **PAMS** — Precision abrasion mass spectrometry
12. **PDMS** — Plasma desorption mass spectrometry
13. **PID-FTMS** — Particle induced desorption Fourier-transform mass spectrometry
14. **REMPI-MS** — Resonance enhanced multiple photon ionization mass spectrometry
15. **SIMMS** — Secondary ion microprobe mass spectrometry
16. **SSIMS** — Static secondary ion mass spectroscopy
17. **TOF-MS** — Time-of-flight mass spectrometry
18. **TOF-SIMS** — Time-of-flight secondary ion mass spectrometry
19. **TSQ-MS** — Triple-stage quadrupole mass spectrometry
20. **UMPA** — Universal microprobe mass analyzer

3. Röntgenmethoden

3.1 Einzelbezeichnungen

1. **ED** — Energiedispersiv
2. **F** — Fluoreszenz
3. **R** — Röntgenstrahlung
4. **WD** — Wellenlängendispersiv
5. **X** — X-ray, Röntgenstrahlung

3.2 Kombinationen

1. **EDX**   Energy dispersive-X-ray fluorescence spectrometry
2. **JCPDS**   International Centre for Diffraction Data (computer evaluation programmes for XRD)
3. **RBA**   Röntgenbeugungsanalyse (Röntgendiffraktrometrie)
4. **RFA**   Röntgenfluoreszenzanalyse
5. **WDX**   Wave length dispersive X-ray fluorescence
6. **XRD**   X-ray diffraction spectrometry auch: (RBA)
7. **XRF**   X-ray fluorescence spectrometry
8. **TR-RFA**   Totalreflektierende Röntgenfluoreszenzanalyse

4. Atomspektrometrie

4.1 Allgemeines

1. **AA**   Atomabsorption
2. **AE**   Atomemission
3. **AS**   Atomspektrometrie
4. **ES**   Emissionsspektrometrie
5. **TS**   Transmissionsspektrometrie
5. **S/B**   Signal-to-background ratio
7. **S/N**   Signal-to-noise ratio (auch SNR)
8. **SNR**   Signal-to-noise ratio
9. **SLS**   Sequential linear scan
10. **SSS**   Sequential slew scan

4.2 Atomabsorptionsspektrometrie

4.2.1 Einzelbezeichnungen

1. **AAS**   Atomabsorptionsspektrometrie
2. **AAC**   Atomic absorption (spectrometry) with continuum source and flame or non-flame atomizer
3. **AAL**   Atomic absorption (spectrometry) with linear source and flame or non-flame atomizer
4. **EDL**   Elektrodenlose Entladungslampen
5. **ET, ETA**   Elektrothermische Anregung
6. **F**   Flamme(nanregung)
7. **Feststoff-** Instrumentelle AAS für Feststoffe, ohne vorherige chemische Probenvorbereitung
8. **G, GF**   Graphitrohr(methode) (auch GF)
9. **GF**   Graphitrohrofen(methode) (auch G)
10. **HCL**   Hohlkathodenlampe
11. **HIL**   Hochintensitäts-Hohlkathodenlampe
12. **Hydrid-**   Hydrid(methode)

13. **NF**   Nichtflammen(methode)
14. **VDL**   Gasentladungslampe
15. **Z**   Zeeman-Kompensation

4.2.2 Kombinationen

1. **ET-AAS,**   Atomabsorptionsspektrometrie mit elektrothermaler
   ETA-AAS   Atomisierung
2. **F-AAS**   AAS mit Flammenatomisierung
3. **Feststoff-AAS** AAS mit Direktinjektion von festen Stoffen
4. **Feststoff-ZAAS** AAS unter Verwendung der Zeeman-Kompensation
5. **G-AAS**   AAS mit Atomisierung im Graphitrohr
6. **NF-AAS**   Nichtflammen-Atomabsorptionsspektrometrie
7. **STPF-AAS**   Stabilized temperature platform furnace atomic absorption spectrometry
8. **Z-AAS**   AAS mit Zeeman-Kompensation

4.3 Atomemissionsspektrometrie (AES)

4.3.1 Einzelbezeichnungen

1. **OES**   Optische Emissionsspektrometrie,
                    (Aufteilung in AES und MES ist präziser)
2. **AES**   Atomemissionsspektrometrie
3. **F**   Flamme
4. **AC**   Wechselstrom
5. **DC**   Gleichstrom
6. **RF**   Radiofrequenz
7. **AC spark**   Wechselstrom-Bogenentladung
8. **DC arc**   Gleichstrom-Funkenentladung
9. **RF spark**   Radiofrequenz-Bogenentladung
10. **DCP**   Direct current plasma, Gleichstromplasma(-anregung)
11. **ICP**   Inductively coupled plasma, Induktiv gekoppelte(s) Plasma(anregung)
13. **MWP**   Micro wave plasma
14. **L**   Laseranregung
15. **CSN**   Conductive solids nebulizer

4.3.2 Kombinationen

1. **CSN-ICP-AES** Atomemissionsspektrometrie mit ICP-Anregung und Feststoffzerstäubung für elektrisch leitende Feststoffe
2. **DC-AES**   Atomemissionsspektrometrie mit DC-Bogenanregung
3. **F-AES**   Flammen-Atomemissionsspektrometrie
4. **ICP-AES**   Atomemissionsspektrometrie mit ICP-Anregung
5. **L-AES**   Laser-Atomemissionsspektrometrie
6. **MINDAP-AES**   Microwave-induced nitrogen discharge at atmospheric pressure

4.4 Atomfluoreszenzspektrometrie (AFS)

4.4.1 Einzelbezeichnungen

1. **AFS**   Atomfluoreszenzspektrometrie
2. **AFC**   Atomic fluorescence (spectrometry) with continuum source
3. **AFL**   Atomic fluorescence (spectrometry) with (ion source) and flame or non-flame atomizer
4. **F**   Flamme(nanregung)
5. **HCL**   Hohlkathodenlampe(nanregung)
6. **ICP**   Atomisierung durch induktiv gekoppeltes Plasma
7. **L**   Atomisierung oder Anregung durch Laser

4.4.2 Kombinationen

1. **F-AFS**   Atomfluoreszenzspektrometrie mit Atomisierung in der Flamme
2. **F-HCL-AFS**   dsgl., mit Atomisierung in der Flamme und Anregung durch Hohlkathodenlampe
3. **F-L-AFS**   dsgl., mit Atomisierung in der Flamme und Laseranregung
4. **HCL-AFS**   dsgl., mit Hohlkathodenlampenanregung
5. **ICP-HCL-AFS** dsgl., mit Atomisierung im ICP-Plasma und Anregung durch Hohlkathodenlampe
6. **L-AFS**   dsgl., mit Laseranregung

5. Molekülspektrometrie (MS)

5.1 Allgemeines

1. **MS**   Molekülspektrometrie (aber auch: Massenspektrometrie)
2. **MAS**   Molekülabsorptionsspektrometrie
3. **MES**   Molekülemissionsspektrometrie
4. **MFS**   Molekülfluoreszenzspektrometrie
5. **MLS**   Moleküllumineszenzspektrometrie
6. **MAL**   Molecular absorption (spectrometry) with line source
7. **MTS**   Molecular transmission spectrometry

8. **VUV**   Vakuum-Ultraviolett (spektrometrie)
9. **UV**   Ultraviolett (spektrometrie)
10. **VIS** Sichtbarer Bereich (spektrometrie)
11. **IR**   Infrarot (spektrometrie)
12. **NIR** Nahes Infrarot (spektrometrie)
13. **FIR** Fernes Infrarot (spektrometrie)

14. **DRS** Diffuse reflection spectrometry
15. **ERS** External reflection spectrometry
16. **FTS** Fourier-transform Spektrometrie

17. **FFTS** Fast Fourier transform spectrometry
18. **HTS** Hadamard-transfer Spectrometry
19. **FHTS** Fast Hadamard transfer spectrometry
20. **IRS** Internal reflexion spectrometry
21. **PAS** Photoacoustic spectroscopy
22. **VPIR** Vapor phase infrared (spectrometry)

5.2 Fourier-transform-Spektrometrie

1. **FT** Fourier transform
2. **FTS** Fourier-transform Spektrometrie
3. **FFTS** Fast Fourier-transform spectrometry
4. **FT-IR** Fourier-transform Infrarot(spektrometrie) (dsgl. FT-NIR, FT-FIR)
5. **ATR-FTIR** Abgeschwächte totalreflektierende Fourier-transform(spektrometrie) im IR Bereich
6. **DR-FTIR** Diffuse reflection Fourier transform (spectrometry) IR region
7. **EMIR-FTS** Electromodulated infrared reflectance Fourier transform spectrometry
8. **FFT-ICP-AES** Fast Fourier transform, inductively coupled plasma atomic emission spectrometry
9. **FT-IR-CIR** Fourier transform infrared cylindrical internal reflection spectrometry
10. **SNI-FT-IR** Substractively normalized interfacial Fourier-transform infrared spectroscopy

5.3 Molekülfluoreszenz-Spektrometrie

1. **MFS** Molekülfluoreszenz-Spektrometrie
2. **MLS** Molekülumineszenz-Spektrometrie
3. **MLC** Molecular luminescence (spectrometry) with continuum source
4. **MLL** Molecular luminescence (spectrometry) with line source
5. **LIMFS** Laser induced molecular fluorescence spectrometry
6. **CES-FS** Constant energy synchronous fluorescence spectrometry

5.4 Ramanspektrometrie

1. **RS** Ramanspektrometrie
2. **RRS** Resonanz-Ramanspektrometrie
3. **UV-RRS** UV-Resonanz-Ramanspektrometrie
4. **RAMA** Raman micro analysis
5. **RAMP** Raman micro probe
6. **CARS** Coherent anti-Stokes raman spectrometry
7. **SERS** Surface enhanced Raman scattering spectrometry
8. **SERRS** Surface enhanced resonance Raman scattering (spectrometry)

5.5 Weitere Molekülspektrometrien

1. **CD** Circulardichroismus
2. **RD** Rotationsdispersion
3. **ESR** Elektronenspinresonanz-Spektrometrie

4. **LIBS**   Laser induced break down spectrometry
5. **MCD**   Magnetic circular dichroism
6. **MORD**   Magneto-optische Rotationsdispersion
7. **NMR**   Kernmagnetische Resonanz (-Spektrometrie)
8. **NQR**   Kernquadrupolresonanz (-Spektrometrie)
9. **SOM**   Scanning optical microscope
10. **TDLAS**   Tunable diode laser absorption spectrometry

6. Thermoanalyse

1. **TA**   Thermoanalyse
2. **TG**   Thermogravimetrie
3. **TMA** Thermomechanische Analyse
4. **DTA** Derivative Thermoanalyse, Differentialthermoanalyse
5. **DTG** Differentialthermogravimetrie
6. **DIE** Direkt-Injektions-Enthalpiemetrie

7. Chromatographie

7.1 Chromatographiearten

1. **C**   Chromatographie
2. **AC**   Affinitätschromatographie
3. **DC**   Dünnschichtchromatographie
4. **GC**   Gaschromatographie
5. **IC**   Ionenchromatographie
6. **LC**   Flüssigchromatographie
7. **PC**   Papierchromatographie
8. **SC**   Säulenchromatographie
9. **CCC**   Counter current chromatography
10. **DCCC**   Droplet counter current chromatography
11. **GPC**   Gel permeation chromatography
12. **FIA-IC** Flow injection analysis ion chromatography
13. **HPLC**   High performance liquid chromatography
14. **HS-CCC** High speed counter current chromatography
15. **HS-GC**   High speed gas chromatography
16. **ISRP-LC** Internal surface reverse phase liquid chromatography
17. **LDLC**   Low dispersion liquid chromatography
18. **SCIC**   Single column ion chromatography
19. **SFC**   Super critical fluid chromatography

7.2 Detektoren für Chromatographie

1. **EC**   Elektrochemischer (Detektor)
2. **ECP** Electron capture (detector)

3.	**FID**	Flammenionisationsdetektor
4.	**FTIR**	Fourier-transform Infrarot(detektor)
5.	**HECD**	Hall electrolytic conductivity detector
6.	**IR**	Infrarot(detektor)
7.	**ITD**	Ion trap detector
8.	**MF**	Molekülfluoreszenz(detektor)
9.	**MS**	Massenspektrometrie(detektor)
10.	**PCD**	Photo column derivativization
11.	**PCD**	Photo conductivity detection
12.	**TCD**	Thermal conductivity detector (WLD)
13.	**WLD**	Wärmeleitfähigkeitsdetektor (TCD)

7.3 Kombinationen

1.	**GC-FTIR**	Gaschromatographie mit Fourier-transfer infrarotspektrometrischer Kopplung
2.	**GC-MI-FTIR**	Gaschromatography with matrix isolation and Fourier transfer infrared detection
3.	**GC-MS**	Gaschromatographie mit Massenspektrometriekopplung
4.	**LC-ES**	Flüssigchromatographie mit elektrochemischer Detektion
5.	**LC-MS**	Flüssigchromatographie mit Massenspektrometerkopplung
6.	**IC-PCD**	Ionenchromatography with post column derivativization
7.	**PPINICI-MS**	Pulsed positive ion-negative ion chemical ionization gas chromatography-mass spectrometry (!)
8.	**Thermospray-LC-MS**	Thermospray liquid chromatography mass spectrometry

8. Oberflächenanalyse *

AEAPS	Auger Electron Appearance Potential Spectroscopy
AEM	Auger Electron Microscopy
AEMA	Auger Electron Microanalysis
AETM	Analytical Electron Transmission Microscopy
AES	Auger Electron Spectroscopy
APFIM	Atom Probe Field Ion Microscopy
APS	Appearance Potential Spectroscopy
ARXPS	Angle-Resolved X-Ray Photoelectron Spectroscopy
ASLEEP	Automated Scanning Low Energy Electron Probe
BIS	Bremsstrahlung Isochromate Spectroscopy
CL	Cathode Luminenscence
CTEM	Conventional Transmission Electron Microscopy
DAPS	Dis-Appearance Potential Spectroscopy
DIMA	Direct Imaging Mass Analysis
DSIMS	Dynamic Secondary Ion Mass Spectrometry

EBIC	Electron Beam Induced Current
ECP	Electron-Channeling Pattern (KLP)
EDA	Energy Dispersive X-Rax Microanalysis (EDX)
EDAX	Energy Dispersive Analysis of X-Ray (EDX)
EDS	Energy Dispersive Spectrometry (EDX)
EDX	Energy Dispersive X-Ray analysis
EDXRF	Energy Dispersive X-Ray fluorescence
EDXS	Energy Dispersive X-Ray Spectrometry (EDX)
EELS	Electron Energy Loss Spectroscopy (ELS)
EEM	Emission Electron Microscopy
EEVS	Elektronen-Energie-Verlust-Spektroskopie
EID	Electron-Induced Desorption
EIM	Emission Ion Microscopy
EIS	Electron-Impact Spectroscopy
ELS	Energy Loss Spectroscopy (EELS)
EM	Electron Microscopy
EMPA	Electron Microprobe Analysis (ESMA, EPMA)
ESCA	Electron Spectroscopy for Chemical Analysis
ESD	Electron-Stimulated Ion Desorption (ESID)
ESDN	Electron-Stimulated Desorption of Neutrals
ESID	Electron-Stimulated Ion Desorption (ESD)
ESMA	Elektronenstrahl-Mikroanalyse (EMPA, EPMA)
ESP	Electron Spectroscopy (ESCA)
EPMA	Electron Probe Microanalysis (EMPA, ESMA)
EPXMA	X-Ray Microanalysis Electron Probe
EXAFS	Extended X-Ray Absorption Fine Structure Spectrometry
FAB	Fast Atom Bombardment
FDM	Field Desorption Microscopy
FDMS	Field Desorption Mass Spectrometry
FEES	Field Electron Energy Spectroscopy
FEM	Field Electron Microscopy
FES	Field Electron Spectroscopy
FIM	Field Ion Microscopy
FIMS	Field Ion Mass Spectrometry
FIS	Feldionenspektroskopie (FES)
HEED	High Energy Electron Diffraction
HEIS	High Energy Ion Scattering (RBS)
HREM	High Resolution Electron Microscopy
HVEM	High Voltage Electron Microscopy
ICEED	Inelastic Low Energy Electron Diffraction
IEE	Induced Electron Emission (ESCA)
IETS	Inelastic Electron Tunneling Spectroscopy
IEX	Ion-Excited X-Rays
IIL	Ion-Induced Light Emission (IIR)

IIR	X-Ray Emission Induced by Ion Bombardment (IIX)
IIX	Ion-Induced X-Ray Spectroscopy
ILEED	Inelastic Low Energy Electron Diffraction
ILS	Ion Loss Spectroscopy
IMA	Ion Microprobe Mass Analysis (IMPA), Ionenstrahl-Mikroanalyse
IMP	Ion Microprobe
IMPA	Ion Microprobe Analysis (IMA)
INMS	Ionized Neutral Mass Spectrometry
IMXA	Ion Microprobe X-Ray Analysis
INS	Ion Neutralization Spectroscopy
IRRS	Infra-Red Reflection Spectroscopy
IS	Isochromate Spectroscopy
ISD	Ion-Stimulated Desorption
ISMA	Ionenstrahl-Mikroanalyse (IMPA)
ISS	Ion Scattering Spectroscopy
ITS	Inelastic Tunneling Spectroscopy
LAMES	Laser Micro Emission Spectroscopy
LAMMA	Laser Microprobe Mass Analysis
LAMMS	Laser Micro Mass Spectrometry
LDMS	Laser Desorption Mass Spectroscopy
LEED	Low Energy Electron Diffraction
LEEIXS	Low Energy Electron-Induced X-Ray Spectroscopy
LEELS	Low Energy Electron Loss Spectroscopy
LEEM	Low Energy Electron Microscopy
LEIS	Low Energy Ion Scattering (ISS)
LEMBS	Low Energy Molecular Beam Scattering
LM	Light Microscopy
LMA	Laser Micro Spectral Analysis
LMP	Laser Microprobe
LRMA	Laser Raman Microanalysis
LOES	Laser Optical Emission Spectroscopy
MBRS	Molecular Beam Reactive Scattering
MBSS	Molecular Beam Surface Scattering
MEED	Medium Energy Electron Diffraction
MEM	Mirror Electron Microscopy
MOLE	Molecular Optical Laser Examiner = Raman Microprobe
MOSS	Mössbauer Spectroscopy
MRS	Micro Raman Spectroscopy
NIRMS	Noble Gas Ion Reflection Mass Spectrometry
NIRS	Neutral Impact Radiation Spectroscopy
PDMS	Plasma Desorption Mass Spectrometry
PES	Photoelectron Spectroscopy
PESIS	Photoelectron Spectroscopy of Inner Shells (XPS)
PESM	Photoelectron Spectral Microscopy
PESOS	Photoelectron Spectroscopy of Outer Shells (UPS)

PHEEM	Photoemission Electron Microscopy
PIR	Photoneninduzierte Röntgenstrahlung
PIXES	Particle-Induced X-Ray Emission Spectroscopy
PXA	Primary X-Ray Analysis
QMS	Quadrupole mass spectrometry
RBS	Rutherford Backscattering
RED	Reflection Electron Diffraction
REELS	Reflection Electron Energy Loss Spectroscopy (auch HRELS)
REM	Raster-Elektronenmikroskopie (SEM)
RFA	Röntgenfluoreszenzanalyse (XRF)
RHEED	Reflection High Energy Electron Diffraction
RIBS	Rutherford Ion Backscattering (RBS)
RIM	Raster-Ionenmikroskopie
RMA	Röntgenmikroanalyse (ESMA, EMPA)
SAD	Selected Area Diffraction
SAEM	Scanning Auger Electron Microscopy
SAES	Scanning Auger Electron Spectroscopy
SAM	Scanning Auger Microprobe
SCANIIR	Surface Composition of Neutral and Ion Impact Radiation
SDLTS	Scanning Deep Level Transient Spectroscopy
SEM	Scanning Electron Microscopy (REM)
SESD	Scanning Electron-Stimulated Desorption
SEXAFS	Surface Extended X-Ray Absorption Fine Structure
SHEED	Scanning High Energy Electron Diffraction
SIIMS	Secondary Ion-Imaging Mass Spectroscopy
SIMA	Sekundärionen-Mikroanalyse
SIMMS	Secondary Ion Microprobe Mass Spectrometry (IMMA)
SIMS	Secondary Ion Mass Spectrometry
SIPS	Sputter-Induced Photoelectron Spectroscopy
SLAM	Scanning Laser Acoustic Microscope
SNMS	Secondary Neutralization Mass Spectroscopy
SNMS	Sputtered Neutral Mass Spectroscopy
SPIXE	Scanning Proton-Induced X-Ray Emission
SRS	Surface Reflectance Spectroscopy
SSIMS	Static Secondary Ion Mass Spectroscopy
STEM	Scanning Transmission Electron Microscopy
SXAPS	Soft X-Ray Appearance Potential Spectroscopy
SXPM	Scanning X-Ray Photoelectron Microscopy
SXR	Scanning X-Ray Radiography
SXS	Soft X-Ray Spectrometry
TDS	Thermal Desorption Spectroscopy
TED	Transmission Electron Diffraction
TEELS	Transmission Electron Microscope Electron Loss Spectroscopy
TELESCA	Transmission Electron Loss Spectroscopy for Chemical Analysis

TEM	Transmission Electron Microscope
THEED	Transmission High Energy Electron Diffraction
TRIX	Total Rate Imaging of X-Rays
UMPA	Universal Microprobe Mass Analyser
UPS	Ultraviolet Photoelectron Spectroscopy
WDAX	Wavelength Dispersive Analysis of X-Rays (WDX)
WDS	Wavelength Dispersive Spectrometry (WDX)
WDX	Wavelength Dispersive Analysis of X-Rays (WDAX)
XAES	X-Ray-Induced Auger Electron Spectroscopy
XEM	Exoelectron Microscopy
XES	Exoelectron Spectroscopy
XPMA	X-Ray Photoelectron Microprobe Analysis (ESCA)
XPS	X-Ray Photoelectron Spectroscopy
XRD	X-Ray Diffraction
XRF	X-Ray Fluorescence (RFA)
XUPS	Extremely Ultraviolet Photoelectron Spectroscopy

---

* Nach H. Hantsche, Microscopia Acta, *87* 271 (1983) und M. Grasserbauer, K.F. Heinrich, G.H. Morrison, Pure & Appl. Chem., *55* (2023) 1983

2.3. Alphabetische Auflistung der Methodenabkürzungen

A	1.1.1	**ESR**	5.5.3
AA	1.2.1; 4.1.1	**ET, ETA**	4.2.1.5
AAC	4.2.1.2	**ET-AAS**	4.2.2.1
AAL	4.2.1.3	**ETA-AAS**	4.2.2.1
AAS	4.2.1.1	**ETV-ICP-MS**	2.2.2
AC	4.3.1.4; 7.1.2		
AC sparc	4.3.1.7	**F**	3.1.2; 4.2.1.6; 4.3.1.3; 4.4.1.4
AE	4.1.2	**F-AAS**	4.2.2.2
AES	4.3.1.2	**FAB-FTMS**	2.2.4
AFC	4.4.1.2	**FAB-MS**	2.2.3
AFL	4.4.1.3	**F-AES**	4.3.2.3
AFS	4.4.1.1	**F-AFS**	4.4.2.1
AS	4.1.3	**FDMS**	2.2.5
ATR-FTIR	5.2.5	**Feststoff-**	4.2.1.7; 4.2.2
		Feststoff-AAS	4.2.2.3
C	7.1.1	**Feststoff-ZAAS**	4.2.2.4
CARS	5.4.6	**FFT-ICP-AES**	5.2.8
CCC	7.1.9	**FFTS**	5.2.3; 5.1.17
CD	5.5.1	**FHTS**	5.1.19
CES-FS	5.3.6	**F-HCL-AFS**	4.4.2.2
CSN	4.3.1.15	**FIA-IC**	7.1.12
CSN-ICP-AES	4.3.2.1	**FID**	7.2.3
		FIMS	2.1.2
DC	4.3.1.5; 7.1.3	**FIR**	5.1.13
DC-AES	4.3.2.2	**FL-AFS**	4.4.2.3
DC arc	4.3.1.8	**FMS**	2.1.3
DCCC	7.1.10	**FS**	5.1.9; 5.6.1
DCP	4.3.1.10	**FT**	5.2.1
DIE	6.6	**FT-IR**	5.2.4; 7.2.4
DR-FTIR	5.2.6	**FT-IR-CIR**	5.2.9
DRS	5.1.14	**FT-MS**	2.2.6
DSIMS	2.2.1	**FTS**	5.1.16; 5.2.2
DTA	6.4		
DTG	6.5	**G**	4.2.1.8
		G-AAS	4.2.2.5
E	1.1.5	**GC**	7.1.4
EC	7.2.1	**GC-FTIR**	7.3.1
ECP	7.2.2	**GC-MI-FTIR**	7.3.2
ED	3.1.1	**GC-MS**	7.3.3
EDL	4.2.1.4	**GF**	4.2.1.8; 4.2.1.9
EDX	3.2.1	**GPC**	7.1.11
EMIR-FTS	5.2.7		
ENAA	1.2.5; 1.2.12	**HCL**	4.2.1.10; 4.4.1.5
ERS	5.1.15	**HCL-AFS**	4.4.2.4
ES	4.1.4		

HECD	7.2.5	MINDAP-AES	4.3.2.6
HIL	4.2.1.11	MIP	4.3.1.12
HPLC	7.1.13	MLC	5.3.3
HS-CCC	7.1.14	MLL	5.3.4
HS-GC	7.1.15	MLS	5.1.5; 5.3.2
HTS	5.1.18	mono	1.1.7
Hydrid-	4.2.1.12	mono,$\bar{\text{I}}$	1.1.9
		MORD	5.5.6
I	1.1.6	MS	2.1.1; 5.1.1; 7.2.9
IC	7.1.5	MSMS	2.2.9
ICP	4.3.1.11; 4.4.1.6	MTS	5.1.7
ICP-AES	4.3.2.4	MWP	4.3.1.13
IC-PCD	7.3.6		
ICPDS	3.2.2	**N**	1.1.2
ICP-HCL-AFS	4.4.2.5	NAA	1.2.2
IENAA	1.2.9	NAA, mono	1.2.10
IMMA	2.1.4	NAA, rel	1.2.11
INAA	1.2.6	NF	4.2.1.13
INMS	2.2.7	NF-AAS	4.2.2.6
IR	5.1.11; 7.2.6	NIR	5.1.12
IRNAA	1.2.8	NIRMS	2.2.10
IRS	5.1.20	NMR	5.5.7
ISRP-LC	7.1.16	NQR	5.5.8
ITD	7.2.7		
ITNAA	1.2.7	**OES**	4.3.1.1
L	4.3.1.14; 4.4.1.7	**PAMS**	2.2.11
L-AES	4.3.2.5	PAS	5.1.21
L-AFS	4.4.2.6	PC	7.1.7
LAMMA	2.1.5	PCD	7.2.10; 7.2.11
LC	7.1.6	PDMS	2.2.12
LC-ES	7.3.4	PID-FTMS	2.2.13
LC-MS	7.3.5	PPINICI-MS	7.3.7
LD-FTMS	2.2.8		
LDLC	7.1.17	**QMS**	2.1.8
LIBS	5.5.4		
LIMA	2.1.6	**R**	1.1.4; 3.1.3
LIMFS	5.3.5	RAMA	5.4.4
LMS	2.1.7	RAMP	5.4.5
		RBA	3.2.3
MAL	5.1.6	RD	5.5.2
MAS	5.1.2	rel	1.1.8
MCD	5.5.5	REMPI-MS	2.2.14
MCP	2.1.13	RF	4.3.1.6
MES	5.1.3	RFA	3.2.4
MF	7.2.8	RF spark	4.3.1.9
MFS	5.1.4; 5.3.1	RNAA	1.2.4

RRS	5.4.2	TMA	6.3
RS	5.4.1	TMS	2.1.12
		TNAA	1.2.3
SC	7.1.8	TOF-MS	2.2.17
SCIC	7.1.18	TOF-SIMS	2.2.18
SERRS	5.4.8	TR-RFA	3.2.8
SERS	5.4.7	TR-XRF	3.2.7
SFC	7.1.19	TS	4.1.5
SIB	4.1.6	TSQ-MS	2.2.19
SIMMS	2.2.15		
SIMS	2.1.9	**UMPA**	2.2.20
SIN	4.1.7	UV	5.1.9
SLS	4.1.9	UV-RRS	5.4.3
SNI-FT-IR	5.2.10		
SNMS	2.1.10	**VDL**	4.2.1.14
SNR	4.1.8	VIS	5.1.10
SOM	5.5.9	VPIR	5.1.22
SSIMS	2.2.16	VUV	5.1.8
SSMS	2.1.11		
SSS	4.1.10	**WD**	3.1.4
STPF-AAS	4.2.2.7	WDX	3.2.5
		WDL	7.2.13
T	1.1.3		
TA	6.1	**X**	3.1.5
TCD	7.2.12	XRD	3.2.6
TDLAS	5.5.10	XRF	3.2.7
TG	6.2		
Thermospray-		**Z**	4.2.1.15
LC-MS	7.3.8	Z-AAS	4.2.2.8

Sachverzeichnis

Abfall, radioaktiver 8, 9, 635, 641, 643, 645
Abkürzungen (Akronyme) für Analysenmethoden 751f
Abrieb
– von Skalpellen 153, 154f
Abtrennung: siehe Trennung
Abwasser, Analysen von städtischem VIII, 643
– mit Voltammetrie 434
– mit ICP-AES und RFA 643
Ahrland-Chatt-Davies-Klassifizierung 462
Akronyme (Abkürzungen) für Analysenmethoden 751f
Aktinidenelemente 98
– Analyse mit Alphaspektrometrie 98
– in-line Analyse mit RFA, Wiederaufarbeitung von Kernbrennstoffen 645
Aktivierungsanalyse VIII, 18, 32, 109, 123, 141, 153, 163, 169, 177, 179, 558, 570, 577, 603
– Neutronenaktivierung, siehe dort
– geladene Teilchen, Aktivierung mit 179
  – Analyse leichter Elemente in schwerer Matrix 179
  – Matrixeinflüsse 179
  – Oberflächenuntersuchung 180
– Verluste bei Bestrahlung mit Bremsstrahlung hoher Energien 183
– Algen, Analyse von mit totalreflektierender Röntgenfluoreszenzanalyse (TXRF) 282
Allgegenwart der Elemente 8
Alphaspektrometrie IX, 19, 23, 24, 28, 98f
– Anwendungen 98
  – Kernbrennstoffabbrand 106
  – Aktinidenisotope 106
  – natürliche Alphaaktivität 107
– hochauflösend 29, 98
– Ionisationskammer 102
– Low-level Spektrometrie 27, 98
– natürliche Alpharadioaktivität 98, 99
– Oberflächensperrschichtdetektoren 98
– Probenvorbereitung 98, 99
  – Laser-Mikrobohren 99
  – Vakuumverdampfung 101
– Spektrenauswertung 103
– Spektrometer 28
– unabhängig von Beta- und Gammastrahlung 105
Alphastrahlung 18
– Energie von 19
Amortisierung von Geräteanschaffungskosten 47
Analog-Digital-Wandler 23, 87, 724
Analyse, chemische (Materialanalyse) 3
– produktionsbegleitend 79
– Teilschritte 3, 4
– zerstörungsfrei 5, 577
– instrumentell 5, 23, 577
Analysendienst, chemischer VIII, 647, 677
– Laser-Massenspektrometrie 201
– Neutronenaktivierungsanalyse 123f
– simultane ICP-AES 647

Analysendurchsatz pro Zeiteinheit 677, 690
– ICP-AES von Fluß- und Bachsedimenten 570, 677
Analysenfunktion 17
Analysengeschwindigkeit, siehe Arbeitskapazität
Analysenkapazität, pro Zeiteinheit, geologische Proben 690
Analysengang 3
– Teilschritte 3
Analysenmethoden 3, 677
– Kombination, von zwei 62
  – LC/MS 62
  – GC FT-IR 62
  – ICP-AES-/RFA 677
  – MS/MS 62
– Kurzbezeichnungen (Akronyme) 751f
Analysenproben 8, 9, 11, 12, 13, 14
– eines Analysendienstes 9, 647, 677
Analysenschritte 3, 4, 39
Analysensignale 3, 17, 18, 22, 23, 65
– Arten von 19
– Bruttomessung 21
– Energie des Analysensignales 18, 19, 20, 21
– Intensität 21
– Messung durch Spektrometrie 21
Analysenvergleich (intercomparison run) 491
Analytik
– produktionsbegleitend 76
Analytische Chemie 57, 58, 59
Anionenbestimmung, Oligo- 15, 137, 415
Anpassung von Analysenverfahren 3
– an prozeßbedingten Datenfluß 508
Anregungsfunktionen
– für Kernreaktion, in der NAA 109, 134
Anregungsquelle 23, 24
Anreicherung, siehe Vorkonzentrierung 459, 485
Arbeitsbereiche
– RFA, wellenlängendispersiv, von Bach- und Flußsedimenten 563
– geologische Proben 683
Arbeitskapazität, pro Zeiteinheit
– geologische Proben 677f, 690
– Umweltproben, mit Voltammetrie 429
Archäologie
– Anwendung chemischer Analysen 57
– ICP-AES Anwendungen 381
– Instrumentelle Elementanalysen 657
Archäometrie, Anwendungen der Multielementanalyse VIII, 459, 657, 667, 669
Atomabsorption
– Energie von 19
Atomabsorptionsspektrometer, handelsübliche 728; Abb.: 740, 741
Atomabsorptionsspektrometrie (AAS) VIII, 28, 47, 65; Abb.: 740, 741
– biologische Proben 535, 608
  – Leber 614
– Elemente, Klassifizierung der bestimmbaren 412
– geologische Proben 535, 690
  – Eruptiv- und Sedimentgesteine 717

Sachverzeichnis

- Graphitrohr, Einheitsverfahren 411
- Linienarmut 24
- Meerwasser 413
- Multielementanalyse 411
  - Einheitsverfahren als Voraussetzung 411
  - geologische Proben 535, 690
  - geothermale Wässer 524
- Nahrungsmittel 623

Atomemission
- Energie von 19

Atomemissionsspektrometer, handelsübliche 727
- DCP-Anregung, Echellegitter; Abb.: 733
- Funkenanregung 728; Abb.: 743
- Glimmentladungsanregung 728; Abb.: 741
- ICP-Anregung 727, 728; Abb.: 731, 736, 737, 738, 740, 744

Atomemissionsspektrometrie (AES)
siehe auch ICP-AES, MIP-AES, kohärente Vorwärtsstreuung 19, 24, 33, 36, 65, 77
- **Anwendungen,** siehe auch: AES
  - Archäologie 657, 658, 660
  - Keramik 661
  - Obsidian 662
- **Bogen-AES** IX, 26, 33, 36, 45, 46, 47, 689
- **Elektrothermale-AES** 19
- Fehlerdiskussion 713
- **Flammen-AES** 19
- **Funken-AES** 19
- **Glimmentladungs-AES** IX, 19
- **Heiße Graphit-Hohlkathoden-AES** 38
- **Laser** 19
- Metalldämpfe von Festproben 82
- **Matrixeffekte**
  siehe auch unter ICP-AES 337, 344
  - Vidiconsystem 342
  - Vielkanaldetektion 342
  - Zerstäubereffekte 344
- spektrale Störungen
  siehe auch: spektrale Störungen ICP-AES 337, 339
- Untergrundkorrektur 339
  siehe auch: Untergrundkorrektur
  - Wellenlängenmodulation 340
  - Zweikanalmethode 342
- Vorkonzentrierung 476

Atomfluoreszenzspektrometer, handelsübliche
- ICP-Plasma, Hohlkathodenlampe, 12 Elemente sequentiell 729; Abb.: 735

Atomfluoreszenzspektrometrie (AFS) VIII, 19, 28, 65, 397
- elektrothermale Atomisierung 19
- Flammenatomisierung 19
- Hohlkathodenlampenanregung 397f
- ICP-Atomisierung VIII, 19, 38, 48, 397f
- **ICP-Hohlkathodenlampen-Atomfluoreszenzspektrometrie (ICP-HCL-AFS)** 397f
  - Anwendungen 397
    - Multielementanalyse, schnelle sequentielle 405f
    - Wasser, Stahl, Blut, Öl, Edelmetalle, S, P 397f
  - Arbeitsbedingungen 398
  - Hohlkathodenlampenmodule 397, 398
  - Lichtstreuung 398
  - Monochromatormodul 400

- Nachweisgrenzen 401, 403, 409
- organische Lösungsmittel 403
- Propanzusatz zum Plasma 397f
- Refraktärelemente 402
- Reproduzierbarkeit 409
- Sauerstoffzusatz zum Plasma 397
- spektrale Störungen 399
  - Überlappung von Spektrallinien 399, 400
- Ultraschallzerstäuber 409

Atomstrahlung 18

Aufschlüsse, zur Multielementanalyse 459
- Boraxaufschluß 652
- Fluß- und Bachsedimente 570
- geologische Proben (HF/HClO$_4$, HF/H$_2$SO$_4$, HNO$_3$/HClO$_4$) (HF/HClO$_4$,/HNO$_3$) 535, 682
- Gesteine (HF/HClO$_4$; NaOH), Seltene Erden-Analyse 551
- Klärschlamm 651
- Königswasser 654
- Litiummetaborataufschluß 713
- Seltene Erden-Minerale, Lithiummetaboratschmelze, HNO$_3$/HCl, Spektrosolv 549
- wasserfreie H$_3$PO$_4$/HNO$_3$ 652
- Wein 633

Auftraggeber VII, 50, 677

Auger-Elektronenspektrometrie (AES)
- hoch radioaktiver Partikel 635

Ausbeute von Detektoren, für Kernstrahlenspektrometer 89

Automation VIII, 41, 43, 48, 51, 80, 677
- von Aufschlüssen 80
- von Probenahme 41, 50

Bakterien, Analyse von
- Mikroanalyse von Metallen in Hydrogenase 283

Baumstruktur von Analysenmethoden 4, 16
Baustufen der Materie 4
Begleitelemente 79
Bestimmungsmethoden 85ff
Betaspektrometrie 19, 24, 28, 558
- Anwendungen
  - Bestimmung von K-40, P-32 558
  - INAA für Geo- und Kosmochemie 558
- Spektrometer 28, 558
Betastrahlung 18, 558
- Energie von 19
Betriebsprodukte, technische 8, 9
Bildanalyse 61

Biologisches Material, Analyse von 57, 58, 63, 153
- AAS 535
- Algen, mit TXRF 282
- Blindwerte 153
- Blut
  - mit ICP-MS 234
  - mit ICP-AES 406
  - Veraschung mit HNO$_3$/H$_2$O$_2$ 234
- Enzyme von Bakterien 283
- Haar, mit kohärenter Vorwärtsstreuung 388
- ICP-AES 535, 647
- Kaninchenleber, mit PIXE 298
- Kollagen, mit NAA 158
- Leber, menschliche 614f
  - Vergleich von AAS, IDMS, NAA, VOL 616
- Meerwasser, mit E-AAS 413

- Milch 609
- Muskel, tierischer 148, 149
  - von Kaninchen, mit NAA 153, 160
- Nahrungsmittel
  - mit Indikator-NAA, 14 MeV-Neutronen 177
- Neutronenaktivierungsanalyse 137, 171
- Voltammetrie 436
- Wundheilung, mit NAA 153

Biotechnologie 59
Blindwerte 461, 469
Blutserum, geforderte Nachweisgrenzen 10
Böden 8, 9, 678f
- ICP-AES 647
- Nachweisgrenze, geforderte 10
  - NAA 123f
  - mit 14 MeV Neutronen 174
- RFA 488
- uranhaltig VIII, 25, 26, 45, 46
Bogenanregung in AES 27
siehe: AES

Chemikalien, Analyse von 8, 9
Chemometrie 48, 531
- Auswertungssystem 531
- Clusteranalyse 588
- Datenbanksystem 531
- Dendrogramme 588
- Informationssystem 531
- Simplexmethode 466
- Variationsdiagramme (Ce/La, K/Rb) in geologischen Proben 688

Clerici-Lösung, zur Schweretrennung von Mineralen 566
Computer
- Auswertung mit 60
- Grafik 62
- Computerisierung VII, 57
- Computerwissenschaft 59, 63

Datenauswertung VII, 60
Datenbeurteilung VII, 4, 43, 505, 517, 526
- Einhaltung von Grenzwerten 513
- geologische Proben 517, 552, 711, 717
- industrielle Multielementanalyse 505f
- Plausibilität, hinsichtlich 511
- von Probengüte und Analysenqualität 509
- prozeß- und produkttechnische Gesichtspunkte 506, 514
- Qualitätskontrolle 4, 43, 505, 517, 677
  - von Großserien 517, 677
  - Probensynthese 519
- Selektivität und Reproduzierbarkeit 519
- Sicherstellung von Analysenergebnissen 43, 514, 517
- Stahlanalysen 506
- Überwachung von Betriebsstoffen und Produkten 506, 514
- Unstetigkeiten der Materialzusammensetzung, Auskunft über 572
- zeitaufgelöste Untersuchungsabläufe 511
  - in der Oberflächenanalyse 512

Datenkonzentrierung 43, 66, 517, 577
- Rechnerprogramm 60
Datenreduktion siehe: Datenkonzentrierung
Datenverarbeitung 43, 60
- spektrometrischer Meßsignale 510

Detektoren, in Kernstrahlenspektrometrie VII, 23, 87, 724
- Halbleiterdetektoren 28, 29, 87f
  - Ge(Li), Reinst-Ge-Detektor 28, 87, 724
  - Si(Li) 29, 87
- Ionisationskammer, für -Strahlungsmessung 27, 29
- Szintillationsdetektoren 87
  - NaJ(Tl)-Detektor 87
Differentialpulspolarographie, siehe Voltammetrie
Dynamischer Konzentrationsbereich 47, 65
- ICP-AES 47, 713
- AAS 47
- Hohlzylinderplasma MIP-AES 366
- MIP-AES, Dreifachplasma 365
- Totalreflektierende RFA (TXRF) 262

Eichung 81, 83
Eichkurve 47
Einheitsverfahren 50
Einkristall 7
Eisenlegierungen, Analyse von
- Gußeisen 242f
- Kohlenstoff in Stahl und Gußeisen, mit RFA 241
  - Eichkurven 243
  - Richtigkeit 244
- Stahl, mit ICP-AES 80
- Stahlstandards, Analyse mit FMS und IV-FMS 199

Elektrochemie 60
- Kalmanfiltertechnik 60
Elektroden
- biosensitive 61, 62
- enzymionensensitive 62
- ionensensitive 62
- Radelektrode, Funken-AES 82
Elektromagnetische Strahlung 18
Elektronikmaterialien, Analyse von VIII, 603
Elektronenhülle, von Atomen 7
Elektronenmikrosonde
- Analyse einzelner Luftstaubteilchen, mit EDAX 599
- in Lebenswissenschaften 608
Elektronenspinresonanz 60
- Energie von 19
Elementanalyse 5
Elemente, chemische
- allgemein 7,8
- biologisch-toxikologische Einteilung 14
  - essentielle 14, 607
  - toxische 14, 607
- in kerntechnischen Lösungen 636
Elementkombinationen, bestimmbare 32-39
- AES mit heißer Graphithohlkathode 33, 38
- Bogen-AES 33, 36, 37
- Differentialpulspolarographie 33, 38, 415f
- Flammen-AES, 33, 36
- Funken-AES 33, 37
- Funken-MS 33, 34
- Geologische Proben, geforderte 683
- Gravimetrie 33, 39
- ICP-AES 33, 37, 38
- ICP-AFS 33, 38
- ICP-Massenspektrometrie 227

Sachverzeichnis 771

- Inversvoltammetrie 33, 38, 39, 415
- Neutronenaktivierungsanalyse 33, 35, 109, 123
- Röntgenfluoreszenzanalyse 33, 35
- Titrationen 33, 39

Elementverteilung, Messung 211
- Ionenmikrosonde SIMS 211
  horizontale Elementverteilung, vertikale Tiefenprofile

Energieforschung, Analyse in der 57, 58

Fällungen 16
- in nichtwässriger Lösung 16

Fehler, systematische 41, 42, 43
- Datenauswertung 43
- Flammen-AES 36
- Gleichstrombogen-Atomemissionsspektrometrie 36
- ICP-Atomemissionsspektrometrie 37
- Probenahme 39, 40, 41
- Röntgenfluoreszenzanalyse 35

Fernerkundung 61

Ferrolegierungen, Analyse von 80

Festkörperforschung, Analyse in der 58

Feststoffanalyse 81, 359
siehe auch ICP-AES und G-AAS

Fichtennadeln, Analyse von 489

Flüssigkeiten
- superkritische 63

Forschungsplanung 49

Forschungspolitik VIII

Fourier-transform (FT) Spektrometrien 60
- FT-IR-Spektrometrie 60
- FT-Massenspektrometrie 60
- Algorithmus 60

Fragestellung, definierte
- Voraussetzung jeder Analyse 4, 460

Fremdträger 16

Funkenanregung 27

Funkenmassenspektrometrie, siehe Massenspektrometrie

Gammaspektrometer 28, 90, 91
- Anticompton-Detektor 92, 93
- automatisiertes Vielzweck-Gammaspektrometer 93
- CAMAC-Grundlage, rechnergesteuert 90, 91, 92

Detektoren 88f
  - Energieauflösung (resolution) 88
  - Ge(Li)-, Reinst-Ge-, Si(Li)-, planarer Ge-, low-energy Ge(LEGe)-, reverse-electrode Ge(REGe)-Detektor 88
  - Geometrie des Detektors 88
  - NaJ(Tl)-Detektor 87
  - Zählausbeute (efficiency) 89
- Fernanalyse 87
- Software zur Spektrenauswertung 92
- Vielkanalanalysator 90

Gammaspektrometrie VIII, 18, 19, 23, 24, 28, 77, 87
- Abtrennung von P-32 zur Untergrundreduzierung 163
- Anwendungen 94
- Computerauswertung 87, 98, 123, 577, 583
- hochauflösende 28, 87, 311
- Low-level 25
- Multiradionuklidanalyse 87, 604

- Nachweisgrenzen 150
- prompte 19
- Rechnerauswertung, siehe Computerauswertung
- Totzeitkorrektur 169

Gammaspektrometrie, Anwendungen 94
- Abfallbeseitigung 94
  - Abtrennung von P-32, zur Reduzierung des Untergrundes 163
- Archäologie 94
- biologisches Material 607
- Eruptiv- und Sedimentgesteine 714
- Geo- und Kosmochemie 94, 717
- Kernbrennstoffzyklus 94
- Medizin 94
- Nuklearmedizin 94
- Strahlenschutz 94
- Technik 94

Gammastrahlung 18
- Energie von 19

Geladene Teilchenstrahlung 19

Geochemische Proben, Analyse VIII
- AAS 535
- Betaspektrometrie 558
- Bach- und Flußsedimente 680
  - mit RFA 563, 680
  - mit ICP-AES 570, 680
- FMS mit Isotopenverdünnungsanalyse 195
  - Standardgesteine BCR-1, W-1 199, 200
- geologischer Standard AGV 1, mit Laser-MS 206
- Gesteine (Eruptiv- und Sedimentgesteine) 717
- Gneise, mit INAA 565
- ICP-AES 381, 535
- ICP-AES und RFA, Vergleich 677
  - Flußsedimente 677
  - Böden 677
  - Gesteine 677
- INAA von Silikatkugeln in Sedimenten 574
- Methodenvergleiche an geologischen Proben
  - ICP-AES, RFA Böden, Gesteinen, Sedimenten 677
  - ICP-AES, NAA, SSMS 711
  - ICP-AES, DC-AFS, Spex-Standard 713
- Multielement-Isotopenverdünnungs-Funkenmassenspektrometrie (ID-SSMS) 195
  - Gesteinsstandard, Stahlstandard 199
- Nachweisgrenzen 10
- Neutronenaktivierung, Vergleich 123f
  - geologische Standards 141
- Qualitätskontrolle 517
- SPEX-common-element-standard, mit DC-AES, ICP-AES 713

Geothermalforschung, Analysen in 516

Geräteausstellung VIII, 723

Gerätehersteller, Anschriften 746

Gerätekontrolle durch Computer 60

Gesteine, Analyse von 8, 9, 679
siehe auch: Geochemie
- ICP-AES 551, 647, 679
- ICP-AES und RFA 679f
- K, P durch Betaspektrometrie 558
- Nachweisgrenzen 10
- Vulkanische und Sedimentgesteine mit MS-Isotopenverdünnung 717
- RNAA von Gesteinen und Meteoriten 553

Gitter 23

Gitterspektrograph 22
- Typ Czerny-Turner 38
- Typ Wadsworth 36

Glas, Analyse von VIII
- Bor 247, 248

Glasfasermaterial, Analyse von
- Ausgangsstoffe, mit INAA 605
- INAA 604

Gravimetrie 5, 39

Großgeräte VIII, 723

Großserien
- geologische Proben mit ICP-AES und RFA 677

Gruppentrennungen 10, 15, 16

Gute-Analytische-Praxis 514

Halbleiterdetektoren, für Kernstrahlenspektrometrie 88
- Ge(Li)-Detektor 88
- Koaxialer Ge-Detektor 88
- Niedrigenergie-Ge-Detektor 88
- planarer Ge-Detektor 88

Halbleitermaterialien, Analyse von 603
- Ionenmikrosonde (SIMS) 211
  - Analyse von ZnS-Phase 217

Halogene, analytische Bestimmung 137

Hardware 60

Hauptbestandteile
- Definition 8, Abb. 4

Hochtemperaturforschung, Analyse in der 63

Homogenisierung 41, 153, 459
- Dismembrator 158

Homogenität 81, 158

Homogenitätstest 41

ICP-Atomemissionsspektrometrie (ICP-AES) XI, 19, 33, 37, 45, 46, 47, 78, 80, 311, 381
- **Anwendungen** zur Multielementanalyse
  - Abwasser 643
  - Analysendienst eines Forschungszentrums 647
  - Archäometrie 657, 669
  - biologische Proben 535
  - Fluß- und Bachsedimente 570
  - geologische Proben 535, 677f, 694
    - Gesteine, Seltene Erden-Analyse 551
  - Kerntechnik, hoch-radioaktiver Abfall (HAW) 641
  - Kunstwerke, Bilder 667, 669
  - Lebenswissenschaften 608
  - Metallegierungen 649
  - Minerale 551
  - Nahrungsmittel 622, 629, 633
  - Umweltforschung 697
- Hydriderzeugung 357
- Probenzufuhr 347
  - Gasgenerator, für Halogene 357
  - Graphitrohr, direkt 347
  - Graphitrohr, elektrothermal 347
  - Hydriderzeugung, für Hydridbildner 357
  - pneumatischer Zerstäuber 347
- Signalerzeugung 347f
- Untergrundcharakterisierung 347
  - Optimierung der Zerstäuberparameter 347
  - Optimierung des Spektrometers 347
- Vergleich ICP-AES mit NAA, TXRF, für geologische Proben 694, 697

- **Feststoff-ICP-AES** mit Funkenerosion 359f
  - Instrumentierung 359f
  - Matrixeffekte 361
  - Nachweisgrenzen, Nachweisstärke 359, 360
  - nichtleitende Pulver 361
- **hochauflösende ICP-AES** 311f
  - Anwendung auf linienreiche Spektren 319
  - Auflösung 315f, 324
    - mit Vordispersion 311
  - Echelle-Monochromator 313f
  - historische Entwicklung 312
  - ICP-Plasma 315
  - Kalibrierung 571
  - linienreiche Spektren 311
  - Nachweisgrenzen 317
  - Optimierung der Spaltbreite 315
  - Signal-Untergrund-Verhältnis 315
  - spektrale Störungen 311, 319f
  - Vorteile 326
- Lichtleiterkopplung 373
  - Optimierung 375
  - Anwendung in der Stahlanalyse 374
- **Matrixstörungen** 337f, 344
  - Aerosolerzeugung 339
  - Anregung und Verdampfung 339
  - Plasmageometrie 339
- nachweisstarke Linien 538
- **Optimierungskriterien** für ICP-AES 311, 377
  - Einfluß auf Nachweisgrenzen 377
  - Einfluß auf Reproduzierbarkeit 377
- **Spektrale Störungen** 37, 48, 337, 537, 538
  - Bandenstörungen 338
  - Intensität wahrer Linien 338f
  - Linienstörungen 337
  - Störungen am Spektrometer 338
  - Untergrundkorrekturen 340
- Stöchiometriebestimmung 650
- **Verbrauchsarme ICP-AES** 329f
  - Bornitridrohr 331
  - Langzeitstabilität 334
  - Matrixstörungen 334
  - Miniaturisierung der ICP-Fackel 330
  - Nachweisgrenzen 334
  - Plasmafackel
    - niedriger Argonverbrauch 329, 330
    - niedrige Energie 329

Integrierte Schaltkreise 57

Interelementeffekte, -störungen 49, 337

Impulshöhen
- analysator 23
- spektrum 87

Informationsexplosion 63

Informationskapazität 65

Informationssysteme 76

Infrarotstrahlung
- Energie von 19

Ingenieurwissenschaften, Analyse in den 63

Interferometer 23

Inversvoltammetrie (anodic stripping) siehe auch Voltammetrie 32, 33, 38, 48, 415

Ionenchromatographie 445f, 544
- **Anionen-**Chromatographie
  - Anionenaustausch-Chromatographie 447, 544
  - Detektion mit Leitfähigkeit, RP, UV 448, 449, 544

Sachverzeichnis 773

- Hohlfaser-Suppresor 447
- Ionenpaar-Chromatographie 448
- Multielementanalyse 449, 544
- **Kationen**-Chromatographie 449, 544
- Kationenaustauscher mit niedrigerer bzw. höherer Kapazität 450

Ionenmikrosonde 211

Ionenquelle, für Feststoff-Massenspektrometrie 30, 185
- Gleichspannungs-Bogenquelle 186
- Glimmentladungs-Ionenquelle 30, 31
- Hochfrequenzfunken-Ionenquelle 30, 186
- Ionenbeschuß-Ionenquelle (SIMS) 30, 187, 188
- Laser-Ionenquelle 30, 188, 203
- Niederspannungsentladungs-Ionenquelle 30
- Thermionenquelle 30
- Vakuumentladungsquelle 30

Ionenseparator, für Massenspektrometrie 31
- Bahnstabilitätsspektrometer 31
- doppeltfokussierendes Spektrometer 31
- Energiebilanzspektrometer 31
- Flugzeitstreckenspektrometer 31
- Quadrupolmassenspektrometer 31

Gitterionisationskammer, für Alphaspektrometrie 29

Ionisationspuffer
- Flammen-AES 36
- Gleichstrombogen-AES 36
- ICP-AES 37

Ionisierende Kernstrahlung 18

Kalibrierung 3, 48, 83
- geologische Proben 684

Kalmanfiltertechniken
- in der UV/VIS-Spektrometrie 60

Kathodenstrahlung 19

Kationentrennungsgang, klassischer 10, 15, 16
- Anionen 10, 15
- Kationen 10, 15, 16
- in nichtwässriger Lösung 16, 17

Keramik, Analyse von 657
- Archäologie, in der 661
- Ionenmikrosonde (SIMS) 211

Kernbrennstoff, Analyse von 98, 635
- Hochtemperaturreaktor 98
- abgebrannter Kernbrennstoff 98
- Brennstoffkerne und -kugeln 99

Kernmagnetische Resonanz 60
- Energie 19

Kernstrahlungsspektrometrie
siehe Alpha-, Beta-, Gamma-, Röntgen-Spektrometrie

Kernstrahlungsspektrometer, handelsübliche 723f; Abb.: 730f
- Detektoren
  - für Alphaspektrometer 724
  - für Betaspektrometer 724
  - für Gammaspektrometer 724; Abb.: 733
  - Low-level Detektor mit Anticomptonabschirmung; Abb.: 735
- **Vielkanalanalysatoren** 28f, 723; Abb.: 733, 737, 739
  - Analog-Digital-Wandler (ADC) 28, 724

Kerntechnik
- Aktiniden, in Kernbrennstoffwiederaufbereitung 645

- Mikro-, Spuren- und Multielementanalysen in 635
- hoch radioaktive Abfall-Lösungen, ICP-AES 641

Kohärente Vorwärtsstreuungs-Atomspektrometrie 65, 385, 391
- Dichroismus und Dispersion, Anteile 391
- Dynamischer Bereich 389
- Meßanordnung 385, 386
- Multielementcharakter 386
- Nachweisgrenzen 389
- Natrium-D-Linie 393

Kohle, Analyse von, mit RFA 246
Komponenten einer Analysenprobe 9, 17
Kompromiß, bei Versuchsbedingungen 16

Kontamination
- Neutronenaktivierungsanalyse 154
- Oberflächen 153
- Quarzampullen, von 159
  - Reinigung, von 159
- Wundheilungsuntersuchungen 154

Konzentrationsbereich, nutzbarer 65
Konzentrierung, siehe: Vorkonzentrierung

Korngrenzen
- Elementanreicherung in 81

Korngrößeneffekte
- Röntgenfluoreszenzanalyse 35

Kosmochemie, Analyse in der
- Betaspektrometrie 558
- Iridium, Indikator für extraterrestrische Herkunft 575
- Meteorite 199, 553, 575
  - Seltene Elemente 553
  - Stein- und Eisenmeteorit, mit ID-SSMS 199
  - Multielement-Isotopenverdünnungs-Funkenmassenspektrometrie 159f, 199
- Probenaufschluß
  - HNO_3/HCl, HF/HNO_3/HCl 196

Kosten, von Analysen 45, 47, 50, 679
Kosten-Nutzen-Abschätzung 460, 677
Kristallgitter 7

Kunstwerke, Analyse von VIII, 667, 669
- Flächenuntersuchungen mit Strahlung (IR, UV, X, γ, n, e⁻) 667
- Pigmente mit RFA, NAA, PIXE, AES, ICP-AES, LMA, MS 667
- Punktuntersuchungen mit Mikrosonden, PIXE 667
- Veränderungen von Kunstwerken, mit SEM, EPMA, IPMA, SIMS 667

Laser 59, 61, 203
- Hochenergielaser 61
- Ionisierungshilfe (LEJ) 65
  - monochromatische Strahlungsquelle 65
- Massenspektrometrie 201
  siehe: Massenspektrometrie, Laser
- Neodym-YAG-Laser 61, 99, 204
- Ramanstreuung, für 61
- Verdampfung von Materialien, durch 65, 203
  - Plasma, entstehendes 203

Lagerstättenprospektion, Analyse in der 458
Landwirtschaft, Analyse in der 58
Lebensmittelanalytik, siehe Analyse von Nahrungsmitteln

Lebenswissenschaften, Analyse in den, siehe auch: biologische Proben; Medizin VIII, 607, 621
- Analysenmethoden, verwendete
  - Aktivierungsanalyse 608
  - Atomabsorptionsspektrometrie 608
  - Funkenmassenspektrometrie 608
  - Probenahme 608
  - Röntgenfluoreszenzanalyse 608
- Analysenproben
  - medizinische Proben 611
  - Nahrungsmittel 609, 621f
  - Serum 608
  - Umweltprobenbanken, Material von 611
- essentielle Spurenelemente 607

Leuchtdioden, Analyse von
- INAA von III-IV-Verbindungen 603

Linienreichtum
- Bogen-AES 49
- Funken-AES 49
- Funkenmassenspektrometrie 49
- ICP-AES 37, 48, 49

Luftstaub, Analyse von
- Multielementanalyse und -überwachung von Luftstaub 577, 693
  - synchrotronstrahlungsinduzierte RFA (SYRFA) 301
  - totalreflektierende RFA 276
  - TXRF, INAA, ICP-AES 703

Managementlehre 63

Massenspektrometer, für Festkörper, handelsübliche 30, 725f
- Funkenanregung 725
- Glimmentladungsanregung 725
- ICP-Anregung 227, 725; Abb.: 742, 745
- Laseranregung 185, 201, 725
- Massenseparator 23
- ortsauflösende Rasterung, Ionenmikrosonde (SIMS) 729

Massenspektrometrie IX, 20f, 77, 185f, 192, 211
- **Anwendungen**
  - Eruptiv- und Sedimentgesteine 717
  - geothermale Wasser 524
  - Kunstwerke, Pigmente 667
  - Lebenswissenschaften 618
  - Standardgesteine 711
- Detektion
  - Fotoplatte 30
    - Laser-MS 205
  - elektrische 30
    - Laser-MS 205
- **Funkenmassenspektrometrie (SSMS)** IX, 18, 31, 33, 34, 45, 46, 48, 185, 608
  - Anwendungen
    - geologisches Material 717
  - **Multielement-Isotopenverdünnungsanalyse** 195, 717
    - Reproduzierbarkeit 197
    - Richtigkeit 198
  - **Niedervolt-Bogenentladungsionenquellen-MS** 26, 31, 33, 34, 45, 46, 186
    - Nachweisgrenzen 34
  - Vergleich
    - mit RFA, AAS, (Gesteinsproben) 718

- **Hochpräzisions-Massenspektrometrie** durch Multielement-Isotopenverdünnung (ID-SSMS) 195
  - Anwendungen 199
    Standardgestein BCR-1, Stein- und Eisenmeteorit, Stahlstandard NBS-1161 199
  - Massenspektrometer, mit Photoplatten- und elektrischer Detektion 196, 197
  - Probenvorbereitung 196
    - Auflösungstechnik 196
    - Graphitspiketechnik 196
  - Reproduzierbarkeit 197
  - Richtigkeit 198
  - Störungen 197
- **ICP-Massenspektrometrie** IX, 227
  - Dissoziation der Probe 228
  - Extraktion der Ionen aus Plasma 229
  - Historisches 229
  - Hybridtechnik 229
  - Ionisationsunterdrückung von Linien 234
  - Ionisierung der Probe 228
  - Isotopenverhältnisse 235
  - Kondensierung von Feststoffen 234
  - Nachweisgrenzen für Lösungen 232
  - Probeneinführung 228
  - Probenvorbereitung, Blut mit HNO_3/H_2O_2 227
  - Quadrupolmassenspektrometer 231
- **Ionenmikrosonde (IMMA)** IX, 185, 211
  siehe auch unter Ionenmikrosonde
  - Ionenquellen 185, 211
- **Lasermassenspektrometrie** IX, 48, 201f, 207
  - Anwendungen
    - geologischer Standard AGV 1 206
    - Übersichtsanalysen, Volumen- und Mikroanalyse 201
  - Laserquelle am doppeltfokussierten MS, für Volumenanalyse 204
  - Laserquelle am Flugzeit-Massenspektrometer, für Mikroanalyse als Mikrosonde 204
  - Lasermassenspektrometrie-Mikrosonde (LAMMA)
    - einzelne Luftstaubkörner 598
    - Lebenswissenschaften 608
- **Sekundärionen-Massenspektrometrie (SIMS)** IX, 185, 211
  - Elementtiefenprofile (ZnS) 220
  - Mikrobereichsanalyse von Elementspuren (ZnS-Phase in Abscheidung) 218
  - Mikroverteilungsanalyse von Elementspuren in AlSi 218
  - Multielementspurenanalyse im Volumen(bulk), Ta-Draht 215
  - Oberflächenverteilungsanalyse von Spurenelementen, Cr-Schicht auf Cu 222
  - Quantifizierung 213
  - Signalerzeugung 212
  - Ultraspurenbulkanalyse 215
- **Tandem-Massenspektrometrie (MS/MS)** 62
- **Verdünnungsanalyse** in der Massenspektrometrie (ID-MS)
  - in Geo- und Kosmochemie 195
  - in Lebenswissenschaften 608

Mathematik, Anwendung in der Analytik 63

Materialforschung, Analyse in der 57, 58, 603f

Materialien der **Nukleartechnik** 8, 9, 635, 641, 643, 645
Materialmodelle zur Spurenelementanreicherung 468
Materie, Baustufen der 6
Materie, chemische Zusammensetzung nach Haupt-, Neben- und Spurenbestandteilen 3, 11, 12, 17, 79
– Aluminium 12
– Basalte 11
– biomedizinische Proben, Erwachsene 13, 14
  – Einteilung nach Hauptbestandteilen 14
– Böden 11
– Erde 11
– Flußwasser 11
– Granite 11
– Luft 11
– Meerwasser 11
– Metallurgisches Silicium 12
– Mond 11
– Thulium-Aluminium, $TmAl_2$ 12
– Phosphatdünger 12
– Polyethylen 12
– Regenwasser 11
– Rinderleber 12
– Säugetierblut 12
– Spinat 12
– Spurenbestandteile 12
Matrixabhängigkeit 717
Matrixlösungen 49, 79
Medizin, Analyse in der 57, 58, 607, 611
siehe auch biologisches Material
– mit INAA 607f
– mit Voltammetrie 436
– essentielle Spurenelemente 14, 607
– toxische Spurenelemente 14, 415
Medizinische Diagnostik
– Implantate 611
– Multielementmethoden 459, 611
Meerwasser 10
– chemische Zusammensetzung 11
– Nachweisgrenzen 10
– ICP-Massenspektrometrie 234
– totalreflektierende Röntgenfluoreszenzanalyse 279
Mehrphasensystem 6
Meßpräparateherstellung 3
– für synchrotronstrahlungsinduzierte RFA 302
Metalle, Analyse von
– Artefakte in der Archäologie 658
– Chromschicht auf Kupfer, mit SIMS 223
– Edelmetalle mit ICP-AFS 408
– Kupferhüttenprodukte 147
– Sintermetalle, mit Ionenmikrosonde 211
– Tantaldraht, gesintert, mit SIMS 223
– Tiefenprofile von Cr, Cu, B in Chromschicht auf Kupfer 223
– Toxische Metalle mit Voltammetrie 415f
Metallurgie, chemische Analysen in der 58
Meteorite, Analyse von
– mit RNAA und radiochemischer Trennung 553
Meteorologie
– Anwender analytischer Daten 57
Methodenvergleich 677
– Eruptiv- und Sedimentgesteine mit MS, RFA, AAS, NAA 717

– geochemische und biologische Proben mit ICP-AES, G-AAS, F-AAS 607
– geochemische Großserien mit ICP-AES und sequentieller RFA 677
– geologische Standardreferenzproben mit ICP-AES, NAA, SSMS 717
– ICP-AES und RFA (wellenlängendispersiv und sequentiell) 677
– Leber, mit AAS, IDMS, NAA, VOL 616
– NAA 123, 139
– Umweltproben, mit TXRF, NAA, ICP-AES 693f
– SPEX-Multielementstandard mit DC-AES, ICP-AES 713
– Wein, mit ICP-AES, GC, DPASV 634
Mikroanalyse
– Elektronenmikrosonde 599, 608, 729
– Ionenmikrosonde (SIMS) 211, 729
– in der Kerntechnik 635f
– Lasermassenspektrometrie 203, 207, 729
Mikroelektronik, chemische Analyse in der 603
Mikrosonden, zur Analyse von Elementverteilungen
– Charged particle activation analysis, CPAA 212
– Lasermassenspektrometrie-Mikroanalyse LAMMA 212
– Particle induced X-ray emission spectroscopy (PIXES) 212
– Secundary ion mass spectrometry (SIMS), als Mikrosonde: IMMA 212
Mikrosonden, handelsübliche
– Elektronenmikrosonde (RFA) 729
– Hersteller 746
– Ionenmikrosonde (SIMS) 729
– Lasermikrosonden-Massenspektrometrie (LAMMA) 598, 729
Mikroprozessoren 60
– am Gerät, zur Datenauswertung 48
Mikrowellen 20
mikrowelleninduzierte Plasma-Atomemissionsspektrometrie (MIP-AES) 363f
– Faden-Plasma-AES (F-MIP-AES) 363f
  – Dreifaden-Argon-MIP (3-F-MIP-AES) 364
  – Nachweisgrenzen 365
– Hohlzylinder-Argonplasma-AES (HC-MIP) 363
  – Nachweisgrenzen 366
– Sauerstoff/Argon-MIP (O_2/Ar-MIP-AES)
  – Bestimmung von Hg in organischen Verbindungen 369
  – on-line Detektor für HPLC 368
Milch
– Bestimmung von Jod 140
Minerale, Analyse von 8, 9, 679
– Apatit 568
– Bastnäsit 549
– Monazit 549, 568
– Obsidian, in der Archäologie 662
– Turmalin 568
– Zirkon 568
– Xenotim 568
Modell-Lösungen
– zur Entwicklung von Analysenverfahren 458
Modellwunden
– Wundheilung und Elementspuren-Kombination 157
– Spurenelementbestimmung 163

Molekülabsorption, Energie von 19
Moleküle 7
Molekülspektrometrie
- IR-Spektrometrie in der Archäometrie 669
  - Untersuchung von Kunstwerken 667, 669
Monostandardmethode der NAA (Comparator-, Monoelementstandard- oder k_o-Methode), siehe auch Neutronenaktivierungsanalyse 35, 109, 123, 133
Mosaikkristall 7
Moseley'sches Gesetz 20, 292
Multielementanalyse 3, 4, 5, 6, 17f, 65
- Auswahl von Methoden 45
- Anwendungsbereiche 45
- Aufgaben, in der Stahlindustrie 506
- Großserien 517, 677
  - Qualitätskontrolle 517
- industrielle Prozeßüberwachung und Charakterisierung von Betriebsabläufen und Produkten 506
- Kombination mehrerer Methoden, siehe dort
- tägliche Praxis 75
Multielementanreicherung, chemische 457, 459, 460, 461, 463, 468
- Anreicherung von Seltenen Erden, Uran, Thorium 485
- Artgleichheit von Probe und Modell 468, 470
- Extraktion 460, 463
- Fällung 460, 463
  - Gruppenreagentien 15, 465
  - Emulsionsbildung bei Extraktion 466
  - Fehlerquellen 466
  - Kollektorfällung 462, 463
  - partielle reduktive Matrixfällung 464
  - Spurensorption an Gefäßwandungen 466
  - Übersättigung 466
  - Verzögerte Niederschlagsbildung 466
- Flotation 464
- Ionenaustausch 460
- Kalibrierung 466
  - Anreicherungsausbeute 466, 467
    - Bestimmung mit Radiotracern 466
  - Kalibrierstandardproben 467
- Klassifizierung von Ahrland-Chatt-Davies 462
- Matrixeinfluß 461
- Spurenkonzentrate 470f
  - Aktivierungsanalyse 474
  - Röntgenspektrometrie 472
- Verflüchtigung 460
- Zonenschmelzen 460
Multielementanalyse, instrumentelle VIII, 5, 23, 45, 76, 386, 505f, 677f
- Analysenkosten 679
- Archäometrie 657
  - Keramik 657
  - Metalle (Ag, Cu, Pb) 657
  - Obsidian 657
- Arbeitsbereiche, **Eisen und Stahl** 507
  - metallische Werkstoffe 507
  - oxidische Werkstoffe 507
- einsatzbereite Methodenkombinationen 47, 677
- **geologische Proben** 11, 12
  - Bach- und Flußsedimente, Großserien 678
  - Bewertung von Rohstoffvorkommen 678
  - Bodenproben 678

- Durchschnittsanalysen 678
- Elementverteilungen 12, 13, 678
- Erzprospektion 678
- Exploration 678
- geochemische Prospektion 678
- Großserienanalytik (siehe auch dort) 678
- ICP-AES und RFA 678f
- in-situ Analyse 678
- Klassifizierung von Gesteinen 678
- Rohstoffprospektion 678
- Umweltforschung 678
- geothermale Wässer 523
- **Großserienanalytik** 517, 678
  - Doppel- und Mehrfachanalytik 517, 678
  - Kalibrierung 517
  - Probensynthese 519
  - Qualitätskontrolle 43, 505, 517, 678
- **Kerntechnik** 635, 641, 643, 645
  - simulierter hoch-radioaktiver Abfall 642
- Kohärente Vorwärtsstreuung, siehe auch: AES 386
- **Kombination von Methoden** VIII, 47, 677, 693, 711, 713, 717, 723
- Multielementstandards 491, 711
  - Flugasche 596
  - geologische: Albit, Diabas 711
  - NAA 123
  - Qualitätskontrolle 43, 517, 678
  - Wasserchemie 539
Multielementstandardmethode, NAA (Relativmethode) 35, 109f, 123f; siehe auch: NAA
Multikomponentenanalyse 59
Multivariante Statistik 61, 531, 577
siehe auch: Chemometrie

Nachweise, spezifische 16
Nachweisgrenzen 32, 33, 46, 49, 65, 145, 459, 470, 682
- abhängige Nachweisgrenzen, in NAA 145
- Atomabsorptionsspektrometrie 543
- Atomemissionsspektrometrie mit heißer Graphit-Hohlkathode 33, 38
- Bogen-AES 33, 36, 37
- Differentialpulspolarographie 33, 38, 420f
- Flammen-AES 33, 36
- Funken-AES 33, 37
- Funkenmassenspektrometrie 33, 34
- Gammaspektrometrie, in NAA 20, 21, 32, 86
- geologisches Material 682
- Gravimetrie 33, 39
- ICP-AES 33, 37, 38, 377, 379
  - direkte Zufuhr aus Graphitrohr 350
  - Feststoff-ICP-AES 360
  - hochauflösende 317
  - verbrauchsarme ICP-AES 334
  - Wasseranalyse 543
- ICP-Atomfluoreszenzspektrometrie 33, 38, 401, 403
  - ICP-HCL-AFS 401, 403, 409
- ICP-Massenspektrometrie 232
- Inversvoltammetrie, 33, 38, 39, 420f
siehe auch: Voltammetrie
- Materialien
  - atmosphärische Aerosole 585
  - geologische Proben, geforderte 682, 683
  - geothermale Wässer 526
  - ICP-AES in Wein 633

– mikrowelleninduzierte Plasma-AES (MIP-AES) 363f
  – Dreifaden-Argon-MIP (3-F-MIP) 364, 365, 367
  – Hohlzylinderplasma (HC-MIP) 366, 367
– Neutronenaktivierungsanalyse 33, 35, 145f, 166, 577, 585, 603, 615
  – in atmosphärischen Aerosolen 585
– Röntgenfluoreszenzanalyse 33, 35, 36
  – totalreflektierende (TXRF) 260
  – synchrotronstrahlungsinduzierte (SYRFA) 302, 303
  – teilcheninduzierte RFA (PIXE) 293
– Titration 33, 39

Nahrungsmittel, Analyse von 609, 621, 629
– Analysenmethoden (AAS, Voltammetrie, NAA, ICP-AES, RFA) 622
– Elementgehalte in Nahrungsmitteln 12, 622, 623
– essentielle Elemente 14, 623
– Milch 609
– Mineralstoffe in Nahrungsmittel 621, 623
– Toxische Metalle 14
  – Belastungspfad im Menschen 417
  – Bestimmung mit Voltammetrie 434
– Übersichtsanalysen 622
Nebenbestandteile, Definition 8

Neutronen
– Energie von 19, 109
– epithermische 19, 109, 123
– Reaktorneutronen 19, 109, 123
– thermische 19, 109, 123
Neutronenaktivierungsanalyse (NAA) siehe auch Aktivierungsanalyse VIII, 18, 35, 48, 603
– Absolutmethode 110
– **Aktivierung** mit
  – epithermischen Neutronen VIII, 109f, 124, 137, 584, 603
  – geladenen Teilchen VIII, 179
  – 14 MeV-Neutronen VIII, 133, 137, 173, 696
  – Photonen 594
  – Reaktorneutronen VIII, 32, 109f, 124, 577, 693
  – thermische Neutronen VIII, 109f, 124
– **Anwendungen**
  – Archäometrie 657, 667, 669
  – atmosphärische Aerosole 581
  – Elektronik- und Halbleitermaterialien, INAA 603f
  – geologische Proben 123
    – silikatische Kugeln, INAA 573
    – Obsidian 662
  – Glasfasermaterial 604
  – Halbleitermaterialien 604
    – Integrierte-Schaltkreis-Technologie, Materialien 604
  – Keramik 661
  – Kunstwerke 667, 669
  – Lebenswissenschaften 607f
    – Milch 609
    – orthopädische Implantate 611
  – Luftstaubkörner, einzelne, mit INAA 600
  – Meteoriten 553
  – Nahrungsmittel, INAA 622, 625
  – Optische Fasern 604
  – Pb, Ag, Cu 658, 660
  – Siliziumtechnologie 604
  – Umweltforschung 577, 696
    – 14 MeV Neutronen 696
    – thermische Neutronen 696
  – Umweltprobenbank 619
  – Videomaterialien 605
– Elemente, mit NAA bestimmbar 35, 120, 125f
– **Indikator-Neutronenaktivierungsanalyse** VIII, 177
– k_o-Methode, siehe Monostandardmethode
– Komparatormethode, siehe Monostandardmethode
– **Monostandardmethode** VIII, 109f, 111, 603
  (siehe auch Komparator- oder k_o-Methode)
  – mit epithermischen (epicadmium) Neutronen
    – nach Alian 117
    – indirekter Formalismus 118
  – mit Reaktorneutronen
    – nach Girardi 111
    – nach De Corte 113, 603
    – nach Kim 115
    – nach Simonits 116
  – mit thermischen Neutronen
    – nach Alian et al 120
– **Multielementanalyse** 35, 109, 120, 123, 126, 603
– Multielementstandardmethode (**Relativmethode**) VIII, 109f, 123
– Nachweisgrenzen 35, 127, 139, 145f, 166
  – abhängige Nachweisgrenzen 145
  – 14 MeV-NAA 174
– Radiochemische Trennungen (RNAA) 554
– Rechnerauswertung von Gammaspektren 87, 98, 123, 577, 583
– Relativmethode, siehe Multielementstandardmethode 110
– Reproduzierbarkeit (precision) 127, 128, 135, 138, 146
– Richtigkeit (accuracy) 128, 129, 130, 135, 167
– Störungen durch Neutronenadsorption (Apatit, Monazit, Xenotin, Turmalin) 565
– Vergleiche verschiedener NAA-Methoden 123f, 129, 131, 694
– Verluste von Elementen durch Bestrahlung 183
– Vorhersage optimaler Parameter 141f
– Vorkonzentrierung 474
– zerstörungsfrei 577
– Zählstatistik 145
Neutroneneinfangquerschnitte 109, 124, 135
– effektive 124

Niederschlag, atmosphärischer
– Elementspurenbestimmung mit Voltammetrie 432

Oberflächen, Analyse von 6, 179, 211f, 291, 301
Oberflächensperrschichtdetektoren 9, 102, 723
Ökologie, Analyse in der 415, 577, 594, 598, siehe auch Umweltforschung
– Kontaminationspfade, kritische 415
– toxische Schwermetalle 415
Öl, Analyse von, mit ICP-AFS 406
Oligoelementanalyse 5, 32, 38, 59, 419
Optimierung von Analysenmethoden 508, 677

Partnerschaft, von Analytikern und Auftraggebern VII, 50, 677
Personalkapazität 45
Phase (Einzelphase) 6, 8
Photoelektronenspektrometrie (ESCA) 635
– von hoch radioaktiven Partikeln 635
Photoleiter 27
Photometrie mit Mehrkanaldetektion, Wasseranalytik 542
Photomultiplier 27, 87
Photonenstrahlung, niederenergetische 18
Photonenzählung 65
Phototransistor 27
Photozelle 27
Pimentel-Report 58, 59
Planeten, Analyse von 63
Polarographie, siehe auch: Voltammetrie 5, 31, 32, 33, 415f
– Adsorptionsvoltammetrie (ADPV) 427
– Differenzpulspolarographie (DPASV) 422
– Differenzpulspolarographie 31, 32, 415
– Inversvoltammetrie 32, 415
– Squarewave-Inversvoltammetrie (SWSV) 425
Potentiometrie
– ionensensitive Elektrode 33
  – geothermale Wässer 524
ppb, Definition 8
ppm, Definition 8
ppt, Definition 8
Präzision, siehe Reproduzierbarkeit
Prisma 23
Probenahme 39
– atmosphärische Aerosole, für INAA 581
– Bach- und Flußsedimente 680
– geologische Proben 679
  – biologisch-medizinisches Material 40, 608
  – lithochemische Beziehungen 680
  – Varianz 680
– geothermale Wässer 524
– natürliche Wässer 431
Probenbank, Analyse von Proben in
– Probenahme 608
Probenkapazität von Großserien bei geologischem Material 677
Probenvorbereitung 5, 23, 41, 485
– Alphaspektrometrie 98
  – Laser, Mikrobohren 99
  – Vakuumverdampfung 101
– atmosphärische Aerosole 581
– biologisches Material 40, 485, 607
– Eindampfen von Wässern 525
– Feinmühle, halbautomatische 736
– Gefriertrocknung von Wässern 525
– Geologisches Material 485, 681
  – Bach- und Flußsedimente 681, 695
  – Flußwasser 695
  – geothermale Wässer 524
– Geräte zur mechanischen und chemischen Probenvorbereitung 738
– Lebensmittelanalyse, Bedeutung in der 626
– natürliche Wässer, Analysenschema 431
– Röntgenfluoreszenzanalyse 485, 501
– synchrotronstrahlungsinduzierte RFA (SYRFA) 302
– Targetherstellung für PIXE 295

Proportionalzählrohr, als Detektor für α-Spektrometrie 29
Prospektion, geochemische, Analyse zur 565, 679

Qualitätskontrolle 4, 43, 48, 517, 519, 677, 708
– mit Doppel- und Mehrfachanalytik 517, 677
– über Elementverhältnisse 517
– Geochemische Materialien 517, 677
– nach Großserien 517
– Kalibrierung 517
– Probensynthese 519
Quecksilbertropfelektroden 32, 415

Radioaktive Abfälle VIII, 635, 641, 643, 645
Radioaktives Material, Analyse von VIII, 635, 641, 643, 645
Ramanspektroskopie 61
– kohärente Anti-Stokes Ramanspektrometrie (CARS) 61
– Ramanstreuung, mit Neodym-YAG-Laser 61
Rasterelektronenmikroskop 255
– mit Röntgenfluoreszenzanalyse SEM-EDX 255, 635
  – hoch radioaktive Materialien 635
Rationalisierung 50, 677
Rauschäquivalente Strahlungsleistung 65
Rauschen (von Strahlungsempfängern) 65
Reaktorüberwachung, Analyse zur 8, 9
Recyclingprodukte, Analyse von 458
Referenzmaterialien, siehe: Standards
Reinstmetalle, Analyse von 8, 211
Reinststoffe, Analyse von VIII, 80, 185, 458
Reproduzierbarkeit (precision) 43, 65, 459, 492, 517, 529
– Geologische Proben, gefordert 684f
– Gleichstrombogen-AES 36
– ICP-AES 37, 381
– ICP-AFS 409
– Multielementanalyse in geothermischen Wässern 526
– Neutronenaktivierungsanalyse 35
– Röntgenfluoreszenzanalyse 563
  – Bach- und Flußsedimente 563, 571
  – Leber 616
– Selektivität und Reproduzierbarkeit 529
– synchrotronstrahlungsinduzierte RFA 304
– verbrauchsarme ICP-AES 37
– Wechselstromfunken-AES 37
Resonanzintegrale 109, 123, 138
– effektive 109
Resonanzionisierungsspektrometrie 61
Restverunreinigung, Regel von Zolotov 460
Richtigkeit (accuracy) 43, 65, 83, 167, 459, 492, 517, 519
– Funkenmassenspektrometrie 185
  – mit Isotopenverdünnungsanalyse 198
– geologische Proben 552, 684, 687
– geothermale Wässer 526
– Lasermassenspektrometrie 206
– Probensynthese 519
– Röntgenfluoreszenzanalyse von Bach- und Flußsedimenten 563
– Seltene Erden-Analyse in Gesteinen, mit ICP-AES 552
– synchrotronstrahlungsinduzierte RFA 304, 305
Robotersystem, zur Probenvorbereitung 41

Rohstofforschung, Analyse in der 678
Röntgenbox
- zum Rasterelektronenmikroskop 255
Röntgenfluoreszenz, Energie der 19
Röntgenfluoreszenzanalyse (RFA) VIII, 33, 47, 77, 80, 726
- **Anwendungen**
  - Archäologie 657, 658, 660, 661, 669
  - Bach- und Flußsedimente 563
  - Eruptiv- und Sedimentgesteine 717
  - geologische Proben 485, 563, 690
  - in-line Bestimmung von Aktiniden, Wiederaufarbeitung Kernbrennstoff 645
  - Kunstwerke 667, 669
  - Lebenswissenschaften 608
  - Nahrungsmittel 625
  - radioaktive Stoffe der Kerntechnik 635
  - städtisches Abwasser 643
  - Trinkwasser 297
  - Wässer, geothermale 524
  - Umweltforschung 695
- Detektoren 29
- **energiedispersive RFA** VIII, 257, 524, 635, 690, 726
- **Nachweisgrenzen** 33, 35, 36, 260, 293, 297, 302, 303
- Probenvorbereitung 501, 563
- **sequentielle RFA** 25, 26, 501, 677
- **simultane RFA** 80
- Spektren 25
- Standards auch: Standards 502
- **synchrotronstrahlungsinduzierte RFA** (SYRFA) 301
  - Anwendung auf Luftstaubprobe 308
  - Interelementeffekte 305
  - Meßpräparateherstellung 302
  - Nachteile 308
  - Nachweisgrenzen
    - physikalische 302
    - absolute 304
    - geringere Matrixempfindlichkeit 303
    - optimale Bereiche der Ordnungszahlen 304
  - ortsauflösende Multielementanalysen 305
    - Vorteile 305, 306
    - Abrasterung einer Kohleprobe 307
  - Reproduzierbarkeit 304
  - Richtigkeit 304
  - Spektrenauswertung, halbautomatisch 304
  - Vorteile 307
- **teilcheninduzierte RFA (PIXE)** VIII, 291f
  - Anwendungen zur Spurenelementanalyse 293, 295
    - biologisches Material 298, 485, 608
    - Kunstwerke 667, 669
    - Mikromethode 293
    - Nachweisgrenzen 293, 294
    - Wasseranalyse 296
  - Halbleiterdetektoren, hochauflösend 292
  - physikalische Grundlagen
    - Fluoreszenzausbeute 292
    - Produktionsquerschnitte 293
  - Ionenbeschleuniger, Teilchenstrahlung von 292
  - Target 294, 295
  - Verdampfung 294
  - Vorkonzentrierung 296

- **totalreflektierende RFA (TXRF)** VII, 35, 257, 269
  - Anwendungen 269
    - Austernfleisch 266
    - Flußwasser 267, 695
    - geologische Proben 694
    - Monazitsand 266
  - dynamischer Bereich 262
  - energiedispersiv 257
  - Gerät 258, 260
  - Kalibrierung 263, 264
  - Matrixeinfluß 262
  - Matrixentfernung 263
  - Meßpräparateherstellung 265
  - Multielementanalyse 269
  - Nachweisgrenzen 260
  - Probenvorbereitung 695
  - Spektrenauswertung 263
  - Untergrundunterdrückung 259
  - Vergleich mit INAA, ICP-AES, in geologischen Proben 694
  - Vorkonzentrierung 472
- **weiche und ultraweiche Röntgenstrahlen, RFA mit,** 237
  - Absorptionskanten 240
  - Anwendungen
    - B in Gläsern 247
    - C in Gußeisen 243, 244
    - C in Kohle 246
    - C und B mit totalreflektierendem Monochromator 238, 241
    - O in Chromatfilm auf Al 252
    - O in Kohle 238, 249, 250, 251
    - Seltene Erden, U, Th in Böden, Pflanzen, Monazitsand 485
  - Hochintensitätsmonochromator 237
  - Spektrale Störungen 238
  - Totalreflexion 237, 240
- **wellenlängendispersive RFA,** VIII, 33, 45, 66, 485, 501, 563, 726
Röntgenfluoreszenzspektrometer, handelsübliche 726
- Elektronenmikrosonde 729
- energiedispersiv 726; Abb.: 734, 739
- ortsauflösende Rasterung 729
- totalreflektierend 65, 727; Abb.: 742
- wellenlängendispersiv 726; Abb.: 731, 743, 744
Röntgenfluoreszenzspektrometrie
siehe Röntgenfluoreszenzanalyse
Röntgenspektrometrie 19, 23, 24, 87f
- Spektrometer 28, 29, 87, 237, 257
Röntgenstrahlung, Energie 19
Routineanalytik 45, 77, 79, 269, 603, 647, 677

Schaltkreise, integrierte, Analyse von
- INAA von Si-Wafern 604
- Kontamination 604
Schmieröle, Abrieb 82
Schwebstoffe, Analyse von
- in Flußwasser mit TXFA 270, 271
Sedimente, Analyse von
- Bach- und Flußsedimente, mit ICP-AES und RFA 679
- Flußsedimente, mit TXRF 269
- Neutronenaktivierungsanalyse
  - mit 14 MeV Neutronen 174
Sekundäranregungen, RFA 35

Selektivität in der Multielementanalyse 65, 492, 529
Sequentielle Analysenmethoden 6, 65, 75, 77, 78, 79
– ICP-AES 381, 681
  – geologische Proben mit RFA 681
  – Wasseranalytik 541
– ICP-AES 397
Signal, analytisches 4, 17-21, 65
– zeitkonstant 65
– zeitveränderlich 65
Signal/Untergrundverhältnis 49, 78, 87, 145, 377, 381
Signalunterdrückung, ICP-AES 633
Silizium, Analyse von 181
Simultane Analysenmethoden 6, 65, 75, 77, 78, 79, 681
– ICP-AES 321
  – geologische Proben 681
– ICP-AFS 397
– Massenspektrometrie 30, 185
– Neutronenaktivierungsanalyse, Gammaspektrometrie 87, 123
– Wasseranalyse 541
Software 60
Spaltprodukte, Analyse von 98
Speciation 5, 17, 617
Spektralbereich 19, 20, 65
– Linien 65
– Detektor 65
Spektrale Störungen, in AES 49, 337
– Bandeninterferenzen 338
– Linieninterferenzen 337
– Spektrometerstörungen 338
– Untergrundintensitäten 338
– wahre Nettolinienintensitäten, Bestimmung 338
– Matrixeffekte, Einfluß auf Nettolinienintensität AES 338
  – Aerosolbildung 339
  – Anregung 339
  – Kalibrierung mit Standardaddition 339
  – Matrixentfernung 339
  – Plasmageometrie 339
  – Verdampfung 339
Spektraler Untergrund 65, 337
Spektralphotometrie 5, 19
Spektroelektrochemie 62
Spektrometer 65
Spektrometer, handelsübliche 22, 23, 360, 723f; Abb.: 730f
siehe auch: AAS, AES, AFS, MS, RFA
Spektrum 22
Spezifität 65
Spurenbestandteil, Definition 8
Spurenelementanalyse, Grenzen 10, 212, 457
Spurenkonzentrat, durch Vorkonzentrierung 459, 471
Sputtern, Bestimmung von Sputterraten 182
Strahlungsempfänger 65
Standards und Referenzmaterialien 11, 12, 18, 491f
– Kalibrierung, mit 48
– **Multielementstandards** 491f, 501
  – biologisches Material 495
  – Blut 499

– geologisches Material 711, 718
– Metalle/Legierungen/Oxide 492f
– Oxidpulver, synthetische 501
– Röntgenfluoreszenzanalyse 501, 718
– Seesediment 493
– Umweltmaterialien 492
– Vergleich
  – mit ICP-AES, NAA, SSMS 711
  – Spexstandard, mit DC-AES, ICP-AES 714
– Wasser 498
– Probleme bei Standardherstellung 496
– Sekundärstandards 520
– Standardreferenzmaterialien 8, 11, 12, 18, 83, 123, 491f
– Standardisierung durch Probensynthese 519
– Zertifizierung von Standards 18, 491f, 580
Standardadditionsmethode 337
Stöchiometriebestimmung
– von Sonderlegierungen und Kristallen mit ICP-AES 647
Strategie des Einsatzes von Analysenmethoden VIII, 45, 49, 57
Stromspannungskurven 18, 21, 31, 415
Struktur, chemischer Analysenmethoden 3, 4
Stahl, Analyse von
– ICP-AES, Lichtleiterkopplung 374
– ICP-AFS 405

Technische Anleitung Luft (VDI) 275
Temperatur 65
– Anregungstemperatur 65
– Normtemperatur 65
Tetrabromäthan, Schweretrennung von Mineralen 566
Tiefseeforschung, Analyse für 63
Titration 5, 17, 39
Totzeitkorrektur, bei Gammaspektrometrie 87, 169, 173
Trennungen 3, 15, 16, 163, 457, 551, 553
– Schweretrennung von Mineralen 566
Trinkwasser, Analyse von
– kohärente Vorwärtsstreuungs-Atomspektrometrie 387
– Voltammetrie 433

Übersichtsanalysen, über alle Elemente 45, 201
– Bogen-Atomemissionsspektrometrie 202
– Glow-discharge Atomemissionsspektrometrie 203
– ICP-Atomemissionsspektrometrie 202
– Laser-Massenspektrometrie 201
– Massenspektrometrie 203
– Neutronenaktivierungsanalyse 123
– radiochemische Analyse 202
– Röntgenfluoreszenzanalyse 202
Umweltforschung, Analysen in der VIII
siehe auch: Biologie, Medizin, Geologie
– Anwendung chemischer Analysen VIII, 57, 68, 577, 614
– Anwendung der INAA 577, 614
– wichtige Metalle 579
– geologische Proben 679
– ökotoxische Metalle, Analyse von 417
– Richtigkeit von Analysenergebnissen 580
Umweltproben 8, 9, 11, 12, 415, 523, 539, 558, 577, 580, 581, 594, 596, 598, 607, 677

- ICP-AES 381
- INAA
  - atmosphärische Aerosole 581
  - Staub aus Eisengießerei 585
  - Luftstaub
    - Photonenaktivierung 594
    - einzelne Luftstaubkörner 598
  - NAA mit 14 MeV Neutronen 174
- Voltammetrie 415f
  - Adsorptionsvoltammetrie 426
  - Differentialpulsinversvoltammetrie (DPASV) 422
  - Squarewave-Inversvoltammetrie (SWSV) 425

Umweltprobenbank, Materialien für
- INAA 607, 614
- Leber 614
  - Kombination von AAS, NAA, IDMS, Voltammetrie 614

Umweltüberwachung, Einsatz von Multielementmethoden 459
- Auswahl zu überwachender Elemente 578
- bestimmbare Elemente 578
- Luftstaub 703
- NAA, Anwendungen 580
- Richtigkeit von Analysenergebnissen 580
- Umweltbereiche 578

Untergrund in Gammaspektren 87, 163
- durch Bremsstrahlung 163
- durch P-32 in biologischem Material 163

Untergrund, spektraler (AES) 337
siehe auch: spektrale Störungen
Untergrundkorrekturen, spektrale (AES) 337
- Vielkanaldetektion mit parallelem Eintritt 342
- Wellenlängenmodulation 340
- Zweikanalverfahren 342

Vakuumultraviolettstrahlung (VUV) 19
Veraschung 3, 41, 42, 457
- Naßveraschung 41, 42, 164, 457
  - Abwasser (städtisches) mit HNO_3/H_2O_2 434, 643
  - Blut mit HNO_3/H_2O_2 234
  - Hefe, mit H_2O_2/HNO_3 bzw. HNO_3 unter Druck 633
  - HNO_3/H_2O_2 234
- vollständige, für Voltammetrie 434
- Umweltproben 434
  - Abwasser 434
  - medizinische und biologische Proben ($HNO_3/HClO_4$, HNO_3) 436
  - Nahrungsmittel: $HNO_3/HClO_4$ 434
  - Wein, Fruchtsäfte: H_2O_2/UV 436
  - Wein mit $H_2O_2/H_2SO_4/HNO_3$ 633

Verbundverfahren, -analyse 5, 75, 77, 459
Verbindung, chemische 7
Verdampfungsquelle 24
Verhältnismethode
- zur Vorhersage und Optimierung von INAA-Daten 141

Verteilung von Datenkollektiven 43
- Normalverteilung nach Gauß 43
- logarithmische Verteilung 43
- logarithmische Normalverteilung 43

Vielkanalanalysatoren 29, 87, 723
- für Alphaspektrometrie 103

Videobänder, Material für, Analysen von
- INAA 605

Voltammetrie VIII, 21, 31, 33, 48, 60, 415f, 419, 420
- Adsorptionsvoltammetrie 426
- Anwendungen
  - Abwasser 434
  - atmosphärische Niederschläge 432
  - Wasseranalytik 545
  - biologisches Material 436, 608
  - Lebenswissenschaften 608
  - medizinisches Material 436, 608
    - Leber 614
  - Nahrungsmittel 434, 608, 625
  - natürliche Wässer 431
  - Trinkwasser 433
  - Wein, Methodenvergleich 634
- Bestimmungsgeschwindigkeit 429
- Differentialpulsinversvoltammetrie (DPASV) 422, 545
- Differentialpulsvoltammetrie (DPV) 420f, 545
- Nachweisgrenzen 428f
- Oligoelementanalyse von Umweltproben 415f
- Reproduzierbarkeit 428f
- Veraschung, vollständige 434

Voltammetrie, handelsübliche Geräte
- Hersteller 746
- Wechselstrom-Impedanz-Meßsystem Abb.: 734

Vorkonzentrierung 3, 457, 480, 485
- geothermale Wässer, von 524
- Gesteinsanalyse 551
- hängende Hg-Elektrode, an 38
- von Seltenen Erden, U, Th für RFA 485

Vorwärtsstreuung, kohärente VIII, 385, 391
siehe auch: kohärente Vorwärtsstreuung

Wasser, Analyse von VIII, 8, 9, 523, 539f, 541
- Anionenanalyse 544
- Analysenmethoden, verwendete 541
- Arten von Wässern 540
- Bestandteile 540
- Differentialpulsinversvoltammetrie (DPASV) 422, 545
- Estuarienwasser, mit TXRF, NAA, ICP-AES 700
- Flußwasser (Elbe), mit TXRF, NAA, ICP-AES, 697f
- Gaschromatographie 545
- Gewässerschutz 539
- Güteberteilung 539
- Haupt- und Nebenbestandteile 540
- ICP-AFS 405
- ICP-AES 541, 543
- Ionenchromatographie 544
- Instrumentelle Multielementanalyse 539f
- Meerwasser, mit G-AAS 413
- Metalle 541
- Multielementanalyse 539
  - von Anionen 544
  - von Kationen 539
- natürliche Wässer
  - Analysengang 431
  - Speciation 432
- Niederschlag, atmosphärischer 432
- Photometrie, mit Multikanaldetektion 542

Semiquantitative Instrumental and Simultaneous Multielement Analysis of All Elements in One Sample

Methods: Combination of sparc source mass spectrometry (SSMS), optical emission spectrometry (OES), instrumental neutron activation analysis with reactor neutrons (INAA), supplemented by monoelement values from X-ray fluorescence (RFA), atomic absorption spectrometry (AAS)

Sample: Uranium ore, mixture of subsamples from one body. Sample weight: 10g

Laboratory: Central Department for Chemical Analysis, Kernforschungsanlage Jülich GmbH, 1981/82

B. Sa March 1981

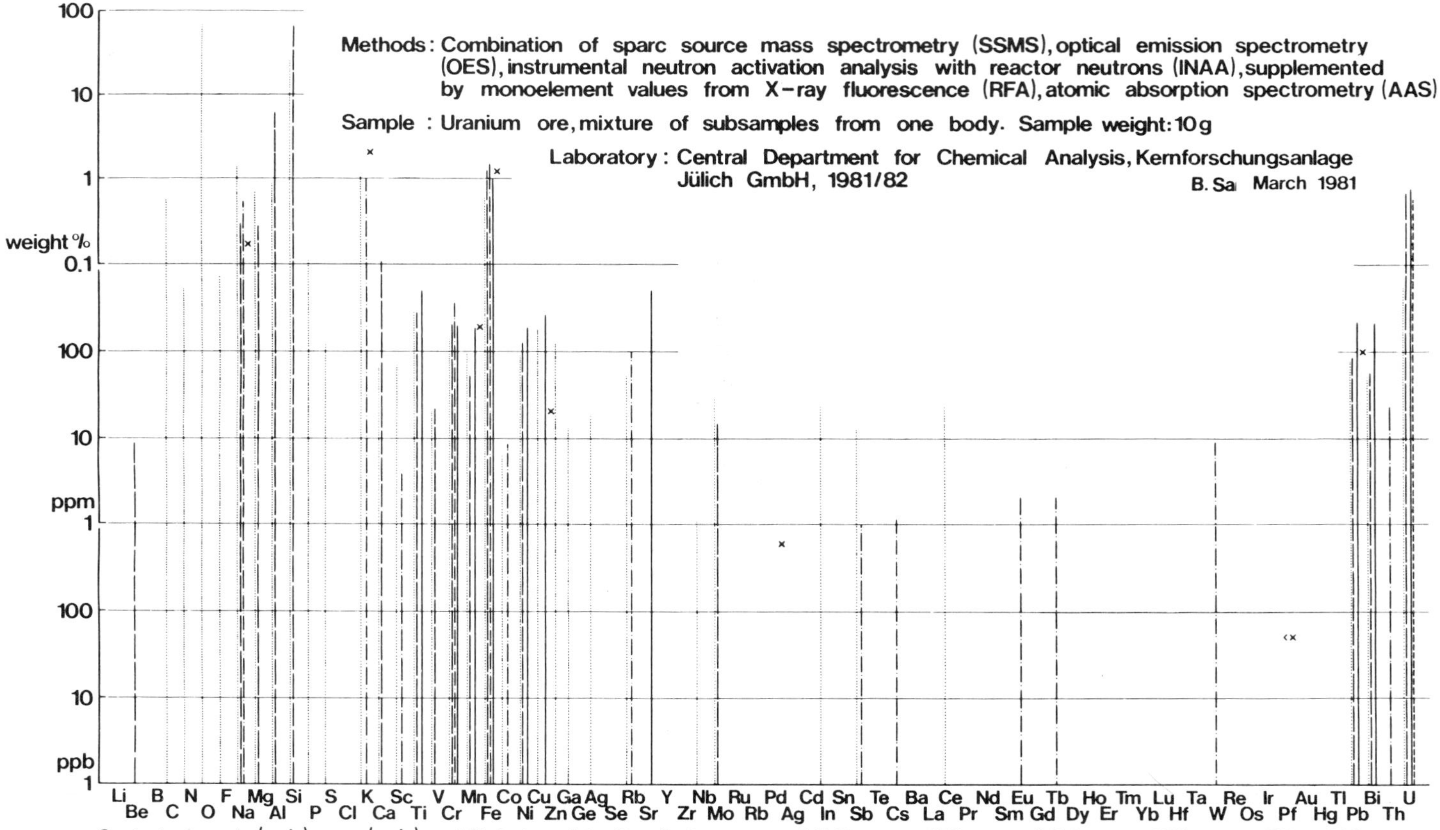

Contents in ppb (ng/g), ppm (µg/g) and %; below detection limits: <; ——:SSMS; — — —:OES; – – –:INAA; ······:RFA; ———:ICP; x:AAS